Earth, Water, Ice and Fire: Two Hundred Years of Geological Research in the English Lake District

Society Publication reviewing procedures

The Society makes every effort to ensure that the scientific and production quality of its books matches that of its journals. Since 1997, all book proposals have been refereed by specialist reviewers as well as by the Society's Books Editorial Committee. If the referees identify weaknesses in the proposal, these must be addressed before the proposal is accepted.

Once the book is accepted, the Society has a team of Book Editors (listed above) who ensure that the volume editors follow strict guidelines on refereeing and quality control. We insist that individual papers can only be accepted after satisfactory review by two independent referees. The questions on the review forms are similar to those for *Journal of the Geological Society.* The referees' forms and comments must be available to the Society's Book Editors on request.

Although many of the books result from meetings, the editors are expected to commission papers that were not presented at the meeting to ensure that the book provides a balanced coverage of the subject. Being accepted for presentation at the meeting does not guarantee inclusion in the book.

Geological Society Publications are included in the ISI Index of Scientific Book Contents, but they do not have an impact factor, the latter being applicable only to journals.

More information about submitting a proposal and producing Society Publications can be found on the Society's web site: www.geolsoc.org.uk.

GEOLOGICAL SOCIETY MEMOIR No. 25

Earth, Water, Ice and Fire: Two Hundred Years of Geological Research in the English Lake District

BY

DAVID OLDROYD

School of History and Philosophy of Science
The University of New South Wales, Australia

2002
Published by
The Geological Society
London

THE GEOLOGICAL SOCIETY

The Geological Society of London (GSL) was founded in 1807. It is the oldest national geological society in the world and the largest in Europe. It was incorporated under Royal Charter in 1825 and is Registered Charity 210161.

The Society is the UK national learned and professional society for geology with a worldwide Fellowship (FGS) of 9000. The Society has the power to confer Chartered status on suitably qualified Fellows, and about 2000 of the Fellowship carry the title (CGeol). Chartered Geologists may also obtain the equivalent European title, European Geologist (EurGeol). One fifth of the Society's fellowship resides outside the UK. To find out more about the Society, log on to www.geolsoc.org.uk.

The Geological Society Publishing House (Bath, UK) produces the Society's international journals and books, and acts as European distributor for selected publications of the American Association of Petroleum Geologists (AAPG), the American Geological Institute (AGI), the Indonesian Petroleum Association (IPA), the Geological Society of America (GSA), the Society for Sedimentary Geology (SEPM) and the Geologists' Association (GA). Joint marketing agreements ensure that GSL Fellows may purchase these societies' publications at a discount. The Society's online bookshop (accessible from www.geolsoc.org.uk) offers secure book purchasing with your credit or debit card.

To find out about joining the Society and benefiting from substantial discounts on publications of GSL and other societies world-wide, consult www.geolsoc.org.uk, or contact the Fellowship Department at: The Geological Society, Burlington House, Piccadilly, London W1J 0BG: Tel. +44 (0)20 7434 9944; Fax +44 (0)20 7439 8975; Email: enquiries@geolsoc.org.uk.

Published by The Geological Society from:
The Geological Society Publishing House
Unit 7, Brassmill Enterprise Centre
Brassmill Lane
Bath BA1 3JN, UK

(*Orders*: Tel. +44 (0)1225 445046
Fax +44 (0)1225 442836)
Online bookshop: http: //bookshop.geolsoc.org.uk

British Library Cataloguing in Publication Data
A catalogue record for this book is available from the British Library.

ISBN 1-86239-107-6
ISSN 0435-4052

Typeset by Type Study, Scarborough, UK
Printed by Alden Press, Oxford, UK

Distributors
USA
AAPG Bookstore
PO Box 979
Tulsa
OK 74101–0979
USA
Orders: Tel. + 1 918 584–2555
Fax +1 918 560–2652
E-mail *bookstore@aapg.org*

India
Affiliated East-West Press PVT Ltd
G-1/16 Ansari Road, Daryaganj,
New Delhi 110 002
India
Orders: Tel. +91 11 327–9113
Fax +91 11 326–0538
E-mail *affiliat@nda.vsnl.net.in*

Japan
Kanda Book Trading Co.
Cityhouse Tama 204
Tsurumaki 1-3-10
Tama-shi
Tokyo 206–0034
Japan
Orders: Tel. +81 (0)423 57–7650
Fax +81 (0)423 57–7651

Contents

It is recommended that reference to all or part of this book should be made in the following way:

OLDROYD, D. 2002. *Earth, Water, Ice and Fire: Two Hundred Years of Geological Research in the English Lake District.* Geological Society, London, Memoirs, **25**.

Preface and Acknowledgments

This book was begun in 1996, and work on it continued through to 2001, with five summer seasons spent in the English Lake District and in British libraries. I called on and taped interviews with a considerable number of geologists (including three in Australia) who have worked in the Lakes or who continue to do so. My inquiries were financed by a grant from the Australian Research Council, to whom I am most grateful, in that it has indulged my penchant for mixing business with pleasure. I am also most grateful to the Geological Society for permitting me to publish such a book with their Publishing House; and to Diana Swan and Angharad Hills for attending to the editorial processes. Bernard Leake and Nigel Woodcock were sympathetic and most helpful referees, making it possible to weed out a number of solecisms, in addition to furnishing much valuable advice and information. I owe them a lot.

Interviews were conducted, or valuable conversations had, with the following persons: Peter Allen, Andrew Bell, Brett Beddoe-Stephens, Michael Branney, John Boardman, Martin Bott, Bob Chaplow, Richard Clark, Tony Cooper, Jack Soper, the late Sir Kingsley Dunham, Ronald Firman, Godfrey Fitton, Neil Fortey, Stuart Haszeldine, Douglas Holliday, Dennis Jackson, Louisa King, Colin Knipe, Peter Kokelaar, Benjamin Kneller, Colin Knipe, Michael Lee, Kenneth McNamara, Uisdean Michie, David Millward, Murray Mitchell, Stephen Molyneaux, Frank Moseley (who supervised my undergraduate work for a year), the late Robin Oliver, Michael Petterson, Tony Reedman, Barrie Rickards, Eric Robinson, Adrian Rushton, Tom Shipp, Eric Skipsey, Phil Stone, Tony Wadge, Barry Webb, Nigel Woodcock and Brian Young. Mervyn Dodd and Alan Smith were good enough to conduct me to some sites in the field also, and I enjoyed having the opportunity to participate in several excursions organized by the Cumberland and Westmorland Geological Societies, while Jack Soper enabled me to see what he got up to while mapping the Kendal Sheet. Many of the above have also assisted through subsequent correspondence. I am extremely indebted to all these people. Without them there would undoubtedly be no book to lay before the reader. Only three people I tried to contact failed to respond; and one other declined to become involved in the project.

Some others whom I was not able to meet personally have also been extremely helpful through correspondence, phone calls or personal conversations: Jim Briden, Norman Catlow, Andrew Chadwick, John Cope, Rudolf Daber, Robert Dott, Kenneth Glennie, Paul Green, David Holmes, Alan Hooper, Stephen Horseman, Richard Hughes, Peter Jeans, Eric Johnson, Bernard Leake, Cherry Lewis, Brian McConnell, Christopher McKeown, Klaus Michels, Richard Moore, Ted Neill, Michael Nutt, Helen Oliver, Vic Parsons, Jack Preston, Brinley Roberts, David Roberts, Eric Robinson, Celal Sengör, Vivien Simpson, David Skevington, David Smythe, Helen Reeves, Pauline Taylor, John Temple, Lorna Thomas, Chris Thompson, Jonathan Turner, Otfried Wagenbreth, Harry Wilson, Derek Woodhall and Kejian Wu. I am much indebted to these persons too. My colleague, David Miller, provided his usual judicious comment on various portions of my text.

The assistance of Misha Au-Yeung and Harshi Gunawardena, who helped with bibliographical matters, is gratefully acknowledged. For references on Lakeland geology, the two bibliographies of Alan Smith have been indispensable. He was also good enough to take some photographs for me in the Hollows Farm area.

Ruth Banger, Simon Bennett, Moira Bent, Lilian Bew, Alan Bowden, Diana Chardin, Jon Clatworthy, Jeffrey Cowton, Hazel Davidson, Paul Davis, Michael Dorling, Peter Eyre, Jackie Fay, Stephen Hewitt, Richard Gillanders, Andrea Fazackerley, Brian Houghton, Martina Koelbl-Ebert, Bob McIntosh, Kate Manners, Mark Nicholls, Peter Nockles, Kate Perry, Jane Pirie, Emma Robinson, Mary Sampson, Claire Slater, Michael Smallman, Jonathan Smith, Pauline Taylor, Chris Terrey, Anne Thompson, Hugh Torrens and Michael Walton have provided valuable assistance in the matter of archival sources, information about Lakeland geologists and related matters, permissions, photographs of specimens, references, and library services. In this last regard, I should *particularly* like to thank the staff of the interlibrary loan service at The University of New South Wales (mostly anonymous), Wendy Cawthorne of the Geological Society's library, and Graham McKenna at the British Geological Survey for their invaluable assistance. Steve Preece, Janet Latham and Jane Oldroyd gave valued assistance with the preparation of the topographic maps. Special mention should be made of the late John Thackray, formerly Honorary Archivist at the Geological Society, who, during his last illness, made the effort to send me a copy of his transcription of an interview he had recorded with Tressilian Nicholas.

At the Geological Society Publishing House, Angharad Hills and Diana Swan have been of the greatest assistance to me, and I should also thank all those, with names unknown to me, who did the work necessary to turn my text and illustrations into a very presentable book. The work of Robert Holdsworth as overseeing editor of this book is also most gratefully acknowledged, as are the comments and advice of Bernard Leake and Nigel Woodcock as referees.

My wife, Jane Oldroyd, has been patient (and sometimes rightly impatient) with me over several years, and has assisted in all sorts of ways in the matter of computing problems.

The hospitality of numerous proprietors of bed-and-breakfast establishments in Britain is recorded with thanks. Among them, it is nice to remember Margaret Lamb at St Bees, whose nine-year-old daughter Harriet entertained with enchanting tap-dancing on the flags of their warm kitchen at breakfast time, not to mention playing the harp, violin and recorder for my benefit. Margaret herself (an itinerant music teacher) produced a cello from somewhere, and we were able to read through the great Brahms Piano Quartet, opus 25, with musician friends of hers in the village. At Keyworth, Anne Carnell made it possible for me to make repeated pleasant visits to the British Geological Survey. I almost became a member of the Carnell family. Graham McKenna effected the introduction.

In a book published in 1990, I thanked the middle fingers of my left and right hands for their stalwart work in typing my manuscript. Matters remained the same in 2002, and I should like to thank them both again.

A note on abbreviations, units, and maps

SS, Skiddaw Slates.
BVG, Borrowdale Volcanic Group.
CPT, Causey Pike Thrust.
'Otley I' is used as an approximate synonym for the Skiddaw Slates.
'Otley II' is used as an approximate synonym for the Borrowdale Volcanics Group.
'Otley III' is used as an approximate synonym for the Windermere Group or Windermere Supergroup.
Ma, million years.
Ka, thousand years

Units of length (feet, yards, miles, metres, kilometres) are given in accordance with the usage of the primary sources under discussion.

1 metre = 3.281 ft
1 kilometre = 0.6214 miles
1 yard = 3 ft

In the topographical maps, thin lines represent waterways. Thick lines represent roads (of unequal importance), coastlines or lake margins. Footpaths are not represented. Filled circles represent towns or villages, of unequal numbers of inhabitants. Filled triangles represent hilltops or mountain peaks, not all of equal topographic significance.

Plates

Plate I. First known geological map of the Lake District (*c.* 1814?), attributable to Joseph Fryer. Geological Society archives (LDGSL 1008/29). Reproduced by courtesy of the Geological Society, London.

Plate II. Manuscript geological map of Cumberland by Adam Sedgwick (n.d.), held at the Sedgwick Museum, Cambridge. Reproduced by courtesy of Sedgwick Museum.

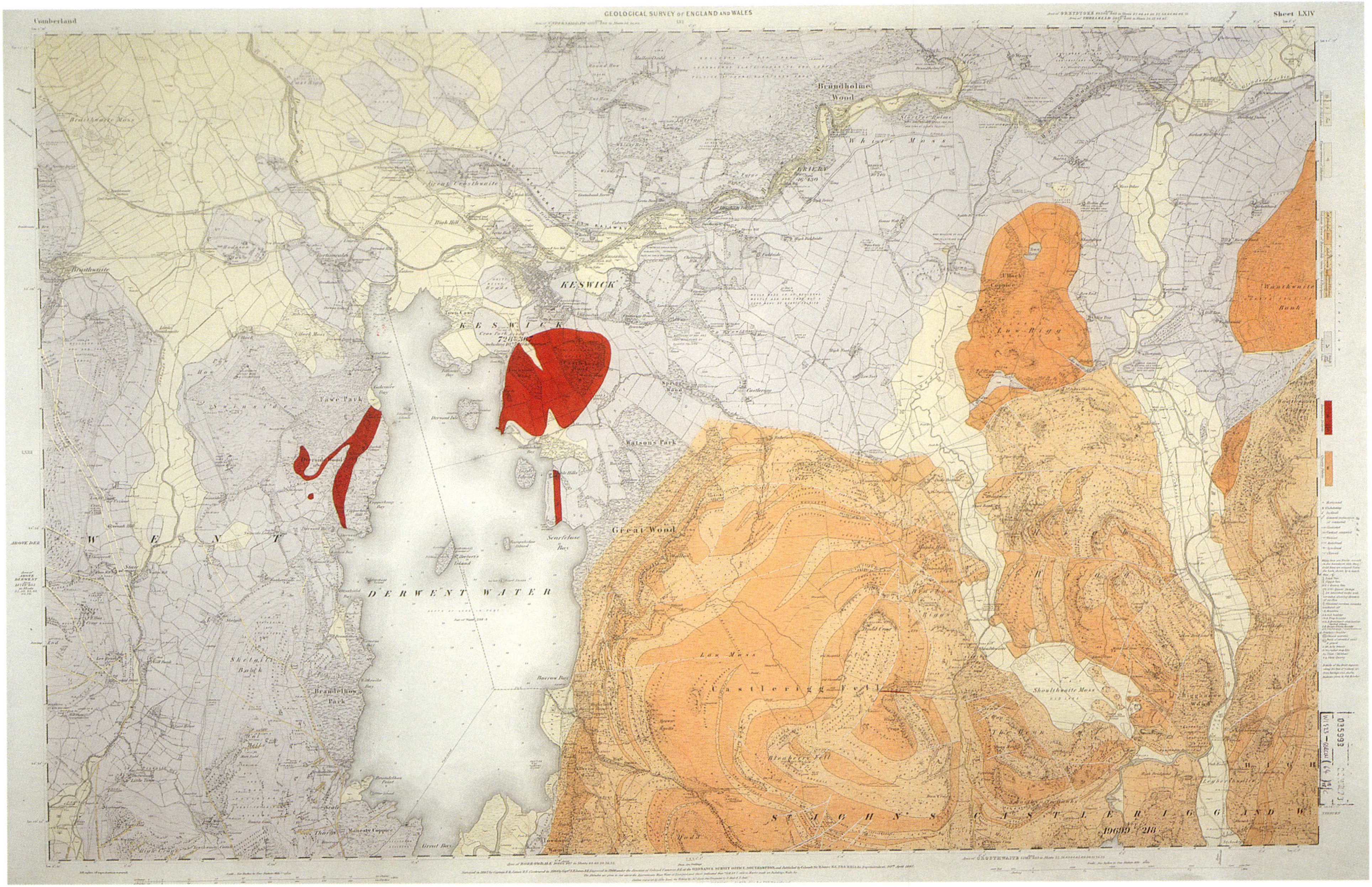

Plate III. Geological Survey of Great Britain, Sheet LXIV. *Borrowdale, Derwent Water, Keswick and environs, six inches to mile.* Surveyed by J. C. Ward under the supervision of W. T. Aveline. Published April, 1873. Reproduced by courtesy of the British Geological Survey.

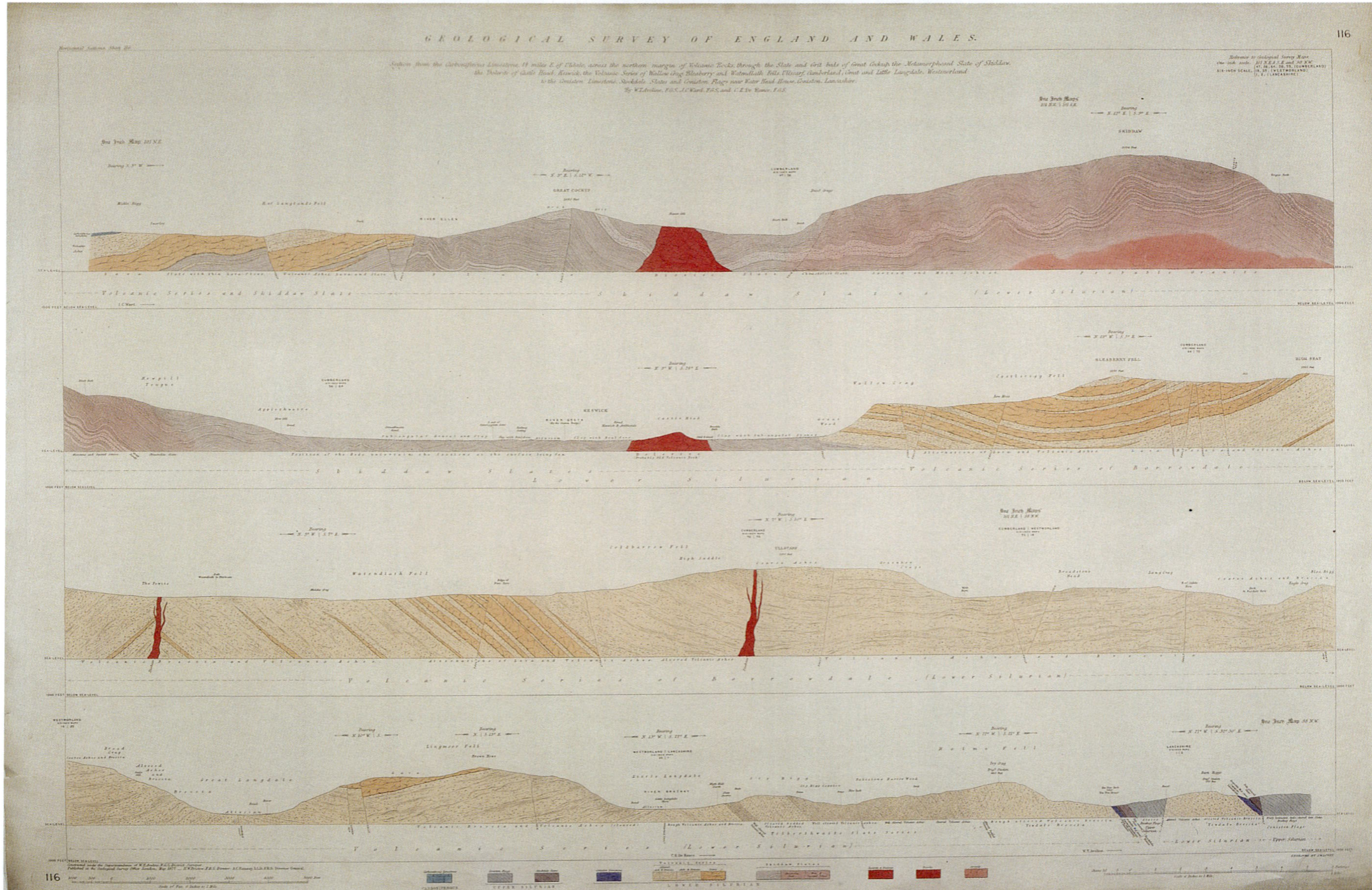

Plate IV. Geological Survey of Great Britain [1877], Horizontal Section 116. *Section from the Carboniferous Limestone, 11 miles east of Ulldale, across the northern margin of Volcanic Rocks, through the slate and grit beds of Great Cockup, the metamorphosed slate of Skiddaw, the dolerite of Castle Head, Keswick, the Volcanic Series of Wallow Crag, Bleaberry and Watendlath Fells, Ullscarf, Cumberland, Great and Little Langdale, Westmorland, to the Coniston Limestone, Stockdale Slates [sic] and Coniston Flags near Water End House, Coniston, Lancashire*. W. T. Aveline, J. C. Ward & C. E. De Rance. Reproduced by courtesy of the British Geological Survey.

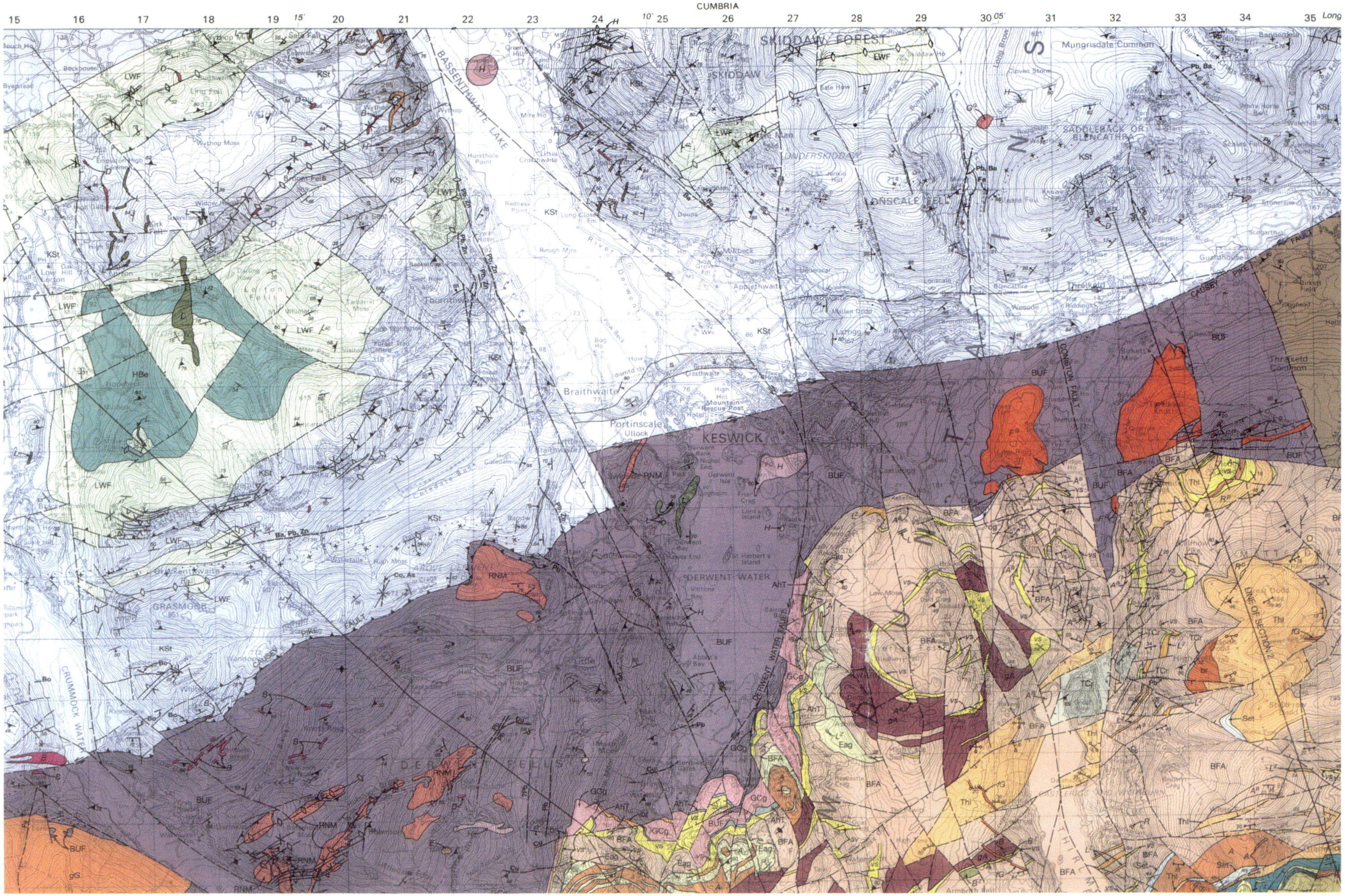

Plate V. Geology of the area of Causey Pike and Keswick. Portion of: British Geological Survey, 1: 50,000 Series, Sheet 29, Solid geology. *Keswick* (1999). IPR/25–3C British Geological Survey, Keyworth. HBe, Hope Beck Formation (green); LWF, Loweswater Formation (light green); KSt, Kirkstile Formation (lilac); BUF, Buttermere Formation (purple); RNM, Robinson Member (brick red); Tarn Moor Formation (brown); BFA, Birker Fell Formation (Lower BVG) (flesh colour); Ennerdale Granophyre (terracotta).

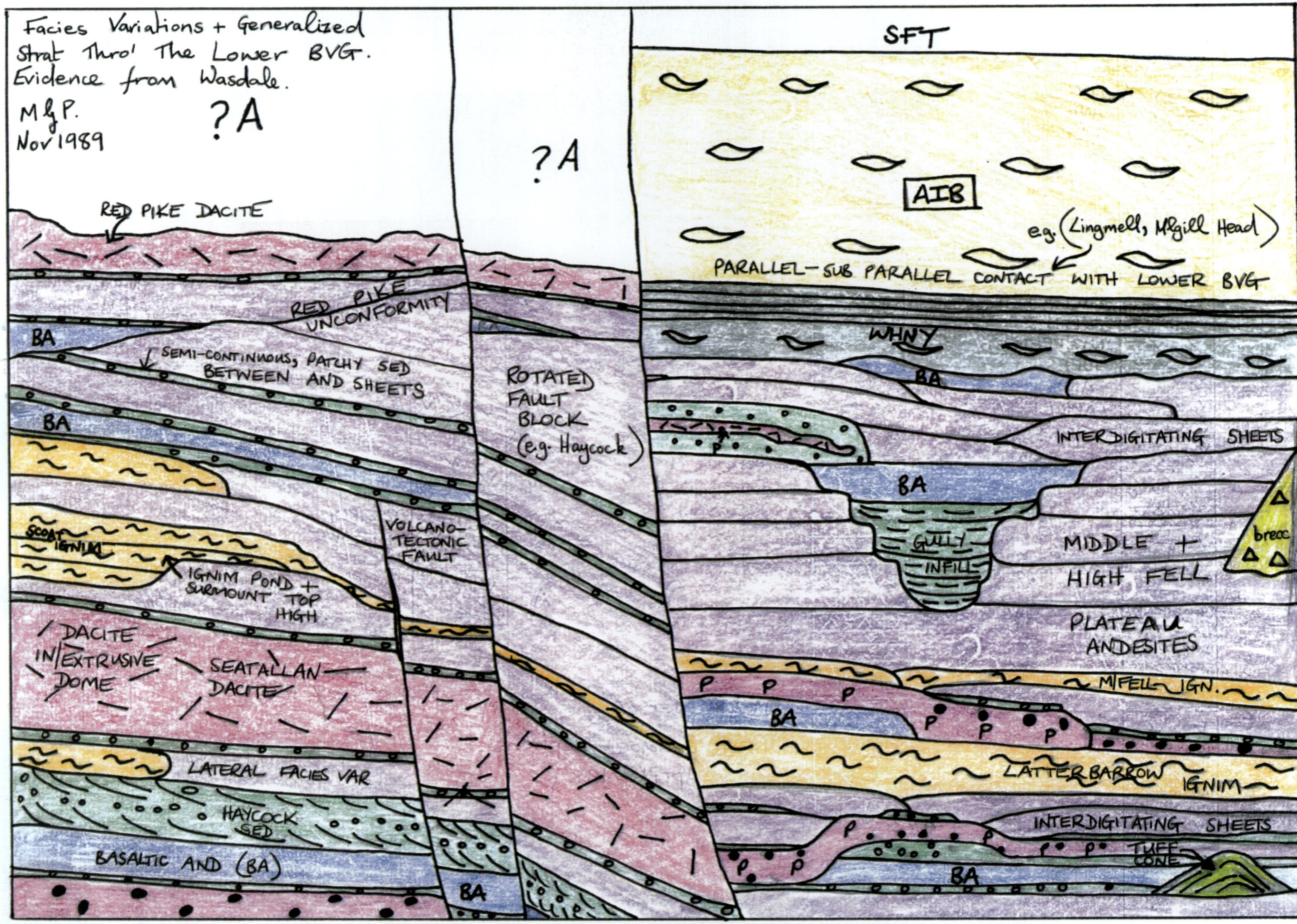

Plate VI. Work-in-progress hypothetical sketch-section of Lower Borrowdale Volcanics, drawn by Michael Petterson, 1989 (AIB, Airy's Bridge Formation; WHNY, Whorneyside Tuff; BA, basalt; the 'fish' signs indicate ignimbrites). By courtesy of Dr Petterson and IPR23/7C British Geological Survey.

late VII. Map depicting parts of the Borrowdale Volcanic Group. Portion of: British Geological Survey, 1:50,000 Series, England & Wales, Sheet 38. Solid eology. *Ambleside* (1996). British Geological Survey, Keyworth. IPR/23-7C British Geological Survey.

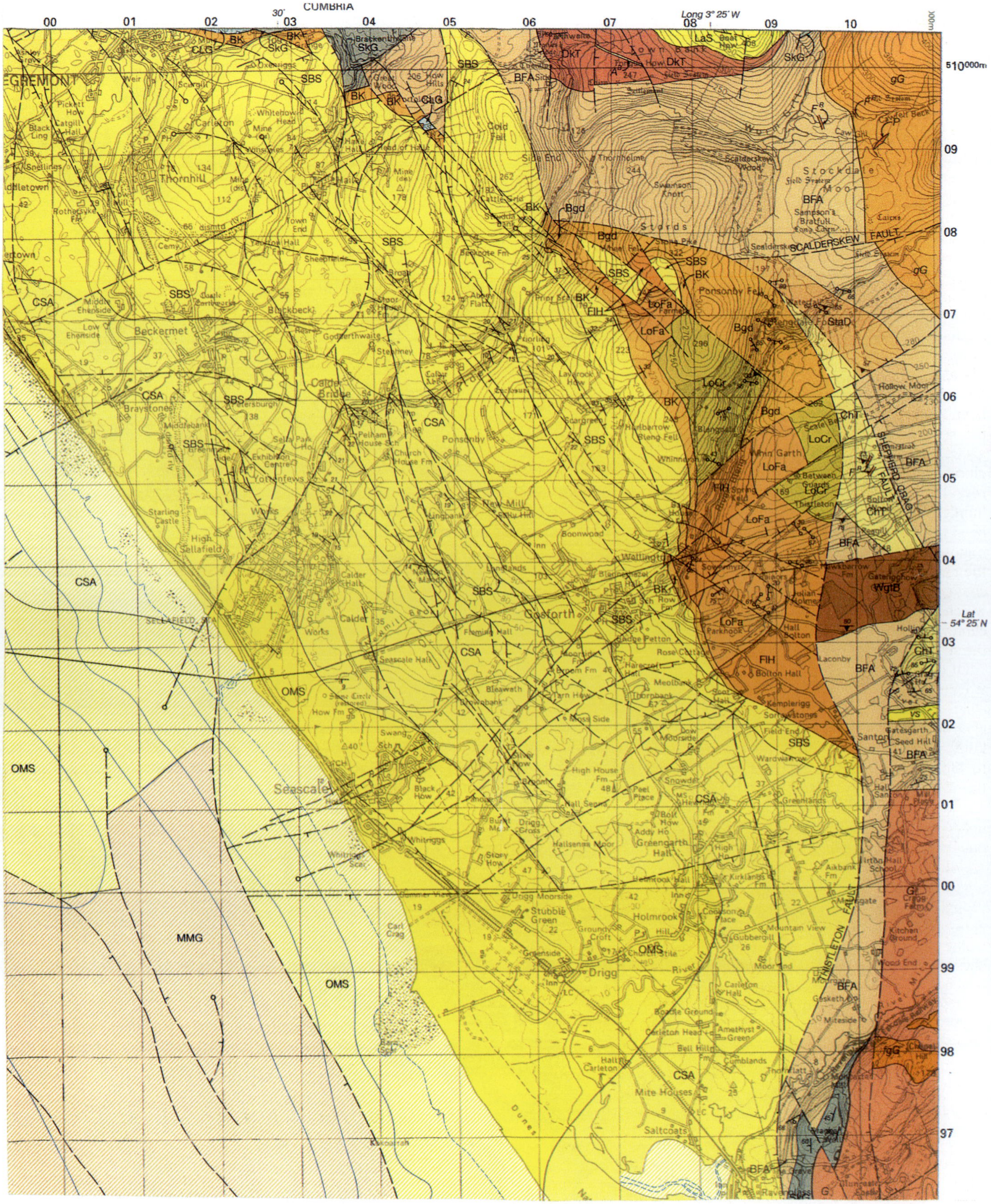

Plate VIII. Portion of: British Geological Survey, 1: 50,000 Series, England and Wales, Sheet 37. Solid geology. *Gosforth* (1999). British Geological Survey, Keyworth. IPR/23–7C British Geological Survey. The Ennerdale intrusion (orange: gG); the Birker Fell Andesites (copper beech: BFA); the ignimbrite (etc.) complex (Blengdale Formation, orange–brown: Bgd; Lowcray Formation, green–brown: LoCr; Longlands Farm Member, light terracotta: LoFa; Fleming Hall Formation, dark terracotta: FlH); and the Permo-Triassic cover to the west (yellows: OMS [Ormskirk Sandstone]; CSA [Calder Sandstone]; SBS [St Bees Sandstone]; BK [Brockram]).

Introduction

Many people – and I am one of them – think that the Lake District is the most beautiful and finest part of England. That is why so many people go there every year. Moreover, although the region is surely a place of great geological interest, its very beauty may well account for the fact that so many geologists have chosen to work there, selecting it for their special area of research. For surely it is a region that one can *love*, as much as study. I candidly acknowledge that I chose to carry out my investigations for the present book because of my devotion to the region. I lived there when I was a small boy during World War II, and always wanted to return, to get to know the place better. This prospect was so attractive that I made it my first and major preoccupation upon my retirement in 1996.

Uncharitable critics may think that my task has been a case of self-indulgence. They may, I admit, be a little bit right. Even so, I think that my occupation has been honourable, and justifiable as a contribution to the history of science. For I like to think that I have opened a new doorway in the study of the history of geology. My aim has been to show how, in all their complexity and detail, the current ideas about the geology of a particular region have been arrived at, right to the 'present' – in this case to the end of the second millennium. A similar task has not, so far as I am aware, been attempted hitherto for any specific region of the globe. And, by concentrating on my chosen region, I wish to reveal much about the history of geology in Britain in the last two hundred years – the time during which geological research has been pursued in a scientific manner in the Lake District. Thus the region provides a kind of lens through which one may examine the history of geology itself, revealing the changing styles of work, changing social relationships, the changing social environment in which geology operates, changing techniques of research, and the changing theories that have been devised in order to understand the structure of the Earth, its 'workings' and its fascinating history.

Of course, such a study must necessarily have a ragged ending. It cannot be one in which all details of the plot are finally made clear, and the story or play achieves its happy conclusion. It is not even like the historical study of a scientific controversy, where we know what the final consensus is; in which case our task is simply to show how that consensus was accomplished. On the contrary, the study of Lakeland geology is anything but complete, and if we are to judge by the continuing flood of publications on the area, not to mention changing notions about the nature of science that one may derive from the literature on the philosophy and sociology of science, or possible future major changes in geological theory, it never will be complete. On the other hand, though the story seems to get ever more complicated as we approach the present, rather than the plot being neatly concluded, it does have a kind of denouement, as we shall see. Thus I have chosen to bring my narrative to a close at the end of the year 2000, which provides a natural ending of a sort.

Of course, for those who want a finished story, my task is inherently unsatisfactory and incapable of adequate accomplishment. Nevertheless, I believe it has been worth the undertaking. First, it allows one to take stock of the present situation, to see how we have reached our present position in the understanding of the geology of this interesting and beloved region. Second, as said, it provides one way, at least, of writing the history of British geology, and more generally of the development of at least some aspects of geological method and theory. Third, it brings into focus the changes that have occurred over the years in the conduct of geology, from being an agreeable amateur avocation to that of a university discipline or national survey, or the domain of the 'consultant', where the cuts and thrusts of politics are at work as people have to battle for funding in a 'corporatized' world. Fourth, it explores aspects of the history of the British Geological Survey (BGS) and the relationship of that organization to the university system, and the different 'interests' of the two. Fifth, it examines a fascinating episode in the recent history of British science, which was overtly political, and in which one can see remarkable contrasts between the conduct of science in a 'courtroom situation' as opposed to within the walls of academe or amongst the usual social community of geologists, whether they be academics or in the 'private sector'. And sixth, it provides an entrée to the study of many interesting and important persons in the community of British geologists, most of whom are but little known except by a small number of specialist historians, or by geologists themselves, alive and walking today.

On the other hand, I do not offer (for example) a gratifying narrative of the relationship between geological science and painting or literature. My story offers nothing for the aesthete, such as the relationship between Goethe's philosophy and his ideas about the Earth, or the relationship between Ruskin's or Wordsworth's ideas about the Earth and the Lakes and their artistic or poetic accomplishments (though when I started this study I thought there might be more to say on such matters than proved to be the case). I offer a much more prosaic product.

One does not even find a roll of celebrated names in the history of geological science in this book. Almost the only person with a leading role whose name is remotely a 'household word' is the parson–geologist and nineteenth-century Cambridge professor, Adam Sedgwick – assuredly one of the 'giants of geology'.[1] A few other geological 'greats' such as William Buckland and Charles Lapworth make brief appearances on the stage. But for the most part I deal with geologists of no wide renown outside their professional circles.

I make no apology for this. Rather, it may be a source of strength in that so many otherwise little-known persons are revealed as playing or having played such important parts in the open-ended drama. It is high time that they be brought forward and presented to the world – for there are many fascinating and distinguished men and women in the cast of hundreds, if not thousands (though necessarily the majority of players have silent roles only, even in this quite detailed story). In the modern world, the geological enterprise is not carried forward chiefly by 'giants'. So, raise the curtain and let the performance begin!

[1] Cf. Carol & Mildred Fenton, *Giants of Geology* (1945).

Chapter 1
In the beginning

First we must put some boundaries around our topic, to contain it as far as possible in space and time. For space, I consider largely the areas shown in Figures 1.1 and 1.2. Looking at the geological map, we see a central region, running from Cockermouth in the NW to Kendal in the SE, which is divided into three main geological regions. This forms the main area of mountains, the central part of the threefold division being the most mountainous. (The southern section is hilly rather than mountainous.) To the west, there are large masses of pink crystalline rock, near Ennerdale and Wastwater, and somewhat similar rock is found on the eastern side at Shap. Dark crystalline rock is found at the northeastern edge of the mountains at Caldbeck Fell, and also round the northern edge of the hills. There is a 'frame' of sandstones, coals and limestones right round the mountain region.

In the present study, I shall be largely concerned with the ancient rocks of the mountainous region of the Lakes, and will not have too much to say about those of the surrounding 'frame'. We must, however, give attention to the rocks near Cross Fell, to the east of Penrith, or to the west of the Pennines, for some of them are similar in kind to those of the main body of the Lake District, and their study has always been intertwined with those of the Lakes proper. In fact, the very first 'scientific' account of Lakeland rocks was written by a local parson who lived in the Cross Fell area. Reference will also be made to 'Lakeland-type' rocks to the south, in the Howgill Fells, and yet further south, rocks in the 'Lakeland-type' 'Craven Inliers' in NW Yorkshire.

As to time, my account deals principally with the two hundred years from 1800 to 2000 (and I am typing the words of the present sentence on Christmas Day, 2000). However, humans have been interested in Lakeland rocks from time immemorial. There is, for example, an impressive prehistoric stone circle in St John's Vale, near Keswick, which shows that the people of that period were at least capable of selecting and cutting stone for their own mysterious purposes, whatever they may have been. There is also a gully near Harrison Pike, Langdale, high in the Central Fells, where a hard 'hornstone' is exposed. It takes an edge like a knife and is known to have been used extensively, perhaps some 5000 years ago, and 'exported' to many parts of Britain (though there has been controversy amongst experts as to whether the stone was carried as boulders by ice and then worked, or carried by human hands after working). There are also many indications of the Roman occupation of the Lake District, and a system of roads was constructed, one 'crazy' one going right along the top of a mountain range – High Street – providing what must have been a miserable, albeit direct, route for travelling between Penrith and Ambleside. More relevant to our purposes, it is believed that the Romans conducted mining operations in the Lake District. And there was mining activity of some importance, chiefly for copper and lead, in various parts of the region over the centuries, long before geology became an established science at the beginning of the nineteenth century. For example, there was quite a large German mining community in Keswick in Elizabethan times. Copper was worked near Coniston, and various metals were obtained from the rocks near Caldbeck. Coal too has been worked for many years around the margins of the Lakes, especially near Maryport, Workington, Whitehaven and Barrow, along the coast of the Irish Sea. There are also large slate quarries and mines, some still in operation. However, there is already a large literature on the history of Lakeland mining (e.g. Postlethwaite 1913; Adams 1995), and I do not propose to add to it in the present study.

In the beginning, then, was Reverend Thomas Robinson, who from 1672 to 1719 was Rector of the little church of St Luke, at Ousby, at the foot of the scarp of the northern Pennines near Cross Fell (see Fig. 1.3). From the manor house/farm/rectory, where Robinson probably resided, one can look westwards over the valley of the Eden, where Penrith stands, and see the fine outline of the hills of the Lakes, with High Street clearly visible. Robinson's parish is in a remote spot today, and it must have been much more so at the beginning of the eighteenth century, but he was evidently an educated man and interested in the natural history of the region. In 1709, Robinson published a treatise, lengthily entitled in the fashion of the day, *An Essay Towards a Natural History of Westmorland and Cumberland, Wherein an Account is Given of their Several Productions, with some Directions how to Discover Minerals by the External and Adjacent Strata and Upper Covers, &c. To which is Annexed, a Vindication of the Philosophical and Theological Paraphrase of the Mosaick System of the Creation, &c.*

Robinson's title, despite its detail, is something of a misnomer, for the author's attention was largely given to the strata in his immediate neighbourhood, near Cross Fell, rather than the Lake District proper. He described (or named and numbered) the separate 'elevations' of the different local strata, and mentioned that there were different kinds of rocks in the valley below where he lived, and in the 'Western Mountains', which 'gradually ascend to Skidday, or Skyday' (Robinson 1709, p. 57). According to his Preface, Robinson had 'been now Thirty Years concerned in the Inspection of Under-ground Projects of several Kinds and Natures'. Thus he was chiefly interested in the small mining operations that were conducted in his area, for coal and metals. His book is interesting to us, however, in that it had a theory of strata – as the title announced. It was that 'after [the] first division of water from land, the Almighty caused particles to juxtapose by a 'secret magnetism' so as to produce ores, *etc.*' (Robinson 1709, preface). This 'hypothesis' was, of course, a speculation, and one that complied with the usual physico-theological efforts of the day to weave together the ideas of science and those of revealed religion. Nevertheless, it was a general idea, applied to the Lakeland strata; and so far as I am aware it was the only one specifically applied to the Lake District before the advent of scientific geology, in the nineteenth century.

Thus, Robinson's book stands alone. For the rest of the eighteenth century, travellers and naturalists came to the Lakes in increasing numbers; topographic maps were made; and numerous quite elaborate topographic descriptions were composed, along with illustrations of the scenery (see Bicknell 1990); and there were artistic representations of the scenery aplenty (Victoria and Albert Museum 1984). However, there was little or no scientific description of the rocks, even if they were known all too intimately as a result of the arduous efforts of the miners and quarrymen, and the men who built the fell walls.

There is, however, one notable exception to this statement. The great Scottish geologist, James Hutton (1726–1797), accompanied by his friend John Clerk of Eldin, visited the Lakes in 1788 on his way home to Edinburgh from the Isle of Man (Hutton 1795, vol. 1, p. 330) ('man' is a Cumbrian word for a cairn marking a hilltop or for a prominent point of a mountain[1]). They found a limestone quarry east of Windermere (near a place called Low-wood Inn) and some included fossils, which they thought were 'entrochi' or crinoid remains. The find was important to Hutton, as he recorded, for the 'schistus' (slate) of such areas as the Lakes was

[1] Nigel Woodcock has informed me that, so far as the Isle of Man is concerned, the name derives from Manannan Mac Lir (Manannan, Son of the Sea), a Celtic Sea God.

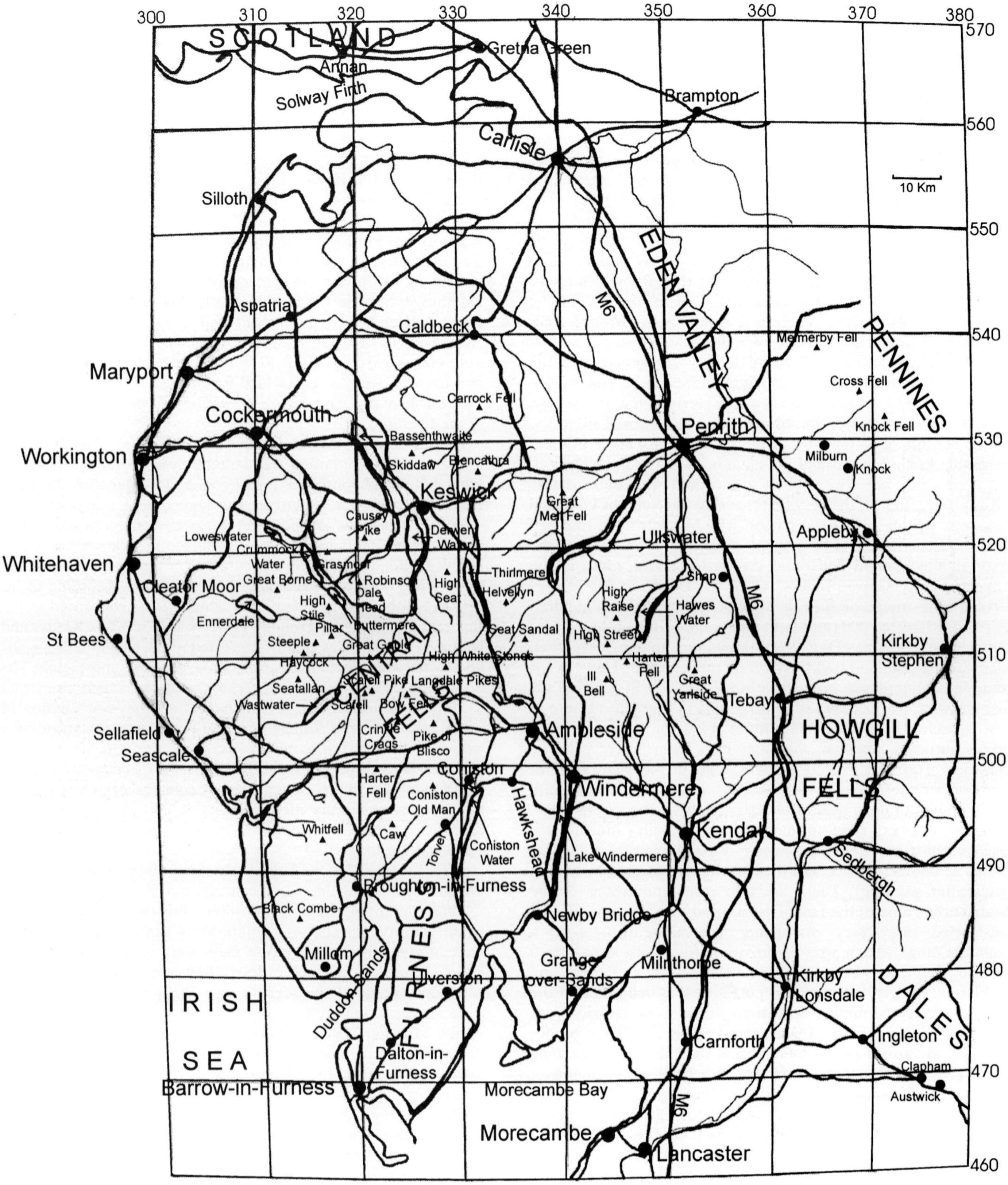

Fig. 1.1. Sketch-map of the geography of the Lake District.

thought by many of his contemporaries to be a 'Primitive' rock – providing a vestige of the Earth's crust as it had been first formed. Such rocks were thought to be devoid of organic remains, according to the theories of the German 'geognost' Abraham Werner, whose ideas, influential at the time, were contested by Hutton. The discovery of fossil remains (to which were added shells provided by Hutton's landlord) showed, in Hutton's view, that the Cumbrian 'schistus' was *not* 'Primitive'. (This meshed with his general theory that *no* rocks observable today were ones dating back to the Earth's first formation.)

In 1791, Hutton's geological friend and biographer, the Edinburgh mathematician John Playfair (1748–1819), visited the Lakes and also searched for fossils near Windermere. In a letter to Hutton, published in his *Theory of the Earth* (Hutton 1795, vol. 1, pp. 332–334), Playfair described how, with the assistance of local guides, he had found fossils in the limestone to the west of

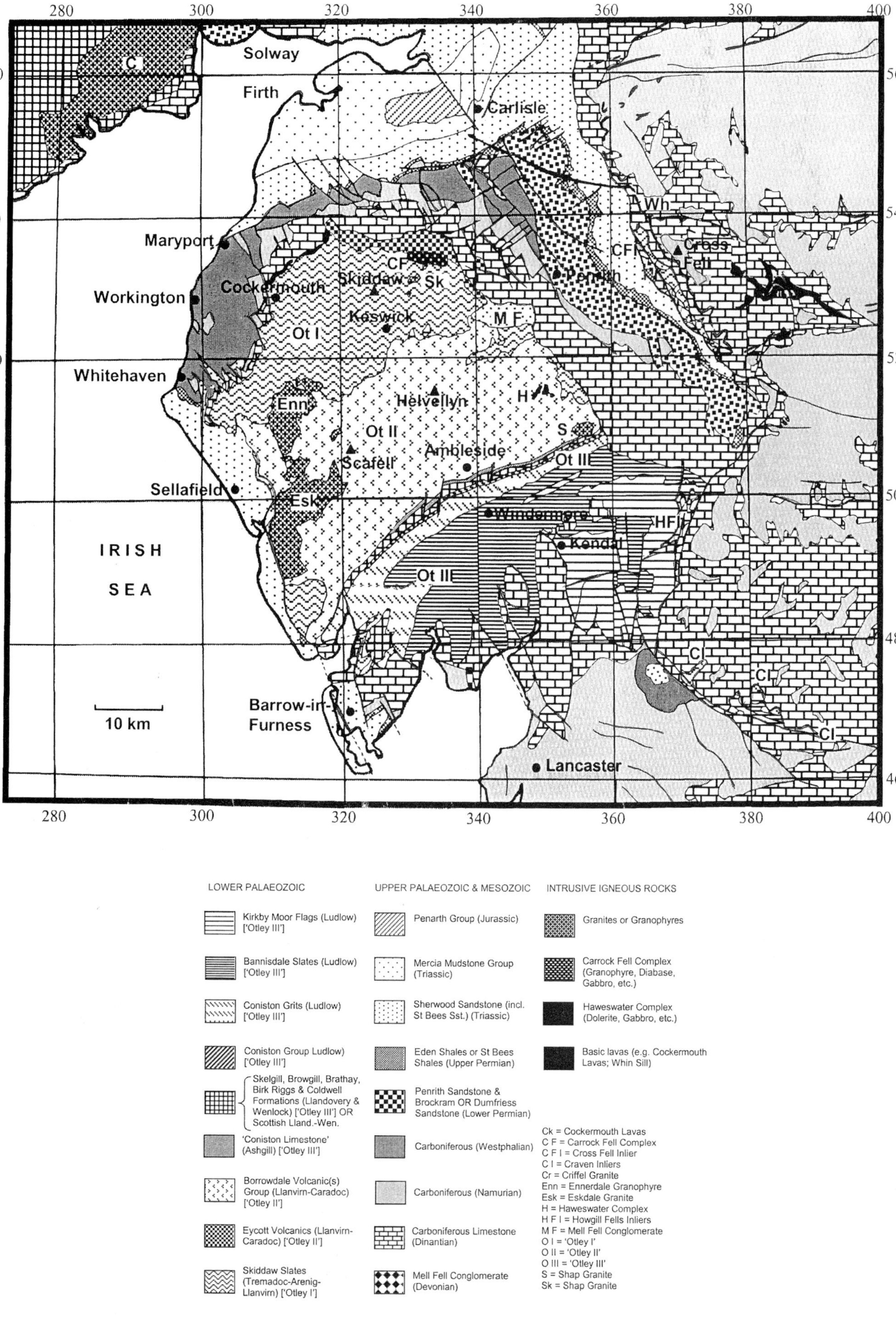

Fig. 1.2. Sketch-map of the geology of the Lake District, based on the British Geological Survey map: *Lake District 54N 04W (S) Solid Geology. Scale 1: 250 000*, 1982. IPR/25-3C British Geological Survey. © NERC. All rights reserved.

Fig. 1.3. Topography of the area to the west of Cross Fell. Waterways are depicted by thin lines, and roadways (not all of equal importance) are shown by thick lines.

Windermere, between Ambleside and Coniston, and had also found some bivalve shells in the associated 'schistus'. He mentally connected exposures of limestone to the east and west of the lake, envisaging the existence of a continuous stratum. Subsequently, Playfair (1802, p. 164) stated that he had extended his observations to the head of Coniston Water, finding additional organic remains. The fossil evidence indicated that although the rocks were early ('Primary' strata), they were not primaeval ('Primitive').

A slight indication of geological interest in the Lake District is to be found in a book, entitled *A Descriptive Tour, and Guide to the Lakes,* published in 1800 by one John Housman (1764–1802), son of the gardener of Corby Castle, near Carlisle (Housman 1800). The author was concerned with the soils of NW England, and the book contained a soil map that contains just a hint of a tripartite division of the Lakes. This was not a scientific work, but slate and a little limestone were noted near Ambleside; also the existence of lead and copper ore, and 'black lead' ('wad': plumbago or graphite) elsewhere in the Lakes. The Lakeland soil was said to be mostly dry gravel, and there is little indication that Housman had much interest in the rocks lying below the surface detritus. However, a letter dated 17 March 1823, viewed at the library of the Wordsworth Trust, Grasmere in 2001, from John Farey (see p. 7) to Jonathan Otley (see p. 8), refers to some notes made by Housman sent by Otley to Farey. The letter implies that Houseman had a significant amount of geological knowledge.

Much more geological in character than Housman's book was a *Treatise of a Section of the Strata . . .* (1809), by the mineralogist and mining surveyor Westgarth Forster (1772–1835). His section ran from Newcastle on the east coast through to Cross Fell; but stopped short before it reached the Lakes. Forster gave tables of strata for Yorkshire and Derbyshire, and displayed extensive knowledge of the rocks adjacent to our region. Surely, had he attended to the Lakeland area, he would have provided the first geological account of the district. However, he cannot be credited with this distinction – regrettably, given that his section was said to be intended to 'amuse the mineralogist, and assist the miner in his professional researches' (subtitle of the *Treatise*).

While Forster's literary work stopped short before the Lakes, another practical man, Joseph Harrison Fryer (1777–1855), surveyor, geologist and mining engineer, came over from Newcastle, spending part of each year from 1808 in the Keswick area. He eventually settled there, marrying a lady from a well-to-do family, successively residing in quite large houses at Lyzzick Hall, Ormathwaite Hall and Braithwaite Lodge. In 1826, Fryer went off to South America to develop mining interests in the southern continent. These failed, however, and he returned two years later and settled in the Newcastle district, where his scientific interests were renewed (Torrens 1994). We need not, however, follow the latter part of his career.

Fryer was son of John Fryer, a Newcastle schoolmaster, surveyor and mathematician, who played an active part in the Industrial Revolution in NE England. His mother too came from a family with scientific and technical interests. Joseph Harrison Fryer was active in the work of the Newcastle Literary and Philosophical Society, where he was friend and correspondent of Nathaniel Winch (1768–1838), the leading northeastern naturalist of the day and author of an important early paper on the geology of Northumberland and Durham (Winch 1817). So far as I am aware, Fryer was the first person to do serious geological work in the Lakes.

As a resident of Keswick, with scientific and technical interests, Fryer was sought out by several geologists who visited the Lakes in the early years of the nineteenth century. As we shall see, he was visited by Adam Sedgwick in the 1820s. However, before then he had given assistance and information to men like George Bellas Greenough (1778–1855), founder and first President of the Geological Society of London, to whom he reported in 1814 that he had prepared geological maps of Durham, Northumberland and Cumberland. He also guided Greenough and the redoubtable Oxford Reader in Mineralogy, the Reverend William Buckland (1784–1856), around the Cross Fell area in 1814 (Torrens 1994, pp. 32–33). The geological consultant and author, Robert Bakewell (1768–1831), was likewise assisted, and his geological section across northern England from the Irish to the North Sea (Bakewell 1815) was claimed to be plagiarized from Fryer, though Bakewell denied the charge.

The mention of a geologically coloured map of Cumberland by Fryer is particularly interesting. In fact, it is probably a map held at the Geological Society, coloured onto a topographic map of the county, published by the cartographer John Cary in 1793. Of course, the date at which the colouring was undertaken cannot be determined, but given that Fryer sent such a map to Greenough, and given that there is at the Geological Society a map that would fit the bill, this could well be it (see Plate I). Fryer, in 1814, might well have had to use a 1793 map, *faute de mieux*.[2]

Already in this early map, the granite exposed over a small area in the valley to the east of Skiddaw and again in a river bed near Caldbeck (which, though small in outcrop, had presumably been known to many generations of Lakelanders) was indicated, and round it was represented an area of 'clay slate'. The rest of the slates of the northern Skiddaw region of the Lakes were represented as 'Grauwacke'. The granite of Eskdale, etc. (omitted in the early maps of William Smith; see p. 11) was called a 'syenite', in keeping with the usage of that period, but there seems to be no clear distinction between the northern 'Skiddaw Slates' and the central 'Borrowdale Volcanics' (to use anachronistic terminology). The 'framing' sandstones, coal and limestone are all marked in unambiguously. From the map, it would appear that Fryer – assuming he was the colourist – was not very familiar with the rocks of SW Lakeland.

There is an anonymous article, entitled 'A Geological Sketch of a Part of Cumberland and Westmorland' in the *Philosophical Magazine* (Anon. 1816), which Hugh Torrens (1994) has shown, by comparison with the original manuscript held in Newcastle, to have been authored by Fryer. It likewise offers no subdivision of the rocks of the main ranges of the Lakes, but states that the author had observed granite at Skiddaw 'about twelve years ago' (1804?). Most attention was given to the 'grau-wacke slate' forming 'small conical hills' immediately to the west of the limestones, etc., of the Cross Fell area. The anonymous writer referred to a short paper on the area, published by Buckland in *Annals of Philosophy* (1815). This was a prolegomenon to his full paper on the area published in the *Transactions of the Geological Society* (Buckland 1817), which will be discussed shortly. As already mentioned, Fryer accompanied Buckland and Greenough to the Cross Fell district in 1814.

Interestingly, Fryer stated that 'the counties of Cumberland and Westmorland ha[d] been frequently visited by very eminent mineralogists' (Anon. 1816, p. 41). This may have been something of an exaggeration, but one can think of at least Greenough, the great map-maker and surveyor William Smith (1769–1839), his polymathic friend and co-worker John Farey (1766–1826),[3] Winch, Buckland, and probably the Manchester chemist John Dalton, as candidates to fit this description. Smith and/or Farey would presumably have visited the Lakes in the course of their work for Smith's great geological map of England (1815), but the Smith archives at Oxford give no confirmatory information on this point. In Book 3 of his *Excursion* (1814), William Wordsworth (1994, pp. 788–789) referred to 'He who with pocket-hammer smites the edge/Of luckless rock or prominent stone,/. . . detaching by the stroke/A chip or splinter – to resolve his doubts;/And, with that ready answer satisfied,/The substance classes by some barbarous name,/And hurries on; . . .'. The poet, it seems, did not quite approve of the activities of the 'mineralists'. As I say, we cannot be sure who Fryer (or Wordsworth) was referring to, but certainly there is corroborative evidence that the beginnings of geological investigation were occurring in the Lakes by at least the second decade of the nineteenth century.

The first major publication on 'Lakeland-type' rocks was, then, that of Buckland, in a paper read before the Geological Society on 28 March 1815 (Buckland 1817). So it was that the first substantial study of Lakeland geology dealt with an 'inlier' of ancient rocks that lies apart from the Lake District proper – rocks 'squeezed' between the sandstones in the valley of the Eden (which runs northwards towards the Solway Firth) and the limestones of the northern Pennines. Buckland stated that he visited the Cross Fell region in September 1814, in the company of Greenough, but he failed to mention that he had Fryer as a guide – an interesting instance of the almost systematic exclusion of the work of the practical men such as Fryer from the printed public record.

Buckland's paper was furnished with a map and sections. He noted the occurrence of distinct exposures, running N–S, of 'trap and greenstone' and 'grauwacke slate' between the limestone of

[2] The Geological Society's catalogue has a note about the map saying 'Cumberland geolog. report by Mr. Fryer'. These words were written on the back of the map and can just be seen through the linen on which it was mounted. It was, however, attributed to Greenough in Sheppard's list of maps (TS 191). On the other hand, in the Society's 1860 printed catalogue it does not have an asterisk against it to indicate that it had been 'geologically coloured, more or less, by Mr. Greenough'. I thank Wendy Cawthorne (pers. comm., 2002) for this information.

[3] On Farey, see Ford & Torrens (1989). Doubtless there were others who left no record of their work. In any case, a distinction needs to be made between 'mineralogists' and 'geologists'.

Fig. 1.4. Jonathan Otley. Photograph of portrait in the Keswick Museum. Reproduced from negative held at Kendal Public Library (920 OTLEY 003937) by courtesy of Kendal Public Library.

the Pennines and the sandstones of the Eden Valley, the 'trap and greenstone' being to the west of the 'grauwacke'. He also recorded the appearance of 'shattered and vertical coal, grit, and limestone' to the west of the 'trap and greenstone'. This suggested that the contacts were faulted, but Buckland did not come out clearly with the notion that the 'inlier', as we now call it, had been faulted into its present position. He followed the views of Winch and Forster that a band of 'Old Red Sandstone' could be found running N–S along the base of the Pennine limestones (the 'Mountain Limestone'). This would appear to have been what are today regarded as 'Basement Beds' to the Carboniferous, consisting of unfossiliferous fluviatile sandstones and quartz conglomerates (Burgess & Wadge 1974, pp. 40–46). Buckland's interpretation of the gypsum-bearing sandstones of the Eden Valley as New Red Sandstone has found subsequent favour. The 'trap and greenstone' and the 'grauwacke slate' corresponded to two of the three Lakeland units referred to above, soon to be recognized in the Lake District proper by Jonathan Otley (see below). Today we call them the 'Borrowdale Volcanics' and the 'Skiddaw Slates' respectively.

So far as William Smith is concerned, one must suppose that he visited the Lakes, albeit briefly, before the publication of his first geological map of England and Wales (Smith 1815); yet it is possible that he used second-hand information for at least the first issue of the map. As was shown long ago (Eyles & Eyles 1938), there are various versions of the famous map, and the surviving copies can be arranged in approximate chronological order by looking for their gradual elaboration of geological detail and for the presence or absence of certain numbers and letters. Smith had rather limited knowledge of the lower parts of the British stratigraphic column, and the rocks of the Lake District were, in 1815, merely designated 'killas', the term used by the Cornish mining fraternity to designate schist or clay-slate, sometimes containing metalliferous veins. The exposure of Skiddaw Granite near Caldbeck was marked on Smith's map, as was a narrow band of limestone extending across the Lakes from the Duddon Sands in the SW to the Troutbeck area, east of Windermere, previously noted at some parts of its outcrop by Hutton and Playfair. A somewhat later version of Smith's map included granites at Eskdale and Buttermere, but the outline of the limestone was in a way less satisfactory than the earlier version, for it was now represented as a smooth continuous band, whereas the 1815 map showed the outcrop as discontinuous, which is truer to the way things actually are in the field.

As said, I am uncertain how much first-hand knowledge Smith had of the Lakes in 1815. His surviving archives, held at Oxford, give no indication that he went to the area before 1821. That year, he was in the Ulverston region on the Cumberland coast, looking at copper mines on behalf of a mining entrepreneur and landowner, Colonel T. R. G. Braddyll.[4] According to his nephew and assistant, John Phillips (1800–1874) (1836, p. xi), the two were in the same district in 1822, and also at Hesket Newmarket to the NE of the Lakes. Moreover, they spent some considerable time exploring the area around Kirkby Lonsdale. In 1823, Smith was again in the north, doing survey work for the Carrock Fell mining region (see Fig. 12.1). For this purpose, he prepared a topographical map of this important mining area, showing the locations of mine shafts and smelting mills, but there was no indication of geological information as such.[5] A geological map was produced for the Bewcastle region immediately to the east of the Lakes,[6] but this need not concern us here.

As is well known, Smith's 1815 map was soon followed by that of George Greenough and his co-workers at the Geological Society (1820), at a scale of six inches to the mile. The main point of interest to us is whether or not Greenough recognized the important distinction between the rocks of the northern Lakes (Skiddaw Slates) and the central Lakes (Borrowdale unit). He did not. All were mapped as 'Talc slate, Chlorit slate, Hone [whetstone], Clay slate, Porphyry'. A unit to the west of Derwent Water was mapped in a separate colour (green) as 'Sca Fell Cumberland: Compact feldspar or Hornstone. Hornstone porphyry. D° with Chlorit', but it is by no means evident what rocks Greenough had in mind for this.

An outcrop of limestone, with argillaceous slate, was shown on the Greenough map running from the SW to the NE across the main mass of slates, as Smith had it in his map. To the south of this, down to Morecambe Bay, a region was tinted yellow, but was un-keyed. Presumably, it was what Greenough called 'Argillaceous slate, Greywacke, Culm'. He also had sandstone mapped to the NW of Ullswater (corresponding to what is today regarded as an outcrop of 'Old Red' conglomerate at Mell Fell), and the general Lakeland 'frame' of Carboniferous Limestone was clearly indicated. On the whole, the Greenough map made little advance on that of Smith so far as the Lake District was concerned.

We may now consider the important work of Jonathan Otley (1766–1856) (see Fig. 1.4), often called the 'father of Lakeland geology'.[7] He came from a working family at Loughrigg, at the lower end of Langdale, not far from Ambleside. Though

[4] See Smith Archives, University Museum, Oxford, Box 8, Folder 4.

[5] See Smith Archives, University Museum, Oxford, Box 16, Folder 1(5).

[6] See Smith Archives, University Museum, Oxford, Box 16, Folder 1(3).

[7] On Otley, see Ward (1877), Rownsley (1894, pp. 123–135), Challinor (1948), Shackleton (1963), Smith (1998–1999 [2000]), Smith (2000), Lietch (n.d.), Wilson (n.d.). Some papers of Otley were recently sold at auction. They are currently in private hands but at the time of writing were located at the library of the Wordsworth Trust at Grasmere.

apparently well educated by the standards of his time and place, with study of both mathematics and Latin, Otley was early employed as a menial basket maker, watch repairer and engraver. In 1791, he moved to Keswick where he established a modest business as watch and clock repairer and maker. Remaining a bachelor, he interested himself in natural history, particularly geology and meteorology, and did topographic survey work in the district. He supplemented his income by working as a guide, and by 1817 had engraved a topographical map of the Lakes, which he published for tourist use in 1818. Work of this kind was enlarged with the publication of Otley's famous *Guide* (1823),[8] which contained an important chapter on the geology of the Lake District, being based on a brief paper that he published in 1820 in the short-lived Kendal journal *The Lonsdale Magazine*.[9]

Through his work as a guide, Otley became intimately familiar with the rocks of the Lake District, and was soon recognized as the local authority on the geology of the region. Eminent men of science such as the geologists Adam Sedgwick and John Phillips (of whom more anon), the Manchester chemists John Dalton and William Henry, and the Astronomer Royal, George Airy, became his friends and correspondents, and Dalton and Sedgwick at least accompanied Otley on field trips, being glad to have him as a well-informed guide.

Otley lived to a great age, but in 1853, when he was 87 years old, he auctioned his books, tools, and scientific instruments, leaving himself a profit of a mere £20. In 1877, the geologist Clifton Ward (see p. 41) published some letters of Otley and a brief memoir of his life. At that time, the Otley letters and papers were in the hands of a nephew, J. Otley Atkinson of Kendal. Their present whereabouts are not known to me (but see Note 7), only a few items being preserved at the Keswick Museum, along with some of the old guide's surveying instruments; and some papers are preserved at the County Archives in Kendal. Unfortunately, these give little information as to how Otley arrived at his important scientific ideas, or precisely where he went or what he did or thought. We thus have to rely chiefly on his small number of publications. It was recorded by one who met him that he was 'an interesting specimen of native genius'; '[a] modest, sensible man, with a head that would make a fine study for the Phrenologist' (Bryant & Baker 1934, p. 81).[10]

Otley made two major theoretical contributions, which may or may not have originated in his own mind. He distinguished between bedding (or 'stripe'), cleavage (or 'bate'), and jointing – which may easily be confused in some Lakeland rocks. He also subdivided the Lakeland rocks into three main units, which I shall refer to from time to time as Otley I, II, and III. In 1820, this was done according to the geographical distribution of the rocks and their physical appearance. Later, in the *Guide* (1823, and later editions), the units were called 'Clay-slates', 'Greenstone' or (following Sedgwick 1826–1833 [1832], p. 400) 'Green slates and Porphyries', and 'Greywacke'. These corresponded with what later came to be called the Skiddaw Group or Skiddaw Slates, the Borrowdale Volcanics Group, and the Windermere Supergroup. Otley also recognized the existence of a mass of granite underlying and apparently altering the 'clay-slates'.

Otley was not well versed in palaeontology, as he himself admitted, writing in a letter to Sedgwick in 1847: 'you know I am very ignorant on the subject of fossils' (Ward 1877, p. 156). So Otley's subdivision of the Lakeland hills was made according to their general appearance and to the lithologies of the constituent rocks. In fact, the three divisions are quite evident in the Lakes, once one thinks to look for such distinctions. The 'clay-slates' form, as Otley said, the northern mountains of Skiddaw, Saddleback (also called Blencathra), Grisedale Pike and Grasmoor, and cut across Crummock Water to the lower part of Lake Ennerdale.[11] The mountain peaks are generally smooth and dark, with poor vegetation. The rocks are, in Otley's words, 'of a dark colour, inclining to black, and generally of a slaty structure', being 'shivered [fragmented] by the weather in[to] thin flakes' (Otley 1820, p. 433). They had lead and copper workings in some places.

The succeeding unit, which like the preceding one appeared to be unfossiliferous, consisted of varied slaty rocks, blue-grey or greenish, with various porphyries, 'sienites' and granites apparently belonging to the unit. They made up the highest mountains of the central Lakes, such as Coniston Old Man, Scafell and Helvellyn, and the ranges around Eskdale, Wasdale, Borrowdale, Grasmere, Patterdale, Martindale, Mardale and so on (see Fig. 1.1). As Otley subsequently wrote, they form the 'fine towering crags' and 'bold colossal features' of the Lakeland mountains, and 'most of the cascades of the Lakes fall over it' (Otley 1843, p. 146). One might add that they created the most wild and beautiful topography of the Lakes, with surfaces of a characteristic 'knobbly' appearance. These rocks sometimes effervesced weakly with acid; and they contained mineral veins in some abundance, particularly the copper deposits of the Coniston region. They also yielded the famous green slates used for buildings in many parts of the Lakes.

Otley's third division formed the southern part of the Lakes. It started with a bed of 'dark blue limestone', which ran right across the region, from the Duddon Estuary in the SW, past the lower slopes of the Old Man of Coniston, across the head of Lake Windermere, and across the valleys of Troutbeck and Kentmere (see Fig. 3.6). (Later, he extended mention of the outcrop to Long Sleddale (Longsleddale), the next valley beyond Kentmere, to the east.) Then came a series of dark flags and slates, quarried particularly at Brathay near Ambleside. Mention was made of the fossils to be found in the limestone, but no discussion followed. It was suggested that the limestone belonged to the 'transition' series 'of some geologists' (Otley 1820, p. 434).[12] Thus our watch repairer and guide evidently had some familiarity with the theoretical literature of geology, as well as local knowledge.

Otley recorded two other most interesting observations. First, he remarked on the granite to be found in some of the valleys in the northern region of the 'clay slate', as in the bed of the River Caldew, calling it the 'primary granite, or foundation rock' (Otley 1820, p. 433). He also recorded the occurrence of mica, 'dark spots' (of hornblende), and the white mineral 'chyasatolite', cruciform in cross-section, in the exposures of the clay-slate adjacent to the granite. The terminology of 'foundation rock',

[8] This book, of 130 pages, was published by Otley in Keswick and by A. Foster & J. Richardson in Kirkby Lonsdale and London respectively, and sold for 4/6d. It seemingly went through nine editions, the last appearing in 1857, though Bicknell (1990, p. 128) was apparently unable to locate a copy of the last edition.

[9] The paper also appeared in *The Philosophical Magazine* the same year.

[10] I thank Hugh Torrens for this reference.

[11] Here, and elsewhere, modern names will be used for features of Lakeland topography, unless otherwise stated. Otley wrote 'Grisdale' and 'Cromack'.

[12] Here Otley was referring to the theory of the eighteenth-century German geologist, A. G. Werner (1749–1817), who thought that the rocks of the Earth's crust were deposited successively from a hypothetical universal ocean. The first deposits (primary, so-called) were crystalline in character (as in granite). Then came a 'transition series' that consisted partly of crystalline and partly of mechanically derived material. Third, came the 'layered rocks', or materials of sedimentary (mechanical) origin. More detailed subdivisions were provided, and it was supposed that there was a general and universal order amongst the strata, even if not all units were universally present. The theory was deployed quite widely in Britain in the early years of the nineteenth century, particularly as a result of the efforts of Robert Jameson (1774–1854), Professor of Natural History at Edinburgh University. The relative merits of the Huttonian (Vulcanist–Plutonist) and Wernerian (Neptunist) theories were hotly debated round the turn of the nineteenth century, and particularly so in Edinburgh.

used to refer to the granite, suggests that Otley, in 1820, was thinking according to the theory of Werner (see note 12). Sedgwick followed him in this for a time. Today we ascribe the presence of the mica, hornblende and chiastolite to the action of a hot intrusion of granitic magma altering the sediments into which it was intruded. Clearly, this was not what Otley had in mind; but it would appear that he was familiar with the rocks that are today interpreted as a 'metamorphic aureole' around the Skiddaw Granite. They would, however, have long been known to Lakelanders, the outcrops being easy of access.

Otley's second major generalization was that the upper slates of his second unit were inclined to the north in Borrowdale, while they inclined to the south in Langdale – 'as though the mountain ridge dividing the counties of Cumberland and Westmorland had acted as a wedge in separating them' (Otley 1820, p. 434). Otley also noted the ferruginous conglomerate at Mell Fell, near Ullswater, on the NE side of the region, forming part of the frame of sedimentary rocks round the central slate region; and the further deposits of sandstone and limestone around the Lakes. The limestone, he wrote, 'does not conform to the direction of the strata beneath it in a regular order of succession' (Otley 1820, p. 435). So the unconformable relationship between the sedimentary 'frame' around the Lakes and the slates of the central mountains was evidently known and understood.

Subsequently, John Phillips (1856, p. 369) stated that it was Otley who first understood the distinction between bedding (stratification) and cleavage, a distinction that needed to be known and understood in map-making and in comprehending the geological structure of a region. Sedgwick seemingly recognized the distinction later, but independently, for he wrote to Otley in 1836 that in his first year in the Lakes (1822) he was 'puzzled to death at first by the appearance of three parallel systems of planes . . . joints, *cleavage planes*, and *beds*'; and he only found the bedding planes to be indicated by 'the stripe and other appearances not easy to describe' (Ward 1877, p. 150), i.e. by variations in colour and perhaps grain size in different sedimentary layers.

In fact, in the entry for 3 October 1822, during a period when he was in west Cumberland, Sedgwick's field notebook (No. 10, Sedgwick Museum, Cambridge) states that he observed a rock with 'a slaty bait [bate or cleavage] almost perpendicular to the plane of stratification'; and in his draft autobiographical notes, held at the Cambridge University Library (Sedgwick papers, Add. 7652. III H2, pp. 44–45), he wrote rather specifically:

> [I]n 1822 in Cumberland I learned . . . after vainly puzzling myself a long time with joints and cleavage planes, not then distinguished from bedding – I found out that the striped structure of the slate was the true key to stratification, and in that year 1822 I had learned to distinguish the slaty cleavage from the joints & bedding.

However, this was written in Sedgwick's old age when his memory was failing and he may have been mistaken as to the date, for he went on to say that 'this was published in the year 1822', which is, to my knowledge, incorrect. Sedgwick first met Otley in 1823. Whether Sedgwick learnt something from Otley on this matter, and whether Otley learnt it as just a practical distinction from quarrymen, is therefore not certain.

It may be mentioned further that there was some earlier published recognition of the distinction between bedding and cleavage before either Otley or Sedgwick, as for example in Robert Bakewell's *Introduction to Geology* (1813, p. 28), the third plate of which clearly figures a unit displaying both bedding and cleavage. Perhaps Otley gained his knowledge of the distinction from this source – or from quarrymen. In any case, he did not ascribe cleavage to the action of pressure (and neither did Bakewell). Sedgwick (1835*c*) published an extended discussion of the distinction between bedding, cleavage and jointing in a well-known paper presented to the Geological Society on 11 March 1835. William Smith knew about the distinction in 1821 and sketched a piece of slate in such a way as to show the difference (Phillips 1844, p. 99), but he was at that time in contact with Otley. In fact, the distinction *was* known to quarrymen in Britain, being recorded by Farey (1811, p. 155), who wrote concerning the rocks of Charnwood Forest in Leicestershire: 'it [the appearance of stratification] is the *stratula*, or Folia, or what the Masons call the beat [bate] or grain of the stone, which here determines its fracture into blocks or slates, and not the lamina of its stratification'. So the concept of foliation was present 'in embryo' there too. Phillips (1836, p. 2) wrote that '[t]he 'bate' may be looked upon as the most secret laminar structure of the stone, which is occasionally developed by weathering to an obvious degree'.

Regardless of the uncertainties on this point, one must surely agree that Otley's contributions were remarkable, as a first effort to unravel the geology of the district, both theoretically and empirically, through actual survey. Other guidebooks of the same decade, such as that of Parson & White (1829), described and subdivided the Lakeland rocks in much the same manner as Otley, though Otley's first unit (now called the Skiddaw Slates) was by then called the 'Transition slate formation' (Parson & White 1829, p. 84). Otley's work stood as the best published information on Lakeland geology for more than a decade.

Did Otley make a geological map of the Lake District? In a letter to Sedgwick dated 24 February 1847 (Ward 1877, p. 156), Otley stated that some ten years after he met John Farey in the Lakes he had been encouraged by him to prepare a coloured geological map of the region, using his published topographic map of 1818[13] as a base. It is known that Farey was in Borrowdale in 1818, preparing a report on the 'black wad' (black lead or graphite) mine there, and that he met Otley in the Lakes (Ford & Torrens 1989). Apparently, Otley did attempt a few colourations and sold copies to Greenough and Buckland for five shillings a copy (Ward 1877, p. 156), but the date of Otley's preparation of the maps cannot have been about 1828, as implied above, since Farey died in 1826. So the details of the date must have escaped Otley's memory.

In any case, I have not been able to locate either of the two early geological maps known to have been made by Otley. The Geological Society does possess a geological map of the Lakes, coloured onto a copy of the 1837 edition of Otley's topographic map.[14] Given that this post-dates the publication of Sedgwick's geological map of the Lakes (see p. 14), the Otley map is not such a remarkable accomplishment. However, it is of considerable interest, none the less. Its theory seems to go back to the early days, 1820–1823, for the outcrop of the 'transition limestone' is not shown extending northeastward beyond Longsleddale; yet Sedgwick showed in 1823 that it extended beyond Longsleddale to Wasdale Head near Shap (see Fig. 3.7), though he only published this observation belatedly (Sedgwick 1831*d*, p. 211). It is also remarkable that Otley did not have regard, in his post-1837 map, to the fact that the long outcrop of the 'transition limestone' was repeatedly ruptured (or displaced by faults), a point that Sedgwick had discussed in detail in his important Lakeland paper read to the Geological Society in 1831 (Sedgwick 1835*d*), which showed the discontinuities in outcrop in its accompanying map (see Plate II). One must conclude, then, that Otley did not keep up with the latest developments in Lakeland geology as he advanced in years. His post-1837 map did, however,

[13] Or perhaps the second edition of 1827.

[14] 'The District of the Lakes': Catalogue LDGSL 1003: Otley 1837. The map has the words 'Cumberland, Westmorland &c' pencilled on its reverse, possibly in the hand of Sedgwick.

bear a stratigraphic column, written in his impeccable surveyor's hand: [15]

12. Red Sandstone
11. Grey Sandstone
10. Coal
9. Mountain or Carboniferous Limestone
8. Old Red Sandstone. Conglomerate
7. Dark coloured Slate and other Rocks, Greywackes
6. Limestone interstratified with Slate
5. Greenstone, and Green or pale Blue Roofing Slate
4. Clay Slate, Skiddaw Slate, Chiastolite in places
3. Imperfect Gneiss, Mica Slate, Hornblende Slate
2. Reddish Granite, Sienite, and Porphyry
1. Grey Granite of Quartz, White Felspar, Black Mica.

The three main units of 1820 correspond to units 4, 5 and 6–7 respectively.

We do not know exactly how this map was compiled – whether it was done from general knowledge (for Otley did really know the Lakes like the back of his hand), by systematic mapping, or by drawing in part on the work of Sedgwick, Smith, Fryer, Farey, Greenough or whoever.

According to John Phillips (1800–1874), William Smith's biographer and an important geologist in his own right who eventually became Professor of Geology at Oxford, Smith made various traverses in and around the Lakes in 1821 (Phillips 1844, p. 98). He met with Otley in Keswick in 1821 and the two compared notes. Apparently they found themselves in general agreement (Phillips 1844, p. 98). Not only that, in September that year they coloured in a geological map of the Lakes, using a sheet map of John Cary as a base (Phillips 1844, p. 99). Unfortunately, I have not been able to locate this map.

It may be, however, that there was some initial rivalry or 'distance' between Smith and Otley. Ward (1877, p. 138) recorded an ambiguous extract from a letter from John Farey to Otley, dated 17 March 1823, now held at the Wordsworth Trust in Grasmere (see Note 7). Ward's transcription was not exactly correct. It should read:

> Mr. Smith's conduct, on your first meeting with him in summer 1821, affects me much more with shame and *regret*, than with surprise: had he met me there, *or any one* else, who he knew, *had laboured* at the "knotty parts," he was examining, he probably would, alike have shunned us.

Is this saying that Smith was not, on their first meeting, very friendly towards Otley – in fact shunning him? It does seem so. Farey went on to say that Smith (his 'untractable Friend') had for eighteen years from 1792 been in advance of others, but in the following years others had got '*far beyond him*' in the knowledge of '*particular districts*'. Otley would have been one such, and it would seem that, for whatever reason, Smith did not desire to meet him.

Of course, Smith was not long out of debtor's gaol in London, and had withdrawn to the north of England. The year 1821 was some time before he achieved the recognition that was his due. We can well understand why he might have been curmudgeonly and avoided Otley. Farey recommended that if Otley should find errors in Smith's maps for Cumberland and Westmorland when they were later published he should not hesitate to 'contradict him on any material misrepresentation he may make through not conferring with you'. This 'exposure must be made if he [Smith] will persist in the extraordinary conduct I have alluded to'. To my knowledge, Otley did not choose to confront Smith in print, or privately, but it is clear that what might have been a fruitful collaboration never occurred.

As mentioned above, Smith was in the Lake District again in 1823, this time doing survey work in the Carrock Fell region. Also that year, he studied the geology of the limestone districts round Kirkby Lonsdale in northern Lancashire, and while staying in that town he wrote some interesting notes that throw light on his thoughts about Lakeland geology. The rocks were, he said, 'destitute of those organic remains which enable us to trace the beds which refer to one rock from [sic] another . . .'. Indeed, he continued, the rocks were so different from the ones usually met with that they formed 'almost a separate branch of Geology dependent on mineralogy and crystallography [rather than palaeontology]'.[16] Thus Smith could not deal with the Lakeland rocks in his accustomed manner, using the general principle that each kind of stratum was characterized by its own particular types of organic remains.

The apparent absence of fossils from at least the first two Otley units certainly made life difficult for a Smithian stratigrapher in the central part of the Lakes. In fact, no fossils were found in the Skiddaw Slates for many years, a point noted by Phillips (1849, p. 210) in his general survey of Lakeland geology. Yet there is a curious statement made by Smith in his Kirkby Lonsdale manuscript:

> The existence of [some[17]] shells in the Cumberland Welsh and Cornish Districts and probably of plants, in the Chyastolite slate of Skiddaw, proves that these parts of the Earth are not altogether totally different from the other.[18]

Might this mean, perhaps, that Smith had some knowledge of graptolites, mistaken as plants, in the Skiddaw Slates? Or some shelly fossils?

Smith's last contributions to Lakeland geology that need be noticed here were in the form of a manuscript cross-section (1824) and his 'county maps' of Cumberland and Westmorland, coloured onto Cary maps (also of 1824). The manuscript section[19] forms part of a larger cross-section from Whitehaven on the Cumberland coast through to the Yorkshire coast. But for the Lakeland mountains (labelled 'Schist Mountains'), Smith did no more than show them in profile without geological colouration. He had the 'frame' of sandstone, coal and limestone, but the section suggests that Smith did not have much to say about the geology of the hill country. This was hardly surprising. Smith never had the opportunity to study the region carefully, and it lacked the fossils necessary for 'Smithian' geology.

However, the maps show (I suggest) influence from Otley. They distinguished between 'Dark soft schist of Skiddaw Saddleback and Grisdale Fells', 'Grey slate of Buttermire [sic] and Borrowdale, associated with hornstone, amygdaloid & argillaceous porphyry characterized by bold towering crags', and 'Schistose rocks of nearly uniform composition forming hills of inferior altitude producing dark slate in Kentmere'[20] – equivalent to Otley's three units. Otley's 'Transition Limestone' was also delineated between the second and third units, as 'Thin beds of limestone alternating with schist and containing madrepore and

[15] The identification of the handwriting is based on a known Otley topographic map held among the Sedgwick archives at the Sedgwick Museum, Cambridge: 'A Map of Skiddaw, Saddleback, and Caldbeck Fells; with their Environs. Drawn by J. Otley Keswick 1827' (uncatalogued). The verso of this topographic map bears the heading 'Skiddaw' in Sedgwick's hand.
[16] Smith archives, Box 35, Folder 1(iii), 19 April 1823, Kirkby Lonsdale.
[17] Word added in pencil to the ink MS.
[18] Smith archives, Box 35, Folder 1(iii), 29 April, Kirkby Lonsdale.
[19] Smith archives, Box 16, Folder 5.
[20] One of the smaller Lakeland valleys, to the SE of the Lake District. See Figure 2.2.

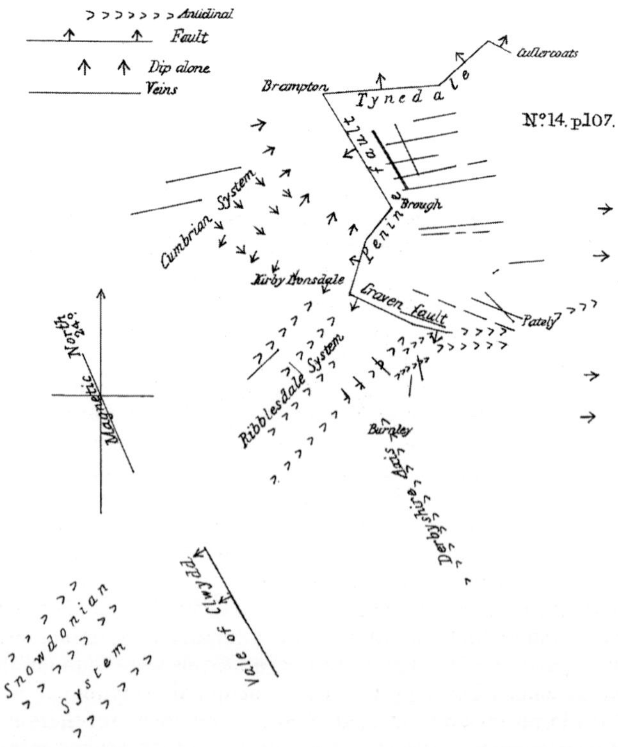

Fig. 1.5. Pattern of faults to the east and SE of the Lake District, according to Phillips (1836, plate 24).

producti'.[21] Interestingly, part of the important boundary between Otley's first and second units was marked in as being the outcrop of a 'Mottled red argillaceous rock near Keswick', running diagonally across Derwent Water from NE to SW. This 'Otley I'–'Otley II' (Skiddaw Slates–Borrowdale Volcanics) boundary can be discerned in a general way merely by looking at the shape and appearance of the hills that are formed of the two kinds of rocks. We shall refer later to this 'mottled rock', which makes its appearance, for example, at the shore of the lake, near a place where a stream called Cat Ghyll (or Gill) runs into the lake (see pp. 46–47). So we approach from a distance for the first time a question that was to provide one of the major sources of controversy for Lakeland geology: was the boundary a line of fault, or unconformity, or simply a change of rock type without angular unconformity?

This interesting question was not treated further by any of the geologists mentioned in the preceding pages. Further progress in understanding Lakeland geology depended in the first instance on a more exact knowledge of the outcrops of the various rocks and the production of a more satisfactory geological map of the region. The provision of such knowledge was soon undertaken by the arduous investigations of the Cambridge geologist, Adam Sedgwick, in the years 1822 to 1824. We turn to an examination of Sedgwick's work in the following chapter. It may be remarked, however, that according to Phillips (1844, p. 103), Smith and Sedgwick met rather briefly in Kirkby Lonsdale (it was probably in 1822), and walked together a few miles up the road to Kendal. We may reasonably assume that they had much to discuss, and that information passed in both directions. Sedgwick, at that time, was still something of a neophyte, whereas Smith was the experienced surveyor, map-maker, stratigrapher and field geologist.

It should be remarked also that Phillips made important contributions to the study of Lakeland geology by his investigations of the fault system to the east, i.e. the several boundaries with the Pennine region. He recorded (Phillips 1836, p. x) that in 1821 he made a careful survey of the Craven Fault, which runs SE from Kirkby Lonsdale (see Fig. 1.2); and the region was re-examined with William Smith in 1822. In 1827, Phillips took an extended walk through the Lakes, and made an examination of the various kinds of slates to be seen in the district. In December that year and the following January, he read a paper in two parts on the results of his fieldwork (Phillips 1827–1828, 1829), and his *Geology of Yorkshire* contained a sketch of the line of interconnecting faults (Phillips 1836, fig. 14) (see Fig. 1.5). Phillips's main interest was in Yorkshire rocks, for he was preparing his volumes on the geology of that county. Thus while his paper of 1829 gave a general account of the 'Slate Series of the Lakes', what he had to say about the Lakes proper had mostly been said previously by Otley. It is relevant, though, that Phillips (1829) observed and mapped some modest-sized inliers of rocks of the 'Slate Series' in the valley areas of Ribblesdale and Chapel-le-Dale, and in the Barbon Fells north of Kirkby Lonsdale, recognizing them as similar in type to the rocks of the southern Lake District ('Otley III'). Thus, from early on, Lakeland rocks were recognized outside the Lakes both in the Cross Fell area and in the so-called 'Craven Inliers' of NW Yorkshire.

These matters need not, however, be pursued further here. Let us now retrace our steps a little to look in some detail at the major contributions of Adam Sedgwick to Lakeland geology, building on, but massively extending, the contributions of Jonathan Otley and the other early observers. Further reference to Phillips's diagram of 1836 will be made in our discussion of Sedgwick's ideas.

[21] 'Madrepore' is an old name for a fossil coral, or an organism resembling a coral. *Productus* is a well-known Upper Palaeozoic brachiopods genus; but the term 'producti' was used by Smith more generally, for what would today be called brachiopods.

Chapter 2
Adam in Lakeland

As is well known, one of the great founders of British geology was the Cambridge professor, Adam Sedgwick (1785–1873). Aspects of his work have been intensively studied by Rudwick (1985) and Secord (1986), regarding his role in the establishment of the Devonian and Cambrian Systems respectively. Earlier, Sedgwick was the subject of a two-volume *Life and Letters* by Clark & Hughes (1890), and there is a short non-technical biography by Speakman (1982). Elsewhere, I have already written on his early work in the Lakes (Oldroyd 1998–1999 [2000]) and about his personal character and beliefs (Oldroyd 2002). Some parts of the present chapter draw on my previous publications, but Sedgwick's studies in the Lake District as a whole have not been analysed previously in any detail.

Sedgwick came from Dent in the Yorkshire Dales, to the SE of the Lake District. His father was the local parson, and the family seems to have been reasonably well-off. It is interesting that parts of the floor of his father's church of St Andrew's are crammed with fossils, and one may wonder whether these curious enclosures may have turned the boy's interests towards the rocks of the Earth's crust and their origins and history. Adam was the third of seven children. He attended a nearby school of some reputation, Sedbergh, and went on to Trinity College, Cambridge. There he studied mathematics and theology with great diligence, and in 1808 he was placed fifth in the University for mathematics (5th Wrangler). This led to a fellowship (by further examination) at Trinity in 1810, and then to the geology chair in 1818. The present geology department at Cambridge has its museum named in Sedgwick's honour, and his field notebooks and some manuscript maps are located there; also some items of his field equipment. Sedgwick's correspondence is held in the Cambridge University Library.

The story of Sedgwick's appointment has often been told. According to a statement that Sedgwick himself originated (Clark & Hughes 1890, vol. 1, pp. 160–161), there were two applicants for the professorship, and although Sedgwick knew no geology he got the position since what the other candidate knew was wrong! This tale cannot be entirely true, however. Sedgwick's fragmentary autobiography, preserved at the University Library, tells us that as a lad he became intrigued by the rocks and fossils that could be seen near Dent and he realized how the structure of the strata cropping out along the sides of the valley could be understood. Sedgwick suffered a breakdown in his health in 1813 and it is surely relevant to the present inquiry that he recuperated by taking a walking holiday in the Lake District. He was actively geologizing on the Continent in 1816, and was 'introduced' at the Geological Society that year. He was elected Fellow of the Society in 1818[1] and FRS in 1821, with Sir John Herschel heading the list of 12 who nominated him.[2] Doubtless the Cambridge chair smoothed the path to the latter fellowship.

Immediately after being elected professor, Sedgwick began fieldwork in a serious way. He went to Derbyshire, Somerset and Cornwall in 1818–1819. He went to Wiltshire, Somerset and Dorset in 1820, to the Yorkshire coast and Teesdale in County Durham in 1821, and in 1822 he worked his way north through Nottinghamshire and Lancashire, and into Cumberland, thus initiating his Lakeland work proper. In 1823, he went again to Yorkshire and to Cumberland (where he first met Otley); and in 1824 he spent the whole season in the Lake District. He returned in 1833, 1835, 1845, 1851 and 1857, but his major work there was done in 1822–1824, when the greater part of the region was given its preliminary survey.

Lakeland geology is difficult compared with that of southern Britain. The terrain is tough, and the rocks are very confused – in most places quite different from the nicely layered fossiliferous strata of the Yorkshire Dales where Sedgwick had grown up. So how did Sedgwick go about his task, assuming that he already had some familiarity with Lakeland-type rocks from his schooldays in Sedbergh and from earlier excursions into the Lakes?

First let me say something about Sedgwick's financial situation, as background to his work. His income in 1822 from his chair and his College fellowship was £407 6s 8d, but, depending as it did on Trinity's financial well-being, it declined in the following two years to £283 1s 0d and £232 15s 8d,[3] although he probably had additional income from private pupils. I have no evidence that Sedgwick obtained separate financial support for his fieldwork, but being unmarried and living in college he could have had sufficient money to be self-financing for travel and board. He was in the field from about the beginning of June until mid-October each year.

In the early 1820s, Sedgwick was undoubtedly very tough physically, despite his earlier period of illness, and the troubles, doubts and conflicts that characterized so much of his later life. An idea of his character in his early days as Cambridge professor can perhaps be had from a cartoon of him, sketched by one of his students (see Fig. 2.1). Sedgwick travelled by carriage, on horseback or on foot. His obituarist John Phillips stated (1873, p. 257) that in fact Sedgwick mostly went on horseback during his field trips, as was usual in the early nineteenth century, but this cannot have been the case for many of the localities that he visited in the Lakes. He travelled with a portable writing desk. He had leather specimen bags, and an assortment of hammers, some of which are now on display in the Sedgwick Museum, Cambridge. He also used a portable laboratory, though his notes say little or nothing about the results obtained with it. No doubt he had an acid bottle.

For accommodation, Sedgwick stayed with friends or at local inns, one of those he patronized being Lowood Inn, near Ambleside (see Fig. 2.2). There is no record of his having made arrangements in advance of his arrival, so I cannot say whether he just 'turned up', or made elaborate prior 'bookings', but one is struck by the way everything slotted into place so far as accommodation was concerned, despite the fact that he seemed to leave little time to organize his domestic arrangements. He was in the field all day and every day, except for Sundays, when he attended church. Often Sedgwick was in the field for 14 hours or more. There is a record that when he was visited in the Lakes by his colleague William Whewell in 1824 both were in 'breathless haste' (Todhunter 1876, vol. 1, p. 32).

Just as Sedgwick's accommodation arrangements worked smoothly, so too did his transport. There always seemed to be a carriage waiting at an appropriate spot, or a horse, for which purposes he may have had a servant with him. Further, it is not clear whether he planned everything carefully beforehand, or altered arrangements from day to day according to the weather and the exigencies of fieldwork. However, one thing is certain: he frequently used local people as guides, and Phillips (1873, p. 257)

[1] Sedgwick was proposed on 5 June 1818 and elected on 6 November that year. His admission certificate was signed by Thomas Harrison FRS, James Burton, Samuel Solly, James Cumming FRS (Professor of Chemistry at Cambridge), and John George Children FRS, none of whom apart from Children were leading lights in the Society. (I am indebted to the Society's former archivist, the late John Thackray, for the list of names and the dates.)

[2] I thank Mary Sampson, archivist at the Royal Society, for this information.

[3] I thank Jonathan Smith of the Trinity Library for this information.

Fig. 2.1. Cartoon of 'Sedgwick when a young man' by one of his students, H. A. Jukes. Held at the Sedgwick Museum, Cambridge. Reproduced by courtesy of the Sedgwick Museum.

recorded that when he met with Sedgwick by chance in the field in Teesdale in 1822 he had a miner's boy as guide, who had him 'in tow'.

Sedgwick seemed to be remarkably little bothered by the weather. There were occasional days when he did not attempt to go out because of rain, but he never mentioned problems due to cloud, and he was infrequently driven inside by rain, despite the Lake District's high rainfall. However, 1824, the year that Sedgwick made his greatest effort in the Lakes, had an exceptionally dry summer.

The question of maps available to and used by Sedgwick is important. There was a published map of the Lakeland region by Thomas Donald (second edition published in 1810, but originally surveyed in 1771), and Sedgwick's copy still exists in the Sedgwick Museum. However, at two miles to the inch Donald's map lacked detail and Sedgwick mentions numerous places and topographic features in his notebooks that do not appear on Donald's map, or any other published map available at that time. Sedgwick complained in his notes on several occasions about the inaccuracy of 'the map' and once he mentions Donald by name. One can see, by comparing Donald's work with modern survey maps, how inaccurate his map was.

Significantly improved topographic maps for Cumberland and Westmorland were published in 1823 and 1824 by C. & J. Greenwood, the ground having been surveyed in 1821–1822 and 1822–1823 respectively. However, even these did not give all the detail that Sedgwick mentioned in his notebooks. Therefore, much of his topographic information must have been supplied by local guides.

Sedgwick's geological map of the southern Lakes was only published as late as 1835 (Sedgwick 1835*a*, plate 4), when the geologist was already 50 years old. His manuscript geological map of Cumberland (only), entered on an edition of Donald (1802), is preserved at the Sedgwick Museum (see Plate II). It is interesting to consider how and when this was coloured in.

On his days off in the field, due to rain or the Sabbath, Sedgwick frequently mentioned writing up his journal ('journalise', he wrote), but he said little about maps. He probably synthesized his Lakeland work some time in the years 1825 to 1830, and the manuscript Cumberland map probably dates from that period. Interestingly, an annotation in the hand of George Greenough[4] to a published map of Westmorland by John Cary (1811), held at the Geological Society, states that 'a map of Westmorland on a large scale was published by Hodgson at Lancaster 1828 coloured geologically by Prof. Sedgwick'. I have not located Sedgwick's Westmorland map, but the annotation may give a clue to the approximate date of the colouration of his surviving Cumberland map. He did, however, refer in a later paper (Sedgwick 1841, p. 551) to a manuscript map, apparently covering parts of southern Lakeland, prepared about 1822, which must have been

[4] I am indebted to the late John Thackray for this identification.

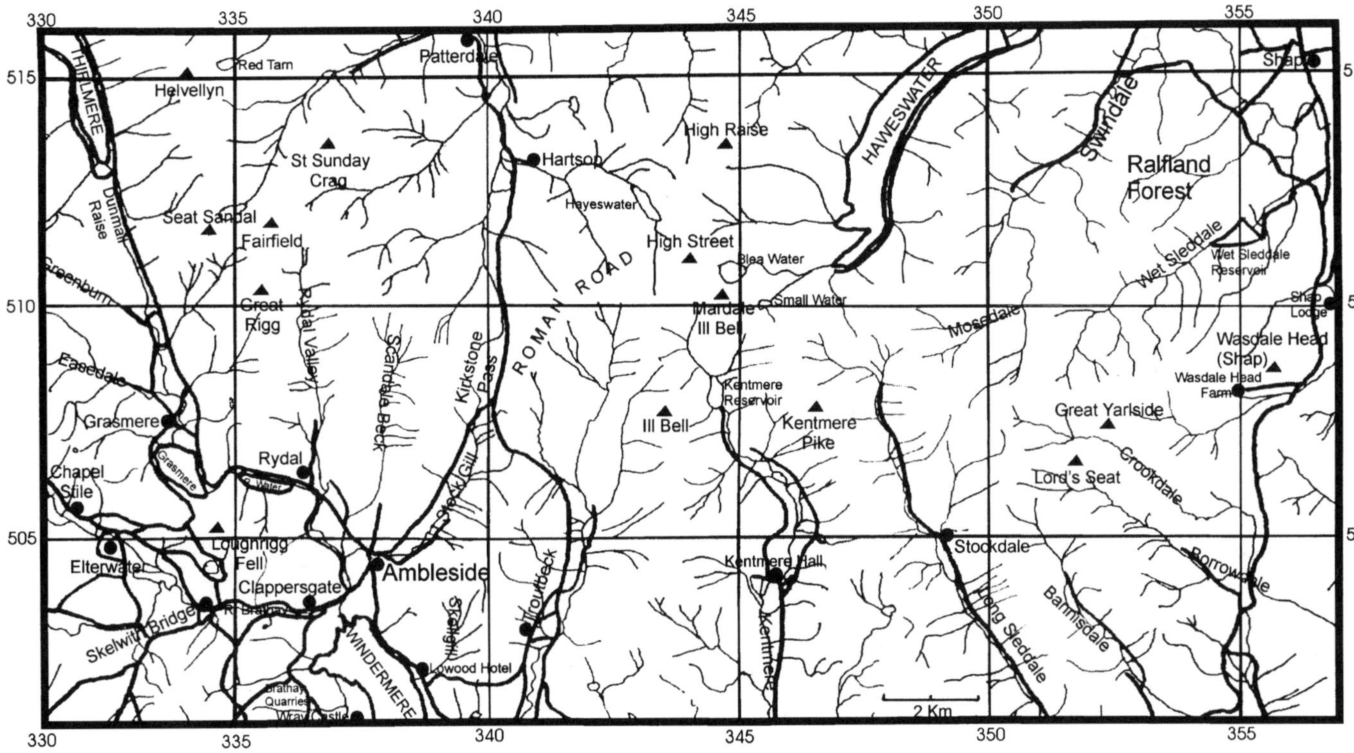

Fig. 2.2. Topography of the area round Ambleside and Grasmere and the valleys and hills of SE Lakeland.

his first effort, in which case he must have prepared draft maps from the outset, as we might expect. Indeed, his journal for 28 August 1823 contains the entry: 'Miserable rain. Colour maps *pour passer le temps*.' His only published coloured geological map of the Lakes was for the southern part, mentioned above (Sedgwick 1835*a*).

Sedgwick's journals for 1822–1824 name 51 persons in the Lakes with whom he journeyed, stayed, dined or conversed, or who acted as guides. These included farmers, miners, the Lakeland poets Southey and Wordsworth, and local naturalists.[5] His most frequent and important scientific contacts were Fryer and Otley. Sedgwick undoubtedly gleaned much empirical information from these two men, as well as how to find his way around in the mountains. Most importantly, he utilized Otley's division of the Lakeland strata into three main units, with granite underlying the Skiddaw region.

After transcribing Sedgwick's Lakeland 'journals', I have been able to sketch where and when he went onto modern maps. In 1822, starting only in August, he worked chiefly in the western to southwestern side of the Lakes, seeking to map particularly the granitic outcrops in Eskdale and round the foot of Wasdale. On 14 September, he ascended Scafell from the SW. Sedgwick was also much interested in the outcrops of limestone and slates in the region of the Duddon Estuary.

In 1823, Sedgwick started in the southeastern area of the Lakes, from Kendal up to Shap, then looked at Longsleddale and Kentmere and worked up to and around Penrith. Thereafter he crossed over to Ambleside (with a look at Haweswater) and examined the rocks of Langdale and Coniston, a month later pushing over Wrynose Pass into the upper Duddon Valley (see Fig. 16.1). He then spent time with Wordsworth in the Grasmere area. Thence he returned to Penrith and worked round the NE boundary of the Lakes, looking at the mining area of the Caldbeck district and the unusual rocks of Carrock Fell on the way (see Fig. 12.1). From Keswick, where he met Otley, Sedgwick examined the slate rocks between Skiddaw and Blencathra (Saddleback), and the underlying granite. The surrounding altered rocks were observed to contain chiastolite.

Sedgwick's most energetic year was 1824 (which, as mentioned, had a good summer). He began in June, near Kendal in the SE, and walked round the Grasmere district again with Wordsworth, making huge journeys from that centre in all directions. Then he went across to Ullswater and made excursions from there, using Pooley Bridge as a centre (see Fig. 9.11). In this district, at Mell Fell, he encountered the important red conglomerate that forms the lowest unit of the rocks lying unconformably around the Lakeland mountains. From Ullswater, Sedgwick went round again to Keswick, re-examining the mining area near Carrock Fell. He then began to explore the NW of the Lakes, meeting up with his Trinity College colleague, the professor of mineralogy William Whewell (1794–1866), to visit the coal regions of the Cumberland coast. In July, he got up into the high and difficult country of the

[5] Wyatt (1996) suggests that Wordsworth had a considerable interest in geology, and that some kind of geological understanding underlay a number of his Lakeland poems, particularly the Duddon Sonnets. I doubt, however, whether Wordsworth made any systematic study of geology, though his circle of acquaintances certainly included some men of science, of whom Sedgwick was the most important for present purposes. The first edition of Wordsworth's famous *Guide to the Lakes* appeared in 1810 as an introduction to the Reverend Joseph Wilkinson's *Select Views in Cumberland, Westmorland, and Lancashire*, and contained just a few remarks of a geological nature. Wordsworth (1810, p. v) referred to the rocks of the central mountain region as being made of 'schist', and added that this gave way to limestone 'as you approach the plain country'. He saw ferruginous matter as the 'principle of decomposition' for the rocks making up the mountains, perhaps thinking of, say, the severely weathered Red Pike near Buttermere when he made this comment. Subsequent editions of the *Guide* contained gradually greater geological content, until in the edition of 1842 he added three 'letters' by Sedgwick (1842) on geological matters (to which were added two further letters in the editions of 1846 and 1853, the last being a *cri de coeur* from Sedgwick about his contest with Murchison). Wordsworth had some interesting ideas about geomorphological matters, such as the unexpected course of the river near the head of the Duddon Valley. For details, see Wyatt (1996).

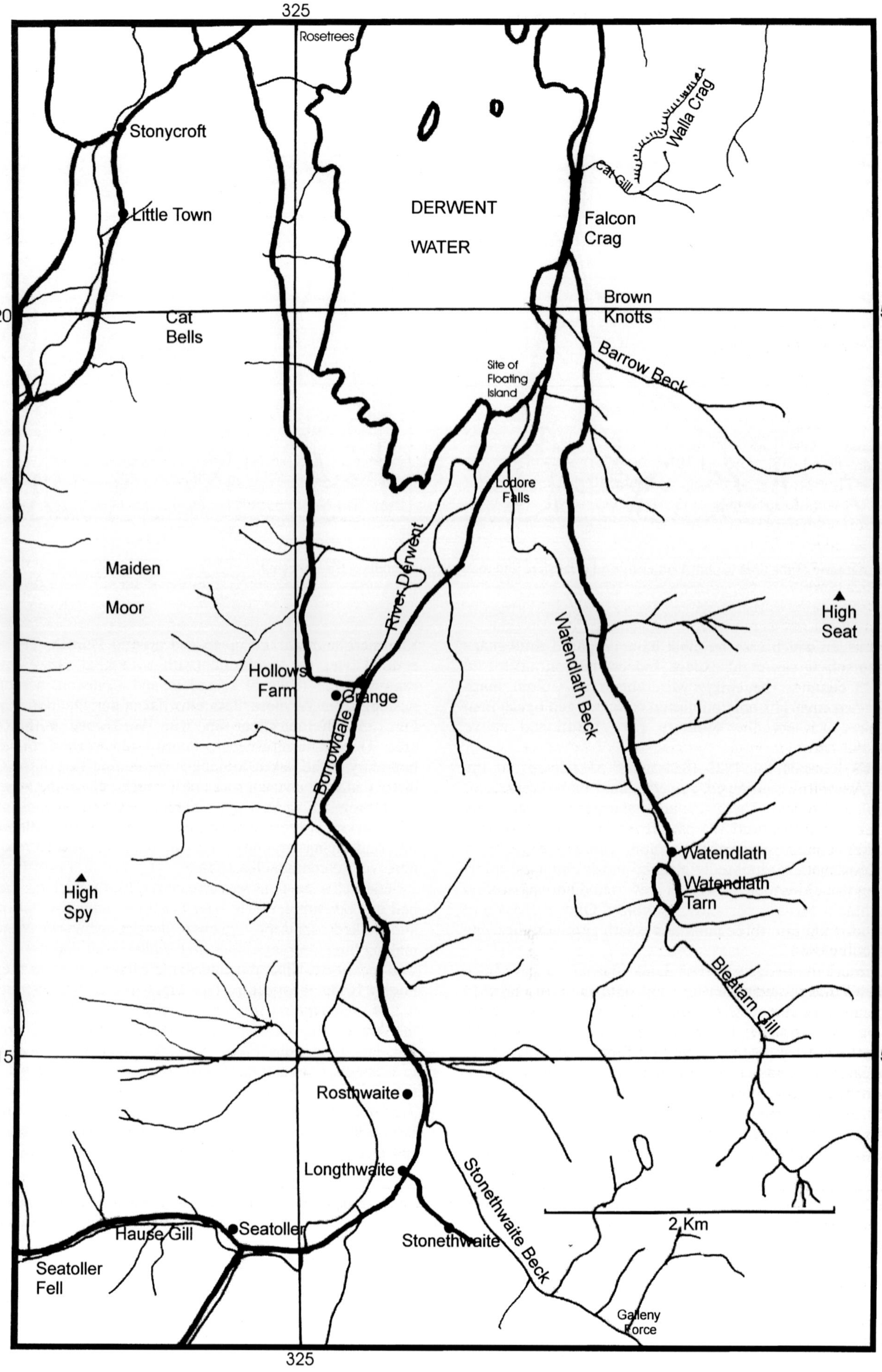

Fig. 2.3. Topography of the area of Derwent Water and southwards, showing Walla Crag.

Central Fells – Scafell, Gable, and the fells round Ennerdale, Buttermere and Crummock (see Figs 7.4, 16.1 and 16.2). In August, the area round Keswick was again visited with Otley, and particular attention was given to the rocks round Derwent Water and the junction between Otley I and II (see Fig. 2.3 and Chapter 9). Skiddaw was visited, and also the area about Bassenthwaite. Finally, Sedgwick worked his way round the western coastal region of the Lakes, revisiting Wasdale, and getting down to Broughton-in-Furness and Ulverston, before cutting across to Kendal on 9 October, to be back in Cambridge for the new academic year.

At first glance, Sedgwick's journals do not suggest that he was working to any systematic plan; perhaps he was not. Nevertheless, when his routes are marked onto modern maps one can see how, after three years' work, he had gone over a great deal of the ground, at least in a superficial manner. And what may at first seem aimless rambles were, in fact, routes that allowed good coverage of the ground with the minimum climbing. He would have seen most of the hills, at least from a distance, walked up most of the major valleys, and climbed the principal hills. The main area that seems to have escaped his attention was the southern district round the lower parts of Coniston Water and Windermere, and Dunnerdale. In 1823, he did, however, get to Ash Gill near Torver, west of Coniston Water, an important geological site that will attract our attention in later chapters (see Fig. 5.2). The fact that he made a special point of going there, and then returning directly, indicates that Sedgwick was specifically directed to the spot by someone with local knowledge. He did not happen on the locality by chance.

One cannot say precisely how far Sedgwick walked on average each day, as we do not know when he was and when he was not on horseback, and we do not know how many little detours he made to hammer interesting outcrops. Perhaps he journeyed on average 20 miles per day, six days a week, from June until October. It was tough work, but he revelled in it, and in later years his health always improved when he was in the field.

In his published description of Lakeland geology, based on his fieldwork of 1822–1824, Sedgwick (1835*a*)[6] described the outcrop of Otley's 'Transition Limestone' (a term that Sedgwick himself deployed in his notebooks) as if he had traced it in the field from west to east. In the published version, he described how the outcrop would continue steadily for a while, and then be displaced to the north or south for perhaps up to three miles. Sometimes a lake or valley could be found associated with the line of displacement.

On the basis of his published paper, it would appear that Sedgwick followed the outcrop of limestone from west to east, using it as a marker band. However, his notebooks reveal that this was not so. Rather, he would stay in one spot for three or four days and range out from there in several directions. When his observations were transferred to his map, perhaps back in Cambridge, the outcrop of the 'Transition Limestone', and other important features such as the granites near Eskdale and Skiddaw, or Otley's three types of slates, would have emerged as the various areas of the map were coloured in. In fact, not all the limestone's outcrop was covered even in the same year. It is true that for some days Sedgwick would become particularly interested in the limestone and he was evidently following its course; but he did not do so consistently. Actually, he very likely knew what to expect, for even in his short paper of 1820 Otley had indicated in words the general outcrop of the 'Transition Limestone'. However, Sedgwick's careful attention to the 'range' of the narrow outcrops of limestone revealed how the stratum had been repeatedly fractured in some way.

Sedgwick's notebooks contain much information about different rock types, their areas of outcrop and the boundaries between them, and, most particularly, information about dips and strikes ('ranges' as he called them); also the bearings of principal topographic features. His interest was evidently lithological and structural, and in his early Lakeland work he did not give much attention to fossils. Indeed, as Sedgwick wrote years later to Robert Harkness (see p. 27), he did not initially expect to find fossils in the Lakeland rocks, thinking that they 'were all below the region of animal life'.[7]

There is so much numerical information about dips and strikes in Sedgwick's journals that one wonders how he remembered it all. Were rough notes made during a day's excursion, and then entered up in the journal in the evenings? Or did he just have a phenomenal memory? I have no means of answering this for certain, but I believe that there *must* have been rough notes used to compile the surviving notebooks. So we have several stages of knowledge filtration and construction: (1) field observations; (2) writing of rough notes; (3) composition of the 'journal' and the destruction of the original notes; (4) preparation of provisional geological maps from the 'journal' or perhaps in the field; (5) composition of a paper for public presentation and evaluation, perhaps in the light of higher geological theory or some ongoing theoretical controversy; (6) defence of the paper at the meeting of some learned society (usually the Geological Society); (7) revision of the paper, and its refereeing; (8) further revision and eventual publication.

Sedgwick's extant notebooks contain few sketches and those there are unimpressive. His procedure seems to have been to cover as much ground as possible in a theoretically neutral or positivistic way, and fill in a map. His categories for this were simple, but that was doubtless to his advantage in dealing with such a difficult area in pioneering fashion. However, by contrast with the impoverished petrographic categories, there was a wealth of detail regarding dips and strikes of beds and of cleavage. The notebooks contain few sections (in contrast to the notebooks of his later friend, but eventual enemy, Roderick Murchison).

As said, Sedgwick's petrographic vocabulary was impoverished. He often referred in his journal to a rock type by its locality or approximate appearance. Thus we have the 'Wallow Crag' type ('Walla rock'), the 'Barrow rock', 'the concretionary',[8] 'the porphyry' and even 'the blue'. Interestingly, some of these terms appear in Otley's writings, so Sedgwick probably got them from him. However, he did not use them in his published work. Wallow (or Walla) Crag was a hillside east of Derwent Water (see Fig. 2.3). Sedgwick did not say in his notes what this rock looked like, but we find it described in Otley (1823, p. 99) as a composite greenstone,[9] with veins of quartz, calcareous spar, chlorite and epidote.[10] There are indeed greenish andesites on Walla Crag.

Back at Cambridge, Sedgwick's positivism dropped away to

---

[6] The paper, which was delivered on 5 January 1831, was published in brief in the *Proceedings of the Geological Society* and *The Philosophical Magazine* for that year (Sedgwick 1831*a*, *b*) – but without the map. The version published in 1835 was refereed by Henry De la Beche (Wendy Cawthorne, pers. comm., 2000).

[7] Sedgwick to Harkness, 28 August 1856, Sedgwick papers, Cambridge University Library, Add 7652/IVA 1.

[8] This may correspond to what are today called 'accretionary lapilli'. For example, Sedgwick's notes for 1824 refer to a 'concretionary slate' on Loughrigg Fell, near Ambleside. This could have come from what are today called the Seathwaite Fell or the Lincomb Tarn Formations, both of which contain welded lapilli tuffs. Some modern geologists may use quite informal nomenclature when in the field, as did Sedgwick.

[9] A field term, now obsolete, for more or less metamorphosed basaltic or doleritic rock, the green colour being due to the chlorite or epidote (derived from German, *Grünstein*.)

[10] Interestingly, Wordsworth (1820, p. 46) referred to an outcrop in the Duddon Valley as 'Wallow-Barrow Crag', and he pointed out that the name 'occurs in several places to designate rocks of the same character'.

some extent as he developed grander theoretical pictures. For example, when his map was finished it was obvious from its outcrop that the 'Transition Limestone' had been affected by massive faults that cut across the outcrop, wrenching it into a number of severed fragments; from which it appeared that the whole region of the Lakes had been upheaved and fractured. Then the great valleys, such as those of Windermere or Coniston, had seemingly been carved out along the lines of weakness produced by the faults. Later, when he published his work, Sedgwick (1835*a*) suggested that the movements and fractures were caused by the intrusion of the great masses of igneous rock, such as the granites near Eskdale or under Skiddaw, which had also produced an anticlinal structure for the Lake District as a whole.

It is interesting that Sedgwick did not observe or report folds and contortions in the central volcanic rocks – Otley's second division, the Green Slates and Porphyries, as Sedgwick was later to call them – though they showed clear evidence of dislocation, as did the 'Transition Limestone'. He remarked that the unstratified porphyries seemed to be contemporaneous with the stratified deposits of 'chloritic slate'. So, he suggested, the igneous and aqueous causes somehow seemed to have acted together: 'the porphyries . . . [being] produced by some modification of submarine igneous action – and the chloritic slates . . . [being] deposited from the waters in the same region, and in the same periods of time . . . the first operation supplying, at least in part, the materials for the second' (Sedgwick 1835*a*, p. 55). This was a prescient suggestion. It did not mean that he was necessarily trying to provide a synthesis of so-called Vulcanist and Neptunist theories (see p. 4). In fact, Sedgwick was dealing with what is today regarded as the product of a great system of volcanoes in the Lake District, the outpourings of which gave rise to what came to be called the Borrowdale Volcanics Group (BVG). We can, therefore, construe his porphyries as being the products of lava flows, while the 'chloritic slate' would be the product of masses of volcanic ash, deposited as layers of sediment in water. It is, of course, inappropriate to read Sedgwick's observations through the eyes of twentieth-century theory. But it is interesting to see how he referred to the rocks the way he did; and we can thereby 'make sense' of what he wrote.

Besides noting the dislocations of the 'Transition Limestone', Sedgwick was also interested in the apparent fractures to the east and southeast of the Lakes. John Phillips (1829) had recognized what came to be called the Craven Fault (see Fig. 1.5 and Phillips 1837–1838, p. 182), running SE from near Kirkby Lonsdale, which brought up the great masses of limestone of the Pennines to a higher altitude than the rocks of the upper Carboniferous and New Red Sandstone. Sedgwick had traced the dislocation NE towards Brough and Stainmore on the western side of the Pennines, and thence in a NNW direction, past Cross Fell (to the east of the fault) and up to near Brampton (see Fig. 1.1). Thus, 'the entire carboniferous zone of the Lake Mountains has been nearly cut off from the central chain with which it must undoubtedly have been once continuous' (Sedgwick 1831*c*, p. 285). The limestone that lapped the eastern side of the Lakeland rocks unconformably, as near Orton for example, was evidently repeated by faulting, the Pennine limestones apparently having been upheaved by some gigantic force (which also upheaved the ancient rocks of the Cross Fell inlier, previously examined by Fryer, Buckland and Greenough). Phillips (1837–1838, pp. 183–184) sought to relate the observations to the 'catastrophist' tectonic theory of the Cambridge mathematician and geologist William Hopkins (see pp. 243 and 255).

Interestingly, Sedgwick's great Pennine fault ran close to his old home village of Dent, where it brought into juxtaposition the limestones of the Pennines and the flagstones of Otley's third Lakeland slate unit (or one of its subdivisions, nowadays called the Brathay Flags[11]), such as occur near Ambleside. Today, a nature trail has been established near Dent at which the so-called Dent Fault, where the Brathay Flags (named after the Brathay River and Brathay slate quarries near Ambleside; see Fig. 2.2) and the Carboniferous Limestone adjoin, can be conveniently inspected (Rickards n.d.). Whether Sedgwick knew of this contact when he was a boy I know not,[12] but even if he did, I doubt that he would have realized its geological significance. Knowledge of a wider range of country is required for that.

Since Sedgwick made no effort to squeeze the history of the Lake District into a Biblical time-scale, there was no problem about the erosion of the lines of fracture taking an indefinitely large time. However, he thought the fractures were produced by what came to be called (by William Whewell) 'catastrophic' Earth movements, acting for a relatively short time, before the subsequent unconformable deposition of the New Red Sandstone round the margins of the Lake District. Thus the arguments of Reijer Hooykaas (1970), that catastrophism could be scientific and empirically based, are well supported by the example of Sedgwick. His catastrophism was not just a form of physico-theology. It emerged from, or was grounded in and warranted by, an enormous effort of fieldwork.

It may be noted that Sedgwick freely used the term 'diluvium' to refer to the deposits of boggy ground. The term had its origin in the notion of such deposits having been laid down at the time of the Deluge; but Sedgwick did not consider this issue in his journals. In fact, so far as Lakeland geology was concerned 'diluvium' was just a blanketing deposit that had to be walked over – a geological nuisance, obscuring the underlying solid rocks and making mapping difficult.[13]

Over and above Sedgwick's catastrophism, he sought to account for what he had seen in the Lakes with the help of the overarching theory of Élie de Beaumont (1829–1830, 1831), though with some caveats (e.g. Sedgwick 1831*c*, p. 289).[14] The French geologist had the idea of a cooling and shrinking Earth, which formed 'wrinkles' in its crust as it cooled. These supposedly gave quite a regular pattern of mountain ranges, and the suggestion was made that each range (or 'wrinkle') corresponded to a particular episode of mountain building. In agreement with the Frenchman's doctrines, Sedgwick asserted that the main lines of mountain chains in the Southern Uplands of Scotland, the Lakes, Wales, the Isle of Man and Cornwall, were approximately parallel, and were thus generated at the same epoch. The elevation of the Pennines, running N–S, could be attributed to a different (later) epoch.

Said Sedgwick in his Presidential Address to the Geological Society in 1831: '[t]he investigation of the faults and dislocations interrupting the continuity of our secondary deposits is becoming, daily, a subject of increasing importance; and we are now called upon, not to regard them as solitary phenomena, but to trace

---

[11] Sedgwick himself used this term for a time, and then changed to the use of the term 'Coniston Flags'. The old name was resurrected by Marr (1878).

[12] Much later, Sedgwick wrote that he had been familiar with the fault for many years, as it 'crosse[d] my native valley of Dent, about half a mile below the village, dislocating and setting on edge all the lower limestone beds' (Sedgwick 1852a, p. 36).

[13] 'Diluvium' also included material that later came to be thought of as being of glacial origin, such as morainic matter and 'boulder clay'. Sedgwick subsequently lost faith with the idea that all 'diluvium' could have a single causal explanation. He stated (Clark & Hughes 1890, vol. 1, p. 371) that he gradually abandoned the idea through additions to his stock of empirical knowledge (partly acquired during a visit to Scotland with Murchison in 1827), which suggested that there was so much evidence of 'local' diluvial action – or multiple inundations. Sedgwick publicly abandoned his adherence to the notion of a single Deluge being responsible for the occurrence of 'diluvium' in his Presidential Address to the Geological Society in 1831 (Sedgwick 1831*c*).

[14] Élie de Beaumont, then an engineer with the Department of Mines in France, visited Britain in 1822 to examine Greenough's map and matters connected therewith. Whether he met with Sedgwick in Britain I do not know.

them through whole regions, and to examine their relations to each other' (Sedgwick 1831*c*, p. 289). Further, the uplift and fracture of central Lakeland might, Sedgwick suggested, be an exemplification of the theory of 'craters of elevation' of Alexander von Humboldt (1880–1883, vol. 5, p. 313) and Leopold von Buch (1810). Thus what started off with observations about the obscure outcrops of calcareous rocks in southern Lakeland – which rocks fizzed with acid and which did not – could, with the help of the maps constructed on the basis of Sedgwick's fieldwork, turn into grand speculations about the history of the globe. Thus it was that the endless notes about dips, strikes ('ranges') and cleavage were used to underpin grand theoretical models. Ultimately, it seems, *this* was the point of Sedgwick's preoccupation in the field with all the structural details of the Lakeland rocks. A visual impression of the Sedgwick–Élie de Beaumont theory, as applied to the areas of the Pennines to the east of the Lakes, and in NW Yorkshire, subsequently appeared in John Phillips's *Geology of Yorkshire* (see Fig. 1.5).

Sedgwick's enterprise was in keeping with the trend in geological theory at Cambridge at that time. In the Newtonian tradition, Cambridge was renowned for its mathematics, and it will be recalled that mathematics had been Sedgwick's forte when he was an undergraduate. It was his expertise in the subject that had secured him his fellowship at Trinity; and this led on to his chair. Sedgwick had been one of those most active in the establishment of the Cambridge Philosophical Society in 1819 (Clark & Hughes, 1890, vol. 1, pp. 205–208), which was in practice concerned with science (or natural philosophy), not philosophy. And the science practised in the Society was generally as mathematical as it could be made (Smith 1985).

So one can see Sedgwick's early Lakeland geology as part of a programme intended to provide a mathematical theory of the Earth. In practice, however, this goal was not held permanently in view, and Sedgwick's later work in the Lakes became more concerned with palaeontology and with stratigraphic correlations. Nevertheless, his Lakeland self-tuition in mapping and 'hard-rock' geology, with correlations based on lithological considerations, left its mark on Sedgwick's 'style' as geologist. It subsequently enabled him to tackle the difficult terrain of North Wales, though, as is well known, Sedgwick's initial failure to provide a palaeontological basis for his Cambrian System led him into difficulties and controversies (Secord 1986).

Sedgwick was slow in getting the results of his Lakeland work into print. Perhaps, at first, he could not see the wood for the trees, by which I mean that he was overwhelmed by the mass of his own observations until he found them falling into some kind of pattern with the help of Élie de Beaumont's theory. Whole barrels of rock specimens were delivered to Cambridge, and often, it seems, they went unopened for several years. But Sedgwick also needed to widen his geological experience, to see how his Lakeland observations fitted in with those in other parts of the world. He had already been in Cornwall in 1819, and he went there again in 1828, 1836 and 1851. In 1827, he was with Murchison in the north of Scotland, and in 1829 the two geologists had a grand season in the Alps. In 1830, Sedgwick was in the Cheviots of Northumberland, and he also made investigations in the Buttermere area and the mining regions of the Cumberland coast with a mining engineer, Williamson Peile (Sedgwick & Peile 1835), which led to a discussion of the outcrop and subdivision of the Mountain Limestone and the Coal Measures around the Lake District. There were further visits to the Lakes in 1833 and 1835, and again in 1844, 1845, 1851 and 1856. However, for much of the middle part of his career – especially in 1831, 1832, 1842, 1843 and 1846 – Sedgwick was caught up with fieldwork in Wales, becoming more and more tangled in controversy with Murchison about the placement of the boundary between the Cambrian and Silurian Systems (Secord 1986). Apart from trying to establish a stratigraphic arrangement within the Lakes there was also the necessity to establish correlations with other areas that seemed to have analogous rocks, such as North Wales and Cornwall.

The first effort in this line was revealed by William Fitton (1780–1861), in his Presidential Address to the Geological Society in 1829 (Fitton 1834), in which he surveyed the progress in geology for the previous year. The threefold division of the Lakeland rocks was given, with Otley's name mentioned (though not specifically credited with the establishment of the three units). It was suggested that the 'grauwacke system' of Cumberland (the slates of southern Lakeland) might be correlated with similar rocks of Somerset, Devon and Cornwall; the 'green-slate of Cumberland' had no analogue in SW England but might be related to part of the Snowdonian formation in Wales; and the 'clay-slate' (Skiddaw Slates) might be correlated with the 'metalliferous killas' (see p. 8) of Cornwall. These correlations were probably suggestions of Fitton as much as Sedgwick. Fitton saw Sedgwick's Cumberland work as helping to connect 'the several groups in the distant parts of England, in a series of similar and probably contemporaneous formations' (Fitton 1834).

In a paper presented to the Geological Society on 1 February 1832, Sedgwick (1836) described the outcrops of the New Red Sandstone in the Vale of Eden to the east of the Lakes (see Fig. 1.3) and on the Cumberland coast near St Bees (see Fig. 20.4). He also mentioned the great system of faults between the Lake District and the Pennines. On 16 May in the same year, Sedgwick (1832) developed the subdivisions of the principal Lakeland units (originating with Otley) and introduced some new terminology (Unit 5 being uppermost):

5. *Greywacke and greywacke-slate* (Windermere Supergroup in modern terminology; 'Otley III' in the present book)
4. *Green slate and porphyry, &c.* (equivalent to what were later known as the Borrowdale Volcanic Group; 'Otley II' in the present book)
3. *Skiddaw slate* (with the same meaning as at present; 'Otley I' in the present book)
2. *Crystalline slaty rocks*
1. *Granite*

Thus we have the introduction of the terms 'Skiddaw Slates' and 'Green Slates and Porphyry' (later 'Porphyries') as lithological units – ones that would play such an important role in the history of Lakeland geological studies. Sedgwick thought that the 'Green slate and porphyry' was equivalent to the 'Snowdonian series', that the Coniston Limestone (the lowest member of Unit 5) corresponded to the Bala Limestone of Wales, and that parts at least of the slates of Unit 5 corresponded with the contorted slates of South Wales. Only the top part of Unit 5 (a reddish flagstone that cropped out north of Kendal) was manifestly equivalent to the 'tilestone' of what came to be called the Silurian System. However, at the time not all his fossils were unpacked and examined; when that was done he 'revoked' the correlation of the Coniston and Bala limestones, linking the former to a higher Welsh unit (the Llansaintffraid Limestone) (Sedgwick 1848, p. 218). The Survey geologists were to maintain the correlation, however.

The term 'Green slate and porphyry, &c.' is interesting. Most of the Lakeland volcanics, others than those of the northern volcanic belt (later called the Eycott Volcanics), are not markedly porphyritic in the sense of having large phenocrysts in a fine-grained matrix. However, Sedgwick did not then regard the central and northern types as distinct units (and a temporal distinction was not made until the 1970s), so he may have had these northern types in mind when he coined the term as a lithostratigraphic unit. On the other hand, Sedgwick used the term 'porphyry' in his notes from his first year in the Lakes (1822) in reference to rocks in western Lakeland, describing one of his specimens as a *dark* porphyry. He recognized the volcanics as a separate unit when he was with Otley, when they both accepted Wernerian theory. I am informed by Professor Otfried Wagenbreth of Freiberg (pers. comm., 2000) that in the old German terminology 'porphyry' meant 'old volcanic rock' (say Palaeozoic) while the term 'liparite' was used for 'young volcanic rock' (say Tertiary). Subsequently 'porphyry' came to mean acidic volcanics such as rhyolites. Given the vast amount of rhyolite and ignimbrite (see p. 99) in the central Lakes, *either* of the German meanings would suit reasonably well; so perhaps there was a residual Wernerian sense in Sedgwick's terminology of 1832. On the other hand, he may just have been thinking of the Lakeland andesites, and probably was. To my knowledge there are no dark rhyolites in the western Lakes.

In a further paper, read on 6 November 1833, Sedgwick (1833–1834) stated briefly that he had, with further fieldwork that summer, discovered that the outcrop of the 'Transition Limestone' (part of Unit 5 above) could be extended eastwards to Shap Wells, where it could be seen passing under the Old Red Sandstone and Carboniferous Limestone. The accompanying slates appeared to have been metamorphosed by the action of the heat of the intruded Shap Granite.[15]

By 1838, Sedgwick was endeavouring to fit his old 'Otley' schema into the new stratigraphic column being developed – with increasing rancour – in Wales. In a paper presented at the Geological Society on 21 March and 23 May 1838 (Sedgwick 1838), it was suggested that the Skiddaw Slates and the Green Slate and Porphyry might together be located in the 'Lower Cambrian', though neither unit had been found to contain fossils. The Greywacke was divided into a lower fossiliferous calcareous slate, passing into fine roofing slates,[16] and an upper division of flagstones and coarse slates with imperfect cleavage and few fossils. Sedgwick put the richly fossiliferous lower division into his Upper Cambrian and the poorly fossiliferous upper division into the Silurian system, thus subdividing Otley's old lithological unit. The fossiliferous lower division was still at this stage envisaged as analogous to the Bala Limestone, which was deployed as an important marker-band in Wales and was regarded by Sedgwick as the beginning of the Upper Cambrian. Thus Sedgwick had no very clear indication in the Lake District of the boundary between the Cambrian and the overlying Silurian: it occurred somewhere within Otley's 'third slate'.

In a subsequent paper read in November 1841, following a brief visit to Westmorland on a return journey from Scotland, Sedgwick (1841) maintained that some fossils from the 'lower division' were recognizable as Lower Silurian (e.g. the corals *Porites* and *Favosites*; brachiopods such as *Leptaena*, *Orthis* and *Atrypa*; and trilobites such as *Asaphus* and *Isotelus*), being equivalent to the then (as later became apparent) ambiguously defined 'Caradoc Sandstone' (see Whittard & Simpson 1960, pp. 5–6, 22–24; Whittard & Simpson 1961, p. 42; Murchison 1839, vol. 1, pp. 216–222); and a good list of fossils from the 'upper division' was determined by the conchologist J. C. Sowerby as Upper Silurian. No clear boundary between Upper and Lower Silurian could be found in Westmorland, though another section had been traced southwards from Shap through the Howgill Fells towards Sedbergh.

Over on the west, the sequence appeared to be as follows:

5. Upper flags and roofing slate (unfossiliferous)
4. Calcareous slates with Lower Silurian (as originally defined by Murchison) fossils
3. Kirkby Ireleth[17] slates
2. Quartzose flagstone, coarse pyritous shale and slate
1. Calcareous slates of Caradoc age (Lower Silurian)

This meant that the Coniston Limestone was no longer regarded as an equivalent of the Bala Limestone, but as an analogue of the so-called Caradoc Sandstone. As Sedgwick (1852b, p. 137) later stated, he did not at the time of forming this view have all his old specimens to hand, but he did have some Coniston specimens collected that year, together with some others supplied by an amateur geologist, James Marshall, FGS, of Kendal.

In the third of his famous letters to Wordsworth, composed in less than a week in May 1842, and published in the later editions of his poet friend's *Guide to the Lakes*, Sedgwick made subdivisions as follows:

[Old Red Sandstone]
Upper Group
- Red flagstone (like the 'tilestones' of Herefordshire)
- Purple, grey, greenish-grey and blue flagstones (like Murchison's Ludlow rocks)
- Hard, grey siliceous rocks with imperfect slaty cleavage, fossiliferous earthy bands, and impure calcareous beds
- Hard thick beds of various colours, with few fossils

Middle Group
- Hard, thick, siliceous beds with striped flagstones (well developed in Howgill Fells; see Fig. 1.1)

Lower, or Ireleth Slate, Group
- Roofing slates, as at Kirkby Ireleth[18] (Wenlock), with several fossiliferous, calcareous, slaty bands, the lowest being the Coniston limestone at the base of the Group

Sedgwick thought that, according to Murchison's original terminology, and the evidence of the fossils, the Upper Group belonged to his Upper Silurian (Ludlow). The fossils of the Lower Group suggested Murchison's original Lower Silurian (Wenlock), and the rocks of the Middle Group contained too few fossils for their status to be determined.

A general indication of the geological structure of the Lakes, as understood by Sedgwick in 1842 (Letter 2), is shown in Figure 2.4.

Another geologist, Daniel Sharpe (1806–1856),[19] an important amateur in the Geological Society, also looked at the strata of southern Lakeland, doing fieldwork there in 1841 and 1843, and published two papers on the stratigraphy of the region (Sharpe 1842, 1843*a*). His stratigraphic subdivisions (1842) of Otley's third unit differed from those of Sedgwick, being as follows:

5. Old Red Sandstone
4. Ludlow Rocks (grey shales)
   Unconformity
   Upper division (hard unfossiliferous, little cleaved greywacke) [Coniston Grits]
3. Windermere Rocks Middle division (argillaceous slates, shales, grits, often banded) (Series) Lower division (schistose grits and argillaceous shales with thin limestone bands)
2. Blue Flagstone Rock (no fossils found, but may have been lost during imposition of the slaty cleavage)
1. Coniston Limestone (dark blue, slaty limestone, resting on brown shale)

Sharpe had the Windermere Series as Upper Silurian and the Coniston Limestone as Lower Silurian. It is interesting that Sharpe in a sense introduced the term 'Windermere Group', though the modern usage is wider than that contemplated above. Thus the 'Windermere Group' in the modern sense would run from Sharpe's Coniston Limestone to his Ludlow Rocks in

[15] A famous building-stone much used for tombstones, bank façades, etc., containing large pink crystals of feldspar, which will receive frequent mention in the present book. The rock is so distinctive that it was much used later in the century for tracing the movements of glaciers (see Chapter 18).

[16] Later termed the Brathay Flags.

[17] A village near the Duddon Estuary, to the SW of the Lakes.

[18] Ireleth is now a part of Askam-in-Furness. The name Kirkby Ireleth appears to refer to both Kirkby-in-Furness and Askam-in-Furness. See Figure 6.3.

[19] Sharpe was a business man, antiquarian and philologist, as well as a geologist. He did a considerable amount of geological work in Portugal, as well as in Wales, where he became involved in the Cambrian–Silurian debates, which attracted his attention after he had formed views somewhat different from those of Sedgwick concerning the Lakeland rocks (see Secord 1986).

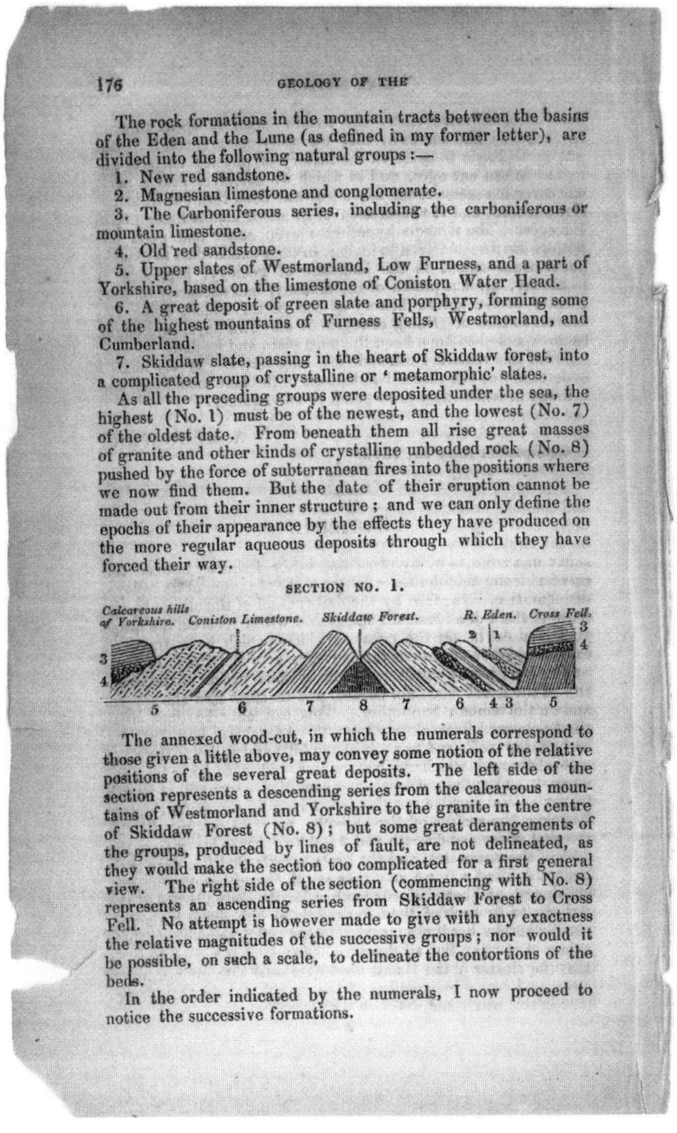

176 GEOLOGY OF THE

The rock formations in the mountain tracts between the basins of the Eden and the Lune (as defined in my former letter), are divided into the following natural groups :—

1. New red sandstone.
2. Magnesian limestone and conglomerate.
3. The Carboniferous series, including the carboniferous or mountain limestone.
4. Old red sandstone.
5. Upper slates of Westmorland, Low Furness, and a part of Yorkshire, based on the limestone of Coniston Water Head.
6. A great deposit of green slate and porphyry, forming some of the highest mountains of Furness Fells, Westmorland, and Cumberland.
7. Skiddaw slate, passing in the heart of Skiddaw forest, into a complicated group of crystalline or 'metamorphic' slates.

As all the preceding groups were deposited under the sea, the highest (No. 1) must be of the newest, and the lowest (No. 7) of the oldest date. From beneath them all rise great masses of granite and other kinds of crystalline unbedded rock (No. 8) pushed by the force of subterranean fires into the positions where we now find them. But the date of their eruption cannot be made out from their inner structure; and we can only define the epochs of their appearance by the effects they have produced on the more regular aqueous deposits through which they have forced their way.

SECTION NO. 1.

The annexed wood-cut, in which the numerals correspond to those given a little above, may convey some notion of the relative positions of the several great deposits. The left side of the section represents a descending series from the calcareous mountains of Westmorland and Yorkshire to the granite in the centre of Skiddaw Forest (No. 8); but some great derangements of the groups, produced by lines of fault, are not delineated, as they would make the section too complicated for a first general view. The right side of the section (commencing with No. 8) represents an ascending series from Skiddaw Forest to Cross Fell. No attempt is however made to give with any exactness the relative magnitudes of the successive groups; nor would it be possible, on such a scale, to delineate the contortions of the beds.

In the order indicated by the numerals, I now proceed to notice the successive formations.

Fig. 2.4. General arrangement of Lakeland strata, according to Sedgwick in his Second Letter to Wordsworth (1842). 1, New Red Sandstone; 2, Magnesian Limestone and conglomerate; 3, Carboniferous Series; 4, Old Red Sandstone; 5, Upper Slates of Westmorland, Furness, and parts of Yorkshire, with Coniston Limestone at base; 6, Green Slate[s] and Porphyr[ies]; 7, Skiddaw Slate[s]; 8, Granite and other unbedded crystalline rocks.

the scheme above, being equivalent to what I have called 'Otley III'.

In 1843, Sharpe's arrangement was developed further, as follows:

[9.] Mountain Limestone
[8.] Old Red Sandstone
[7.] Ludlow Rocks
[6.] Windermere Rocks
5. Flagstones & Slates of Kirkby Ireleth – top of Lower Silurian (placed below Blawith Limestone by Sedgwick)
4. Blawith Limestone (Sedgwick's 'Second band of calcareous slate')
3. Grey Slaty Grits (Lower division of Windermere Rocks) (1842)
2. Slates, Shales and Flagstones
   - f. Shear Bed (Sheerbate?; see p. 31)
   - e. Indurated shale
   - d. Blue flagstone
   - c. Indurated brown shale
   - b. Dark blue slate
   - a. Brown shale
1. Coniston Limestone – bottom of Lower Silurian.

This might seem innocuous, but it had already produced knock-on effects when Sharpe got to work in Wales, where he did fieldwork in 1842; for there he identified the Bala Limestone with the Caradoc (Sharpe 1843*b*). And if the Coniston Limestone were an analogue of the Welsh Bala Limestone, and if the Lakeland limestone was the lowest fossiliferous unit in the north of England, this did not bode well for Sedgwick's Cambrian in Wales.

Sedgwick's next Lakeland paper (1845*a*) was again trying to effect correlations between the rocks of Otley's third unit and those of Wales. It was based on Sedgwick's fieldwork in the southern part of the Lake District in the summer of 1845 (Notebook 40). Now aged 60, he was accompanied on the trip by one John Ruthven (1793–1868), originally a cobbler from Kendal, but a man who seems to have gone up in the world, to judge by a portrait of him held at the Kendal Museum, which presents him in the clothes of a middle-class gentleman. Ruthven also accompanied Sedgwick on field excursions in Wales, and did some independent fossil collecting for him in southern Scotland. He became an active member of the Kendal Natural History and Scientific Society, which was founded in 1835 with the active support of Sedgwick.[20]

Sedgwick stated that he had now recognized a fault near Dalton-in-Furness that produced a repetition of the main limestone band, and that when this was allowed for, the several strata fell into place, as it were. He was now paying much more attention to fossils than he had done in his preliminary survey of 1822–1824, and with the help of palaeontological and lithological evidence he now proposed the following subdivisions of Otley's third unit:

6. A fossiliferous group parallel to the Upper Ludlow (i.e. top of the Silurian of the Welsh Border region)
5. Coarse slates, flags, grits: a coarse development of the Ireleth Slates (called by Daniel Sharpe the 'Windermere Rocks'; see p. 20)
4. Ireleth Slates
   - (g) Upper Ireleth Slates (no fossils discovered)
   - (b) Calcareous slate with limestone concretions (Upper Silurian fossils)[21]
   - (a) Lower Ireleth Slates (no fossils discovered)
3. Coniston or Furness Grits (no fossils discovered, but probably equivalent to grits of the Lower Denbigh Flags of Wales)
2. Coniston Flagstone: unfossiliferous, but perhaps equivalent to the lower flags of Denbighshire (equivalent to the Wenlock Shale or Lower Denbigh Flagstones; Upper Silurian)
1. Coniston Limestone (Lower Silurian)

Thus Sedgwick was now placing the division between the Lower and Upper Silurian immediately above the Coniston Limestone.

Sedgwick returned to the problem in a paper presented to the

[20] See the Society's Minute Book, 1835–1852, p. 98 (Public Record Office, Kendal). Ruthven became one of the Society's Curators, and subsequently published his own geological map of the Lakes (Ruthven 1855). He was one of the most active amateur naturalists of the district, and became in part a professional, being employed on occasions by Sedgwick, and collecting and selling fossils on a substantial scale. Sedgwick described him as a 'famous fossil collector' and a 'geologist whose fame will last longer than the stoutest shoe that ever came off his ancient last' (Clark & Hughes 1890, vol. 2, p. 91). Sedgwick was elected President of the Society in 1840 and regularly gave lectures in Kendal about the progress of his Lakeland work and other matters.

[21] These had been described as Lower Silurian in his Letters to Wordsworth.

Geological Society on 7 and 21 January and 16 December 1846, following his Lakeland fieldwork of 1845 (Sedgwick 1846). He now absorbed his fifth unit into the fourth, making it a part of the Ireleth Slates; and the sixth unit was subdivided into two units:

(b) Arenaceous slates, grits and flags, without cleavage
(a) Ditto, with cleavage

He now had some palaeontological evidence (e.g. in the form of the lamellibranch, *Cardiola interrupta*, a Ludlow type) warranting the placement of the Coniston Flags in his Upper Silurian.

Units 5 and 6 of the previous year were now united, but were given three subsidiary subdivisions:

c. Red flagstone (Tilestone)
b. Coarse flagstone (Upper Ludlow)
a. Finer flagstone

The Ireleth Slates (4) were now divided into

d. Coarse slate and flags (Lower Ludlow)
c. Upper or great Ireleth slates
b. Upper limestone
a. Lower Ireleth slate

The following three units remained essentially unchanged, but were given some palaeontological backing, in part by the assistance of the emerging palaeontologist, John Salter (1820–1869). Only the lowest unit, the Coniston Limestone, was deemed to be Lower Silurian. This put Sedgwick at odds with Sharpe (see p. 21), who regarded the Ireleth Slates as Upper Silurian. Sedgwick's paper was supported by a considerable number of sections for the rocks of southern Lakeland.

Interestingly, Sedgwick contended that the sequence in Westmorland was more complete than that in North Wales; and in neither area was it the same as in Murchison's 'Siluria' i.e., the Welsh Border country. Sedgwick also thought that the appearances of limestone at horizons below the huge masses of the Carboniferous Limestone were everywhere irregular and mere local phenomena, so that exact correlations between sub-Carboniferous limestones of different areas was not to be expected. Thus one could not, for example, hope to make a simple comparison of the Bala Limestone of Wales with the Coniston Limestone of the Lakes – which might mean that if the Coniston Limestone comprised the oldest fossiliferous rocks in Lakeland this did not necessarily have undesirable implications for his Cambrian in Wales.

These later Lakeland papers of Sedgwick were beginning to become heavy with palaeontological argument, and it should be remarked that simple identification of particular strata by means of unique fossil content was not possible. Rather, one had to use a kind of statistical argument, showing changing proportions of the various fossils in the different units. This was what might be expected from the perspective of Lyellian geology, or from the standpoint of Darwinian transmutationism. Sedgwick did not accept many of the arguments of either of these theorists (Darwin's ideas were published later, of course); yet a kind of statistical stratigraphy was deployed without apparent difficulty or incongruity. (Interesting statistical arguments in relation to the geology of Somerset, Devon and Cornwall had also recently been deployed by John Phillips; see Knell 2000, pp. 240–249).

In December 1846, Sedgwick again returned to the problems of the subdivision of the upper Lakeland rocks, but more particularly in an attempt to effect a correlation with the ancient rocks of Wales (Sedgwick 1847). His correlations were given diagrammatically, as shown in Figure 2.5. However, as will be seen, the location of the boundary between the Cambrian and the Lower Silurian in Westmorland was excessively vague. The fossils were not behaving as precisely as might have been wished. It seemed that the lamellibranch, *Cardiola interrupta*, had a greater vertical range in South Wales than in Westmorland.

Sedgwick again tackled the problem of the subdivision of the lower Lakeland rocks in a paper read to the Geological Society in February 1848 (Sedgwick 1848). This, one would suppose, would

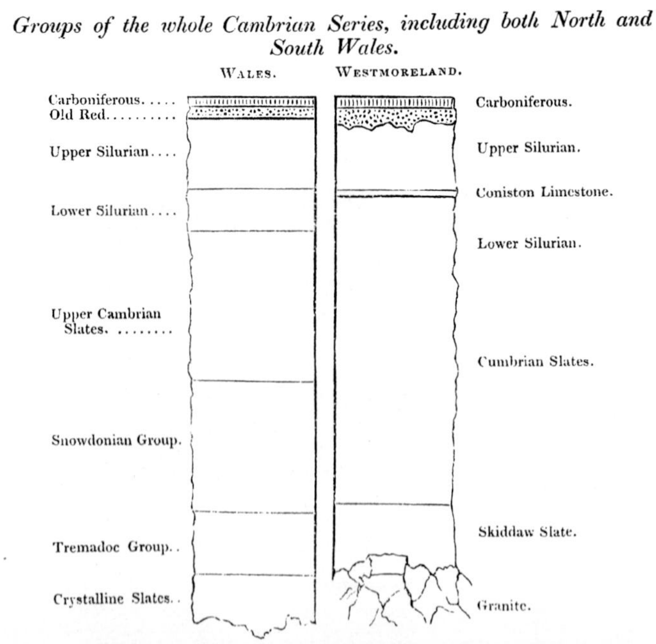

Fig. 2.5. Correlation of strata in Wales and Cumbria, according to Sedgwick (1847, p. 156).

necessitate travel into the northern Lakes. In fact his journal (Notebook 40), which seemingly represented his latest fieldwork for the Lake District, indicates that he himself had only visited the southern part of the region in 1845. However, in 1847 Sedgwick had requested Ruthven to search for fossils in the Skiddaw Slates, looking at sites that his 1822 notebook suggested might be fossiliferous by virtue of the presence of carbonaceous matter. This was really the nub of the problem for Sedgwick. He wanted life forms below the bottom of his Lakeland Lower Silurian (Coniston Limestone), which would give support to his desire to have a palaeontologically warranted Cambrian system (both in the Lakes and in Wales). The Green Slates and Porphyries could not be expected to yield fossils. So the place where they needed to be found and where one had to look was clearly the Skiddaw Slates.

To Sedgwick's pleasure, Ruthven had some success in two trips, finding some fucoidal remains and several species of Skiddaw Slate graptolites, which were displayed to the Geological Society at its meeting of 2 February 1848. The newly discovered fossils, which were discovered in the general area of Lamplugh Fell between Loweswater and Ennerdale on the NW margin of the Lakeland mountains (see Fig. 20.1), were described, figured and named by Sedgwick's Cambridge palaeontologist, the Irishman Frederick McCoy (1823–1899). The discovery was extremely important for Sedgwick in his controversy with Murchison, for it now appeared that he had information about some of the very early and simple forms of life – perhaps even the earliest. These Lakeland fossils could be claimed as Cambrian, and, Sedgwick suggested, they might be even older than the claimed Cambrian rocks of North Wales. Then in 1855, in an introduction to a volume on the palaeontological collections at Cambridge (Sedgwick & McCoy 1855, p. iv), Sedgwick proposed that the Skiddaw Slates might be 'physically co-ordinate' with the unfossiliferous Longmynd rocks in the hills above Church Stretton, Shropshire, which were of somewhat similar physical appearance.

We have seen above that Sedgwick initially thought that the Coniston Limestone might be the Lakeland equivalent of the Bala Limestone in Wales; but he gave up this idea in 1848 in favour of the suggestion that it was analogous to the Llansaintffraid Limestone. Yet in 1851, he was having second thoughts on this question: perhaps the Coniston Limestone did not 'overlie, but . . .

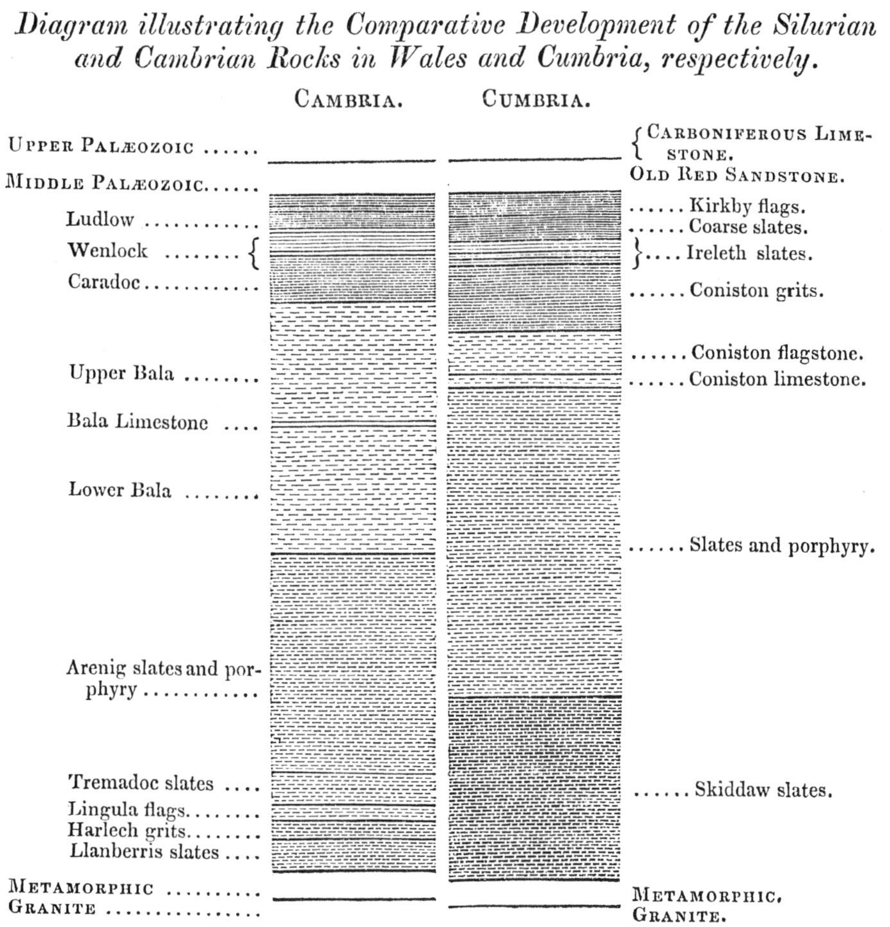

Fig. 2.6. Correlation of 'Cambrian' and 'Silurian' strata in Wales and Cumbria, according to Sedgwick (1852*b*, p. 150).

[was] actually interlaced with, the great central group of Cumberland, or the equivalent of the Cambrian' (Sedgwick 1852*a*, p. 54). Were this so, the Coniston Grits might be correlated with the controversial Caradoc Sandstone of Shropshire and the Ireleth Slates might be equivalents of the Wenlock Shales (Silurian). Sedgwick undertook brief excursions in the Lakes with Ruthven in 1851 and 1852 to look into this possibility (Sedgwick 1853*a*, p. 70).

The idea about the 'relocation' of the Coniston Limestone was developed further in one of Sedgwick's last papers on Lakeland stratigraphy (Sedgwick 1852*b*), as he retracted his ideas of 1841. This was the famous paper that slipped through the refereeing process at the Geological Society, with the help of the Cambridge President William Hopkins, who was sympathetic to Sedgwick in his tournament with Murchison (Secord 1986, p. 232). The paper had been favourably reviewed by Phillips, who had, however, recommended that its polemics against Murchison be toned down. But it went forward unchanged, and caused consternation amongst the Council members of the Society, to the extent that an attempt was made to withdraw the publication. By then it was too late, and the attempted censorship only served to inflame Sedgwick's anger, to the point that he never thereafter published with the Society.

Sedgwick now (1852*b*) placed emphasis on the fact that the Coniston Limestone was substantially different in lithology from the 'Caradoc Sandstone', and having begun to think (from as far back as 1845, he claimed) that the Coniston Limestone, at its southwestern end, was 'interlaced' with the Green Slates and Porphyries ('Otley II'), he was again pushing the Coniston Limestone down to a lower horizon. That is, it was being represented as Cambrian, for which claim Sedgwick held that McCoy – by then a professor at Trinity College, Dublin – was willing to vouch on the basis of fossil evidence. The status of the overlying Coniston (Brathay) Flags and Coniston Grits was less certain, however.

Sedgwick felt strengthened in his view that a re-classification was warranted when, as he recounted, he received a letter from Salter in November 1851 saying that he had identified fossils from the Brathay Flags, somewhat above the Coniston Limestone, as being of Llandeilo or Bala age (which had by then become *Cambrian* units by Sedgwick's book). Sedgwick's Lakeland stratigraphic sequence (1852*b*) was thus

10. Carboniferous Limestone
9. Old Red Sandstone
8. Kirkby Moor Flags[22] (Upper Ludlow: Silurian)
7. Coarse slates, flags, grits (approx. Lower Ludlow: Silurian)
6. Ireleth Slates (Wenlock: Silurian)
5. Coniston Grits (equiv. Caradoc Sandstone: Upper Cambrian)
4. Upper Cumbrian
    3. Coniston Flagstone (Upper Cambrian)
    2. Coniston Limestone (Bala: Upper Cambrian)
    1. Green slates and porphyries[23] (Upper Cambrian)
3. Skiddaw Slate (Lower Ca[u]mbrian)
2. Metamorphic slate
1. Granite

Comparisons between Welsh and Lakeland rocks were also made, diagrammatically, as is shown in Figure 2.6. It will be remarked from the list above that the old well-defined lithological arrangement of Otley's day was now being seriously compromised.

[22] There is a 'Kirkby Moor', close to Kirkby Ireleth, near the Duddon Estuary, i.e. towards the SW end of the line of exposure of 'Coniston rocks'. However, there are no exposures there of Kirkby Moor Flags, a name that has survived into the modern geological vocabulary. This unit crops out in the Kendal–Kirkby Moorside locality, and the name may have referred to hill country nearby, formerly called Kirkby Moor. Such a name does not appear on modern topographic maps, however.

[23] Sedgwick did not actually call them this, but gave a list of the component rocks of Otley's second unit.

However, the demands of palaeontology, the desire for coherence with the stratigraphic arrangements of other parts of Britain, and the demands for a palaeontologically grounded Cambrian System, were necessarily leading to this result. In his polemical attack, Sedgwick (1852*b*, p. 168) made much of the claim that his Cambrian system was not restricted to Wales; it was 'general and for all England'. One may doubt, however, whether Sedgwick would have suggested his stratigraphic reclassification of the Lakeland rocks if he had not been influenced by the controversies that were then going on about Welsh geology. Interestingly, he contended that the application of Smithian stratigraphy could not be undertaken *just* on the basis of fossils (which had not, in fact, been Smith's practice). Rather, physical sections had to be made out first, and then fossils could be used to determine where those sections fitted into the stratigraphic column.

In 1852, Sedgwick had made a visit with McCoy to 'Silurian Country' (i.e. the Welsh Border region), and the two had become convinced that the so-called 'Caradoc Sandstone' in different areas contained two different suites of fossils, the upper being of Wenlock vintage while the lower was apparently of Bala age. They gave the name May Hill Sandstone to the upper of these two groups, and proposed to correlate it with the Coniston Grits. It was this change, down south, that was driving the placement of a major stratigraphic boundary within 'Otley III'. However, for Sedgwick it had the great attraction of presenting the Cambrian with its own distinctive shelly fossils – something better than the still-obscure graptolites of the Skiddaw Slates.

Sedgwick's Lakeland work was now coming to a close. He was ageing, busy with ecclesiastical and University work, but still battling with Murchison about the placement of the Cambrian–Silurian boundary in Wales (Secord 1986). Having withdrawn from effective communication with the Geological Society, he maintained his arguments (and his rage) in unlikely publications such as a fifth letter to Wordsworth's *Guide to the Lakes* (Sedgwick 1853*a*), and the extraordinarily polemical introduction to the *Synopsis of the Classification of the British Palaeozoic Rocks . . . in the Geological Museum of the University of Cambridge* (Sedgwick & McCoy 1855).

The ire in this last publication was stirred up by the publication the previous year of Murchison's treatise, *Siluria* (Murchison 1854), which endeavoured to carry down the Silurian System to the first appearances of organic life in the Earth's crust. In fact, Murchison tried to have his cake and eat it. On the one hand, he complained that the rocks from the Skiddaw Slates up to the base of the Coniston Limestone were largely unfossiliferous, and in particular lacked the little brachiopod, *Lingula*, which was found low down in the flags of North Wales, and *might* there serve as a mark of Cambrian strata. On the other hand, he referred to Ruthven's discovery of graptolites and fucoids in the Skiddaw Slates, which gave them organic content, albeit limited. So, with some audacity, Murchison claimed the Skiddaw Slates, and all the rocks up to the Coniston Limestone, as Lower Silurian (Murchison 1854, pp. 146–147). The rich Coniston Limestone fauna he also had as Lower Silurian, being allocated to his Llandeilo Formation.

Hardly surprisingly, Sedgwick was incensed. He objected to the imperious claim that rocks must be Silurian simply because they contained graptolites. And the correlation of the unfossiliferous lower part of the Skiddaw Slates (with no base visible) with the lower part of the Longmynd rocks (again without visible base) was to him preposterous (Sedgwick & McCoy 1855, pp. lxxxiii–lxxxiv), though he had earlier countenanced such a correlation. In fact, it was now Sedgwick who wanted to emphasize palaeontological correlations, though, broadly speaking, during the Cambrian–Silurian controversy he was somewhat more reliant on lithologies and structure than was Murchison, who chiefly tried to use Smithian fossil-based stratigraphy. Sedgwick acknowledged that he had not seriously examined the Skiddaw Slates since 1824; but Murchison, he pointed out, had never examined them at all.

Sedgwick went once again to the Lake District in 1856, in the company of John Ruthven, and Thomas Gough of Reston Hall near Kendal, Curator of the Kendal Natural History and Scientific Society. Their time was chiefly spent in the Lancashire part of the Lakes, that is in the SW area around the lower reaches of the Duddon River, and Ulverston (see Figs 1.1 and 5.2). There they met one of the most remarkable amateur collectors of the nineteenth century, John Bolton (1791–1873),[24] and examined his extensive fossil collection. Sedgwick met with Bolton a second time the following year, and did two days of fieldwork under Bolton's guidance.

Following this work, Sedgwick, with Bolton's assistance, achieved a better understanding of the faults that afflicted the Coniston rocks in this southwestern region, seemingly as a result of the earth movements that had caused an inlier of Skiddaw Slates to be faulted into position (as was then thought) in the hills around Black Combe, north of the Duddon Estuary (see Fig. 6.3). It now appeared that, contrary to what was supposed in 1852, the Ireleth Slates were not an independent stratigraphic unit on a particular stratigraphic horizon, but were merely a lateral equivalent of the Coniston Flagstone, which was earlier thought to be a lower unit. However, further east in Bannisdale[25] there appeared to be slates that definitely overlay the Coniston Grit. So the unit Ireleth Slates disappeared from the books, and the Bannisdale Slates made their first appearance in their stead (Sedgwick 1843–1863 [1857], p. 188).[26] In addition, the Coniston Grit was tentatively correlated with the May Hill Sandstone, down on the

[24] See Deardon (1968), Shackleton (1974), and Kelly (1998–1999 [2000]). Bolton began his working life as a weaver at Ulverston at the age of nine, but studied natural history in the amateur style even in his teens. At 18, he moved to Barnsley, Yorkshire, to further his weaving skills; and on seeing one of the Jacquard looms he set to to design an improved version, which was soon in use in the Barnsley district. In 1842 he returned to Ulverston, where he set himself up as a land surveyor, and began serious fossil collecting, especially from the exposures of Coniston Limestone in his district, but also further afield. Later he engaged in archaeological excavations too; and some minor papers were published, even by the Geological Society. In 1869, at the age of 79, he published his remarkable *Geological Fragments*, which gives a splendid account of his life as an amateur geologist, and his fieldwork in the Lakes. (He was such an enthusiast that not infrequently he slept in the open on the fells in what he called 'Rock Hotel'.) Many good itineraries were described, and a general account of the geology of the region as it was known in the late 1860s. The topographic description of the Furness district to the SW of the Lakes was especially impressive. Some time before 1866 he moved to Swarthmore, naming his home 'Sedgwick Cottage'! In his old age, Sedgwick used his good offices to obtain a small pension for Bolton from the Queen's Bounty Fund, and he recorded that Bolton made laborious collections of fossils and recorded their localities accurately. It was Bolton's work that enabled Sedgwick to 'correct some of the blunders . . . [he] had made in the arrangement of the groups which appear in Low Furnace, and enabled . . . [him] . . . to interpret the gigantic faults which break up the stratas on the shores of the Duddon estuary, and throw the whole mass of the Country between Dalton and Ulverston, 4 or 5 miles out of their normal position' (letter from Sedgwick to un-named recipient, 18 January, 1873, facsimiles of which are bound into a few copies of Bolton's *Geological Fragments*, e.g. one held at the public library, Barrow-in-Furness).

[25] A small valley, parallel to and NE of Longsleddale, running NW from the old A6 main road between Kendal and Penrith. See Figure 3.7.

[26] It is stated in the *Lexique Stratigraphique* for the British Silurian (Whittard & Simpson 1961) that the term 'Bannisdale Slates' was used by Sedgwick in the field or in his notebooks, but was not published by him. He did indeed mention 'the slate beds in Bannisdale' in his journal for 15 August 1823. The name was also mentioned in print in his paper of 1843–1863 (1857), though he only said that the term was appropriate when a limestone that appeared to lie within the unit in some places was absent.

western side of the Malvern Hills, which unit, as we have seen, Sedgwick (1853*b*) had recently introduced as an upper division of Murchison's controverted 'Caradoc Sandstone'.

Sedgwick was a giant in the early years of Lakeland geology. For thirty years and more he, and to some extent Sharpe and Phillips, were virtually the only geologists of note who did substantial research in the district. He published no less than 23 papers that dealt with the area, or made substantial reference to the Lakes in the context of his other theoretical or fieldwork. Sharpe, in fact, did not pursue his Lakeland work, so when the geologists of the next generation took over in the late 1850s (see Chapter 3) they picked up from Sedgwick, leaving aside Sharpe's investigations. One or two other amateurs did stratigraphic work (e.g. Marshall 1840; Binney 1848) but their contributions were even less influential. Sedgwick's Cambridge ally, the mathematician William Hopkins (1793–1866) (1842, 1848), used the results of Sedgwick's survey work, and indeed Sedgwick's own ideas about the 'catastrophic' scale of earth movements in the Lake District (as evidenced by the rupture of the Coniston Limestone and associated strata) in an interesting way, seeking to link it to his theory of how sudden uplifts of the Earth's surface might occur, so as to give rise to topographic features such as the Lakeland mountains.[27] However, this work was tangential to the unravelling of the stratigraphic sequence in the Lakes, which was the first and most pressing task for geological research in the region.

Earlier accounts of Sedgwick's work, other than those of Rudwick and Secord, which tell much about his work as a geological controversialist, have perhaps suggested that grand questions of a theological or metaphysical nature were central to Sedgwick's work (e.g. Gillispie 1951). It is certainly the case that Sedgwick was at first a Wernerian, and in his younger days also believed in the idea that the Biblical Flood might account for numerous superficial or 'diluvial' deposits that he observed in the field. Indeed he was always a 'catastrophist', and wrote vehemently against transmutationist (evolutionary) ideas. Moreover, he averred in his only major book: '[g]eology, like every other science when well interpreted, lends itself to natural religion' (Sedgwick 1833, p. 22). 'It proves that a pervading intelligent principle has manifested its power during times long anterior to the records of our existence' (Sedgwick 1833, p. 23).

So principles of natural theology assuredly underpinned, or acted as a protopostulate to, Sedgwick's geological work. But first and foremost he was a field geologist and stratigrapher, and, we might say, an early structural geologist. His aim was to know the rocks, sort them out into their correct stratigraphic order, and determine the structures of the strata of the regions that he investigated, and the sequence of events that might have led to the formation of the rocks that he observed in the field. He wanted to discover the law-like pattern of regularities of structure that he thought he could discern amongst the rocks, and then he sought a physical theory to account for the pattern. These were his geological goals. They were accomplished to a large extent single-handed so far as the generalizations were concerned, though, as we have seen, he was assisted in finding important exposures and fossils by amateurs such as Otley, Ruthven and Bolton.

It should be said again that Sedgwick's stratigraphic work in the Lake District was to a considerable degree driven by his controversies in Wales and the Welsh Border area. The several reclassifications that he made, as described above, were not just the product of his considerations of Lakeland geology. He was trying at first to accommodate himself to the demands of Murchison's Silurian empire building. Then, when the great rupture with Murchison occurred, and Sedgwick struck out on his own to reassert his Cambrian domain, subdividing Murchison's Caradoc Sandstone and introducing the boundary between the Cambrian and the Silurian into the middle of that unit (Secord 1986, p. 245), this had repercussions in the Lake District, such that the Cambrian–Silurian boundary was now required to be located within Otley's third unit. As in all geological (and other scientific) research, results had to be made to cohere with one another as best they could, giving the closest fit with observations that could be achieved. But so far as Lakeland stratigraphy was concerned, the fit was not so very comfortable for Sedgwick, even at the end of his career.

On the other hand, Sedgwick's stratigraphic column, given on p. 23, is recognizable to modern geological eyes:

Sedgwick	*Modern*
Carboniferous Limestone	Carboniferous Limestone
Old Red Sandstone	Old Red Sandstone
Kirkby Moor Flags	Kirkby Moor Flags
Coarse slates, flags, grits	Underbarrow Formation?
Ireleth Slates	Bannisdale Slates
Coniston Grits	Coniston Grits
Coniston Flagstone	Brathay Flags
Coniston Limestone	Coniston Limestone Series (obsolete term; includes modern Ashgill)
Green slates and porphyries	Borrowdale Volcanics Group
Skiddaw slate	Skiddaw Slates
Metamorphic slate	Metamorphic aureoles round Skiddaw Granite, etc.
Granite	Granites or granophyres of Skiddaw, Eskdale, Ennerdale, Shap, etc.

Thus Sedgwick had a grip on the lithostratigraphy of the Lakes, and his units, mappable in the field, provided the framework for the Surveyors' work that was soon to follow (see Chapter 4). Obviously, there have been numerous subdivisions of these units since Sedgwick's day, and we shall endeavour to follow the development of Lakeland stratigraphy in our subsequent narrative, particularly in Chapters 5, 8 and 13. However, although much was to follow, by the end of his career Sedgwick had achieved a remarkably good grasp of Lakeland lithostratigraphy.

Mention may also be made here of an unpublished sketch section made by the out-of-work palaeontologist John Salter, who had resigned from the Survey in 1863, following a row with Thomas Henry Huxley (Secord 1985). In 1864, when Salter was hungry for employment, he was getting some assistance from Sedgwick (doubtless willing to assist any refugee from the Murchison citadel in London) and did some palaeontological work for him at Cambridge, eventually published in Sedgwick & Salter (1873). Whether Salter's troubles at the Survey had anything to do with the Sedgwick–Murchison dispute is not known, but the manuscript sketch (see Fig. 2.7) gives a Sedgwickian view of the relationships between the stratigraphy of the Lakes and Wales, at the same time showing how they would appear from a Murchisonian perspective. The diagram is worth reproducing here as, to my knowledge, Sedgwick produced no similar figure from the same period. The provenance of the figure is also interesting: it turned up in 1999 among the McKenny Hughes papers, deposited that year at Cambridge by one of his descendants (see Chapter 4, Note 3). Hughes joined the Survey in 1861 and began mapping in the north of England, moving towards the Lakes, in 1866. He is known to have met with Sedgwick in north Lancashire at that time (Anon. 1906), and was almost certainly influenced by him, as he shortly thereafter published a somewhat Sedgwickian paper (Hughes 1867; see Chapter 4). Whether Salter made the sketch for Hughes in 1864 in order to explain to him the issues in contention in the Sedgwick–Murchison debate, or whether Salter made it to help himself clarify his ideas, and Hughes later got hold of it, cannot now be determined, but it may have made Hughes lean towards

[27] On Hopkins, see Smith (1989), and Chapters 18 and 19.

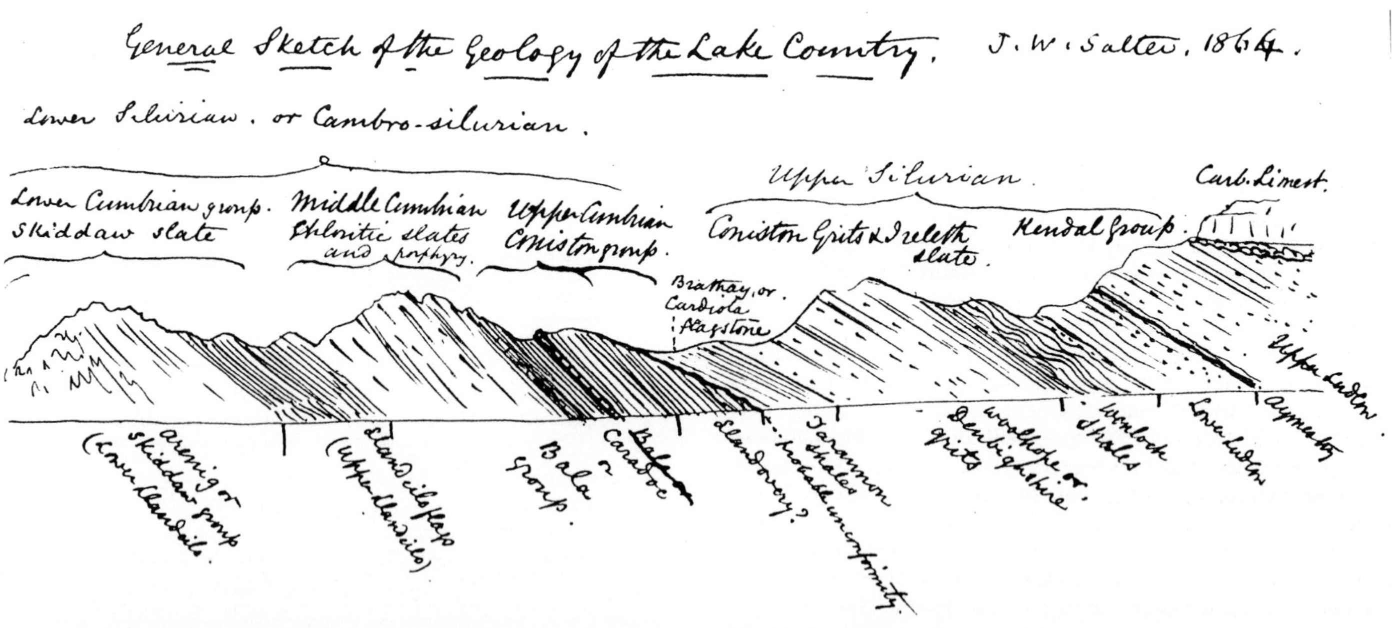

Fig. 2.7. Sketch-section of the geology of the Lake District, according to John W. Salter (1864). Cambridge University Library (Add. MSS 9557). Reproduced by permission of the Syndics of Cambridge University Library.

Sedgwick's position. And it interests us here as it gives what is reasonably interpretable as the elderly Sedgwick's view of Lakeland stratigraphy.

Here, then, we take leave of Adam Sedgwick in the Lake District. By heroic endeavour, he had accomplished much conceptually, almost single-handed (though with considerable practical assistance from local collectors). The problem of the awkward placement of the boundary between the Cambrian and Silurian that he proposed was not resolved in his lifetime. There was much more to be done on the subdivision and classification of 'Otley III'. The Skiddaw Slates needed closer investigation. Hardly anything had been done to unravel the mysteries of Otley's second unit with its 'volcanic silt' or 'chloritic slate', and a host of unusual igneous rocks and complex sedimentary structures. This story begins rather than ends with Sedgwick's heroic investigations.

Chapter 3
Robert Harkness, Henry Alleyne Nicholson and Charles Lapworth

For some forty years after Adam Sedgwick took to the field in the Lakes in 1822, he was virtually the only major geologist to make the region one of his main objects of study. We have seen how he made considerable efforts to sort out the stratigraphy of the southern Lakes, relating his work there to his investigations in Wales, but to a large extent he left the central volcanic region alone; likewise the Skiddaw Slates – apart from claiming them for his Cambrian System on the basis of their apparently meagre fossil contents, as collected by Ruthven.

The man who took up from where Sedgwick left off was Robert Harkness (1816–1878) (see Fig. 3.1).[1] His family came from Ormskirk, near Liverpool, but he attended school in Dumfries in SW Scotland, his father's home town. From there he went to Edinburgh University, where he studied under the geologists Robert Jameson and James David Forbes, and the chemist Thomas Charles Hope. In his early twenties Harkness was making geological investigations in Lancashire, particularly among the coalfields and the rocks of the New Red Sandstone.

In 1848, the family moved to Dumfries, from which centre Harkness, being of independent means, began geological investigations among the rocks of the Southern Uplands, including the region near Moffat, which, rich in graptolites, later became a classic site through the work of Charles Lapworth (1842–1920) (Hamilton 2001). Thus early in his career, Harkness became familiar with these Lower Palaeozoic fossils. In fact, his interest in them was stimulated by Sedgwick, who exhibited examples of the organisms at the British Association in Edinburgh in 1850 (Shackleton 1966*a*). Sedgwick introduced Harkness to John Ruthven and also put him in touch with John Salter. Harkness subsequently submitted his finds to Salter for identification.

Harkness was particularly interested in comparing the Triassic rocks with which he was familiar in Lancashire with those of the Solway Basin and the Carlisle district. His work there attracted favourable attention, and he obtained a chair in geology at Queen's College, Cork, in 1853. He was supported by a number of eminent geologists, including Murchison, who, when informing him of the success of his candidature, referred to him as a 'good Silurian' (Goodchild 1882–1883, p. 153). The following year, with the support of Henry de la Beche, Harkness was offered a chair in India but he declined the opportunity. In 1865, feeling the strain of his teaching load in Ireland, which included lecturing in geography, botany and zoology, as well as geology, Harkness made an attempt to move to Edinburgh, but being unsuccessful, he remained at Cork. He retired in 1877, to settle with his married sister in Penrith, but in October 1878, on a journey back to Ireland to examine some of his former students, he died suddenly of heart disease, premonitions of which had prompted his retirement.

Harkness's first visit to the Lakes was in 1849, and thereafter, having relatives in Penrith (New Red Sandstone country), he made repeated visits to the region. He made important comparisons between the rocks of Ireland and those of Cumbria, as well as the Old and New Red sandstones of Scotland and northern England.

Though, as already said, Harkness was (in 1853) a 'good Silurian', he had cordial relations with Sedgwick, and from an important letter that the old geologist wrote to the rising man in August 1856 (Cambridge University Library, Sedgwick Papers, Add. 7652/IVA13), it appears as though he were handing over his Lakeland baton to Harkness. At any rate, Sedgwick gave him advice as to how and where to undertake his researches. He suggested that he call on Ruthven in Kendal, Jonathan Otley and another guide Charles Wright in Keswick, and James Garth Marshall of Coniston who would direct him to good localities there. Sedgwick recommended Otley's guidebook, and his own 'letters' in Wordsworth's *Guide*, as introductions to Lakeland geology.

Fig. 3.1. Robert Harkness. Copy of engraving accompanying his obituaries (Anon 1878; Goodchild 1882–1883).

Evidently, Harkness was particularly interested in the fossils that might or might not be found in the Skiddaw Slates (SS), and Sedgwick recommended that a good place to look might be the slate hills to the NE of Crummock and Buttermere (up Mill Beck, I presume; see Fig. 12.3), where he had in fact himself described fossils (Sedgwick 1848). He also recommended investigation of the isolated block of SS in the SW corner of the Lakes at Black Coomb (Black Combe) (see Fig. 6.3). It was, he said, 'of Skiddaw Slate, brought up by enormous dislocations & its ravines are of good promise'. Interestingly, Sedgwick emphasized that the observations did not indicate a simple relationship between cleavage and pressure or stress. Indeed, he wrote, 'you have to acct. for a second cleavage plane among beds that are by no means crystaline [*sic*]'. This was, perhaps, the first indication of the idea that the SS had been affected by more than one episode of stress and consequent cleavage. Or it may have been no more than an indication of the importance to Sedgwick of distinguishing bedding from cleavage. It may be remarked that in fact very few fossils have subsequently been found in the Black Combe area.

With considerable energy, then, Harkness set out to determine more accurately the total outcrop of the SS in the Lakes, their structural features, the extent to which they displayed signs of metamorphism, and their fossil contents. The results of his labours were presented to the British Association in 1857 (Harkness 1858)

[1] On Harkness, see Anon. (1878), Goodchild (1882–1883), Shackleton (1966*a*), Fox (1998–1999 [2000]).

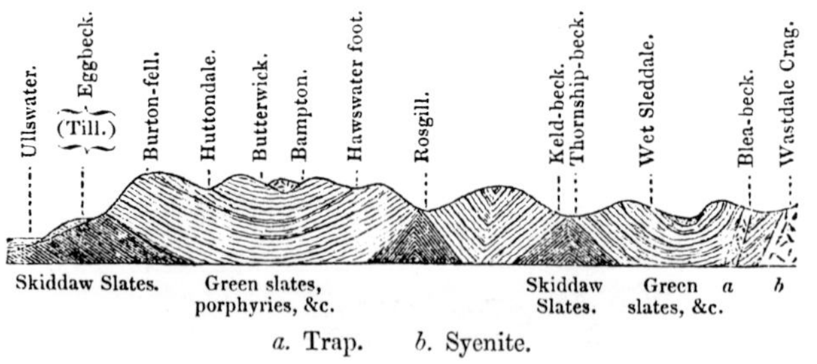

Fig. 3.2. The structure of the eastern Lake District, according to Robert Harkness (Harkness 1863, p. 127).

and then to the Geological Society in a paper read on 17 December 1862 (Harkness 1863). Harkness examined particularly the region between Derwent Water and Crummock (see Figs 12.1 and 12.3) – or from Newlands Valley northwestwards, perpendicular to a series of ridges and valleys, and parallel to the west side of Bassenthwaite, extending the traverse to Sunderland (a small village NE of Cockermouth; see Fig. 8.6). From the fell road that runs from the Newlands Valley to Buttermere, his traverse took him across what appeared to be an anticline running parallel to Coledale Beck and a parallel syncline, roughly perpendicular to Bassenthwaite, through Wythop parish (see Fig. 12.1).

Harkness was now able to describe a number of fossil localities, yielding chiefly graptolites found in various screes along the line of his section. This was not particularly satisfactory for the accurate identification of strata according to their fossil contents in a Smithian fashion. However, it did bring the Skiddaw Slates indubitably into the zone of life – whereas in his early investigations of the 1820s Sedgwick had not seriously looked for fossils in the oldest slates, since, as he put it in his letter to Harkness of 1856, he had not then shaken off his 'Wernerian nonsense', thinking that the Skiddaws had been deposited before life appeared on Earth. However, for Harkness, even what he took to be the lowest SS from near Threlkeld, east of Keswick and below Blencathra, seemed to have indications of worm traces (Harkness 1858). (By the time of which we speak, upper Skiddaw fossils had already been added by Sedgwick to the Woodwardian Collection of fossils at Cambridge.)

Harkness supposed the fossils (graptolites) – which occurred in the occasional coarser beds of a 'flaggy nature' – indicated that the SS belonged to Murchison's 'Lower Llandeilo' subdivision of his Lower Silurian, and were analogous to the fauna of the Quebec beds of Canada, the graptolitic strata of which had been described by the American geologist James Hall. (Analogies with Australian types, found near Melbourne by McCoy, who had taken up a chair in the Antipodes, were also mentioned.) The new Lakeland specimens were described for Harkness by Salter, who also figured 15 types from the new Skiddaw assemblage. Indeed, he referred to the SS as the 'metropolis of the Graptolites'. It appears that Harkness did not himself do all the collecting. He mentioned that a Keswick mineral dealer, Joseph Graham, was the first to discover the fossil localities near Keswick, as at Outerside, Barf (close to the western shore of Bassenthwaite), Skiddaw and Longside. Other collectors mentioned were a mineral and fossil dealer, W. West of Wimpole Street, London, and another dealer, Bryce Wright of Great Russell Street.[2]

Examining the SS as a whole, Harkness decided that there was insufficient differentiation, either palaeontological or lithological, to warrant their subdivision. They appeared to be about 7000 ft thick and the overlying 'greenish-grey rocks' (Borrowdale Volcanics) might be some 14 000 feet thick. Importantly, though he did not discuss the issue specifically in his paper, Harkness represented the contact between the SS and the overlying volcanics as *conformable* in all his figured sections, which were for the Newlands–Embleton Valley–Derwent Valley area; Matterdale–Uldale; Ullswater to Wasdale Crag (near Shap) (see Fig. 3.2); and Black Combe. The Coniston Limestone was accepted as an equivalent of the Bala Limestone of Wales.

Harkness thought that the Lakeland graptolites had affinities with those he had discovered in the Southern Uplands; and the rocks' similarity of strike suggested to him (in the style of Élie de Beaumont) that the two sets of hills were uplifted at about the same time. All the main Lakeland rocks (i.e. Otley's three units) were seemingly uplifted together, before the unconformable deposition of Upper Old Red Sandstone rocks. So the elevation might have occurred in Lower Old Red Sandstone times. Harkness's fossils included a shrimp-like simple crustacean, *Caryocaris*, and the several graptolites. He averred that the existence of such complex organisms in ancient rocks counted against the 'progression theory' or the theory of evolution (Harkness 1864).

Harkness next had regard to the rocks of the Cross Fell Inlier, producing a simple map and section of the area (see Fig. 3.3) (Harkness 1865). His attention was drawn to an important fossil locality (still a classic site) at Pus Gill (or Pusgill), to the NE of the conical hill, Dufton Pike, by William Wallace, resident of the village of Dufton and author of *The Mineral Deposits of Alston Moor* (1861) (see Fig. 1.3). Similar richly fossiliferous deposits were found in stream beds nearby, with crinoids, corals, brachiopods and trilobites. Henry Alleyne Nicholson (1844–1899), who was soon to become Harkness's co-worker and co-author, was also collecting in the area at about the same time.[3]

A little further to the SE of the Pus Gill beds Harkness reported the occurrence, near Keisley (see Fig. 1.3), of a fossiliferous limestone that had been worked in several quarries for many years. The fossils appeared to be of Bala age, and the strata in which they occurred seemed to be approximately in line with the outcrop of the Coniston Limestone in the main body of the Lake District. So he proposed that it was an eastward extension of that well-known unit. Further analogy was suggested with some strata in Kildare, Ireland. A little south of the Keisley Limestone, Harkness found what appeared to be a faulted contact, bringing back the SS. So one might say that the rocks of the Cross Fell Inlier were a kind of microcosm of the larger mass of rocks in the Lake District proper.

If Sedgwick passed his baton to Harkness, one could also say that Harkness passed it to Henry Alleyne Nicholson (1844–1899) (Fig. 3.4), though the two undertook collaborative research in the Lakes before Harkness retired from the field. Nicholson was born in Penrith, son of an Oriental scholar, Dr John Nicholson. After schooling at Appleby Grammar School, he studied under the Göttingen University zoologist and palaeontologist, Wilhelm Keferstein. Nicholson then moved to Edinburgh University where he read medicine from 1862 to 1867, taking a first in all subjects and gaining the University's Gold Medal. It may seem extraordinary today, but Nicholson's graduation thesis was an essay on the geology of the Lake District. It was published the following year (Nicholson 1868*a*), and constituted the first book to be published on the geology of the region. It was dedicated his 'friend and teacher' Robert Harkness.

Nicholson's career soon flourished further. After a period as medical practitioner, he turned to his real interest: natural history. He was swiftly appointed to chairs at Toronto (1871), Durham

[2] To this day, one hears of fossil dealers ransacking Lakeland sites, much to the annoyance of geologists proper. One site at Longsleddale has been fenced off, but the depravations are said to continue. In the nineteenth century, collectors such as John Bolton amassed huge amounts of fossils, for the joy of the specimens, probably much to the disadvantage of subsequent scientific work.

[3] On Nicholson, see Lapworth (1899), Whitaker (1899), Hinde (1899), Anon (1903*b*), Marr (1905*b*), Shackleton (1969), Dickins (1998–1999 [2000]). The importance of professional collectors and dealers of fossils to nineteenth-century British geology is explored in detail by Knell (2000).

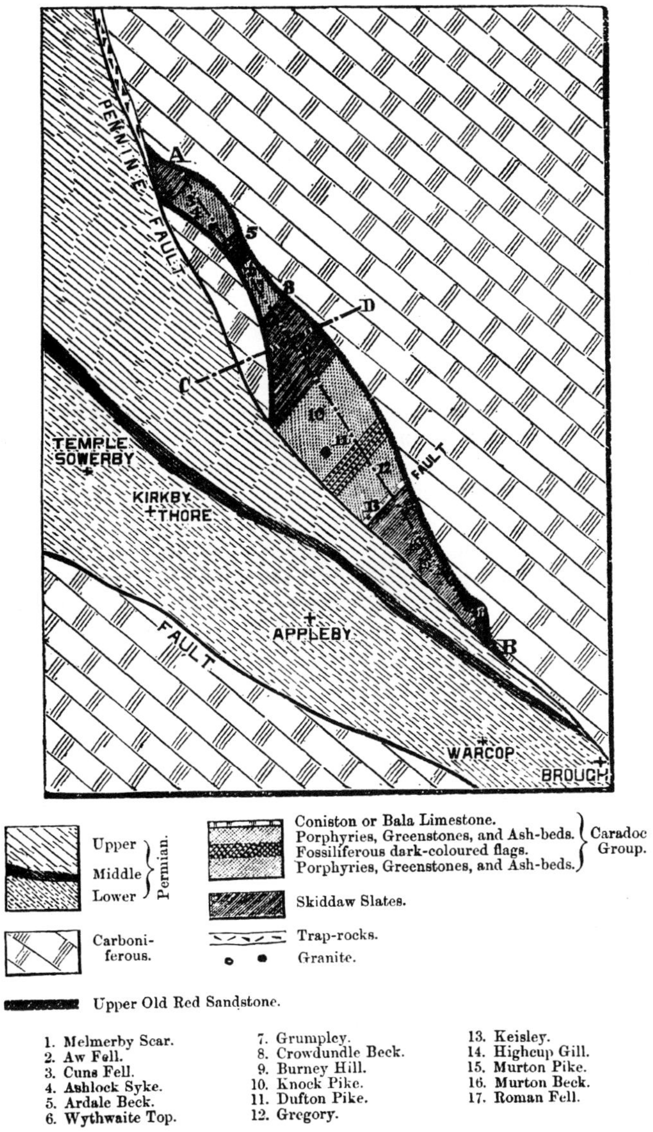

Fig. 3.3a. Geological map of area of Cross Fell Inlier, according to Harkness (1865, p. 236).

(1874), St Andrews (1875), and Aberdeen (1882–1899), where he was Regius Professor of Natural History; and he was elected FRS two years before he died. Nicholson's teaching was in natural history not medicine, and his specialist work was in palaeontology and stratigraphy. He missed out on a chair at Edinburgh – it is said because of his support for evolutionary doctrine (Hinde 1899, p. 140).

In Canada, Nicholson gained experience with Lower Palaeozoic fossils, especially corals, and he later developed knowledge of the anatomy of such organisms by preparing thin sections and examining them microscopically. He likewise became a leading British authority on graptolites (especially through his *Monograph of the British Graptolitidae* (1872*a*) and on stromatoporoids.[4] Besides his early collaborative work in the Lakes with Harkness, Nicholson also carried out major stratigraphic work in the district with J. E. Marr (see p. 59).

Presenting a paper to the Geological Society on 9 May 1866, on the results of their fieldwork for the previous summer, Harkness & Nicholson (1866), referred to their investigations of the rocks and fossils of the Cross Fell Inlier. In Eller Gill, a stream running up to the Pennine scarp east of Milburn, Harkness found a trilobite (named *Agnostus morei* by Salter) from the Skiddaw beds, which could be correlated with rocks in Shropshire, and which appeared to be of Lower Llandeilo age, i.e. close to the bottom of Murchison's Silurian System. (The actual horizon for the Silurian base was, of course, hotly disputed at that time, and the Cross Fell rock would have been Cambrian by Sedgwick's reckoning at that time.) The unfossiliferous upper beds of Skiddaw rocks in the Inlier were tentatively placed in the Upper Llandeilo. So the geologists were beginning to 'get a handle' on the stratigraphy of the most ancient rocks then known in the north of England.

Regarding the relationship between the SS and the overlying 'greenstones, porphyries, and ash-beds' ('Otley II'), Harkness & Nicholson (1866) asserted that the two units were conformable to one another. They did not reveal the empirical evidence (or the geographical locality) on which this important claim was based, which is unfortunate since it was an issue that was to form a central focus of attention in Lakeland geology until relatively recently. Perhaps Nicholson took Harkness's word for it without examining the issue closely. Investigations were also reported for the rocks of southern Lakeland around Windermere and Coniston, and over to the west near Broughton and Ulverston. In

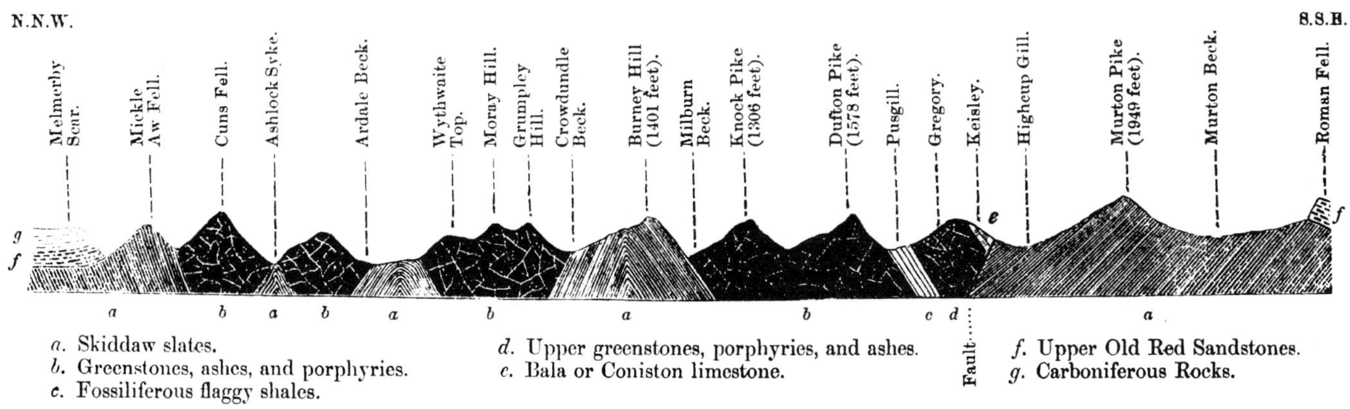

Fig. 3.3b. Horizontal section of Cross Fell Inlier, according to Harkness (1865, p. 239).

[4] Laminated calcareous organisms, now extinct, and perhaps related to the hydrozoans.

Fig. 3.4. Henry Alleyne Nicholson. Undated photograph in Geological Society archives (LDGSL P46/10).

that area, they departed from Sedgwick, identifying his Ireleth Limestone with the Coniston Limestone, the outcrop being repeated by faulting, but they accepted the correlation of the unit with the Bala Limestone of Wales. Significantly, they claimed that there were no Llandovery or Wenlock rocks in the Lakeland area, referring the Coniston Flags and Coniston Grits to the Upper Caradoc. The existence or otherwise of Llandovery rocks[5] in that part of the world subsequently became a topic that exercised geologists' attention (see p. 43). Finally, Harkness & Nicholson discussed the systems of faults in the Lake District, pointing out that the several lakes might be aligned according to different systems of faults of different ages. Nicholson (1868*a*, p. 11) subsequently credited Harkness with the discovery of the occurrence of 'two great systems [of fault] which cut each other at right angles', though Phillips had earlier had ideas on this question.

By this time, members of the Geological Survey were beginning to work their way up from Yorkshire, through north Lancashire, and towards (but not yet into) the Lake District. Some small inliers of 'Lakeland' rock were encountered on the way, near Casterton and Kirkby Lonsdale, in the lovely Lune Valley running up from Lancaster below the Pennine scarp towards Sedbergh; and somewhat to the east, in valleys to the south and west of Horton-in-Ribblesdale. (As mentioned in Chapters 1 and 2, they had previously been described in a preliminary way by Phillips (1829), and Sedgwick (1846, 1852a) used them to support his revised belief that the Coniston Limestone should be regarded as Bala rather than Caradoc.) In particular, Thomas McKenny Hughes (1832–1917), then with the Survey (which he had joined in 1861), was at work in the district. Son of the Bishop of St Asaph, Hughes had studied under Sedgwick at Cambridge. In 1873, he succeeded to the Woodwardian chair on which his old master had sat for so long, and years later he co-authored Sedgwick's *Life and Letters* (Clark & Hughes 1890). Hughes was always one of the strongest supporters of the great pioneering geologist. The two met in the Sedbergh region in 1866 (Anon. 1906, p. 3).

In a paper of 1867, Hughes said that he opposed the opinion of Harkness & Nicholson (1866) that the Coniston Flags belonged to the Caradoc, and thought that there had been some mistake in the collection of fossils from the Flags. Salter, Hughes said, had stated in an unpublished letter that there might have been two sets of flags, from which the confusion might have arisen (Hughes 1867, p. 347). Hughes went on to describe in some detail his observations to the SE of the Lakes in the Ribble Valley, in the Doe Valley to the west (running up from Ingleton to Chapel-le-Dale), and in the intervening Crummack Dale (see Fig. 3.5), all with their ancient 'Lakeland' rocks, with Carboniferous Limestone overlying them unconformably (broadly to the NE); and bounded to their SW by the Craven Fault(s) (see Fig. 1.5), which bring Upper Carboniferous strata against the ancient 'Silurians'.

Hughes contended that there was evidence of unconformity *within* the ancient rocks. The upper part, he said, consisted of grits, slates and flags (Coniston Flags), with a basal conglomerate evident in some outcrops. Below, there were some cleaved slates, the Coniston Limestone and 'green slates' (equivalent to Sedgwick's Lakeland 'Green Slates and Porphyries'). The upper unit had graptolites in its lower beds and the fossils *Cardiola* (a mollusc) and *Acidispus* (a trilobite) in its upper sandy slates, the latter fossil indicating a Wenlock age. (The lower beds contained few fossils, and those that had been found did not pinpoint the stratigraphic horizon precisely.) Hughes claimed that the upper unit did not always lie on the same beds of the lower unit: hence an unconformity. The fossils in the Coniston Limestone indicated a Bala age. So, following his teacher, Sedgwick, Hughes was inclined to link the Coniston Limestone (Bala) with the underlying 'Green Slates and Porphyries', and have Coniston Flags lying above unconformably, though of somewhat indeterminate age. However, the evidence for unconformity was not clear-cut. Hughes, like Sedgwick, was breaching Murchison's Silurian, with the Coniston Flags at the base of the upper division. It was recorded (Anon. 1906, p. 3) that this move was 'received with some disapprobation on the part of the Director-General of the Geological Survey [Murchison]'. The issue will be discussed further in Chapter 4.

When Nicholson published his doctoral thesis (*An Essay on the Geology of Cumberland and Westmorland*, 1868*a*), a remarkable work of synthesis for one so young, he was not inclined to follow Hughes in this matter, leaning at that time towards Murchisonian views (as were most of the members of the geological community outside Cambridge).

Following Hughes's work, Harkness and Nicholson spent the summer season of 1867 in the Lakes, giving a preliminary account of their results at the British Association in Dundee that year – which was not, however, published – and a fuller version in March 1868 at the Geological Society (Harkness & Nicholson 1868). They were able to find numerous graptolites in the rocks above the Coniston Limestone, particularly at Skelgill and Longsleddale (see Figs 2.2 and 3.6), which appeared to be of Caradoc type; and an apparently uninterrupted conformable sequence of strata, extending from the 'Green Slates and Porphyries', through the Coniston Limestone, graptolite-rich mudstones (which formed an important marker band in Lakeland geology), grits, the Coniston Flags,[6] to the Coniston Grits, all regarded as *Lower* Silurian (in

[5] The unit that Murchison envisaged as forming the 'transition' between his Lower and Upper Silurian, and which today forms the lowest of the main subdivisions of the Silurian.

[6] The mudstones, grits and flags were all taken as belonging to the Coniston Flags.

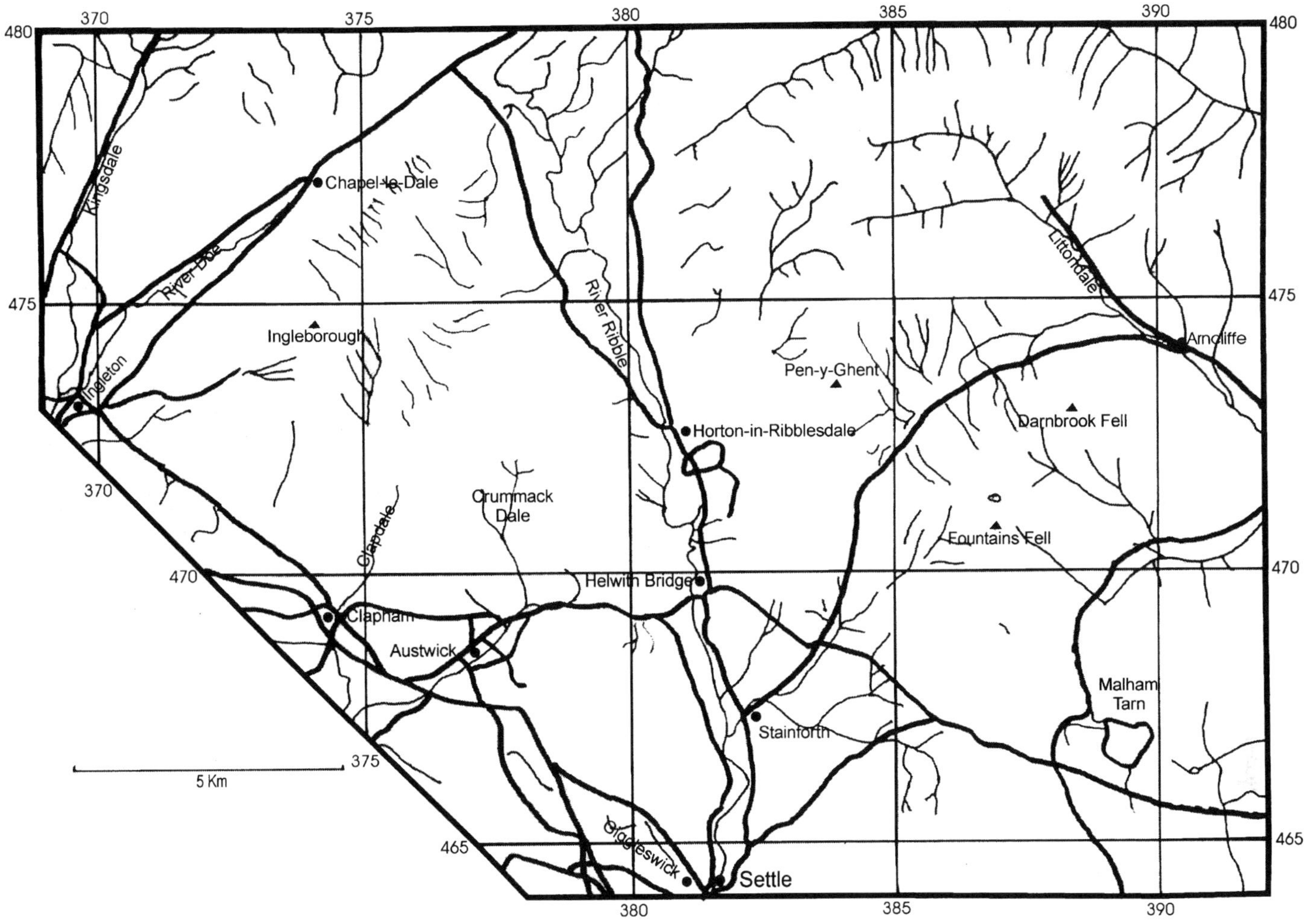

Fig. 3.5. Topography of the area of the Craven Inliers, NW Yorkshire.

Murchisonian terms) – though Nicholson acknowledged that the Coniston Grits appeared to contain some admixture of Upper Silurian fossils.[7] However, the argument was based on what the two geologists had observed in southern Lakeland, not down in Ribblesdale (Yorkshire), etc., where Hughes had been working.

Regarding 'Otley II' (or Sedgwick's 'Green Slates and Porphyries'), Harkness & Nicholson reported their discovery in 1865 of a fossiliferous band of shales just below the last 'porphyry' of the unit, on which lay the Coniston Limestone.[8] These shales cropped out at a locality called Style [*sic*] End Grassing between Longsleddale and Kentmere (see Fig. 3.6), and also by the road between Ambleside and Coniston, near Sunny Brow. The fossils included the forms designated *Petraia sub-duplicata*, *Orthis vespertilio* and *Stenopora fibrosa*, all deemed to be Bala types. Like remains had also been found in the Cross Fell Inlier by Harkness. Thus it appeared that the 'Green Slates and Porphyries' (as then delimited) were, in their upper beds at least, of Bala age. Nicholson thought that the Coniston Limestone was *conformable* to the 'Green Slates and Porphyries'; and since the underlying SS were, on the basis of their graptolite contents, regarded as Lower Llandeilo, it seemed reasonable to regard the unfossiliferous lower parts of the 'Green Slates and Porphyries' as Upper Llandeilo. However, the relationship between the SS and the 'Green Slates and Porphyries' was not then certain – and was to remain contentious for many years. Nicholson (1868*a*, p. 33) simply said that it was 'a question which admits of doubt, though data are wanting to arrive at a definite conclusion'.

In 1868, Nicholson read two papers to the Geological Society (Nicholson 1868*b*, *c*) in which he formally described the graptolites of the SS and then those of the Coniston Flags (approximately equivalent to what were subsequently called the Brathay Flags), which he subdivided into: (1) black mudstones or shales alternating with grey grits; (2) cleaved flags; (3) 'sheerbate' flags.[9] The various graptolite types were also figured and the localities of the specimens stated (see Figs 3.7 and 3.8). It should be noted that many of the 'Flag' types were figured from Scottish specimens, indicating the analogy between the Lakeland rocks and those of the Southern Uplands. For the Skiddaw types, strong analogies were noted with graptolites of the Quebec Group from Canada, being of Lower Llandeilo age; and the Coniston Flags were – by North American and Bohemian analogies – of Caradoc/Bala age. Further types were described and figured the following year (Nicholson 1869*a*).

As mentioned, Nicholson left open the question of the relationship between the SS and the 'Green Slates and Porphyries' in his doctoral thesis (1868*a*), but in November that year he revisited the Lakes to re-examine the junction between the two units in the northern Lakes, publishing his results in two papers in March 1869

[7] Today, they are located in the Ludlow group (Upper Silurian).

[8] The discovery was reported at the meeting of the British Association in 1865, but was not published. (See Nicholson 1868*a*, p. 35.) Publication occurred in Harkness & Nicholson (1866) and later in Harkness & Nicholson (1877). The unit was subsequently known as the 'Stile End Beds'. 'Grassing' = pasture.

[9] 'Sheerbate' was an old quarrymen's term for flags or slates in which the plane of cleavage coincided with the bedding (though the quarrymen did not think of the matter in that theoretical fashion), so that it parted with a hackled fracture.

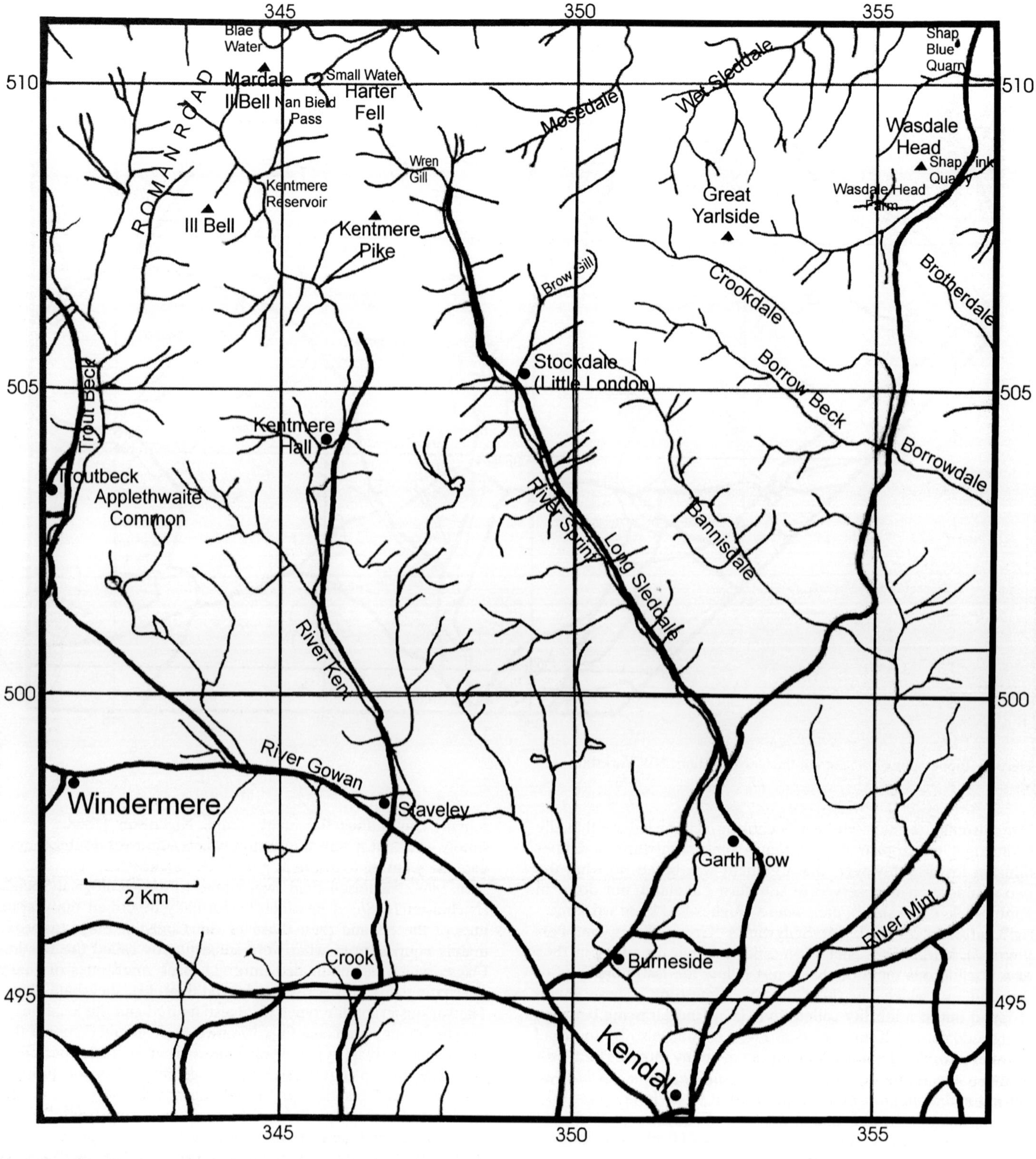

Fig. 3.6. Topography of the area of Kentmere and Longsleddale, southeastern Lake District.

(Nicholson 1869*a*, *b*). He made a careful examination of the whole line of junction in the Lakes, so far as there were manifest exposures, except for the region of Black Combe to the SW of the region. It appeared to Nicholson that, on the basis of differences of dip and strike, and because the 'Green Slates and Porphyries' appeared to lie sometimes on the upper and sometimes on the lower parts of the SS, there could be no doubt that there was a substantial unconformity between the two units. There were, however, certain problems with his argument. For the purpose of the problem in hand, the upper and lower SS were merely differentiated on the basis of lithologies, not fossils. Also, Nicholson took the distinctive, sometimes strongly porphyritic, igneous rocks of Eycott Hill (to the east of the main mass of the northern part of the Lakeland hills) to belong the 'Green Slates and Porphyries', though they are rather different from the lavas of the main body of the Lakes. Furthermore, he could not find a clear contact that, in itself, displayed unconformity. (For example, the visible contact at Hollows Farm, in Borrowdale – see pp. 129–132 – did not 'speak clearly'.) However, in an addendum to one of the papers it was claimed that a good exposure had been found in the Vale of

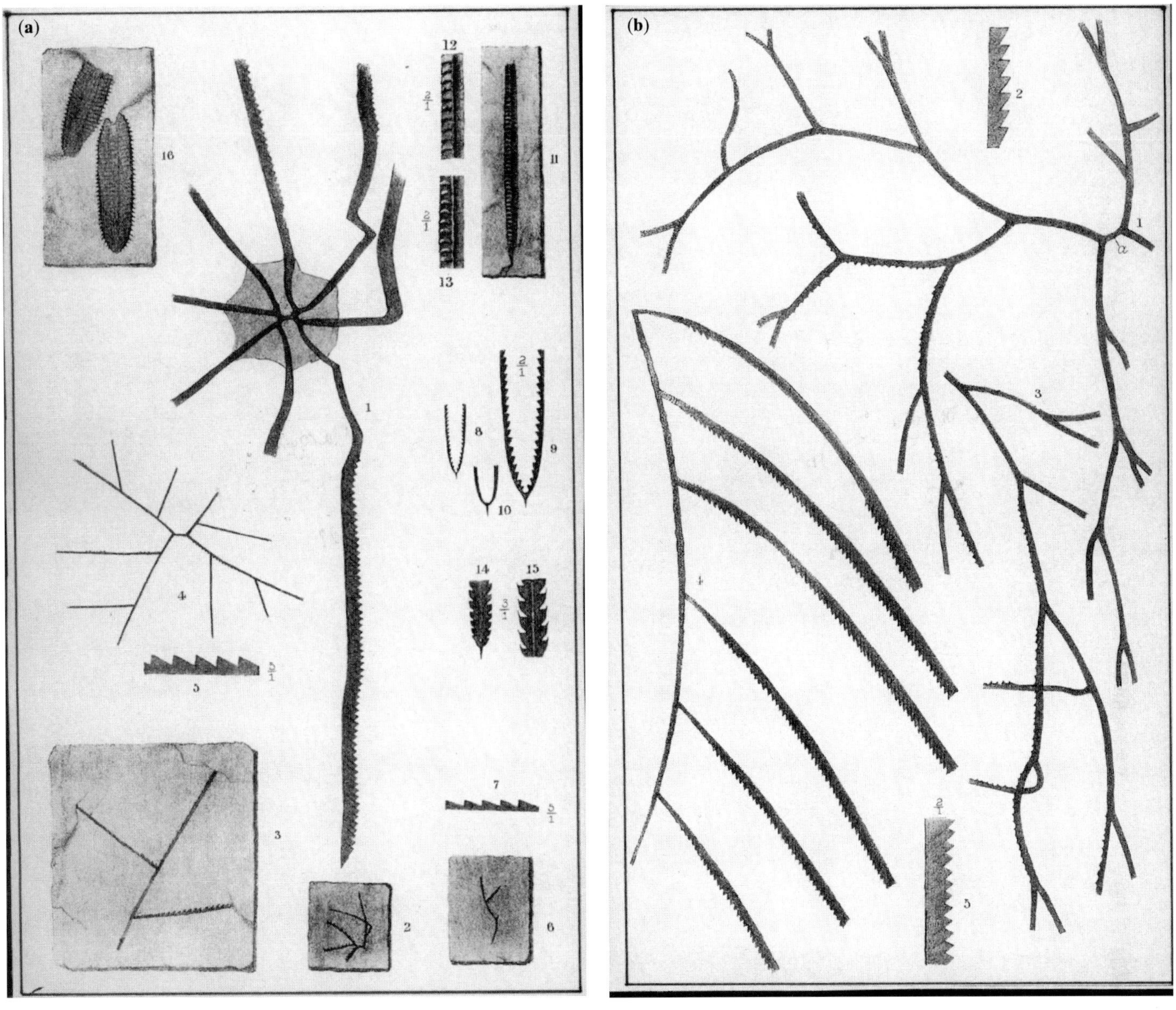

Fig. 3.7. Graptolites from the Skiddaw Slate Series, as figured and named by Nicholson (1868*b*, Plates V & VI). (**a**) 1–2, *Dichograpsus octobrachiatus*; 3–5, *D. reticulatus*; 6–7, *Dendrograpsus hallianus*(?); 8–10, *Didymograpsus geminus*; 11–13, *Diplograpsus teretiusculus*; 14–15, *D. pristiniformis*; 16, *Phyllograpsus typus*. (**b**) 1–3, *Dichograpsus multiplex*; 4–5, *Pleurograpsus ? vagans*.

St John, about two miles SE of Keswick, where the Green Slates could be seen lying directly over the dipping SS and Porphyries (Nicholson 1869*a*, p. 173). (It is not well exposed today.)

In another paper (Nicholson 1869*c*), an important fault was proposed on the western side of Derwent Water, responsible for a supposed repetition of the SS and the lowest part of the Green Slates and Porphyries; and another fault was said to run N–S approximately through the middle of the lake, below Wallow Crag and Barrow, where Sedgwick had made a number of his observations (see Figs 2.3 and 3.9). However, Nicholson's supposition of an E–W fault at the north end of the lake was based on erroneous mapping. The igneous rock near Rosetrees is an extension of an intrusion on the east side of the lake near Keswick, and does not mark a repetition of the rock at Hollows Farm due to faulting, as Nicholson supposed, and indicated in one of his sections.

In June the following year, Nicholson added more detail to his earlier account of the relationship between the 'Green Slates and Porphyries' and the underlying SS. He examined a number of key sites and provided horizontal sections of the localities concerned, each time representing the strata in such a way that the relationship between the two stratigraphic units appeared unconformable.

Also in 1870, Nicholson (1872*b*) presented a paper to the Edinburgh Geological Society, in which he examined the possible correlations between Lakeland and Scottish strata. The SS were now represented as either Llandeilo (in terms of Murchison's stratigraphy) or Arenig (Sedgwick's). The 'Green Slates and Porphyries' were now called the 'Borrowdale Series', which was a term that came to be used (with some variants) until J. F. N. Green (1920) introduced the name 'Borrowdale Volcanic Series', underscoring the acceptance of the idea of a volcanic origin for the unit. The 'Stile End' fossils at the top of the Series were still represented by Nicholson as Caradoc; and the overlying Coniston Limestone was regarded, as usual, as an equivalent of the Bala Limestone of Wales. However, the overlying 'Graptolitic Mudstones' appeared to be Upper Llandeilo in age, with many fossils in common with those of the 'anthracitic shales' of

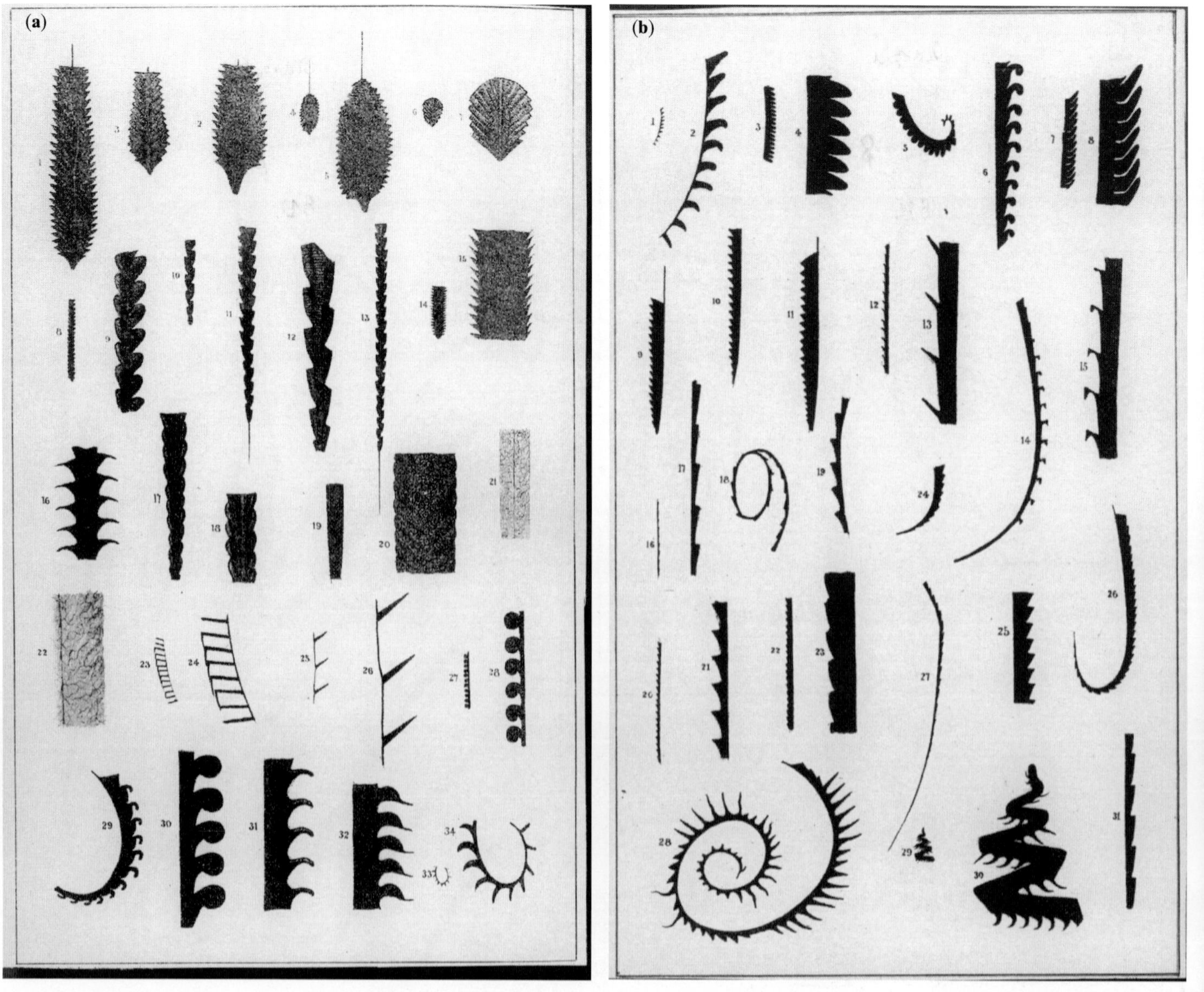

Fig. 3.8. Graptolites from the Coniston Flags [or Brathay Flags], as figured and named by Nicholson (1868c, Plates XIX & XX) (including some specimens from Southern Uplands).
(**a**) 1, *Diplograpsus palmeus*; 2, same enlarged; 3, ditto; 4, *D. folium*; 5, same enlarged; 6, *D. folium*; 7, same enlarged; 8, *D. angustifolius*; 9, same enlarged; 10, base of *D. tamariscus*; 11, *D. tamariscus*; 12, fragment of *D. tamariscus* enlarged; 13, ditto; 14, fragment of *D. confertus*; 15, same enlarged; 16, fragment of *D. mucronatus*; 17, *D. putillus*; 18, portion of same enlarged; 19, fragment of *Retiolites geinitzianus*; 20, same enlarged; 21, fragment of *R. perlatus*; 22, portion of same enlarged; 23, *Rastrites peregrinus*; 24, same enlarged; 25, fragment of *R. linnaei*; 26, same enlarged; 27, fragment of *Graptolites lobiferus*; 28, same enlarged; 29, fragment of base of *G. lobiferous* enlarged; 30, fragment of *G. lobiferous*; 31, fragment of *G. sedgwicki*. 32, fragment of *G. sedgwicki* enlarged; 33, fragment of = *G. sedgwicki* base; 34, same enlarged.
(**b**) 1, fragment of *Graptolites sedgwicki*; 2, same enlarged; 3, fragment of *Graptolites fimbriatus*; 4, same enlarged; 5, base of *G. fimbriatus* enlarged; 6, fragment of *G. priodon*; 7, *G. priodon*; 8, same enlarged; 9, *G. colonus*; 10, fragment of *G. colonus*; 11, *G. colonus*; 12, fragment of *G. discretus*; 13, same enlarged; 14, *G. discretus*; 15, fragment of *G. discretus*; 16, fragment of *Graptolites nilssoni*; 17, same enlarged; 18, *G. nilssoni*; 19, same enlarged; 20, fragment of *G. nilssoni*; 21, same enlarged; 22, fragment of *G. bohemicus*; 23, same enlarged; 24, base of same enlarged; 25, fragment of *G. sagittarius*; 26, *G. sagittarius*; 27, base of *G. sagittarius*; 28, *G. sedgwicki*; 29, *G. terriculatus*; 30, same enlarged; 31, fragment of = *G. tenuis*.

Dumfriesshire. Thus the strata seemed to be out of order according to the standard Welsh sequence.[10]

To deal with the problem Nicholson had recourse to a theory used quite widely at about that time when stratigraphic anomalies appeared. That is, he utilized the 'theory of colonies' of the French–Bohemian geologist Joachim Barrande (1799–1883), the great authority on the Lower Palaeozoic rocks of Bohemia. According to this theory, there might in some areas be certain 'precursorial' forms, in isolated geographic pockets in a given area, preceding the main advent of fossils of that type in the

[10] Ludlow
Wenlock
Llandovery
Bala–Caradoc
Llandeilo
Arenig

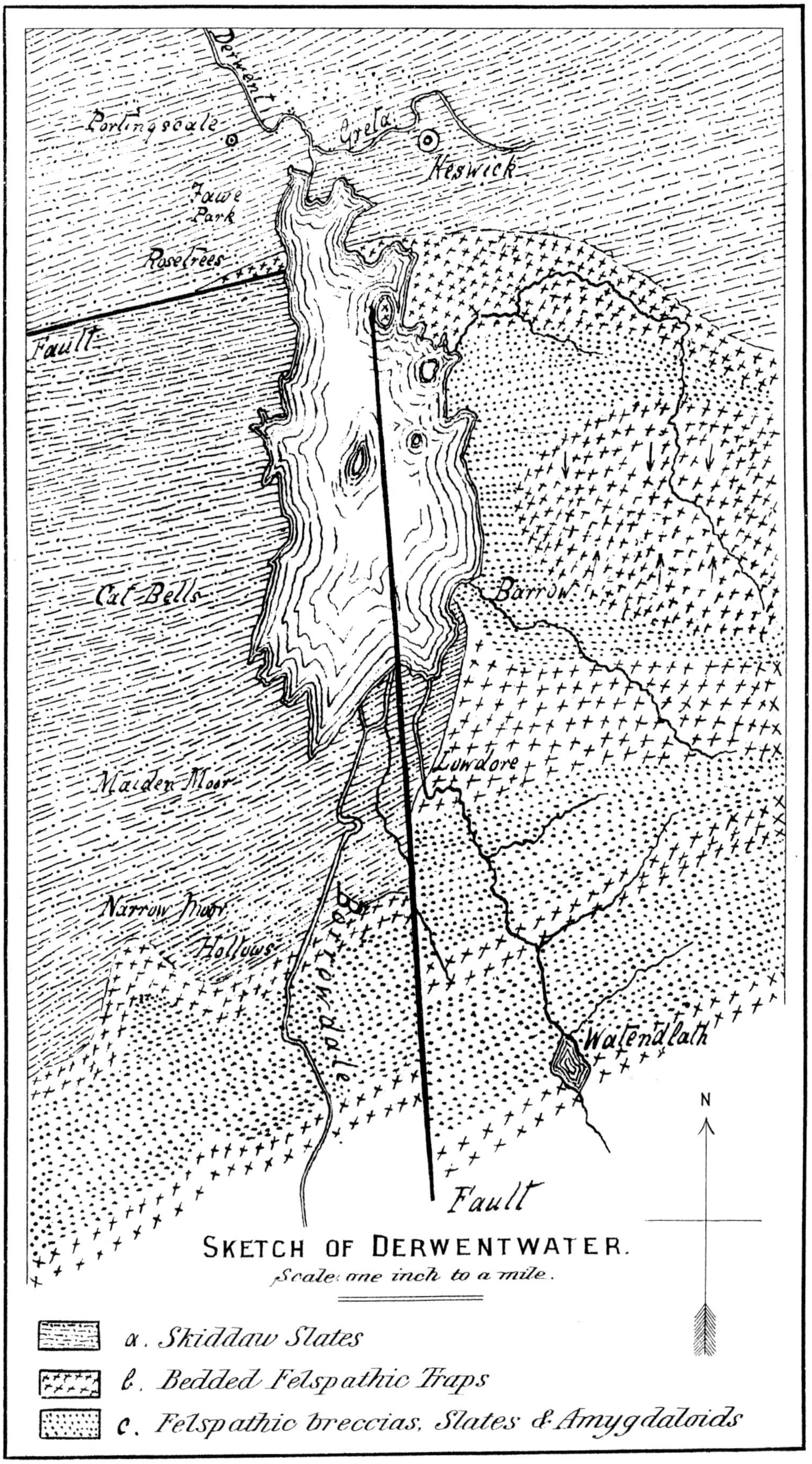

Fig. 3.9. Sketch of the geology of the Derwent Water area, according to Nicholson (1869*e*, Plate IX).

stratigraphic column at that locality.[11] This was not an impossible notion; but it played havoc with Smithian stratigraphy, and was not immediately reconcilable with the emerging Darwinian theory, though that was an extremely flexible construct. In fact, the ‘anomalies’ that Barrande's theory sought to explain (in Bohemia) were subsequently accounted for by recognition that

[11] For more on Barrande's theory, see p. 59.

certain strata might be repeated by faulting, as suggested by Marr (see Chapter 5). Such structural repetition might not always be apparent without careful mapping and, at the time, Murchison and other members of the official Geological Survey were willing to entertain the essentially *ad hoc* doctrine of 'colonies' (which was later linked to a significant misunderstanding of the geology of the Southern Uplands, and made it necessary for much of the surveying to be revised for that part of Scotland).

Above the Graptolitic Mudstones came the Coniston Flags, not rich in fossils; and above these came the Coniston Grits, with associated slaty bands that Nicholson termed the Bannisdale Slates. This name had recently been used by the Survey, who were by now working in the district (see Chapter 4), in their map for the Kirkby Lonsdale area (Sheet 98/49, 1869); and in the accompanying memoirs (Aveline *et al.* 1872; Aveline & Hughes 1872). Sedgwick had passed this way in 1823 and later introduced the name (see p. 24). Nicholson was by now persuaded, on palaeontological evidence furnished by Hughes, that the Coniston Grits might be regarded as Upper Silurian, not Lower Silurian as he had supposed in his doctoral thesis (Nicholson 1868*a*, p. 68), at which time he doubted that the Coniston Flags and the Coniston Grits were distinct units.

Finally, at the top of the Silurian sequence, there came what Nicholson called the 'Kendal Rocks' (Ludlow; later called the Kirkby Moor Flags), about which there was little palaeontological ambiguity, though the Lakeland Ludlows seemed to be thicker than those in their type area in the Welsh Border region.

Analogies could be made with the Southern Uplands. There Nicholson had 'Bottom or Axial rocks' near Hawick – of uncertain age, but possibly analogous to the SS. The Moffat Series (then being researched by Lapworth, who had not yet, however, fully published his results) might be Upper Llandeilo. A limestone near Girvan on the Ayrshire coast, and perhaps also the Wrae Limestone of Peebleshire, were regarded as possible equivalents of the Bala and Coniston Limestones. The Coniston Mudstones (later called the Stockdale Shales) appeared to have no analogue in Scotland; but their position was palaeontologically anomalous in the Lakes. The Coniston Flags and perhaps some of the Coniston Grits appeared to be equivalent to Lapworth's Gala Group near Galashiels. The Balmae Beds of Kirkudbrightshire might be analogues of the Coniston Grits. And some Ludlow rocks from the Pentland Hills near Edinburgh might be related to the Ludlow strata of the Kendal district. So by and large, there seemed to be similarities between the rocks of southern Scotland and NW England. Nicholson (1874) reasserted essentially the same views in 1873 in a paper to the Geologists' Association (founded 1858).

Nicholson and Lapworth became acquainted with one another as early as 1869, when Lapworth, a schoolmaster at Galashiels, was beginning his stratigraphic work based on graptolites in the Southern Uplands and wrote to Nicholson for advice. They promptly entered into a cordial relationship, and Nicholson and Harkness went to visit Lapworth in the field to view his Southern Uplands sections (Lapworth 1899). Then in 1875, the careers of Lapworth and Nicholson came together more closely in St Andrews, where Lapworth was appointed to a secondary teaching position at Madras College, and Nicholson took up his chair of natural history at the University. The two were then able to enter into detailed scientific conversations, and they worked together in the Lake District and the Cross Fell Inlier that summer, thus making possible the closer integration of English and Scottish Palaeozoic stratigraphy. They presented the results of their collaborative work at the British Association, which met in Bristol that year. In combination, the two geologists had the greatest expertise available in Britain at that time on graptolites and their use in stratigraphy, and between them they were able to devise a significantly improved and clarified stratigraphic sequence for the part where there had previously been the greatest uncertainty (Nicholson & Lapworth 1876):

A. Coniston Mudstone Series [*Middle* Silurian: Llandovery]
   b. Knock Beds[12] [graptolitic shales and grits, equivalent to the Gala and Hawick Beds of southern Scotland]
   a. Skelgill Beds [graptolitic mudstones, equivalent to Lapworth's 'Birkhill Beds' near Moffat]
B. Coniston Limestone Beds [equivalent to limestones at Kildare, Ireland, and Girvan, Scotland]
   c. *Trinucleus* Shales [a local group found in the Sedbergh district]
   b. Coniston Limestone
   a. Dufton Shales [fossiliferous shales, only found in the Cross Fell Inlier]

Above the Coniston Mudstone came the Coniston Flags (*Upper* Silurian), equivalent to the Denbighshire Flags of North Wales and the Balmae and Riccarton Beds of southern Scotland. However, the palaeontological evidence for these allocations and correlations was not published.

Harkness and Nicholson's last stratigraphic paper on Lakeland geology was presented to the Geological Society on 21 March 1877 (Harkness & Nicholson 1877). The SS were now assigned to the Arenig, where they mostly remain to this day. The fossil-bearing beds near the top of the Borrowdale Series were termed the 'Stile-End Grassing Beds', and were apparently of Bala age (taken to be Lower Silurian). Above the Borrowdales occurred the sequence as described above (Nicholson & Lapworth 1876), and the whole was depicted as shown in Figure 3.10. Copious palaeontological evidence was provided for the proposed stratigraphic subdivisions, but the Graptolitic Mudstones were relegated to the *Lower* Silurian (high Bala or Llandovery) – an interesting point, suggesting that the choice of *Middle* Silurian in 1875 was chiefly the idea of Lapworth (and may have been foreshadowing his introduction of the Ordovician System in 1879). It should be remarked that the relationship between the Borrowdale Series and the Coniston Series was envisaged as being one of conformity; but the Coniston Limestone was coming to be regarded as a complex unit, inviting subdivision.

Yet the question of the boundary between Upper and Lower Silurian was evidently causing trouble. In fact, Harkness and Nicholson – not choosing to deploy the subdivision of Middle Silurian – could not decide whether to place the Dufton Shales at the top of the Lower Silurian or the bottom of the Upper Silurian; or possibly as some kind of 'passage beds'. If anything, they inclined toward the Upper Silurian alternative.

Interesting discussion followed the presentation of the paper, with the Survey geologists (hereinafter 'Surveyors') by no means convinced of Harkness and Nicholson's findings. Hughes stated that in the Craven district to the south he had found a basal conglomerate at the bottom of the Coniston Flags (see p. 31) and no equivalents of the Knock Beds or the Graptolitic Mudstones, so the generally conformable sequence suggested by Harkness and Nicholson's vertical section (Fig. 3.10) might be doubted. In North Wales there appeared to be two 'pale slates', only one of which was represented by the Knock Beds. The Graptolitic Mudstones did not, Hughes maintained, always lie on the same unit in different places. The fossil evidence for the placement of the Graptolitic Mudstones was challenged by both Hughes and the Survey palaeontologist, Robert Etheridge. The Surveyor, Charles de Rance, said that he had spent two years mapping the Borrowdale Series, and that he disagreed with the idea that the fossiliferous 'Style-End Grassing Beds' were in fact part of the Series at all, but belonged, rather, to the overlying Coniston Series. He also questioned the suggestion that the Coniston Series was conformable to the Borrowdale Series.

Thus there was substantial disagreement emerging between the

[12] Knock is a village in the area of the Cross Fell Inlier. The Knock Beds were later renamed the Browgill Beds by Marr & Nicholson (1888).

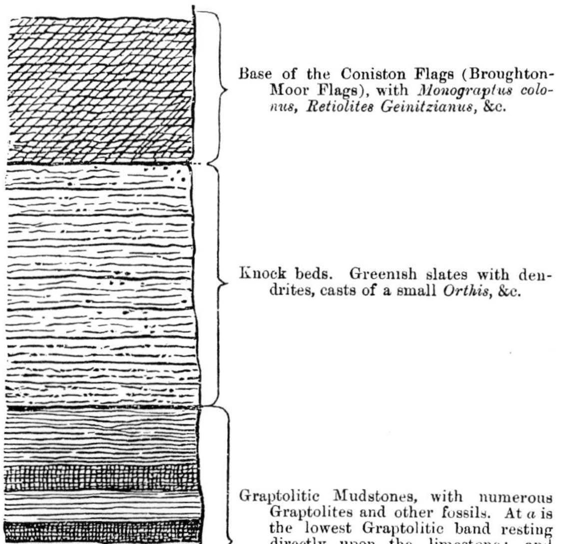

Fig. 3.10. Vertical section of strata at Skellgill Beck, near Ambleside, according to Harkness & Nicholson (1877, p. 472).

Surveyors and the 'amateur' geologists such as Lapworth, Harkness and Nicholson. The Surveyors, of course, could spend long months in the field, and prepare detailed maps. However, their work in the Lakes was by no means complete by 1877, when the disagreements outlined above were beginning to come seriously into view, and they certainly did not have the monopoly of ability. But to examine the work of the first Government Surveyors requires a fresh chapter. Harkness's fieldwork in the Lakes was closed by 1877. Nicholson's certainly was not; but when he re-entered the field in the 1880s it was with another partner, John Marr. So we shall look at Nicholson's later Lakeland work in conjunction with the early investigations of Marr. We shall also revisit his work on the Skiddaws in Chapter 8.

Before considering Nicholson's investigations with Marr, however, it should be mentioned that he undertook some work on his own in the Lakes in the early 1880s. There are letters from Nicholson to Lapworth in the Lapworth archive in Birmingham that throw an interesting light on the relationship between the two geologists and the attempts that Nicholson was making to sort out the stratigraphy of the region and the relationship between the Skiddaws and the volcanics around 1882–1883. This was the time when Lapworth was hard at work unravelling the structure of the rocks of the Northwest Highlands of Scotland, a period of work that was so physically and mentally exhausting that he suffered a breakdown in 1883. Thus, although Nicholson wrote imploringly that year to Lapworth, asking him to come and give him a hand in the Lakes, Lapworth never came (though he apparently had a look at the rocks near Millom in March or April that year and gave a lecture in Keswick at some stage; see Marr 1916, p. 75). Nicholson never published the ideas expressed in his correspondence, probably doubting that they were secure enough to lay before the public single-handed. It is interesting, however, to see what ideas Nicholson was trying to develop.

In the next chapter, we shall describe the work of the Surveyors in the Lakes, and will see that the most important contributor to the fieldwork was James Clifton Ward; and in Chapter 5 we shall look in some detail at the work of John Marr, who collaborated with Nicholson in the late 1880s. One of Ward's achievements was to describe a 'standard section' for the Borrowdale Volcanics (at least in their lower part) at Falcon Crag on the eastern side of Derwent Water (see Figs 2.3 and 4.4). In 1887, Nicholson and Marr described for the first time a fossiliferous inlier at 'Drygill' to the NW of Carrock Fell (which they construed as belonging to the general sequence of the Eycott Volcanics, which crop out nearby). It is interesting to see, from Nicholson's letters to Lapworth, how he was endeavouring to synthesize this knowledge in his fieldwork at least as early as 1883. According to Nicholson (Nicholson to Lapworth, 17 April 1883), Ward thought he could find Skiddaws interstratified with Borrowdale Volcanics at Eycott Hill (to the NE of the main Lakeland hills) and at Drygill (see Fig. 12.1). There Nicholson found light-coloured unaltered shales or mudstones that Ward had mapped as belonging to the top of the Skiddaws (but see p. 52–53). Nicholson found them to be crammed with fossils (brachiopods and trilobites, not graptolites). He took them to be at the bottom of the 'Eycott Hill Series'.

Nicholson (letter to Lapworth, 17 April 1883) made a somewhat unwarranted reading of Ward's Derwent Water section and construed the ashes as being interbedded with mudstones. So he took this part of the section to be equivalent to the beds at Drygill, and assumed that the 'traps' observed by Ward at Falcon Crag by Derwent Water were equivalent to the volcanics at Eycott Hill at the NE edge of the Lakes (Fig. 12.1). Nicholson thought he could see fossil cavities in the beds near Keswick, which made him begin to doubt the accuracy of Ward's work, in that he had missed the supposed mudstones at Falcon Crag and the numerous fossils at Drygill (the latter indeed being a reasonable complaint). This meant that Nicholson was bringing the Eycott into the Borrowdale Volcanics, and low in that series too. Indeed, by 12 May (Nicholson to Lapworth, 12 May 1883) he contemplated the Eycotts running all the way through to Honister Pass, and was looking forward to finding them at Millom at the southwestern corner of the Lakes!

A letter to Lapworth of 22 April gives further clues about Nicholson's ideas and the correspondence suggests that he and Lapworth were trying to work out a joint theory for the structure and stratigraphy of the Lakes. Nicholson wrote of a '*great Pebidian nucleus*' to the region, evidently thinking of the theory that was being developed by Henry Hicks (1837–1899), the gifted but cantankerous amateur medico-geologist from St David's, Pembrokeshire, who had been trying to develop the idea of islands of plutonic or metamorphic Precambrian ('Archaean') rocks in various parts of Wales, the Malverns and Shropshire, contrary to the views of the Survey officers (Oldroyd 1991, 1992, 1993, 1994). The 'amateur' geologists formed a group pressing this idea contra the Survey, and evidently there was a hope that it might also be applied in the Lakes. However, precisely what the Lakeland Pebidians consisted of was unclear, both to Nicholson and the present writer. Were they the various Lakeland granitic masses or Ward's 'altered ash-series' of central Lakeland?

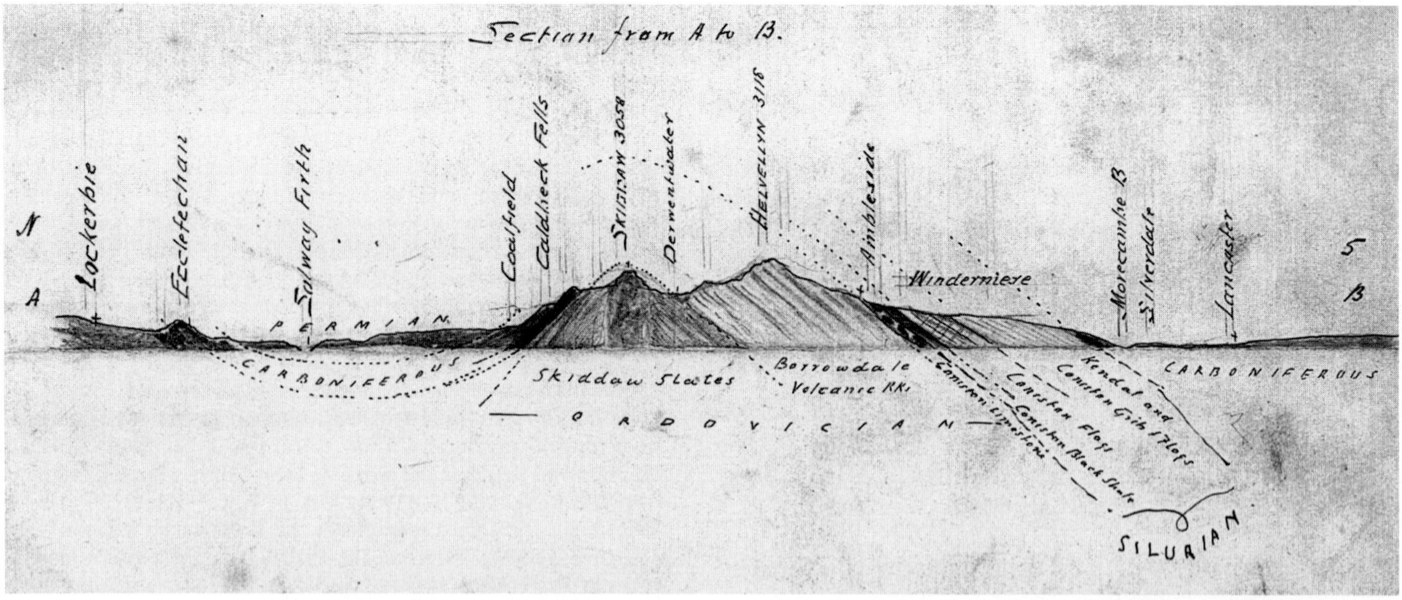

Fig. 3.11. Horizontal section of the Lake District, according to Charles Lapworth. Undated, but thought to be early 1880s. Item 82 in the Lapworth Archive, Lapworth Museum of Geology, The University of Birmingham (coloured in original). Published by courtesy of the Curator of the Lapworth Museum.

To my knowledge, the only surviving evidence of a Lakeland Pebidian series exists in the Nicholson–Lapworth correspondence. The idea did not get published. As mentioned, Nicholson several times urged Lapworth to come into the field with him in the Lakes, but Lapworth had too many other things on his plate at that time, and so far as I am aware he did not do much, if any, Lakeland fieldwork in the 1880s. Without his support, Nicholson was apparently unwilling to publish his ideas, which included the concept of an 'Eycott Series', consisting of a limestone (top), felsites, basic traps and breccias, and the fossiliferous Drygill Shales (bottom). He was further trying to establish correlations all the way from the Cross Fell Inlier, to Longsleddale, to the strata of the Coniston area and, as mentioned above, across to Millom. Perhaps the Keisley Limestone of the Inlier was equivalent to the Coniston Limestone in the Lakes? However, as Nicholson himself admitted, he was having difficulty in making sense of it all. He was not able to distinguish all the volcanic rocks satisfactorily, and he was unclear how many limestones he was dealing with. In 1884, he admitted that he could not find rhyolites at Falcon Crag (Derwent Water) and at the entrance to Borrowdale, as he hoped would be the case if there were analogy with the strata to the east in the Cross Fell Inlier.

Yet the model evidently had some attraction to the 'amateur party'. Nicholson's correspondence shows that he wanted to keep his ideas from local amateurs such as John Postlethwaite (see Chapter 6) and Kinsey Dover, presumably to avoid being pre-empted; yet he sought collaboration with the influential amateur geologist, the Wellington schoolmaster, Charles Callaway (1838–1915), and the two did some fieldwork together in the Lakes in 1886 (Nicholson to Lapworth, 11 August 1886 and 24 October 1886). One can see how this collaboration would have had appeal. Callaway had first-rate experience in unfossiliferous ancient rocks, and played an important role in the 'Highlands Controversy' and the 'Archaean Controversy' (Oldroyd 1990, 1992, 1993, 1994); and he too would have been glad, presumably, to see the ideas of the anti-Survey 'Archaean' group applied in the Lakes. But apparently he and Nicholson did not hit it off, and probably Callaway did not think Nicholson's hypotheses satisfactory. Be this as it may, their plans to publish something together came to nothing. When we consider the various stratigraphic implausibilities that Nicholson was entertaining this is by no means surprising, and I can believe that Callaway thought Nicholson was getting out of his depth in 'hard-rock' geology in the Lakes.

Things were different with John Marr, however. The two were introduced by Lapworth, and did much fieldwork together. Marr wrote to Lapworth (10 April 1887), 'I like Nicholson very much. He is great fun and I am very grateful to you for bringing us together'. Thus a major collaboration of two Lakeland geologists was initiated. However, both of them were better at the stratigraphy of fossiliferous rocks than with the complexities of igneous rocks. Their collaborative work will be discussed in Chapter 5.

Regrettably, we know little about Lapworth's views on Lakeland geology, since only one side of his correspondence with Nicholson is extant, and there is no evidence that Lapworth did any substantial work in the Lakes after his early investigations with Nicholson. There is a hint about his ideas in a section (Fig. 3.11) that he is thought to have prepared for a lecture in Keswick, 'The Silurian System in Britain'. This evidently refers to an unpreserved map. The date must be 1879 or later, as it refers to the Ordovician System, and it may well be from about the time of the Lapworth–Nicholson correspondence that we have been discussing. Certain features are to be noted: the supposed conformity of 'Otley I', 'II' and 'III'; the hint at a separate unit between the Skiddaws and the Caldbeck Fells, which might be the Drygill Shales; the placement of the base of the Silurian *below* the Coniston Limestone; the fourfold division of the Windermere Silurians; the possible compatibility of the model with the 'Archaean' hypothesis, with sedimentary rocks lapped around a granitic island (but the formation of a grand anticline by the action of a granitic intrusion, as envisaged in the Surveyor's maps and sections – see Chapter 4 – is also possible); and the implicit correlation of the SS with the rocks of the Southern Uplands of Scotland. Regrettably, we have no further clues as to Lapworth's ideas, and as has already been said, he never seriously devoted himself to Lakeland geology.

So now let us turn our full attention to the work of the Survey in the Lakes, to which it has already been necessary to make mention.

Chapter 4
The first surveys

From the time that the Geological Survey was established in 1835, its officers tramped the country, entering information on topographic maps, and colouring them according to the determined rock types, so that – through the preparation and publication of maps – the essential tools for elucidation of the geological structure and history of the United Kingdom were prepared (or constructed), and much information was gathered of economic significance. By and large, the team of surveyors worked systematically across the country from the SW northwards, though some preference was given to regions of commercial significance, with important mineral deposits. The task was a large one, which even now is not complete. It posed difficulties in that geological theories, and stratigraphic subdivisions, were altered during the course of the work; and sometimes there were differences of factual, methodological or conceptual opinion between the Survey officers themselves and in relation to the ideas of 'amateur' geologists – which according to the accepted terminology of the time might include anyone from impecunious collectors such as John Bolton to university professors such as Sedgwick, Nicholson or Lapworth. This terminology may seem odd today, but certainly it was the Survey staff who earned their money by the making of geological maps; and they were in the field for much longer stretches of time each year than were the 'amateurs'. Indeed, one way or another, they were at it all the time. So they regarded themselves as the 'professionals'. On the other hand, they might sometimes be less familiar with the details of specific areas than 'amateur' geologists who happened to reside in those areas or who made them their special areas for scientific research. Certainly the word 'amateur' was not intended as one of opprobrium or condescension. Lapworth in particular liked to refer to himself as an 'amateur' well after he was Professor at Birmingham.

The primary survey began in the Lakes in 1866, and was largely complete for our region by 1887, though the last map was not issued until 1893. During this period, the Directors General of the Survey were Sir Roderick Murchison, until his death in 1872; Andrew Ramsay (knighted in 1881), from 1872 to 1882; and Archibald Geikie (knighted in 1891), from 1882. Interestingly, all three hailed from Scotland. As I have shown elsewhere (Oldroyd 1990), Geikie was a protégé of Murchison, and Ramsay generally went along with his predecessor's stratigraphic subdivisions. So the Survey was very much part of Murchison's 'kingdom', and followed his ideas about the subdivision of the Lower Palaeozoic rocks, contra Sedgwick. As is well known, a compromise between the conflicting views of Murchison and Sedgwick was eventually proposed by Lapworth (1879), with his idea of a new geological system, the Ordovician, to occupy the contested stratigraphic territory between Sedgwick's Cambrian and Murchison's Silurian.[1] Yet this solution to the controversy was not accepted by the Survey until the end of Geikie's reign as Director General, and the appointment of his successor, Jethro J. H. Teall in 1901. So the Survey maps were still labelled according to the Murchisonian nomenclature until the beginning of the twentieth century.

The different hand-coloured maps of the Lakeland region for the first Survey are listed in Table 4.1, along with the dates of publication, and the names of the surveyors responsible for the work. Hand-coloured six-inch maps were also produced. The original field slips (six-inch) are preserved at the Survey office in Edinburgh. The primary field surveying of England was essentially completed, apart from minor revisions, in 1884, so the Lake District was one of the last areas to receive the Survey's attention.

Amongst the geologists listed in Table 4.1, the leading figures were William Talbot Aveline (1822–1903) and James Clifton Ward (1843–1880), both of whom warrant more detailed remarks than those given for their colleagues in Note 2. Aveline was an early member of the Survey, joining the staff as a mere teenager under Henry De la Beche in 1840. He was put to work in the Mendip Hills and then worked under Ramsay in South Wales. Later he traced the important Silurian Tarannon Shales in North Wales, which were at first thought to overlie Bala and Llandovery rocks unconformably, although later, after Lapworth had worked out the Silurian succession in the Southern Uplands with the help of graptolites, it was found that the Tarannon Shales near Conway contained a type, *Monograptus exiguus*, characteristic of Lapworth's Gala Group (Upper Llandovery).

Aveline (see Fig. 4.1) was described by Geikie, quoting Ramsay, as 'a tall, dark, silent, big-booted man who strode with gigantic steps over the hills; whose eyes seemed always directed towards the front, but never let anything escape them; who wrote like a schoolboy, but was the ablest field-geologist on the staff' (Geikie 1904, p. lxvii). He was a man of few words, probably taciturn. According to one anecdote, 'he spent a whole day among the Welsh hills, and his conversation was said to have consisted of only two words. In the morning, as he passed a crag of rocks, he tapped it with his hammer, and remarked 'Grits'. In the evening, on the way homewards, he had to chip another block, and again broke the silence with 'more Grits' (Geikie 1904, p. lxvii). Yet Aveline was said to be gentle, kindly and modest, and a favourite with his colleagues. In 1867, he was promoted to District Geologist and took charge of the Lakeland mapping, residing at Kendal until his retirement to Somerset in 1882.

Aveline gave his life and soul to the Survey, and was responsible for some of its toughest early work in England and Wales. It was, then, somewhat shaming that the organization tried to squeeze him out in the closing years of his career, as is shown by letters between him and the Survey's Senior (or deputy) Director, Henry Bristow, held in the institution's archives at Keyworth. By 1880, most of the survey work was complete in the Lake District, and several of the staff had been transferred to North Wales, at the behest of Ramsay, who wanted more information for the second edition of his memoir on the geology of North Wales (Ramsay 1881). Some work was still going on in the Lakes, and understandably Aveline was reluctant to move before his project there was completed. However, Ramsay and Bristow desired his presence in Wales and Aveline was instructed to move to the Vale of Clwyd in Denbighshire. He protested, and an angry exchange of letters between Bristow and Aveline followed (British Geological Survey archives, IGS 1/1273 and 1/1284).

Issues became heated, with Aveline writing to Bristow on 17 May 1880 that his last two years in the Survey had been 'utterly miserable'. The dispute became increasingly acrimonious, so much so that it was suggested to Aveline in 1881 that it might be time for him to retire. But he was not yet 60, so that were he to resign he would forfeit his pension. In the event, matters were not pushed to a showdown, and Aveline was permitted to resign in 1882, his pension intact. He retired to his family farm in Wrington, Somerset, and so far as is known lived there contentedly as a country squire. However, the incident made an unfortunate conclusion to a dedicated career. Aveline was a tough man, his maps and accompanying sheet memoirs being his chief bequest to posterity. He was not an active theoretician, at least in print.

[1] Interestingly, Lapworth himself thought that the term 'Ordovian' would be better than 'Ordovician', as it would be analogous to 'Silurian'. See Lapworth to Hughes, 19 August 1880 (Hughes papers, Cambridge University Library, Add MSS 9557, Packet F). However, to my knowledge, the term was never published.

Table 4.1. *The primary mapping of the Lake District: the Surveyors and dates of first publication of the Lakeland one-inch maps*[2]

Maryport **101 NW (1892)** *Aveline* Russell Holmes	**Cockermouth** **101 NE (1885)** *Ward* Russell Holmes Colvin	**Penrith** **102 NW (1893)** *Aveline* Dakyns Russell Burns Hebert Clough
Whitehaven **101 SW (1892)** *Aveline* Russell	**Keswick** **101 SE (1875)** *Aveline* Ward [Hebert]	**Appleby** **102 SW (1893)** *Aveline* Hughes Dakyns Tiddeman Ward Russell Dalton Goodchild Lightfoot Hebert
	Coniston/Ambleside **98 NW (1882)** *Aveline* Ward Wollaston Rutley Cameron Hebert	**Kendal** **98 NE (1871)** *Aveline* Hughes Goodchild Rutley Wollaston Lightfoot Dakyns Hebert
	Dalton in Furness **98 SW (1877)** *Aveline* Cameron	**Kirkby Lonsdale** **98 SE (1869)** *Aveline* Barrow Hughes Tiddeman
		Mallerstang **97 NW (1889)** *Aveline* *Howell* Hughes Goodchild Dakyns Tiddeman Russell Clough De Rance Strahan Barrow[3]

Fig. 4.1. William Talbot Aveline. British Geological Survey archives (GSM 1/639/5). Photograph reproduced by courtesy of the British Geological Survey.

Some indications of Aveline's weariness with Survey work in his latter years can be discerned in his field-slips for his Lakeland work (mostly done in the area of the Duddon Valley and in southern Lakeland), which show signs of perfunctory effort, with only a few outcrops coloured in. On the other hand, a later commentator, J. F. N. Green (see Chapter 6), regarded Aveline's maps of the Duddon area as 'meticulously accurate' (Green 1919, p. 154) and his cartography beautiful. Also, I found (during a day in the field in 1999) that the modern geologist, Jack Soper (see Fig. 14.4), doing contract mapping for the Survey in the Kendal area, used copies of Aveline's field-slips to take him unerringly to spots where outcrops might be found. Even so, the archived slips

[2] The supervising geologists are indicated by italics. Information about the less important subordinate staff is given as follows. George Barrow (1853–1932) never really worked in Lakeland proper and was presumably just giving a hand from his main Yorkshire area of Cleveland. His best-known work was done in Scotland, where he studied metamorphic (Barrovian) zones. He was a constant critic of his colleagues' ideas and eventually was moved south to England again. In Chapter 5 we shall encounter him giving support to another iconoclast, J. F. N. Green. David Burns (1843–?) was with the Survey as Assistant Geologist from 1867 to 1880. Alan Charles Grant Cameron (1844–?) was appointed Assistant Geologist in 1868 and was promoted to Geologist in 1893. He retired in 1900. Charles Thomas Clough (1852–1916) studied science at St John's, Cambridge, and was appointed Assistant Geologist in 1875. Initially working in northern England, he was transferred to Scotland in 1884, where he was acclaimed for his work in the Northwest Highlands and the Hebrides. He was promoted to Geologist in 1896 and District Geologist in 1902. Alexander Colvin was appointed to the Survey as Assistant Geologist in 1874 and resigned in 1897. Born in the West Indies, John Roche Dakyns (1836–1910) studied science at Trinity College, Cambridge. He joined the Survey in 1862 and was promoted to Geologist in 1868. Having surveyed large tracts of northern England, he was transferred to Scotland in 1884, and then to South Wales in 1894, retiring in 1896. Dakyns was a political radical and a friend of Marx and Engels (Robinson 1991). William Herbert Dalton (1848–1929) joined the Survey as Assistant Geologist in 1867 and was promoted to Geologist in 1883. Two years later, he resigned to enter a career as consulting geologist, studying oil-fields in many parts of the world. Of French parentage, Charles Eugene De Rance (1847–1906) attended King's College School, London, and then trained as an engineer. He was with the Survey as Assistant Geologist from 1868 to 1898, when he resigned to become a consultant water and mining engineer at Blackpool. (His position in the Survey had become untenable because of his 'drink problem'.) John George Goodchild (1844–1906) was

Fig. 4.2. James Clifton Ward. Geological Survey archives (GSM 1/639/91). Photograph reproduced by courtesy of the British Geological Survey.

reveal that Aveline left most of the work in the high fells to his younger colleague Ward. Soper found, incidentally, that Aveline was not always exact in his determinations of dips and strikes.

Ward (who used Clifton as his first name) (see Fig. 4.2) was a man of more intellectual bent than Aveline. He was born at Clapham Common, son of a schoolmaster, and studied at the Royal School of Mines in Jermyn Street, London, from 1861 to 1864, where he gained the Edward Forbes Medal. The School was effectively a tertiary college where one might study geology; and in the manner of the British education system in the nineteenth century, it did not have a strong practical or engineering component, its name notwithstanding. However, in the 1860s it was the best place in Britain to obtain a thorough grounding in geology, especially in its practical aspects. In 1865, Ward was appointed Assistant Geologist in the Survey and began work in the Millstone Grit and Lower Coal Measures of Yorkshire. In 1869, he was transferred to Keswick and was subsequently promoted to Geologist. His next eight years were spent in survey of the north Lakes, with the productions of maps, as listed in Table 4.1; also beautiful horizontal sections, produced in collaboration with Aveline and various other geologists named in Table 4.1. As will be discussed below, Ward made particular studies of the Borrowdale Volcanics, which were examined in detail in an important memoir (Ward 1876*a*).

Unlike Aveline, Ward did not choose to remain with the Survey throughout his career. He was deeply religious and of romantic temperament (as can be seen from jottings in his field notebooks, which are preserved along with his maps and sections at Keyworth). Having lived in the Lakes for several years, he did not want to move on to some other district. So he studied to become a clergyman, and was ordained in 1878, the year after his marriage to the daughter of a Cockermouth solicitor (who was doubtless a factor in binding Ward to the Lakes), becoming assistant curate at St John's, Keswick. Thus he departed from the Survey at the end of 1878, though he continued do some work for Ramsay, naming and arranging specimens of Lakeland rocks for display in the Jermyn Street Museum. In 1880, Ward was appointed to the beautiful parish of Rydal, in Wordsworth country, but he died that year, probably of pneumonia, leaving two young daughters. He was only 37 years old.

Yet Ward achieved much in his few Lakeland years. He played an important role in Lakeland intellectual life, helping to establish the Cumberland and Westmorland Association for the Advancement of Literature and Science, and lecturing extensively in the district on geological and archaeological topics. It is recorded that he displayed geological maps and models at the Keswick Museum, and two of the latter have recently been rediscovered and restored and are now on public view. The geological model consists simply of his Keswick six-inch geological map, pasted over a three-dimensional plaster model of the area. Much more than Aveline, Ward was interested in developing general theoretical ideas about the Lakeland rocks.

The Survey's Annual Reports provide information about the progress of the survey through the District (see Table 4.2).

Regarding official publications, the main products from the primary survey of the Lakes – apart from the maps and horizontal sections – were Aveline & Hughes (1872, 2nd edn 1888), Aveline *et al.* (1872), Aveline (1873), Ward (1876*a*) and Dakyns *et al.* (1897). The surveyors also entered into some debates in the

appointed Assistant Geologist in 1867 and did most of his fieldwork in NW England. Suffering from a heart problem, in 1887 he moved to the Head Office in London and in 1899 he took on the position of museum curator for the Scottish Branch of the Survey, but died before retirement. Goodchild played an active role in the Cumberland and Westmorland Association for the Advancement of Literature and Science, and played a prominent part in the development of the geological collections at Carlisle Museum. William Gunn (1837–1902) joined the Survey as Assistant Geologist in 1867, was promoted to Geologist in 1884, and became District Geologist the year before he retired in 1902. His work was chiefly with the Carboniferous rocks of northern England, with the ancient rocks of NW Scotland, and with the Tertiaries of the Hebrides. E. J. Hebert was Assistant Geologist from 1875 to 1880. Thomas Vincent Holmes (1843–1923) came from Kirklington Hall, Cumberland, and studied science at King's College, London. He was Assistant Geologist with the Survey from 1868 to 1879, following which he resigned and moved to Essex, where he maintained an active interest in geology, becoming President of the Geologists' Association in 1889–1891. Henry Hyatt Howell (1834–1915) was appointed Assistant Geologist in 1850 and rose to become Scottish Director in 1882 and Director for Great Britain in 1888 (i.e. assistant director of the Survey), retiring in 1899. Specializing in economic geology, he played no substantial role in the fieldwork in the Lakes. George Herbert Lightfoot was appointed Assistant Geologist in 1867 but resigned in 1869. Robert Russell (1842–1900) was trained as a civil engineer and joined the Survey in 1867. His main work was done in the Yorkshire and Whitehaven coalfields. Frank Rutley (1842–1904) studied at the Royal School of Mines and joined the Survey in 1867, beginning his fieldwork with Aveline and Ward. He soon became the Survey's leading authority on mineralogy and petrology, and in 1882 took up a lectureship in mineralogy at the Royal College of Science. His textbooks on mineralogy and petrology dominated the English-language field for many years. Richard Hill Tiddeman (1842–1917) joined the Survey as Assistant Geologist in 1864 and was promoted to Geologist in 1870, retiring in 1902. He was President of the Yorkshire Geological Society in 1914, having done much work in the cave deposits of that county. George Hyde Wollaston (1844–1926) was Assistant Geologist from 1868 to 1871, and then took up a teaching career at Clifton College.

[3] Russell, Goodchild and Strahan were authors of the sections of the accompanying Memoir (Dakyns *et al.* 1891) concerned with Lower Palaeozoic rocks, and Hughes did the initial mapping of the relevant parts of the sheet.

Table 4.2. *Progress of first Lakeland survey*

Year	*Area surveyed (square miles)*	*Counties or regions examined*
1866	720	Northumberland; Durham; Yorkshire; Westmorland; Derbyshire; Cheshire. Lakeland Silurians
1867	692	Durham; Northumberland; Westmorland; Yorkshire; Lancashire.
1868	2152	Northumberland; Durham; Westmorland; Yorkshire; Lancashire. Work proceeding from Kendal north towards Appleby
1869	183	Cumberland; Westmorland; Yorkshire; Lancashire. Much of Westmorland completed
1870	1145.5	Cumberland; Westmorland; north Lancashire. Work spreading into south Cumberland
1871	1196	Northumberland; Durham; Cumberland; Westmorland; Lancashire. Work beginning south of Whitehaven
1872	948	Northumberland; Durham; Cumberland; Westmorland; Yorkshire; Lancashire. Whitehaven coalfield begun
1873	737	Northumberland; Durham; Cumberland; Westmorland; Yorkshire; Lancashire. Whitehaven coalfield
1874	288.5	Cumberland; Westmorland; Lancashire. Carboniferous; Permian; volcanics
1875	546.5	Cumberland; Westmorland; Lancashire. Skiddaw Slates and associated igneous rocks. Whitehaven coalfield
1876	604.5	Cumberland. Silurians and associated volcanics. Whitehaven coalfield; Vale of Eden. Lake District almost completed
1877	345	Cumberland; Westmorland. Whitehaven coalfield completed
1878	164.5	Cumberland
1879	222	Cumberland; Cheshire. Specimens collected for Geological Museum
1880	18.5	Cumberland. Revision work, with help of new railway cuttings and colliery workings
1881	336	Cumberland; Westmorland. Revisions
1882	198	Cumberland; Westmorland. Revisions
1883	335	Cumberland; Westmorland. Revisions (mostly Westmorland)
1884	173	Cumberland; Westmorland. Revisions
1885	142	Cumberland; Westmorland. Revisions (mostly Westmorland)
1886	33	Cumberland. Revisions
1887	59	Cumberland; Westmorland. Resurveying
1888	25	Westmorland. Revisions

journal literature, notably those to do with glaciation (see Chapter 19), the question of whether or not Llandovery rocks might be found in the Lake District, and the issue of the nature of the boundary between the SS and the overlying volcanics (see Bibliography, and Smith n.d.). For unpublished sources, the most complete information in the Survey archives is available in the notebooks of Ward and De Rance.

A major problem in southern Lakeland was a satisfactory classification of the several 'Coniston' beds, and an acceptable placement of the boundary between the Lower and Upper Silurian. We have already encountered these difficulties in Chapter 3, where differences of opinion had cropped up between the 'amateurs', Harkness, Nicholson and Lapworth, on the one hand, and surveyors such as Hughes and De Rance on the other. Hughes himself was not in full agreement with his Survey colleagues, who, as mentioned, were largely Murchisonian, and, as is well known, he resigned to take up Sedgwick's vacated chair at Cambridge in 1873.[4]

[4] Details of the circumstances of Hughes' departure from the Survey have only become apparent through the recent (1999) deposit of his correspondence at Cambridge University Library (MSS, Add 9557) by one of his descendants, Jane Fawcett. He was evidently not happy in the Survey and his superiors were not happy with him. Murchison wrote a memorandum (19 August 1867) stating his concerns about the publication of the paper by Hughes in the *Geological Magazine* (Hughes 1867). Ostensibly the issue was Hughes' unacknowledged use for private publication of material obtained during the course of official work; and Murchison issued an edict (19 August 1867) that in future all 'private' work should be submitted to Surveyors' superiors before publication (a practice that continues to the present). At the time the real issue was surely the fact that Hughes' paper proposed stratigraphic ideas about the subdivision of the Silurian that were 'Sedgwickian' in flavour and contrary to Murchison's views. Hughes (30 August 1867) protested to Ramsay that others had previously published privately without official objection, and it was, in his opinion, a good thing that unorthodox ideas should be made available for

In the first published sheet memoirs for the Lake District, which described Map 98 for the southeastern part of the region (Aveline *et al.* 1872; Aveline & Hughes 1872), Aveline subdivided the strata as follows:[5]

Upper Silurian	Kirkby Moor Flags	Upper Ludlow
	Bannisdale Slates	Lower Ludlow or Wenlock
	Coniston Grits and Flags	Wenlock (equivalent to Denbighshire Grits and Flags)
	Stockdale Shales with Graptolite Shales	Tarannon Shale or Pale Slates
Lower Silurian	Coniston Limestone and Shales	Caradoc or Bala
	Green Slates and Porphyry	Volcanic lavas, ashes and breccias

The Stockdale Shales were a newly named unit and the Bannisdale Slates had scarcely been mentioned previously in print (see p. 24), though both had previously been identified on the key of the map (Sheet 98), which appeared before the Memoir. The Bannisdale Slates were described as sandy mudstones, with thin bands of sandstone, roughly cleaved and poorly fossiliferous. They contained the brachiopod *Rhynchonella navicula*, known from the Wenlock Shale, near Llangollen in Wales. However, there being no Lakeland equivalent of the famous Wenlock Limestone of Shropshire in the Lakes, exact correlation with the Welsh or Shropshire subdivisions was not possible. Aveline stated that part of the unit might be Ludlow in age, and in fact this is the place where the Bannisdale Slates have subsequently come to rest.

The Stockdale Shales occurred as a thin dark band, swarming with graptolites, below paler overlying shales. By palaeontological evidence, they appeared to be equivalent to the Tarannon Shales of Montgomeryshire, North Wales. They were named after the tiny hamlet of Stockdale on the east side of the upper end of the valley of Longsleddale in SE Lakeland, also called Little London (see Fig. 3.7). The beds had been traversed many years before by Sedgwick in 1823, but he was not at that time much on the lookout for fossils, and certainly not for graptolites, and he had not made anything of them then. In the 1880s, they became an important area for study by Nicholson and Marr, as will be discussed in Chapter 5. Aveline drew the base of the Upper Silurian, as then understood, at the bottom of these shales, as did Lapworth. (Today they are regarded as belonging to the Llandovery, at the bottom of the modern Silurian.)

The Coniston Flags were described as a blue sandy mudstone, while the Grits consisted of beds of tough sandstone with associated shale and slate. One band of the Grits near Sedbergh, apparently well below the usual Ludlow horizon, contained abundant Ludlow fossils, conflicting with the usual association of the Grits with Wenlock fossils. As was often the case at that time when such difficulties were encountered, Hughes had recourse to Barrande's theory of colonies (Aveline & Hughes 1872, p. 11).

From the stratigraphic subdivisions outlined above, and also for those described for Harkness and Nicholson in Chapter 3, it will be observed that the rocks forming the unit known as the Llandovery Series, between the Bala Series and the Wenlock Series in Wales, were apparently absent in the Lake District. Yet there was no apparent significant unconformity marking the gap. This issue attracted discussion in the pages of *The Geological Magazine* in 1876.

In April 1876, Henry Hicks (see p. 37) published a table of the divisions of the Cambrian and Silurian Systems, as he understood them, for different parts of Britain, Europe and North America (Hicks 1876*a*). For the Lakes and Scotland, and their Welsh equivalents, he had (in part) the correlations shown in Table 4.3.

As can be seen, the line of division between the Lower Silurian and the Upper Silurian was placed above the Hirnant and Coniston Limestones, or at the base of the Lower Llandovery, as warranted by the opinion at that time of Sir Charles Lyell, and in accordance with the seeming existence of a band of limestone through Britain at the close of the Bala period. (The Bala Limestone in Wales does resemble the Coniston Limestone in some outcrops, and has fossils in common, but it is generally thicker, and, in some exposures I have seen, more fossiliferous.)

Hicks had not worked in the Lakes. He was merely putting together correlations, as seemed to him appropriate, on the basis of the published work of Sedgwick, Harkness, Nicholson, Lapworth, and some of the Survey officers themselves (but not Aveline and his colleagues in the Lakes). So when Hicks published correlations differing from their own, they were not likely to be attractive to the Lakeland fieldsmen, who had been tramping the Cumbrian fells for nearly a decade. Aveline swiftly (and uncharacteristically) took up his pen and sent a letter from Kendal to the *Geological Magazine*, baldly asserting that Hicks had got things wrong: there were no Llandovery rocks in the Lakes; and a stratigraphic break occurred between the Coniston Limestone and the Green Slates and Porphyries, rather than above the Limestone (Aveline 1876*a*).

Hicks was not a man easily put down. He quickly responded, saying that the distinguished Swedish geologist, Gustav Linnarsson, held, with Harkness and Nicholson, that the Coniston (graptolitic) Mudstone was not Upper Silurian (Hicks 1876b) (though the unit had been represented as such in his table above). Hicks envisaged a mobile, changing geographical environment, such that the same conditions did not necessarily apply in different regions at the same time: some areas, such as the Lakes, might be above sea level while at the same time other areas, such as South Wales, might be undergoing continuous sedimentation. However, the overall order of deposition should be the same for the two separate areas.

Aveline (1876*b*, *c*) responded by asserting the importance of stratigraphic survey, rather than uncritical reliance on fossil

public discussion. Further, he objected that the first instruction on the matter should be in the form of a censure. This was the beginning of further tiffs, and it seems likely that Hughes thenceforth dragged his feet somewhat. Ramsay (29 December 1868) complained about the number of days Hughes spent indoors, and it appears that he was trying to arrange things so that he spent more time in London than in the field. Ramsay subsequently wrote to Hughes (7 February 1870) reprimanding him for his slow rate of progress. There was further trouble about Hughes's inflated travel allowances. (He claimed a shilling a mile for horse hire at one stage.) On the other hand, at the time when Aveline was having his own difficulties with the management he wrote to Hughes that Murchison's Report for 1869 stated that only Hughes and Ward were worth anything. Rutley was 'not worth anything all the year' and Cameron 'did his work so badly that [for] at least a quarter [of his time] he was resurveying' (Aveline to Hughes, 17 November 1870, Hughes papers, Cambridge University Library, Add. MSS 9557, Packet O). Even so, it is unsurprising that Hughes applied for Sedgwick's post when it became vacant in 1873. He received support from Lyell (whom Hughes had helped considerably in the revision of later editions of the *Principles of Geology*), Ramsay, Thomas Davidson, Joseph Prestwich, Warington Smyth, Henry Bristow, Henry Woodward, Searles Wood and David Forbes. The support from the Survey may indicate that they wished to be rid of an irritant, and the Cambridge authorities could have been pleased to appoint someone of good family with Sedgwickian proclivities and the practical experience of a Survey man. So Hughes got the job, even though his publication record was then meagre. He proved to be a good appointment. For a cartoon of Hughes with his students at Cambridge, see Figure 4.3.

[5] The different sections of this memoir had separately signed authors. It may be noted that the stratigraphic subdivisions were largely followed by Lapworth in his section of Figure 3.12.

Fig. 4.3. Cartoon of Hughes leading a party of Cambridge students on a field excursion in the Lake District (1 July 1882). Archives of the Sedgwick Club, Cambridge University. Reproduced by permission of Claire Slater, President 2001, on behalf of the Committee of the Sedgwick Club.

Table 4.3. *Correlations of Silurian strata in Britain, according to Hicks (1876a, facing p. 156)*

Formations	*Wales*	*South Scotland and Cumberland*
Ludlow (Upper Silurian)	Sandstone and shale Limestone (Aymestry) Shale (Lower Ludlow)	Grits, flags and shales (Kendal Group)
Wenlock (Upper Silurian)	Limestone (Wenlock) Shale (Wenlock) Limestone (Woolhope)	Grits, flags, and shales (Ireleth Group)
Llandovery (Upper Silurian)	Conglomerates, sandstones, and shales (Upper Llandovery and May Hill) Sandstones and shales (Lower Llandovery)	Grits (Coniston, etc.) Flags etc. (Coniston, etc.) Shales (Coniston, etc.) (Grapt[olitic] Mudstones)
Bala (Lower Silurian)	Limestone (Hirnant, etc.), sandstones and shale, with contemporaneous trap Limestone, sandstones, and shales (with trap)	Limestone (Coniston) Slates, etc. (Green-Slates and Porphyries)
Llandeilo (Lower Silurian)	Limestones, sandstones and shale (with contemporaneous trap) Limestones and slates Slates and flags (with contemporaneous traps)	Traps Slates and flags (Upper Skiddaw, etc.) Contemporaneous trap
Arenig (Lower Silurian)	Shales, slates and flags Flags, sandstones and slates Slates and flags	Slates and flags (Lower Skiddaw)
Tremadoc	Slates and flags Flaggy sandstone, etc.	

evidence. He pointed out that the geologist might sometimes argue in a circle: one might assume a certain bed to be, for example, Lower Silurian, so that the fossils found therein were taken to be Lower Silurian; and then the bed might be regarded as Lower Silurian because it contained Lower Silurian fossils. However, stratigraphic survey might reveal that the bed was 'connected with' Upper Silurian beds; so the fossils that were previously taken to be characteristic of Lower Silurian strata might have to be reconsidered. He was indeed correct: it could well be a kind of chicken-and-egg situation. The way out of the problem was to define some units as characteristic of a particular system, series or other subdivision. But mechanism for enforcing adherence to such definitions was not then in place. So even when no recourse was made to Barrande's theory of colonies, nineteenth-century stratigraphy could become racked with controversies of this kind, as different proponents of different type sections or boundaries vied with one another to have their preferred sections and subdivisions accepted. It was not really possible to follow the biological typologists, and work simply by precedent. Sections do not have 'replicas' as do members of biological species; they have analogues.

Such difficulties continue to the present, though there are mechanisms, through the activities of the various commissions of the International Union of Geological Sciences, to try to achieve consensus and compliance. The beginnings of such developments occurred through the work of the International Geological Congresses (Ellenberger 1978), but the first of these was not held until 1878. Before then, and for long after, stratigraphy was plagued by disagreements, though perhaps never in quite such an overt manner as occurred in the Murchison–Sedgwick dispute about the placement of the Cambrian–Silurian boundary. As we shall see, stratigraphic controversy has been ceaseless in the Lakes, up to the present – which is one reason why a study of the history of research in the region is illustrative of the general problems that arise in geology and the manner in which they are tackled.

Aveline was supported by his young subordinate Ward (1876*b*), but he did no more than pour scorn on Hicks's arguments on the grounds that he had never actually worked in the Lakes. This was true, and doubtless the argument appealed to the geological community, who placed such emphasis on the virtues of fieldwork. Yet it did not touch the central conceptual and procedural problems that were manifesting themselves in stratigraphy at that period.

Hicks (1876*c*) wrote again, referring to the authority of the sources used in his table, but more importantly to the recent discovery by Harkness and Nicholson in the Coniston Mudstones of fossil shells that were received as Upper Bala or Llandovery in Wales (published in Harkness & Nicholson 1877). So the Mudstones could not, Hicks maintained, be so high in the stratigraphic sequence (Ludlow) as Aveline supposed. Hicks also maintained on the basis of palaeontological evidence that the Tarannon Shales of Wales did not, as Aveline supposed, overlie the Llandovery, but were in fact a member of the Llandovery. He was asserting the old virtues of Smithian stratigraphy, whereas the Surveyors were having recourse to lithologies while stressing the importance of stratigraphic work in the field.

Charles Lapworth (1876) then entered the correspondence, pointing out that the fossils discovered by Harkness and Nicholson, on which Hicks placed reliance, were brachiopods and trilobites; but one might gain more certain results by means of graptolites, which nobody at that time in Britain apart from Lapworth himself was using to much stratigraphic advantage. On the basis of the graptolites, Lapworth believed that the Graptolitic Mudstones of the southeastern Lakes were indeed (Lower) Llandovery in age. (For discussion of Lapworth's procedures at about that time using graptolites, see Chapter 8; and Hamilton 2001.)

As we have seen in Chapter 3, Lapworth was never a major player in Lakeland geology, but he had been there with Nicholson in 1875 (Nicholson & Lapworth 1876), and he could certainly see the potential for stratigraphic work with graptolites in the region. First, he noted (Lapworth 1876) the similarity, both lithological and palaeontological, between the Coniston Mudstone and the Birkhill Shales – which had emerged as the top of Lapworth's 'Moffat Series' where he was working at Dob's Lin in the Southern Uplands; and similar units could, he claimed, be found in Ireland and Thuringia. Lapworth argued further that the graptolite fauna of the Coniston Mudstone was essentially different from that of the overlying Coniston Flags, which were regarded as

Upper Silurian. Below the Mudstones were the Coniston Limestone of (then) agreed Bala age, with a possible break between them, as argued by Aveline. However, continued Lapworth, there was a faunal gap between the Coniston Mudstones and the Coniston Flags, as all the graptolites of the former had seemingly died out by the time the latter were deposited. So, he concluded, the Mudstones must be Lower Llandovery (which is where they remain to this day!); and the chief time-break must be between the Mudstones and the Flags, not between the Mudstone and the Limestone. In Lapworth's view, then, Hicks was right. The Welsh sequence could, if suitably applied, work for the Lake District.

This all sounds satisfactory; but Aveline (1876*b*) was by no means satisfied. He pointed out that if there was to be a significant time-break at the top of the Coniston Mudstones, this should reveal itself as some kind of physical break or sharp junction – but he had found no such break, and neither have subsequent geologists. We have, then, an example of ongoing debates about the relative merits of rocks and fossils as stratigraphic determinants, long after Smithian principles were enunciated and refined. We shall see how the investigations proceeded in relation to this matter when we consider the work of Nicholson and Marr in Chapter 5.

Ward's work, by contrast, was chiefly concerned with the igneous rocks of 'Otley II' – Sedgwick's 'Green Slates and Porphyries'. A clear picture of Ward's day-to-day work is revealed in his notebooks. In the years 1869 to 1876 he surveyed 377 square miles of territory and 1011 miles of boundary. His diaries or notes show him constantly engaged in the field, preparing and revising his maps and sections, and writing memoirs, but because of his early resignation from the Survey and premature death, only the explanatory memoir on Sheet 101 SE was actually published by the Survey (Ward 1876*a*), though this was a work of considerable importance, much more substantial than the other early sheet memoirs for the Lakes. We also see Ward liaising with the other surveyors, writing notes on their publications, and interesting himself in the work of his chief, Ramsay, in Wales – which was important as there was always a wish to achieve correlations with the 'standard' stratigraphy of the Lower Palaeozoic in Wales.

Early on (September 1869), the experienced Ramsay set Ward going, instructing him to note carefully the distinction between bedding and cleavage, to look for fossils in the SS, to look for evidence of metamorphism round the Skiddaw Granite, to distinguish ash and breccia beds from what the surveyors called 'close flinty lavas' (felstones – or ignimbrites: see Chapter 7), to distinguish between volcanic necks and volcanic 'products', to map-in moraines, to look for scratches on rock surfaces indicative of glaciation, to try to recognize evidence for silted-up lakes, and to note whether the fragments of ash beds were rounded or angular.[6] Points to consider included such questions as whether the SS were cleaved before the deposition of the Green Slates and Porphyries; the relationship between cleavage and foliation; whether the Skiddaw Granite was in some manner metamorphic, its age relative to the adjacent rocks, and its geometrical relationship to them; whether the SS might be regarded as Llandeilo; whether they were or were not conformable to the Green Slates and Porphyries; whether the Green Slates were deposited under water or slopingly round volcanic vents; why purple ash beds prevailed at the lower part of the series; the proportions of the ash beds and the lavas; the present rate of denudation of the region and the silting of lakes.[7] These, then, were the questions of the day. Ward certainly had plenty to get on with! And we find that the questions that he dealt with in his memoir of 1876 were along just these lines.

An immediate problem for Ward was to try to understand the kinds of rocks he was dealing with in the 'Green Slates and Porphyries' (or the 'Volcanic Series of Borrowdale' as he preferred to call them, following Nicholson 1872*b*)[8], and for this purpose he focused on a good section in 'Cat Gill' (or Ghyll) – a stream that runs into the eastern shore of Derwent Water, about halfway between Keswick and the head of the lake (see Figs 2.2 and 4.4). There, in ascending the stream, one passes conveniently over or through (in a gorge) alternating beds of volcanic ash and lavas, with purple breccias also present. Ward thought that the breccia-tion might be due to the lava having rolled over a cool surface, breaking into clinkers as it did so (Ward 1876*a*, p. 14). The ash beds displayed both bedding and cleavage, and contained some small garnets, which were also found in the lavas, as we shall see: for subsequently the source of such garnets became a matter of ongoing argument in Lakeland geology. Ward cut thin-sections of the different rocks and examined them under the microscope, obtaining the assistance of Samuel Allport (1816–1897) and the Survey's mineralogist and petrologist, Frank Rutley (1842–1904). The thin-sections were drawn and reproduced in colour in the 1876 memoir (Plates I–III). However, Ward did not provide petrological identifications of the various rocks in the Cat Gill section (though some chemical analyses were given). At the bottom of the sequence, near the lake-side, he found what appeared to him to be a fault between adjacent SS and the lowest purple breccia (see Fig. 4.5). He made no mention of mudstones, such as Nicholson was to suppose he saw there a few years later (see p. 37). Ward's six-inch map of the area is reproduced in Plate III, which illustrates the mapping style of the primary survey.

Ward then tried to use the Cat Gill rocks as a standard for the 'Volcanic Series of Borrowdale', but in doing so he was extrapolating excessively, for the rocks near Derwent Water only represent the lower part of the total volcanic succession. Nevertheless, Ward sought to extrapolate, considering supposed variants on the type section. Over at Eycott Hill, on the north-eastern edge of the Lakes, near Carrock Fell – outside the main mountain mass – he described a handsome porphyritic rock with large crystals of plagioclase feldspar in a fine-grained or glassy groundmass; other variants rich in magnetite were found. Further variants were found at St John's in the Vale, east of Keswick, at Scarf gap, and near Grange in Borrowdale (see Figs 2.3 and 12.1). In general, Ward thought the lavas might be said to belong to the dolerite family, but also having some similarity to 'felstones' – both vague categories. In any case, there was much variation; and the Cumberland rocks did not entirely resemble their ancient counterparts in North Wales.

Regarding the ashes, they too exhibited considerable variation. For the most part, they were not much water-worn, though some, as at Bleaberry Fell, the hill behind Cat Gill and Falcon Crag (see Fig. 2.2), contained what looked like conglomerate material. At Latterbarrow in western Lakeland (see Fig. 20.1) there appeared to be rolled inclusions of Skiddaw Slate. The ashes (in the Derwent Water region) were often strongly cleaved, but did not usually form good slates. An important variety was what Ward called 'concretionary ash'; or in his field notes he called it 'close-blue-flinty'.[9] It was very likely the rock that Sedgwick had called 'the concretionary' in his field notes back in the 1820s, a good example being found at Grisedale Hause (a col at the southern end of the Helvellyn range). Ward also found it in the highest mountains of the Lakes, as round Green Gable for example (see Fig. 16.1). Sometimes, this kind of rock displayed crude columnar structures. Suggestions of bedding could also be discerned, but sometimes the boundaries of the 'concretions' mysteriously cut

---

6 Ward, Notebook 4, pp. 27–28.

7 Ward, Notebook 4, pp. 28–30.

8 Nicholson (1872*b*, p. 107) used the term 'Borrowdale Series', mentioning that this was the name favoured by himself and Harkness.

9 Ward, Notebook 2, 1871, p. 37 and *passim*.

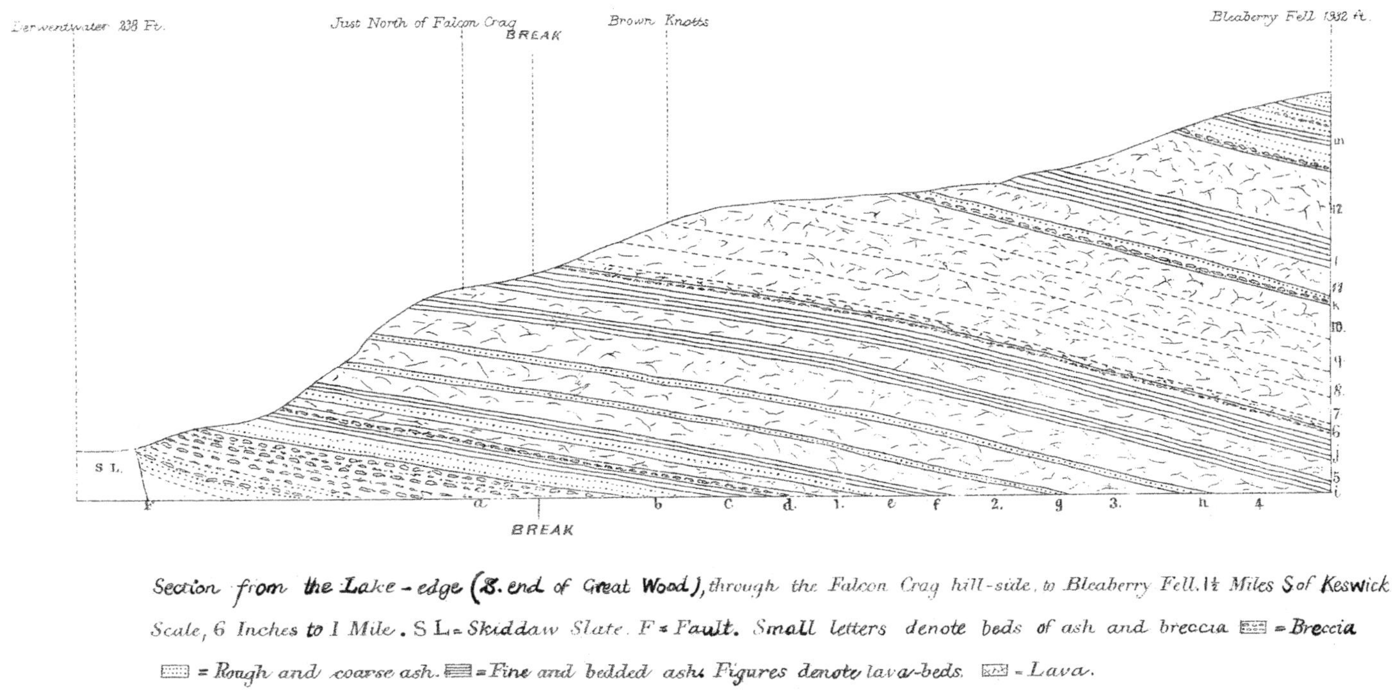

Fig. 4.4. Ward's 'standard section' for the (Lower) 'Volcanic Series of Borrowdale', Cat Gill (running eastwards from the shore of Derwent Water to Falcon Crag and Bleaberry Fell; Ward (1876*c*, Plate V). Reproduced by courtesy of the British Geological Survey.

Fig. 4.5. Contact between Skiddaw Slates (left) and the 'purple breccia' of the 'Volcanic Series of Borrowdale' (right), with fault gouge between, at Cat Gill, eastern lake-side of Derwent Water. Photographed by the author (1996).

through the larger fragments of ash. The texture was sometimes flinty, and sometimes the concretions were oval in form along the apparent lines of bedding. Sometimes, the exterior of the rock appeared to exhibit 'streaky lines'. Sometimes 'well-marked fragments . . . [were] discernible on the outside, with streaky lines more or less bending round them, or the streaky appearance . . . [was] combined with a truly bedded one (Ward 1876*a*, p. 25). Such mysterious, apparently altered, rocks ('doubtful' as Ward termed them) kept turning up all over the place, as Sedgwick had found. Indeed much of Map 101 (SE), around Thirlmere and Helvellyn, was mapped as 'Highly altered volcanic rocks mostly ash', and included the 'doubtful' rock – though needless to say they were not labelled 'doubtful' on the map.

The 'doubtful' rock was compact and trap-like, though definite outlines of fragments were plainly discernible on its weathered exterior. In thin section, the rock revealed a granular matrix, but the granules were 'frequently collected along lines which flow round and among the separate fragments' (Ward 1876*a*, p. 27). Ward supposed that the granular flow of chloritic material had somehow been produced subsequent to the rocks' formation and by the same process (whatever it was) that had converted the separate fragments into a trap-like rock that did not display a fragmentary structure on freshly broken surfaces. Yet on microscopic examination, the granules seemed to form patches or streaks that went around small fragments, which themselves seemed to be altered.

According to Ward's field notes, Ramsay thought that the rock might be a 'contemporaneous trap', but it did not seem to have a definite upper boundary, which made Ward doubt his chief's hypothesis.[10] His notes show him struggling with the problem, and leaning to the view that the rock was a kind of 'altered ash', the altering cause seemingly lying somewhere to the west, possibly having to do with the Eskdale granite; for the thicknesses of these rocks seemed to increase the nearer they were to some possible intrusive igneous centre. So Ward suggested that the rock had originally been bedded, and was then somehow altered so as to produce much chloritic matter and the appearance of a flow structure, the flow being in the direction of least resistance, which corresponded to the original bedding. He did not suggest that the rock was metamorphic and had acquired its peculiar state as a result of pressure. Though resembling the Welsh felstones in general appearance, analysis showed that the chemical composition was significantly different, an analysed Welsh sample being much richer in silica.

Altogether, the rock was quite a mystery; and in fact there was no suitable petrographic category available at that time into which Ward could pigeon-hole it. Many years later, in the 1950s (see p. 100), the rock type was construed as a welded tuff (or ignimbrite) – a rock formed when a mass of red-hot rock fragments is emitted from a volcano as a *nuée ardente* or 'fiery cloud'. As the fragments settle they adhere to one another, partly fusing together, but also flowing somewhat before the material finally consolidates. Then columnar jointing may occur as the solid rock cools and contracts. Ward had no knowledge of such a phenomenon, so he had to struggle to provide an interpretation with the categories then available to him. It seems to me that Ward did well under the circumstances. A sample of the rock, from Langdale, collected by Ward, is held at the Tullie House Museum, Carlisle (see Fig. 4.6a), with a modern photograph of rock of a similar type for comparison (Fig. 4.6b).

Attempting to answer some of the other questions posed by Ramsay, we find Ward reading Darwin's and Sedgwick's writings on cleavage and foliation, and suggesting that both types of structure might be produced when rocks are deeply buried. Also, for the area surrounding the granite near Skiddaw, he seems to have had the idea of 'progressive metamorphism'. Cleaved slates might become spotty on the cleaved surfaces. Spots might develop as hornblende crystals. Those crystals become more prominent and arranged in layers agreeing with the cleavage. Eventually, the hornblende slates become micaceous and gneissose. Such changes could be discerned the closer one approached the Skiddaw Granite. In general, the theory invoked to account for the Skiddaw Granite and its environs was much like that of the eighteenth-century Scottish geologist James Hutton (Oldroyd 1996, pp. 92–100). It is well shown in several of the sections that the Survey published, illustrative of the geology of the Lakes. An example is shown in Plate IV.

Ward did not attempt subdivision of the SS on the basis of their rather rare fossil contents (see Chapter 8), but by looking for differences in lithologies and evidence of bedding and cleavage he was able to discern something of their structure, as shown for example in Plate IV. He was also much interested in the contact between the Slates and the overlying volcanic series. As we have seen, he believed that the contact at the outlet of Cat Gill into Derwent Water was a fault. Likewise, the contact at Eycott Hill appeared to involve a fault, but Ward thought that the Eycott volcanics were a low member of the larger 'Volcanic Series of Borrowdale' unit, for it appeared that they contained some interbedded SS.

In general, Ward differed from the earlier geologists such as Harkness and Nicholson, or even earlier surveyors working in the district such as Dakyns (1869*a*, *b*), by thinking that the slates–volcanics contact in the main body of the Lakes, as near Derwent Water, was one of a system of faults, not an unconformity. In fact, this became a point of difference amongst the surveyors themselves, which was unusual in that they generally maintained a united front in public. As we have seen, Nicholson had done fieldwork in the Lakes early in the winter of 1868 (November), and the following year he had described his results for the region around Derwent Water (Nicholson 1869*c*). Nicholson thought that the lowest trap of the Volcanics overlay the SS *unconformably* at Hollows Farm near Grange (see Fig. 2.3). By contrast, Aveline thought there was a fault at this locality. However, Dakyns (1869*b*) reported that he had been in the district in December 1868, and had made a traverse from Keswick to Buttermere, passing between Robinson and Hindscarth, and from the vantage point of the watershed between these mountains it appeared from a distance as if, at Honister Crag on the opposite side of the Gatesgarth Valley, there was an unconformity between the Green Slates and Porphyry and the underlying SS. On a second walk, from the Borrowdale Valley, up behind Hollows Farm to Eel Crags of High Spy (see Fig. 9.7d), Dakyns again claimed to be able to see evidence for unconformity, with successively higher beds of the Volcanic Series abutting the SS.

It is interesting that Dakyns said that he was accompanied by another geologist during the first walk, but he did not feel at liberty to say who it was, as they disagreed about the interpretation of the contact. The second trip, which supposedly vindicated his views, was made alone. We know that Dakyns and Nicholson were in the same general area at about the same time, but the unnamed geologist is more likely to have been his supervisor Aveline, since Dakyns and Nicholson broadly agreed that the contact was one of unconformity. In June 1869, Aveline (1869) wrote to the *Geological Magazine*, saying that he had doubted Dakyns' claim, so he (Aveline) and Ward had gone over the ground between Derwent Water and Buttermere together, and had come to the conclusion that the contact was a fault, or rather a whole set of faults. This was the interpretation that Ward put forward in his memoir (1876*a*), and which thus became the official Survey view at that time, appearing in the maps (which were, as said, largely the work of Ward for the Central Fells). The issue was important, as one of the most basic questions for Lakeland geology was to establish the nature of the relationships between Otley's three main units; and the nature of the junction between the SS and the Volcanic Series needed to be understood in order to get the relative time relationships right, and to effect satisfactory correlations with the Welsh and Scottish rocks.

Favouring the idea of faulting, then, Ward reckoned that he could see evidence of 'slickensiding' and in Matterdale Beck (see Fig. 12.1), the Slates and overlying ashes (tuffs) appeared 'smashed together', though sometimes producing the appearance of conformity thereby (Ward 1876*a*, p. 49). (For further discussion of this site, see pp. 132 and 137.) However, an interbedding of Skiddaw Slate and rocks of the Volcanic Series was recorded as having been found by Aveline in the Black Combe district in southwestern Lakeland (Fig. 6.3), and claimed interbedding of the volcanic rocks with Skiddaw Slate in of the Eycott Hill area of NE Lakeland (Fig. 12.1) likewise suggested to Ward that the change in rock type did not necessarily involve a long time-gap and unconformity, but simply a change in the nature of the deposits. Certain rocks, such as those of Friar's Crag, on the NE shore of Derwent Water (Fig. 2.2), close to Keswick, appeared to Ward to represent volcanic plugs, and might have marked the source whence came the ashes and lava flows of (say) Falcon Crag and Cat Gill. Ward thought that most of the ashes and lavas were deposited subaerially. This might explain the absence of fossils, though the circumstances of extended and severe eruption conditions would hardly have been conducive to life.

For correlations with the Welsh rocks, Ward (1876*a*, pp. 46–47) proposed the parallels shown in Table 4.4. They were, he well knew, to a significant extent only guesses.

Ward (1876*a*, p. 68) thought that the SS and the Volcanic Series

[10] Ward, Notebook 2, p. 39.

Fig. 4.6. (**a**) Specimen of 'felstone (altered ash)' collected by Clifton Ward. Tullie House Museum, Carlisle: Acc. No. 1979.189.1). Photograph reproduced by permission of Tullie House Museum. (**b**) Rock of similar type, photographed by the author (1997), by a track from Langdale valley to Stickle Tarn (see Fig. 16.1).

Table 4.4. *Stratigraphic Comparisons for Lakeland and Welsh rocks, according to Ward (1876*a)

Unit	*Localities*	*Welsh equivalents*
Coniston Limestone	Coniston, etc.	Bala Limestone
Volcanic Series of Borrowdale (about 12 000 ft)		
9. Bedded, mostly fine, flinty ash	Great End, Esk Pike, Allen Crags	
8. Unbedded coarse ash and breccia	Broad Crag, Long Pike	
7. Bedded and rough ash	Scafell Pikes, Glaramara, Ullscarf	
6. Partially bedded, fine, flinty ash	Base Brown, Rossthwaite Fell	Llandeilo and Arenig Series
5. Well-bedded ash	Seathwaite	
4. Contemporaneous traps	Watendlath Fell, High Seat, Bleaberry Fell	
3. Breccia and bedded ash	Brund Fell, Watendlath	
2. Contemporaneous traps	Honister, Dale Head, Gate Crag, Falcon Crag [Cat Gill]	
1. Purple breccia		
Transition Beds		
Interbedded volcanic strata and Skiddaw Slates		
Skiddaw Slates (about 10 000 to 12 000 ft; base not visible)		
5. Black slates	Skiddaw	Arenig
4. Gritty beds	Gatesgarth, Latterbarrow, Tongue Beck, Watch Hill, Great Cockup	Caernarvonshire Grit and Stiperstones (Shropshire)
3. Dark slates		Tremadoc
2. Sandstone series	Grasmoor, Whiteside	*Lingula* Flags
1. Dark slates	Kirk Stile, between Loweswater and Crummock	*Lingula* Flags

had been cleaved by lateral pressure at the same time; but with the softer SS there was much crumpling, whereas with the more resistant Volcanics, strengthened by the lava bands, the strata were merely thrown into 'a series of low curves'. His generalized sections (Ward 1879, p. 54) reveal his ideas about the structures of the Skiddaw and Borrowdale rocks (see Fig. 4.7). The large anticline and syncline for the Langdale area and north past High Raise to Borrowdale are noteworthy.

A convenient summary of Ward's ideas on the geological history of Lakeland can be found in a letter from him to Ramsay, dated 10 March 1877 (KGA/Ramsay/8/724; reproduced here in part by courtesy of Imperial College Archives):

1. [oldest] The Skiddaw Slate sea with continued depression of seabed.
2. Volcanic Period, volcanic deposits[,] at first submarine, afterwards subaerial.
3 Depression of the area with its piled up volcanic material, and denudation of the same.
 [Substantial unconformity]
4. Submergence of the area beneath the Con. L. and U. Sil. sea with continued depression of seabed.
5. Elevation of the sea and enormous denudation throughout Old Red Period.
6. Carbonif. rocks formed around the early nucleus of the Mountain district.
7. Post-Carbonif. subaerial sculpture of the rough-hewn(?) block.
8. [youngest] Mid-glacial submergence, the country having previously acquired its present configuration.

This sequence presented Ward's matured thoughts on the geological history of Lakeland, but did not differ substantially from the sequence suggested in his north Lakes Memoir. It would allow for the possibility of the Coniston Limestone Series, etc., being laid down right over the top of the Borrowdale Volcanics and the Skiddaws, as J. F. N. Green later envisaged (see p. 82), but this was not explicitly stated in Ward's letter to his chief, and it was not shown in Ward's generalized section of the Lakes (Ward 1879, plate II).

With his training at the Royal School of Mines, Ward was familiar with the examination of rocks in thin-section, according to the methods pioneered by Henry Clifton Sorby (1826–1908), and using a technique of Sorby (1858) he sought to find out something about the conditions of depth and pressure under which the Skiddaw granite had been formed. The idea was to study the fluid-containing cavities found within crystals of some minerals, and the mobile vacuoles within the fluid cavities.[11] If v were the ratio of the 'sizes' of the vacuoles in the cavities to the 'size' of the cavities, V their relative size at zero external pressure and 0 °C, and p the pressure 'beyond that equal to the elastic force of the vapour' at a temperature, t, then according to a dubious calculation made by Sorby (1858, pp. 463–464):

$$p = 369\,000\,[(V - v)\,/\,(1 + V)] \qquad \text{(Ward 1875}a\text{, p. 575)}$$

The formula depended on the idea that with an increase in pressure, a vacuole would eventually disappear. (It had come into being, by contrast, as the crystal had cooled and the pressure had decreased.) Sorby had estimated, from examination of Italian granite from Ponza, that $V = 0.3$ and Ward used as the formula for his Lakeland granite:

$$p + 4000 = 369\,000\,[(0.3 - v)\,/\,(1 + 0.3)] \qquad \text{(Ward 1875}a\text{, p. 576)}$$

He could measure average values of v for specimens seen under the microscope in thin-section, and thence values of p were calculated, supposedly representing the pressure of the rock (measured in feet of rock) necessary to compress the liquid sufficiently to make the vacuoles disappear at 360 °C – the temperature at which granite would glow in the dark with a dull red heat. (The source of the number 4000 in the empirical formula was not stated.)

So, Ward suggested (1875*a*, p. 588), one could think of the total pressure at the granite's formation as being due to overlying rock ('downward pressure'), plus some additional 'surplus or outward

[11] Sorby noted that his attention had been drawn to such phenomena by Alexander Bryson, whose petrological and mineralogical collection he had examined in Edinburgh. Bryson had discovered fluid cavities within minerals of Aberdeen granite. On the early history of the study of fluid inclusions, see Wiesheu & Hein (1998).

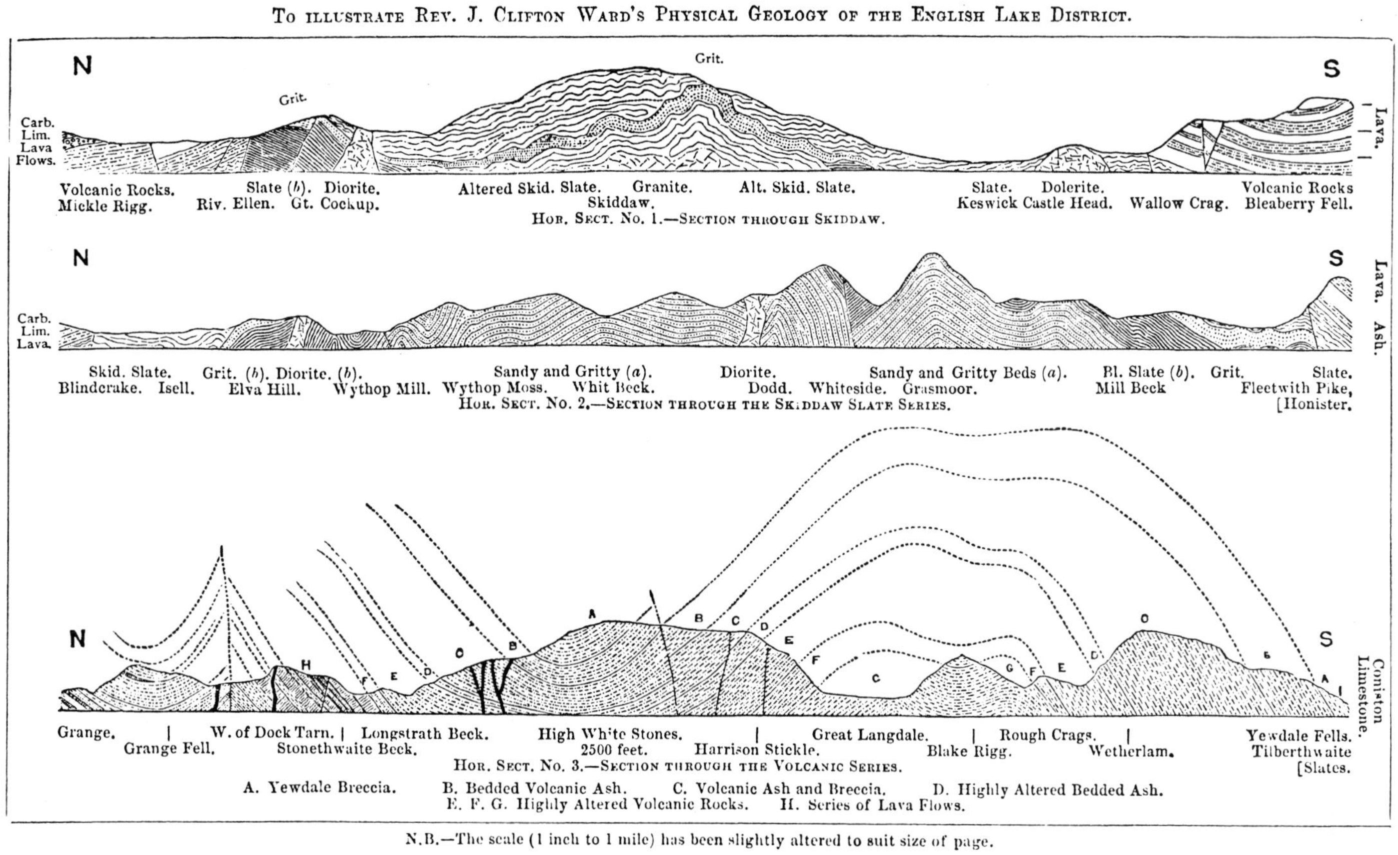

Fig. 4.7. Horizontal sections of the Lake District, according to Clifton Ward (1879, p. 54).

pressure' which could give rise to the forces producing elevation, contortion, metamorphism, etc. If the calculated pressure (from examination of the cavities and vacuoles) was about the same as the pressure expected according to the thickness of overlying rock, inferred from field evidence, then it would be likely that the 'outward pressure' had been relieved by volcanic action. On this basis, Ward estimated that the Skiddaw and Eskdale granites of the Lake District, formed at depths of about 30 000 and 22 000 feet respectively, and had been at a pressure such as to be able to effect elevation, contortion and metamorphism, there being a subsequent relief of pressure as the granite was exhumed; but the measurements and calculations for the Shap granite suggested that it was formed at lesser depth with less 'outward pressure', and so might have been an immature volcanic vent.

The arguments used by Ward were insecure, as it seems to me, but they represented a remarkable first effort to deduce something about the circumstances of formation of the Lake District's igneous rock by means of physical calculations based on microscopic examination of thin-sections. The great excess of pressure indicated for the Skiddaw granite suggested that the igneous mass hypothesized beneath the mountain (see Plate IV) had produced an upheaval and contortion of the overlying strata, followed by denudation. This could have occurred during Old Red Sandstone times, after the massive deposition of sediments during the Silurian. Indeed, the uplift could account for the cessation of deposition. The Skiddaw Granite might be a product of the fusion of Skiddaw Slate with added silica (Ward 1876*c*, p. 30). The granite or syenites (granophyres) of St John's in the Vale, Buttermere (see Plate IV), and Ennerdale might represent transition beds between the Volcanic Series and the SS, metamorphosed *in situ* (Ward 1876*c*, p. 28). The more basic crystalline rocks of Carrock Fell might have been produced by metamorphism of the lower part of the Volcanic Series (Ward 1876*c*, p. 28).

Interestingly, then, like some twentieth-century authors also, Ward thought that there might be various different origins of granite, including metamorphism of rocks such as clay slate in 'aqueo-igneous fusion' if a suitable admixture of requisite additional elements was available. The existence of liquid cavities in the quartz veins commonly found penetrating the slates suggested that the silica and water necessary for metamorphism into granite might indeed have been supplied from below. He further thought that metamorphic processes occurring among the many different rock types of the Volcanic Series could give rise to a variety of products, as: hard flinty-looking felstones; white-weathering felstones; banded felstone-like rock, trap-like within; amygdaloidal rocks; and fine-grained traps; massive rock with metamorphic fragments apparently welded together. So, although the metamorphic agencies acting over a given area might have been more or less uniform, the specific effects could vary considerably. In presenting such ideas to the Geological Society, Ward was bringing to the attention of British geologists some of the work of Continental 'neo-Wernerians' such as Gabriel Auguste Daubrée (1860).

In the reported discussion following the presentation of Ward's (1876*c*) paper, his suggestions received a mixed reception. His colleague Rutley thought his ideas about metamorphism feasible; but John Judd, Professor at the Royal School of Mines and not a particular friend of the Survey, did not think that the calculations based on Sorby's equation, and the necessarily approximate measurements for the sizes of the cavities and their vacuoles, could be relied on, and (as well one might!) he questioned the whole basis of Ward's arguments.

It is interesting that Ward's chief, Andrew Ramsay, expressed some sympathy with the idea of granite being of metamorphic origin (A. Geikie 1895, p. 326) and it is not impossible that Ward picked up such ideas from Ramsay's lectures at the Royal School of Mines. Though the Survey during Ramsay's directorship (1867–1881) supported Ward's Lakeland idea that what were later called andesitic lavas were altered tuffs, Ramsay's successor Archibald Geikie dissociated himself from such notions. From

Geikie's field notebooks (held at Haslemere Educational Museum) it is known that Geikie passed through the Shap area on his way north in 1873 (Notebook N), and he may have seen Ward then. Four years later, the Surveyors held a Conference at Kendal and journeyed through the Lakes so that all could get an idea of how the work was progressing. Geikie recalled that he and Ward had discussions of Ward's slides one evening – the occasion when Ramsay made the famous remark: 'I don't believe in looking at a mountain with a microscope' (A. Geikie 1895, p. 343), which was rather what Ward was doing.

In 1890, Geikie (see Notebook LL) made a more extended visit to the Lakes in company with the Survey petrologist Frederick Hatch (1864–1932), collecting information that he was to include in his Presidential Address to the Geological Society in 1891 (Geikie 1891) and which passed virtually unchanged into his *Ancient Volcanoes of Great Britain* (Geikie 1897). Geikie visited the region of Longsleddale and looked at the Stockdale (or Yarlside) Rhyolite nearby (see Fig. 3.6), which has an outcrop with strike parallel to the sediments of the Stile End Grassing Beds or the Coniston Limestone, and is distinct and separate from the underlying 'Green Slates and Porphyries'. Geikie preferred to call the rock a 'quartz-felsite' rather than rhyolite. He next proceeded to the area SW of Coniston to examine the rocks of the Appletreeworth area, where the Coniston Limestone (as then called) and associated sediments could be seen striking unconformably and almost perpendicularly across the outcrops of the Borrowdale Volcanic rocks of the Dunnerdale Fells (see Fig. 5.2). Following Ward's ideas, the volcanics had been shown on the Survey maps as tuffs or altered tuffs, but Geikie saw them as andesitic lavas, not tuffs (Notebook LL, p. 11). Geikie also went up to the copper mining area in the hills to the west of Coniston (Fig. 5.2), where he saw cleaved volcanics almost resembling shales or slates, along with coarse volcanic breccias, brecciated tuffs, felsites, and vesicular andesites. Next he proceeded to Elterwater, into Langdale (Figs 2.2 and 16.1), and to Loughrigg and Grasmere (Fig. 2.2), observing well-bedded tuffs *en route*. In Langdale he saw 'felsitic-looking materials'.

From Grasmere, Geikie drove past Thirlmere to Keswick and remarked that rocks marked on the Survey map (by Ward) as 'ash' seemingly contained lava bands. It was arranged that Hatch should section them. From Keswick, the granitic rock of St John's was examined, and an unsuccessful search was made for possible volcanic vents in the area. Significantly, Geikie could not find some of Ward's mapped faults. Examining Ward's 'standard' section at Cat Gill and Falcon Crag, Geikie could not distinguish between the lavas and tuffs as Ward had done, though, being on a quick visit, he did not say categorically that Ward was mistaken.

From Keswick, Geikie went up Borrowdale and over Honister Pass to Buttermere (Figs 2.3 and 7.4). Again he saw evidence of what he took to be lavas as well as ashes. Importantly, he could not see evidence for the complex of faults that Ward, with Aveline's concurrence, had mapped for the boundary between the Volcanics and the SS. Geikie then looked at the Slates of Skiddaw and Saddleback (Fig. 12.1) and then went over to the NE to look at the volcanics of Eycott Hill. He remarked (Notebook LL, p. 32) that Ward had correctly mapped them as lavas but he could see no essential difference between these rocks and others that Ward had mapped as ashes. Also, Ward's fault mapped at the base of the Eycott Volcanics was in a 'peaty plain' and was therefore 'speculative'. Geikie noted (Notebook LL, p. 33) that he could not 'understand Ward's incessant invocation of a faulted boundary'.

Returning to Borrowdale, Geikie (still with Hatch) examined the area to the east of Grange, and further south around Rosthwaite, but unfortunately he did not get to visit the classic exposure of the Skiddaw–Volcanics contact at Hollows Farm to the west of Grange; or no observations were recorded if he did go there.

Geikie referred to Ward's work in his Presidential Address and in his *Ancient Volcanoes*. In the latter, he described Ward as his 'late lamented friend' and mentioned that he had been over the ground with him, learning about his ideas on the nature of the rocks and the structure of the region (Geikie 1897, vol. 1, p. 233). But, said Geikie, much of what had been mapped as tuffs appeared to him to be lavas. Likewise, Geikie expressed his disagreement with Ward about the boundary between the Volcanics and the Skiddaws as being a complex series of faults. On both these points Geikie had differed from his colleague when they had been in the field together (presumably in 1877) but had then deferred to his opinion in view of Ward's greater field experience in the region. However, Geikie's reservations about Ward's ideas were strengthened as a result of his fieldwork of 1890. Thus Ward's interpretations of Lakeland geology did not stand long, even among the members of the Survey.

J. F. N. Green (1919, p. 154) (see Chapter 6) later made much the same complaints against Ward's work, and went so far as to say that since Ward 'regarded most of the lavas as altered ashes', 'stratigraphical mapping became impossible', for the succession effectively became continuous for much of the Volcanics. Like Geikie, Green also objected that Ward was prone to insert faults without justification.

Though one may doubt whether Ward's colleagues were in a position to make an informed judgement of his work, its apparent completeness – judged by his field-slips, which had no blank spaces – was at least temporarily persuasive and it became embodied, along with his theories, in the official maps, with the sanction of Ramsay. From the slips it would appear that Ward's only companion in the high fells was E. J. Hebert, a junior surveyor who was only with the Survey from 1875 to 1880, and about whom little is recorded. Some of the preserved slips are marked 'Aveline's Copy' and many of these are largely blank. It should be remarked that several of the Lakeland maps listed at the beginning of this chapter remained the only 'official' ones covering the main body of the Lakes until the last two decades of the twentieth century, following the remapping of the area inaugurated in 1982 (see Chapter 14), and the maps for the Ullswater–Haweswater–Appleby, and the Kendal and Kirkby Lonsdale areas, are only to appear in the third millennium.

As we have seen, doubts were expressed about the maps of the Central Fells by the Director-General as early as the 1890s. A good deal of Ward's work was carried out without the benefit of the presence of another pair of well-informed eyes, and his theory was suspect, so 'his' maps were never well regarded. The more experienced Aveline, by contrast, perhaps did more precise mapping (in easier country), but many of his field-slips were blank. Thus the situation for the central Lakes was by no means satisfactory, though to the south, amongst the 'Silurians', or the world of 'Otley III', better, palaeontologically controlled, mapping was possible.

Ward certainly covered much ground that no previous geologist had examined with any attention, but there were deficiencies in the young man's work (or that of his co-workers). An important example would be the dreary valley of Drygill, to the NW of Carrock Fell, for which he and his co-worker for the area, Robert Russell, produced the relevant field-slip (47 SE (E)) (see Fig. 12.1). The Drygill outcrop was outlined, coloured grey, and labelled 'slate', but its special significance does not seem to have been realized. In a published map, Ward vaguely mapped the shales there as belonging to the SS (Ward 1876*c*, p. 16),[12] but neither he (nor Russell) can have looked at them closely, for their extensive and definitely non-Skiddaw fossil deposits were not discussed, and it was not until 1887 that Nicholson and Marr reported fossils there (though Nicholson had found them there earlier; see p. 37). In 1892, John Marr regarded the fossils as being

[12] The sketch-map shows the area of Drygill mapped as 'Skiddaw Slate in Volcan. Ser.' The Carlisle–Cockermouth map (Sheet 101 NE, 1890) does not indicate the distinctive rocks of Drygill.

indicative of the Coniston Limestone Series. This, as we shall see, was a highly important piece of evidence in favour of Marr's tectonic theory for the Lakes; and for later theorists too.

Slip 47 SE (E) (which is annotated in Ward's hand, not Russell's) is of special interest as evidence of what Ward was or was not able to accomplish. Like his other field-slips, the whole is coloured in, suggesting that he had covered the entire ground.[13] But he cannot have gone up the Drygill Valley with any care (if at all) or he would surely have noticed its richly fossiliferous shales, with types (Caradoc, according to modern determinations) quite unlike those (rarely) found in the SS. The Drygill Shales are also weathered brown, rather than the dark colour characteristic of the SS. No fossils from the Drygill area are held in the Survey's collections dating earlier than the 1930s, and there are no references to the area in Ward's notebooks other than mention of work to be done on Sheet 47 (Pauline Taylor, pers. comm., 2000). Thus we must conclude that Ward overlooked some vital evidence, and his fully-coloured field-slips – unlike those of Aveline which mostly show dabs of colour where rocks were actually observed at outcrop – give a misleading impression of his accomplishment.

But who would blame Ward? The vast bulk of the work in the high Lakeland fells was cruelly left to him alone. In any case, he clearly did do some work in the Carrock Fell and Drygill area, for his mapped pattern of outcrops there, and his suggested lines of fault, on his Cockermouth sheet are, in general, compatible with the modern Cockermouth sheet (1997).

This said, and though modern geologists still use the field-slips of the first Lakeland surveyors to help them find exposures in the field, finding them extremely useful for this purpose, the initial surveying was not, perhaps, done to the highest standards. Indeed, it came in for criticism even in its own day from those in the mining community who wanted to use the official maps to assist them in their work. In particular, there was criticism, both private and public, of the Whitehaven map by the mining engineer John Dixon Kendall (1848–1929?), of Kendall and Main, Mining and Civil Engineers, 58 Roper Street, Whitehaven (BGS archives, GSM 1/29, p. 91 (5 April 1894); Kendall 1920*a*, p. 58).[14] Kendall (1920*a*) told the world that in consequence of his representations made in 1894 the Whitehaven map (1892) had to be withdrawn and revised. Aubrey Strahan (1852–1928) (who was Survey Director from 1914 to 1920) did some resurveying in 1894, and the map was reissued in 1895, but his field-slips reveal that he made rather few observations. Kendall placed his notes at Strahan's disposal, and he was then (according to Kendall) able to achieve in a few weeks what his predecessors had failed to do in more than twenty years! Goodchild did not take up Kendall's offer to show him round some of the exposures and the mine workings. When the memoir on the haematite ores was eventually published by Bernard Smith (1919) (see below) it still attracted criticism: there were many factual errors according to Kendall, sometimes relating to matters that would have been known if mines, by then disused, had been examined when they were working in the nineteenth century. Moreover, according to Kendall, Smith had not even taken the trouble to examine old mine records as he should. So a second edition had to be got out only five years later. Even then Kendall (1925) found that it contained errors. It would seem that the Survey was not particularly good at 'industrial' geology at that time.

Following the completion of the first survey and publication of the memoirs, maps and sections, there was a pause for a few years. However, perhaps in response to criticisms such as those of Kendall, work was subsequently resumed to the west and north of the Lakes and into western Lakeland, a local office being established at the Oddfellows Buildings, Queen Street, Whitehaven, in 1920, with Bernard Smith (1881–1936) in charge. (At the same time, an office was opened in Newcastle-upon-Tyne, headed by Robert Carruthers.) Work continued in Cumberland until the 1930s, the chief surveyors besides Smith being Ernest Dixon (1876–1963), Tom Eastwood (1888–1970), James Maden (1894–1927), Frederick Trotter (1897–1968), Sydney Hollingworth (1899–1966), and Colin Rose (1908–1991), with Kingsley Dunham (1910–2001) working down in the Furness Peninsula.[15] The staff of

---

[13] I thank Richard Gillanders for information on these points concerning the field-slip.

[14] Kendall, who was born in Ulverston, was elected Fellow of the Geological Society in 1874, being supported by George Dixon, James Dees and John Taylor. His nomination paper describes him as a 'civil and mining engineer' from 'St Bees – Carnforth'. (I am indebted to Ms Wendy Cawthorne for this information.) According to Mr Peter Eyre of the Cumbria Archive Service, Kendall was residing in Whitehaven in 1894. However, his company had apparently gone out of business by 1910, not being listed in the Trade Directory for that year. Ms Andrea Fazackerley of the Barrow-in-Furness Library provided Kendall's date and place of birth.

[15] Ernest E. L. Dixon obtained a first in geology at the Royal College of Science in 1897 and joined the Survey in 1897, remaining with the organization (apart from a period of war service) until his retirement in 1938. He was awarded the Geological Society's Murchison Medal in 1936. In pre-War days Dixon did important work in structural geology in South Wales, and also studied the evidences of glaciation in the Midlands. He was with the Cumberland group at Whitehaven from its inception in 1920, and worked on Carboniferous geology, glacial geology, and sedimentary structures in the Skiddaw Slates, as well as developing ideas on the origin of the Cumbrian haematite deposits (see Eastwood 1963; Hollingworth 1964). He helped train Trotter and Hollingworth. Bernard Smith obtained a first in geology from Sidney Sussex College, Cambridge, in 1903. After a period as Demonstrator he was appointed to the Survey in 1906 and worked in the Midlands and North Wales before the War. During the War he worked on economic projects, especially the Cumberland haematite deposits, and this led to his being placed in charge of the Whitehaven office when a Survey branch was established there in 1920. After 1930, he became a senior member of the Survey staff. He was elected to the Royal Society in 1933 and appointed Survey Director in 1935, but sadly he died from cancer only a year later (see Wills 1937). Tom Eastwood was a Lancastrian who studied geology under William Watts at the Royal College of Science, from where he joined the Survey in 1911. He worked first on the Midland coalfields, and then, during the War, on reports on British mineral resources. After service in France, he was appointed to the Whitehaven office in 1920, and took charge of the Cumberland group in 1930. Eastwood was promoted Assistant Director of the Survey in 1937, and thereafter concentrated chiefly on economic geology, retiring in 1949 (see Edmonds 1970–1971; S. B[owie], 1971). Little is known of Maden, who died young. J. Kidd (see Fig. 4.10) is not even named as a Survey staff member in Flett's (1937) history of the Survey. Frederick Trotter studied geology and chemistry at Armstrong College, Newcastle-on-Tyne, with a period for war service, during which he was severely wounded (see Fig. 4.10). He joined the Survey in 1921, being posted to Whitehaven. He worked on the stratigraphy of the Carboniferous rocks of the region, on the haematite deposits, on the Skiddaw Slates, and most importantly on the glacial history of Cumbria, though he was actually a specialist in coal geology. After 1933, he worked in South Wales, and took charge of the Survey's Manchester office after the Second World War, rising to the position of Assistant Director of the Survey in 1956 and retiring in 1962 (see Taylor 1968–1970; R[ose] & D[unham] 1969; Rose 1969.) Born and schooled in Northampton, and after study under Marr and Harker, Sydney Hollingworth (known to colleagues as Holly) obtained a first in geology at Clare College, Cambridge, and worked with the Survey from 1921 to 1946, his first assignment being at the Whitehaven office, where he was stationed for ten years, and from where he mapped on the Brampton, Whitehaven, Gosforth and Cockermouth sheets. Though working for much of the time on the Upper Palaeozoic rocks, Hollingworth eventually became involved with the study of the Carrock Fell complex in NE Lakeland, the Skiddaw Granite and its metamorphic aureole, and the Borrowdale Volcanics and Skiddaw Slates. He was also interested in Lakeland geomorphology and in evaporites. Hollingworth took up the Chair of Geology at University College, London, in 1946, and rose to be President of the Geological Society in 1960–1962. In his later years he gave much attention to engineering geology (see Wells 1967; and also Robinson 1998–1999 [(2000)]). Colin Rose (1908–1991) (see Dunham 1991) attended John Lyon School, Harrow (which is of interest to me as my first job was as a chemistry teacher at that institution), from whence he proceeded to University College, London, to read geology.

Fig. 4.8. Staff of the Whitehaven office of the Geological Survey, 1927. Left to right: Trotter, Eastwood, Smith, Kidd, Dixon, Hollingworth. British Geological Survey archives, GMS 1/641. Photograph reproduced by courtesy of the British Geological Survey.

that time are recorded in two official Survey photographs, apparently taken at the same time, one of which is shown in Figure 4.8. The Whitehaven office closed in 1927, with the staff and stores moving to headquarters in London, but fieldwork continued in the north for several years. The photographs were likely taken on the occasion of the office's closure.

Thinking back to those pre-War days, the former Survey chief, the late Sir Kingsley Dunham (record of interview, 10 June 1998), recalled the 'Cumbrian room' in London in the 1930s, where the maps were collated. There were then four staff at work at four desks in the same room, plus the District Geologist (Eastwood, in Dunham's day), who had his own den. They would meet for tea in the 'London room' and there were particularly animated discussions about the origin of the Cumbrian haematite deposits. Trotter was noted for his ability to jump onto a table in a pub without a run. During periods of fieldwork, they went out 'about nine'. Transport in the field was chiefly by bicycle. It appears to have been a fairly relaxed life. Thus Eastwood (1963, p. 140) recalled that his colleague Dixon 'had no notion of time, whether it was his own or other people's. . . . [H]e was always much more

Footnote 15 *continued*

After an expedition to Lake Tanganyika, he took his MSc in 1930, joined the Survey, and was assigned to the Cumberland unit. He worked on the Cockermouth sheet and on the Monmouth Sheet when the unit moved to South Wales. He then went north again to do detailed work on the Cumberland haematite deposits with Kingsley Dunham, mapping at 1: 2500, and later examined the Lake District metalliferous resources. From 1941, Rose worked on a series of jobs related to the War effort, and in 1948 he transferred to the Civil Service, being involved with the mining and petroleum industries. Then, quite late in life, he went back to complete the mapping of the Lower Palaeozoic rocks in the Furness area. I have been told that he had some difficulty in picking up the threads of his pre-War work. Nevertheless, he was the senior author (with Dunham) of the Memoir on the Cumbrian haematite deposits (1977); and from 1973 to 1978 he represented the Department of Energy at UN conferences on the Law of the Sea. Dunham obtained a first at Durham under Arthur Holmes (as his second student there) and did his PhD on the genesis of the North Pennine ore deposits (1932), following which he held a Commonwealth Fellowship at Harvard for three years, working on the geology of the Organ Mountains, New Mexico. Returning to Britain, he was appointed to the Survey in 1935, joining the Cumbrian unit, which, however, had by then been moved to Monmouthshire to work on the Forest of Dean coalfield, where he was initiated into Survey work by Rose. Then, because of the approaching war, the two were moved to Barrow and Dalton-in-Furness to map the Cumberland haematite deposits. Dunham recalls (record of interview, 10 June 1998) that the six-inch map of the area by Dakyns and Tiddeman was a 'remarkably feeble effort' in that they wholly ignored the haematite. After finishing his Furness work, Dunham returned to the north Pennines to re-examine the ore-bodies there; and after the War he transferred to the London office, becoming Chief Petrographer (1948), where his work included the examination of specimens sent down from the Lake District. In 1950, he was appointed to the Geology chair at Durham, where he remained for 16 years, regularly taking students to Westmorland for map-work. In 1966, Dunham returned to the Survey as Director, reigning over one of its most successful periods. He was elected FRS in 1955, served as Vice-President from 1971 to 1976, and was knighted in 1972. He retired to the position of Professor Emeritus at Durham in 1975, laden with honours, and was still active when I met him in 1998, though afflicted by blindness. He died on 5 April 2001. For an outline of Dunham's career, see McQuillan *et al.* (1986); Anon. (2001*a*); Johnson (2001).

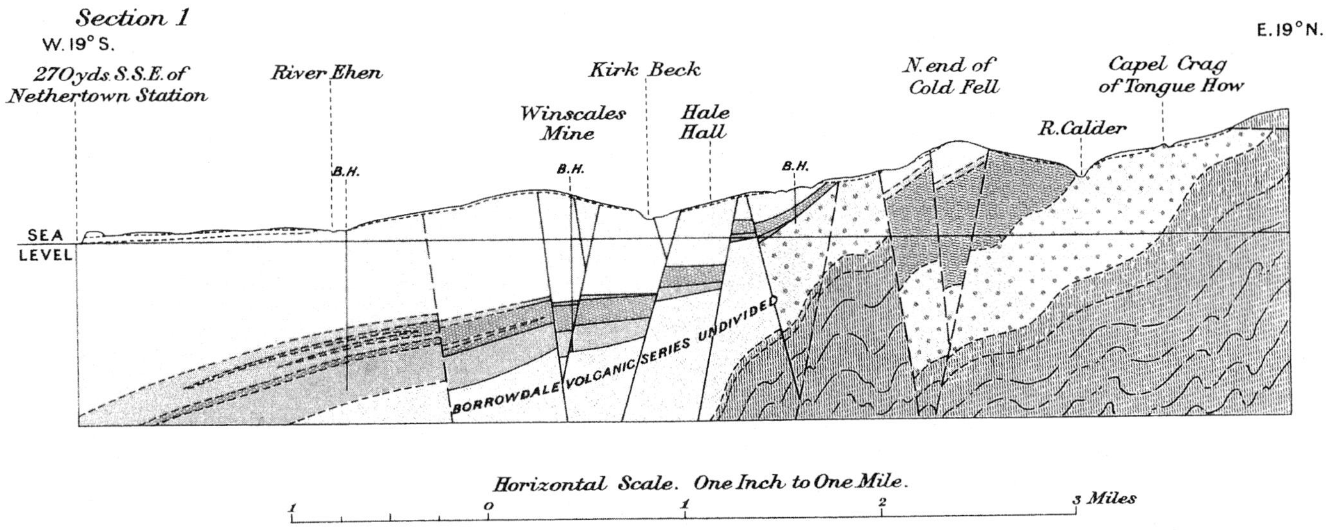

Fig. 4.9. Section accompanying one inch to the mile geological map of the Gosforth district (Sheet 37) (1937, reprinted 1952) (coloured in original). IPR/23–7C British Geological Survey. © NERC. All rights reserved.

interested in things over the hill than under his feet'. Perhaps this attitude to life was reflected in the leisurely production of maps.

Thus the following one-inch maps were eventually published, from the six-inch 'standards':

Whitehaven (Sheet 28)	1929
Maryport (Sheet 22)	1930
Gosforth (Sheet 37)	1937
Cockermouth (Sheet 23)	1960

Ulverston (Sheet 48) only appeared in 1977, from the work of Rose and Dunham, but at a scale of 1: 25 000. Rose and Dunham's Dalton-in-Furness map (Sheet SD 27) appeared the same year and at the same scale, but only following further post-war work by Rose in 1970–1972, and with additions by W. B. Evans in 1974. Given that the Whitehaven office opened in 1920, this was not an impressive rate of progress, but interruption of work by World War II must be taken into account. Moreover, with all the mine workings in west Cumberland to be considered, along with a vast amount of borehole data (it being at that time the most bored region Britain), there was a tremendous amount of work to do. Also, work further north in the Carlisle–Solway region was undertaken, but this lies outside our present concern.

In addition to the maps, annual reports and memoirs were published, as follows:

Smith 1919; Flett 1921; Smith 1921; Smith 1922; Dixon 1922; Smith 1923; Smith 1924*a*, *b*; Smith 1925; Smith 1926; Dixon *et al.* 1926; Smith 1927; Eastwood 1928; Dixon 1928; Smith 1928*a*, *b*; Smith *et al.* 1929; Smith 1930; Eastwood 1930; Eastwood 1931; Smith 1931; Eastwood *et al.* 1931; Eastwood 1932; Trotter & Hollingworth 1932a; Eastwood 1933; Elles 1933; Trotter *et al.* 1937; Eastwood 1938; Eastwood 1939; Dunham and Rose 1941; Eastwood *et al.* 1968; Rose & Dunham 1977.

There were also quite a number of papers in academic journals, reports of field excursions, etc. Regardless of the rate of progress, there was much to do. The terms of reference for Dunham and Rose, who began their main mapping of the Duddon Estuary and Furness areas in 1937 (Dunham & Rose 1949, p. 14), required them to examine every extant mine plan and record of boring. Such a task could not be completed overnight.

Regarding the work of the 1920s and 1930s, it should be noted that much of it had to do with the geology of the Cumberland coalfields and the deposits of Carboniferous Limestone, and the Permo-Triassic rocks of the coastal region, which largely lie beyond the scope of the present study. The work also had much to do with glacial geology, which was clearly of relevance to the regional agriculture but also to high-level geological theory. (Discussion of the Pleistocene geology of Cumbria will be deferred to Chapter 19.) However, the eastern outcrops of the Skiddaws were surveyed near Ennerdale, and the relationship to the Ennerdale granophyre was examined. The SS were subdivided into the following lithological units, chiefly at the suggestion of Dixon (Smith 1925, p. 70):

Mosser Striped Slates[16]
Loweswater Flags
Kirk Stile Slates[17]
Blake Fell Mudstones[18]

Subdivision of the Skiddaws on the basis of graptolite zonation was also undertaken, using the services of the Cambridge specialist Gertrude Elles; but this topic will be deferred until Chapter 8.

The surveying of the 1920s and 1930s did not reach the mountainous volcanics of the Central Fells, but in revising the Cockermouth map the officers worked their way round the north of the Lakes, examining the Eycott Volcanics and the rocks of the Carrock Fell complex, both of which were regarded as distinct from the regular tuffs and andesites of the Borrowdale Volcanics. The metamorphic aureole round the Skiddaw granite was re-examined by Hollingworth, who also studied the sections made available by the drilling of the tunnels in relation to the establishment of the Haweswater reservoir in eastern Lakeland.

However, I should like to focus attention here on the geological structure proposed for the land between the Lakeland mountains and the sea, as revealed in the section appended to the Gosforth map (see Fig. 4.9), as this has an important bearing on the discussions that developed in the 1990s, when the area was being considered for a nuclear waste repository (see Chapter 20). It will be remarked that the surveyors supposed that the rocks of the Borrowdale Volcanics dipped westwards beneath the Carboniferous

[16] Mosser Fell lies between Loweswater and Lorton Vale, which runs north from Crummock Water. See Figure 12.1.

[17] The name Kirk Stile seems only to survive in the name of the Kirkstile Inn at Loweswater village. (I am grateful to Chris Thompson for this information.)

[18] Blake Fell is approximately on a line between Loweswater and the northwestern end of Ennerdale.

and Permo-Triassic cover, all being substantially faulted. The map itself showed the mountains of the Lake District proper meeting the rocks of the coastal plain at a fairly simple system of faults, running approximately N–S. This was subsequently dubbed the Lakeland Boundary Fault, and has received much attention in recent geological investigations, as we shall subsequently see.

From the list of publications given above, we may also focus on Eastwood's paper on the 'Cockermouth Lavas'. These are a series of dark, basaltic lavas, sometimes amygdaloidal, lying below the Carboniferous Limestone, with the same dip and strike, and in places over a thin conglomerate, which lies unconformably on the SS. The two units cropped out to the north of Cockermouth and ran eastwards, parallel with the Derwent River. They were surveyed by Eastwood, who initially supposed that they belonged to the Borrowdale Volcanics, and were written up as such for the Survey *Progress Report* for 1927. However, specimens were sent to London for examination by the Survey petrographer, H. H. Thomas, who reported that they contained much olivine and were quite unlike the typical Borrowdale rocks. This opened up a number of theoretical possibilities, as is shown by correspondence in 1927 between Smith and Thomas and others, held in the Survey archives (GSM2/519): they might be a particular unit belonging to the Carboniferous; they could be a variant of the (Ordovician) Borrowdales; or they might be related to the basic rocks of Carrock Fell to the east, and perhaps also to the Eycott-type rocks, exposed along the northern margins of the Lakeland mountains. If the last hypothesis were true it could mean that the Eycotts and the Carrock Fell complex might all conceivably be Carboniferous. Additional investigation was evidently necessary, and Eastwood's contribution to the 1927 *Report* was temporarily withdrawn, while the rocks were investigated further and compared more closely with the Eycott types.

In the event, the Cockermouth Lavas were deemed by Thomas to be different from the Eycotts, which lacked olivine and were of andesitic composition. The Cockermouth Lavas overlying the conglomerate had previously been noticed by Ward and Goodchild in the primary survey work, and regarded as basal Carboniferous, whereas the Eycott lavas were represented as Borrowdales. They accounted for the situation by introducing arbitrary faults between the conglomerate and the lavas. The conglomerate was noted by Eastwood as containing pieces of Latterbarrow Sandstone (from the Skiddaw Slates; see pp. 81 and 199).

As said, Eastwood's first thought had been that the lavas were of Borrowdale age, and that the conglomerate consequently represented quite a large period of erosion. However, after more detailed survey, and petrological examination by Thomas, he concurred that the lavas were indeed extruded rocks, not sills. Having petrological analogies to the Carboniferous lavas in the Midland Valley of Scotland, the so-called 'toadstones' of Derbyshire, and certain lavas in Shropshire interbedded with Carboniferous Limestone, they might reasonably be regarded as Carboniferous, in agreement with their relationship to the Cumbrian Carboniferous Limestone (Eastwood 1928). The Cockermouth Lavas are unique in the Lake District and though not of wide extent are interesting from the perspective of modern geology in that they would appear to represent a period of modest crustal extension in northern England during the early Carboniferous.

We should also mention the Surveyors' ideas on the origin of the haematite ores of west Cumbria. Within the Carboniferous Limestone of the coastal region, there are large cavities called 'sops' or 'flats' according to their shape, containing rich deposits of haematite (kidney iron ore). These had been quarried or mined since the Middle Ages, and were much used during the nineteenth and early twentieth centuries as a source of iron, serving as the basis of the shipbuilding industry at Barrow (and the characteristic railings around fields on some of the Lakeland farms). The earliest views of Sedgwick (1836), Edward Binney (1847, 1855) and William Brockbank (commenting on Binney 1868) were that the iron was of volcanic origin, and of pre-Permian age. Binney thought that chloride of iron was emitted during volcanic eruptions and could have been accumulated in adjacent oceans. A volcanic vent at Greenscoe near Dalton-in-Furness (see Fig. 1.1) gave some credence to this view, for the iron deposits formed an approximate ring to the south and east of the vent. John Kendall (1875) suggested that the iron might have been derived from the Carboniferous. Later (1881–1882) he suggested that the ore could have been derived from ferruginous solutions associated with the local volcanic activity. Later still (1920*a*), he suggested that the haematite bodies might be the product of the outer zone of mineralization of the copper–lead–zinc province of Lakeland orebodies.

By contrast, the surveyor Goodchild (1889–1890) thought iron compounds were 'washed down' from the overlying ferruginous sandstones of Permo-Triassic rocks such as the St Bees Sandstone, an idea supported by Bernard Smith (1919, 1924*a*, 1928*a*) in the second round of surveying. However, Goodchild failed to take up an offer (*c*. 1890) from Kendall to view the haematite nodules in west Cumberland and compare them with ones of the brockram (i.e. cemented breccias of pieces of limestone, being a Cumberland miners' name for breccia) of the Eden Valley (Kendall 1920*a*, p. 60). In Kendall's opinion Goodchild was 'almost a stranger to the iron-ore fields'. Yet he had been a co-author of the revised Whitehaven map (1895). Differing from Goodchild, Dixon (1928) liked the idea that the source of the downward percolating water was igneous and hydrothermal, having risen up through the Loweswater Flags (SS). Otherwise, he did not see how metalliferous veins could occur in the Skiddaws, where they had no ferruginous Permo-Triassic cover. Smith, however, doubted Dixon's belief that the iron–manganese veins were necessarily of the same age as the lead, copper and zinc veins in the Skiddaws; and he queried how it was that the solutions supposedly only rose where there was a cover of impermeable rocks and descended where the cover was absent. Interestingly, the opposed views were both published in the same progress report (1928) – something that would not occur today! From borehole evidence, Trotter (Trotter *et al.* 1937, p. 70) noticed that where the limestone was covered by the impervious St Bees Shale (which underlay the pervious and ferruginous St Bees Sandstone) it contained negligible iron, which suggested that the haematite was being brought down from above by water of meteoric origin. Later, Trotter (1945) showed that there were some bodies of iron ore that crossed post-Triassic faults without displacement, and were therefore also post-Triassic. He ended up with the view that magmatic water rose into the 'brockram' deposits at the base of the Permian and subsequently percolated down into the sops in the limestone below.

The late, indeed posthumous, paper by Kendall (1929) makes interesting reading. In it, the elderly mining engineer gave expression to long-standing grievances about the work of both Goodchild and Strahan and that of the subsequent Whitehaven Survey group under Smith. It appears that Kendall gave substantial assistance to the nineteenth-century surveyors, so much so that parts of the Whitehaven sheet (as revised by Goodchild & Strahan 1895) were, according to his claim, substantially his work. Yet this went unacknowledged and no copy of the map was sent to Kendall. On complaining to Archibald Geikie, he was told that due acknowledgment would be made in the accompanying Memoir, when published (Smith 1919, 1924*b*), but this didn't happen either. A letter to Smith apparently brought no response, so Kendall ceased providing the Survey with information. He must, therefore, have been amused to see the opposing viewpoints published in the 1928 *Report*. Moreover, he asserted that both parties were wrong. Smith (following Goodchild's idea) was wrong because there were many beds, at different depths, below the ferruginous sandstones that were *not* stained red or iron bearing. Dixon's views were closer to Kendall's, and were indeed

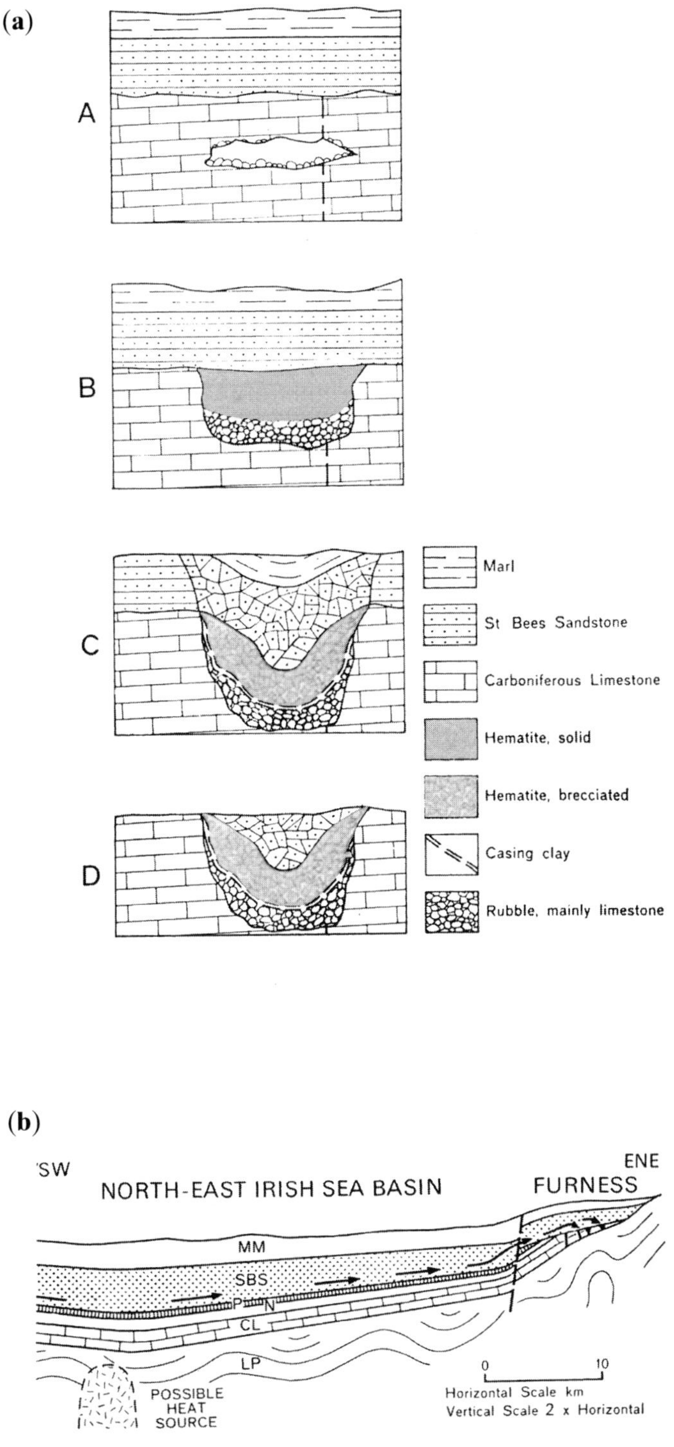

Fig. 4.10. (**a**) Sketch cross-sections to illustrate stages in the formation of the sops. (**b**) Sketch cross-section to illustrate the possible provenance and migration of ferriferous formation waters. Dunham's conception of the formation of the Cumberland haematite deposits, as depicted in Rose & Dunham (1977, p. 91). IPR/23-7C British Geological Survey. © NERC.

probably derived in part from him. But in Kendall's view Dixon was inconsistent, for he envisaged both a magmatic theory and one of downward percolation; and he did not sufficiently relate the formation of the haematite to the formation of the mineral veins in the Lake District proper. In this episode we see that the Survey could be a bit lordly, not always acknowledging for assistance as it should.

Dunham himself, as he told me (record of interview, 10 June 1998), suggested a compromise solution to the 'Dixon–Smith' problem, incorporating both the idea of water bringing iron solutions upward from below *and* the idea that it percolated downwards through the Permo-Triassic. His idea was that there had been, offshore, down-dip, at considerable depth, a centre capable of generating hot water, and that this had travelled up the cliff (fault-scarp) of the Permian sandstone, collecting iron as it went; but when it encountered large holes under the sandstone it went down into them and mineralization of the cavities occurred. In discussing the matter in print (in Rose & Dunham 1977, pp. 89–92), years after his survey work with Rose before and during the war, Dunham showed that there were plenty of modern instances of volcanoes generating ferruginous solutions and sediments, but there was no adjacent volcano of the right age that might have acted thus. Also, there being negligible iron in the lead–zinc–copper ore bodies of the Lakes proper, it was unreasonable to suppose that they had the same source as the haematite of SW, Cumbria.

Dunham's model can be represented by his own diagram, as shown in Figure 4.10. Here it is supposed that solutions have been driven up-dip by a heat source situated somewhere in the Irish Sea region, have leached ferruginous matter from the St Bees Sandstone through which the solutions passed and then deposited it in cavities in the Carboniferous Limestone. The source of iron in the St Bees Sandstone is taken to be laterites developed in hot wet conditions at the end of Carboniferous times.

It is not clear precisely when and how Dunham formulated his theory. It was not present in his paper with Colin Rose (Dunham & Rose 1949), which simply had the idea of ferruginous matter percolating into the sops from the overlying strata but the authors were agnostic as to the actual source of the iron. New ideas were developed in the late twentieth century to account for the Cumbrian haematite deposits and matters relating to tectonic events in NW England and the area of the Irish Sea during the Tertiary. These will be discussed in Chapter 18, where the problem of the source of the heat also receives mention; and the topic recurs in Chapter 20. Meanwhile, in Chapter 5 we return to the nineteenth century and early twentieth century to expound the ideas of John Marr and Alfred Harker, whose Lakeland investigations overlapped those of the primary and the 'Whitehaven' surveys.

Chapter 5
John Marr and Alfred Harker

As we have seen, the main 'professional' mapping of the Lakes was more or less completed by the 1880s, though the Survey continued to publish memoirs on the region for some time after finishing the fieldwork, before returning to the region in the twentieth century to revisit some of the old questions; and then it undertook a major resurvey after 1982 (see Chapter 14). So it was that much of the detailed research from the 1880s to the outbreak of World War I was undertaken by 'amateurs'. Of these, the most important were Nicholson, whose ideas were discussed in Chapter 3, John Postlethwaite (see Chapter 6), John Edward Marr (1857–1933) and J. F. N. Green (see Chapter 6). There are good archival sources on Marr, in that all 65 of his field notebooks and an unnumbered 'journal' survive at the Sedgwick Museum, together with a few letters.[1] The notebooks reveal that from 1874 until 1927 Marr spent a substantial part of almost every year in the Lakes; and he wrote what became the standard book on Lakeland geology for many years (Marr 1916).[2] To my knowledge, rather little on Green has survived beyond his published works.

Marr's family came from Bolton-le-Sands in Lancashire, and he attended school at Lancaster Grammar. While there, he made the acquaintance of the Surveyor Richard Tiddeman, then working in the district, and accompanied him on several field-trips. Marr's unnumbered journal (for 1874–1876) shows that he was examining human remains in caves near Settle, looking at fossiliferous Lakeland strata, and interesting himself in glacial phenomena. Palaeoanthropology, Lakeland stratigraphy and Pleistocene geology were to become some of his chief objects of attention for the rest of his life, though he was also interested in structural questions and sedimentology.

Marr was also meeting the right people while still young. He was with Thomas G. Bonney (1833–1923), the Cambridge petrologist, on a field excursion in the Lakes in September 1874 when only 17 years old, and in December that year he met Clifton Ward through the good offices of Tiddeman. Not surprisingly, Marr decided to read geology under Bonney at St John's, the college which for many years was the University's leader in the study of the subject. He obtained an exhibition in 1875, and Bonney personally sent a telegram informing him of the result: '[y]ou have done very well'.

After graduating, Marr worked as a university extension lecturer for four years and began to develop an interest in geology and scenery. He was stationed in Penrith for some of the time, enabling him to prosecute his Lakeland work. A newspaper of unknown provenance (1881?) reporting on a field excursion led by Marr described him as a

> learned but youthful lecturer (only a few years out of his teens), who . . . has won golden opinion from his numerous hearers, not only [by] giving indubitable proofs of the perfect mastery of his subject, [but] also by his striking modesty, total absence of pretension, pedantry, or dogmatism, as well as by his engaging affability and desire to give full information to everyone desirous of having any difficulty or obscurity elucidated.[3]

After a brief period at Leeds, Marr gained a fellowship at St John's in 1881 and was appointed lecturer in geology in 1886. In 1917, he succeeded McKenny Hughes in the Cambridge chair and held the position until his retirement in 1930. Marr was elected FRS in 1891, and had other important distinctions. Thus he was one of Britain's leading geologists in the late nineteenth and early twentieth centuries. According to geologists' folklore he was a 'prim and proper' man. For example, he could not bring himself to use the name 'Backside Beck' for an important fossil locality in the Howgills, NE of Sedbergh (David Skevington, pers. comm., 2000) (see Fig. 13.2)!

As a young man, Marr did fieldwork in both South and North Wales, and in 1879–1880 he travelled to Scandinavia and Bohemia, visiting the latter region to examine Barrande's theory of 'colonies' (see p. 60). The idea was (as Marr explained in his Notebook IV) that there might, in the Bohemian area, have been depressions of a southern Precambrian barrier, allowing Palaeozoic graptolites to migrate into basins from equatorial regions; but they died out in their basins when sedimentation was renewed there. This theory, if true, might mean that 'precursorial' forms could appear early in a given region, and then disappear from that region for a time, before the subsequent main appearance of those forms in that area, occurring when there was a more general subsidence and marine transgression. Such a suggestion allowed a disturbance of the normal stratigraphic order for a given region.[4]

Probably because he found the Bohemian evidence incompatible with the ideas about graptolite zonation recently developed in the Southern Uplands by Lapworth,[5] Marr (1880b) was dissatisfied with Barrande's ideas and did not see how they could apply in the area of Britain, which seemingly had open ocean at the time. Though shown round the Bohemian sections by Barrande's secretary, and later meeting with the celebrated palaeontologist himself, Marr thought that the sections had been

---

[1] The notebooks' pages are unnumbered. Reference will therefore be made by notebook number and date.

[2] For biographical material on Marr, see Anon (1916), Anon. (1932–1935), Double (1932–1933), Harker (1933), King (1934), Garwood & Thomas (1934), Oldroyd (1998–1999 [2000]). Portions of the present chapter are drawn from my earlier publication.

[3] Notebook VI, 7 April 1881.

[4] The theory had excited considerable controversy. Besides being supported by Murchison and some Survey officers, it was deployed on the continent by such geologists as Heinrich Georg Bronn (1859). The case of Lyell is interesting. In the first volume of his *Principles of Geology* (Lyell 1830–1833, vol. 1, p. 123), there is a famous passage in which he contemplated the possibility of iguanodons and pterodactyls re-appearing in presently temperate localities if appropriate environmental conditions returned, as was possible according to Lyell's thinking. So he, like Bronn and Barrande, was willing to see a disruption of the regular single-path stratigraphic sequence. If this proposal had been taken seriously by nineteenth-century geologists as a whole there would have been great confusion. Lyell visited Barrande in Prague in 1856, and was impressed by the evidence shown to him. On his return, he gave a somewhat confusing account of the idea, apparently supporting the hypothesis on the one hand (envisaging shifts of organisms with the collapse of land barriers between oceans), but at the same time offering a diagram of a synclinal structure for Central Bohemia that required no 'colonial' hypothesis (Lyell 1857, pp. 32–37). Edward Forbes (1854, p. xxxiv), on the other hand, had thought that the stratigraphic problems found in Bohemia arose from physical disruption of strata – the view subsequently taken by Marr in the 1880s.

[5] According to an undated document by Lapworth in his archive at Birmingham, being his commentary on a paper on the Stockdale Shales (Marr & Nicholson 1888), Marr, utilizing the methods and results of Lapworth's Moffat work (1878), was testing the graptolite zones he had worked out in the Lake District in Bohemia, and found that they were at odds with Barrande's hypothesis. Marr himself wrote to Hughes (18 July 1880): 'At first sight they [Barrande's colonies] look all right, but one begins to suspect something when you find a Brathay Flag fauna in beds of Arenig age! There is a Graptolite Mudstone fauna in beds of Upper Bala age! & not graptolites only but higher organisms in each case, brachiopods, lamellibranchs, trilobites, etc.' (Hughes Papers, Cambridge University Library, Add. MSS 9557, Packet K).

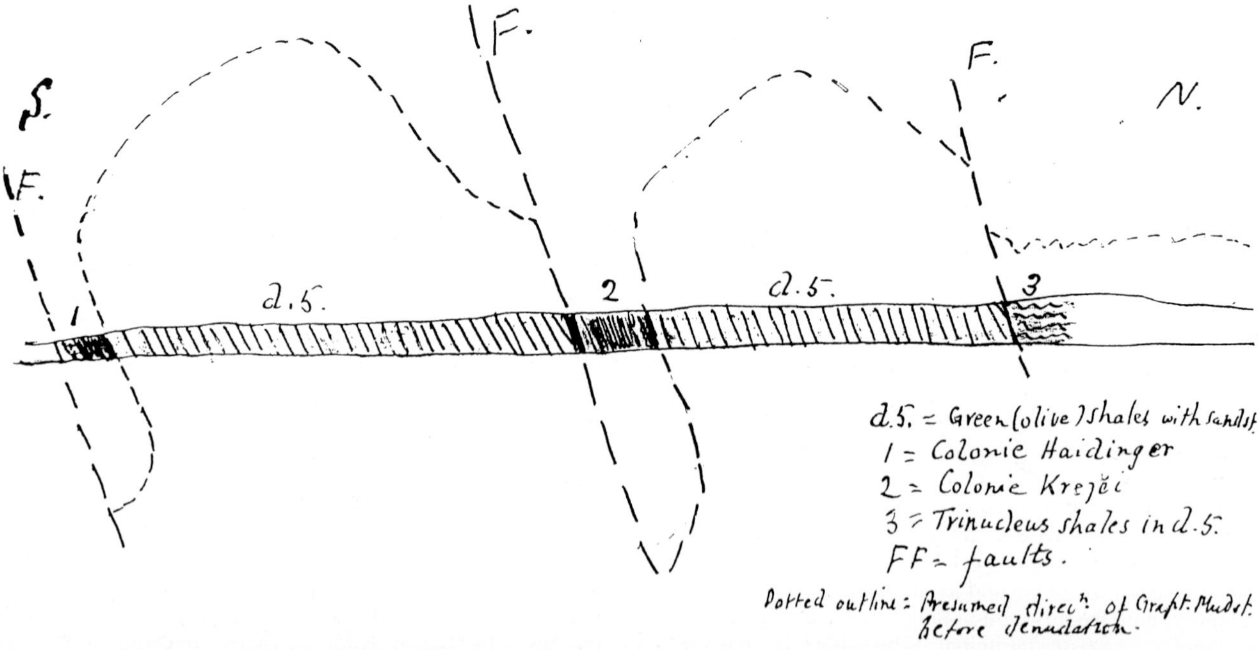

Fig. 5.1. Structures in Bohemian rocks, according to Marr, Notebook IV, Sedgwick Museum, Cambridge, 3 July, 1879. Reproduced by courtesy of the Sedgwick Museum. Previously published by Oldroyd 1998–1999 (2000)b, p. 363.

misconstrued. He claimed that the seeming 'precursorial' colonies of fossils were the result of younger rocks having been faulted into older strata (Notebook IV, 3 July 1879), as shown in Figure 5.1. So the Bohemian evidence provided no reason to discount William Smith's principles, or recent ideas on the use of graptolites as zonal criteria for stratigraphic purposes. This was an important message, which Marr followed in his subsequent Lakeland work. Marr's paper before the Geological Society (Marr 1880*a*) effectively gave the *coup de grâce* to the colony theory in Britain and made a considerable reputation for the young geologist.

Later, Marr (1883, p. 27) blamed Murchison for the erroneous theory that Barrande had propounded. Murchison (and Survey officers such as Archibald Geikie) had been inclined to regard Llandeilo rocks as the only ones containing graptolites; or conversely that all black shales containing graptolites belonged to the Llandeilo. Accepting this, Barrande had supposed that certain rocks that might otherwise have been thought younger than Llandeilo (by the evidence of their fossil contents) *were* Llandeilo simply because they contained graptolites. Hence he developed the idea of precursorial forms in 'colonies'. Incidentally, Barrande used to name his various colonies after geologists who criticized his theory, and one of his colonies in Bohemia came to be known as *Colonie Marr*! According to Perner (1938, p. 518), Barrande was considerably troubled (and annoyed) by Marr's suggestions and made an effort to counter them in his *Défense des Colonies*, using arguments based in part on writings of earlier British geologists. It should be noted that Marr was by no means Barrande's only critic. Other opponents were Lapworth, d'Archiac, Cotta, Haidinger, Lipold and Tullberg.

As a Lancashire man, and following in the steps of Sedgwick and Hughes, it is unsurprising that Marr should have selected the Lake District as the locality for his chief geological investigations – or that he should have adopted a version of the Sedgwickian view in the matter of the Cambrian–Silurian controversy,[6] being influenced by Hughes's views about the placement of the boundary between the two systems (Hughes 1867).

Marr's first Lakeland paper appeared in 1878, being communicated to the Geological Society by Hughes on 19 June. It was only that year that Marr obtained his first at Cambridge, with Hughes as one of his examiners. Marr carried out his fieldwork for his paper in 1876–1877, though he had been geologizing in the Lakes since at least 1874. He often worked there in mid-winter, though later he usually started in March and was away from Cambridge for long spells each year.

Five years before, Sedgwick and Salter had published their *Catalogue of the Collection of Cambrian and Silurian Fossils Contained in the Geological Museum of the University of Cambridge* (1873), which, besides containing Sedgwick's great diatribe against Murchison about the Cambrian–Silurian boundary in the Introduction, also sought to present a 'Sedgwickian' view of Palaeozoic stratigraphy, based on the fossil collection that had been amassed at Cambridge.

In arranging the fossils, Sedgwick and Salter proposed a tripartite division of the Bala Group (Lower, Middle and Upper), all

[6] Marr did this most strongly in his prize-winning Sedgwick essay (Marr 1883) – perhaps not surprisingly, given the context. The classification expounded was highly Sedgwickian:

Downtonian	
Salopian (Wenlock and Lower Ludlow)	Silurian
May Hill (Valentian)	
Upper Bala	
Middle Bala	
Lower Bala (Llandeilo)	Cambrian
Llanvirn	
Arenig	
Tremadoc	

Thus Lapworth's Ordovician was not accepted by Marr (or other Cambridge geologists) in 1883.

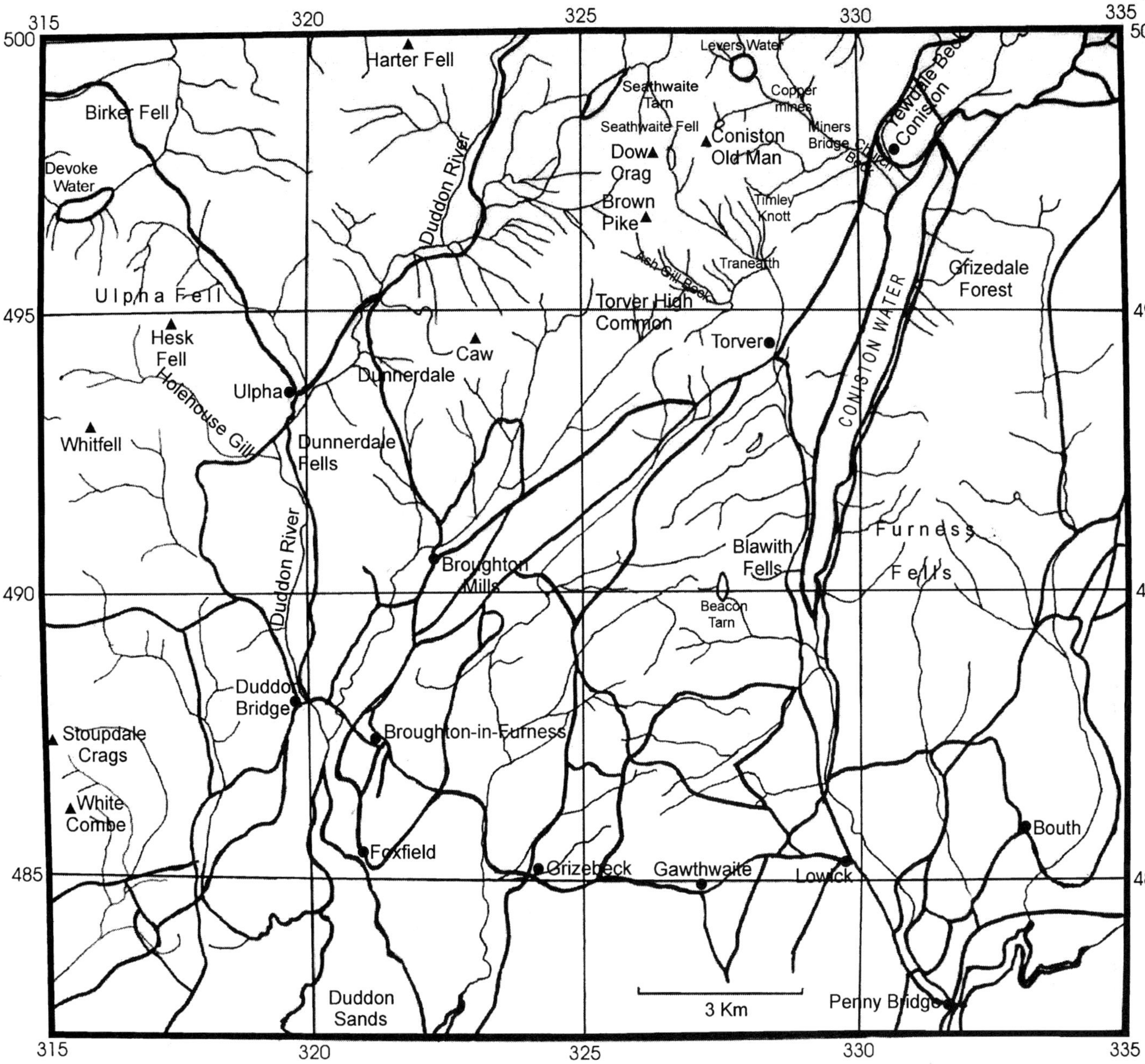

Fig. 5.2. Topography of the Coniston area, Ash Gill, and the Duddon Valley.

allocated to the *Upper* Cambrian. The Coniston Limestone was classified as *Middle* Bala, while the Bala Limestone of Wales was put in the Middle Cambrian (Sedgwick & Salter 1873, pp. 26 and 72). For the Upper Bala, further subdivision was proposed into (1) Hirnant Limestone of Merionethshire (above the Bala Limestone), and 'Ash Gill Slates, &c., above the *Coniston Limestone*', and (2) Llandovery Rocks (called Lower Llandovery by the Survey). This was the first hint of the introduction of the famous term 'Ashgillian' or 'Ashgill',[7] though obviously Salter did not at the time know that he was introducing a word that would come to be used as one of the major subdivisions of the Ordovician.

However, although Salter is interesting for having given the first intimation of the Ashgillian in print, his introduction of the term 'Ash Gill Slates' was by no means felicitous. He made mistakes in placing some of the fossils and got some of the strata misplaced too (Marr 1907, pp. 60–61), but he was ill at the time and worked from the museum and library without examining the Lakeland rocks *in situ*.

Ash Gill is a modest beck that runs over moderately level ground to the SE of the Old Man (= Hill) of Coniston, and ends up in Coniston Lake (see Fig. 5.2). At one point it flows through a small gorge, and nearby there is a quarry where fossils have formerly been found in abundance (see Fig. 5.3). Salter, then, on the basis of the fossils available to him at Cambridge, distinguished the 'Ash Gill Slates' from the Coniston Limestone. They were above the limestone, but below the Brathay or Coniston Flags, which occurred in quarries near Ambleside. Sedgwick and Salter put the Ash Gill rocks in their Upper Cambrian and the Brathay (Coniston) Flags in the Silurian.[8]

[7] Later, Marr (1907, p. 61) stated that he had reason to believe that the name was suggested to Salter by Aveline, who wanted the Ash Gill beds separated from the Coniston Limestone.

[8] Sedgwick & Salter (1873, p. 91) referred to 'Coniston or Brathay Flag (upper part)' as being Lower Wenlock.

Fig. 5.3. View of Ash Gill on Torver Common, west of Coniston Water (waterfall at locality of section studied by Marr [see Fig. 5.6]; quarry to rear of fall, in middle distance). Looking northeast. Photographed by author, 1996.

Marr concurred – as was hardly surprising for one working for his finals under Hughes. In fact, he stated (Marr 1907, p. 61) that it was Hughes who had first pointed out to him, in 1876, that there were rocks near Sedbergh resembling those at Ash Gill, with fossils in common, so that the unit at Ash Gill was not merely a local entity. But Hughes did not publish on the matter. Marr (1878, p. 871) now added detail to the knowledge of the fossils of the area, and suggested that the main divisions of the Lakeland Palaeozoics were (for the most part) characterized by a particular species of the trilobite *Phacops*: [9]

Kirkby-Moor Flags	*Phacops caudatus*
Bannisdale Slates	*Phacops downingiae*
Coniston Grits	*Phacops downingiae*
Coniston Flags	*Phacops obtusicaudatus*
Stockdale Shales	*Phacops* sp.
Ashgill Shales	*Phacops mucronatus*
Coniston Limestone	*Phacops macroura* and *P. conophthalamus*
Green Slates	No fossils
Skiddaw Slates	*Phacops nicholsoni*

A more extensive list of fossils was provided for each of the foregoing units, and a clear demarcation of the Ashgill Shales (now so named) from the Coniston Limestone was made (whereas previously Salter had merely mentioned the distinction). The lithological distinction between the two units is fairly obvious in the field, but Marr now pinned it down by the use of palaeontological evidence.

It was also by fossil evidence that Marr claimed that the top of the Ashgill Shales marked the top of the Upper Cambrian (Upper Bala), as then understood by Sedgwickians. Moreover, Marr noted that the Ashgill Shales appeared to be of very variable thickness along their outcrop. Indeed, he maintained, with Hughes, that there was an unconformity at the base of the overlying Stockdale Shales[10] – as might be expected for a major point of subdivision of the stratigraphic column. This would account for the rapid thickening and thinning of the unit along its outcrop. So it was at *this* horizon that the latter-day Cambridge Sedgwickians determined that the Cambrian–Silurian boundary should be situated. Marr's work was based on observations in the Skelgill area, to the east of the northern end of Windermere; and in the Coniston area – Torver and Ash Gill. It may be noted that the boundary was taken as the base of the 'Upper Silurian' by the Murchisonian Surveyors, not the base of the Silurian (see Note 10). So in placing the boundary between the Stockdale and Ashgill Shales, and putting the latter at the top of the Cambrian, Marr was enormously enlarging the Cambrian at the expense of the Silurian, as originally 'envisaged'[11] by Murchison.

Only a year later Lapworth (1879) proposed his solution to the controversy about the position of the Cambrian–Silurian boundary, which was eventually to meet with general acceptance (though only at the beginning of the twentieth century, after Geikie's retirement, for the Survey). As is well known and as has been mentioned previously, Lapworth's proposal was to establish a whole new system, the Ordovician, which would occupy the contested territory that Sedgwick wanted to be Upper Cambrian and which Murchison and the Survey designated Lower Silurian. Lapworth pointed out that to all intents and purposes a tripartite division of the Lower Palaeozoic had already been established, based on palaeontological principles. In Bohemia, Barrande had recognized three distinct Lower Palaeozoic faunas, and the application of this threefold division to the British scene provided, Lapworth suggested, a neat solution to the problems that had beset British geology in this part of the stratigraphic column since the 1840s. On Lapworth's scheme, then, the Lower Llandovery would be at the bottom of the Silurian, which meant that for the Lakes the Ashgill Shales would be at the top of the Ordovician.

Marr's next fieldwork was undertaken in the Dee Valley of North Wales, and in 1880, now a Fellow of the Geological Society (elected 1879), he presented a paper (Marr 1880*b*) in which he

[9] Following subsequent taxonomic work, *Phacops* is now treated as an order (*Phacopida*), not a genus.

[10] So named by Aveline & Hughes (1872, p. 7) in their sheet memoir for the Kendal, Sedbergh, Bowness and Tebay area, for the unit found by Stockdale Farm, to the east of Longsleddale, but also exposed to the SE of Ambleside in Holbeck Gill. They placed the unit at the base of the 'Upper Silurian' in the Lakes. Lapworth, however, thought they belonged in the Llandovery.

[11] One cannot say that the boundary was yet 'defined'. Murchison's boundaries for the Silurian were forever expanding; and so too was Sedgwick's Cambrian.

Table 5.1.

Epoch	Formation	Lake District	Dee Valley
S	Upper Ludlow	Kirkby-Moor Flags	–
I	L. Ludlow–U. Wenlock	Bannisdale Slates	–
L		Coniston Grits	–
U		Upper Coldwell Beds	Dinas-Bran Beds
R	8. Wenlock continued	Middle Coldwell Beds	Grits above Penyglog Quarry
I		Lower Coldwell Beds	Ditto
A		Brathay Flags	Flags of Penyglog Quarry, etc.
N	7. Tarranon Shales	Pale Shales	Tarranon Shales
	6. Graptolitic Mudstones	Graptolitic Mudstones	Graptolitic Mudstones
	5. May-Hill Group	Basement beds of Silurian	Corwen Grits, etc.
C			
A			
M			
B	4. Upper Bala	Ashgill Shales	Hirnant Limestone (?)
R	3. Middle Bala	Coniston Limestone	Bala Limestone + overlying shales
I	2. Middle Bala in part, Lower Bala	Borrowdale Series	Shales and andesitic ashes of Berwyns
A	1. Arenig	Skiddaw Slates	Beds of Taihirion and Arenig
N			

sought correlations between the rocks of North Wales and those of the Lakes. His stratigraphy was now further developed and Welsh correlations were offered (see Table 5.1).

Interestingly, Marr did not deploy Lapworth's term 'Ordovician'. Just as the Survey did not adopt Lapworth's suggestion in its maps until after Archibald Geikie's reign as Director-General, on the grounds that Murchison's terminology had priority in the scientific literature, the Sedgwickians at Cambridge did not immediately jump to Lapworth's suggestion either. They too had a position to uphold.

In 1883, Marr was awarded the Sedgwick essay prize at Cambridge, to which reference has already been made, wherein he noted a grey limestone band at the lower part of the Ashgill unit, which contained (along with other fossils) a particular trilobite: *Staurocephalus clavifrons* (Marr 1883, p. 58). Thus the Ashgill was beginning to be subdivided.

Marr was introduced to Nicholson by Lapworth in 1886, and they had their first week in the field together the following April (Lapworth archive, Birmingham University: Nicholson to Lapworth, 24 October 1886; and Marr to Lapworth, 10 April 1887), thereafter doing a considerable amount of collaborative work in the Lakes. Important results were soon published. It will be recalled that the 'Green Slates and Porphyries', or the rocks of the Borrowdale Series ('Otley II'), had hitherto yielded no fossils. But Nicholson & Marr (1887) reported the discovery (first made by Nicholson, I believe) of quite abundant fossils in some shales – brown, or of bleached appearance – just NW of Carrock Fell, in northeastern Lakeland. The shales – which Marr described (10 April, 1887) as 'very queer' and nothing like anything he had seen previously in the Lakes – were found in a dreary valley, sometimes dry, appropriately called Drygill, near the top of a col between Carrock Fell and another hill, Great Lingy, to the west (see Fig. 4.9). What came to be called the (previously mentioned) Drygill Shales appeared to be intercalated with breccias and bands of quartz felsite, which Nicholson and Marr took to belong to the lavas and ashes of the Eycott Series, named after the outcrop of volcanics at Eycott Hill close by. These volcanic rocks – today called the Eycott Volcanics – form a substantial outcrop running along the northern and northeastern edges of the Lake District, appearing between the Skiddaw Slates and the surrounding 'rim' of Carboniferous Limestone.[12] We have previously noted that Ward and Russell failed to discover the Drygill fossils in the primary survey, and the rocks were mapped at SS.

In the 1880s, the Eycott Volcanics were generally believed to belong to the Borrowdale Series[13], so finding a rich fossil lode at Drygill was an important achievement. It appeared that, at long last, palaeontological evidence pertaining to the Borrowdale Volcanics might be emerging. Nicholson and Marr claimed that the Drygill fossils and the shales themselves were quite different from those of the Skiddaw Slates; and also differed from any of the units above the 'Volcanic Series'. However, it was – implausibly we recall – suggested they were somewhat like those that Ward had described at Cat Gill, to the E of Derwent Water. They were also somewhat similar to the Dufton Shales in the Cross Fell Inlier. A 'Llandeilo–Bala' age was proposed, and Lapworth's notion of an Ordovician System was deployed as if it were uncontroversial. Very likely, contact with Lapworth's friend Nicholson had caused Marr to shift from his earlier Sedgwickian position. The proposed analogy with rocks near Derwent Water was almost certainly Nicholson's suggestion (see p. 37), not Marr's.

Marr was 14 years younger than Nicholson, and it was unsurprising that he should have been the second author in the paper of 1887. But he moved to the senior position for a paper published the following year – one of the most important and influential of the early publications on Lakeland geology: 'The Stockdale Shales' (Marr & Nicholson 1888).

The Stockdale Shales are named after the tiny hamlet of Stockdale (called 'Little London' in Sedgwick's notes, when he was there in 1823) on the eastern side of the valley of Longsleddale, just before it enters the rough country of the Borrowdale Volcanics to the north (see Fig. 3.7 and cover illustration). They consist of dark shales at the base with paler and sometimes purplish shales above, and are well exposed in the stream banks and beds in several sites. There are some good cliff sections up the hill. At one

[12] However, in earlier correspondence with Lapworth (22 April 1883) Nicholson seemed to think that Eycott-type rocks were all over the place: in the Cross Fell Inlier; at Shap Wells; in Longsleddale; and even at Millom in SW Lakeland! In the same letter, he referred to 'Drygill Shales', though the term did not appear in print until Nicholson & Marr (1887). See p. 37.

[13] Subsequently, from the work of Downie & Soper (1972), with the help of microfossils, the Eycott Group came to be regarded as the product of eruptions preceding the Borrowdale Volcanics.

time they were swarming with graptolites,[14] which naturally made them attractive to Nicholson, a leading authority on such organisms; and since the remarkable work of Lapworth (1878), geologists had come to recognize the great value of graptolites for stratigraphic purposes. As free-floating organisms[15] they were often widely distributed. Moreover, they seemed to have evolved rapidly, making them excellent as stratigraphic markers. Marr and Nicholson specifically acknowledged Lapworth's assistance in their study of the fossils from the Stockdale Shales.

Though Sedgwick undoubtedly saw these rocks in 1823, he was not then much on the lookout for fossils, being chiefly concerned with dips and strikes, as we have seen. Harkness & Nicholson (1868) had first distinguished the shales as a significant unit, forming part of what Sedgwick had called the Coniston Flags. They called the lower part the 'Graptolitiferous Mudstones' and the upper part 'Grey Grits'. The rocks were named the Stockdale Shales by Aveline and Hughes in their memoir for Sheet 98 NE (1872, pp. 3 and 7) on the Kendal area, and divided into 'Graptolite Mudstones' and 'Pale Slates' ('Upper Silurian'). Nicholson & Lapworth (1876) had used the terms 'Skelgill Beds' and 'Knock Beds', relating them to the Upper Llandovery and the Tarranon Shales of North Wales. Aveline (1876*b*) had the Stockdale Shales as Tarranon, while Hicks (1876*a*, *b*, *c*) had them as Llandovery. Lapworth (1876, 1878) put the Skelgill Beds in the Lower Llandovery, whereas Harkness & Nicholson (1877) had them as Upper Bala or Lower Llandovery. The situation was evidently fluid, and complicated by the ongoing battles about the location of the Cambrian–Silurian boundary, which was at the root of the whole trouble. Now, a decade later, it was possible to return to the problem with the help of the concept of the Ordovician System. It was Marr and Nicholson who did this.

After several years' work (Marr was at 'Browgill' in 1883 and recorded a 'beautiful exposure of pale shales' in his Notebook XIII), Marr & Nicholson (1888) distinguished two main stages of the Stockdale Shales: the Skelgill (or Skelghyll) (lower) and the Browgill (upper). The former was given its name from the beck in which such rocks were exposed near High Skelgill farm, SE of Ambleside. The stream runs down from the east of Wansfell Pike to Low-wood Hotel, on the shores of Windermere, an inn that Sedgwick sometimes patronized, and was visited by John Playfair. The Browgill Beds were so named after Browgill – one of the tributaries of Stockdale Beck, which runs past Stockdale Farm and into the River Sprint, the main river of Longsleddale (see Fig. 3.6) – where the relevant rocks crop out. The two units corresponded to the earlier Graptolite Mudstones and Grey Grits or Pale Shales. The lower one appeared at Skelgill, with a small portion of the upper unit, while both were well exposed at Stockdale.

At Skelgill near Ambleside, the two geologists had Ashgill Shales at the bottom of their section, then Lower, Middle and Upper Skelgill Beds, and a small exposure of Browgill Beds. The Lower Beds were allocated two graptolite zones; the Middle received six; and the Upper four. Detailed descriptions were given of the outcrop localities, the rock types, the thicknesses of the beds, and the fossil contents. It was claimed that the Skelgill Beds were *conformable* to the underlying Ashgill Shales. The bottom of the lowest unit of the Middle Skelgill was not seen, owing to the existence of a strike fault. The rationale for the graptolite zones used was not stated, but it appears that they were based on those established by Lapworth (1878) in his work at Moffat.[16] Unfortunately, Marr & Nicholson (1888) did not supply a map, but they provided a 'composite' section, based on the exposures visible at various spots along the line of the stream. The conformity with the underlying Ashgill Shale should be noted, for this was the boundary that was supposed to mark the division between the Ordovician and the Silurian. As noted above, Marr (1878) had previously thought there was an unconformity above the Ashgill Shales in the exposures to the west near Ambleside and Coniston.

At Stockdale and Browgill,[17] Marr and Nicholson followed much the same procedure as that used at Skelgill, except that this time they could focus more attention on their upper unit of the Stockdale Shales, namely the Browgill beds; and at a cliff face in Browgill they were able to find an almost complete section, at a place they called 'The Rake'.[18] It has become one of the classic Lakeland sections. Their diagram of the stratigraphic section with its various zones may be compared with the actual cliff face visible at 'The Rake' (see Figs 5.4 and 5.5). In the section, the Skelgill beds were represented by Ab_1, etc., and the Browgill beds by Ba_1, etc.[19] It will be noticed that by using the graptolite zoning method it was possible to recognize that certain zones were missing from the sequence. Much tighter stratigraphic control was now possible with the use of zonal technique.

Having established their type sections at Skelgill and Stockdale, Marr and Nicholson then examined similar sections all the way along the outcrop of these beds from Shap Wells in the east to Poaka Beck near the Duddon Estuary to the west, and over at Knock (Swindale Beck) in the Cross Fell Inlier; also down south in the Palaeozoic inliers near Sedbergh; and even in Teesdale. Comparison with Lapworth's zones for the Moffat area gave the following correlations:

Lake District	*Southern Uplands*
Browgill Beds	Gala Group
Skelgill Beds	Birkhill Shale

with the graptolite zones for the two units being nicely relatable to one another. This evidently pleased Lapworth, for in commenting on the paper (undated document, Lapworth archive, Birmingham University) he pointed out that Marr and Nicholson had made excellent use of the 'zonal method' for stratigraphy, by means of graptolites, helping to show the generality of the technique and its applicability to widely separated localities.

Marr and Nicholson concluded by asserting that there was a marked faunal break between the Ashgill Shales and the Stockdale Shales (i.e. between the Ordovician and the Silurian), but that the two systems were conformable along the line of

[14] During a visit to the locality in 1996, I was informed by a local resident that the fossils had suffered great depradations by the activities of amateur collectors and dealers, the latter coming to cart them away by the truckload. The area is now fenced off, but the damage is already done. Subsequent discussion in 1998 with the owner of the land indicated that the area had been fenced off at the owner's initiative as he had become perturbed by the seeming absurdity of an endless succession of students hunting for similar fossils in the same place. As a result of the fencing, the area is now wholly overgrown and neither commercial collectors, students, geologists, nor historians of geology can find anything there. See Oldroyd (1999*a*).

[15] Not all early graptolite investigators supposed that the organisms were free-floating. Reinhard Richter (1871), working on Thuringian graptolites, thought that the sicula was a kind 'foot' or 'haft-organ' by which a graptolite could be secured to the seabed.

[16] Marr did not visit Lapworth's classic site in the Southern Uplands, Dobs Lin, until 1892 (Notebook XXIII), but Nicholson knew Lapworth's field area. There was also reference to the work of the late Swedish geologist, Sven Tullberg, and the Norwegian Theodor Kjerulf, and their zone work may also have been used.

[17] In the earlier literature, this stream had been variously termed Ancrowgill, Iron Crow Gill and Arncoside Beck.

[18] A word having much the same meaning as a small couloir – a steep gully, rut or groove on a mountainside. It might occur at a joint plane, a fault, or a weathered-out dyke or vein of ore.

[19] Ba = Lower Browgill; Bb = Upper Browgill.

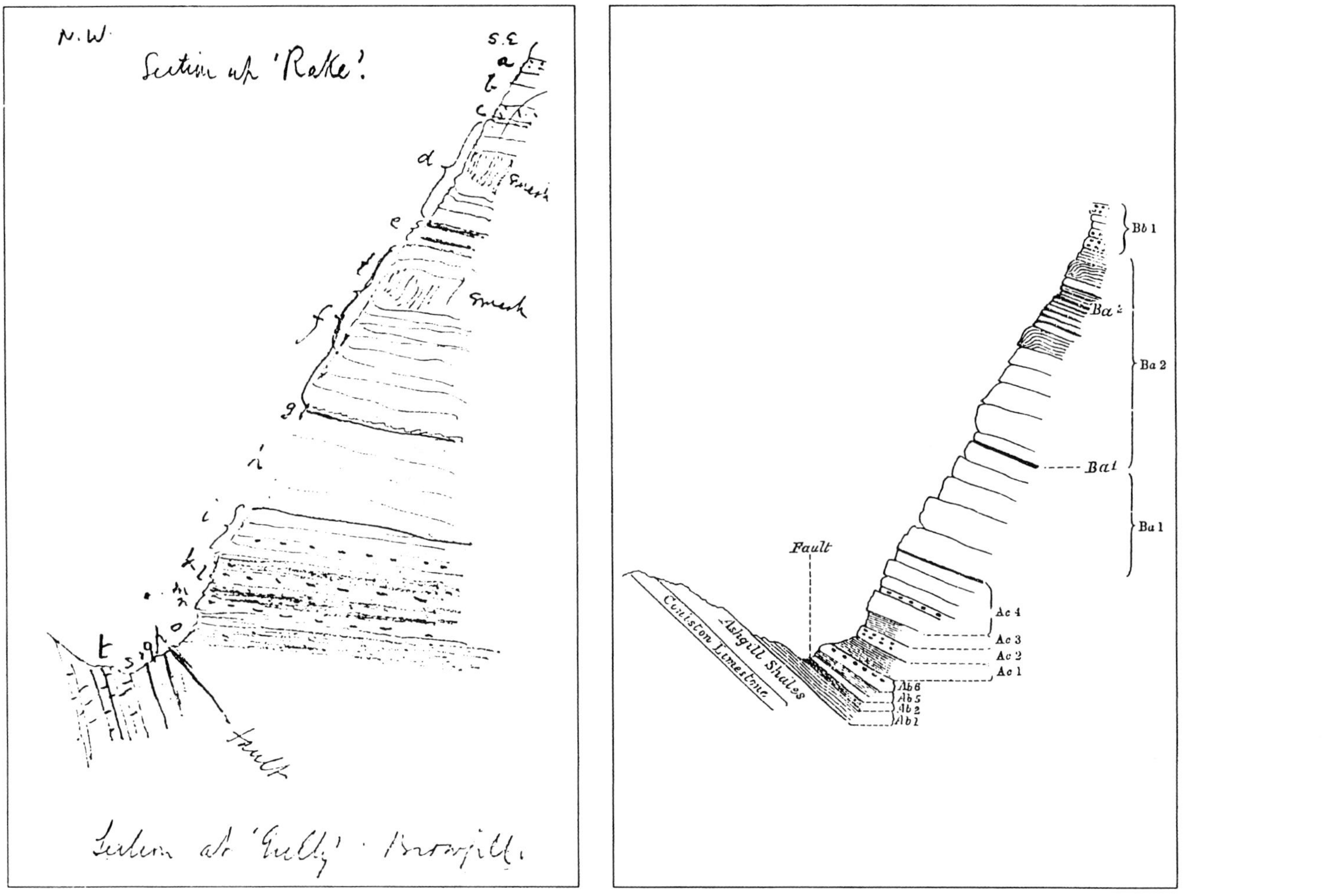

Fig. 5.4. Section at 'The Rake', Browgill, near Stockdale, Longsleddale, according to Marr & Nicholson (1888, p. 675); with figure from Marr's Notebook XVI, 20 Sept., 1886 (Sedgwick Museum archives) for comparison. Marr's manuscript figure reproduced by courtesy of the Sedgwick Museum.

outcrop. This marked a change of viewpoint on the part of Marr, presumably under the influence of Nicholson.

The argument was not entirely *ad hoc*, but was consonant with the lithologies of the rocks concerned. It will be recalled that the Coniston Limestone is affected in several places by substantial dip faults that cause large 'jumps' in the line of outcrop. Yet these large faults seemed to die out quite quickly to the south. Marr and Nicholson suggested, then, that there was a significant strike fault at about the horizon of the top of the Lower Skelgill Beds. It was suggested that the soft Skelgill Beds (black shales) lay between the harder Ordovician rocks below and harder Browgill rocks above. So they had suffered crushing and stretching, with the Lower Skelgills adhering to the Ashgill Shale and the Upper Skelgills adhering to the Browgill Beds. In this way, the rocks had adjusted to the tectonic forces, so that the several N–S faults fracturing the Coniston Limestone, etc., were 'accommodated'. The faults did not have to run right around the globe! They could die out in a relatively short distance. Above the Browgill Beds, the sequence appeared to run conformably into the Coniston Flags.

Marr's last collaborative work with Nicholson described their investigations in the Cross Fell Inlier (Nicholson & Marr 1891). A map of the area was provided, along with the following stratigraphic sequence:

New Red Sandstone	
Carboniferous	
Coniston Grits	Silurian
Coniston Flags	
Stockdale Shales	
Ashgill Shales	Ordovician
Staurocephalus Limestone	
Keisley Limestone	
Dufton Shales	
Corona Beds	
Rhyolitic Group	
Eycott Group	
Skiddaw Slates	

As will be seen (and this was a process that always seems to occur as more detailed stratigraphic investigation occurs), the number of units recognized was increasing. The 'Corona Beds' comprised a series of calcareous ashy shales and mudstones containing the brachiopod *Discina corona*. The fossiliferous Keisley Limestone had previously been noted by other observers and its fossils listed, but was now given its own name. Its fossils were much the same as those of the Dufton Shales though the latter was, as the name suggested, a more argillaceous deposit.[20]

[20] Later, the fossils were closely examined by F. R. Cowper Reed (1896, 1897), who showed that the Keisley Limestone belonged to the Ashgill, and marked the top of the Ordovician in the Cross Fell Inlier. Marr (1906*a*) acknowledged this and responded by putting the *Staurocephalus* Limestone *within* the Keisley Limestone, not *above* it.

It was suggested that the Corona Beds were of Llandeilo or Caradoc age, as compared with the Welsh deposits. The Dufton Shales were thought to be equivalent to the Coniston Limestone or the Bala Limestone; and the Keisley Limestone was seemingly also much the same as the Coniston Limestone, though the fossils were not quite the same. The Ashgill Shales were, however, like those in the Lake District proper.

The 'Ashgill Series' was formally proposed by Marr as the top of the Ordovician in his Presidential Address to the Geological Society (Marr 1905*b*), and the finishing touches to the placement of the Ashgillian at the top of the Ordovician were, so far as Marr was concerned, published in 1915, following his fieldwork at Ash Gill and some other Lakeland localities in 1913 and 1914 (Marr 1915).[21] It was still not an entirely easy problem, and back in 1886 Marr's fieldwork (Notebook XVI) had shown that there were structural complications (both folding and faulting) in the locality of Ash Gill and Ash Gill Quarry. He now examined the strata above the waterfall closely (see Fig. 5.2), producing the rough sketch shown in Figure 5.6, showing that he had formed a definite idea of the succession there.

Thus his fine subdivisions were as follows:

Dimorpho[*graptus*] beds
Ashgill Shales
[*Dalmanites*] *Mucronatus* beds (upper and lower)
Upper White Limestone
Ash
White limestone
Phillipsinella beds
Chasmopses beds [Caradoc]

These units were crammed into just a hundred feet. However, Marr showed that essentially the same sequence could be discerned at Cautley, near Sedbergh, so matters seemed satisfactory. Earlier (Marr 1907), he had argued persuasively for the existence of the Ashgill Series as a widespread unit of more than local, Lakeland, extent and significance, having already – with another Cambridge geologist Thomas Roberts – shown a distinction between Caradoc and Ashgill fauna in rocks in South Wales (Marr & Roberts 1885). Ashgill rocks, Marr claimed, had their own distinctive trilobites; for example, the form *Trinucleus* dying out at the top of the unit. However, at that date (1907) there were still uncertainties as to where the upper boundary should be drawn in different countries, so the Ordovician–Silurian boundary was still fuzzy, even then.

Aspects of Marr's work on Lakeland glaciation and geomorphology will be discussed in Chapter 19. Here it may be mentioned that his field notebooks contain one or two interesting details of a personal nature. When first looking through the notes, I was struck by the frequency with which he visited the Shap Wells district in the early 1890s. Given that he and Harker produced a paper on the Shap Granite in 1891, this might seem unsurprising. But the enthusiasm for research in the area of the Shap Wells Spa Hotel seemed somewhat excessive. The explanation appears in Notebook XXIII (1892–1893), where we find a newspaper announcement of Marr's marriage to Amy, youngest daughter of the Hotel's proprietor, in January 1893. Marr does not, however, seem to have been entirely enthusiastic about the prospect of marriage. While his notes for 15 July 1892, bear an arrow-pierced heart, they also contain the following remarks:

> Temper bad. Drove to and from Morland [a village near Shap] in silence. Reached Shap at 7 p.m. Temper slightly improved – owing to a proposal I made (received). My companion in the lowest depth of despair having been decidedly '*let in*' to her intense delight horror. Unfortunately there is no way out of it. A grief like this must be endured.

Fig. 5.5. View of 'The Rake', Browgill. Photographed by author, 1996.

What this tells us about Marr, or Victorian marriage ambivalence, I do not here inquire. The couple duly married at Orton, not far from Shap, and spent the first night of their honeymoon at St Pancras Hotel, where Marr noticed that their balcony was made of Shap Granite and had orthoclase crystals with biotite inclusions. Perhaps he thought he was trapped in a Shap marriage crystal! According to reports, however, Marr's wife was well liked by her husband's students, whom she used to entertain at Cambridge. A photo in the Marr archives at Cambridge, dated 1 April 1893, shows her looking quite cheerful by a carriage departing from some Lakeland farm. Presumably she accompanied her husband at least for a time on his field excursions, and certainly his geologizing was not curtailed by his marriage.

Another interesting little vignette is provided by Notebook XXII (1890–1891), which shows a picture of Marr enjoying himself getting his feet wet in the field (Fig. 5.7) on 1 April 1891. The photographer was Edmund Garwood (1864–1949), who published papers on the Carboniferous rocks round the Lake District in the early twentieth century, and was a student of Marr's at Cambridge, also doing numerous geochemical analyses for him.[22] They were in the Lakes together again in 1904 (Notebook XXXVIII).

[21] Notebook LIII (1914) reveals that Marr was accompanied by his young son, Alleyne, named after Henry Alleyne Nicholson. Young Alleyne was sometimes sent off to look for fossils in places where his father, now 57, did not feel like walking. The lad was killed in World War II.

[22] Garwood (1864–1949) was a distinguished geologist, explorer and alpinist. He became Professor of Geology at University College, London in 1901, in succession to Bonney. In 1891, he had recently been appointed a Cambridge University Extension Lecturer in geology and had the opportunity, therefore, for fieldwork with Marr. See C[hubb] (1949, 1950).

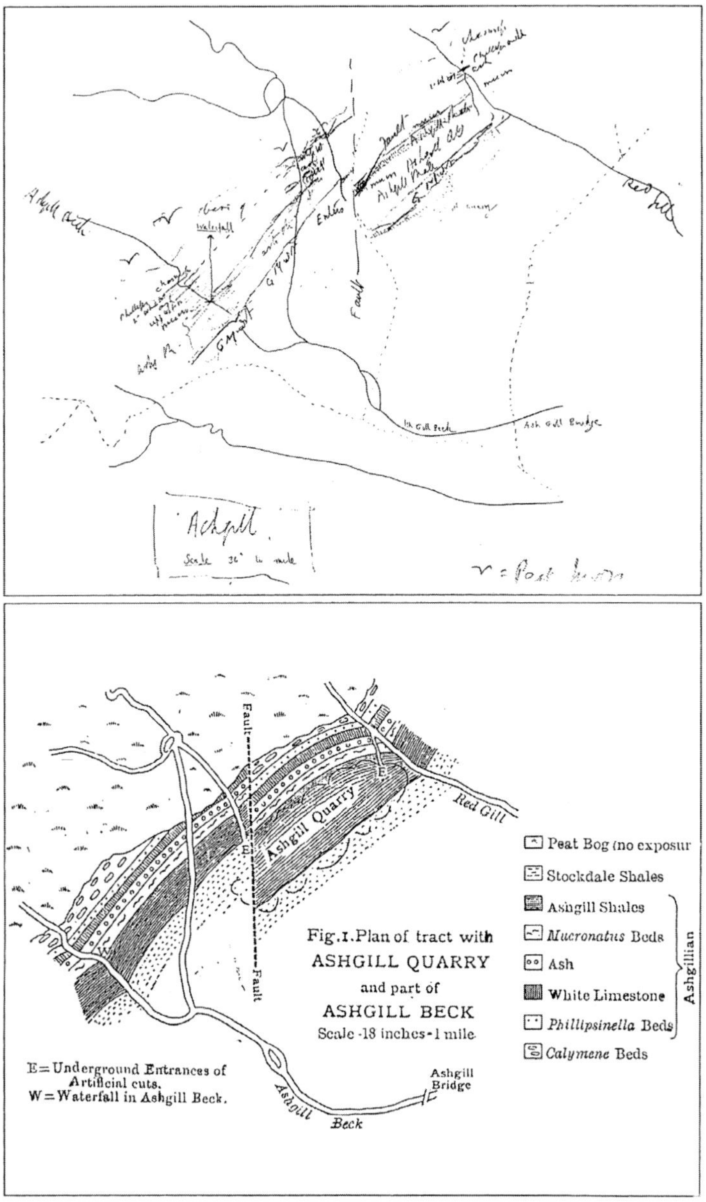

Fig. 5.6. Sketch of the area of Ash Gill quarry: in pocket of Marr's Notebook LIII, 8 Sept., 1914, Sedgwick Museum archives, Cambridge. Reproduced by courtesy of the Sedgwick Museum; with published version for comparison (Marr 1915, p. 190). Previously published Oldroyd (1999*a*, p. 20).

Fig. 5.7. John Edward Marr crossing a Lakeland beck, 1891.Marr archives, Sedgwick Museum, Cambridge. Photographed by Edmund Garwood. Reproduced by courtesy of the Sedgwick Museum.

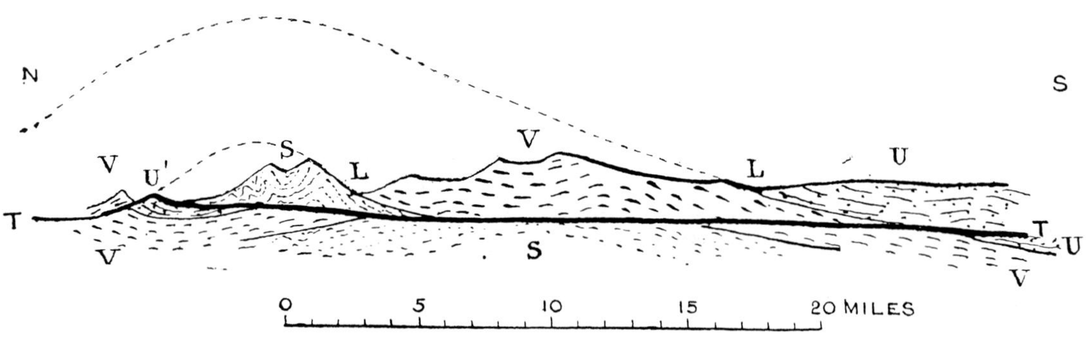

FIG. 3.

S. Skiddaw Slates.
V. Volcanic Series.
U. Upper Slates.
U′ Upper Slates below Thrust Plane at Drygill.
TT. Thrust Plane.
LL. Lag Planes.
For clearness sake, the amount of thrust has been represented as much less than the probable actual amount.

Fig. 5.8. Structure of Lake District, according to Marr's theory of thrust- and lag-faults (Marr 1900*a*, p. 468).

It was noted above that Marr repudiated Barrande's theory of colonies, on the basis of structural considerations. The success of the work of Lapworth and Charles Callaway in the early 1880s, elucidating the structure of the Northwest Highlands of Scotland in terms of thrust faulting, as was subsequently ratified by the officers of the Survey (Oldroyd, 1990), showed the need to take account of reverse or thrust faults when considering the structures and stratigraphies of complex regions. I suggest there could perhaps have been an over-enthusiasm for such ideas in the late nineteenth century, and Marr was, I think, caught up in this.

As mentioned above, in his paper with Nicholson (1888) Marr made use of the idea of the differential movement of beds under faulting, according to their relative resistance to tectonic forces. He developed this idea, along with notions of thrust faulting, in an important paper delivered to the Geologists' Association in 1900, propounding ideas that were given further expression in Marr's influential synthesis of Lakeland geology, *The Geology of the Lake District* (1916). Marr (1900*a*) pointed out that there were already three views on the table, so to speak, regarding the relationship between the Skiddaws and the Borrowdales. The Surveyor Dakyns (1869) thought that the contact was unconformable; Surveyor Ward (1876) thought it was faulted, with a connected line of criss-crossing high-angle normal faults; and some geologist (Marr could not locate the reference) had apparently envisaged a thrust fault. Marr's view – with which his colleague Harker concurred, he said – was that there had been a pushing of the rocks from the south, but because of the different natures of the rocks of Otley's three units the underlying Skiddaws had been pushed furthest; then came the Borrowdale Volcanics; and then the Upper Slates ('Otley III'). Put another way, the Borrowdales had 'lagged' behind the Skiddaws; and the Windermeres had 'lagged' behind the Borrowdales. Thus Marr introduced (so far as I am aware) a new kind of fault into the geological literature: 'lag faults'. These, like thrust faults, were low-angled; but there was no inversion of the strata during their formation, as was the case with thrust faults, and as envisaged by Lapworth and others for the Northwest Highlands of Scotland. Because of the supposed differential south-to-north movement – unequal at different points from east to west – there could also be high-angle 'tear faults', striking approximately N–S and producing erodable shatter belts (fracturing the outcrop of the Coniston Limestone, for example) as had long been known, since the time of Sedgwick and Otley.[23]

To make such a differential 'lagging' movement geometrically possible there had to be compensating thrust faulting. Marr had no direct evidence for such a type of fault in the Lakes, but the occurrence of the Drygill Shales in the northern Lakes provided a 'way out' of his problem. His students Gertrude Elles and Ethel Wood (1895) (see Chapter 7) had suggested on palaeontological evidence – especially that of graptolites – that the Shales were of the same age as the Coniston Limestone in the southern Lake District. Marr's model is shown in Figure 5.8 and repays close study. Note that, like Nicholson, Marr regarded the northerly (Eycott) Volcanics as part of the Borrowdale Volcanics, despite their petrological differences.

What we have here is the idea that Ward's zig-zag outcrop of high-angle normal faults was conceptually representable by low-angle lag faults (L) between the Skiddaws and the Borrowdale Volcanics; and Marr himself had seen evidence of faulting near the boundary of 'Otley II' and 'Otley III', near Torver for example (though they are not obviously low-angle faults there). This differential movement was supposedly compensated by the movement along the large thrust fault (T–T), which was hypothesized as underlying the whole mountain range (but being beneath the Skiddaws was invisible). Put another way, above the thrust plane there had supposedly been a general northward movement of the main sequence (Skiddaws, Borrowdales and Upper Slates). Of these, the Skiddaws had travelled further than the Borrowdales, which in turn had travelled further than the Upper Slates. But a portion of the Upper Slates (the Drygill Shales) was to be seen further north than the Skiddaws, supposedly being a small exposure of the youngest of the three main units. These rocks had not been pushed forward by the thrusting but had been there all the time, and were now 'peeping through' where erosion had cut down below the level of the hypothetical thrust. This was surely ingenious, but we should note that the Shales crop out near the top of a hill rather than at the bottom of a substantial valley.

It is, I think, exceedingly difficult to envisage a system of forces that would give rise to the foregoing state of affairs, but by the early years of the twentieth century Marr's position within the geological community was such that he could command assent. He took members of the Geologists' Association with him to the Lakes in August 1900, and proposed to them the idea of a major thrust fault (Marr 1900*b*, p. 530). Whether it was received positively I do not know. The sketch map produced for the benefit of the excursionists (Marr 1900*b*, plate XIII) is virtually impossible to reconcile with the section of Figure 5.8 above. However, as we shall see in Chapter 7, Marr's ideas held the field for some 20 years. It will be noted that his structure (Fig. 5.8) rightly allowed for a syncline in the high fells around Scafell; and his idea was that the main body of these fells was produced by a kind of 'tectonic accumulation'. The ashes and lava flows had, so to speak, been

[23] Regarding lag faults, Sir Edward Bailey, working in Scotland, referred to fold-faults as 'slides', of which there were supposedly two categories: overthrusts and lags. The notion of 'lag faults' is still found in the literature of structural geology, but such faults are rather rare creatures. They are not formed according to the system of forces envisaged by Marr, but are associated with extensional forces. For an example, see Soper & Anderton (1984).

gathered together where they now stood by means of tectonic activity (not by the accumulation of volcanic debris from a volcano in the locality of the highest mountains). This idea was odd too, even for its time.

In fact, I suggest, Marr himself had difficulty in making sense of the whole business. There is a little poem in his Notebook XXII (24 September 1890) relating to his work in the region between Haweswater and Ullswater, which is worth reproducing here:

A plague upon lavas and ashes,
Agglomerates also be banned,
Away with contortions and smashes;
Such games I don't understand.

Let thrusts be consigned to the devil,
May 'tears' go along with them too;
Let imps of Beelzebub revel
In rocks which are twisted askew.

Accurst be the lavas of Stanah;
The rocks of Galleny be blowed;
Whilst as for the Eycotts, how can a
Man tell where the mischief they're stowed.

The White Stones agglomerates, drat 'em,
As also the tuffs of Bowfell:
I dedicate every atom
To inmost recesses of h[ell].

The rocks that occur on Torpenhow[24]
Make any geologist swear:
Oh! send the whole lot to Gahena,
To fuse and solidify there.

In Hades there may be a Johnny
Who'll venture such rocks to descry,
For instance, our underground B[onney]
Might work his experienced eye.

For my part I hold the volcanic
Deposits are rather too much;
Beds furnishing relics organic
Alone, in the future I'll touch.

There followed a hypothetical sketch section from High White Stones, through Rough Crag, Rigginside, to Randale Beck, i.e. a SW–NE section of the ground at the western side of the south end of Haweswater. Marr had in his section *two* supposed thrust faults, with a 'Stanah Group' thrust over Eycotts. (Stanah is a farm near the NE of Thirlmere.) The poem shows, I suggest, that Marr had been trying, with difficulty, to make sense of the immensely complicated tangle of volcanics in the region between Haweswater and Ullswater, and across to Thirlmere, such as can be seen depicted on the recently published Keswick sheet (1999), which displays complicated folds (but not thrust faulting). To my knowledge, Marr never published a section like the one in his notebook and the name Stanah Group did not appear in the literature except for a mention by John Postlethwaite (1889–1890, p. 45), which stated that rocks of Eycott type were to be found at Stanah. (The name Stanah also occurs in Harker's notebooks for the period, and his notes mention a thrust fault in the Haweswater valley; Harker, Notebook 17, 1890, p. 103).

As for 'Galleny', that presumably refers to Galleny Force, a small waterfall on the eastern side of the Stonethwaite Valley in Borrowdale (see Figs 2.3 and 7.6). Ward's one-inch Survey map (1875) showed a contact between 'highly altered volcanic rocks, mostly ash', and 'contemporaneous trap' near the waterfall. The modern Keswick map (1999) shows outcrop of ignimbrite (Airy's Bridge Formation; see pp. 98–99) and andesite, with faulting nearby. We can appreciate Marr's difficulties, then, and perhaps understand why he preferred to work in rocks where the stratigraphy could be determined on the basis of palaeontological evidence. Marr was later to put his student Edward Walker to work in the Langstrath area, doubtless with a view to getting some light thrown on the problem (see p. 75).

As for the thickness of the Borrowdales, Marr (1916, pp. 7 and 17) estimated that they might be of the order of 10 000–20 000 feet, with the higher figure being more likely. For while the rocks were supposedly thrown into broad folds, they were not held to be repeated by faulting; so the apparent thickness was the real one, so to speak.

Marr's thrust and lag fault 'theory' of the Lakeland structure was, I suggest, rather extraordinary, and it said nothing about the presence of Skiddaw-type rocks in the Black Combe area of SW Lakeland. Perhaps these were rocks that had escaped the northerly thrust or push? But with Marr's strong position at Cambridge and in the Geological Society his ideas carried weight,[25] especially when presented to the Geological Society in Marr's Presidential Address (Marr 1905*b*, p. lxxvi), in the Jubilee Volume of the Geologists' Association (Marr 1910, pp. 642–645), or when embedded in his *Geology of the Lake District* (1916, pp. 75–89).

It is interesting to consider whether Marr's views on this matter were partly derived from his association with Alfred Harker (1859–1939). Born at Kingston upon Hull, son of a corn merchant, Harker studied at St John's Cambridge from 1878 to 1883, reading mathematics and physics, but also taking geology and petrology. He rapidly moved on to posts as demonstrator, lecturer and reader in petrology in the Cambridge department. Harker is best known for his petrological studies of magmatic hybridization, but he was also an excellent fieldsman, doing fundamental map-work among the igneous rocks of Skye.

In 1882, Harker was a member of an undergraduate field excursion to the Lakes and Cross Fell, under Professor Hughes, with Marr in the party as demonstrator. (He must be one of those represented in the cartoon of Fig. 4.3.) Harker wrote down the stratigraphy of the region in his notebook as then understood by Marr; [26] and he was with him in the Lakes on a number of subsequent occasions until about the turn of the century. There is a reference in Harker's notebooks to thrust faulting in the Haweswater valley (Notebook 17, 1891, p. 103), but nowhere do we see any mention of lag faults or the remarkable structure for the Lakes envisaged by Marr (Fig. 5.8). Harker's chief concern was the igneous rocks of the Carrock Fell area (just the thing for him, with its neighbouring gabbro and granite), and he was also evidently interested in the Eycott volcanics.

Harker and Marr went to NE Lakeland in 1893 (Harker, Notebook 20), but, as said already, Harker's notebooks contain no indication of Marr's theory. By contrast, Marr's notes for 1890–1891 are full of thoughts about thrusting, though they contain no overall schematized figure such as that of Figure 5.8. Perhaps his first great triumph – the dispatch of Barrande's theory of colonies by the invocation of thrust faulting – made Marr sympathetic to the concept as a general explanatory tool. Helped out by his new idea of 'lag faults', he felt able to tell a tectonic 'story' about the geological structure of the Lakes. But the theory was suspect, and, as we shall see in Chapter 6, it attracted criticism from another Lakeland aficionado, J. F. N. Green. Incidentally,

[24] A hamlet on the northern margin of the Lakes (see Fig. 12.1). Marr was presumably puzzled by the northern outcrop of the Volcanics, which were difficult to reconcile with his thrust and lag theory, though things could be 'made to fit', as shown in Figure 5.8.

[25] He was Secretary from 1888 to 1898 – an unusually long period; Vice-President for several periods; President from 1904 to 1906; a Member of the Society's Council for a total of 34 years; a Fellow of the Royal Society (Council Member 1904–1906); and President of Section C (Geology) of the British Association (1896).

[26] Harker, Field Notebook No. 2, 1882, Sedgwick Museum Archives, p. 59.

the theory would appear to have been largely Marr's, not Harker's. Marr (1900*a*) stated that his work was carried out in conjunction with Harker, and that he concurred with the ideas, but his colleague never supported the theory in print.

However, Marr's forte was biostratigraphy, physiography and sedimentology, rather than structural geology or igneous petrology. We shall return to his views on physiography in Chapters 18 and 19, and will discuss his ideas on graptolites and their evolution in Chapter 8. Here one should say a few more words about his work on the stratigraphy of the Ordovician and Silurian rocks, which, with physiography, was his Lakeland speciality. And then something will be said on his studies of the igneous rocks of Lakeland, conducted with Harker.

We have seen that Marr, first with Nicholson and then alone, got the stratigraphy of the Ashgill Shales (Ordovician) sorted out at Ash Gill, and the overlying Lower Silurian succession was elucidated at Skelgill near Ambleside and at Stockdale, Longsleddale. But as things turned out, the best Ashgill succession was not at Ash Gill, Coniston. Rather, a thicker and seemingly more complete succession was found near Sedbergh to the SE of the Lakes, near the foot of the Howgill Fells (see Fig. 5.9). (This area had, of course, been mapped in the primary survey, and Hughes had mentioned it to Marr.)

Marr examined three domes of Ordovician and Silurian rocks that cropped out in the area of the valley that runs from Sedbergh to Kirkby Stephen. This is generally called the Cautley district, after the principal topographic feature of the area: Cautley Spout, a substantial waterfall. The exposures had been revealed during the primary survey by a strong team of surveyors (Dakyns, Tiddeman, Russell, Clough and Strahan; see Dakyns *et al.* 1891), with good outcrops being found in the several stream beds of the area, and by the Rawthey River, which runs down to Sedbergh. Marr thought he could recognize fossils of Caradoc and Ashgill age (Ordovician) and Stockdale Shales, Wenlock beds, Lower Ludlow beds, beds equivalent to the Middle and Upper Coldwell Beds and the Bannisdale Shales (Silurian) of the Lake District. During an excursion in 1912 (accompanied by a number of Cambridge students, including Bernard Smith (see Chapter 4), Tressilian Nicholas and W. B. R. King, whose names will appear in later chapters), Marr found that he could correlate the Cautley rocks with those in the Cross Fell area, and the succession in South Wales. In fact, he decided that the Cautley area provided the most complete Ashgill succession that he had seen, and he therefore recommended that the exposures near Cautley, rather than at Ash Gill itself, be taken as the type locality for the Ashgill Series of northern England.

Today, such a change would need ratification by some appropriate committee or international commission, but such formality was not required in Marr's day. Even so, it was probably gratifying to him to hear the Survey Assistant Director, Aubrey Strahan, say in the discussion of Marr's paper at the Geological Society that the term Ashgillian, initially recommended by Marr, had 'proved convenient', and had been adopted in Survey publications as far away as South Wales. Marr (1913, p. 11) also mentioned localities in Backside Beck (the name he found so difficult to utter) and Watley Gill in the Westerdale Inlier – the northeastern of the Palaeozoic domed inliers of the Cautley region – where contacts between Ashgill Shales (Ashgill) and the Skelgill beds of the Stockdale Shales (Llandovery) could be observed. However, none of these sections has survived into modern geology as the Ordovician–Silurian stratotype boundary. This privilege is reserved for Lapworth's famous site at Dob's Lin (modern spelling) near Moffat in the Southern Uplands, studied by him in the 1870s (Cocks 1985, 1988; Hamilton 2001). Even so, Marr's Cautley work was important in that it gave geologists a clear idea of where to look for the best exposures of Ashgill Shales, as well as fossil lists and an indication of where to seek for further specimens. It should be remarked that he did not chiefly use graptolites for this work, though some specimens were reported. In recent times (1996), Barrie Rickards (see p. 176) has been re-examining at Marr's sections at Watley Gill and his studies suggest that the area contains fossils demonstrating a lower zonal position (below the *Dicellograptus anceps* zone) than had previously been thought. However, that is another story, arising from the discovery of more and better palaeontological evidence, and need not be pursued here.

It was mentioned above that Marr's forte was not hard-rock geology. Even so, he did work with Harker on Lakeland igneous rocks – especially near Shap (where there was the romantic inducement). They published a major paper on the area (Harker & Marr 1891) and a follow-up (Harker & Marr 1893). The first of these provided an account of the Shap Granite, which served as the starting point for all future investigations of that important rock. The division of labour cannot be determined with certainty, even with the help of their respective notebooks, but Marr probably provided the local stratigraphy and the regional setting for the large mass of granite, while Harker examined the granite's petrology and that of its surrounding metamorphic aureole. He cut many thin-sections, which are still to be found in the famous Harker slide-collection at Cambridge and have been utilized by subsequent investigators, such as Ronald Firman (see Chapter 7).

The two John's-men examined the granite itself, of course, and what they took to be the changes in the adjacent rocks wrought by the granite, at different distances from the igneous body. They did so in the context of the idea, then current in the Survey under the influence of Ward (1875*a*, *b*), that granite was produced by some kind of 'aqueo-igneous fusion', or what one might call a quasi-metasomatic process. But this idea they rejected, preferring the more orthodox 'Huttonian' idea that the granite was simply emplaced as a melt, interacting with and altering the surrounding rocks in various ways according to their nature and proximity.

The Shap Granite clearly affected all the adjacent Silurian rocks, and pebbles of it could be found in the overlying basal conglomerate of the Carboniferous; so it was evidently Devonian. Inspection of thin-sections showed that the quartz had crystallized before the conspicuous and characteristic pink orthoclase crystals. Ward (1875*a*, pp. 571 and 575) had thought that the fluid inclusions in the quartz suggested its formation at a depth equivalent to 46 000 feet of strata, whereas the actual thickness of overlying Silurians was perhaps 14 000 feet. Harker (I think we can leave Marr out of the discussion here) thought that this indicated that the quartz had perhaps formed at the greater depth, but had subsequently been brought up along with the otherwise still-fluid magma. Thus he was querying the conclusions made by Ward about the depth at which the granite as a whole might have been formed. At best, Ward's argument would only apply to the quartz crystals.

Harker examined certain darker than usual patches in the granite and found that they were similar in character to the rock found in dykes and sills adjacent to the granite. Hence he formed the hypothesis that there might have been a more basic magma underlying that which formed the main mass of granite, and this basic matter got carried up to form the dark patches or 'clots' in the granite. The granite appeared to occur at an intersection of two sets of disturbances: one that gave the general Lakeland line of strike (ENE–WSW) and one that gave the nearest Skiddaws (to the west of Shap village) a NW–SE strike.

A point of considerable interest was the fact that Harker and Marr could *not* find evidence for any significant zonation in the metamorphic aureole. The metamorphism just seemed to die away gradually in all the rock types as one receded from the granite. Also, there appeared to be an absence of minerals such as andalusite (or chiastolite), staurolite, and garnets (other than lime-garnets) such as might be indicative of metamorphic zonation round a granitic mass. The idea of metamorphic zones was then in its infancy, but it is interesting that Harker and Marr were on the lookout for such phenomena.

Also interesting was the character of some of the adjacent

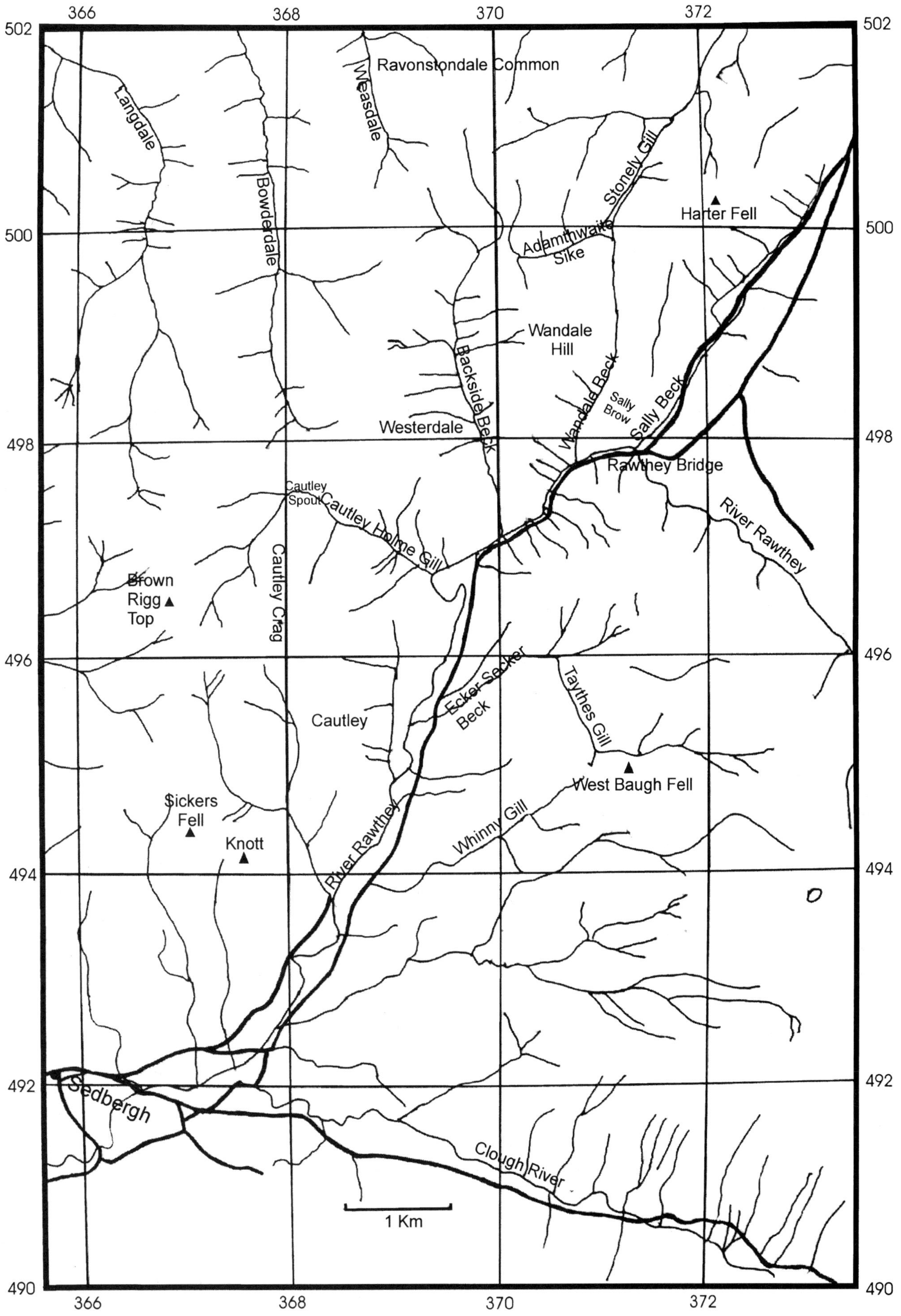

Fig. 5.9. Topography of the Howgill region, northeast of Sedbergh.

'rhyolites', which appeared to have undergone silicification; but the change was not apparently related to the presence of the granite boss (it did not, for example, increase as one approached the granite). Harker noted that the London University geologist, Catherine Raisin, who had done (and was to do) work with Bonney in North Wales, had remarked similar rock in Caernarvonshire, which she ascribed to the percolation of siliceous solutions in a kind of solfataric action (Raisin 1889). Harker thought that something similar might have occurred near Shap. Perhaps he was referring to the 'Yarlside Rhyolite',[27] which crops out in the vicinity and is today thought by some to be an ignimbrite (see Chapter 7; Millward & Lawrence 1985; Branney 1986) – a type of rock that was to cause much soul-searching amongst Lakeland geologists. The invocation of solfataric action may well have been picked up by J. F. N. Green, whose work will be discussed in the following chapter.

Two years later, Harker & Marr (1893) made some minor modifications of their Shap paper, the main difference being that they had now found that the igneous rocks to the north of the granite were more basic than previously supposed. This suggests that they had earlier given these rocks somewhat cursory attention – perhaps because Marr's attentions were earlier focused more particularly on the Shap Wells locality where dwelled his beloved, to whom he proposed in July 1891. Be this as it may, the 1893 paper reaffirmed the view that the thermal metamorphism produced by the Shap Granite did not produce major changes in the bulk chemical composition of the adjacent rocks. The amount of metasomatism was thus insignificant. This now placed the authors firmly in the 'magmatist' camp (as it would have been called in the twentieth century) so far as the Shap Granite was concerned.

The Shap Granite stands at the eastern edge of the Lake District, close to the surrounding rim of Carboniferous rocks, and most non-geologists would not regard it as part of Lakeland. But it is embedded in Lakeland-type rocks and, as one of the chief granitic intrusions of NW England, it has an important part in our story. As will be seen, the Lakeland granites likely played a role of major importance in the geological development of the region and have posed ongoing problems for theorists of Lakeland geology in a number of ways. However, it was not until the 1950s that the area was examined again in detail (see Chapter 7). Harker and Marr had done such a thorough job.

I conclude this chapter by saying something about Harker's work in the Carrock Fell area (see Fig. 4.9). As previously mentioned, one of Harker's special interests was the interaction of acidic and basic magmas, and the resultant production of 'hybrid' rocks. In later years, his investigations of the rock at the mountain of Marsco, near the region of contact between the acidic Red Cuillins and the basic Black Cuillins in Skye, became particularly famous. The investigation of Carrock Fell was a valuable introduction to this later work.

The unusual rock of Carrock Fell had been noticed as far back as the 1820s by Otley and Sedgwick, and there was much earlier mining in the area, but surprisingly little geological work was done on the area until the 1870s. Ward said little about Carrock Fell – a modest E–W aligned hill, running parallel with and adjacent to the River Caldew in the NE corner of the Lakes – in his Lakeland Memoir, though he discussed the area in a paper published the same year (Ward 1876*c*), which provided a simple sketch-map, indicating the outcrops of the following rock types: spherulitic felsite, diorite, hypersthenite, granite, 'bastard granite', all on the same modest-sized hill. There was no such combination of rocks anywhere else in the Lakes. By hypersthenite, Ward 'should' (according to the usage of the 1870s) have meant simply a basic rock in which the main constituent was hypersthene; but by his description he evidently meant something else, as he stated that it also contained plagioclase, with some quartz, hornblende(?), altered augite, and titanium or iron oxide. In fact, such a rock might have passed as a gabbro; but rocks today named gabbros used to be called hypersthenites in the early years of the nineteenth century.

Anyway, a specimen collected by the petrographer J. Jethro H. Teall (1849–1924) (who later became Survey Director in 1901) was passed to the mineralogist and petrologist Charles Trechmann (1851–1917) for examination. He announced (Trechmann 1882) that the supposed hypersthene was diallage (a variant of augite), so that the rock should indeed be identified as a gabbro, not a hypersthenite. This suggestion was accepted by Teall and enshrined in his *British Petrography* (Teall 1888, pp. 178–180). Ward, it may be remarked, had supposed that the rock showed 'bedding', but for Teall it was an example of 'banding'. Indeed, it was a type of banded gabbro such as much interested him at that period. So by the time Harker examined Carrock Fell, the hill was thought to contain both granite and gabbro in close proximity, an unusual state of affairs. Such adjacent acid and basic rocks were of great interest to nineteenth-century geologists, as there was an ongoing debate as to whether there could be more than one kind of magma in the Earth's interior, or whether the different types could be generated by differentiation from a single magma source (Oldroyd 1996, pp. 214–216).

Harker examined the rocks in the years 1882–1883, at times in company with Marr, but elucidation of Carrock Fell was a task for a specialist in igneous petrology, and the publications that flowed from the fieldwork had Harker as sole author. The Carrock granite he now represented as a granophyre, and the investigation was widened somewhat to include an outcrop of 'greisen'[28] (where mining had long been undertaken) down in the Caldew River valley, SW of Carrock Fell, close to the place where small outcrops of the Skiddaw Granite could be seen in the river bed.

According to Harker (1894*a*, *b*, 1895), the rocks of Carrock Fell (largely the gabbro and the granophyre) had somewhat analogous chemical compositions, significantly different from that of the Skiddaw granite and greisen, which suggested that the two represented distinct episodes of igneous activity. Also, Harker wholly rejected Ward's idea that the Carrock rocks might represent some kind of alteration product of the Eycott Volcanics. As to the gabbro–granophyre complex, perhaps the first theoretical question was: which was intruded first, the gabbro or the granophyre, or were they the product of some process of differentiation? From the very geometry of the situation, they did not appear to have differentiated by gravity settling. Rather, considering the geometries of the igneous masses, the rocks' densities, and their variations near their margins, Harker concluded that the gabbro was intruded first and showed signs of differentiation such that it became more basic at its margins. But before the gabbroic matter was intruded, solidified and fully cooled, there came a second intrusion of more acidic rock giving rise to the granophyre. There was some chemical interaction between the two (the basic material still being hot and chemically active). However, there was apparently also some brecciation at the interface due to mechanical interaction, and at some places a quite abrupt change in silica content could be demonstrated across the boundary. Some apophyses of granophyric material into the gabbro could be discerned; yet there also appeared to be some incorporation of gabbroic matter into the granophyre near the junctions; and some 'impregnation' of the gabbro by granophyre. Harker did not offer a firm opinion as to why or how the basic and acidic matter was formed at depth but he supported the view that they were derived

[27] Yarlside is a complex of hills to the SW of the Shap Granite. A band of rhyolitic rock extends SW from the granite, through Yarlside, to Longsleddale (but is arguably an ignimbrite).

[28] An old Saxon mining term for a variety of granite, deficient in feldspar, now thought to be produced by pneumatolysis. Greisens can contain various accessory minerals, especially tin, and serve as useful ore bodies.

from a common source that had undergone differentiation prior to emplacement.

From the discussions recorded as having taken place after the presentation of his paper, Harker's findings met with the approval of Sir Archibald Geikie, which is hardly surprising if we consider the timing of the events. Geikie had for some time been engaged in a furious row with the petrologist and Professor of Geology at the Royal College of Science, John Judd (1840–1916), as to whether the granophyres or the granites of Skye were the older, and Geikie had recently delivered a knockout blow at the Geological Society to Judd on this very question (Oldroyd & Hamilton 1997), demonstrating that the rounded acidic Red Cuillins were *younger* than the jagged basic Black Cuillins.[29] Harker's demonstration of a similar state of affairs at Carrock Fell must have been highly pleasing to Geikie, whose authority as *the* British authority on volcanoes and igneous rocks had been challenged by Judd. By contrast, though Judd's presence at the Harker's paper was recorded, and he apparently spoke on some point(s), what he said was not deemed sufficiently important to be reported.

Given this subtext to Harker's work, it is not surprising that he received a commission from Geikie the same year (1895) to conduct a thorough mapping of southern Skye. This immensely difficult task was carried out with signal success and led to the publication of several major publications on Skye of great theoretical interest. Harker's maps of Skye stood for almost a century. But that is another story. It is interesting, however, to see how Harker's career moved forward from the Lake District to Scotland; and it is not surprising that he played no further major role in the history of Lakeland geology. However, the Carrock Fell investigations would have been an ideal introduction to the problems at Skye in much more remote and difficult country, and on a much larger scale. Thereafter, Harker was beguiled into transferring his affections to Scotland.

[29] Harker's paper was read in January 1895. Geikie's had been read in February 1894.

Chapter 6
Edward Walker, Robert Rastall, Frederick Green, John Hartley, George Mitchell and their work on Lakeland volcanics and structure

Following the work of the first Surveyors, and the Cambridge geologists Marr and Harker, a number of amateurs such as John Postlethwaite[1] and Matthew Stables of Barrow were active in the second half of the nineteenth century and around the turn of the century, the amateurs concentrating particularly on mineral and fossil collection and studies of glacial phenomena (see Chapter 18). There were also visits to the Lakes by members of the Geologists' Association in 1881 (Hudleston 1881*a*, *b*) and in 1900 (Marr 1900*b*). It was during the latter visit that Marr, acting as excursion leader, made known his ideas on lag faulting and the general structure of the Lakes. Postlethwaite acted as an 'unofficial' excursion leader before and after Marr took over guidance of the party.

One of the places visited by the Geologists' Association in 1900 was Langstrath (see Figs 2.3 and 16.1), the eastern branch of Borrowdale, and specifically a locality called Blea Crag, where Marr (1900*a*, p. 530) reported that 'Mr E. E. Walker, B.A., who is occupied with their study, explained what he had learnt about' the intrusive garnet-bearing rocks that formed the Crag. Such rocks had been noticed as far back as the time of Otley and Sedgwick, but were only now becoming a source of particular interest. Edward Eaton Walker (1878–1903) was a product of Bradford Grammar School and Trinity College, Cambridge, where he took up a major scholarship in 1897. A student of Harker and Marr, Walker graduated with first-class honours in both parts of the Natural Science Tripos, with a distinction in geology in Part II in 1900. This led to the award of a Harkness Scholarship, which was intended to be used for research on the Borrowdale Volcanics in the Lake District (Anon. 1903*a*, *b*).

Walker made good progress with the study of these rocks in Langstrath – which were part of what later came to be called the Seathwaite Fell Tuffs (Oliver 1954*a*). So Walker was beginning the modern attack on the understanding of the upper part of the sequence of the Borrowdale Volcanics in the Central Fells. Tragically he did not live to complete his work. Before finishing his Borrowdale investigations, he took a post as geologist to the British East African Protectorate in 1902, and died there of blood poisoning from ulcerated wounds the following year. However, he had one Lakeland paper almost ready for publication before he left England, and Marr presented it to the Geological Society in 1903 (Walker 1904).[2]

Walker mentioned that he had been looking over the Lakeland rocks where garnets occurred and had found them near the Eskdale Granite and the Ennerdale Granophyre particularly. However, he described first the exposure at Blea Fell in Langstrath, where he observed a fine-grained dark rock associated with a coarse-grained pink rock, veins of the latter penetrating the former; but with indications of the one shading into the other at some contacts. This was the kind of phenomenon that interested Harker. Walker, assuredly under Harker's influence, thought that the exposures indicated a brief time interval between the basic and acidic materials' emplacement, and prior magmatic differentiation before emplacement.

Much of the rest of the paper was concerned with petrographic descriptions, the work being directed towards the observation and analysis (chemical and microscopic) of garnet-bearing rocks. Over and above that, Walker also offered a new taxonomy and stratigraphic sequence for the rocks of the Central Fells, namely

Banded Ashes and Breccias
'Streaky' Rocks
Eycott Lavas, and associated ashes and breccias
Falcon-Crag Lavas and Ashes (some garnet-bearing)

Like Ward, it is evident that Walker had been noting the Borrowdale Volcanics near Falcon Crag, adjacent to Derwent Water (Fig. 2.3); but unlike Ward he did not propose that the curious streaky higher-level Borrowdales were altered ashes. Some of the rocks of the upper part of the sequence *were* evidently ashes (as seen well on Glaramara or Bowfell for example; see Fig. 16.1), but there were also the curious 'streaky' rocks, such as Ward had collected from Langdale (see Fig. 4.4), and, like Nicholson, Walker thought that some of the rocks of central Lakeland were of Eycott type.

The chief interest attaches to what Walker had to say about the 'streaky' rocks, which he found commonly associated with garnets – the source of which was apparently his main concern. It appeared, on microscopic evidence, that the garnets were surrounded by coronas of feldspar, the latter having replaced the former in still-liquid magmas. He thought that in general such 'streaky' rocks might be formed either

1. by infiltration of substances like quartz, calcite, chlorite, epidote and ilmenite along bedding planes;
2. by lamination in fine fragmental rocks;
3. by the flow of igneous material; or
4. as a result of dynamic action on included fragments.

Walker found so much variation in the 'streakies' that he thought that they might not all have had the same origin. Moreover, it was not always obvious which of the four possibilities above was the most appropriate. That he should have difficulty with such rocks is unsurprising, given that such things had hardly been seen being formed by modern geologists. A type exposed on Rossthwaite Fell, Walker called a 'tuff-porphyroid'. Other such rocks seemed to have a rhyolitic and/or andesitic matrix, or andesitic fragments occurred within a rhyolitic matrix. The 'streaky' rocks seemed to form a ring around the highest part of the district. Marr, he noted, had put them on the same footing as other garnet-bearing rocks in the Lakes, and had supposed that they were intrusive in the Volcanic Series. However, while Walker made a useful contribution in examining the range and contents of the garnetiferous and 'streaky' rocks of Lakeland he did not develop any firm opinions or general theory as to their formation

[1] Postlethwaite (1840–1925), FGS, was an employee of the Cockermouth, Keswick and Penrith Railway and honorary curator of the Keswick Literary and Scientific Society. Having had a few years' experience underground at the Brandlehow Mines in his youth, he wrote a valuable history of mines and mining in the Lakes (Postlethwaite 1877, 1889, 1913), the third edition of which contained also a good general account of Lakeland geology, based on his earlier separate work, *The Geology of the English Lake District* (1897). This little volume provided a useful black-and-white geological sketch map of the region (which is also useful for its delineation of the network of railway lines in the district in bygone times). The third edition of the mining volume, reprinted several times since 1976, contains a biographical note on Postlethwaite by another important amateur Lakeland geologist, E. H. Shackleton. Postlethwaite became friendly with Clifton Ward, and a good deal of his account is based on Ward's ideas, especially with regard to the Borrowdale Volcanics. Postlethwaite made significant contributions by his fossil collections, especially of graptolites and trilobites in the SS, on which topics he published several papers. Such palaeontological work will be discussed in Chapter 8.

[2] Marr recorded that Walker did the body of the work in 1901 and that he gave his manuscript to him when he left for Africa early in 1902.

or emplacement – scarcely surprising given his youth and lack of experience and the fact that he was not able to complete his investigations. Marr (1905*b*, pp. lxxiii–lxxiv) suggested that there was an identifiable streaky garnetiferous series occurring below the layered ashes of Scafell and he called this the Sty Head Group (see Fig. 16.1). He had earlier (Marr 1900*a*) thought that they might be intrusive (a point on which he and Harker had disagreed at the time) but now, following Walker's work, he was persuaded that they were contemporaneous, or formed part of the volcanic succession. Years later (Oliver 1954), the unit became known as the Airy's Bridge Group with its ignimbrites (see p. 98).

There is published mention of maps made by Walker, which were passed to Marr, and were shown by him to Green (1915*a*, p. 201), but regrettably I have not located them.

Also early in the twentieth century, Marr's Cambridge colleague Robert Rastall (1872–1950) began a re-investigation of the granitic rocks of Lakeland: the Buttermere–Ennerdale Granophyre and the Eskdale and Skiddaw Granites, which had been examined early on by Sedgwick, and had of course been mapped during the course of the first Lakeland survey and were described in Ward's Memoir. Rastall was born in Whitby, east Yorkshire, but his family had estates in Eskdale, to which he moved in 1942, following his retirement in 1937 (Bulman 1949); so it is perhaps not unexpected that Rastall interested himself in Eskdale rocks. Initially, he studied agriculture, and he taught at Tamworth Agricultural College, but in 1899 he decided to read for a science degree at Cambridge. He obtained a first in geology in 1903, which led to a fellowship at Christ's College (1906) and eventually to a University Lectureship (1919). Rastall became well known for his several textbooks. His chief research interests were petrological and related to the study of metalliferous deposits.

When Rastall (1906) published his first Lakeland paper, the idea of granite being formed by some kind of metasomatism or metamorphism, as envisaged by Ward, was on the decline in Britain (though it was later to make a comeback in the 1950s in the work of Herbert Read at Imperial College). Rastall gave it short shrift. As might be expected, he first mapped the granite or granophyre exposures and closely examined the contacts with the Borrowdale rocks. It is interesting in this regard that he construed the blocks of granophyre on the narrow neck by Little Dodd, between Red Pike and Starling Dodd, as having been formed by the weathering out of underlying solid rock, not material moved by deluges or glaciers. (These were the boulders that Sedgwick, long before, thought were emplaced by some catastrophic inundation; see p. 255.) By considering the form of the mass of granophyre rocks, Rastall inferred that their intrusion seemed to be 'intimately connected' to the junction between the SS and the Volcanics; but he suspected the existence of a series of laccoliths, or perhaps a 'cedar-tree' laccolith. The rock of High Stile and Red Pike formed a smaller branch, while that running from Starling Dodd southwards towards Ennerdale and beyond formed the larger (see Fig. 7.4).

Immediately south of Buttermere, and again associated with the Eskdale Granite near Wastwater, as well as at other localities, Rastall described the occurrence of more basic intrusions (previously reported by Walker), which chiefly occurred near the margins of the acidic rocks. Such an association of mixed acid and basic rocks interested Harker of course, and it is not surprising that they also attracted Rastall's attention. He thought that they might be explained by the ideas of the Norwegian geologist Waldemar Brøgger (1851–1940) (1890), according to which melts in a magma chamber might differentiate as a result of differential movement between hotter and cooler regions of the melts (Soret's principle). Then, after differentiation, the more basic material might be intruded first, followed by the acidic matter. Rastall thought that the differentiation might have occurred in two stages: in the deep magma and subsequently within the laccolith, as Brøgger had envisaged. However, it seemed to Rastall that there had also been some hybridization of the previously differentiated magma. The Merseyside geologist Arthur Dwerryhouse (1896–1931) (1909) also investigated the Eskdale Granite and concluded, with Rastall, that it was intrusive, rather than metamorphic as Ward had supposed. Dwerryhouse's chief interest, however, was in glacial theory. (He was subsequently appointed to the chair at Belfast.)

Rastall (1910) also investigated the granite that had long been recognized as lying beneath Skiddaw, and which, as the nineteenth-century geologists had noted, was exposed in just a few spots in the valleys on the eastern side of the mountain at two localities in the Caldew Valley and at Sinen Gill, a small upper branch of the Glenderaterra Beck, which runs southwards from the Skiddaw massif (see Fig.12.1). Exposure is poor around Sinen Gill except in the stream bed (where an excellent contact may be seen), but Rastall thought that the porphyritic granite there suggested that it was part of an apophysis of the main mass of granite; as was the exposure at Grainsgill (a tributary to the River Caldew; see Fig. 12.1), where the intrusion was made up of both granite and greisen. For the main mass of granite he postulated a laccolith, as he had suggested also for the western Lakeland granites.

Though he did not undertake chemical analyses of the Skiddaw Granite, Rastall made numerous thin-sections and concluded that the main mass was an alkali granite, with frequent perthitic or micropegmatitic texture. However, the phenocrysts were not well formed and the texture suggested that they were produced concurrently with the final solidification of the rock, i.e. phenocrysts were forming along with the crystallization of a eutectic mixture.

Regarding the structure of the overlying SS, Rastall proposed a general anticlinal arrangement, with a southwesterly strike; but apparent repetition of fossiliferous beds suggested complex folding and an anticlinorium (plunging to the SW) rather than a simple anticline. The intrusion of the granite might have been along the anticlinorial axis, either accompanying the folding or following on it closely. By analogy with the Shap and Eskdale granites, this emplacement could, suggested Rastall, have been late Silurian or early Devonian. Rastall accepted the palaeontological evidence for some (at least) of the Slates being Ordovician (Arenig) age (see Chapter 7), but he thought that the unfossiliferous older SS might be Cambrian and the grits of Watch Hill (to the east of Cockermouth, i.e. to the north of the main Skiddaw massif; see Figs 8.6 and 12.1) – which, he thought, were similar to Torridonian, Longmyndian or Ingletonian grits – might even be Precambrian.

Rastall's most significant contribution was his analysis of the metamorphic zones in the aureole around the Skiddaw Granite. The classic study of a metamorphic aureole was that of the German petrographer Harry Rosenbusch (1836–1914) who investigated an area of contact metamorphic rocks (the so-called *Steiger Schiefer*) in the region of Barr–Andlau and Hohwald in the eastern Vosges. There he distinguished three zones in an aureole: spotted slates or phyllites, with occasional contact minerals, chiefly chiastolite; spotted schists; and hornfelses (Rosenbusch 1877). In the 1890s, the somewhat maverick Surveyor George Barrow (1853–1932), who gave his name to the metamorphic (Barrovian) zones that he described in the Cairngorms, but whose ideas became so much at odds with those of his colleagues that he was eventually withdrawn from the Scottish team, had developed the idea that different minerals were formed during the processes of metamorphism, according to the different degrees of heat and pressure that rocks had endured during metamorphism (Barrow 1893). Thus in some cases it became possible to map unfossiliferous metamorphic rocks according to their characteristic accessory 'metamorphic' minerals (biotite, garnet, staurolite, kyanite, sillimanite). This method, and that of Rosenbusch, held much promise and zonal mapping was applied to the rocks around the Skiddaw Granite in Rastall's (1910) paper, Skiddaw being potentially favourable for such work, given that granite was intruded into the relatively homogeneous SS.

However, the metamorphic minerals around the Skiddaw granite were not the same as those that Barrow had found in the Highlands, since the degree of metamorphism was substantially less. In fact, Ward had previously noted evidence of zonation: spotted slate; chiastolite slate; spotted (andalusite) schist; mica schist; granite (Ward 1876*a*, pp. 5–12). This had been queried by Rosenbusch (1877, p. 211), who claimed that Ward had not got the order correct, since chiastolite should be associated with spotted slates. However, in Rastall's view the two investigators were not referring to the same entities, for at Skiddaw there were two spotted zones, with the first not being developed everywhere; and Rastall averred that, for the Skiddaw area, Ward's version was more satisfactory than that favoured by Rosenbusch.

Rastall sought to improve on Ward's account, however, by considering the different effects produced by the metamorphism acting on different kinds of sediments: black slates, grey flags or slate, and grey grits. The black slates yielded an outer spotted zone and an inner chiastolite zone (or andalusite when fresh). The degree of metamorphism was slight, as graptolites could be found even within the chiastolite zone. The grey flags yielded spotted slate, chiastolite slate, cordierite gneiss (with andalusite), and mica schist. Microscopic inspection showed that the spotting was chiefly due to biotite. Very close to the granite there were indications of garnet and staurolite. The grey (siliceous) grits appeared to be rather resistant to metamorphism and Rastall reported no clear zonation, though minerals such as biotite and chlorite occurred, and andalusite in some outcrops. There were also slight indications of cordierite close to the granite, but no garnets or staurolite. The greisen was attributed to the action of residual vapours and liquids from the intruding granitic magma.

Rastall noted that cleavage appeared to be absent on the northern side of the supposed laccolith, which suggested that the SS had there been protected from the influence of great thrusts from the south,[3] which had elsewhere affected the Slates. As a generalization, it appeared to Rastall that the SS had been intensely deformed *prior* to the intrusion of the granite. There were no indications of kyanite and sillimanite, characteristic of Barrow's highest metamorphic zones, so presumably the granite was not very hot at the time of its intrusion. In fact, tidy Barrovian zonation was not found – showing that Barrow's Scottish findings could not be applied holus-bolus to other areas.

Rastall noted that he had been encouraged by Marr, and had been in the field with Bernard Smith, whom we encountered in Chapter 4. As we saw, Smith did a considerable amount of work in resurveying western Cumberland in the 1920s, but when he was with Rastall he was still a demonstrator at Cambridge. Harker was not mentioned. Rastall's work was subsequently used in an interesting way, years later, in controversies about the tectonic history of the SS (see Chapter 9).

We should now consider the substantial contributions of the most interesting and controversial of all the Lakeland amateur geologists: John Frederick N. Green (1873–1949) (see Fig. 6.1). Son of a Devonshire clergyman, he attended school at Bradfield College and went up to Emmanuel, Cambridge as a scholar in 1891, obtaining the position of 15th Wrangler in 1894. Thence he went on to Part II of the Natural Science Tripos, gaining a first in geology in 1895. Presumably, he studied geology under Hughes, Harker and Marr. He is known to have been in the same class as Gertrude Elles (see p. 107). In his reply to the citation made on the occasion of his receipt of the Geological Society's Lyell Medal in 1925, Green stated that he owed much to Marr's teaching (Green 1925, p. xlv). Green also stated that he was 'brought to attempt geological work' by the Surveyor, George Barrow. It is not known at what stage in Green's career this occurred, but the connection is interesting. As previously mentioned, Barrow was a

Fig. 6.1. John Frederick N. Green. Geological Society archives (LDGSL P53/57). Photograph reproduced by courtesy of the Geological Society, London.

dissident amongst the Scottish Surveyors,[4] and Green likewise argued for ideas that were contrary to those of 'the establishment' (see below). Green's professional career was that of a senior civil servant in the Colonial Service. However, he devoted much of his spare time to geology and was to become President of the Geologists' Association in 1918–1920 and of the Geological Society in 1934 (the year after his retirement from the Civil Service).

Green's first major investigations were carried out in Pembrokeshire, where, with the help of careful remapping, and digging out a contact, he resolved a long-running controversy between members of the Survey and several amateur geologists such as Henry Hicks, about whether there were or were not Precambrian rocks in the St David's area (Green 1908, 1911*a*; Oldroyd 1991). In 1911, Green turned his attention to the Lake District. Doubting the idea of a long, complexly faulted contact between the Skiddaws and the Borrowdales, as enshrined in the Survey maps, he chose to start his examination of the problem in the area north of the Duddon Estuary, SW Lakeland (see Fig. 6.2), where major intrusions did not seem to complicate the structure. Green probably already doubted Marr's notion of lag faults as the basis of Lakeland structure, though he only attacked it in print at a later date. According to George Mitchell (1956*a*, p. 422) (see p. 85), Green's Lakeland work – I take it in general – depended on 'a critical examination of the six-inch maps of the Geological Survey and numerous traverses', but systematic

[3] Later investigations have suggested thrusting from the north for the northern SS (see p. 199 and Plate V).

[4] There is a remarkable Barrow typescript (BGS, LSA 326), held in the Survey's Edinburgh office, written in his old age, which details his many cogent objections to his colleagues' ideas.

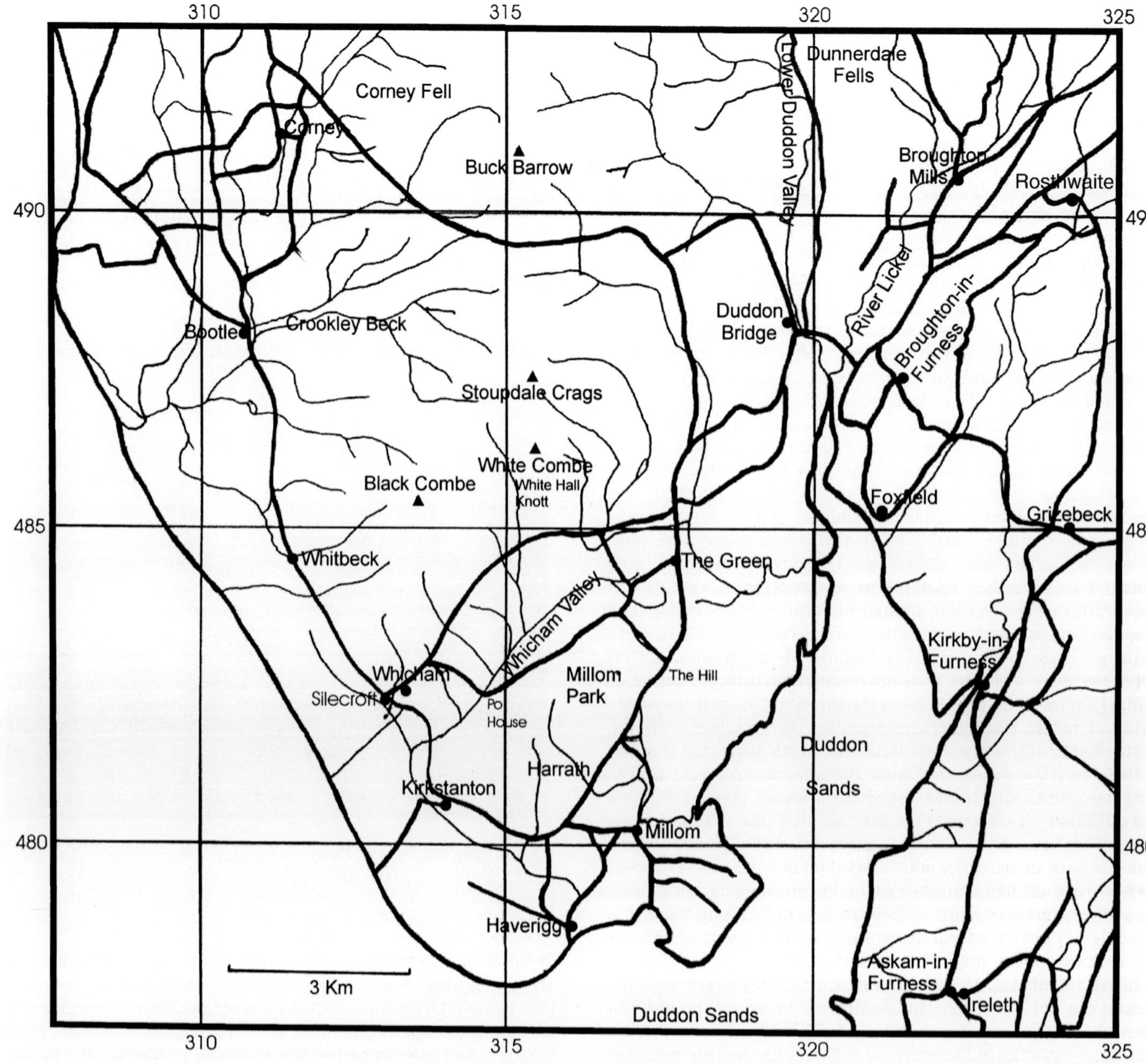

Fig. 6.2. Topography of the area of Black Combe, the lower Duddon Valley, and the Duddon Estuary.

remapping was not undertaken. (He published maps none the less.) Based in London, Green made a close study of the Survey's maps in Jermyn Street, and possibly also the field-slips, but he could only get into the field when his administrative work permitted.

Green read a paper on his Duddon work to the Geological Society (of which he had become a Fellow in 1908) on 15 November 1911. He proposed his own system for subdividing the rocks of the district, as follows (Green 1911*b*, p. 71):

Limestones (Sleddale Group)
Stile End Beds
  Unconformity
Upper Tuffs }
Rhyolites }
Harrath Tuffs } Volcanic Group
Middle Tuffs }
Dog Crag Tuffs }
Mottled shale
Blue shale or (?) Skiddaw Slates – no base seen

The new names, Harrath Tuffs and Dog Crag Tuffs, were coined from localities in the Duddon area north of Millom. Green claimed, on the basis of his mapping and observation of exposed sections, that the shales (SS) passed *conformably* into the Volcanic Group. Further, the rocks appeared to be thrown into folds striking NE–SW, so that the total thickness was less than previously supposed. Moreover, having supposedly determined the succession and structure of the Volcanics, Green's mapping suggested that there was an *unconformity* below the Coniston Limestone, not a lag fault as Marr supposed, for the Limestone lay on different units at different places (which could, however, also be accounted for intelligibly by faulting). Green's idea was confirmed to his satisfaction by trenching, as he had done near St David's. He also postulated a substantial strike fault on the basis of a repetition of the Limestone towards Duddon Bridge. The total thickness of the Volcanics in SW Lakeland was estimated as only about 650 feet (Green n.d. [1913], p. 4). (In fact, later mapping has shown faulting in the rocks near the Volcanics–Coniston Limestone contacts.)

In the ensuing discussion, Green was partly supported by

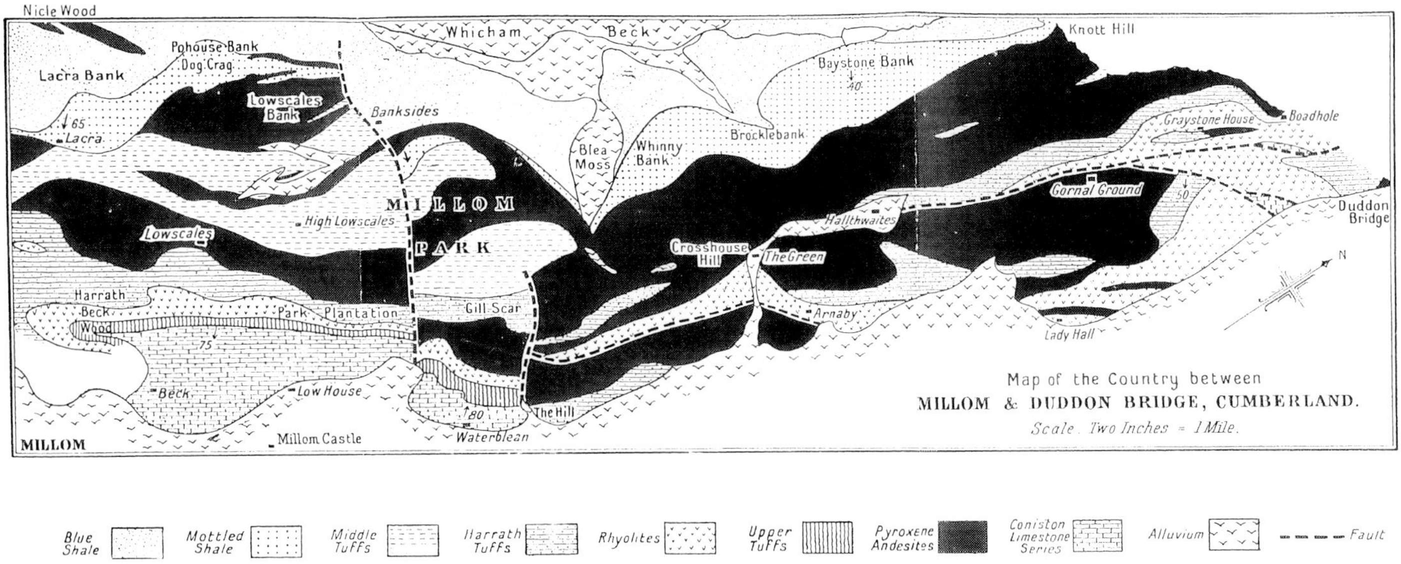

Fig. 6.3. Geological map of the area of the Duddon Estuary, accompanying Green (n.d. [1913]).

Barrow, who saw an analogy between Green's work and that of Lapworth in the Southern Uplands in the 1870s, where patient mapping aided by graptolite zonation had revealed that the sequence was not as thick as the surveyors supposed. The apparent great thickness arose from repetition of beds by folding. Indeed, Barrow was very likely correct, in that Green's proposed structure for the southwestern Lakes was plausibly beholden conceptually to Lapworth's work in the Southern Uplands.

However, the details of Green's mapwork were criticized by Surveyor Bernard Smith, who, as mentioned, had been in the Lake District with Rastall. It was the vexed question of the nature of the Skiddaw–Borrowdale contact that agitated Smith. He claimed that in the Wicham Valley (which runs approximately E–W, north of Millom) the Volcanics could be shown to be striking against the Slates (Green's 'Shales'). He also queried Green's mapping of a number of faults. Marr was not recorded as having participated in the discussion.

In the event, Green's paper was not published in full by the Geological Society. According to the records of the Publications Committee (Society archives COM/P3/2), the referees were Harker, Marr and William Watts of Birmingham University – a reasonable choice, given that the Survey had a kind of vested interest in the issue.[5] But so too did Marr, given his published views on Lakeland structure, with his system of lag faults, and with a fault rather than an unconformity at the base of the Coniston Limestone. The referees' reports are not preserved (they were presumably hand-written and posted to Green). Whether Green was offered the chance to revise his paper is not known, but in the event he withdrew his contribution and published it privately (Green n.d. [1913]), very likely in a huff. (He stated in a footnote: 'Publication declined' by the Geological Society.) The Society only published an abstract of the paper (Green 1911*b*).

The full paper contained a sketch-map (Fig. 6.3) that is almost impossible to reconcile with any other map known to me, ancient or modern,[6] but it would appear that Green's general thesis of there being low-angle folds with repetition of beds may have arisen from the fact that he did not achieve a precise characterization and differentiation of the various units in the volcanic sequence. On the other hand, Green recognized different lenticular andesite sheets within the sequence; and his idea that there was a kind of 'transition' of conditions from the SS to the Volcanics was worthwhile and has attracted later support.

Undeterred by his rebuff by the Cambridge–Geological Society–Survey establishment, Green turned next to the more difficult area of the eastern Lakes (around Haweswater), attempting to use the sequence that he had established in the area north of the Duddon Estuary as a model. His results were published by the Geologists' Association in a lengthy paper, which was furnished with a coloured map and sections (Green 1915*a*). This started with a useful sketch of the history of ideas about Lakeland structure, but, as is the case with some historical work where the ideas of a participant in events are recounted, Green's work was presented in a somewhat favourable light, as representing a return, from Marr, to an earlier and better interpretation.

Green first mentioned the Lakeland andesites, which, he pointed out, frequently had a brecciated structure. This, he suggested, could be attributed to the shattering of a cooling and solidifying crust on a flowing lava. So the brecciation would be expected to occur on the upper surface; but he also envisaged the possibility that it might occur more rarely in the emplacement of sills. By contrast, Ward had, as we know, conflated the andesites with the tuffs in his mapping, and was inclined to suppose that the unbrecciated lavas were metamorphosed tuffs, but it is clear from his Memoir (Ward 1876*a*, pp. 24–29) that he was moved in this direction by the 'streaky' rocks, for which there was at that time no satisfactory 'uniformitarian' or 'actualist' explanation.

Deploying his Duddon Valley (western) sequence of Volcanics in his understanding of the geology of the eastern Lake District (round Haweswater and SW to the Shap area and south as far as Stockdale), Green supposed that he saw a kind of rim of SS and conformably above them mottled tuff and (lower) andesite round the eastern and northern edges of his area. In the centre were tuffs, (upper) andesites, and rhyolites. His mapping was broad-brush, but his theory was clear, and evidently beholden to his work near Millom. The region, he supposed, had been thrown into a series of recumbent isoclinal folds (see Fig. 6.4), again, I suggest, reminiscent of structures envisaged for the Southern Uplands. The rocks of the Coniston Limestone Series lay unconformably on the Volcanics at their southern margin, as at Stockdale, Longsleddale. The complexity of the folding naturally led one to expect a complicated and rather confusing set of outcrops. Green's theory

[5] I thank the late John Thackray for information on this point.

[6] One may suspect that Green's mapwork was done very hurriedly in the single season of 1910.

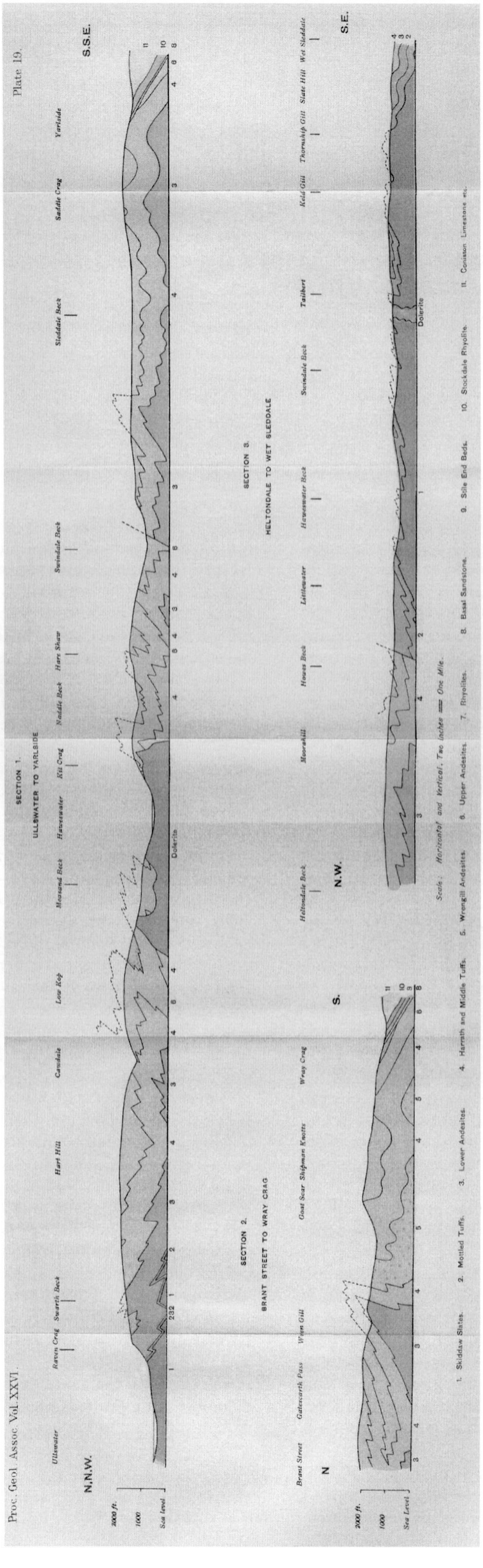

Fig. 6.4. Sections for eastern Lakeland, according to Green (1915*a*, Plate 19) (coloured in original).

led to his sections, I suggest, for it is not clear to me that he could have *deduced* his system of folds from his map of eastern Lakeland (though field evidence for overfolding was suggested, for example between Cawdale and Haweswater). However, Green stated that the main idea for his folded structure was to be found in the work of Harkness (1863). Green envisaged two main foldings, and a third gentle one that also affected the overlying Carboniferous rocks. The first fold was supposedly responsible for unconformity below the Coniston Limestone Series; the second one was that which had produced the anticlinoria, post-Silurian.

Green was dismissive of Marr's lag faults, saying (correctly) that it was a 'task of great difficulty to conceive the peculiar system of mixed pressures and tensions apparently required to produce lag faults in association with a great thrust in rocks that are not behaving as viscous fluids' (Green 1915*a*, p. 222). He was also correct in saying that the various experimental modellings made for foldings and thrustings by people such as Gabriel Daubrée, Henry Cadell or Bailey Willis did not produce Marr's structures. Green also complained that Marr's streaky, garnetiferous Sty Head Group was not to be found in the Haweswater region, suggesting that Marr had not got the general Lakeland volcanic sequence right. The rocks of the Central Fells are, of course, different from the Volcanics round Haweswater, and Marr need not have been dismayed by Green's objection on this issue.

In a second paper in the same year, Green (1915*b*) gave attention to the ongoing problem of garnets and 'streaky rocks'. Contrary to Walker, Green thought (on the basis of observations of crystals seen in thin-section) that the garnets were formed by alteration of feldspars from within. He suggested that the 'streaks' were 'to be found in infolds of rhyolite' (e.g. quite large structures such as the Cawdale Infold); but, contrary to modern opinion, he did not think that they causally related to 'eutaxitic' (the term he used) flow-structures, though there was seemingly some correlation. The 'streaks' appeared to be associated with cracks but not with shearing or foliation. From the dips, it appeared that the 'streaks' had originally been horizontal. Green (1915*b*, p. 217) speculated that the cracks had been formed during cooling and that there had been subsequent infiltration associated with some kind of solfataric action – solutions circulating under pressure 'during a solfataric stage of the Borrowdale volcanic episode'.

Regarding the garnets, Green doubted that they were formed by direct crystallization from a magma, but thought that they might also have been formed by solutions circulating under pressure during the supposed solfataric stage of the vulcanicity.

Two years later, Green (1917) published a paper on the large intrusive masses of the Lake District, such as the Eskdale Granite, the Ennerdale Granophyre, the Carrock Fell complex, etc., and he again gave attention to the general structure of the region. As before, he endeavoured to utilize his old Duddon classification, but now he also brought observations from the eastern Lakes and applied them to the west. Thus, for example, a claimed andesite unit, extending from Honister Pass to Pillar, via Kirk Fell (see Fig. 7.4), was mapped as Wrengill Andesite (so named by Green 1917, plate XXVIII), a unit found in the upper reaches of Longsleddale (see Fig. 3.7). The Latterbarrow Sandstone by the Calder River (west Lakeland) (see Fig. 20.1) was associated with the Mottled Tuffs earlier identified near the Duddon Estuary and was said to lie conformably on, and interbedded with, the underlying SS.[7] The volcanic rocks extending round from Eycott Hill to the Uldale Fells of northern Lakeland were mapped as Lower Andesite; and the inlier at Drygill was represented as belonging to the Coniston Limestone Series, as were rocks of the northern fells east of Cockermouth at Watch Hill, Bewaldeth, Great Cockup and Great Sca Fell (see Fig. 12.1). Green's mapping was exceedingly broad-brush and rather amateurish in appearance. None the less, he had interesting and important ideas, based on a thorough knowledge of the work of his predecessors as well as his own fieldwork. Further, he set out the nature of his problem clearly, namely: what was the age of the several major intrusions relative to the other Lakeland units?

First Green discussed that old question of the relationship between the three main (Otleyan) units of the Lake District. Using the analogy of the exposures at Po House, Wicham Valley near Millom (see Fig. 6.2), Green suggested that when a lava flowed over mud the latter could get squeezed up into, and incorporated in, the flow breccia and into vesicles in the flow. So a variety of structures and textures could be manifest at such contacts, with fragments of mudstone being found within the lower margin of the volcanic material. This, he maintained, was what was to be seen around the Lakes at the Borrowdale–Skiddaw contacts, and there were no indications of faulting. So structures that Marr (1916, pp. 84–85) thought were indications of what might be expected on a larger scale during his postulated thrusting were for Green nothing more than the result of a lava spreading over a muddy sediment![8]

Green (1917, p. 16) thought that he could see evidence at Whitfield Cottage (3236 5349) near Binsey Fell in the northern Lakes (see Fig. 12.1) of overfolding of the SS from the north, such that the slates at that locality lay on andesite; but he also claimed a tongue of lava pushing into the sediment. So even where the succession was apparently inverted this could be ascribed to later Earth movements, and the main idea of a conformable succession could still be upheld. Green mentioned that 'unusual faults' had been postulated for the Borrowdale–Skiddaw boundary for some 40 years; but firm published evidence was lacking. Ward's interpretation was dismissed as a product of his 'peculiar views on metamorphism, long abandoned by all geologists'. Marr had thought that the purple breccia at Falcon Crag and at the edge of Derwent Water (see Figs 4.5 and 2.3) might be a fault breccia; but for Green it was an explosion breccia.

Regarding the Watch Hill Grits NE of Cockermouth (thought possibly Precambrian by Rastall), Green had them as unconformable to the Skiddaws, rather than being part of them, and – supposedly containing fragments of Skiddaw, Borrowdale and Carrock Fell rocks – belonged to the Coniston Limestone Series, which, then, could supposedly be recognized in the northern as well as the southern Lakes. Indeed, Green suggested, the Grits might be linked to the Drygill Shales, which, as we have seen, Marr had ascribed to the Coniston Series (then regarded as 'Bala' in age). All three types seemed to have association with rhyolitic rocks, e.g. the Stockdale Rhyolite in southern Lakeland, or a felsite found associated with the Watch Hill Grits.[9] So these folded Watch Hill rocks were, for Green, post-Carrock Fell, and of Coniston–Bala age. The folding itself into anticlinoria and synclinoria supposedly occurred during the Devonian.

In fact, Green (1917) turned Rastall's suggestion that the Grits were analogous to the supposed Precambrian rocks at Ingleton down in Yorkshire on its head, suggesting that *they* were of the same age as the Coniston Limestone Series. (In saying this, he was adopting ideas similar to those of the nineteenth-century surveyors such as Goodchild (1886)). Indeed Green thought he could find a number of Lakeland rocks of agreed Bala age, such as sandy slates in Longsleddale or sandstones near Millom, that seemingly resembled those at Ingleton. So, he agreed with Rastall that there was a similarity between the Watch Hill Grits and the Ingleton rocks; but for Green that was evidence of Bala-age rocks in the northern Lakes. Green's geology was diverging considerably from 'orthodox' opinion.

[7] The *Lexique Stratigraphique International* for the British Ordovician (Whittard & Simpson 1960) credits Dixon (1925, p. 72) with being the first to recognize this unit; but Green had clear priority for usage of the term, if not its appropriate application.

[8] Green misrepresented Marr's views somewhat here, suggesting that Marr thought that he had seen such structures near his thrust planes, but in fact he said that such structures were such as might be expected to be associated with thrusts or lags.

[9] Years later, this was confirmed as a sill: Hughes & Kokelaar (1993).

Green contended that his (earlier) demonstration of an unconformity below the Coniston Limestone Series allowed a new estimate of the age of the Borrowdales, regarded as Llandeilo by the Survey officers. He claimed that the Borrowdale Volcanics everywhere lay on the *Didymograptus bifidus* zone of the Skiddaws (see Chapter 8), and, by analogy with exposures in Wales and Ireland, it appeared to Green that the Borrowdale episode (which could be relatively brief) occurred in the Middle Llanvirn. The large igneous masses such as that at Eskdale, and the complex at Carrock, were supposedly emplaced late in the Borrowdale episode, but before a closing solfataric stage, to which, as we have seen, he attributed the occurrence of the 'streaky rocks' and perhaps the garnets. Such an idea was not necessarily implausible, a relatively short Borrowdale eruption being compatible with his modest thickness for the volcanic material.

The suggestion of there being 'Bala' rocks in the northern Lakes was important. To pursue the matter further, Green (1918*a*) next turned his attention to the unfossiliferous Mell Fell Conglomerate, exposed at the northern end of Ullswater and forming two rather isolated rounded hills at the northwestern side of that lake (see Figs 1.1 and 9.11). The unit had been noticed as far back as the time of John Playfair (1802, p. 219) at what he called Dun-Mallet (Dunmallard) Hill, just west of Pooley Bridge. The prevailing opinion of Green's time followed that of Ward (1876*a*, p. 75), who had tentatively referred the unit to the Lower Carboniferous. The Conglomerate was obviously lying unconformably on both the Skiddaws and the Borrowdales, yet it contained scarcely any fragments of these.[10] Rather the chief constituents were pebbles of material similar to the Silurians down in southern Lakeland and the Howgill Fells, notably the Coniston Grit. Ward (1879) had speculated that the Coniston rocks had been drifted northwards, perhaps collecting in some kind of fiord. By contrast, in one of his two papers on Lakeland geology, Richard Oldham (1858–1936) (1900), on leave from his work with the Indian Geological Survey, suggested that the conglomerate represented some kind of torrential deposit, formed on dry land near a range of hills in a region of generally arid climate – the sort of thing he knew in India.

Green claimed that the problem could be dealt with by reference to his postulated unconformity at the base of the Coniston Limestone Series, supposedly exhibited by the Watch Hill Grits in northern Lakeland. By geometric argument, he sought to extrapolate the plane representing the base of this unit to the east and south and estimated that it would have passed comfortably over Skiddaw and Saddleback (Blencathra) and might have been about 1400 feet above the bottom of the valley in which the Mell Fell Conglomerate is now situated. The Coniston Grits might, then, have been about 4500 feet 'up in the air'. So Green envisaged the Conglomerate as being part of a gigantic fan-debris from mountain torrents, probably deposited in the Devonian. The small apparent northerly dip observable in the Conglomerate was ascribed to gigantic false-bedding. The overall (implicit) conclusion was, then, that the rocks of southern Lakeland (and the Howgills) might once have extended right over the main Lakeland mass, leaving northern relics in the Watch Hill Grits, the Drygill Shales, and the Mell Fell Conglomerate.

Shortly after the publication of Green's paper on the Mell Fell Conglomerate he produced a paper on the Skiddaw Granite and the metamorphism that it had produced in its neighbourhood (Green 1918*b*). Details need not be given here, but the gist of his argument was that the granite had produced a metamorphic aureole, and that cleavage had been induced on the whole by subsequent Earth movements. An anticlinal axis was represented as being oriented NE–SW and passing through the centre of the Granite, exposed in the upper reaches of the Caldew River; and an approximately parallel synclinorial axis was represented as passing through the northern end of Bassenthwaite Lake and the Skiddaw massif, extending to the Carrock Fell complex at Grainsgill. Green (like Ward and others before him) supposed that the anticline was produced by the action of the emplacement of the granite, and that this occurred at about the same time as the production of the Borrowdale Volcanics. So, he suggested, near the close of the Borrowdale episode there was folding associated with the emplacement of the Skiddaw Granite and the basic rocks of Carrock Fell. Then everything was worn down by erosion, and the Silurians were deposited *over the whole Lakeland region*. In the Devonian, there was severe compression, with the Skiddaws Slates supposedly being caught between the crystalline masses of the Skiddaw Granite and the Carrock Fell complex, with consequent folding and crumpling (of which there is seemingly much evidence in the Caldew Valley south of Carrock Fell; see Fig. 12.1). Green thought there was some kind of thrust fault at the contact. This suggestion has not been upheld as a result of more recent mapping, though there is the Watch Hill Thrust within the SS to the west of Carrock. The modern map does not, however, support the notion of a straightforward anticline and syncline (or anticlinorium and synclinorium) such as Green envisaged; and once again it is difficult to reconcile his observations with modern maps. On the other hand, his striking suggestion of the Silurians having formerly overlain the area of the Lakes as a whole has received later approval.

Green's last main Lakeland papers appeared in 1919 and 1920, but need not be discussed in detail here, since the main elements of his thinking were provided in his earlier publications. The 1920 paper is important, however, in that Green gave diagrammatic expression to the ideas that he had been developing in the previous decade (see Fig. 6.5). In this figure, the upper part reveals his concepts of a conformable relationship between the SS and the Borrowdales; (Ordovician) folding of both; a subsequent period of planation, followed by the unconformable deposition of the Coniston Limestone Series; and then the Silurian sequence. The unconformity below the Coniston Limestone Green called the 'Bala unconformity'. The lower part of the figure shows the whole diagrammatically, subsequent to the Devonian compression, which yielded a set of anticlinoria and synclinoria. As mentioned, the Coniston Limestone Series was deemed to have been deposited right over the top of the main body of Lakeland rocks. Green described the fold that affected the Coniston Limestone Series from Shap Wells to the Duddon Estuary as a *monocline*. His (very sketchy) map (Green 1920, plate 8) showed the two periods of folding postulated in Green's theory.

In his final Lakeland publication, a report of the Geologists' Association's excursion to the Lakes in 1921 (led by Green, with his wife accompanying), Green (1921, p. 123) suggested that his Volcanics sequence was repeated by folding some ten times between Keswick and Coniston, so that the total real thickness was perhaps no more than 3500 feet – significantly less than Marr's approximation of about 20 000 feet. Green also insisted on unfaulted contacts at both the top and bottom of the Borrowdale Volcanics. Unfortunately, however, he did not publish the evidence for his tenfold repetition, and after 1921 he went off to do work in the Scottish Highlands and then in southern England. Interestingly, Green, a persistent critic of Marr's structural theories for the Lake District, was President of the Geological Society from 1934 to 1936 – and Marr died in 1933.

Through his several publications, Green was also persistently critical of Ward's theories; but he spoke favourably of Aveline's work on a number of occasions, especially in his 1920 paper. Green was insistent that the relationship between the Coniston Limestone Series and the Borrowdale Volcanics was one of unconformity, and his target in making this point was presumably Marr. In his work with Harker near Coniston, Marr had discerned strike faults near the boundary, and there are indeed such to be

[10] Later investigations have revealed some Borrowdale Volcanics material in the western part of the Mell Fell Conglomerate.

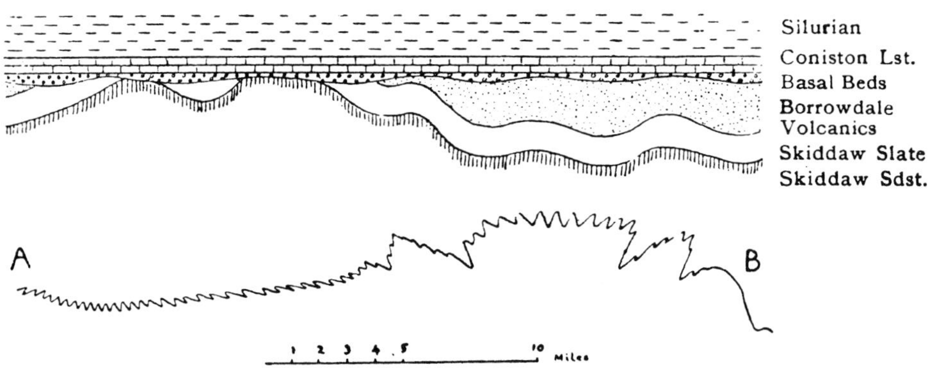

Fig. 6.5. Disposition of Lakeland rocks, pre-Devonian, according to Green (1920, p. 111).

found, as for example near Ashgill quarry (Torver Common) and further east at Browgill. In 1910, Marr (1910, p. 629) situated the Coniston Limestone in the Ordovician and the overlying Stockdale Shales in the Silurian – which suggested that the break would be above, not below, the Coniston Limestone. In his *Geology of the Lake District*, Marr (1916, p. 34) acknowledged the discordance of strike between the Limestone Series and the Borrowdales, but hedged his bets as to whether this was due to unconformity or a fault. The unqualified acceptance of an unconformity below the Coniston Limestone Series (or between 'Otley II' and 'Otley III') did not come until the 1920s.

The old surveyors such Aveline and de Rance had thought that the relationship was one of unconformity, but no Coniston memoir was published, and the matter had not really been pronounced on by the Survey. De Rance opted for an unconformity in discussion of a paper by Harkness & Nicholson (1877, p. 483), saying that he held this view on the basis of two years of mapping in the area. Even so, the matter was still open for discussion in the early twentieth century. It is interesting, then, that the surveyor Bernard Smith (1924*a*) reported his examination of some old notes by Aveline and published a few extracts, which showed conclusively that Aveline *had* thought that the relationship was one of unconformity. The Cambridge Sedgwick Club visited the Lakes in 1923, with Marr and Tressilian Nicholas (see p. 108) as leaders, and finally reached the conclusion (perhaps in response to Green) that there was an unconformity. Whether this persuaded Marr to give up his old theory of Lakeland structure in the 1920s is not known. He was ageing, going blind, and devoting most of his energies to Quaternary studies in the Lakes and East Anglia. It should be noted that Green stated (1920, p. 118) that Aveline had asserted an unconformity before 1880, but had not published on it. Green said that he had only recently got to know of Aveline's views, presumably by reference to the notes eventually published by Smith. Anyway, little was heard of Marr's lag fault theory after about 1920. Nicholas stated in 1925 that Marr had by then fully accepted the notion of an unconformity below the Coniston Limestone Series (Hartley 1925, p. 224).

In summary, Green's work was both interesting and influential, despite the fact that it ran counter to the views of the powerful Cambridge school (Harker, Marr, Rastall) and also to that of the Survey as expressed by Ward's maps and publications – though Green generally spoke well of Aveline's work. Green's mapwork was decidedly amateurish, but he could only get to the Lakes during his vacations. He worked out a general theory of Lakeland geology, starting from his observations and classification in the Millom–Duddon Valley area and applying them wholesale. For the most part, his stratigraphy was not taken up by others, and Marr, while mentioning him several times in *The Geology of the Lake District*, did not deign to argue the toss with Green. Green's theory of the 'streaky rocks' (as being produced by solfataric action) did not attract favour (though solfataric action was invoked by other geologists to explain such phenomena as the emplacement of ore bodies). His placement of the Watch Hill Grits in the Coniston Limestone Series ('Bala') was not widely adopted either. Nevertheless, his idea of having the southern unit formerly reach over into the northern part of the Lakes has received subsequent assent, as have his ideas on the brecciation of the andesitic lavas. It is not known what his personal relations were with the rest of the geological community.

As will be discussed in Chapter 7, Lower Palaeozoic work was pursued in the 1920s chiefly from a palaeontological and stratigraphical point of view by the Cambridge geologists Gertude Elles and Tressilian Nicholas; but this field did not interest Green much. He was concerned to work out the relationships and structure of the three Otleyan units, particularly the Borrowdale Volcanics.

The next person to take up Lakeland volcanic work was not a Cambridge man but John Jerome Hartley (1886–1959) (see Fig. 6.6). Son of a Wesleyan minister from Keighley, Yorkshire, he studied engineering at Liverpool University and was then employed on the Canadian Pacific, working on its celebrated spiral tunnels, and the South-Eastern Railways. He studied at McGill (Montreal) in 1912. Following war service in the Royal Engineers, Hartley took employment as a lecturer at Finsbury Technical College and studied part-time at Birkbeck and Imperial Colleges, taking a geology degree in 1920. He then worked on a London External MSc, with a thesis (of 39 pages and two maps) that was completed in 1924, entitled 'The geological structure of the Borrowdale Volcanic Series as developed in the area lying between Ambleside, Grasmere and Coniston'. To my knowledge, this was the first academic dissertation for a higher degree that was devoted to Lakeland geology. Hartley merely called it a 'paper', which in essence it was.

Hartley was in charge of the Engineering Laboratory at Finsbury from 1920 to 1926 and conducted classes in surveying, which presumably complemented his geological researches. In 1927, he was appointed Demonstrator in Civil Engineering at Queen's University, Belfast, and the following year he moved to a post of the same rank in Geology. After a decade there, he briefly rejoined the Civil Engineering Department, but was appointed Junior Lecturer in Geology in 1939. In 1942, he was promoted to Lecturer, and to Reader in 1948, in the small department of two geologists. Thus Hartley had a steady, albeit unspectacular, career. He was described by his professor, John Charlesworth (1960), as 'a pleasant colleague, modest and unassuming, a gifted lecturer and unstinting in his help to students'. But Dr Harry Wilson (pers comm., 1999), who studied under Hartley in Belfast, has remembered him as a 'very shy character, with curiously leonine features', keeping but one step ahead of the students in his lectures. He was said by the departmental technician (Sam Montgomery) to have private means derived from the Hartley Jam family. Hartley (called 'Baps' by his students) had the misfortune to be knocked down by a US Jeep outside the University during the War, after which he became absent-minded and mildly eccentric, but was a 'well-known and well-loved character around the University' (H. E. W[ilson] 1960). He was an untidy bachelor for most of his life, and surprised his colleagues when, quite late on, he married his landlady's daughter (J. Preston, pers. comm., 1999). Much of Hartley's later work in Ireland was concerned with

Fig. 6.6. John J. Hartley *c.* 1941. Copy of photograph formerly in the possession of the late E. M. Patterson and supplied by John Preston.

the older igneous rocks of Tyrone and on the hydrogeology of the Belfast area.

As we shall see, Hartley published a number of papers on the Borrowdales, and he also did a considerable amount of work on Irish geology. John Preston (pers. comm., 1999), who succeeded Hartley at Belfast in 1951, has kindly provided a photo of him in his later years, taken in Tipperary (Fig. 6.6). It appears that Hartley's specimens have not survived, but some of his field-slips were still held in Belfast in 1999. In his old age, Hartley assisted a Leeds PhD student, Lewis Clark, in the early stages of his dissertation on 'The geology and petrology of the Ennerdale Granophyre, its metamorphic aureole and associated mineralization' (Clark 1963), but this was not actually completed until four years after Hartley's death.[11]

Considerable interest attaches to Hartley's little-known thesis, as it specifically acknowledged Green's assistance in the field. In his 1925 paper, Hartley also acknowledged Green's help in the examination of microscope slides and identification of hand-specimens. Very likely, Hartley attended some of Green's later papers presented to the Geologists' Association and was almost certainly a member of the group of over 80 members that Green led to the society's Lake District meeting in 1921 (Sweeting 1958, p. 42; Green 1921), at which he interpreted Lakeland geology along the lines described above. I suggest, then, that it was through Hartley that some of Green's ideas became disseminated in the geological literature in the 1920s and 1930s. There was no acknowledgement of any supervisor for Hartley's thesis, and possibly Green acted informally in this capacity, there being no other London geologist at that time specializing in Lakeland geology. In favourably discussing Hartley's 1925 paper, Green stated that he had 'been over much of the ground in his company' (Hartley 1925, p. 223).

Hartley's thesis (the essence of which was published in Hartley (1925)) was devoted to the Borrowdales around Grasmere, Elterwater and Ambleside. He provided an intelligible map of the area, with faults clearly delineated and reconcilable with the modern Ambleside map (1996). By now, the old Survey idea advocated by Ward that the andesites of the Borrowdales might be construed as altered tuffs was dead and buried, and consequently more precise differentiation and delineation of different units could be achieved. Hartley's proposed stratigraphic sequence was as follows:

Basal Sandstone of the Coniston Limestone Series
Rhyolites
Acid Andesites with thin tuff bands
Felsitic and Basic Tuffs
Augite Andesites with tuff bands
Bedded Tuffs with andesite at or near base

The fact that Green's work was underpinning Hartley's observations is evident in his statement that '[d]etailed mapping shows that the rocks are folded into a series of anticlines and synclines, the axes of which approximately ENE to WSW. The folds usually pitch to the eastward, and large folds are generally corrugated by smaller folds of the same type and direction' (Hartley 1925, p. 204). This was 'pure Green'. Thus Hartley claimed to be able to relate his subdivisions to those proposed by Green for the Haweswater district; and also to those of Marr and Harker for the Shap area (though the terminologies were different). Hartley was

[11] In his acknowledgments, Clark (1963) thanked Hartley for 'suggesting the field of research, for his constant advice in the field and laboratory, and for his critical reading of the manuscript of this thesis'. It is not known when Clark started his work, and it is not clear whether Hartley was formally his supervisor at any stage. I thank Ms Lilian Bew for supplying copies of some pages of Clark's thesis.

not precise about the matter, but he seems, like Green, to have envisaged an Ordovician folding (NNE–SSW) and a Devonian(?) one (ENE–WSW).

It should be noticed (which my American correspondent Robert Dott did before me!) that Hartley (1925, p. 206) used the truncations of current bedding, observed in rocks near Chapel Stile, in Rydal Park, and near Sweden Crag, to determine the 'way-upness' of beds, employing ideas suggested by Charles Leith of the University of Wisconsin school of geology in his *Textbook of Structural Geology* (1913, p. 132). It was not until the late 1920s that Edward B. Bailey, working in the Scottish Highlands, began, with the help of North American colleagues, to make significant use of such way-upness criteria in Britain.[12] In fact, Hartley was probably one of the first British geologists to exploit the American ideas in his fieldwork. However, the idea had already been used in Scotland by Green (1924) (also from Leith), and it is likely that Hartley picked it up from Green. (Hartley and Green both gave the date of Leith's book incorrectly as 1914.)

Work on the Borrowdale Volcanics was further developed by George Hoole Mitchell ('Mitch') (1902–1976) (see Fig. 8.2).[13] His father was a schoolmaster who taught chiefly English and geography at Heversham School, Westmorland, and subsequently at Liverpool Collegiate School. The family was accustomed to spending lengthy summer holidays in Langdale, and Mitchell became familiar with the Lakes from an early age. He could speak the local dialect, which in those days was largely unintelligible to people from outside the area. From Liverpool Collegiate, Mitchell went on to Liverpool College and thence to Liverpool University in 1921, where he read geology under the recently appointed professor, Percy Boswell. Graduating with a first in 1924, Mitchell studied the rocks of the Coniston Limestone Series in the Kentmere area, his MSc thesis being submitted in 1925. It is interesting that, according to his son Murray (who obtained his information from his mother) (interview, 24 August 1998), the Cambridge geologists let Mitchell know by various nods and winks that it would not be pleasing to them if he continued his Coniston Limestone work further west, as this was regarded as the special area of investigation of Tressilian Nicholas (see p. 108 and Fig. 8.2), who had mapped the area between Yewdale and Applethwaite Beck[14] but never published his work (though apparently he made his results available to Mitchell (see Mitchell 1940, p. 303).[15] So Mitchell turned his attention to the Borrowdale Volcanics of the Kentmere and Troutbeck areas, working on this topic at Imperial College, London, with a Beit research fellowship. This led to a PhD on the Borrowdale Volcanics at Liverpool in 1927, a position as demonstrator at Imperial college in 1928,[16] and an appointment in the Survey in 1929.

For many years, Mitchell's Survey work was chiefly concerned with the Midland coalfields and Pennine Carboniferous rocks. However, he was so enthusiastic about Lakeland geology that he carried on with it privately during his vacations, extending his investigations of the Borrowdales westwards to the Coniston region and then down the Duddon Valley. He became District Geologist for the Southern Uplands in 1954 and Assistant Director of the Survey in 1959, with responsibility for Scotland. Yet even in his late years in Scotland he still kept up his Lakeland work. He was elected FRS in 1953.

Mitchell's last published works for the Lake District were a paper on the Borrowdales of the Seathwaite Fells of SW Lakeland (Mitchell 1963) and a Geologists' Association guide to the Coniston area (Mitchell 1970). His son recalls that his father had a sense of duty that drove him on to complete his work, even in his late fifties. Mitchell was, Murray told me, incredibly fit even then, and he could hardly keep up with his father on the fells. He did not need maps to find his way around either. He knew the ground so well that he could find his way around in thick mist and featureless terrain, such as the fells behind the Langdale Pikes.

Mitchell's early work was undoubtedly influenced by Green, whom he knew well and with whom he corresponded; and Hartley was also a correspondent. Like Hartley, Mitchell accepted Green's idea of an unconformity below the Coniston Limestone Series, and also his conception of two major fold systems for the Borrowdales: the first ('Pre-Bala' or Ordovician) producing folds that strike NNE–SSW; and the second, ENE–WSW, being Devonian. His main concern, however, was to establish a stratigraphic sequence for the Borrowdales, and to try to reconcile it with the schemes of Green and Hartley.

It is interesting that one of Mitchell's notebooks, preserved in the Geological Survey archives at Edinburgh and headed 'Kentmere Petrography KL7', contains what appears to be a draft of the literature review of his doctoral thesis. He showed how Green applied his own terminology to Ward's standard volcanic sequence at Falcon Crag (Derwent Water) (see Fig. 4.4) and then explained how Green thought this sequence was repeated ten times between Keswick and Coniston. Green himself never published a correlation table for his own stratigraphy for the Volcanics and that of Ward – so perhaps Green issued one as a mimeographed document to the members of the 1921 Geologists' Association Lake District excursion; or perhaps Mitchell got it from Green through correspondence that has not survived, or through conversation.

Be this as it may, Mitchell's notebook gives a useful correlation table for the volcanic stratigraphies of himself (Kentmere area), Hartley (Ambleside–Grasmere area), Green (eastern Lakeland), and Marr and Harker (Shap area) (see Table 6.2). It will be seen that the various geologists were endeavouring to establish a unified scheme for the Lakes *as a whole*. However, they had not, at that time, seriously got to work amongst the Volcanics of the Central Fells, so it was something of an illusion to suppose that they were giving a general sequence for the Borrowdale Volcanics.

Mitchell's paper of 1929, which was accompanied by a clear coloured map of the volcanics between Longsleddale and the branches of the Troutbeck valley and also showed the unconformably overlying Ordovician and Silurian sediments to the south, was essentially the published version of his Liverpool PhD. It contained the results of the analyses of rock samples that he had collected as representatives of his several Borrowdale units, the total thickness of which he put at about 3700 feet (Mitchell 1929, p. 13). His general theory of the Lakes was evidently influenced by Green in several ways, notably the belief in a relative modest total thickness for the BVG; the assumption of two systems of folding – an earlier one NNE–SSW and a Devonian one ENE–WSW; the idea of a repetition of units due to folding and faulting, as may be seen from his published sections for the eastern Lakes (see Fig. 6.7); and recognition of flow brecciation. The Devonian movements supposedly gave rise to the complex folds displayed in Figure 6.7. Mitchell further remarked on an alternation of basic and acidic lava flows, thinning westwards and northwards – but there was held to be a corresponding thickening of tuffs toward the NW.

---

[12] The Irish geologist, Patrick Ganly had realized the possible use if current bedding for determining way-upness back in the nineteenth century (Archer 1980), but his idea was not taken up at the time.

[13] I am indebted to Mitchell's son, Murray, for information about his father's work. My other main biographical source is the detailed obituary by Stubblefield & Dunham (1977). Mitchell was called 'Hoole' at home.

[14] On Applethwaite Common between Troutbeck and Kentmere, not named on the 1: 25 000 topographic map.

[15] Nicholas's maps have not been located.

[16] Part of Mitchell's PhD work was actually undertaken at Imperial College with Boswell and the assistance of W. W. Watts in petrological matters; but the degree was awarded by Liverpool University.

Table 6.2. *Correlation table for Borrowdale Volcanics. Prepared by G. H. Mitchell (n.d.), with the scheme of Mitchell (1929) added for comparison*

Marr & Harker (1900)[17]	*Green (1915)*	*Hartley (1925)*	*Mitchell (Notebook, n.d.)*	*Mitchell (1929)*
Shap Rhyolite	Rhyolites	Rhyolites	Upper Rhyolites	Upper Rhyolites
Shap Andesites	Upper Andesites with Tuff Bands	Upper Andesites	Upper Andesites with Tuff Bands	Upper Andesites
Shap and Scawfell[18] Ashes and Breccias	Harrath Tuffs	Felsitic Tuffs	Coarse Lithic Tuffs	Coarse Tuffs
	Wrengill Andesite	Augite Andesite	Middle Andesites	Wrengill Andesites
			Lower Rhyolites	Kentmere Pike Rhyolites
	Middle Tuffs	Bedded Tuffs	Middle Tuffs	Bedded Tuffs with Andesites
Eycott Hill Series(?)	Lower Andesites	Andesites of Scandale Beck	Augite Andesite (intrusive)	Harter Fell Andesites
	Mottled Tuffs			Froswick Tuffs
Falcon Crag & Bleaberry Fell				Nan Bield Andesites, with tuff-bands
Andesites				
				Base not seen

Material from Mitchell notebook reproduced by courtesy of the British Geological Survey.

Meanwhile, Hartley was moving on from the Grasmere and Ambleside area, to begin the major task of remapping the rocks round the Langdale valleys (Hartley 1932) – the first serious engagement with the rocks of the Central Fells since Ward's primary survey. Hartley's stratigraphic taxonomy was essentially the same as that used in 1925, though in moving from Grasmere and Ambleside to Langdale he found no equivalent for his Acid Andesites and he adopted Green's and Mitchell's term Wrengill Andesites for rocks he had previously referred to as Augite Andesites. In the Langdale area (or more strictly that of the Mosedale Valley running north from the upper Duddon to the west of the Bowfell range; see Fig. 16.1), so-called Mosedale Andesites were proposed as equivalents of Green's Lower Andesites, Hartley's own Scandale Beck Andesites, or what Marr (1900) had called the Eycott Hill Series or Ullswater Group. Hartley claimed that he had connected up his Mosedale Andesites with those mapped as Lower Andesites by Green at the head of Wastwater. So at this stage Eycotts were still regarded as having correlates right in the middle of the Lakeland mountains, as Nicholson had conceived long before.

Hartley's Bedded Tuffs, previously so named in Hartley (1925), were those striking rocks as are found, for example, on Stonesty Pike, Crinkle Crags and Bowfell (see Fig. 16.1) with their remarkable sedimentary structures and concretionary features, such as had been noted by Sedgwick back in the 1820s. Like Green, Hartley was inclined to invoke solfataric action to account for these mysterious 'concretions'. He also referred to various kinds of 'rhyolites' exposed extensively in the area of the bedded tuffs. Some were evidently intrusive, while others seemed to have been emplaced *lit-par-lit*, altering the tuffs to such an extent that the two could not be clearly distinguished. Hartley (1932, p. 48) thought that the altered tuffs had yielded a 'very compact greenish hornstone, sometimes spotted', with frequent epidote. Its formation was attributed to thermal or hydrothermal metamorphism. The entities concerned appear from Hartley's photograph (Hartley 1932, plate 2) to have been some kind of peperite (see p. 219). Hartley thought that there were two major rhyolitic intrusions in the Langdale area, perhaps part of a 'cedar-tree' laccolith or phacolith. Marr's Sty Head Group would constitute the tuffs and the intrusive rhyolites, producing metamorphism at the contacts.

Hartley suggested that the general structure of the Langdale Valley area was anticlinal (as earlier envisaged by both Ward (1879, p. 54) and Marr (1916, p. 18)) with the bifid fold pitching eastwards. But, like Green, he had a set of subsidiary folds superimposed on the major ones envisaged by his predecessors. So far as folding in general was concerned, then, Hartley continued Green's idea (supported by Mitchell) that there were two phases, the broader Ordovician folds, which yielded the anticlinal structure of the Langdale and Ambleside areas, and a more concertina-like Devonian folding.

Hartley (1932) also gave attention to the strange sedimentary structures of the bedded tuffs. He noted evidence for graded bedding (not so called), and also considered the common, contorted bands observable in the bedded tuffs. The fact that the contortions were confined to restricted bands suggested that they were formed when the tuffaceous material was still plastic. Hartley thought that they might be drag folds, i.e., minor folds in incompetent beds formed as part of a larger-scale fold structure, and he related the idea to earlier suggestions of Marr (1916, pp. 84–85) that the structures were, so to speak, incipient faults. He also canvassed another possibility, namely that the folds were not of tectonic origin, but represented slips in the sediments before the tuffs consolidated. He mentioned that Green had suggested to him that sediments collecting on steep subaqueous slopes might have been shaken by seismic action, yielding structures such as those illustrated in Figure 6.8. However, the drag-fold hypothesis was preferred. Hartley mentioned, incidentally, that he disagreed with Green's (1919, p. 163) explanation of drag folds, namely that they could arise from the dragging action of the lower surface of a lava flowing over a still-plastic sediment.

Hartley also attended to cleavage, faulting, jointing and quartz-veining, plotting the dips and strikes of some 2000 joint planes on a six-inch map of the Langdale area. The evidence suggested that the joints were tensional and associated with the folding rather than the faulting. In many cases, the fractures had been subsequently sealed by silica-bearing solutions. From consideration of the way in which these sealed fractures were themselves fractured, it appeared that two tensional episodes could be identified; and there was also fracture seemingly associated with the late pitching of the folds. All in all, Hartley offered a comprehensive analysis of this complex region, arguing ingeniously from the large amount of empirical evidence that he compiled. However, his work has been criticized by the Surveyor Tony Wadge (interview, 17 July 1998), who found that Hartley failed to recognize landslips in the Grasmere area, and thus erroneously mapped Borrowdale Volcanics in the valley floor, adjusting his maps by introducing hypothetical faults. (Hmm!) Also, Murray Mitchell recalls

[17] Actually Marr (1900). Mitchell (1929) attributed the work to Marr and Harker. The fact that Harker's name was not attached to the paper suggests that Harker wished to dissociate himself from his colleague's tectonic ideas.

[18] This has to do with Scafell near Shap, not the major mountain of the same name in the Central Fells.

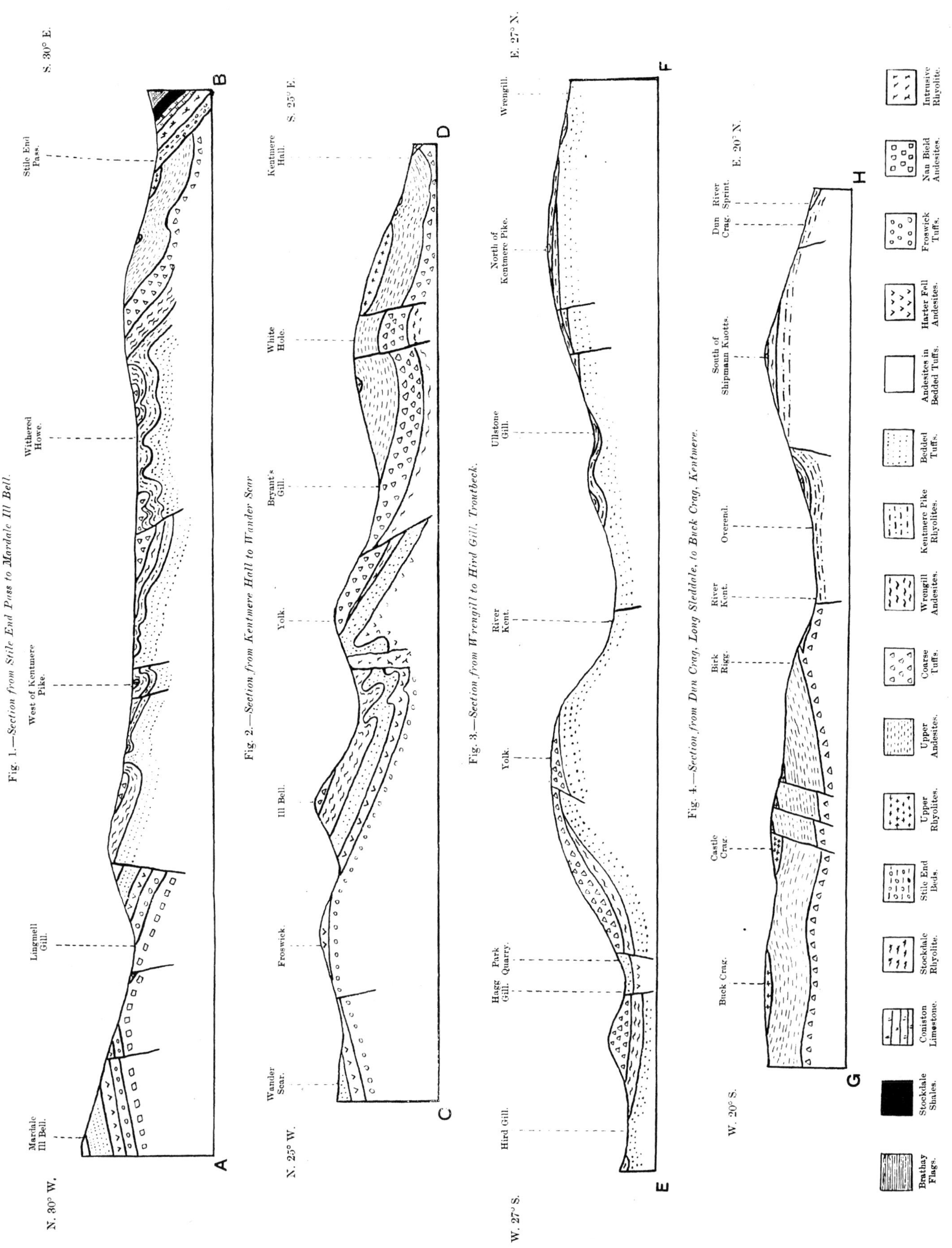

Fig. 6.7. Representative sections for the geology of the Kentmere area, according to Mitchell (1929, Plate V).

A.—CONTORTED BAND IN BEDDED TUFF. THREE TARNS, BOWFELL. View looking E.

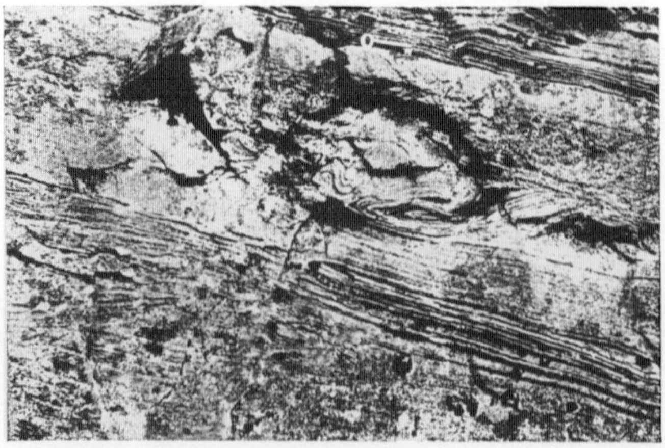

B.—CONTORTED BAND IN BEDDED TUFF. HARRY PLACE, GREAT LANGDALE. View looking E.N.E.

Fig. 6.8. Structures in tuffs, Langdale area, according to Hartley (1932, Plate 3).

(interview, 24 August 1998) walking with his father in the Lakes in the 1940s and that Mitchell could not make much sense of Hartley's mapping of the Borrowdales in the Langdale area.

Mitchell (1934) published another paper on the eastern Borrowdales, focused on the area of Upper Longsleddale, Wet Sleddale, Mosedale, Mardale and Swindale. He and Hartley were in correspondence in the late 1920s, as evidenced by a few letters surviving in the archives of the British Geological Survey; but regrettably we have nothing for the period we are now entering. However, a letter from Hartley to Mitchell, dated 27 October 1927 (GSM 1/393 (ii)), bearing Mitchell's annotations, contains the sentence: 'There are many points I would like to discuss with you especially as to how the E. portion of your Troutbeck area confirms J. F. N. Green's results for the Haweswater district'. Mitchell annotated this with two enigmatic exclamation marks. They might have meant that he suspected that Hartley was trying to scrounge information from Mitchell; or (more likely, I think) that Mitchell was already having second thoughts about Green's structural and stratigraphic ideas. This interpretation is supported by the statement in the 'Previous Literature' section of Mitchell's paper that 'Green published in 1915 an account of the eastern Lake District which included the present area. Some divergences from the views therein expressed have been found necessary by . . . [my] recent six-inch mapping . . .' (Mitchell 1934, p. 421).

This was assuredly an understatement. It is virtually impossible to reconcile Green's crude map of 1915, largely devoid of faults, but indicating large areas of (say) SS where none are directly exposed, with that of Mitchell (1934). In particular, the stratigraphic succession had been radically changed, in line with the table above (p. 86, for 1929), but with Haweswater Rhyolites added below the Nan Bield Andesites, and below that an Andesite of Ralfland Forest, Mottled Tuffs, and SS in unproved succession. Comparison with the two maps suggests that Green's importation of a 'Duddon Valley stratigraphy' (with Harrath Tuffs, etc.) into the eastern Lakes had not been successful. That is, he had not been able to achieve firm boundaries on the ground in the east, using his western stratigraphic sequence. The confusion as to what were tuffs and what were andesites, and which were which, when there was more than one instance of each – the problem that had dogged Ward's mapping – was also profound in Green's work. However, one can see distinct clarifications emerging in Mitchell's work, and in Hartley's too.

For each of his units, Mitchell gave both extended and synoptic descriptions, with mention of good localities and description of the subdivisions, in hand-specimens and as seen in thin-section. However, while differing radically from Green about the stratigraphy, Mitchell retained the main elements of his predecessor's structural theory. That is, there was supposedly a pre-Bala folding with NNE–SSW strike, which did not affect the Coniston Limestone Series lying unconformably on the Borrowdales; and then a more complex Devonian folding with fold axes striking ENE–WSW. This gave long, narrow, boat-shaped inliers of bedded tuffs in Upper Mosedale, etc. Mitchell was also able to take advantage of the section made available by the construction by Manchester Corporation of a tunnel for the aqueduct between Haweswater and Longsleddale, from which it seemed that his (and Green's) previous ideas about the structure of the eastern Lake District were supported, except that Mitchell proposed overfolding to the south, whereas Green had had it to the north (see Fig. 6.9). However, as Mitchell himself acknowledged, while the rocks in the tunnel could be identified with reasonable certainty, the drawing of the section as a whole, and the linking of the underground contacts to those visible in the field, required significant hypothetical extrapolations; so theory could be strongly at work here.

This point was also made in the discussion of Mitchell's paper by Sydney Hollingworth (see p. 53), who had also been into the Haweswater Tunnel, and was familiar with the observations on the basis of which Mitchell re-asserted Green's theory of anticlinorial overfolding. Hollingworth acknowledged that there was evidence in the tunnel of folding, but not necessarily of overfolding as proposed. Thus the Surveyors were pursuing different structural ideas from those envisaged by the amateurs Green, Hartley and Mitchell. (Here I regard Mitchell as an amateur, as his Lakeland work was 'unofficial', though he would naturally have had easy access to the Surveyors' map-revision work.) On the other hand, Hollingworth accepted Mitchell's placement of the Haweswater Rhyolites low in the sequence.

A small additional area of the Borrowdale Volcanics was mapped by George Hadfield and H. C. M. Whiteside (1936), focusing on the ridge, Low Rigg and High Rigg, between St John's in the Vale and Naddle Valley, i.e. north of Thirlmere and east of Keswick[19] (see Fig. 12.1). Low Rigg is made up of a fairly small outcrop of microgranite. In this area, the investigators found the following succession:

Middle High Rigg Andesites
Fine-grained lavas and tuffs
Lowest basic pyroxene-andesites

[19] George Samuel Hadfield (1910–1976) studied at Bootham School, York, Lausanne University, and Imperial College, London, where he took an MSc in geology. He went into his family's sand and gravel business and became managing director of a large hardstone quarry in SW Scotland. See K., J. R. (1977). No secure biographical information has been found concerning Whiteside.

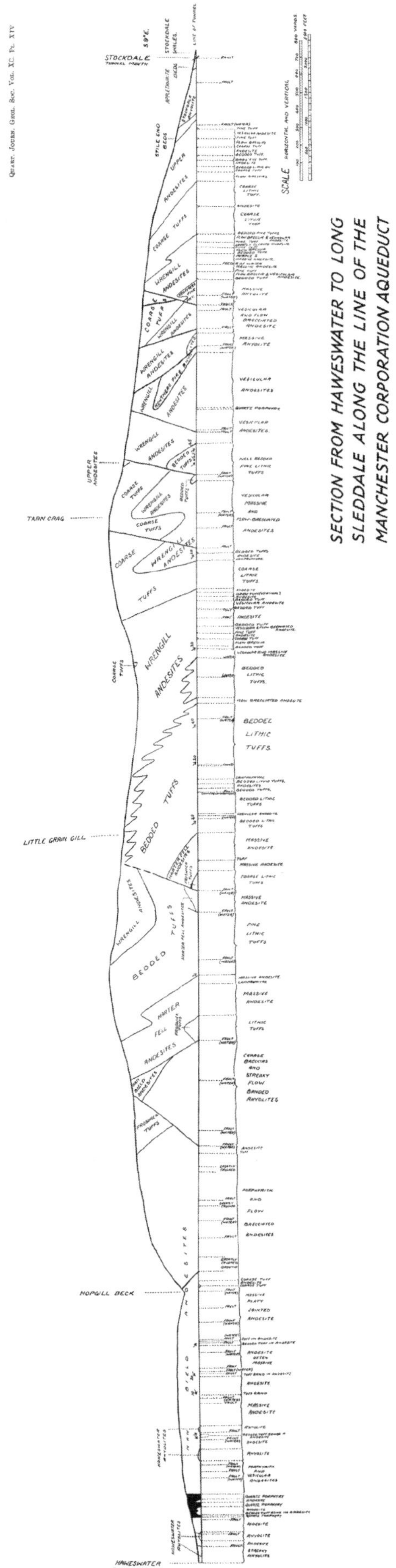

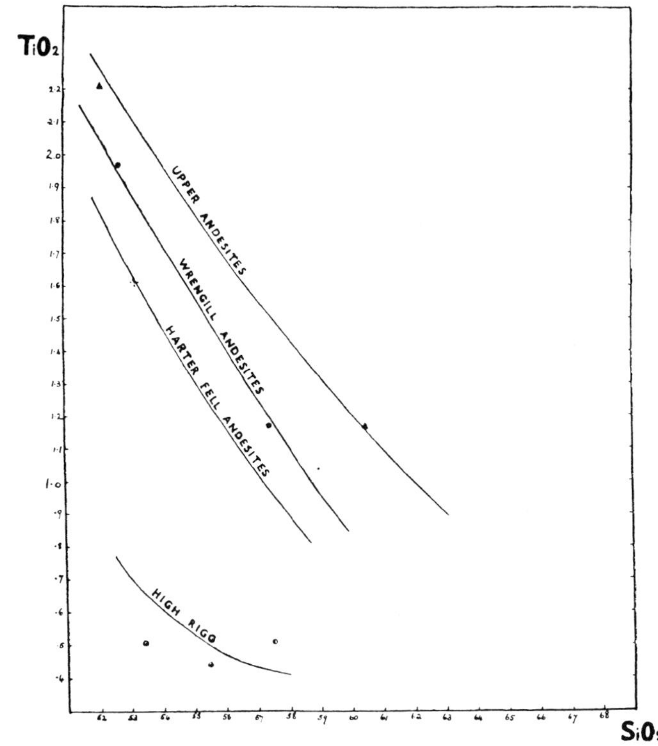

Fig. 6.10. Variation diagrams, plotting TiO_2 versus SiO_2, for various rocks from Borrowdale Volcanics, according to Hadfield & Whiteside (1936, p. 54).

The total thickness was estimated to be about 2400 feet. These units were thought to be low in the Borrowdale sequence, and analogous to those investigated at Falcon Crag, east of Derwent Water, by Ward. The pyroxene-andesites might be similar in their groundmass to the Eycott Volcanics, though they lacked the characteristic plagioclase phenocrysts of the latter. In fact, the authors' interests were chiefly petrological, and a good deal of evidence was brought forward relating to the chemical composition of the rocks concerned, which was correlated with data for Borrowdale rocks from other parts of the Lakes. It was found that when TiO_2 was plotted against SiO_2, the titanium content was inversely proportional to the silica content (see Fig. 6.10); and more interestingly, the different plots appeared to follow a pattern that was relatable to the supposed age of the various Borrowdale units, namely

Upper Andesites
Wrengill Andesites
Harter Fell Andesites
High Rigg lavas

This plot suggested that the proportion of titanium gradually increased as the eruptions of the Borrowdale rocks occurred. Here we have perhaps the first example in Lakeland geology of the use of geochemical data as direct aids to stratigraphic investigation.

Structurally, Hadfield and Whiteside thought that the area was dominated by the supposed Devonian folding, with ENE–WSW strike. Their theory was similar to Green's in a number of other ways: the Volcanics being conformable to the underlying SS; and there being flow brecciation with the garnets arising from solfataric action.

Work was continued in the following years by both Hartley and Mitchell, with both of them publishing in the *QJGS* early in the

Fig. 6.9. Section from Haweswater to Longsleddale, according to Mitchell (1934, Plate XIV).

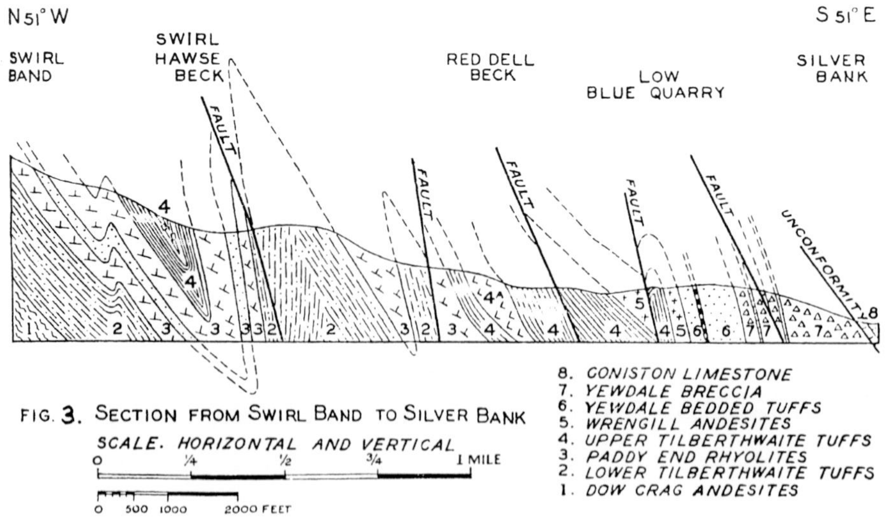

Fig. 6.11. Structure of Borrowdale Volcanics in the Coniston area, according to Mitchell (1940, p. 315).

Table 6.1. *Comparison of Mitchell's and Hartley's successions*

Mitchell	*Hartley's equivalents*
[Upper Rhyolites and Upper Andesites of Kentmere and Ambleside not seen]	
7. Coarse Tuffs (Yewdale Breccia of Ward)	Felsitic and Basic Tuffs
6. Yewdale Bedded Tuffs	Fine tuffs at head of Troutbeck
5. Wrengill Andesites	Wrengill Andesites
4. Upper Tilberthwaite Tuffs	Bedded Tuffs
3. Paddy End Rhyolites[20]	–
2. Lower Tilberthwaite Tuffs	Bedded Tuffs
1. Dow Crag Andesites	Mosedale Andesites

1940s (Mitchell 1940; Hartley 1941). In his paper, Mitchell extended his work westwards, covering the area to the NW of Coniston Water, which included Coniston Old Man, the Coniston copper mines, and important accessible contacts between the Coniston Limestone and the Borrowdale Volcanics, at places such as Timley Knot and in Yewdale (see Fig. 5.2). Introducing some new units, his succession was (with Hartley's units, as known to him, given for comparison) as shown in Table 6.1. The total thickness was estimated to be about 5550 feet.

Mitchell (1940) published a coloured map of the area, a diagram summarizing the chief folds and faults (which as in his earlier work he regarded as having occurred in the Devonian), and an interesting section drawn perpendicular to the general line of strike (see Fig. 6.11).[21] From this, it can be seen that he was now considering the substantial role of faulting as well as folding in order to produce a repetition of beds, and hence, like Green, he had a lesser thickness than that envisaged by the earlier surveyors and Marr. In this, Mitchell's theory was, then, an extension of Green's thinking, though Green did not like to invoke faults.

In Hartley's (1941) paper, he showed how he had extended his investigations into the region around Helvellyn, and across the southern part of Thirlmere to the southern Armboth fells and Dunmail Raise (the low pass through which the main road runs north from Ambleside to Thirlmere and Keswick), thereby linking up with his previous mapwork (Hartley 1932) in the Langdale area. His stratigraphy was adapted from his 1932 paper, with Helvellyn Andesites occupying the same stratigraphic position as the Wrengill Andesite, and the Upper Rhyolites becoming the Steel Fell Rhyolites (Steel Fell being the area immediately to the west of Dunmail Raise). Again, Hartley did not find the Upper Andesites of Green and Mitchell. Hartley marked in faults with confidence, whereas Green had found them almost an anathema. 'Streaky rocks' were recorded near Harrop Tarn to the west of the southern end of Thirlmere, but they seemed to be less common in the area than in Langdale. Hartley thought that the texture might be due to 'autobrecciation and subsequent partial assimilation', the autobrecciation being 'perhaps subsequent to local differentiation in necks' (Hartley 1941, p. 144), whatever that might mean![22] Green's idea of alteration of tuffs due to late solfataric action following their intrusion by rhyolites was revived. Hartley estimated the total thickness of the succession in the Helvellyn area to be about 5000 feet (to which one might add the 2400 feet of Hadfield and Whiteside at High Rigg).

On the important question of structure, Hartley again sided with Green *contra* Marr. He suggested that the outcrop of lava flows that he had mapped on the steep hillside of the southeastern end of the Thirlmere valley was such as to provide an approximation to a horizontal section; from which it appeared that the structure involved isoclinal folding, with folds leaning northwards. Mapping north and NE of Helvellyn also suggested isoclinal folding. There were, besides, broader folds with axes running approximately NE–SW: a syncline to the south of Helvellyn and a monoclinal structure along the northern side of the mountain, and running down towards Glenridding at the southern end of Ullswater (see Figs 2.2 and 9.11). Marr (1916, p. 24) had invoked nearly horizontal strike faults to account for repetition of Lakeland beds, but Hartley stated that he could not find such structures in the Helvellyn area. In fact, he concluded his paper with a warm acknowledgement of the importance for his work of Green's publications.

The Fellows of the Geological Society did not rush to Marr's defence in the discussion following the presentation of Hartley's paper, but Hollingworth and Mitchell doubted that the claimed pneumatolysis could be linked to the intrusive rhyolites and was more probably due to regional changes associated with the emplacement of the Borrowdale Volcanics. Mitchell thought that the Thirlmere rhyolites might be extrusive but with lava also burrowing into underlying tuff deposits, which Hartley conceded as a possibility for the Helvellyn area if not for Langdale. This was Hartley's last published contribution to Lakeland geology, his

[20] Paddy End was part of the mining complex at Coniston. The rhyolites there contained veins of copper ore.

[21] This section evidently runs for somewhat more than two miles across the 'grain' of the hills north of Levers Water towards the general direction of Yewdale Beck (see Fig. 5.2), but Low Blue Bank and Silver Bank are not marked on the modern 1: 25 000 map, nor on Mitchell's geological map.

[22] Hartley said he had in mind Marr's (1916, p. 16) suggestion that a scum of lighter material rose to the top of the liquid rock filling a vent and was then showered out by subsequent explosions.

later investigations largely being confined to Irish geology. He was 55 in 1941 and probably did not feel like further difficult mapwork in the Lakes. Besides, travel to England during the War would have been difficult. Also, he suffered his accident in Belfast. However, as mentioned, he did, in his old age, act as supervisor to a Leeds PhD student, Lewis Clark (1963), on the volcanic rocks to the east of the Eskdale Granite.

Let us now push on to the 1950s to review Mitchell's further work on the Lakeland volcanics, the investigations of the new post-war generation of doctoral students, and an eventual solution to the conundrum of the 'streaky rocks'.

Chapter 7
Granites, garnets, the 'streaky' rocks and Mitchell's later work

We now come to a time when I have been able to meet and get to know some of the protagonists in the story of the history of Lakeland geology. The first two post-war doctoral students working on hard rocks in the Lake District were Ronald Firman (b. 1929), with a Durham University PhD on 'Metamorphism and metasomatism around the Shap and Eskdale Granites' (1953) and Robin Oliver (1921–2001), with a Cambridge PhD on 'The Borrowdale Volcanic and associated rocks of the Scafell area, Lake District (England)' (1953). Though Firman was born before Oliver, his thesis was completed earlier, being referenced in Oliver's, so I shall discuss Firman's work first.

Ron Firman (Fig. 7.1) attended City of Norwich School, where he included geology in his curriculum, becoming interested in the subject through the efforts of his 'first-class' geography teacher (Firman, pers. comm., 1999). Though hindered by the disability of mild spasticity and associated speech impediment, he attended Durham University, where, as a goalkeeper, he captained his college hockey team and regularly contributed to Union debates. Despite this, and his previous experience of fell walking in the Lake District, Professor Lawrence Wager cast doubt on Firman's ability to meet the demands of geological fieldwork, by omitting to invite him, when an honours student, to join an undergraduate field excursion to the Lakes. In the ensuing 'animated' discussion, Firman's response was 'try me'! The trial was surely successful, for Firman participated in all subsequent departmental excursions both official and unofficial, in a Durham University Exploration Society training course, and while a student he made private geological trips to Skye, North Wales and the Malverns. Moreover, as was then normal practice for those who had studied geology at school, Firman completed the four-year honours degree after only three years of study (Firman, pers. comm., 1999).

Stimulated in his final year by Kingsley Dunham, then replacing Lawrence Wager as Professor at Durham, Firman became interested in the question of the origin of granite – at that time a controversial topic – and the processes of mineralization. It was Dunham who suggested that a study of the Shap and Eskdale granite aureoles would make a good PhD problem. Although Firman's supervisor was Frederick Stewart (who later became Regius Professor of Geology at Edinburgh), Dunham continued to take a lively interest in Firman's research, visiting him at Shap with Stewart and encouraging him to lead a field excursion for the Yorkshire Geological Society in 1952, before he had completed his doctorate. George Mitchell, who later acted as Firman's external examiner, was joint excursion leader, and among the participants was Professor W. D. Evans from Nottingham – chance meetings, but ones that turned out to be important for Firman's subsequent career.

Firman duly completed his doctorate in 1953 after only two years and three months, and promptly obtained a research assistantship at Manchester before joining the lecturing staff at Nottingham in 1954. There he stayed until the Department closed in 1989, albeit as a Senior Research Fellow without teaching duties during his last five years before retirement. During his career, he did further Lakeland research and advised the Manchester Corporation Waterworks on geological aspects of their engineering projects in the Lake District. He also interested himself in other aspects of economic and mining geology, doing work in the Midlands, Turkey and Canada. However, he was always particularly interested in processes of mineralization and geochemical problems. He also became interested in the applications of geology and mineralogy to archaeology, and in 2000 was attached to the Archaeology Department at Nottingham.

Fig. 7.1. Ronald Firman in 1954. Copy of photograph supplied by Dr Firman and reproduced by his permission.

J. G. Marshall (1858), Nicholson (1868*a*), and Ward (1875*a*, *b*) thought of the Lakeland granites as being very likely the product of some kind of metamorphic processes acting on adjacent igneous rocks and being consolidated from a state of 'aqueo-igneous fusion', giving rise also to an aureole of altered rock around the granitic centres. As we have seen, Ward's views influenced his Survey colleagues, and thence the ideas presented to the public in the maps and sections resulting from the nineteenth-century Lakeland surveying. Harker & Marr (1891, 1893), we recall, preferred the more traditional theory that the granites were emplaced as molten magmas. They held that the Shap Granite was intruded after the main Lakeland earth movements, and that the Lakeland lavas were affected by these movements before the emplacement of the granites. Following Dwerryhouse (1909), Marr (1916) tentatively had the Shap and Eskdale Granites as approximately the same age, i.e. rather late in the history of the Lakeland region (Devonian). Another worker, D. R. Grantham (1928), had had the idea that the Shap Granite was composite, an earlier somewhat basic type being produced by the assimilation of pre-existing volcanics and then being disrupted and invaded by a more acidic biotite granite. B. Simpson (1934) thought that the Eskdale intrusion had the form of a laccolith, whereas the Survey Memoir for the Gosforth area (Trotter *et al.* 1937), resulting from the resurvey of west Cumberland, envisaged it as an intruded stock.

Firman entertained the view, quite widely held in the 1950s by geologists such as Herbert Read (1957), that granites might be produced by metasomatic changes. However, after carrying out many mineralogical, petrographical and a few geochemical analyses, Firman concluded that he had no petrological criteria for settling the migmatist–magmatist dispute so far as the Shap and Eskdale Granites were concerned. Rather, one had to rely on traditional mapping techniques, showing the shapes of the granite masses and their attendant aureoles. From this evidence, and with the help of his chemical, petrological and mineralogical examination of the aureoles, Firman suggested that following the intrusion of the granitic magmas, there were subsequent hydrothermal stages (first a high-temperature phase and then one at a lower temperature), arising from the action of the hot fluids rising up with and above the magma, causing pneumatolysis and mineralization in 'pre-granite joints'. Firman's (1954) proposal for the sequence of changes at the Shap 'Blue Quarry'[1] (see Fig. 3.7) was, in summary, as follows:

1. intrusion of granite;
2. thermal metamorphism with introduction of Si and K into the country rock;
3. epidotization and formation of garnet;
4 sericitization associated with the formation of pyrite veins;
5. low-temperature hydrothermal mineralization accompanied by the production of a variety of clay minerals.

The garnet-bearing veins typically had outer epidote and inner garnet zones. Their production was attributed to percolating siliceous fluids bearing calcium and iron. Supposedly, the fissures subsequently provided channels for the percolation of the low-temperature solutions, which also appeared to have worked their way into joints and shatter belts that transected the 'metasomatic structures' and therefore were presumably late features.

Firman rejected the rather odd suggestion of Marr (1902) that the 'metasomatic structures' might represent thermally metamorphosed pre-granite metalliferous veins. Detailed chemical analyses were provided in Firman (1957*a*). The manner in which the garnets (andradite) were developed in the central parts of the veins at Shap suggested to him that they were formed from solution. Whether epidote or garnet was formed in the vein depended on the Al/Fe ratio. This was related to the proximity to the wall material of the vein, which had become hornfelsed by the chemical and thermal action. Garnet formation occurred through the introduction of Si, Ca, OH and Fe into the aureole from the granite – or, as was acknowledged in the discussion after the presentation of the paper, Ca could have been derived from pre-existing calcite veins, though such veins had not been found outside the aureole. The garnets in the Shap veins were different from those described by Oliver (1956*a*, *b*) as occurring in the Borrowdale Volcanics (almandine). (Subsequently, in the light of fresh evidence in the form of quartz–calcite–chlorite veins found in excavations for a dam in the valley of Wet Sleddale (see Fig. 3.7), Firman (in Mitchell *et al.* 1972, p. 450) began to think there might be something in the idea that metamorphism of pre-existing veins could have occurred.)

Firman's mapwork task was relatively straightforward around Shap and the northern part of the Eskdale aureole, which he had covered during his PhD. However, after submission of his thesis he examined hitherto unmapped ground, both within and beyond the Eskdale aureole, as far south as the Duddon Valley and thus had to face all the complexities of the Borrowdales for that region. His published map appeared in 1957 (see Fig. 7.2). The map's artistic character is noteworthy, and Firman has informed me that it, and others, were done for him by an old school-friend and artist, James Codd. Firman also told me that some of his maps were burnt in a fire at the Lower Duddon Youth Hostel, and much of the area had to be remapped, but 'thankfully' this geological catastrophe occurred after the submission of his thesis.

Firman (1957*b*) was chiefly concerned with structures and attempting to get a stratigraphic sequence that could be meshed with that of Mitchell (1956*a*), who was then working to the south in the Duddon–Ulpha area (see p. 101). Initially, Firman could find no evidence for a thrust near Devoke Water, earlier proposed by Green (1917, 1920), but later he conceded that there might be minor thrusts in that area (Firman 1960). Firman (1957*b*) agreed with Mitchell about the existence of the Ulpha Syncline, striking ENE, which was attributed to a 'pre-Bala' movement (i.e. before the unconformable deposition of the Coniston Limestone Series and younger rocks; hence Ordovician). Also, Firman followed Mitchell in the idea that the Black Combe inlier of Skiddaw Slates in SW Lakeland was produced by anticlinal folding in 'pre-Bala' times. In addition, there were 'Caledonian' folds, superimposed on the 'pre-Bala' structures, striking northeastwards and also swinging southwards towards the western end of these younger folds.

Did the emplacement of the western granitic rocks of Lakeland cause the folding(s)? Firman took the view that both sets of main folds were there before the granite, but minor monoclinal folds on the northern side of the Ulpha Syncline seemed to have resulted from the pressure of the Eskdale Granite 'shouldering aside' Borrowdale Volcanics. These folds died out eastwards. Considering the form of the aureole, along with other factors, Firman concluded that the Granite was a partly unroofed stock with steep sides. It was, he suggested, intruded after the 'pre-Bala' folding, but before all the Caledonian movements were over, for the granite itself supposedly showed evidence of Caledonian faulting.

However, more data were needed to be sure of this conclusion, and so Firman gathered the help of some 22 undergraduates from the Swinnerton Geological Society of Nottingham University, and interested sixth-form geologists from Norwich, to collect large quantities of information on the structures within the Eskdale Granite, its aureole, and the surrounding Borrowdale Volcanics. With such inexperienced observers it was necessary to standardize the observational procedures precisely, but this was done, and structural generalizations became possible (Firman 1960).

Within the granite, two principal types of joints were recorded: high-angle and low-angle. The former chiefly had a strike approximately to the NW, while the latter struck northeastwards in the northwestern part of the granite and swung round to an approximately southerly or even southeastern strike as one moved west and south. The observed aplite veins did likewise. A similar swing could be discerned in the high-angle joints, the unfaulted granite margin, the margin of the aureole, and also folds and cleavages in the Borrowdale Volcanics further south. So the Eskdale Granite's structures might be related to regional tectonic trends. However, the low-angle joints were cut by the higher-angle joints, and by hydrothermal mineral veins and sheets of microgranite and aplite; and low-angle joints were not found in the neighbouring Borrowdales. So, suggested Firman, the low-angle joints, sometimes slickensided, could be construed as small shear-structures, which was consistent with their being generated as marginal thrusts at the time of the emplacement of the granite. On the other hand, they might have been produced by the central collapse of an intrusion (Balk 1937). Which of these possibilities was correct was to prove an ongoing question for students of Lakeland geology.

Leaving aside here a large number of complexities concerning the joint and cleavage patterns, we can summarize by saying that Firman thought that they could be accounted for by assuming that the granite had been subjected to a major stress directed a little to the east of south, with shearing, wrench or tear faulting occurring in the still-plastic mass, and also rotational strain. The granite and the Borrowdales had seemingly responded differently to the forces acting, for the fractures in the former did not correspond with the folding and cleavage in the volcanics. Firman thought that the granite was emplaced during the Caledonian Orogeny, and that the shear-joints were generated fairly soon after the intrusion. This was consistent with the Devonian age of the Eskdale Granite (383 Ma), determined by John Miller at the Cambridge radiometric laboratory shortly after (Miller 1961).

In the discussion following the presentation of Firman's (1960) paper, Professor W. D. Evans (Nottingham University) pointed out that the author's method allowed him to discriminate between joints produced by contraction in an igneous mass and those produced by 'external' tectonic forces; and it could be accomplished by close examination of only small areas, showing the

[1] There are two main quarries at Shap on the SE corner of Lakeland: the 'Pink' and the 'Blue'. It was at the former, where quarrying is no longer proceeding, that the rock, with its beautiful characteristic large pink feldspars crystals, so frequently seen in tombstones all over Britain, was obtained. The Blue Quarry is still working (1999), the product being chiefly used for road metal.

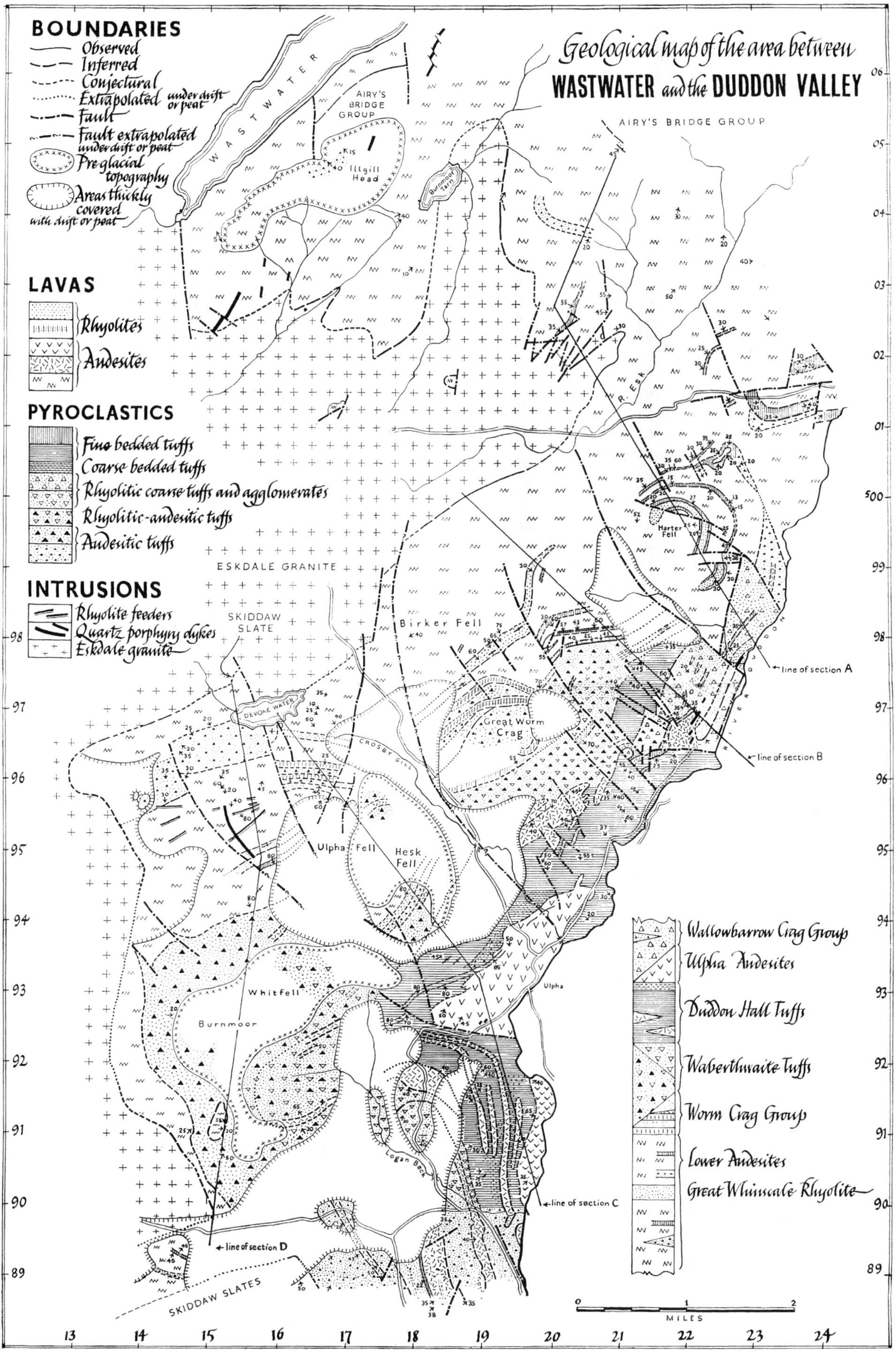

Fig. 7.2. Geological map of the area between Wastwater and the Duddon Valley, according to Firman (1957*b*, Plate 3). Reproduced by courtesy of Dr Firman and the Yorkshire Geological Society.

forces that had formerly been acting in a region. This was a nice boost for the Nottingham efforts.

Firman's contributions to the gravimetric investigations of the area will be discussed later (see p. 156). Let us now turn to consider the work of the late Robin Oliver (Fig. 7.3), who, as already said, was doing his doctoral research in the Lakes at the same time as Firman.

Oliver was a New Zealander, son of the botanist and ornithologist, W. R. B. Oliver, Director of the Dominion Museum, Wellington. He entered Victoria University College, Wellington in 1939, where he took an MSc under the noted geomorphologist Charles Cotton, completing it in 1943. For the remainder of the War, Oliver was stationed at remote Campbell Island, SE of New Zealand, on coast-watch duty. This gave him the opportunity to do some botanical work and prepare the first geological map of the island. Returning after the War, he was briefly employed by the NZ Geological Survey on groundwater studies in Canterbury. During this period, being an adventurous man, he chose (while on holiday) to camp on the edge of the crater of Mount Ruapehu in the North Island, while the volcano was erupting, and got himself charred by hot ejectamenta as a result!

In 1945, Oliver took a job with Shell, and after instruction at The Hague was posted to Venezuela for a three-year contract. This set him up financially, and enabled him to enrol at Cambridge for a PhD. Professor Cecil Tilley, Head of the Department of Mineralogy and Petrology, suggested a thesis topic that involved working on the highest and most demanding terrain that the Lake District could offer, and which had hardly been examined since the days of Ward and Walker. This was a challenge that Oliver welcomed; and although the Lake District has some tough country it is but nothing compared to the mountains of New Zealand. Overall, Oliver told me, he enjoyed his time in the Lakes as he camped in the hills for weeks on end with his young wife, Helen; and as a keen mountaineer and rock climber he took the opportunity to tackle some of the classified routes in the area.

On finishing his degree in 1953, Oliver held an appointment as demonstrator at Oxford for a year under Wager, teaching metamorphic geology. Then, looking for something more remunerative, he joined the Canadian geology and geophysics company, Huntings, and contributed to the mapping of Sri Lanka and parts of Pakistan (where he was briefly kidnapped!), making geological interpretations from aerial photographs taken by Huntings. In 1958, he took up an appointment at Adelaide University, lecturing in sedimentary and metamorphic geology, and developing a special interest in Antarctic geology. When I interviewed him in 2000, he was still working at the University with an honorary appointment. Among other things he was doing consultancy work on accessory minerals – a topic that he started in the Lake District, where he became interested in the problem of garnets in the Borrowdale Volcanics. Oliver suffered a stroke early in 2001, while returning by bus to Adelaide from a geology conference in Melbourne, and died soon afterwards. He had been active to the last, having played cricket and tennis until his mid-seventies.[2]

Oliver's Lakeland publications are chiefly cited today for their solution of the problem of the 'streaky rocks', but while he was engaged on his thesis his principal concern was the question of the origin of the garnets: were they crystallized from a magma or were they the product of metamorphism or metasomatism? Oliver also had to map the Central Fells, and did much geochemical work (for which he was assisted by the use of the spectrometer of Dr S. R. (Bobbie) Nockolds, lecturer at the Cambridge 'Min and Pet' Department). Tilley was Oliver's supervisor, and, Oliver recalled, imparted wise – albeit infrequent – counsel. Oliver visited Mitchell in Edinburgh quite early on and received useful advice and encouragement from that quarter. On the other hand, he did not meet Hartley. The examiners for the PhD were Tilley and Mitchell. Oliver only met Firman on a few occasions, and only once in the field, in the Shap area (Oliver, pers. comm., 2000).

Fig. 7.3. Robin Langford Oliver, in mid-career. Copy of photograph supplied by Mrs Helen Oliver and reproduced by her permission.

Oliver started his three seasons of mapping from Seatoller in Borrowdale (see Fig. 2.3). Then he camped at the head of Eskdale; then at Wasdale Head; and finally in Langdale (see Figs 7.4 and 16.1). For this reason, he did not distinguish the important Whorneyside Tuff, the type locality of which is in upper Langdale, below Bowfell, which subsequent geologists have found so important as a marker band separating the effusive andesitic lavas of the lower Borrowdales and the overlying, often welded, pyroclastics of the upper Borrowdale Volcanic Group (see p. 217). Oliver's early fieldwork involved a study of the rocks around Great Gable (see Fig. 7.4), which are good for rock climbers, but had hitherto deterred close geological investigation.

Oliver hoped that aerial photographs would help him recognize the presence of faults, but the method proved disappointing, only the most distinct and obvious fractures showing up on the photos then available. His practice was to make a tracing of the topography for a day's work from a six-inch map and enter the geological information on this, transferring the information to his manuscript map in the evenings or back at Cambridge. He did traverses across the strike and also followed units laterally. He gauged his altitude with the help of an aneroid barometer. He did not consult the old field-slips of the primary survey.

Regarding the origin of the garnets, there were several competing hypotheses in the literature. Sorby (1880) had thought that they were original constituents in the lavas, or pyrogenic. Marr (1900) took the same view concerning the garnets in the 'streaky rocks', but later supposed that the latter as a whole were of metamorphic origin. Harker (1902) maintained that the garnets

[2] See Mason (2001) and Cooper (2001). I last met Robin Oliver in December 2000, at the 4th International Mineralogy and Museums Conference in Melbourne.

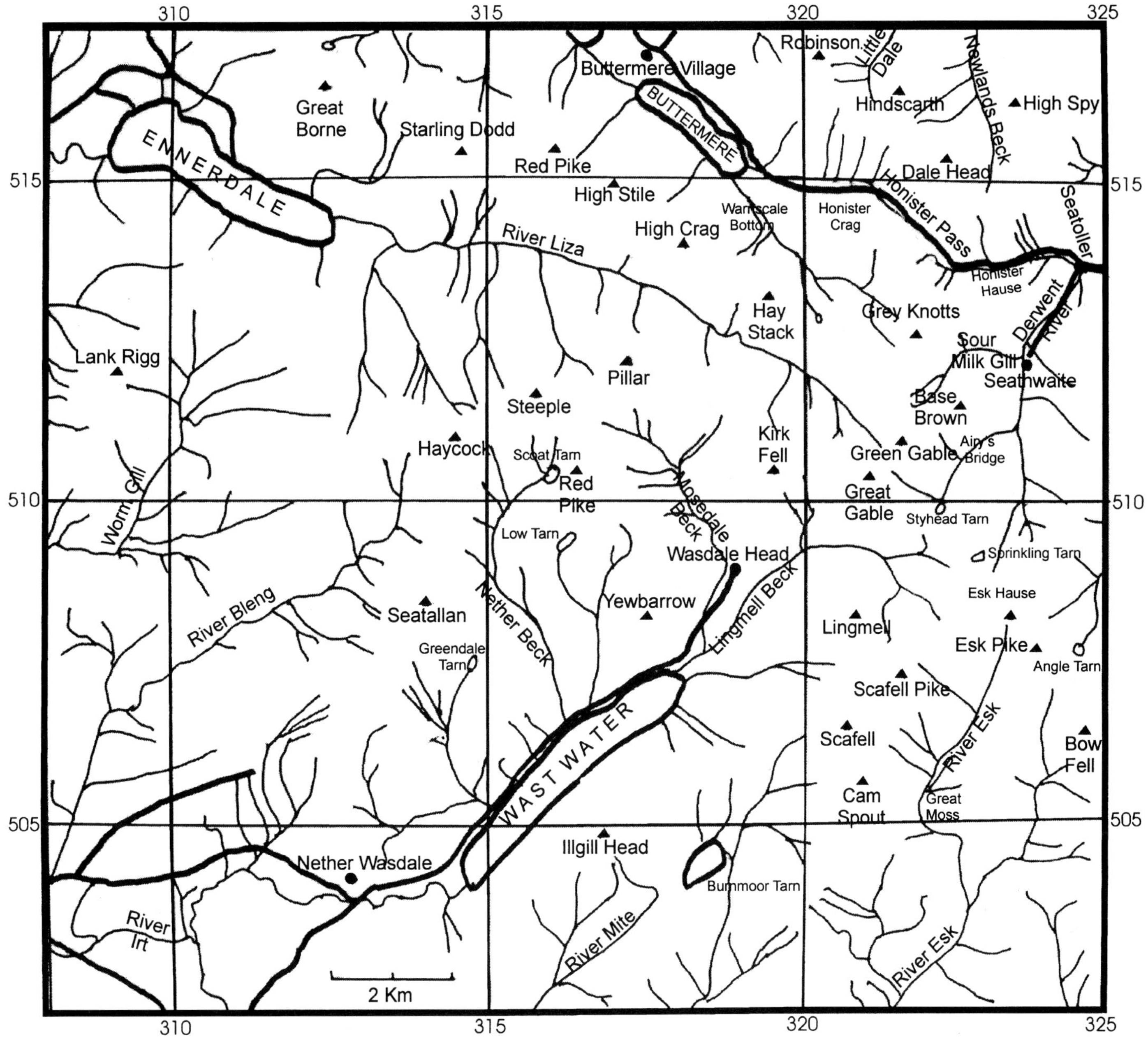

Fig. 7.4. Topography of the western part of the Central Fells, Buttermere, Ennerdale, and Wast Water.

were primary in intrusive rocks but secondary in the tuffs, but (in discussion of Walker 1904) he subsequently entertained the possibility that the garnets were the product of regional metamorphism and he was supported by Hutchings in this view. Walker himself thought the garnets were crystallized from magma. Green (1915), Hartley (1932), Hancox (1934) and Hadfield & Whiteside (1936) had supposed that they were formed by the action of circulating solutions under high pressure, or were of solfataric origin. Garnets were also of interest to Firman, as discussed above. One can readily see, then, why Tilley gave Oliver his research topic.

Oliver's thoughts on the problem were influenced by considerations of microscopic textures particularly. He found a tuff on Rosthwaite Fell (see Fig. 16.1) with a garnet at the edge of an igneous rock fragment, with the fragmentation edge cutting through the garnet crystal, which suggested that the crystal had been formed before the igneous fragmentation process that gave rise to the tuff, rather than by some metasomatic or metamorphic process (Oliver, pers. comm., 1998). Elsewhere, he found garnets surrounded by coronas of plagioclase, suggesting igneous origin. We saw in Chapter 6 (pp. 75 and 81) that Walker and Green gave different explanations of the coronas. Oliver sided with Walker, thinking that if Green were right there would be indications of garnet reaction zones between iron ore and feldspar. For Oliver, the garnets had served as nuclei, around which the feldspar crystals had formed during crystallization from the melts.

Further, Oliver noted indications of flow around garnet crystals, for example in a rhyodacite from High Gait Crags (see Fig. 7.5a). This suggested that the garnets formed before solidification of the lava; and a slide showing garnets in a biotite–quartz porphyry from Scafell (see Fig. 7.5b) suggested that the quartz and garnet crystals had formed simultaneously, since they were euhedral except at the contact between the crystals. Since the quartz was evidently of magmatic origin, it seemed that the garnet was formed similarly.

Many of Oliver's arguments (Oliver 1954*a*, *b*, 1956*a*, *b*, 1961) were concerned with chemical and spectrographic analyses of his rock samples and extracted minerals. Variation diagrams were prepared in which the proportions of major constituents (alumina, magnesia, etc., and trace elements) were plotted against silica, as pioneered by Harker (1909). The data revealed iron enrichment

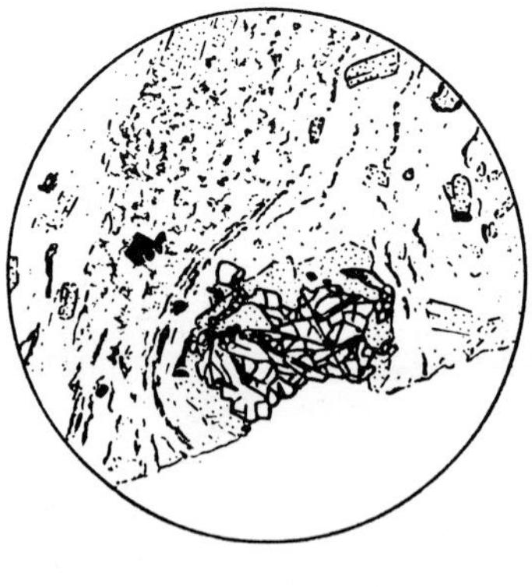

Fig. 7.5a. Garnet in rhyodacite, according to Oliver (1956*a*, p. 127). Reproduced by courtesy of Cambridge University Press.

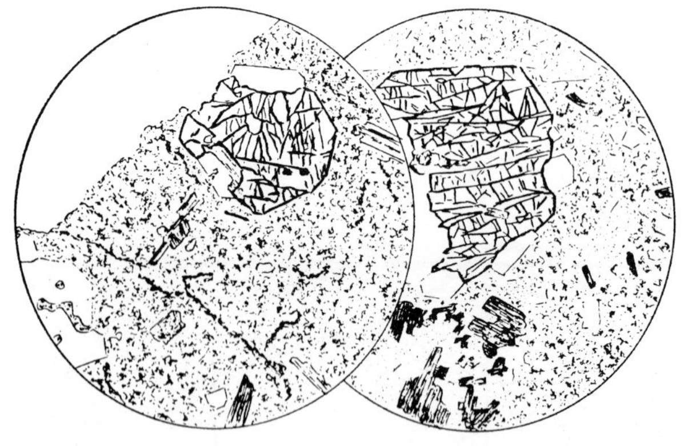

Fig. 7.5b. Garnet and quartz in biotite quartz-porphyry, according to Oliver (1956*a*, p. 128). Reproduced by courtesy of Cambridge University Press.

for the rocks of lower silica content, which suggested magmatic differentiation. An increase in the secondary minerals such as epidote and amphiboles as one neared the Eskdale Granite was thought to reflect the influence of the occurrence of a mainly hydrous phase emanating from the Granite.

Magnesium, iron, manganese and other elements were also determined for the garnets, the composition of which varied somewhat through the Borrowdale succession. Garnets in the lower andesites appeared to have the highest magnesium content, and the magnesium/iron ratio in both the garnets and the containing rock decreased as one ascended the sequence, whereas the manganese concentration increased. Oliver had the idea that these changes might be related to chemical differentiation in the magma chamber and wrote a paper on the topic; but Nockolds thought the evidence insufficient to justify this conclusion and the idea was not pursued in the publications that flowed from the PhD (Oliver, pers. comm., 1998).

On the basis of his own chemical analyses and those of others, however, Oliver's (1961) data showed minor iron enrichment for rocks with lower silica content suggesting that there had been alteration by assimilation, perhaps supplementing magmatic differentiation; and the increase in secondary minerals such as epidote and amphiboles as one neared the Eskdale Granite indicated that this large mass had produced subtle chemical changes in the country rock.

However, Firman was not convinced about the magmatic origin of the Lakeland garnets. In a letter to the *Geological Magazine* (Firman 1956), he claimed that almandine garnets had not been found in modern lavas; that such garnets were extremely rare in any lavas; that they had not been synthesized artificially under conditions that might be expected to prevail in andesitic lavas, or crystallized from melts of andesitic composition; and that the garnets were not found in many Borrowdale andesites that otherwise resembled the ones that did contain garnets. At the time these were arguments that carried some weight. Firman (1956, 1957*b*) suggested that the Lakeland garnets (almandine) (i.e. other than the andradite he had studied in the veins around Shap) might be restricted to formations containing welded tuffs (now coming into 'fashion' following Oliver's publications: see p. 100).

Oliver (1956*b*) responded to Firman's letter of 1956 in a missive from Sri Lanka, where he was by then working. He pointed out that almandine had in fact been synthesized successfully at high pressures. Further, some Australian work by Melbourne University's A. B. Edwards (1936) suggested that almandine might form in a melt if that was 'contaminated' by aluminous material; and in the Lakeland case such matter might readily have been derived from the SS. Such a source was compatible with the 'patchy' distribution of garnets in the Lakeland andesites. As things turned out, later workers have sided with Oliver on the question of the origin of Lakeland garnets (see p. 141), and as will be mentioned later (see p. 105) Firman himself subsequently came round to the view that they were pyrogenic.

As to stratigraphy, Oliver (1953, 1954*a*, 1961) recognized the following units, which are compared here with ones previously suggested by Hartley (1932) and Mitchell (1934):

Oliver	*Hartley and Mitchell*
–	Upper Andesites
Esk Pike Hornstone[3] Lincomb Tarns Formation[4] Seathwaite Fell Tuffs[5]	Coarse Tuffs; Felsitic and Basic Tuffs
–	Wrengill Andesites
–	Kentmere Pike Rhyolite; Intrusive Rhyolite
Airy's Bridge Group[6]	Bedded Tuffs
Birker Fell Andesite Group; Andesites	Dow Crag Andesites/Mosedale
Andesites (Grey Knotts Type[7])	

It should be noted that these successions were established for Central and Eastern Lakeland respectively. So the absence of a unit in the foregoing lists did not necessarily imply unconformity, or conflict of judgement. When Oliver's units were mapped, they yielded what we see in Figure 7.6, which was the same map as that accompanying his thesis, except for the units being represented uncoloured. His estimated thickness for the Borrowdales in the Central Fells was 9160 to 10 210 feet, getting back to the value that Marr had envisaged.

Oliver's (1954*a*) published unit names, as above, have all survived in the 1:50 000 Ambleside map (British Geological Survey 1996; see Plate VII), and in the same order, though naturally there have been subsequent subdivisions of the units in the more recent publications. His published section is reproduced

[3] Esk Pike is on the high ground between Bowfell and Great Gable.

[4] Lincomb Tarn is near the ridge between Allen Crag and Glaramara.

[5] The Seathwaite Fell occupies ground south of Seathwaite, upper Borrowdale, in the angle between Styhead Gill and Grains Gill (see Fig. 16.1).

[6] Airy's Bridge was, in the 1950s, a stone pack-horse bridge across the river along the track from Sty Head Tarn to Seathwaite (at 3224 5103). Today there is only a wooden structure and the name has disappeared from recent maps (see Fig. 7.4).

[7] Grey Knotts is on the ridge SW of Honister Hause towards Brandreth (see Fig. 7.4).

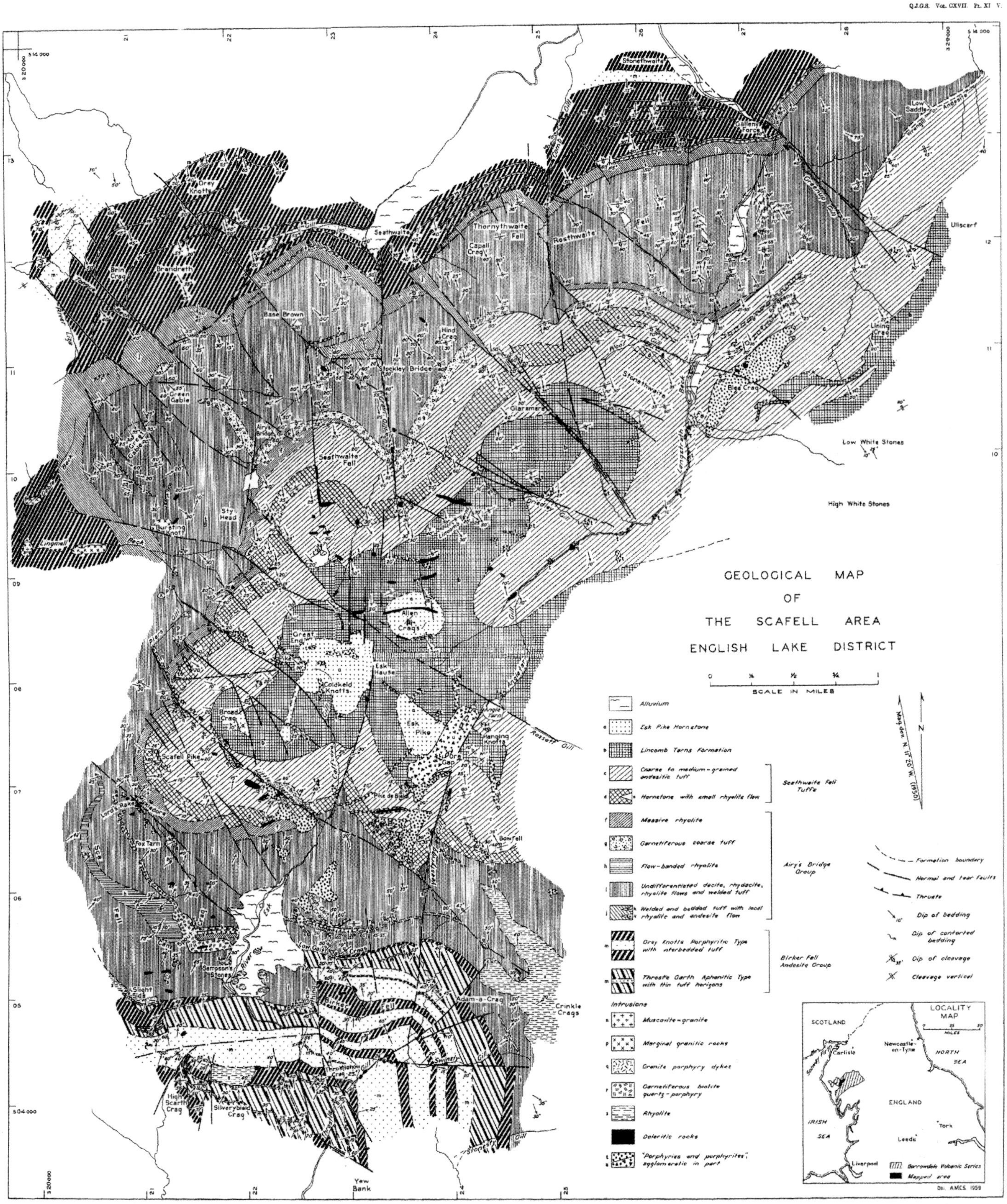

Fig. 7.6. Geological map of the Scafell area, according to Oliver (1961, Plate XIV).

in Figure 7.7, which extends from Stonethwaite in Borrowdale (see Fig. 2.3) SW towards Glaramara (see Fig. 16.1). It was with Oliver's revision and redefinition of the stratigraphic units for the Central Fells in his 1954 paper that the modern period for the study of geology in this important part of the Lakes was initiated.

Oliver (1954*b*) also made a major contribution to Lakeland geology by publishing a paper on the nature of the mysterious 'streaky rocks', such as are particularly common in the Lincomb Tarn and Airy's Bridge Formations. These were construed as *ignimbrites*, that is indurated tuffs consisting of various crystals and

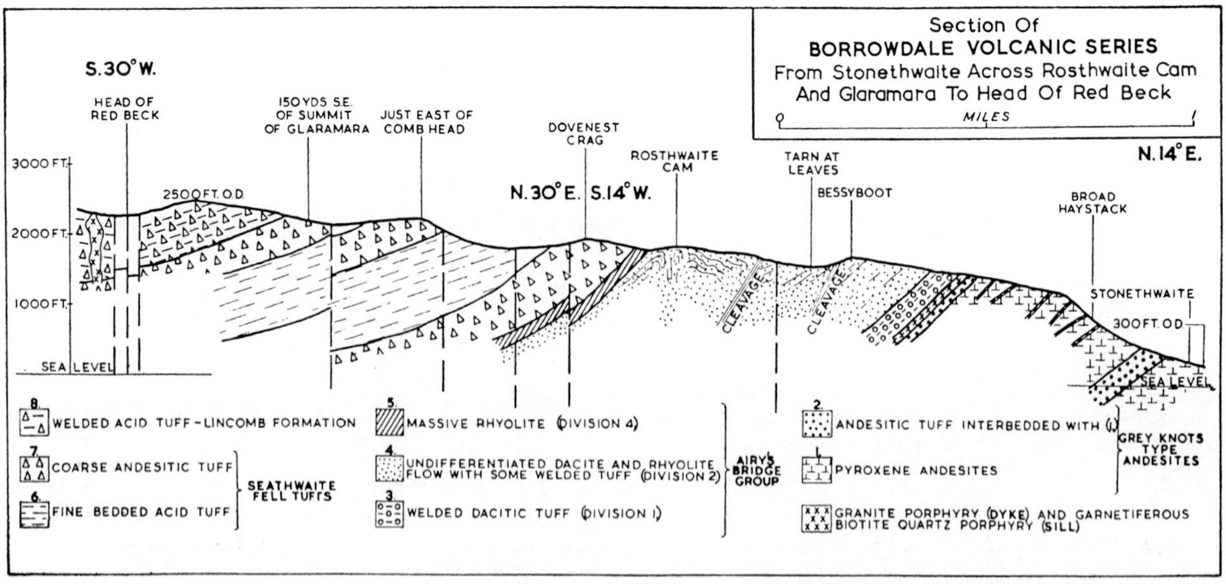

Fig. 7.7. Profile showing the succession of Borrowdale Volcanics, Stonethwaite to Red Beck, according to Oliver (1954*a*, p. 409). Reproduced by permission of Helen Oliver and the Geologists' Association.

fragments embedded in a matrix of glass shards ejected by a *nuée ardente*, or cloud-like incandescent mass of vapours and solid volcanic fragments thrown out of a volcano. As a result of such an eruption, the glowing fragments may partly fuse together while still in motion. Such rocks typically contain lenticular structures (*fiamme*) produced by fragments of pumice being partly flattened and elongated by the weight of the overlying parts of the deposit, and which in cross-section give the impression of flow.[8]

The term *nuée ardente* was first applied scientifically to an eruption that occurred in Martinique in 1903, destroying the capital, Saint-Pierre (Lacroix 1903). Subsequently, the rocks produced by such clouds were described in New Zealand's North Island by Patrick Marshall (1935), and soon became quite familiar to New Zealand geologists. One might suppose, then, that as the first New Zealander to undertake extended work in the Lakes, Oliver would have been the person who recognized the 'streaky rocks' as ignimbrites. But it was not quite like that. Oliver informed me (pers. comm., 1998) that when he first saw these rocks he judged them to be rhyolitic or rhyodacitic lavas, exhibiting flow banding or flow layering. Then one day, perhaps in his second season of Lakeland fieldwork – he could not quite recall the date – he was joined on a field excursion by a fellow New Zealand geologist also at Cambridge, (Maurice) Hugh Battey,[9] and it was he who first remarked on the striking similarity between the Lakeland Ordovician 'streaky rocks' and Marshall's New Zealand ignimbrites.[10] However, it was Oliver (1954*a*, *b*) who researched the literature on ignimbrites and published the details of the new nomenclature.

Thus a major puzzle about Lakeland petrology was solved at last, relatively easily. Problems naturally remained. The 'streaky rocks', if ignimbrites, would have been expected to exhibit rapid changes in thickness, as they might flow down one side of a volcano but not another. So it would not be easy, and sometimes impossible, to trace continuous layers laterally. Association with marine fossils was unlikely. There might be caldera collapses and concomitant faulting. Such issues were not flagged by Oliver, but knowing that the 'streaky rocks' were ignimbrites at least took away their 'mystery' and made possible their incorporation – and that of the Lakeland rocks more generally – into evolving volcanological theory. Naturally such developments took time.

As said, Oliver proposed a new stratigraphic succession for the Central Fells. He also proffered ideas about the structure of the area, giving attention to faults, folds, cleavage and joints. A broad shallow syncline was apparently confirmed, running approximately SW–NE through Scafell, Esk Hause and High White Stones (see Figs 7.6 and 16.1), which linked up with a similar structure figured by Hartley (1941); and several approximately parallel folds were revealed on the northern limb. Oliver followed Mitchell and others in proposing a Devonian age for these structures. Broadly speaking, the cleavage orientation was found to be compatible with the notion that it was produced at the same time as the folding. The orientations of 581 joint planes were determined and their poles plotted on a stereogram (see Fig. 7.8).[11] There was, as can be seen, a preponderance of steeply inclined joints, chiefly striking a little east of north and a little north of east. The gently inclined joints, represented by the contour line near the centre of the figure, were found to be commonly quartz-filled. Oliver suggested that the gently and steeply inclined joints might have been produced by tension and shear respectively, and that the former fractures remained more open and thus able to be penetrated by silica-bearing solutions.

Oliver's paper (1961) was well received at the Geological Society and generated considerable discussion. However, not all members of the audience were persuaded by his argument about the magmatic origin of the garnets. Kingsley Dunham, an authority on ore bodies who, as we have seen (Chapter 4), had done extensive work on the haematite deposits of west Cumbria, suspected that geophysical evidence might one day reveal the presence of a much larger granite mass under the Borrowdales than had previously been proved by mapping (a prescient suggestion) and therefore suggested that the garnets might in fact be of metamorphic origin, not magmatic.

---

[8] Not all ignimbrites display fusion and welding, however, and welding is likely to be more pronounced at the bottom of a flow than at the top, where the pressure is less. There is a range in the degree of welding, but all are thought to be formed from *nuées ardentes*. See p. 220.

[9] Battey later took a post at King's College, Newcastle, where he taught mineralogy and petrology.

[10] Actually, a particular variety thereof, called 'owharoite' by Marshall, from the locality of Owharoa in New Zealand.

[11] For discussion of stereograms, see p. 121.

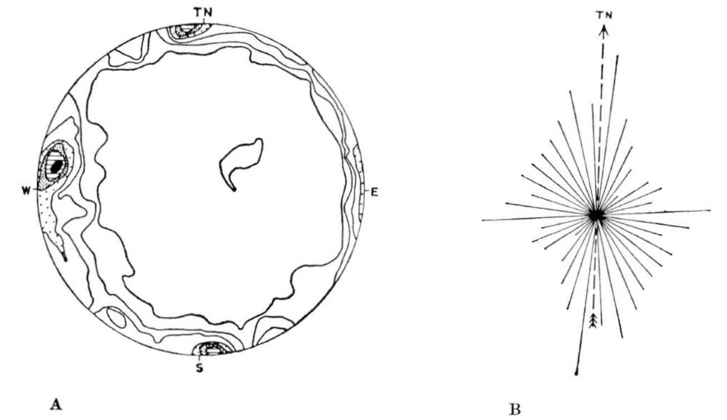

Fig. 7.8. Contoured stereographic diagram showing orientations of joints in rocks of Scafell area, according to Oliver (1961, p. 386).

The mineralogist Roger Strens, then a PhD student at Nottingham University working on 'The geology of the Borrowdale–Honister district (Cumberland) with special reference to the mineralisation' (1962), who later taught 'Min and Pet' at Cambridge, wondered whether the occurrence of two generations of epidote postulated by Oliver, the first due to hydrothermal alteration in the volcanic pile and the second to emanations from the Eskdale granite, might not be manifested in compositional differences in the two generations. Strens himself liked the idea of the formation of epidote and secondary feldspar by the action of soda and potash-rich magmatic liquids acting on anorthite.

Sydney Hollingworth (President) expressed 'surprise' that the large variations in dip over Oliver's map were not manifested in changes of orientation of outcrop boundaries, corresponding to folds.

Ronald Firman, who actually presented the paper for Oliver since he was overseas at the time, expressed views somewhat similar to those of Dunham, and also queried Oliver's proposed correlations from the Central Fells SW towards the Ulpha area in the Duddon Valley (where Mitchell was doing his last major piece of Lakeland mapping, and where Firman had also been mapping, as we have seen).

Oliver responded in writing to the foregoing comments, stating that the grade of metamorphism was too low, overall, to be compatible with the idea that the almandine garnet was of metamorphic origin. His more detailed maps and sections showed how outcrops were related to structure. He acknowledged that his correlations with the Duddon area were based on Mitchell's earlier work and might need revision. He had not done enough work on the epidotes to distinguish two generations on the basis of compositional differences to answer Strens's query. However, he maintained his opinion on the garnet question, and, as mentioned above, the balance of opinion on this question subsequently shifted in his favour.

Let us now focus attention on Mitchell's later work. He published several Lakeland papers in the latter part of his career (Mitchell 1956*a*, *b*, 1957, 1963) and a Geologists' Association guide to the Lakes (Mitchell 1970). Mitchell's son, Murray, told me how his father was determined to complete his mapping of the southern portion of the Borrowdales (pers. comm., 1998). This he did with papers (accompanied by maps) on the Dunnerdale Fells (1956*a*), i.e. in the region between the Coniston–Broughton road and the Duddon Valley and Ulpha; and on the Seathwaite Fells (1963), which carried the 1956 map northwards past Seathwaite Tarn, to the west of Coniston Old Man, and to where the Duddon River divides at Cockley Beck Bridge (see Figs 5.2 and 16.1).

There are certain significant differences between Mitchell's pre- and post-war work. He began to relinquish the goal of establishing a general sequence for the Borrowdales all the way from (say) Kentmere in the east to the Duddon Valley in the west (though indications of this goal remained in his later work). Rather, there was recognition that there might be similar volcanic rocks cropping out at separate localities, which were not necessarily produced by the same volcanic episode(s), and so could not be correlated, despite the similarities. Thus his correlation diagram (see Fig. 7.9) was significantly less precise than the ideas implicit in his earlier work, undertaken under the influence of Green. Mitchell also acknowledged that there could be considerable lateral variation in specific units. So 'repetition' was not necessarily an indication of either faulting or folding, though these were naturally not excluded.

Even so, Mitchell had not, in 1956, wholly lost sight of the wish to establish E–W correlations of the Borrowdale Volcanics across the Lake District, as is evident in a diagram from the very same paper (see Fig. 7.10). His geological map of the Dunnerdale Fells (see Fig. 7.11) lives on to some extent, even in the modern maps such as the Ambleside Sheet (1996; see Plate VII).

Mitchell's last major Lakeland paper (1963) developed the aforementioned tendency of his 1956 papers. Also, some of the previous units now became subdivided groups, so that the proposed succession was as follows:

Wrengill Andesites
Tilberthwaite Group
Lickle Rhyolites
Dunnerdale Group
Cockley Beck Group

As might be expected, further work in a different area generated different and more precise stratigraphic units. The estimated overall thickness for the Borrowdales of the Seathwaite area was now given as 13 550 feet, a value reminiscent of Marr's ideas, re-emerging with the decline of Green's notion of anticlinoria.

Mitchell, of course, continued to address problems of correlation and synonymy. He thought his Cockley Beck Group might correspond to Firman's (1957*b*) Lower Andesites, and to Hartley's (1932) Mosedale Andesites (Mitchell 1963, p. 296). But in the main, Mitchell worked out his own stratigraphic sequence. There was no one else doing comprehensive mapping of the area at the time other than a PhD student of Hollingworth at University College, London, Jacques Konig (1964), who was working on the 'Middle Tuffs' in the Coniston, Tilberthwaite, Kirkstone Pass and Langdale areas, and who obtained assistance from Mitchell and Firman.

Mitchell's correlation difficulties highlighted the problems encountered by Lakeland geologists in the 1960s. Though they met for organized field excursions, and encountered one another in the field from time to time, or on other occasions, they were essentially working single-handed (Firman's use of a quite large team of relatively inexperienced observers being very much the exception rather than the rule); and with such a tangle of igneous rocks it would have been almost miraculous if the different classifications had meshed perfectly. Konig (1964, p. 20) recorded the fact that the maps of Hartley and Mitchell were not linked. As recounted to him by Hollingworth, both geologists were in the field at the same time but neither knew (or took the trouble to ascertain) the extent of the other's proposed mapping limits. Early PhD students such as Firman and Oliver were able to spend continuous periods in the field, but they did not meet in advance of their mapping and agree upon a unified stratigraphy before setting to work (though Firman recalls that he discussed matters of stratigraphy with Oliver when he visited Cambridge to examine Harker's slide collection for his Shap work). A further obstacle to the establishment of a unified stratigraphy lay in the fact that the early students were not supervised by lecturers who were themselves experts in Lakeland geology. It was only as late as 1982, when the systematic Survey revision of Lakeland geology began (see Chapter 14), that a synthetic overview of the geology began to be developed, with regular meetings held between the field

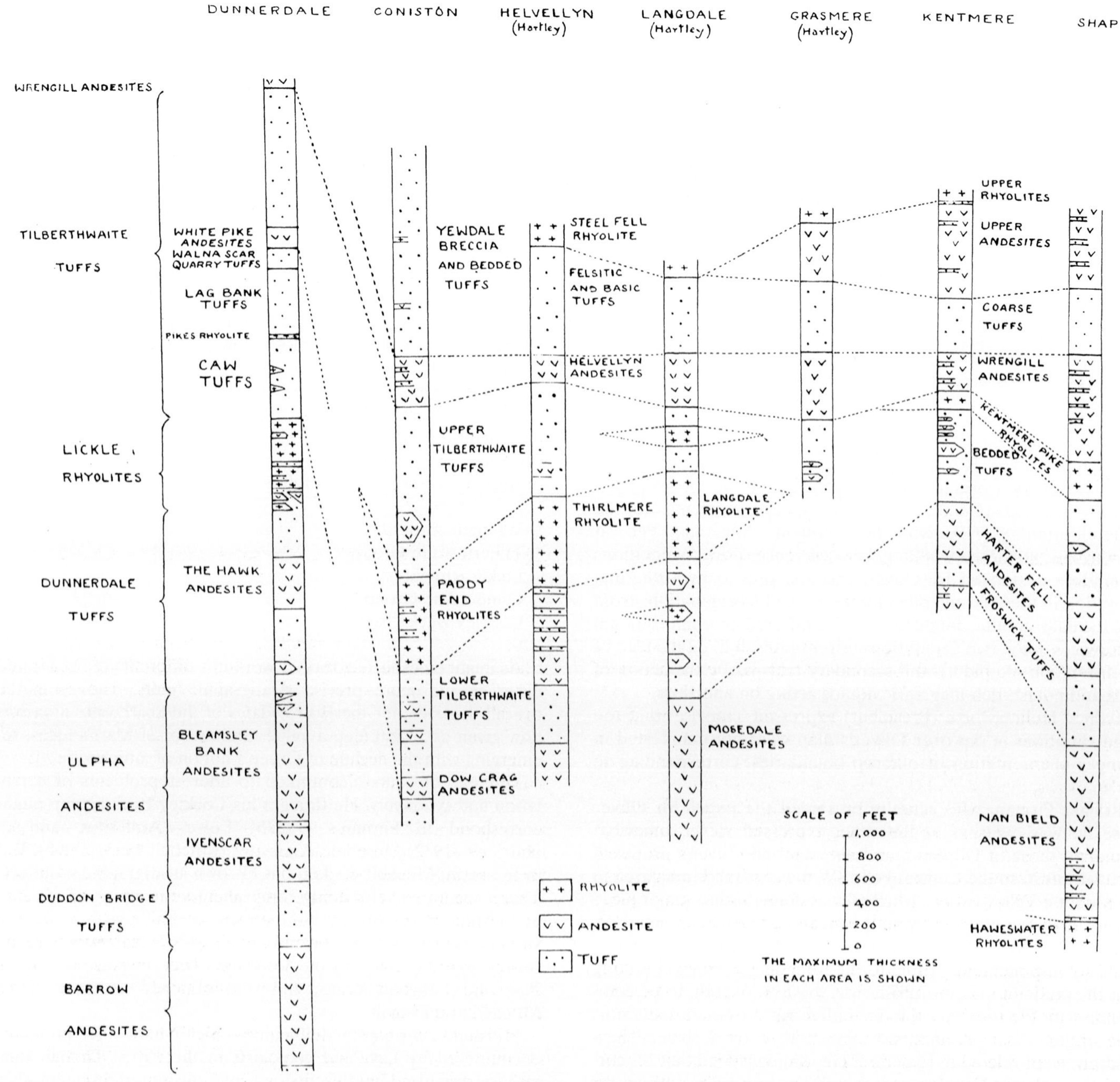

Sections showing suggested correlations of Borrowdale Volcanic Series.

Fig. 7.9. Correlation of Borrowdale Volcanics, according to George Mitchell (1956a, Plate XXXVIII).[12]

officers to iron out problems of correlation, etc.[13] Even at the end of the twentieth century, not everything was settled (see Chapters 13, 15, 16 and 17).

We have little information about the relationships between 1950s workers such as Mitchell and earlier geologists such as Hartley or Nicholas. Murray Mitchell does recall (pers. comm., 1998) that when he was walking with his father in the Langdale area in the 1940s, Mitchell had Hartley's map with him and appeared to be utterly perplexed, being unable to make sense of the Borrowdales of that region, as mapped by Hartley. Perhaps it was the 'liberties' that he took, mentioned by Wadge (see p. 86), that caused the problems.

It is interesting to consider what factors may have led Mitchell to move away from the 'Green' view of the structure of the

[12] In the text of this paper, Mitchell did suggest that the Lickle Rhyolites might be correlated with the Haweswater Rhyolites. The units 'Paddy End Rhyolites', 'Tilberthwaite Tuffs' (Upper and Lower), 'Yewdale Bedded Tuffs' and 'Dow Crag Andesites' had been introduced in Mitchell (1940). In this paper, Mitchell (1940) also used the term 'Yewdale Breccia', though that first appeared in a section in Ward (1879, p. 54) (who referred, incidentally, to 'Tilberthwaite Slates'). Mitchell (1956) gave 'White Pike Andesites', 'Dow Crag Andesites', 'Walna Scar Tuffs', 'Lag Bank Tuffs', 'Caw Tuffs', 'Lickle Rhyolites', 'Dunnerdale Tuffs', 'Ulpha Andesites', 'Duddon Bridge Tuffs' and 'Barrow Andesites' as newly named units, thus substantially revising the older work of Green for SW Lakeland.

[13] However, before then Frank Moseley and his students were beginning to develop a stratigraphic synthesis, and his edited volume (Moseley 1978*a*) evidenced an emerging consensus in the 1970s.

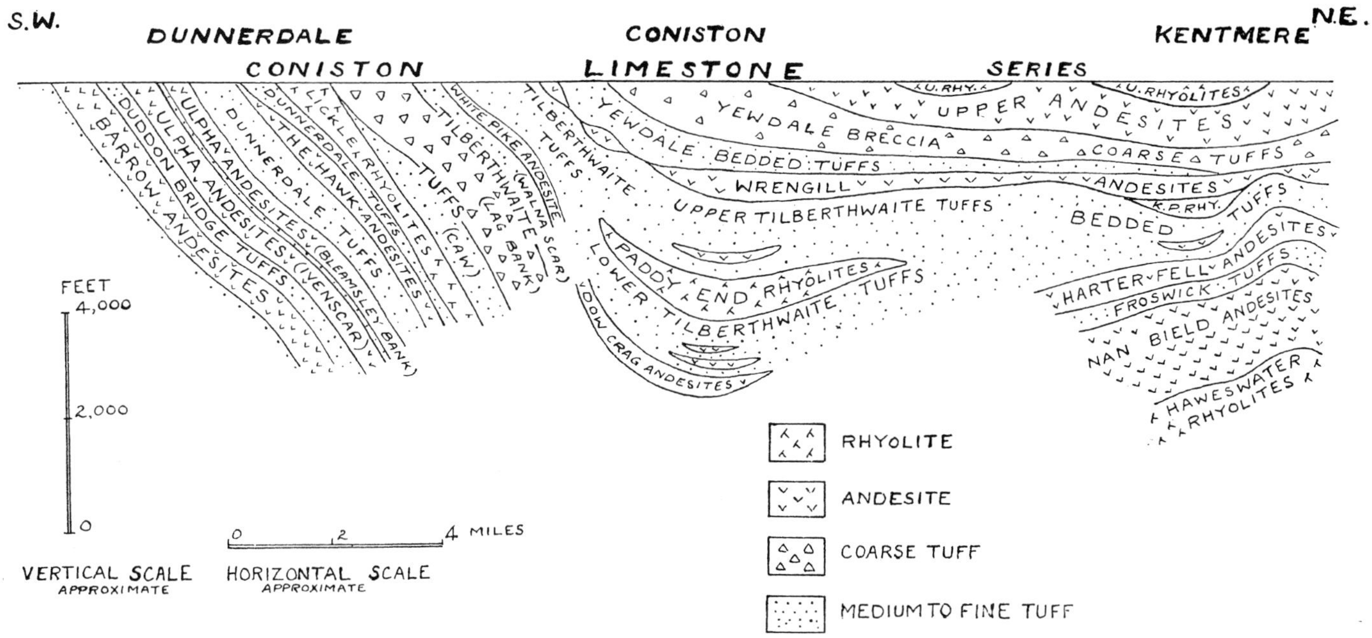

Fig. 7.10. Profile of Borrowdale Volcanics from Kentmere to Dunnerdale, according to Mitchell (1956*a*, p. 447). © John Wiley & Sons Ltd. Reproduced by permission.

Borrowdale Volcanics, as expressed in his earlier papers, such as Mitchell (1940) on the Coniston area. Murray Mitchell recalled to me that he went out in the field with his father in the seasons 1949, 1950, 1951 and 1952, during the mapping of the Dunnerdale Fells and down towards Broughton-in-Furness and Duddon Bridge. He did not remember his father discussing the issue of the folding of the Volcanics, but recalled that the area between Duddon Bridge and White Pike was important to his thinking. Perhaps Mitchell's views changed when he began to see the results of his mapping unfolding (Murray Mitchell, pers. comm., 1999)? G. H. Mitchell discussed the issue somewhat in his survey of Lakeland geology given in his Presidential Address to the Yorkshire Geological Society (Mitchell 1956*b*, pp. 425–426), where he stated that his remapping of the Dunnerdale Fells had not revealed the anticlinorial folding previously envisaged for the Borrowdales by both himself and Green. By the 1950s, then, it was beginning to become apparent that the andesites simply died out at the ends of their flows, rather than disappearing from view by folding. Also, Mitchell acknowledged that he had previously conflated the Lickle and Paddy End Rhyolites at different localities, thinking that it was the same unit repeated by folding. The precise thought processes that occurred as he changed his views cannot now be recovered. The change presumably preceded the contributions of Firman and Oliver, given the dates of Mitchell's post-war Dunnerdale fieldwork. Sir Kingsley Dunham, writing a note on Mitchell's Borrowdale work in Sir James Stubblefield's obituary of Mitchell (Stubblefield & Dunham 1977, pp. 375–378), thought that the change of heart was indicative of Mitchell's theoretical open-mindedness.

Looking forward from the 1950s, it is interesting to examine Mitchell's two published maps for 1956*a* and 1963, and compare them with the 1990s maps of the Ambleside and Ulverston sheets, which cover the area that Mitchell was dealing with in his late papers. Considering Mitchell's map of 1956 and its accompanying section (see Fig. 7.11), we can see that he envisaged an essentially simple structure with units forming part of a syncline plunging to the NE. His overall structure was complicated by faulting and there was some consequent repetition of beds, notably the Tilberthwaite Tuffs. This repetition has been preserved in the modern map, though the two western outcrops of Mitchell's Tilberthwaite Tuffs are assigned to the Caw Tuffs, and the eastern one is placed in the Seathwaite Tuffs of Oliver (1954*a*). It will be noted that the central part of Mitchell's map was an area of complex faulting, and the work of the late twentieth-century survey (i.e. the Ambleside map) differed substantially from Mitchell's single-handed effort, though there was a general similarity thereto. However, as we shall see (p. 228), the field men who made the Ambleside map did not agree among themselves about some aspects of the mapping of this area, even at the end of the twentieth century and after the official map was published. So it is scarcely surprising that one cannot reconcile Mitchell's map with the modern product in some places. There is no disagreement, however, that in this part of the Borrowdales there is a swing in the strike towards the line of the unconformity indicated at the bottom of the map.

In his first 1956 paper, Mitchell referred to 'flow-banding' in the Dunnerdale Tuffs, and mentioned that Oliver thought that such rocks were 'formed by the deposition from a *nuée ardente*'. This was hardly a ringing endorsement of the ideas of Oliver and Battey, though Mitchell was conversant with them, since he had examined Oliver's thesis.[14] However, Mitchell's 1963 map did show a 'Mainly welded tuff' in his Tilberthwaite Group. Comparison with the modern map reveals that he was referring to what Oliver had designated the Airy's Bridge unit, which contains many ignimbrites.

In a second paper Mitchell (1956*b*) gave an important overview of the geology of the whole region, with valuable historical synopses for the different areas and rock types. He suggested that the Caledonian (Devonian) Earth movements involved pressures directed from the south, an idea reflected in his N–S sketch section for the Lake District (see Fig. 7.12) and still favoured today (see Chapters 10 and 17), though some of the Lakeland folds are today thought to be sag rather than compressional structures (see Fig. 16.7), and southwards-directed thrusting is also envisaged for the northern Lakes (see Chapter 15 and Plate V). Figure 7.12 is of special interest in that it suggests an essential unity to the

[14] Firman has pointed out to me (pers. comm., 1999) that the flow-banding in Lakeland lavas may sometimes be difficult to distinguish from elongated *fiamme* in ignimbrites; and reading a thesis is something different from looking at rocks together in the field. According to Frank Moseley (see p. 133), Mitchell was never very enthusiastic about the idea of ignimbrites in the Lakes (pers. comm., 1996).

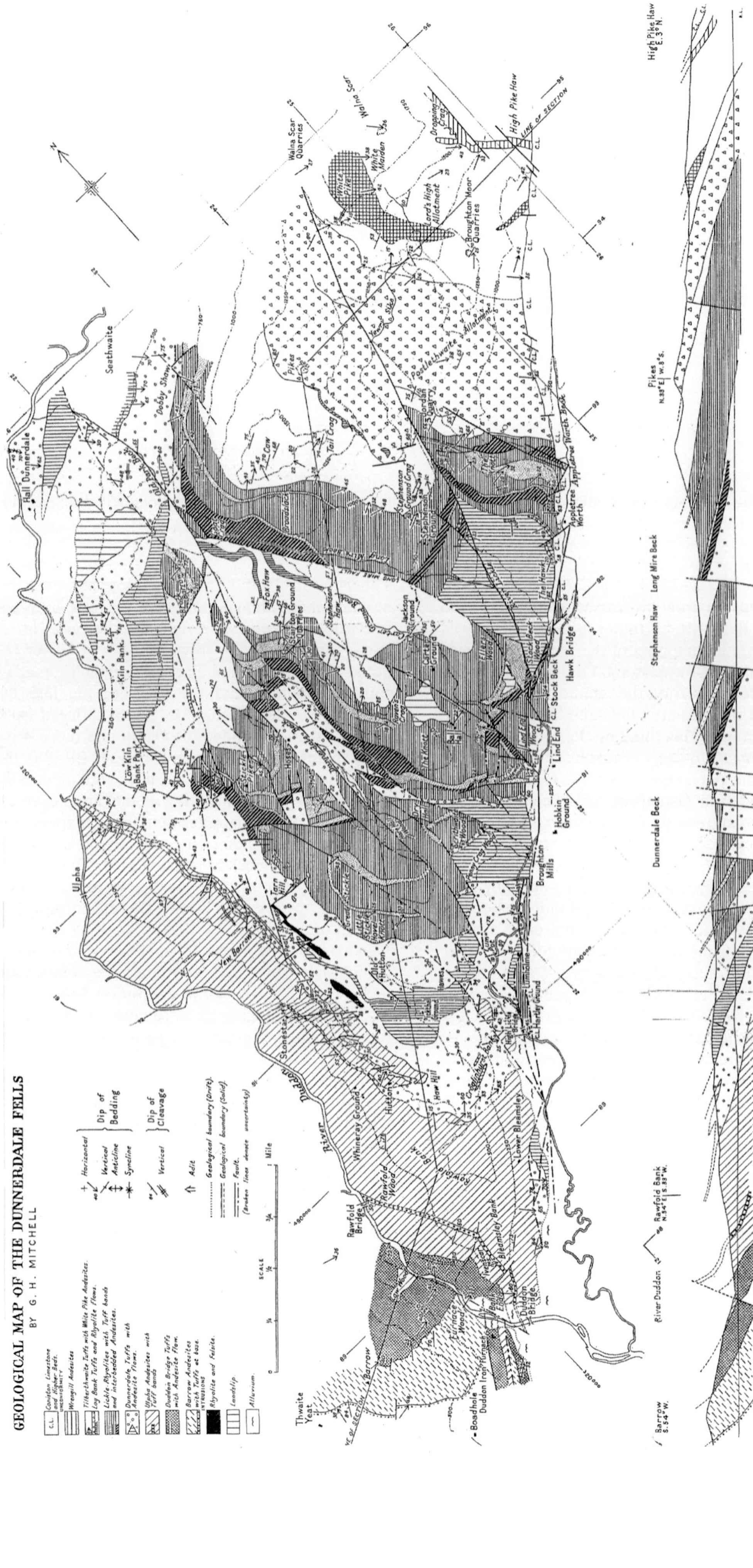

Fig. 7.11. Geological map and Section of Dunnerdale Fells, according to Mitchell (1956a, Plate XXXIX). © John Wiley & Sons Ltd. Reproduced by permission.

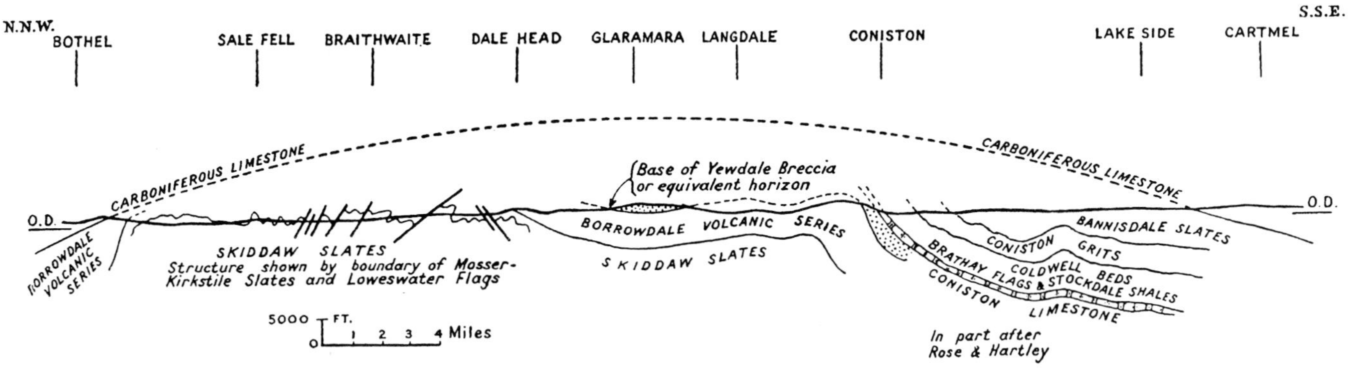

Fig. 7.12. General structure of the Lake District, as seen in N–S section, according to Mitchell (1956*b*, p. 443). Reproduced by courtesy of Murray Mitchell and the Yorkshire Geological Society.

Borrowdale Volcanics, which were only moderately folded, because of their competence, with the unit seemingly extended right across to northern Lakeland, where they appeared in the guise of the distinctive porphyritic Eycott rocks. By contrast, the Skiddaw rocks had become crumpled and faulted, and so too had the rocks of the Coniston Limestone Series, whereas the overlying tough Coniston Grits had formed broad folds, such as the Bannisdale Syncline. There was a further N–S connection in the occurrence of the supposedly downfaulted inlier of Drygill Shales in north Lakeland, thought to be associated with 'Otley III' rocks as Marr had previously suggested (see p. 68). The connection was mentioned in Mitchell's text, though it is not evident in Figure 7.12. It may be noted that the Scafell Syncline (at the centre of the diagram) seemed to have been formed in the Caledonian Earth movements along with the other structures, and was suggested to be complementary to a broad anticlinal structure developed in the SS. Mitchell's figure does not display an unconformity below the Coniston Limestone, but he made it clear that he accepted such a structure, and the horizon was also thought to be the site of extensive strike faulting.

Before we leave the work of these three influential geologists, Firman, Oliver and Mitchell, it may be mentioned that Firman did eventually accept the idea of a pyrogenic, rather than a xenocrystic, metamorphic or metasomatic origin of the Borrowdale Volcanics garnets, though he never published his change of heart. In the 1980s, he had a Pakistani PhD student, Safdar Ansari (1983), doing work on the Eskdale Granite. After the doctorate was completed, it was proposed that Firman and Ansari might write a joint paper on the garnets in and around the Eskdale Granite. In the event, the paper was never completed, but Firman did draft a four-page introduction and he kindly gave me a copy of the first page of this document, probably dated 1984. It states that the xenocrystic hypothesis had been invalidated by the work of Godfrey Fitton (whose contributions and thoughts on this matter will be described in Chapter 10). It also mentions that Green & Ringwood (1968) had shown that material of the same bulk composition as garnet was a likely liquidus phase for magma stored in the upper mantle or lower crust. So it might, under some circumstances, yield garnets in lavas or ash flows.

Let us now leave the three main Lakeland geologists of the 1950s, and have a look at the development of ideas about the Skiddaw Slates.

Chapter 8
The Skiddaw graptolites

When I was at Cambridge in the 1950s the Professor of Geology was Oliver Bulman (1902–1974) who lectured in palaeontology. He was a quiet man, with a quiet sort of interest: graptolites. He did not fraternize much with the undergraduates, but was regarded as a great authority on his subject. He was a wonderful artist, and when we got to dinosaurs in the syllabus his blackboard drawings drew spontaneous applause from the students. I never got to know him, but I do recall that when we were on a field excursion in South Wales a student found some nodules with three-dimensional pyritized graptolites therein. This was an exciting find. A telegram was sent to Cambridge and Bulman was on the next train to Pembrokeshire and soon arrived, full of excitement. He took away the specimens, and I suppose they were eventually described somewhere. I never saw the paper.

Also at Cambridge at that time was a deaf old lady, Dr Gertrude Elles, whose work provides one of the main foci for the present chapter. She was reputed to inhabit a room somewhere in the attic, but we never saw her. Born in 1872, she would have been about 84 when I was a student. She died in 1960. However, she too was said to be a great authority on graptolites. From photographs (see Fig. 8.1), Elles was strikingly beautiful when young.[1]

Fig. 8.1. Gertrude Elles. Copy of photograph in Newnham College archives. Reproduced by courtesy of the Principal and Fellows, Newnham College, Cambridge.

Elles was raised in Wimbledon, daughter of a Scottish businessman father and an English mother; and with her Scottish connections she had a life-long interest in Scottish geology and scenery, though her best-known work was done in Wales and the Lakes. She went up to Newnham College, Cambridge, on a scholarship in 1891, and studied under Hughes and Marr. On graduating with a first in geology in 1895[2] (as did J. F. N. Green the same year), she obtained a further scholarship that enabled her to study in Sweden for a year with Sven Törnquist and others, familiarizing herself with the stratigraphy and palaeontology of the Lower Palaeozoics of Scandinavia. Back in Cambridge, she obtained a College Research Fellowship at Newnham; then a position as University Assistant Demonstrator; then a College Lectureship; and in 1926 she was appointed University Lecturer in Geology. Ten years later she became the University's first female Reader. Elles was Director of Studies in Science and Medicine at Newnham from 1927 to 1936, and Vice-Principal from 1925 to 1936. She served as President of Section C of the British Association in 1923, giving an important address on graptolite evolution; and again in 1939. Elles was awarded the Geological Society's Lyell Fund in 1900 and the Murchison Medal in 1919, though she was only elected a Fellow that year. She was soon on the Council (1923–1927). Elles thus showed what could be achieved by female geologists in Britain in her day, once appropriate institutional backing was available, and if marriage was foregone. She was belatedly awarded a Cambridge ScD in 1949, Dublin having sensibly awarded her a doctorate in 1907. Elles was supervisor to Bulman (Rickards 1999).

While an undergraduate, Elles joined forces with another Newnham student, Ethel Mary Wood (1871–1946), and at Marr's suggestion they made a joint study of the Drygill Shales while undergraduates (Elles & Wood 1895). Wood, who also got a first in geology in 1895, went on to become research assistant to Lapworth at Birmingham, and both she and Elles studied graptolites under his guidance. Assisted by Gwendoline Watney (see p. 177), they undertook a huge joint research project on the literature, museum specimens and field localities of all known British graptolite types, which culminated in the great *Monograph on British Graptolites* (1901–1918), published by Elles and Wood under the editorship of Lapworth. Wood was chiefly responsible for the drawings; Elles provided the descriptions; and Lapworth wrote the historical introduction, which was essentially an annotated bibliography of all prior writings on graptolites.[3]

[1] For further on Elles, see Bulman (1960), King (1961), White (1961), Creese & Creese (1994), Creese (1998), Rickards (1999).

[2] Actually it would only have been the 'equivalent' of a first. Female students could not actually take out degrees in Cambridge at that time.

[3] I have been told by Adrian Rushton (pers. comm., 1998), retired Survey palaeontologist, that the Birmingham geologist William Watts is reputed to have said of Elles and Wood: 'Wood had the brains in that pair'. According to an anecdote transmitted to me by David Skevington (pers. comm., 2000), the division of labour in the *Monograph* arose from the fact that the two young ladies were scarcely on speaking terms at the time of its compilation, both being enamoured of the same physicist at Birmingham – and Ms Wood emerged the victor, getting married to Gilbert Shakespear in 1906, during the monograph's preparation! However, according to what Bulman told Skevington, Wood's microscope had a built-in magnification error, so that there was discrepancy between the rhabdosome dimensions recorded in the text and those obtained by measuring Wood's figures. The existence of discrepancy is supported by Rickards (1999), who states, however, that Wood's drawings were more accurate than Elles's descriptions.

On the question of the Birmingham physicist's marriage choice, I would say that, judging by the portraits of the two young ladies (see Rickards 1999), Shakespear made a good, but difficult, decision. A woman of serene beauty, Ethel Wood was a tennis champion, an outstanding pianist, and active in undergraduate (Liberal) politics to boot. Though gaining a DSc she went into social work, eventually being made Dame for her contributions to the administration of war pensions. I can imagine that she was admired and loved by all, perhaps including Watts? For more on Wood (Shakespear), see Creese (1998, pp. 299–230). Ms Anne Thompson, archivist at Newnham College, has kindly informed me that Wood was daughter of the Vicar of Biddenham, a village near Bedford. Her schooling was at Bedford High School.

While Elles's fame rests chiefly on her palaeontological work it should be noted that in middle age she also put her hand to metamorphic and structural geology in Scotland. She was a keen Cambridge sportswoman too, and adept at fly-fishing. Besides her collegial, sporting and scientific activities she was active in Red Cross work and had musical interests. She is described as having blue eyes and 'a mass of bright corn-coloured hair' (White 1961).

Another Cambridge graptolite aficionado was Mr Tressilian Nicholas (see Fig. 8.2), who was born in 1888, and lived, remarkably, until 1989. Nicholas's chief published work on graptolites was undertaken as a student in North Wales before World War I – in which he served with distinction, taking part in the Gallipoli campaign. He obtained a Fellowship at Trinity in 1912, and worked in the Lakes in the 1920s but did not publish. In fact, he eventually largely dropped out of science, for though he did much teaching in his earlier years he became Bursar at Trinity in 1929, and was able to enjoy the pleasures of the Trinity Senior Common Room for an unprecedented 73 years. By various accounts, Nicholas was a notable success as Bursar, his investments adding considerably to the College's ample coffers. His field notebooks, held at the Sedgwick Museum, reveal that he did try to get back to geology after World War II, but found many of his old sites covered with pine trees. So he gave up Lakeland geology and held court at the Senior Common Room (I guess). However, for many years he led popular field excursions to the Lakes for the Sedgwick Club. I never saw him as a student, and only heard of him relatively recently. The present (2000) leading Cambridge graptolite expert is Professor Barrie Rickards, the Sedgwick Museum Curator and Fellow of Emmanuel, whose name will recur in the present book (see p. 176).

As is well known, though widely distributed horizontally in space, graptolites had rather short vertical distributions in time. So they are invaluable as stratigraphic indicators and are used extensively for marking time-subdivisions in the older fossiliferous rocks, where graptolite zones have gradually been erected, performing somewhat the same role for the Palaeozoic as ammonites do for the Mesozoic. The concept of biozones was chiefly the invention of the French traveller and palaeontologist Alcide d'Orbigny (1802–1857) (1849–1852) and the German stratigrapher, Albert Oppel (1831–1865) (1856–1858). The concept of biozones can be deployed in essentially two ways: either by taking the first appearance of a species as the indicator of the beginning of a zone; or by taking the presence of a characteristic set of fossils as indicative of a zone. That graptolites and ammonites were free-swimming obviously makes them appropriate for biozonation. As previously mentioned, Charles Lapworth, with Nicholson, had pioneered the use of graptolites in British biostratigraphy.

However, the sudden appearance of a species at some locality, marking a new zone, is a puzzle. It could correspond to an artefact of the knowledge of the fossils in an area (or incompleteness of the fossil record); to sudden (punctuated) evolution; or to migration into an area from some part of the world where the creatures had previously been living. The first possibility means that we are not really in a position to begin establishing zones satisfactorily. The second means that we are not dealing with classical Darwinian evolution; and the third means that we do not have a proper grip on time zones. That is, the same fossil species (or fossil assemblage) may indicate rocks of different ages in different faunal provinces – as appears to be the case for the *Didymograptus bifidus* zone for the Lower Ordovician in the Atlantic and Pacific provinces (D. Skevington, pers. comm., 2000). This could suggest that the zone concept does not provide good time-precision for stratigraphic work.

Nevertheless, the concept of biozones has proved essential in stratigraphy and can be used in limited areas without too much unease.[4] Thus if new types appear in the Skiddaw Slates one can use them as markers for mapping purposes, even if we do not know whether the new forms evolved there or migrated from elsewhere. And the actual *nature* of biological change need not be understood for the concept of zones to be applied. The causes of the turnover of fossils did not really concern the early workers such as d'Orbigny and Oppel, so long as the concept was pragmatically useful. In fact, they were pre-Darwinian, non-Lamarckian, palaeontological 'catastrophists', rather than evolutionists. It is worth emphasizing that the fundamental concept of biozones, so widely used by stratigraphers, goes back to pre-Darwinian days, and was and arguably is more compatible with catastrophism–saltationism than evolutionary gradualism.

Although the status of graptolites as the remains of living organisms was initially uncertain, by the early nineteenth century, it became widely recognized that their remains – on occasions somewhat resembling writing – did indeed represent former organisms, though they were at first conflated with cephalopods such as *Orthoceras* (Ernst Friedrich von Schlotheim 1822). Alexandre Brongniart (1828) established to general satisfaction that the remains were of animals, not plants. Around 1830–1835, the Swedish naturalist Sven Nilsson is believed to have expressed the idea that graptolites were *Polypi ceratoporae* or 'horny polyps', but the concept was apparently not published.

Drawing on the work of the Danish naturalist Beck, Roderick Murchison figured three graptolites in his *Silurian System* and the generic name *Graptolithus* was used (Murchison 1839, part 2, pp. 694–696 and part 3, plate 26). As was his wont, Murchison claimed all of them for his Silurian System. In the next few years many reports and descriptions of graptolites were made and numerous synonyms got into the literature. Workers included such men as Hans Geinitz in Germany (who abandoned the cephalopod hypothesis and referred the organisms to 'Zoophytes'), the biozone man Alcide d'Orbigny in South America, James Hall in North America, Joachim Barrande in Bohemia, and Frederick McCoy at Cambridge and later Melbourne. Barrande (1850) studied graptolite anatomies and remarked the existence of the *virgula* and a hypothesized common canal linking the separate *thecae*. He classified them as belonging to the *Polyparia*. Scharenberg in Norway (1851) offered criteria for classification, such as the number of stipes, the manner of arrangement and number of rows of *thecae*, and the 'attitude' of the stipes relative to the *sicula* (which, however, was not described using these terms until the work of Reinhard Richter (1871) on Thuringian graptolites – and he called it the 'foot' or 'haft-organ', since, unlike later workers, he believed that the creatures were attached to the ocean floor). The term *sicula* was introduced by Lapworth (1873).

Hall (1865) published a major work, *Graptolites of the Quebec Group*, which summarized his North American investigations of the previous decade. He had been studying 'Lower Silurian' (i.e. Ordovician) types, while Barrande in Bohemia had worked on 'Upper Silurians' (i.e. Silurian). Studying the specimens in detail, and providing figures drawn by his assistant Robert Whitfield, Hall did not accept the idea of single-stiped types (e.g. *Monograptus*) such as Barrande had described in Bohemia. Hall mistakenly thought that such things were merely the broken-off bits of multi-stiped species. However, his contribution to the growth of graptolite palaeontology was none the less considerable. The single-stiped forms in fact follow the multi-stiped types in the stratigraphic succession.

In the next twenty years or so, many new forms of graptolites were discovered, named and figured, and subdivision into different genera began. Indeed, Adrian Rushton has informed me (pers. comm., 2000) that he once prepared a histogram for the naming, year by year, of the British Caradoc and Llandovery graptolites, and found that by 1880 about 90 out of a total of 150 had already been discovered and named. Hall in America was

[4] Doubtless many biozone boundaries are diachronous, but the time taken for migrations of forms may be small compared with the durations of biozones; so the zones can serve satisfactorily for stratigraphic division and correlation.

Fig. 8.2. Group of geologists in the Lake District, International Geological Congress, 1948. Left to right: Murray Mitchell, George Mitchell, Tressilian Nicholas, W. B. R. King, Sydney Hollingworth. Copy of photograph supplied by Murray Mitchell and reproduced by his permission.

similarly active work on lower units. The field was rife with synonyms, partly because one could not always tell whether one was dealing with a whole or a part of a specimen. In the Survey monograph on the geology of North Wales, the Survey Palaeontologist, John Salter (in Ramsay 1866, pp. 328–329), took issue with Hall's rejection of monograptids. Despite the growing knowledge-base concerning graptolites, their use for establishing stratigraphic zones did not get under way before the 1860s to 1870s. Let us see how this approach was developed for the SS.

As mentioned above, there was little knowledge of graptolites in the early 1820s and Sedgwick neither looked for nor found them in his reconnaissance work of 1822–1824. In the 1820s, then, he treated the SS as a single unit. In his letter on Lakeland geology in Wordsworth's *Guide to the Lakes*, dated 30 May 1842, Sedgwick stated that the Skiddaws contained no organic remains; and in a supplementary letter of 1846 he acknowledged that he had 'not been able to bring [the Skiddaws] into close comparison with any of the great rock formations of North Wales'. The Skiddaws still seemed bereft of fossils, whereas some organic remains had turned up in what were plausibly comparable rocks in Wales.

However, as we saw in Chapter 2, it was most important to Sedgwick that every effort be made to find fossils in the SS as palaeontological succour for his Cambrian System, so we find that in 1847 he was employing his collector, John Ruthven (see p. 21), to search in the ancient slates. According to a later recorder of events, John Postlethwaite (1882–1883, p. 38), fossils were first found in the SS by a local mineral dealer, Joseph Graham, and it was on the basis of this report that Sedgwick sent Ruthven hunting, directing him also to places that Sedgwick had noticed in 1822 as having dark and potentially fossiliferous slates. Ruthven camped on the flanks of Skiddaw for about a week, but the fossils that he eventually found and which were described and published by Sedgwick came from Scaw Gill near the mountain road that crosses the Whinlatter Pass between Keswick and Cockermouth (see Fig. 12.1), not from the Skiddaw massif. There were some 'fucoids' (seaweeds) and two graptolites: *Graptolites sagittarius*, and a new species, *Graptolites latus*, described and named by McCoy, who claimed to have seen the type at Builth in North Wales also (Sedgwick 1848). Sedgwick's specimens were displayed at the British Association in 1850.

Robert Harkness followed Sedgwick's recommendations and contacted Ruthven, exploring some other areas, probably where Ruthven suggested looking (see p. 27). Harkness's collections are preserved at Tullie House Museum, Carlisle, and contain some fine specimens, revealing what could be found in the SS in the mid-nineteenth century, before generations of collectors had been at work. As we have seen, Harkness also began to provide a structural picture of the Skiddaw rocks in the northern Lakes, envisaging a set of anticlines and synclines, running approximately E–W across Bassenthwaite. However, this structure was deduced from dips and strikes, not fossil zones.

The development of a zonal scheme subdivision of the SS began with a paper of Nicholson (1868*d*) entitled 'Notes on the distribution in time of the various British species and genera of Graptolites', following papers read to the British Association and to the Geological Society in 1867 (Nicholson 1868*d*, *e*). Making use of Harkness's material, these papers described the types then known in the Skiddaws. Nicholson stated that graptolites were generally characteristic of Silurian strata (*sensu* Murchison), but that there was also the net-like *Dictyonema*, found in the Tremadoc, which might be Upper Cambrian. Graptolites seemed to be commonest in the 'Llandeilo'; and as previously mentioned this was for a time used as a kind of rule of thumb by the Surveyors in the Southern Uplands: dark shales with graptolites were Llandeilo. This was an assumption which, as later pointed out by Lapworth (1879–1880,

p. 251), caused much confusion in the Survey's work, both in Ireland and Scotland.[5]

Nicholson listed some of the typical forms for the various units of the 'Lower Silurian' as follows: [6]

Caradoc	*Climacograpsus*, *Rastrites*, *Helicograpsus*, *Cyrtograpsus*, *Graptolites*
Llandeilo (Upper)	*Climacograpsus*, *Diplograpsus*, *Dicranograpsus*, *Rastrites*, *Pleurograpsus*, *Cyrtograpsus*, *Graptolites*
Llandeilo (Lower)	*Dichograpsus*, *Tetragrapsus*, *Phyllograpsus*,
Skiddaw Slates	*Diplograpsus*, *Didymograpsus*
Arenig	Not mentioned
Tremadoc	*Dictyonema*

It should be noted that the divisions had been established in the first instance on lithological grounds (though with an attempt to identify a diagnostic fauna for each unit); and then fossils such as the above were found in the beds as indicated. Later, the arguments could be reversed: the presence of certain fossils could be used to determine the unit to which rocks at a certain site belonged. Note that the above fossils are only stated at the level of genera, but actually Nicholson named species. The zone concept was not yet deployed, but graptolites were becoming useful for characterizing Lower Palaeozoic units.

The same year, Nicholson (1868*b*) published detailed descriptions and figures of all the species then known in the SS, again utilizing material collected by Harkness. He said that he did not think that the Skiddaws had palaeontological analogies with other rocks in the United Kingdom, but there appeared to be similarities to the fossils of the Quebec Group in North America, as described by Hall. Four years later (1872*c*), Nicholson began to develop the idea of the SS (Upper Cambrian) as having the oldest British graptolites (other than the retiform *Dictyonema*), there being similar contemporaneous forms in the Quebec Group.

Specifically, Nicholson discussed the supposed former migrations of graptolites. At the end of the period of the SS, most forms seemed to become extinct, but, suggested Nicholson, the few that survived (such as *Climacograps[t]us* and *Diplograps[t]us*) migrated from the Lakes into Wales and Ireland, and into the Southern Uplands and Scandinavia. This occurred either because of uplift in the Lakeland area or because volcanic activity there made it uninhabitable for graptolites. New forms such as *Rastrites* appeared in Scotland at that time (Llandeilo). From there, Nicholson suggested, some forms migrated to North America via Ireland in Caradoc times (Hudson River and Utica Shales), the forms there being different from the earlier Quebec Group graptolites. However, the graptolites also supposedly developed in Scotland and subsequently back-migrated to the Lakes, founding there the graptolite fauna of the Coniston Mudstone (Skelgill Beds, then regarded by Murchisonians as Caradoc, Upper Silurian). While there were no graptolites in the underlying Coniston Limestone (Bala), there were some in the supposedly equivalent Girvan Limestone in southern Scotland. Following the Coniston Mudstone, the Coniston Flags and Coniston Grits were deposited, with different forms appearing and others disappearing.

Some of the later graptolites also supposedly travelled back to Scotland, and thence to Ireland and America, and towards Saxony and Bohemia. However, some of the Bohemian stages seemed to have come from the Coniston area, not Scotland, as particular Scottish types were not present. Such backward and forward migration was theoretically possible, and meshed to an extent with the idea of 'colonies' developed in Bohemia by Barrande, discussed in Chapter 5. All the toing and froing was not helpful to those who wanted to use the stratigraphic principles of William Smith. While the Survey countenanced Barrande's theory, on the whole the Surveyors were doubtful that graptolites served as useful stratigraphic indicators. It should be remarked, however, that Barrande's work was hindered by the fact that Sedgwick had proclaimed the SS as Lower Cambrian (perhaps equivalent to the Longmynd rocks, but, unlike this Shropshire unit, were possessed of fossils). So Barrande was led to believe that graptolites made their first appearance in or immediately below his 'Primordial Zone' of life, much lower in the stratigraphic column than was actually the case.

From the published discussion of Nicholson's paper on graptolite migrations, it would appear that it was not too well received, which is hardly surprising given the number of postulated migrations. There was, I suggest, a risk of arguing in a circle or circles, if one did not know the relative ages of the rocks in different parts of the world except by the evidence of the fossils. Indeed, I would say that the uncertainties and implausibilities revealed by Nicholson's paper (which, of course, assumed that land and water masses had essentially retained their relative positions over time) were such that one can sympathize with those who were sceptical about using graptolites in stratigraphy. The organisms seemed to be wandering all over the place. Were the changes in forms in some given locality due to evolution, or the arrival of forms from other localities? Without independent knowledge of relative ages, who could be sure of the direction in which the migrations were supposedly taking place? Such problems were compounded by the often poor preservation of specimens, difficulties in their identification and synonymy, and the pervasive uncertainty at that time as to whether one should be regarding the rocks in which they occurred as Cambrian or Silurian, according to whether one was a Sedgwickian or a Murchisonian – not to mention the 'fixist' (non-mobilist) assumptions of some nineteenth-century geologists (though Nicholson himself was an evolutionist).

---

[5] The subdivisions of the Lower Palaeozoics envisaged by Murchison and/or Sedgwick at about that time were as follows:

Ludlow (proposed by Murchison 1833)	Upper Silurian
Wenlock (proposed by Murchison 1834)	Upper Silurian
Llandovery (proposed by Murchison 1859)	Transitional between Lower and Upper Silurian (Murchison)
Caradoc (proposed by Murchison 1835)	Lower Silurian
Llandeilo (proposed by Murchison 1835)	Lower Silurian
Arenig (proposed by Sedgwick 1852, Cambrian)	Lower Silurian
Tremadoc (proposed by Sedgwick 1846, Cambrian)	Lower Silurian
Lingula Flags (proposed by Sedgwick 1847, Cambrian)	Lower Silurian
Longmynd (proposed by Sedgwick 1852)	Cambrian

Later, Henry Hicks (1881) introduced the Llanvirn, below the Llandeilo, equivalent to his earlier designated Upper Arenig (Hicks 1875). In 1873, Salter noted 'Ash Gill Beds' in the Lake District, and these, following the work of Marr (see p. 66), came to be accepted as another major subdivision, between the Caradoc and the Llandovery. The Longmynd unit is Precambrian by today's reckoning. Today, the Ashgill, Caradoc, Llandeilo, Llanvirn and Arenig are allocated to the Ordovician. The allocation of the Tremadoc has been controversial, but it is generally accepted as belonging to the Ordovician by British geologists. There are, of course, further divisions of the Cambrian at present, and all the units named above have several stages, though the Llandeilo has become reduced to a single graptolite zone (perhaps less), and is no longer regarded as one of the Ordovician epochs (Fortey *et al.* 1995, 2000). The classification given above was generally used by the Survey in the nineteenth century, the Caradoc being regarded as Lower Silurian.

[6] The modern names end in 'aptus', as *Climacograptus*, *Tetragraptus*, etc.

The same year, Nicholson (1872*b*, p. 107) reported that the Skiddaw graptolites suggested correlation with Hall's Quebec Group of Canada, but had not yielded trilobites such as Barrande thought were characteristic of the lowest ('Primordial') life-zones. So Nicholson placed the Skiddaws in the Lower Llandeilo (or Sedgwick's Arenig). It may be remarked that the Arenig is, for the most part, where they remain to this day, though the unit has a different definition from that which it had in Sedgwick's day.

As previously indicated, the first successful use of zonal techniques in Britain using graptolites was developed by Lapworth in the early 1870s. This was done for rocks near Galashiels where he worked as a schoolmaster in the Southern Uplands (Lapworth 1870, 1873, 1878; Hamilton 2001). He corresponded with Gustav Linnarsson (1841–1881), palaeontologist to the Swedish Survey, who is thought to have been the first to deploy zonal techniques using graptolites. However, regardless of whether he initiated the procedure, Lapworth certainly used it in Scotland with outstanding success. At the celebrated Dob's Lin site, near Moffat in the Southern Uplands, he investigated a small valley where there were graptolite shales folded into a tight anticline, which indicated with reasonable certainty which way up the beds were (they were *not* inverted). Lapworth observed five lithologically distinct bands of rock in a cliff face and collected from each of them. Then, on getting home to his nearby shepherd's cottage (Birkhill) in the evenings, he carefully arranged the fossils in five drawers. There is also a legend, recounted to me by Beryl Hamilton who is writing a biography of Lapworth, derived orally from Lapworth's granddaughter, the late Patricia Lapworth, that Lapworth wore a special coat in the field with five or perhaps more pockets, like a portable filing cabinet, and the separate fossils were slipped into the appropriate pocket when out in the field, and then put in the drawers in the evenings. This sounds verisimilitudinous, and gives an idea of how the zone concept was applied to British graptolitic strata in the early days.

However, in Scotland Lapworth was working above the level of the Skiddaws, in what he called Llandeilo and Llandovery. Later, as we have seen, the top of the Llandeilo was separated off as the Ashgill, now regarded as the uppermost Ordovician subdivision. In fact, as mentioned earlier, Dob's Lin is now the world stratotype for the boundary between the Ordovician and the Silurian – arising from Lapworth's gallant work. But I digress.

In Sweden, Linnarsson proposed six distinct horizons, defined by graptolites, from Cambrian to Silurian, and showed how they had analogues in Britain and in North America. He sent a paper on this topic, and samples of his specimens, to Nicholson, who made the work known to British audiences through publication in the *Geological Magazine*, along with commentary (Linnarsson 1876; Nicholson 1876). He explained how his Swedish colleague had shown that graptolites were 'restricted to certain very definite horizons' and were widely distributed in different countries and were thus appropriate for zonal stratigraphy. The SS and the Quebec Group were apparently analogues of Linnarsson's 'Lower Graptolitic Schists', his second unit, which Nicholson referred to the 'Lower Silurian'. There were also Swedish analogues for the Skelgill Beds, the Coniston Mudstone (Stockdale Shales), the Coniston (Brathay) Flags, and the Coniston Grits, for the upper part of the Lakeland succession.

We know that Nicholson was in correspondence with Lapworth at about this time, and they presented a joint paper on the 'Silurians' of the north of Britain at the British Association in 1875. Then in a long series of papers Lapworth (1879–1880) – utilizing his new System, the Ordovician (Lapworth 1879) – produced a major survey about what was known of graptolites *and* set out a scheme (Table 8.1) for their zonation, naming zones according to certain characteristic graptolites that they contained. Of the zones listed in Table 8.1, Lapworth only recognized Zones 2, 3, 12, 13, 15 and 17 in the Lakes at that time.

Lapworth's table – which referred to Britain, Europe and North America – was based on his vast knowledge of the literature, his correspondence with Linnarsson and Nicholson, and his wonderfully skilful work at Dob's Lin and elsewhere. He placed the SS in the Arenig, and suggested a subdivision into Lower Slates and Upper Slates, each having a characteristic fauna, though the precise boundaries or zones were not clear. Just naming the genera here, however, there appeared to be the following:

Upper
: *Trigonograptus*, *Trichograptus*, *Didymograptus*, *Glossograptus*, *Phyllograptus* and *Diplograptus*

Lower
: *Loganograptus*, *Temnograptus*, *Schizograptus*, *Ctenograptus*, *Dichograptus*, *Tetragraptus*, *Didymograptus*, *Phyllograptus*, *Diplograptus* and *Azygograptus*

Monograptids were now excluded from the SS. What appeared to be such were, it was realized, only found in the Silurian, and seeming Skiddaw 'monograptids' were in fact merely broken-off fragments of some other forms.

Meanwhile, Marr was the up-and-coming geologist in the Lakes. As mentioned in Chapter 5, after graduating he obtained scholarships enabling him to travel to Bohemia and Sweden, where he met Linnarsson and Sven Törnquist, who were working out the graptolitic zonations there; and he could study the Swedish sections, which are, I understand, less disturbed than in Britain.

As we have already seen, Marr (1878) did fieldwork in the Lakes while still an undergraduate, and sought to subdivide the upper succession (above the Coniston Limestone) into 'well-defined life-zones' (using trilobites rather than graptolites). His mention of 'life-zones' in that early paper indicates his intended approach to stratigraphy, but he did not tackle the Skiddaws in this paper. In a paper of 1880, comparing the strata of the Lake District with North Wales, he assigned the Skiddaws to the Arenig (i.e. he located it in Sedgwick's Cambrian) (Marr 1880*a*).

As we have also seen, Marr's next main work in the Lakes was undertaken with Nicholson, on the rocks above the Skiddaws, examining what was now regarded as the top of the Ordovician and the transition from thence into the Lower Silurian (see p. 64) (Nicholson & Marr 1887; Marr & Nicholson 1888).

Table 8.1. *Graptolite zonation, according to Charles Lapworth (1879–1880)*

20.	Upper Silurian	*Monograptus nilssoni*	Lower Ludlow
19.	Upper Silurian	*Monograptus testis*	Wenlock
18.	Upper Silurian	*Cyrtograptus linnarssoni*	Wenlock
17.	Upper Silurian	*Cyrtograptus murchisoni*	Wenlock
16.	Lower Silurian	*Cyrtograptus grayae*	Tarranon
15.	Lower Silurian	*Monograptus exiguus*	Tarranon
14.	Lower Silurian	*Rastrites maximus*	Tarranon
13.	Lower Silurian, Llandovery	*Monograptus spinigerus*	
12.	Lower Silurian, Llandovery	*Monograptus gregarius*	
11.	Lower Silurian, Llandovery	*Diplograptus vesiculosus*	
10.	Lower Silurian, Llandovery	*Diplograptus acuminatus*	
9.	Upper Ordovician	*Dicellograptus anceps*	Caradoc
8.	Upper Ordovician	*Dicellograptus complanatus*	Caradoc
7.	Upper Ordovician	*Pleurograptus linearis*	Caradoc
6.	Upper Ordovician	*Dicranograptus clingani*	Caradoc
5.	Upper Ordovician	*Coenograptus gracilis*	Llandeilo
4.	Lower Ordovician	*Didymograptus murchisoni*	Llandeilo
3.	Lower Ordovician	*Didymograptus bifidus*	Arenig
2.	Lower Ordovician	*Tetragraptus bryonoides*	Arenig
1.	Upper Cambrian	*Bryograptus callevei*	–

But could one see any theoretical pattern in the graptolites, besides using them in a 'positivistic' way for the purposes of zonation? Marr's notebook for 1892 has a little poem that touches on this question, which shows that he was trying to figure out an *evolutionary* history for graptolites (see Fig. 8.3):

How first of all it came about I do not care a fico [fig]; But something turned without a doubt To octobrachiate *Dicho*.	*Dichograptus* [Arenig]
Then 'Evolution once again Came into play', *&c.* And thus we find our old sea-pen Is posing as a *Tetra*.	*Tetragraptus* [Arenig]
Another pair of branches drops As Mother Nature bid em, oh! And in the sea are many crops Of double-branching *Didymo*.	*Didymograptus* [Llandeilo]
Through nearly half a circle sweeps Each stipe from off its fellow And from each hydrotheca peeps The flesh of a *Dicello*.	*Dicellograptus* [Caradoc]
Now trouble comes about and s[d]wells In corpore non sano So back to back the early cells Unite to form *Dicrano*.	*Dicranograptus* [Caradoc]
The distal thecae follow suit Unto the very tip. So! The first Diprionidian brute Emerges as a *Diplo*.	*Diplograptus* [Llandovery]
Each theca finds opposing cell Is to itself a sore foe; So some are dropped, and all is well At first with the *Dimorpho*.	*Dimorphograptus* [Llandovery]
But not for long, and so at last One row of cells is gone oh! Our beast (through many stages passed) Becomes the perfect *Mono*.	*Monograptus* [Llandeilo–Ludlow][7]

This, I reckon, was a clever ditty, and shows more clearly than any published work how the members of the Cambridge group were trying to develop a story about the evolution of graptolites, avoiding all the toing and froing of Nicholson's graptolites, and also Barrande's theory of colonies. In brief, the poem told a 'story' about the progressive decrease in the number of stipes and evolution by a process of the two-stiped forms 'doubling back' on themselves, and then eventually becoming the 'advanced' *Monograptus*. It should be emphasized that most of the evolutionary scheme envisaged in Marr's poem referred to units well above the Skiddaws (Arenig). It foreshadowed some ideas of Gertrude Elles, which we shall look at shortly. I can imagine that she heard this poem in a Marr lecture, and conceivably was inspired thereby to study graptolites and endeavour to work out their evolution properly. Elles was in her first academic year at Cambridge in 1891–1892.

In 1894, Marr published a paper listing the graptolites then known in the Skiddaws, and gave a somewhat more detailed system of subdivision, based on graptolites, than Lapworth had earlier proposed. Marr (1894, p. 128) suggested:

2d.	Milburn Beds[8]	Upper Arenig or Lower Llandeilo
2c.	Ellergill Beds[6]	Upper, with *Didymograptus nanus*
2b.	*Tetragraptus* Beds	
2a.	*Dichograptus* Beds	Lower [with *Azygograptus lapworthi*]
1.	*Bryograptus* Beds	Tremadoc Slates

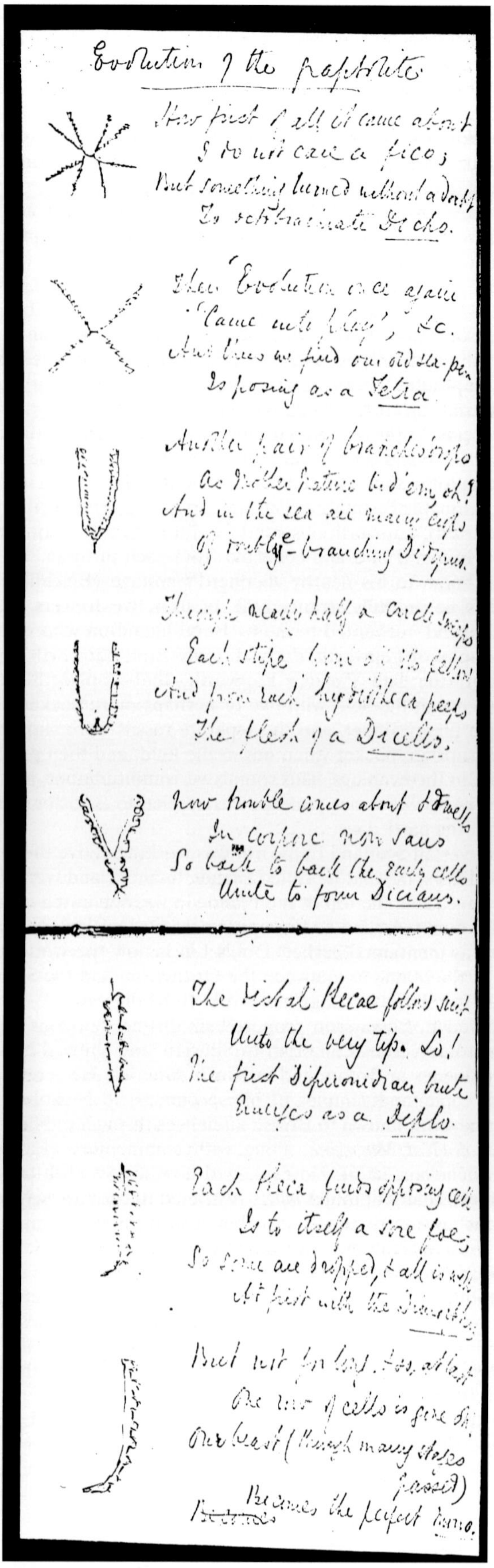

Fig. 8.3. Extract from Marr, Notebook XXIII, 30 April, 1892, Sedgwick Museum, Cambridge. Reproduced by courtesy of Sedgwick Museum.

[7] I interpolate epochs here according to Lapworth (1879–1880).
[8] Found in the Cross Fell Inlier.

This scheme had as much to do with the Skiddaw-like rocks in the Cross Fell Inlier – where Nicholson and Marr had been working both jointly and separately – as the main Skiddaw outcrop in the Lakes. It should be remarked that Marr thought that the Skiddaws went as low as the Tremadoc.

The question of graptolite evolution was, however, more complicated than might appear from Marr's little poem. In 1895, he and Nicholson published a somewhat curious paper (Nicholson & Marr 1895) that reflected some of the varied views on evolutionary theory that were floating about in the late nineteenth century. It invoked the concept of 'heterochronous convergence', a term (*heterochrone Convergenz*) proposed by the Austrian stratigrapher and palaeontologist Edmund Mojsisovics (1839–1907) (1893, part 2, p. 5),[9] and also deployed by the English Mesozoic brachiopod and ammonite specialist Sidney Buckman (1860–1884) (1895).[10]

Nicholson & Marr (1895) considered three quite different graptolite forms: *Bryograptus/Dichograptus*, *Tetragraptus* and *Didymograptus*. The theory of heterochrony was deployed to suggest that these types did not evolve one from the other (as envisaged in Marr's poem), but that all three groups underwent similar changes. That is, it was suggested that the various forms were 'the result of the variation of a number of different ancestral types along similar lines'. On this view, a form such as *Monograptus* might have evolved from a number of different original types, which could account for the great variety of thecae found on monograptid stipes. It could, as we might say today, be a case of 'convergent evolution' or a kind of mimicry. The different monograptids might have different phylogenies. This would not, however, preclude the use of graptolites in stratigraphy, or the use of zonal techniques, for it seemed that the similarly shaped forms occurred at the same time: the homoeomorphic forms were isochronous. It was suggested that insofar as phylogenetic relationships were concerned the most important criteria were the shapes of the thecae and the angle of divergence of the stipes – not the number of stipes as earlier supposed (i.e. slow evolution from the dendroid *Dictyonema* to the single-stiped *Monograptus*).

So, having introduced this chapter by describing the career of Gertrude Elles in a little detail, and having presented something of the background to her work, let us now consider her investigations on Skiddaw graptolites (1898*a*, *b*). Her task involved a survey of all known major collections, including the specimens of various Cumbrian amateurs, many of which had found their way into the Keswick Museum. The specimens were organized into six families (*Dichograptidae*, *Nemagraptidae*, *Dicranograptidae, Diplograptidae*, *Glossograptidae*, *Retiolitidae*) and one of doubtful character; 22 genera; and 58 species. On this reckoning, some of the forms such as *Tetragraptus* and *Didymograptus* had been subsumed under the larger group of *Dichograptus* or *Dichograptidae*. Many of the forms described were clearly figured – and Elles was an excellent draughtswoman, as can be seen from her notebooks at Cambridge (see Fig. 8.4).

In Elles's 1898 papers, the Skiddaws were in the main allocated to the Arenig, with some of the rocks being thought to be older and some younger, a determination that is still sustained in the modern literature, though as we shall see her ideas were challenged from the 1950s to the 1980s. Elles's subdivisions were as follows:

Upper Skiddaw Slates		
(a) Milburn Beds	? Llandeilo	(in Cross Fell Inlier)
(b) Ellergill Beds	Llanvirn	(in Cross Fell Inlier)
Middle Skiddaw Slates		
(a) Upper *Tetragraptus* beds	Upper Arenig	
(b) *Dichograptus* beds	Middle Arenig	
(c) Lower *Tetragraptus* beds	Lower Arenig	
Lower Skiddaw Slates		
(a) *Bryograptus* beds	Tremadoc[11]	
(b) ?	Lingula Flags	

This result was reached on palaeontological grounds, analogies being made with Swedish fossils.

On the question of graptolite evolution and phylogeny, Elles had to contend with the fact that her teacher Marr had endorsed 'heterochronous convergence'. She discussed this idea, referring to Nicholson & Marr (1895) and acknowledging their evidence for the seeming convergence, but she evidently did not embrace it. Elles suggested that there had been evolution from one type to another, with a reduction in the number of stipes – rather as envisaged by Marr in his poem and by earlier writers. Nevertheless, Elles viewed the evolution of the Skiddaw graptolites in terms of derivations from two early types: *Bryograptus* and *Clonograptus*. From *Clonograptus*, in the Lower Skiddaw Slates, there supposedly evolved various forms of *Tetragraptus*; and then *Didymograptus* in the Middle Skiddaws. From the latter in the Lower Skiddaws, came *Loganograptus*, then two kinds of *Dichograptus*, both of which then supposedly evolved into *Tetragraptus* and further into *Didymograptus* in the Middle Skiddaws. Thus, on this view, the various species of both *Tetragraptus* and *Didymograptus* had more than one evolutionary ancestor; so there was some acceptance of Marr's newer ideas. However, Elles's overall scheme was different from her teacher's. In the discussion following the paper, Marr conceded that Elles's ideas on the graptolite evolution were 'more likely to be correct' than the ones that he and Nicholson had put forward not long before.

Elles's work was, of course, done well before the advent of cladistics. However, the arguments as to what were ancestral and what were derived characters were based on a detailed knowledge of the anatomies of the different species; the localities where they were found; and analogies from regions outside the Lakes such as North Wales or Sweden. The evolutionary arguments evidently depended on knowledge as to which rock units were older. But here lay the difficulty, for the Skiddaws did not lie in a nice regular sequence. In many places they were strongly folded and contorted, and although quite a lot of fossils

[9] Mojsisovics distinguished between 'heterochronous' and 'isochronous' convergence, being two kinds of what we might call 'convergent evolution'. In the 'heterochronous' case, one might have *similar* ammonites in (say) the Triassic and the Jurassic. I thank Martina Koelbl-Ebert for checking Mojsisovics' text for me.

[10] According to the modern view of heterochrony, different features of organisms may develop at different rates so that they may be 'mature' in some respects and 'immature' in others (paedomorphism). As a result one may find organisms of different type but similar form living at different times; or ones of different form living at the same time. Adult 'larval forms' of a more recent species may resemble the larval forms of former species. It is not evident that the nineteenth- and twentieth-century concepts covered by the same name are one and the same, or not at least 'causally'. On the basis of his detailed stratigraphic studies in Gloucestershire, Buckman (1895, p. 456) suggested that brachiopods of the same form ('homoeomorphous') might sometimes be 'heterogenetic'. He envisaged a stock that might 'throw off' various homoeomorphic forms; but the sequence in which such 'throwing off' might occur was not always the same; hence there might be similar forms of different ages and of different 'genetic' (Buckman's term) origin. Deploying Mojsisovics' ideas, he suggested that the terms 'heterochronous homoeomorphy' and 'isochronous homoeomorphy' would be useful. His concept was not obviously related to paedomorphism. Nicholson and Marr seemingly found the concept useful to 'tell a story' about the origin of monograptids, but its biological basis was insecure.

[11] Only recently, in the 1980s, has the presence of Tremadoc rocks been confirmed by the discovery of trilobite remains (Molyneux & Rushton 1984), whereas Elles's Tremadoc suggestion had for a long time been refused. On this matter, see further below.

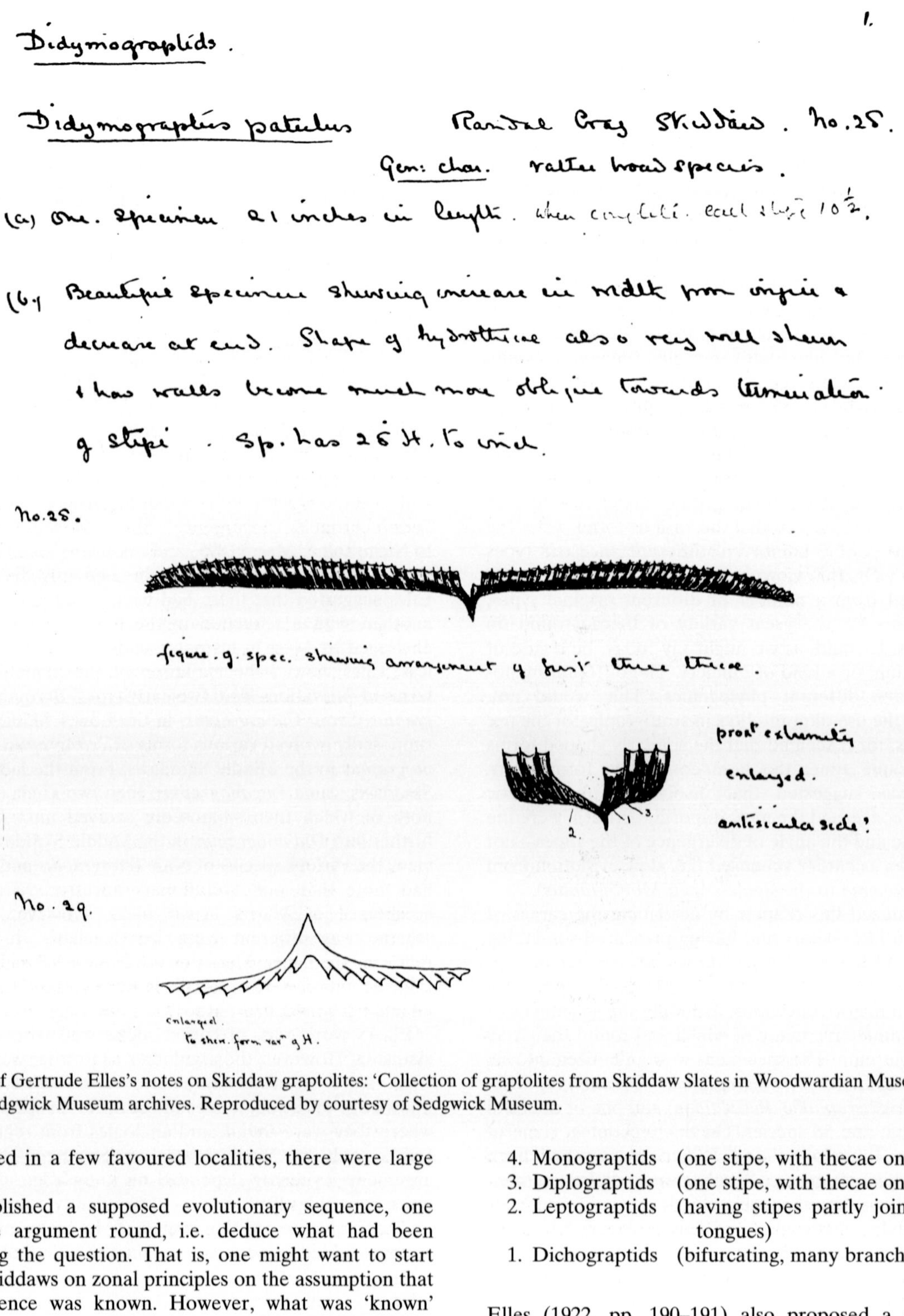

Fig. 8.4. Sample of Gertrude Elles's notes on Skiddaw graptolites: 'Collection of graptolites from Skiddaw Slates in Woodwardian Museum', undated notebook, p. 1, Sedgwick Museum archives. Reproduced by courtesy of Sedgwick Museum.

had been located in a few favoured localities, there were large barren areas.

Having established a supposed evolutionary sequence, one might turn the argument round, i.e. deduce what had been assumed, or beg the question. That is, one might want to start mapping the Skiddaws on zonal principles on the assumption that the fossil sequence was known. However, what was 'known' depended to an extent on theories as to how the graptolite evolution had occurred and the empirical 'validity' of the zones.

In 1922, Elles published an important paper on 'The graptolite faunas of the British Isles: a study in evolution'. She supposed that there were four main aspects to the evolutionary development of the monograptids: change in the direction of growth of the stipes (from pendent to scandent); simplification in branching; elaboration of the type of thecae; and localization of the thickening of the periderm wall. The first two processes supposedly occurred in tandem, leading to the eventual dominance of *Monograptus* in the Upper Silurian. The thecae became increasingly separated from one another, and more lobed or hooked. Changes in the thickening of the wall led (in some cases) to a kind of lattice-like skeletal structure for the organisms. Thus the overall evolutionary sequence was represented as follows:

4. Monograptids (one stipe, with thecae on one side only)
3. Diplograptids (one stipe, with thecae on two sides)
2. Leptograptids (having stipes partly joined – like forked-tongues)
1. Dichograptids (bifurcating, many branched)

Elles (1922, pp. 190–191) also proposed a zonal scheme for Skiddaw rocks:

Two-branched forms	Pendent series	*Didymograptus murchisoni* zone
		Didymograptus bifidus zone
	Horizontal series	*Didymograptus hirundo* zone
		Didymograptus extensus zone
Many-branched forms, becoming simpler		Upper *Dichograptus* zone with horizontal *Tetragraptus* types
		Lower *Dichograptus* zone with rarity of *Tetragraptus*

The paper had as its *pièce de résistance* an elaborate table setting out graptolite phylogenies from the Tremadoc to the Ludlow, according to the evolutionary changes supposedly at work during this long span of time; but the further details of this need not concern us here. Elles always emphasized the importance of zonal assemblages rather than individual index fossils, but she regarded the coming in of new forms in abundance as indicative of passage to a higher horizon or zone. As she put it, the '*coming in of new forms in abundance*, ... usually indicative of a more advanced stage in evolution, ... [is] the basis of modern zonal stratigraphy' (Elles 1925, p. 337).

Then all this had to be applied to the Lake District stratigraphy and mapwork. Elles did not really do this herself, but from Cambridge she acted as a consultant to the Surveyors from the Whitehaven office.[12] In their *Summary of Progress* for 1924 (issued 1925), the Survey proposed the following *lithological* subdivisions for the Skiddaws:

4. Mosser Slates[13] and Watch Hill[14] Grits
3. Loweswater Flags
2. Kirkstile[15] Slates
1. Blakefell[16] Mudstone

In 1931, with the help of Elles's identifications in Cambridge, the Survey correlated these units with the following fossils:

4. Mosser Slates and Watch Hill Grits	*Dichograptus octobrachiatus*, *Tetragraptus headi*, *T. quadribrachiatus*, *Didymograptus extensus*, *D. gibberulus*, *D. hirundo*, *D. extensus*, *D. nitidus*, *Phyllograptus angustifolius*, *Glyptograptus dentatus*
3. Loweswater Flags	*Dichograptus octobrachiatus*, *Tetragraptus amii*, *T. quadribrachiatus*, *Didymograptus extensus*, *D. nitidus*
2. Kirkstile Slates	*Didymograptus extensus*, *D. nitidus*, *D. hirundo*
1. Blakefell Mudstone	Fossils rare; no graptolites

Then in the *Annual Report* of the Survey for 1932, Elles extended her theorizing, defining a zone as a belt of strata containing a definite assemblage of graptolites characteristic of a 'certain stage of evolution' (Elles 1933, p. 94). Using an evolutionary perspective she saw graptolite development in terms of three trends: stipe reduction; change in attitude of stipes; and development of sigmoid thecae. These trends did not always operate sequentially or simultaneously: an earlier trend could supposedly be re-established at some later time. However, a concentration of many-stiped graptolites at a particular horizon was indicative of a particular stage of general evolution. Or, if there was a concentration of four-stiped and then two-stiped types, the latter would be younger than the former. Elles claimed, as an evolutionist would, that many 'transient forms' could be discovered between the various types. So she named such entities as, for example, *Didymograptus protobifidus-bifidus*. On this basis, a supposed scheme for the evolutionary development for the Lower Ordovician graptolites of the Lake District (Skiddaw Slates) was proposed, with the whole linked to Elles's zonal scheme for the British graptolites, with comparisons to Canada, Sweden and Australia. The scheme for Skiddaw zonation was as shown in Table 8.2.

For simplicity, only the zone index fossils, not the full zonal assemblages given by Elles, are named in Table 8.2. Elles claimed her scheme agreed with that being developed in Victoria, Australia. The graptolites were widely distributed indeed (from a pre-mobilist standpoint), were that so.

The localities reported above are not important in themselves for our purposes, but it should be noted that some of the localities such as the Barf scree[17] apparently had several zones, seemingly extending down into the Tremadoc, given the presence of *Bryograptus*.[18] (But, as said above, large areas of the Skiddaws were unfossiliferous.) The apparent occurrence of several zones at one locality, as at Barf, could mean three things: there is a succession at that place that is highly condensed for some reason; the zone scheme is in error; or there has been disturbance of the rocks at that locality.

A point should be made about Elles's method of working. At the time of which we are speaking, she was chiefly a *museum* palaeontologist. She was brilliant at observing, identifying and drawing graptolites, as can be seen from her notebooks. She was often in the field too, of course, and there are legends of her packing up boxes of graptolite-bearing slates at Keswick station, to be sent down to Cambridge for examination. However, a lot of her work was done with other people's specimens, or relied on specimens in local museums or amateur collections, for which there might be uncertainties about localities. Her method of working is also reputed to have been a bit haphazard. I have been told by a long-serving staff member at the Sedgwick Museum, Michael Dorling, that there is an institutional legend that 'Gertie's' room was like a series of strata. She spread all the fossils for a particular investigation on a table. Then when that was finished she laid paper over the whole and started again on the next batch of fossils. Also, David Skevington recalls that, when fresh space was required the 'strata' were moved sideways and specimens sometimes slipped onto the floor, making a scree even in Fenland, one might say! Working thus, I can imagine that specimens sometimes got muddled. Elles may have been a rather vain woman too, as, according to Dorling, she always had her publications on display in a kind of horse-shoe arrangement round the room (or perhaps that was so she could just put her hands on them conveniently).

There is also a tradition in the Sedgwick Museum that Elles was 'dominated' by her evolutionary ideas. Rickards has said (pers. comm., 1998) that she would look at a specimen and think more of its 'evolutionary setting' than its 'place in the rocks'. For

---

[12] Elles's reports were surely useful to the Survey, but Adrian Rushton recalls (pers. comm., 2000) that its palaeontologist Cyril Stubblefield (later Director) used to examine Elles's reports, and on occasion question or challenge her identifications or conclusions.

[13] Mosser: a hamlet south of Cockermouth on the road to Loweswater, situated below and to the west of Mosser Fell.

[14] A hill to the NE of Cockermouth (see Figs 8.6 and 12.1).

[15] Kirkstile: the name survives in 'The Kirkstile Inn' in Loweswater village, between Loweswater and Crummock Water. (I am indebted to Chris Thompson for this information.)

[16] Blake Fell: an area situated between Loweswater and the westward end of Ennerdale.

[17] Barf (see Fig. 12.1) is an interesting place, being a hill of modest size, just to the west of the road that runs up the west side of Bassenthwaite Lake. The hill has a steep scree slope, near the top of which there is a peculiar knob-shaped outcrop of slates, for some reason painted white, and known as 'The Bishop'. The area has for years been a favourite locality for collecting and few fossils can now be found there, as it is readily accessible to the road, where a pub is conveniently placed.

[18] This fossil was named after Theodor Kjerulf, first Director of the Norwegian Geological Survey. Bulman (1941) subsequently renamed the material from Barf *Pseudobryograptus*, taking the view that it was not indicative of a Tremadoc age (see p. 117).

Table 8.2. *Zonal scheme for the graptolites of the Skiddaw Slates, compiled from Elles (1933*a)

Zone	Sub-zone	Index fossil	[Fossil localities]
Didymograptus bifidus	–	*D. bifidus*	Cross Fell inlier, Mungrisdale, Outerside
D. hirundo (with *D. hirundo* transients)	–	*D. hirundo*	Dead Beck, Outerside, Randel Crag
D. extensus	(d) *Isograptus gibberulus*	*I. gibberulus*	Outerside, Hazelhurst, Lord's Seat, Randel Crag, Watches, Whitehouse Fell
	(c) *D. nitidus*	*D. nitidus*	Barf, Mungrisdale, Outerside, White Horse Bent, Undercrag
	(b) *D. deflexus*	*D. deflexus*	Hudson Place Farm, Whitehorse Fell, Undercrag, Barf
	(a) Reclined *Tetragrapti*	Reclined *Tetragrapti*	Great Knott, Barf
Dichograptus (many-branched graptolites)	Horizontal *Tetragrapti*	Horizontal *Tetragrapti*	Barf, Hodgson How, Scaw Gill
		D. octobrachiatus	Randel Crag, Gategill screes [on Saddleback], Carlside, Ullock Pike, Mirehouse, Barf, Braithwaite, Grisedale Pike, Brackenthwaite
Bryograptus kjerulfi	–	*B. kjerulfi*	Barf

example, she would look at a *Dicranograptus* and think of it as having evolved from a *Dicellograptus* and on its way to becoming a biserial form such as *Diplograptus* (see Fig. 8.3). Yet the actual appearance of these forms in the rocks is just the opposite of this arrangement: the biserials appear first. It is remarkable that the example quoted to me by Rickards in 1998 is precisely what one might expect from someone who adhered to the scheme outlined in Marr's ditty!

Perhaps more importantly, Elles did not do systematic mapping in the Skiddaws – she never personally put her zones to rigorous test in the field in the Lakes. Consequently, when the Survey officers tried to do their mapwork with the help of her zonation, as one of them, Sydney Hollingworth, put it delicately: '[i]t was not [found] possible to apply . . . [Elles's] detailed zonal scheme during the mapping of the slates in the Cockermouth sheet' (Hollingworth *et al.* 1954). In other words, the scheme was found not to work for the purposes of detailed mapping.

At the beginning of this chapter, I mentioned the work of Tressilian Nicholas, who did much careful collecting in the Lake District in the 1920s, as his notebooks at Cambridge attest. His fossils were stored away there and are still extant, as are his lists of Skiddaw graptolites and the localities where they were collected. The palaeontologist Adrian Rushton (see Chapter 15) has informed me (pers. comm., 2000) that Nicholas's specimens were always 'well localized', unlike nineteenth-century collections such as that of Harkness at Carlisle. But Nicholas did not publish his work. Whether this was because he was too busy teaching to get his results published (which may well have been the case in the years of staff shortages during the Depression), or whether he was too beguiled by the delights and responsibilities of being Bursar of Trinity, or whether he was unwilling to challenge the ideas of his senior colleague, Elles, I do not know. In an interview recorded in 1975 with the aged geologist, the transcript of which was kindly made available to me by the recently sadly deceased John Thackray (d. 1999), former archivist at the British Museum of Natural History and the Geological Society, Nicholas recalled that Elles had lectured him on graptolites for his third-year geology, and she had helped him by identifying specimens when he was doing his mapping for his thesis on the geology of St Tudwal's Peninsula, North Wales (the only major piece of work Nicholas ever published). They had been on field excursions together when he was a student. She gave him a fishing rod, and they went up to Scotland to try for trout just before World War I; but as soon as the news of the outbreak of War reached them Elles rushed south to fulfil her obligations in the Red Cross. Nicholas described her as a great friend of the family in later years, but she was a person who spoke her mind and did not suffer fools gladly. Maybe he did not want to question his mentor's authority. Anyway, Elles's (1933) work was not seriously challenged in print for many years.

Needless to say, World War II interfered with the progress of geology for at least a decade, and it was only in the 1950s that serious research was taken up there again, with students beginning to do PhDs there. One of them, Dennis Jackson (see Fig. 8.5), now retired, has told me what he did, and how he came to think that Elles's scheme had problems. Jackson was a pupil at Cockermouth Grammar School and read geology at King's College, Newcastle-upon-Tyne. For his undergraduate thesis he decided to map the area immediately to the east of Cockermouth – the hills he had seen out of his schoolroom window, and which he knew intimately (see Fig. 8.6). He initially intended to study the Cockermouth Lavas (Carboniferous) (see p. 56), but he soon got interested in the larger canvas of the Skiddaws. One day, while doing an undergraduate mapping exercise for his degree at Newcastle, he visited a quarry by a lonely church at Setmurthy to the east of Cockermouth (see Fig. 8.6), and found there specimens of *Tetragraptus*. He reported these to his professor, Stanley Westoll, who urged him read Elles (1933) to find out about such matters. As Jackson explained to me, the Setmurthy discovery was a surprise: he had not expected to find rocks there so low in Elles's succession. So, beginning with Barf and Scaw Gill (see Fig. 12.1), he started to examine all the exposures where Elles claimed to have age-dated fossils (see the list above).

For Elles's overlying zone of *Dichograptus* (many-branched) graptolites, she had claimed that it cropped out over a wide belt of Lakeland country, from Saddleback (Blencathra), round the flanks of Skiddaw, and over to Braithwaite and Grisedale Pike. But Jackson could find no such zone. Wherever he looked, he found assemblages that suggested higher zones such as *D. hirundo* or *I. gibberulus*. The situation at Barf was a real mess! Elles had supposed there was a condensed section there, produced by thrusting. Jackson thought otherwise. The rocks on the cliff on the south side of the hill near 'The Bishop' appear to have intense zigzag folding; however, in his view this was not due to tectonic activity but to 'creep' of the beds, which had got shattered during the Pleistocene. The rocks were rich in fossils, with *Dichograptus*, '*Bryograptus*' and *Tetragraptus*, but the specimens represented long-ranging forms and were compatible with younger zones.

Ultimately, Jackson found that he accepted all Elles's zones from *D. deflexus* upwards as far as *D. bifidus*, but in his estimation her zones of *Bryograptus kjerulfi*, *Dichograptus* and the two sorts of *Tetragraptus*, were illusory so far as the Skiddaws were concerned. So Jackson mapped the strata according to his own revised zonal schema. However, for much of the work he, like others, had to rely in part on lithological subdivisions, for there

Fig. 8.5. Dennis Jackson in graduation regalia (1953). Copy of photograph supplied by Dr Jackson and reproduced by his permission.

just were not enough fossils overall to do the job properly on a palaeontological basis. His lithological units were (Jackson 1962) as follows:

4.	Redmain Sandstone (see Fig. 8.6; equivalent to the Latterbarrow Sandstone)	Unfossiliferous
3.	Mosser and Kirkstile Slates	*D. hirundo*, *I. gibberulus*
2.	Loweswater Flags (? equivalent to Watch Hill Grits)	*I. gibberulus*, *D. nitidus*, *D. deflexus*
1.	Hope Beck[19] Slates (new unit proposed by Jackson)	Unfossiliferous

All these, Jackson thought, were compatible with an Arenig age.

Jackson was an indefatigable observer and scoured hundreds of thousands of pieces of slate during his thesis years. He told me (pers. comm., 1998) that he seldom turned over a piece of loose slate. There was too much rock to afford time for that. And he liked to work in the rain, as the graptolites showed up more clearly! He did his own identifications, but made use of existing records such as those of Nicholas. In fact, Nicholas had the evidence, years before, to undermine Elles's Lakeland stratigraphy, but had not revealed it to the world or perhaps did realize its implications.

Jackson's (1956) thesis was examined by Bulman, but – perhaps fortunately for the candidate – not by Elles, who was by then very old. Bulman specifically avoided introducing him to Elles. Jackson told me that they did pass the old lady, but without exchanging a

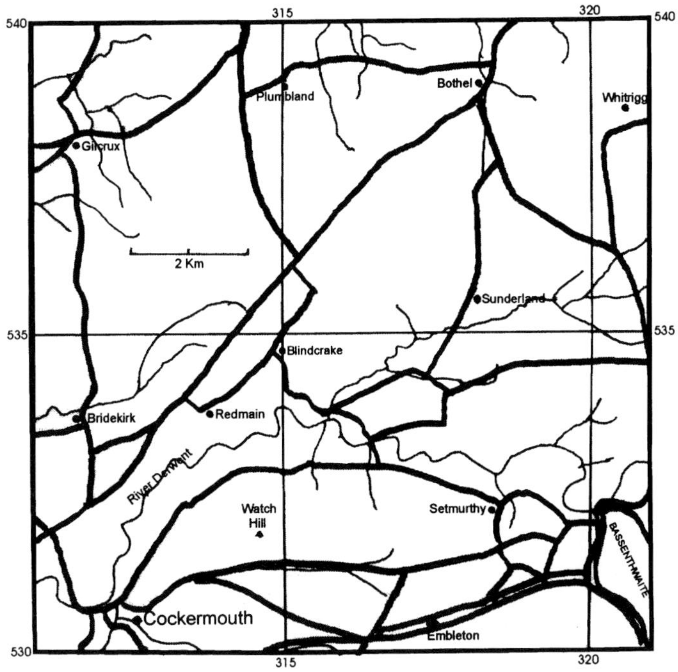

Fig. 8.6. Topography of the country to the north and east of Cockermouth.

word. Be this as it may, though the thesis undermined what had been the received view, it would appear that its findings were not altogether unwelcome. Bulman said: 'I am delighted'; and 'that boy is observant beyond the ordinary'. In fact, there had been disquiet about the SS zonation and mapping for some time. In a paper as far back as 1941 Bulman had taken the view that species such as Elles's zone fossil '*Bryograptus kjerulfi*' were not necessarily indicative of Tremadoc but could range into the Arenig; so the whole of the Skiddaw Group could be Lower Ordovician (Arenig). (Also, two asserted Tremadoc bryograptid species at Barf had been mis-identified.) Even earlier, the Surveyor Ernest Dixon (in Eastwood *et al.* 1931, p. 31) had inclined to the view that the overall fossil evidence for the Skiddaws was 'contradictory', with difficulties arising from the fact that fossils had been found by diverse collectors of 'diverse trustworthiness'. Also, the great thicknesses of zones and lithological units revealed in the general mapping did not lead one to expect a remarkable attenuation and concentration of zones at Barf as Elles supposed was the case. (Dixon's point had, however, been challenged by Elles (1933*a*, pp. 102–103), who thought that the evidence was 'entirely satisfactory' if 'properly interpreted'.) In 1954, Hollingworth's important review of Lakeland geology stated that 'the zonal succession remains somewhat uncertain'; and that the Survey officers (such as Dixon) working in the 1920s and 1930s had not found it possible to apply Elles's ideas successfully.

Bulman, incidentally, was pleased that the young Jackson had not come bothering him, as did so many research students, to ask for help with graptolite identifications. Identification may have to rest on examination of a vast literature in, for example, awkward languages like Swedish. It is by no means an easy task. Back in the 1930s, even the Survey had had to farm out specimen identification to Elles in Cambridge.

Gertrude Elles died in 1960, and shortly thereafter Jackson (1961, 1962) published two papers, the latter of which criticized her work (though he had also done so in his thesis). In his 1962 paper, Jackson provided a map of the northern Skiddaws according to the distribution of their graptolite zones, as he had previously presented them in his thesis. Jackson's last paper on the

[19] See Figure 12.1 for the location of Hope Beck.

topic appeared in a collection of papers on Lakeland geology edited by Frank Moseley (1978*a*). Jackson was revisiting his earlier work, having been away in Canada for a good many years after he finished his PhD. Again he stated his objections to Elles's work, and suggested that her scheme had its origins in Marr's ideas about the evolution of graptolites involving a progressive reduction in the number of stipes; and I believe that this was so. Thus for Elles there *ought* to be a downward sequence:

Didymograptus	(2 stipes)
Tetragraptus	(4 stipes)
Dichograptus	(multi [8]-stipes)

So, for Elles, *Tetragraptus* indicated the occurrence of a Lower Arenig fauna. By contrast, Jackson, looking at a range of types to characterize a zone (although Elles did this too), did not assume that the presence of tetragraptids was conclusive as to age.

All this illustrates several interesting features about the nature of biostratigraphic work, and the social workings of the geological community. It also raises the question of what is going on when graptolites evolve, and the biological theory underlying the zone concept. First, it is evident that one cannot characterize a zone satisfactorily by a single zone fossil – a point realized from the early days by Oppel, and agreed to by all concerned in the story outlined above. However, this leaves open the biological question of where the different species suddenly spring from. In fact, they do not always do that. They may appear, increase in numbers for a time, and then diminish. Nevertheless, they do sometimes seem to emerge from nowhere, so to speak. So graptolites, as found in the rocks, do *not* always support Darwin's classical evolutionary theory, with smooth transitions of forms in the stratigraphic record, Elles's claimed 'transients' notwithstanding. On the contrary, in certain aspects the new species seemingly arrived rather suddenly in a given area, as might be expected by creationist theory, catastrophism, or punctuated equilibrium theory. On the other hand, as Skevington has emphasized to me, the graptolites were colonial organisms, and the number of stipes *had* to be integral. So while there can be, and is, gradualistic evolution of the forms of the thecae, the change in stipe number will necessarily be 'punctuated'.

When I have asked graptolite experts about such matters, their response has normally been that sudden arrivals of new forms result from migration from neighbouring faunal provinces. This is all very well, but it seems to leave the problem of why there appears to be an absence of sequences of graptolites with smooth 'Darwinian' transitions. Of course, one should be taking a 'populational' rather than a 'typological' approach, and apparently the issue does not seem to worry graptolithologists unduly.

Interestingly, if we consult modern texts such as Palmer & Rickards (1991), we find that their 'story' of the evolution of graptolites is not really so very different from that encapsulated in Marr's poem. The animals are thought to have followed the same broad sequence as that envisaged by Marr and Elles, with the suggestion that the early retiform types such as *Dictyonema* were bottom-dwellers, whereas the later forms were free-floating. The passage from the Ordovician to the Silurian supposedly coincided with a glacial epoch, which killed off most of the earlier types, and it was only the diplograptids that made it through to the Silurian, whence were derived the dimorphograptids and the monograptids. Modern studies try to see the development in terms of an evolution towards mechanical and hydrodynamic stability, together with a distribution of thecae on the stipe that maximizes the scope for feeding. This similarity between the nineteenth- and twentieth-century ideas about graptolite evolution suggests that Elles's problem was not so much that she was over-wedded to a false evolutionary sequence, but that she did not do enough careful *fieldwork*. There can be a disadvantage in the museum specialist laying down the law to those who work in the field, though obviously the museum workers make essential contributions through pronouncing on identifications, and publishing formal descriptions of species.

In any case, I should mention that Elles may, in a way, be said to have had the last laugh. As we shall see, resurvey of the Lake District began in 1982 and was approaching completion by the end of the century. Quite early on in this resurvey, rare but indisputable evidence from new trilobite discoveries came to light in exposures of Skiddaw rocks on the western side of the Lakes, indicating a Tremadoc or Lower Ordovician age (see Chapter 15). However, it is believed that Elles did not visit this locality, or others where evidence for Tremadoc rocks were later found, so, on the evidence available to her, she was not really justified in hypothesizing rocks so low in the stratigraphic sequence for the Skiddaws. Her '*Bryograptus*' and *Clonograptus* had longer ranges than the Tremadoc, but the trilobites found in the 1980s have confirmed the occurrence of Tremadoc rocks in the SS and the recent Cockermouth (BGS 1997*a*) and Keswick (BGS 1999*a*) maps both entrench the idea of the occurrence of Tremadoc in the lower Skiddaws.

Elles might even have had the last laugh at the Setmurthy Quarry, had she ever visited the place. We recall that Jackson, as a student, found *Tetragraptus* specimens there, but thought they were compatible with a younger age. Once the Survey's geologists began a systematic re-examination of the SS in the 1980s, and convinced themselves that there was acritarch and trilobite evidence for Tremadoc rocks in the Skiddaws of the northern and western Lakes (see p. 198), then one could look at the Setmurthy fossils in a new light. Jackson provisionally put his tetragraptids in the lowest part of his Skiddaw succession, and suggested that the overlying beds could be low in the Loweswater Flags. However, for the Surveyors, now countenancing the possibility of Tremadoc rocks, it seemed equally possible that the tetragraptids might be below the Hope Beck Slates: old and Tremadoc in age. In fact they now placed them in a new group at the bottom of the Skiddaws called the Bitter Beck Formation, Bitter Beck being a small river in the Setmurthy area (see Fig. 8.6). On the basis of evidence from its microscopic acritarchs and some rare graptolites, this unit was determined as belonging to the Tremadoc, as was the nearby Watch Hill Grits. As we have seen, these grits were allocated to an extraordinary number of horizons by earlier workers – from Precambrian (Rastall 1910) to the SS (Dixon, in Smith 1925, p. 71) to late Ordovician and younger than the Borrowdale Volcanics (Green 1917)! This last suggestion was by no means unreasonable, given that the Watch Hill Grits crop out in some localities as uncleaved turbidites with 'Bouma' features (cf. Bouma 1962), not unlike some of the 'Otley III' strata in southern Lakeland. But the new Tremadoc determination was based on a host of correlative evidence from strata in other parts of the world, such as Australia, Scandinavia and China. The evidence must cohere; and it did.

The uncertainty about the stratigraphic situation at Watch Hill arose in part because the area is, according to the thinking of the modern Survey (see p. 199), structurally more complex than was envisaged by Jackson. The Grits form a ridge along the top of a hill. Below them and to the south are the Bitter Beck Beds, and below them in the Embleton Valley are the Kirkstile Slates, which are now known to be Arenig (younger). To account for this state of affairs, the older Watch Hill Grits and Bitter Beck Beds are suggested to be thrust over the Kirkstile Slates from the north, and are mapped as such on the new Cockermouth map (1997). Thus a tectonic argument can on occasions trump a stratigraphic problem, avoiding the need to invoke palaeontological *ad hoc* hypotheses such as Barrande's theory of colonies. However, it too should have coherence with other evidence and not just be *ad hoc*. The tectonic situation in the Watch Hill area has been incorporated with general theories of Lakeland structure and geological history, as we shall see, but I must say that I have not myself been able to discern any physical expression of the fault on the ground.

For convenience, we may conclude this chapter by presenting (anachronistically) two modern versions of the lithostratigraphy and British graptolite zonation for the SS, remembering of course that the arrangements were arrived at by making extensive use of

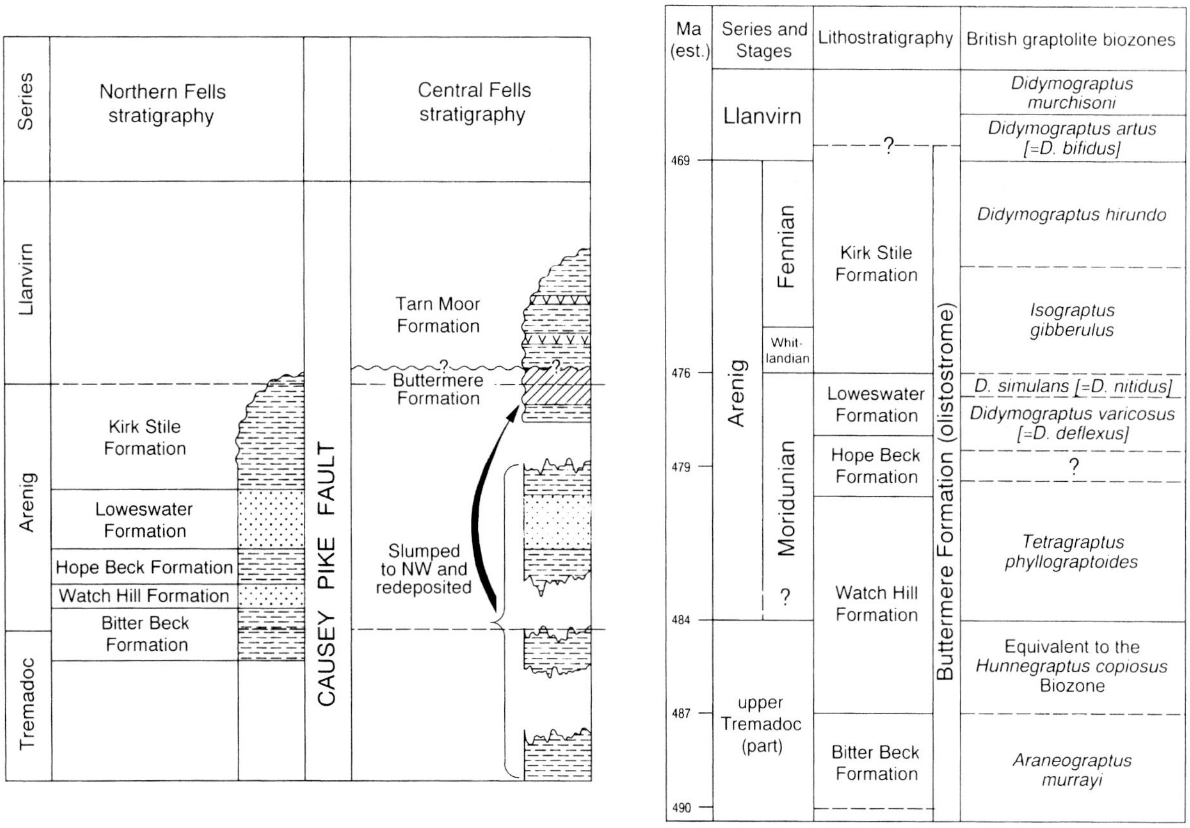

Fig. 8.7. Subdivisions of the Skiddaw Slates, as given in Akhurst *et al.* (1997, p. 39); based on Cooper *et al.* (1995). IPR/25-32C. British Geological Survey. (For discussion of slumping in the Central Falls, see pp. 200–201.)

Table 8.3. *Graptolite zonation of the Skiddaw Slates according to Woodhall (2000)*[20]

Series	Stage	Graptolite biozone	Lithological unit (Northern Fells)
[Llanvirn	Upper	*Didymograptus murchisoni*]	–
[Llanvirn	Lower	*Didymograptus artus*]	–
			.... Unconformity
Llanvirn	Lower	*Didymograptus artus*	Kirkstile Formation
Arenig	Fennian	*Aulograptus cucullus*	Kirkstile Formation
Arenig	Fennian	*Isograptus caduceeus gibberullus*	Kirkstile Formation
Arenig	Fennian	*Isograptus victoriae*	Kirkstile Formation
Arenig	Whitlandian	*Isograptus victoriae*	Kirkstile Formation
Arenig	Moridunian	*Didymograptus simulans*	Loweswater Formation
Arenig	Moridunian	*Didymograptus varicosus*	Hope Beck Formation
Arenig	Moridunian	*Tetragraptus phyllograptoides*	Watch Hill Formation
Tremadoc	Upper	*Araneograptus murrayi*	Bitter Beck Formation
Tremadoc	Lower	No zones defined	Base not seen
Tremadoc	Lower	*R. [Dictyonema] flabelliformis*	Base not seen

palynological data to which neither Elles nor Jackson had access, long after the remapping of these rocks was done in the Survey's 'Lakeland Project', initiated in 1982 (see Chapter 14). First we have a BGS Memoir version (Fig. 8.7) dating from 1997, based on a paper by Surveyor Tony Cooper *et al.* (1995). Then this was soon superseded in a report compiled by Derek Woodhall (2000, p. 6) to accompany the new Keswick map (BGS 1999), which offered a significantly revised and refined graptolite zonation (Table 8.3) developed by Cooper for the then forthcoming Skiddaw Memoir.

It should be remarked that (1) the subdivisions shown in Table 8.3 were based on both acritarchs and graptolites; (2) the international stage names used were imported from outside the Lake District (as might be expected); (3) the acritarch zones were largely given as matching the stage names, but not exactly so in all cases; (4) the graptolite zones did not fit the Whitlandian–Fennian boundary at all; and (5) the lithological subdivisions more or less fitted the graptolite zones albeit not perfectly, the small mismatches not being shown in Table 8.3.[21] Thus are the endless tricky problems in stratigraphy.

[20] Based on Woodhall's table, by courtesy of the British Geological Survey. IPR/25–32C British Geological Survey.

[21] The top of the Hope Beck Formation is given as slightly below the top of the *D. varicosus* zone; and its bottom slightly above the top of the *T. phyllograptoides* zone. The unconformity represented actually occurs within the *D. artus* zone.

This was the end of the story, so far as establishing the biostratigraphic sequence in the Skiddaw Group was concerned in the twentieth century – which is something altogether different from the establishment of the unit's immensely complicated structure. The scheme shown in Table 8.3 will probably not be developed much further in the twenty-first century, as the shortly-to-be-published Skiddaw Memoir should wrap everything up and it will probably be some considerable time before professional stratigraphers revisit the Skiddaw Slates. It may be mentioned, however, that the Survey officer, Tony Cooper (see p. 197), informed me (pers. comm., 1998) that over to the east, at Eycott Hill, patches of mudstones have been found that appear to contain Cambrian acritarchs. It is possible, then, that there are Cambrian slates underlying the whole Skiddaw pile, although this is not specially important conceptually, since the existence of very early Skiddaws would not in itself necessitate a change in structural or theoretical interpretation. There is no reason in principle why sedimentation need have altered radically from the Cambrian into the Ordovician. Skiddaw rocks have also been found further east: in the Cross Fell Inlier, as previously mentioned; and in a small inlier in Teesdale in Northumberland, east of the Pennines (Johnson 1961). The rocks have shown up in boreholes on the eastern side of the Pennines too. Again, that does not come as any particular surprise, though it does raise the question as to why such rocks should be elevated in the Lake District and not so further east.

The foregoing account, then, tells us something of the story of the establishment of the stratigraphy of the Skiddaw Slates by means of macrofossils, with just a mention of microfossils. The complicated question of their structure will be considered in various places in the chapters that follow. Chapter 15 also contains further discussion of stratigraphic matters, and shows how some of the ideas mentioned at the end of the present chapter were developed in the last two decades of the twentieth century.

Chapter 9
The Skiddaw Slates and the Borrowdale Volcanics

In the present chapter, we shall, with a few backward glances, begin to look at the large amount of work done in the Lakes from the 1960s, concentrating on the structure of the Skiddaw Slates, and ideas concerning the stratigraphic relationship of the slates to the overlying Borrowdale Volcanics.

Working out the stratigraphy of the SS through consideration of lithologies and fossil zonation was one thing; examining them from the perspective of structural geology was something different again. To my knowledge, the first person to undertake such work chiefly from the perspective of modern structural techniques was Kenneth Glennie, in a Master's thesis entitled 'The structure of the Skiddaw Slates in the north-west Lake District', at Edinburgh University (1955),[1] the work being undertaken under the supervision of Donald McIntyre and the American Lionel Weiss.[2] Dennis Jackson and Glennie were at work in the Skiddaws at the same time, but their work did not intersect.

Glennie's work, which was focused on the area round Loweswater, Crummock, Buttermere and Ennerdale, was a continuation to the NE of his earlier undergraduate mapping in the area of the hills of Dent and Latterbarrow, SW of Ennerdale, across the Calder Valley and Uldale (see Fig. 20.1).[3] Glennie's empirical investigation was straightforward. He determined the dips and orientations of bedding, cleavage and folds of the slates in his study area, and plotted the data on six-inch maps. His analysis of the data was, however, novel for the Skiddaws, though Oliver and Firman were doing somewhat similar work in the Borrowdales at about the same time. Considering any surface (bedding plane, joint, cleavage plane or whatever), a line drawn perpendicularly to that surface points towards its *pole*; and the pole may be plotted on a stereogram. If we have a folded surface, the plot of the poles to the surface on the stereogram will yield a great circle: the 'π-circle'. The pole of *this* circle, determinable from the stereogram, will reveal the trend and plunge of the fold, represented by the point 'β', on the diagram (see Fig. 9.1). Using multiple data, one could also provide 'contoured' diagrams, representing the averages of many individual empirical determinations (see Fig. 9.2). Such techniques derived from, among others, the work of the Swiss metamorphic and structural geologist Eugen Wegmann (1896–1982) and the Austrian structural geologist Bruno Sander (1884–1979), whose ideas were being introduced into Britain in

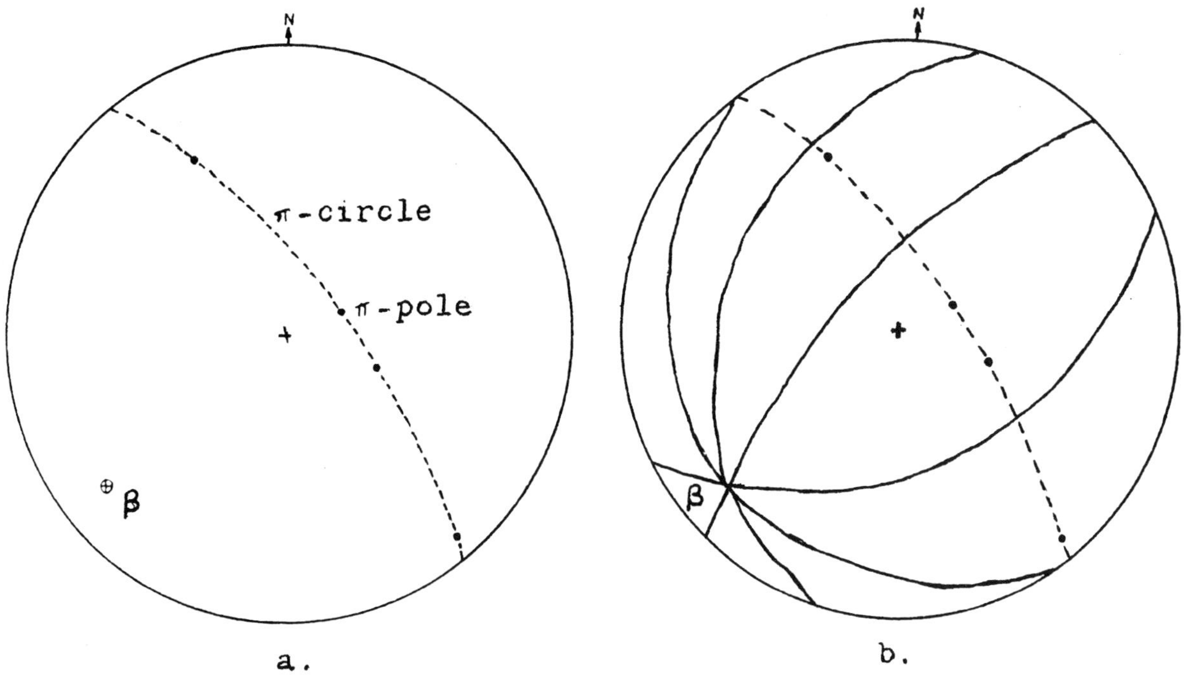

Fig. 9.1. Stereographic representation of a fold, as depicted by Glennie (1955, p. 9). Reproduced by courtesy of Dr Glennie. (**a**) Idealized π-diagram to show the relation between the π-circle and β. (**b**) Idealized β-diagram to show the relation between the π-circle, great circles of s-surface, and β.

[1] Glennie has informed me (pers. comm., 1999) that his work was completed by the end of 1954 and was the first MSc to be awarded by a Scottish university. His thesis research was funded by Shell, and he subsequently became an authority on North Sea oil geology.

[2] McIntyre subsequently left Edinburgh to take a chair at Pomona College, California. At the end of the century, he was back in Scotland in retirement at Perth, active in studies in the history of geology. He played an important role in the early application of computers in the study of structural geology. Glennie has asked me to mention his special indebtedness to his supervisor, for his introduction to him of the potentialities of stereographic methods and their applicability in the Lakes.

[3] Latterbarrow, incidentally, has a somewhat distinctive greenish sandstone. It was thought by Green (1917) to be a unit different from the SS, but overlying them conformably. It was mapped as a separate unit by Hollingworth for the Whitehaven map (Smith 1925, p. 72), but was regarded then as forming the top of the Skiddaw sequence. In the original survey, Ward had conflated it with the Watch Hill Grits, which, as mentioned earlier (pp. 76 and 81), were at different times placed all over the stratigraphic column. In the Gosforth Memoir (Trotter 1937, p. 20), Hollingworth placed the Latterbarrow Sandstone below the Mosser Slates. For its subsequent interpretation, see p. 199.

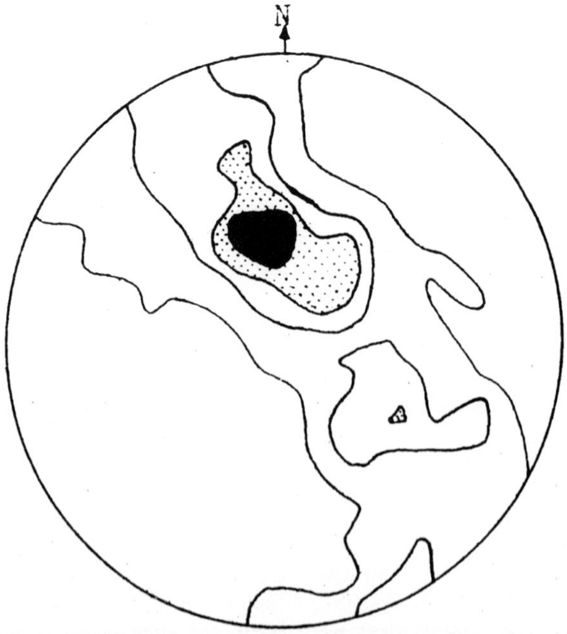

Fig. 9.2. Contoured π-Diagram, displaying a 'girdle' equivalent to an average axial plunge of 10°–20° to the southwest. Based on the determination of the attitudes of 1,400 bedding planes; contoured at 1, 2, 3, and 4% for each 1% of the area examined: Glennie (1955, p. 15). Reproduced by courtesy of Dr Glennie.

the 1950s by McIntyre in Edinburgh (who had studied with Wegmann) and Coles Phillips at Bristol (who later co-translated Sander's work into English).[4]

Glennie also examined 'B-lineations': lines representing the lineations corresponding to the intersections of cleavage and bedding, which are parallel to the directions of the fold-axes. His main conclusion, which was arrived at in effect by the averaging of many subsidiary folds, was that in his area there was a large-scale anticlinal folding, with the axis oriented N240°E and with a plunge of about 15°. The limbs of the fold appeared to be somewhat concave, suggesting that his area was only revealing part of a larger-scale structure.

Glennie (1955, p. 48) noted that there was considerable variation in the directions of the subsidiary fold-axes examined, which *might* be due to repeated (or multiple) deformations. He did not pursue this idea, but as we shall see it was developed with interesting results in the 1960s.[5] It should be remarked that the general system of folding examined by Glennie had been identified since the time of the early Lakeland geologists such as Harkness; but now the evidence was being quantified, using techniques for the averaging of observations by stereographic representation of data. Unlike many students of Lakeland geology, Glennie did not continue his research in that area of Britain but became a petroleum geologist and an authority on North Sea oil.

The next person to undertake a structural study of the Skiddaw Slates, and an important figure in our story, was the Scottish geologist Alexander Simpson (1925–1997). Son of a schoolteacher, he attended Ayr Academy and Glasgow University (where Arthur Trueman was professor), graduating with a first in geology in 1946. After a brief period of research on the Scottish Coal Measures while working as a research demonstrator at Glasgow, Simpson moved to Africa in 1947, where he was on the staff of the Nigerian Geological Survey for six years, employed in mapping the coalfields in the south of that country. In 1953, he was appointed to a lectureship at Birkbeck College, London University, where he spent the rest of his academic career. He was promoted to senior lecturer in 1965, but suffered a stroke in 1979 at the age of 54, which necessitated his taking early retirement in 1984.[6] His former chief, John Temple, remembers him as presenting very carefully prepared lectures, lucidly delivered. They were chiefly on structural geology. Simpson 'was regarded by both students and colleagues with respect and affection, especially for his wry, laconic, sense of humour and his soft Glaswegian accent' (typescript of obituary of Simpson, copy supplied to me by Professor Temple).

Late in 1954, Simpson started mapping in the Isle of Man, for his PhD at Birkbeck on 'The stratigraphy and tectonics of the Manx Slate Series, Isle of Man'. The degree, completed under the supervision of Archibald Dollar, Reader at Birkbeck, was awarded in 1961, though the fieldwork was finished by the autumn of 1958. It was a difficult topic for a unit on which little had been done since the time of George Lamplugh's Survey Memoir (1903), in part because of the sheer difficulty of making anything of the geology there. The Manx Slates have commonly been described as 'monotonous' (but see Woodcock *et al.* 1999), and they contain so few macrofossils that their study would have thrown but little light on the Slates' structure or history.

In his thesis, Simpson developed a lithologically based stratigraphy for subdivision of the Slates, and sought to use the 'way-upness' criteria based upon sedimentary structures, previously deployed by Hartley in the Lakes, Robert Shackleton in Anglesey, and Edward Bailey in Scotland. Simpson also proposed three phases of folding (F_1, F_2 and F_3) for the Slates, his techniques with stereograms being similar in principle to those described above for Glennie's work, derived from the methods of Sander and Phillips. Simpson claimed to discern suites of congruous minor flexures, a generation of b-lineations,[7] and a succession of three axial-plane cleavages defining three surfaces (S_1, S_2, S_3). Such concepts and associated techniques were 'in fashion' at that time, but were difficult to apply with surety. Simpson had a basic synclinal structure (F_1) (like Lamplugh's), approximately parallel to the topographic axis of the Isle of Man; the imposition of a crumpling (F_2) on the foregoing, supposedly resulting from a nearly vertical compression; and a third folding (F_3) due to ENE–WSW compression.

Later work, based on the micropalaeontological study of Manxian acritarchs (Molyneux 1979) has not substantiated Simpson's stratigraphy, for his 'oldest' unit has apparently yielded the youngest organisms; and the 'youngest' has yielded Tremadoc forms (Jackson *et al.* 1995, p. 12). Simpson noted that a *Dictyonema* (Tremadoc) had long ago been found in what he regarded as his uppermost unit. As stratigrapher, Simpson was seemingly a 'splitter' rather than a 'lumper', proposing 11 subdivisions for the Manx Slates. Because he had Tremadoc (then regarded as Cambrian by some) at the top of his sequence, he placed the whole of the Manx Slates in the Cambrian, as had Lamplugh. Subsequently, Ordovician (Arenig) graptolites were found in the Manx Slates (Rushton 1993).

[4] For a synoptic account of the history of structural geology, including this period, with mention of the persons referred to here, see Howarth (1999).

[5] Looking at the matter retrospectively, Glennie (pers. comm., 1999) thought that some of the subsidiary folds he examined may have been due to deformation of sediments at the time of their deposition; and others might be due to 'bed slippage along difficult-to-recognize shear planes'. The former might be a sedimentary feature arising near the depositional margin of the Iapetus Ocean (a theoretical entity unheard of in the 1950s!) and the latter due to Caledonian deformation. He wondered also whether the lower-angle strikes observed in the SE of his area were caused by the emplacement of the Ennerdale granophyre, or the rigidity of the adjacent Borrowdale Volcanics during deformation.

[6] I am indebted to Professor John Temple, formerly of Birkbeck, for information in a typescript of his obituary of Simpson; and Chris Terrey of Birkbeck and Simon Bennett of Glasgow University for additional information.

[7] b = the fabric axis; or the axis perpendicular to the girdle plane of a petrofabric diagram. The term 'b-lineation' was due to Sander.

Towards the end of his thesis, Simpson mentioned that he had already made a preliminary comparison between the Manx and the Skiddaws, from which it seemed that both had undergone polyphase deformation. So following publication of the main results of his thesis in 1963, Simpson turned his attention to the Lakes, examining the suggestion made long before by Harkness & Nicholson (1866) that the Skiddaw and Manx rocks were structurally and lithologically analogous and contemporaneous.[8]

Simpson's daughter Vivien – at the time of writing a teacher domiciled in the United States – has some memories of her father's work in the Lake District, for between the time she was 12 and 15 she spent the summers in the field with him, helping to carry specimens and trekking over many hills (pers. comm., 1998). She said that she rather began to dislike the Lakes! Simpson did much work with a graduate student, David Wedden; and another student, Douglas Helm, who worked on the Skiddaw rocks of Black Combe in SW Lakeland under Simpson's supervision. Ms Simpson recalled that her father and Wedden worked particularly round the Buttermere area and (surprisingly to me) Windermere; also near Keswick and Ullswater. He did not study the Lakeland graptolites so far as she could recall, but did much work with his microscope at home (where he chiefly worked, as Birkbeck lectures were in the evenings) (pers. comm., 1999).

Simpson's Skiddaw work, based on mapping on the 1: 25 000 scale (B. Roberts, pers. comm., 2000), was published in 1967[9] and soon caused a stir in the geological community. Several geologists I have interviewed have expressed surprise, even disbelief, that he covered such a large area – about 126 square miles, more than the whole of the western half of the main Skiddaw outcrop – in such a short time, with the close attention needed to establish a satisfactory stratigraphy and elucidate the structure of such a bewildering set of rocks as the SS. Be that as may, Simpson again showed his leaning towards taxonomic 'splitting', envisaging eight lithological subdivisions for the Skiddaws, and apparently not utilizing Jackson's work:

Sunderland Slates[10]
Watch Hill Grits and Flags
Mosser Slates
Loweswater Flags
Blakefell Mudstones
Buttermere Flags
Buttermere Slates

The Latterbarrow Sandstone was not placed in the SS but at the bottom of the Borrowdales (where it is still located). The total thickness of the Skiddaws was estimated to be about 30 000 feet.

Simpson's published maps were essentially structural and lithological, with his lithostratigraphic map showing similarities to the one he had published for the Isle of Man; and as such his structure was substantially more complex than Glennie's. Simpson determined numerous dips and strikes, some of which were plotted on his second map (see Fig. 9.3), which depicted a system of anticlines and synclines running approximately NE–SW. As in the Isle of Man, he envisaged three generations of folding: F_1, F_2, and F_3. The folds were represented on his first map (not reproduced here) as closing to the SW.

While it is not generally good historiographic practice to compare earlier work directly with modern work, it can sometimes be helpful in understanding what an earlier worker did or did not accomplish. In the case of Simpson's map, we can, by comparison with the modern Keswick sheet (BGS 1999*a*; see Plate V), see that his general outline of what came to be called the Lorton anticline has been sustained; but his clearly marked Buttermere anticline does not appear on the modern map. On the other hand, some of his structural features had long been suggested. For example, the anticline running through the northern end of Crummock Water had been mentioned by Harkness, and still makes an appearance on the modern Keswick sheet. However, some of Simpson's structures, such as the F_3 synforms and antiforms, running perpendicular to the other structures, and his long Crummock synform, were novel and do not appear on the modern map as such, the Crummock synform now being interpreted as a thrust fault (thrusting from the north), called the Causey Pike Thrust. Simpson's Gasgale Slide is now interpreted as a substantial thrust fault system (it forms a wildly fractured hillside on the southeastern slopes of Whiteside); and his Braithwaite syncline only appears as short traces on the modern map. Simpson envisaged a tangle of dips immediately to the NE of Buttermere, but he made no attempt to unravel this structure (which occurs on the mountain Robinson and was to become an important site for investigation in later work; see p. 200). It would appear, then, that subsequent survey has not sustained several features of Simpson's heroic single-handed effort.

With his proposed F_1–F_3 structures, then, Simpson was now deploying the concept of polyphase folding in the interpretation of the Skiddaw Group, which was in itself an important development. His paper is difficult to describe briefly, but I may indicate its essential features by reproducing his summarizing sections (see Fig. 9.4) and accompanying commentary. It will be seen that the nub of his interpretation was that there were two distinguishable phases of deformation *before* the Borrowdale Volcanics were emplaced (but within the Lower Ordovician); that there was a marked unconformity below the Borrowdales; and that the forces that had deformed the Borrowdales had produced a further, relatively mild, deformation of the Skiddaws.

Simpson's arguments were supported by stereograms representing (1) poles to bedding planes; (2) poles to S_1; (3) F_1 linear structures; (4) poles to S_2; (5) F_2 linear structures; (6) poles to S_3; and (7) F_3 linear structures. It would have been difficult to challenge him on these empirical findings directly without going over all the ground again and repeating all the structural observations. However, one may notice that his stereograms (3) and (5) were very similar, even if (2) and (4) were not; so one might wonder whether two distinct foldings (F_1 and F_2) could be distinguished, let alone the 'weak' F_3 folds. It is perhaps not surprising, therefore, that the geological community was sceptical of Simpson's ideas. The notion of two distinct intra-Ordovician orogenies did not mesh with received theory, though they were not of course impossible *a priori*. Moreover, the idea that the Borrowdales and the Skiddaws were conformable still had currency.

Nevertheless, Simpson (1968*a*) pushed ahead boldly, attempting correlations across the Irish Sea between the Isle of Man, the Lake District and Ireland, and further afield into southern Scotland and the Ingleton inliers (where we shall not attempt to follow him on this occasion). Thus he attempted to relate the supposed Lakeland structures (Fig. 9.4) to those envisaged for the Isle of Man (Simpson 1968*a*, p. 144, fig. 3).[11]

---

[8] He also published a paper in *Nature* on polyphase folding in Scotland.

[9] According to Moseley (pers. comm., 1996), the paper was initially submitted to the *Journal of the Geological Society*, but was refused. It was, however, published successfully in the *Geological Journal*.

[10] Sunderland is a village to the NE of Cockermouth, near the last northern outpost of the Lakeland hills. See Figure 8.6.

[11] There is a comment in the literature by Downie & Soper (1972, p. 259) suggesting that Simpson deployed ideas from his student, Douglas Helm, who was working under his supervision on the Skiddaw Group in the Black Combe inlier further south (see p. 126). As I understand the matter, the issue was that Helm had the idea of the deformation of the SS as being prior to the eruption of the Borrowdale Volcanics, thus giving a substantial early Ordovician deformation. *This* idea was first published by Simpson, but the ideas of Helm and Simpson were substantially different in other respects in that Helm had the idea of six phases of deformation in the Black Combe area, whereas Simpson envisaged three phases, as he had previously done in the Manx Slates. See also Note 16.

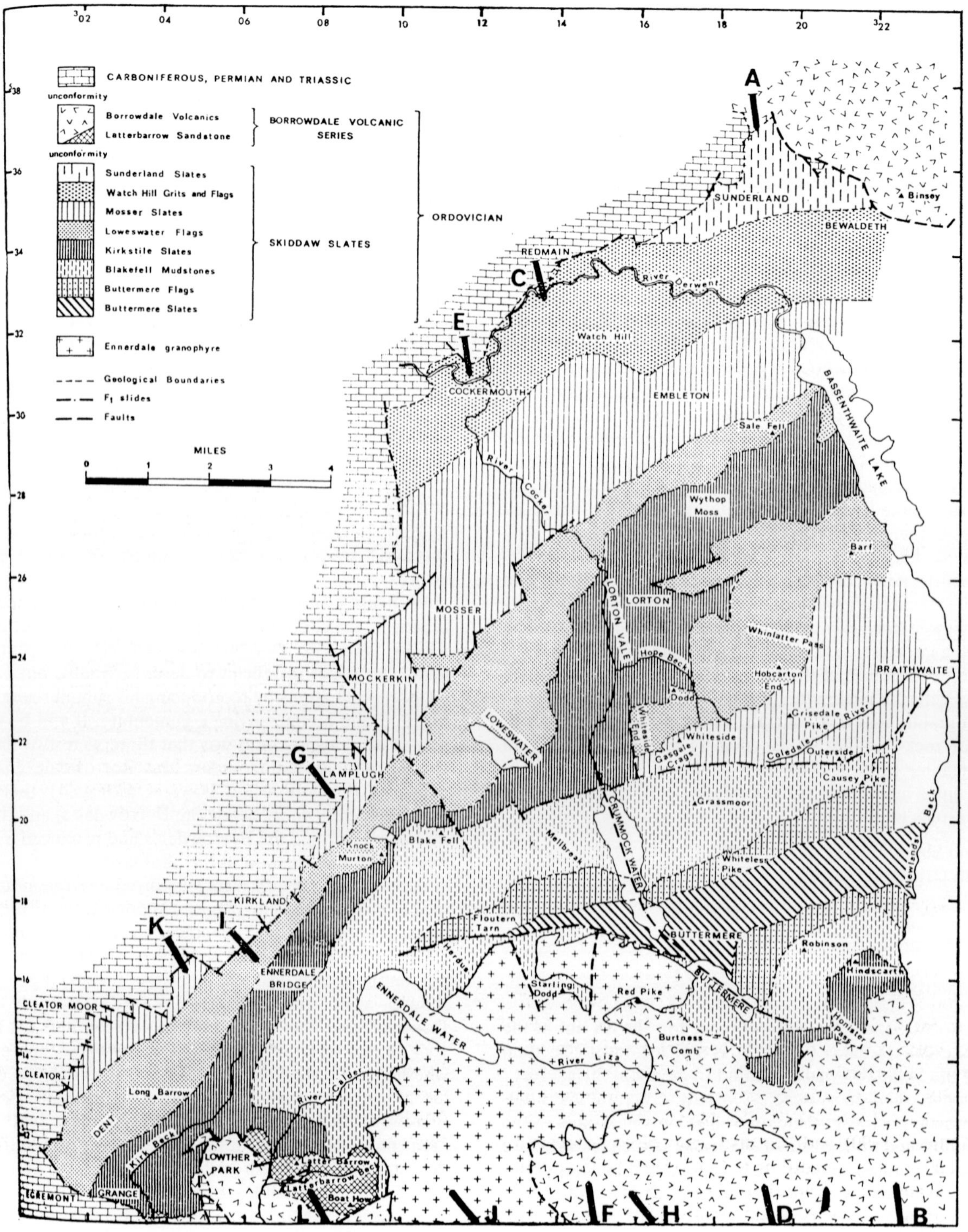

Fig. 9.3. Geological map of northwest Lake District, according to Simpson (1967, plate 18). © John Wiley & Sons Ltd. Reproduced by permission.

As is often the case in science, and particularly in geology, Simpson's work was not simply repeated, holus-bolus, there and then, to check it out. Rather, attention became focused on an issue that seemed readily testable: was there or was there not an unconformity at the base of the Borrowdales? This was, in any case, an issue that had been nagging away in Lakeland geology for over a century. It was a fundamental question for the understanding of the geology of the Lakes. Let us see how the controversy on this issue developed, stimulated by Simpson's claims, returning first to some of the older literature and the extraordinary range of views that had been put forward on this single topic.

Nicholson (1868*a*, p. 33) had supposed that there was an 'apparently conformable' succession, though it was a 'question that admits of doubt'. The idea of conformity was challenged, however, by the Surveyor Dakyns, as we saw in Chapter 4. About the same time, Nicholson (1869*a*) changed his mind and thought that he could see evidence for unconformity at what was to become a famous and controversial locality: Hollows Farm, near Grange in Borrowdale, not far from the upper end of Derwent Water (see Fig. 2.3). Though now receiving Nicholson's support, Dakyns was not defended by his Survey colleagues, for as we saw in Chapter 4, Aveline and Ward went to have a look over the ground in 1869 and concluded that the junction consisted of a whole series of small normal faults, so that the contact could follow the contours approximately. This opinion was embodied in the eventual Survey Keswick map (1875 Sheet 101 SE), and in

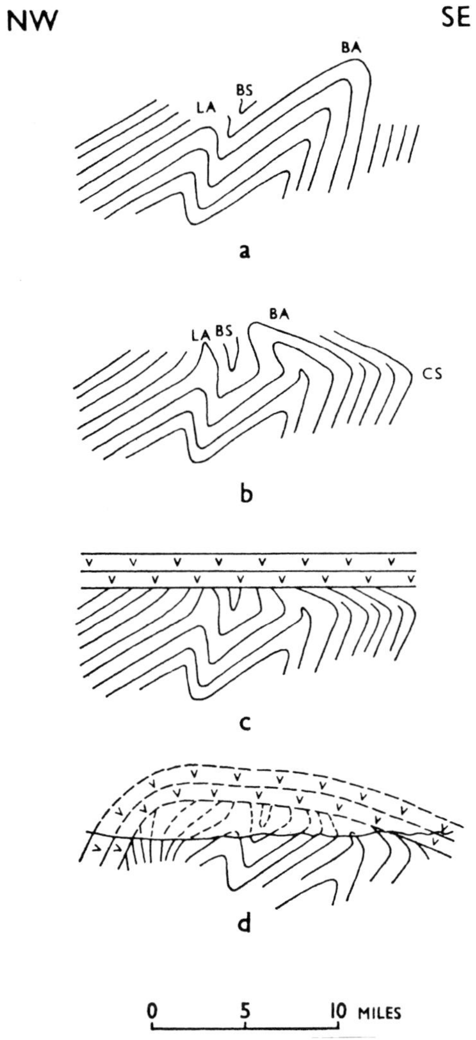

(a) F1 movement-phase. Deformation of the Skiddaw Slates by a southeasterly orogenic compressive stress. Formation of the Buttermere anticline with (in the northeast of the area) the subsidiary Lorton anticline and Braithwaite syncline on its northwestern flank.

(b) F2 movement-phase. Deformation of the F1 fold-system by northward gravitational sliding. The recumbent Crummock synform refolds the Buttermere anticline.

(c) After this dual intra-Lower Ordovician deformation, the folded Skiddaw Slates were truncated by rapid erosion. The Borrowdale Volcanic Series accumulated unconformably on the Skiddaw basement.

(d) During the end-Silurian orogenesis the Borrowdale Volcanic Series (with the overlying younger Ordovician and Silurian sediments) was flexed into the Lake District anticline by northward orogenic compressive stress. Synchronous upward warping and mild F3 deformation of the Skiddaw basement. The sub-Borrowdale unconformity was flexed by the Lake District anticline and the axial plane of the F2 Crummock synform given its existing gentle south-south-easterly inclination.

Fig. 9.4. Sections illustrating tectonic evolution of the Lake District, according to Simpson (1967, p. 412). © John Wiley & Sons Ltd. Reproduced by permission.

Ward's Memoir (1876*a*, p. 43). However, Dakyns & Ward (1875) noted that in Swindale (a valley running parallel to and SE of Haweswater) volcanic ashes of the 'Green Slates and Porphyries' could be seen interbedded with the SS.

In the 1880s, thrust faulting was in vogue, following the exemplary work of Lapworth in the Northwest Highlands of Scotland (Oldroyd 1990) and the discussion of the concept by Archibald Geikie (1884). According to Marr (1916, p. 75), quoting an unspecified newspaper report, Lapworth gave a lecture at Keswick in 1883 in which he suggested that the Borrowdales were older than the Skiddaws, with the implication that the former had been thrust over the latter; but (perhaps fortunately for Lapworth's reputation!) this idea was never published in a geological journal.

The surveyor Goodchild (1885–1886) referred to a meeting of the Cumberland and Westmorland Association at 'Hollas' (= Hollows [Farm]; see Fig. 2.3) close to Grange in Borrowdale in May 1886, led by John Postlethwaite. The party had examined the contact between the Skiddaws and the Volcanics, and had remarked volcanic material interbedded with the Skiddaws close to Grange. Goodchild had the idea, then, that the Volcanics could lie on widely separated parts of the Skiddaws at different places, implying an unconformable relationship; yet at some localities (such as 'Hollas') they might appear to be conformable. His suggestion was that the effusion of volcanic ash could have led to 'local subsidence round the vent', so that some of the later-formed ash might be deposited in a basin-like structure. In consequence, there was the possibility of the appearance of conformity in some localities and unconformity in others. His hypothetical sketch-section showed how this might have occurred (see Fig. 9.5). Postlethwaite (1890–1891, p. 48) thought that there was a 'great fault' between the Volcanics and the Skiddaws, but did not suggest what type it might have been.

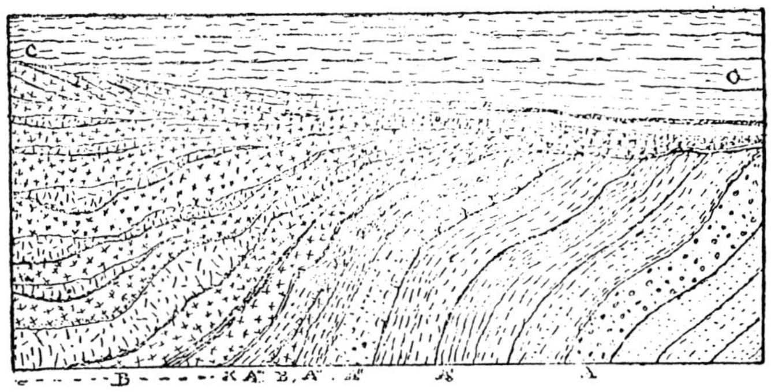

Fig. 9.5. Possible relationship between Borrowdale Volcanics and Skiddaw Slates, according to Goodchild (1885–1886, p. 477).[12]

[12] Note that Surveyor Goodchild used the term 'Ordovician' in this 'unofficial', non-Survey publication.

In his Anniversary Address to the Geological Society, Archibald Geikie (1891) gave an account of volcanicity in the Lakes, and was inclined to think of the volcanics as having been deposited under water and probably conformably to the SS.

In the Survey Memoir for Appleby, Ullswater and Haweswater, John Dakyns suggested that there might be thrust faulting in the Ullswater district (Dakyns *et al.* 1897, p. 16), but the idea was not pursued in any detail, and thrusts cannot reasonably be inferred from the relevant map (Appleby [Penrith], Sheet 102 SW (1893)). Then, as we have seen, Marr (1900, 1916) had the idea of a 'lag fault' between the Skiddaws and the Volcanics, but Green thought that the two units were conformable to one another. Trotter in the 'second survey' thought that conformity was evident in the northern fells near Overwater (see Fig. 12.1) (Smith *et al.* 1929, p. 68) (but he was there dealing with the Eycott Volcanics, not the central Borrowdales). On the other hand, the surveyors (Eastwood *et al.* 1931, p. 43) reported what appeared to be evidence of unconformity near the hamlet of Wilton, east of Egremont and SW of Ennerdale (see Fig. 20.1). Yet again, Rose (1954, p. 406) thought that the contacts were conformable in the area between Keswick and Buttermere; while Mitchell (1956*b*, p. 415) thought that there was evidence of movement in some places (as in the valley of the Lowther River, which runs northward to Penrith, receiving the outflow from Haweswater), along the bedding planes more or less adjacent to the boundary of the two units, with the fault planes perhaps themselves folded by later movement. Hollingworth (following Rose 1954 [*fide* Helm 1970, p. 124]) suggested that the structural dislocation might be located in the Skiddaws at some localities and in the Borrowdales at others, whereas in his well-known summary of Lakeland geology (Hollingworth *et al.* 1954, p. 387) he had accepted a conformable relationship. The Leeds PhD student Lewis Clark (1963, 1964), studying the rocks around the Ennerdale Granophyre in the general area between Buttermere and Wasdale, thought that the Skiddaw Slates–Borrowdale Volcanic Group (hereafter SS–BVG) contact displayed a thrust fault. To put it bluntly, the interpretive situation was a mess! It was Simpson who stimulated geologists to examine the whole matter more fully.

Given the aforementioned 'mess', it is difficult to say what was the 'standard view' of the matter, but a number of geologists were of the long-standing view that the boundary was one of conformity. Among these geologists was Jack Soper (b. 1934) (see Figs 14.3 and 14.5). His name will appear frequently in the pages that follow, as he has taken an interest in Lakeland geology for many years and has contributed greatly to our understanding of the structure of the region and its geological history, and even at the beginning of the twenty-first century was doing contract mapping for the Survey in southern Lakeland and the Howgill Fells; so it is appropriate to introduce him here.

Soper, who, I should mention, has been notably helpful in providing information for the present study, hailed from the West Riding. He attended Woodhouse Grove Methodist School near Bradford and became interested in geomorphology through walking on the moors with his father (pers. comm., 1998). After national service, he read Geology at Sheffield University, where he got a first, doing his honours mapping in Crummack Dale (one of the 'Lakeland-type' inliers of NW Yorkshire; see Fig. 3.5), for which he received the Fearnsides Prize in 1959. Thence he proceeded to a PhD (1963) on the Rogart Granite area and its migmatitic envelope in remote NE Scotland (inland from Brora), his supervisor being Peter Brown, with whom he used to go rock climbing in the Lakes. The doctoral research naturally involved Soper in the later phases of the migmatist–magmatist debate, particularly since he took up a DSIR fellowship at Imperial College even before he had finished his doctorate, and so came in contact with controversialists such as Herbert Read. However, the migmatist school was already in decline by that time. Soper was soon offered a job back at Sheffield, where he stayed until 1988, when he moved to Leeds following the downgrading of his old department.

Jack has been an enthusiastic controversialist. His work has focused on interpretations of the deep structure of Britain and geotectonic models of the British Caledonides in terms of terrane accretion, on the tectonics of the Scottish Highlands and the Lake District, and on the tectonics of Greenland, where he spent 15 seasons. He was an active participant in the BGS–universities collaboration on the remapping of the Lakes and the structural interpretation of the region (see Chapter 14). He moves easily from minute details in the field, to maps, and on to the large-scale theoretical picture. At the end of the twentieth century, he was Adjunct Professor in Structural Geology at Galway.

From Sheffield, Soper was soon at work in the Lakes, and in contact with Simpson and the Birmingham geologist Frank Moseley, who will be appropriately introduced later in this chapter. Soper was reacting to Simpson's idea that there had been three sets of Palaeozoic foldings in the Lakes: the first two constituting a tectonic episode *within* the Ordovician, which had folded and cleaved the SS *before* the deposition of the Borrowdale Volcanics; and another at the end of the Silurian (Caledonian movements).[13] With his former supervisor, Soper wrote to the *Scottish Journal of Geology*, taking issue with Simpson's (1968a) ideas (Soper & Brown 1968*a*). Soper referred to a BVG site called the Greenscoe Vent, near Dalton-in-Furness (SW Lakeland, see Fig. 1.1), where Borrowdale Volcanics could be found in contact with Skiddaw rock. On Soper's view, there was but one cleavage there, which passed through all the considerable variety of rocks present. Fragmental rock in the vent contained Skiddaw material; and examination of thin sections showed that the cleavage 'passe[d] through the slate fragments and volcanic matrix alike'. Bedding could be seen in the slate, but no earlier cleavage. This evidence, if correct, was damaging to Simpson's theory. Further, as Soper pointed out, there are areas of the Skiddaws (such as the Watch Hill Grits, near Cockermouth) that do not display cleavage at all.

Simpson, however, had two London co-workers, and partial allies: Brinley Roberts and Douglas Helm, who had completed a PhD on the Skiddaw inlier in SW Lakeland at Black Combe in 1968.[14] (For the topography of the Black Combe area, see Fig. 6.3.) While Simpson thought that there were three phases of folding, Helm (1968, 1970) went further and claimed that he had evidence for six in the Black Combe district. Polyphase deformations were definitely in fashion. However, the 'northern' geologists were suspicious, particularly because Simpson did not bring out his field maps and show them to other geologists; and he was disinclined to go into the field with them, demonstrate his structures, and debate the issues on the ground (Soper, pers. comm., 1998, 2000).[15]

There was also apparently some trouble in the London camp. Helm & Brinley Roberts (1968) accepted Simpson's notion of two phases of intra-Skiddavian deformation, but disputed that the folds had the same alignments in the Black Combe area and the main Skiddaw inlier. Helm's F_4 was pre-Borrowdale, whereas for

[13] Subsequent work, consequent upon the dating of the Shap Granite, places the later folding, cleavage, etc., in the early Devonian.

[14] Unfortunately, I have been unable to make contact with Dr Helm to hear his account of events. I understand that he subsequently became Head of the Department of Geology at Goldsmiths' College, London University. Brinley Roberts was an approximate contemporary of Helm as a research student at Birkbeck. His PhD (supervised by Frank Fitch) was concerned with the Llwyd Mawr ignimbrite in North Wales and was completed in 1965. However, he participated in field excursions in the Lakes and discussed Lakeland problems with Helm. Roberts went on to join the staff at Birkbeck and worked on low-grade metamorphism in Wales, the Lakes, the Southern Uplands and Antarctica (Roberts, pers. comm., 2000).

[15] Adrian Rushton (pers. comm., 2000) also recalled that the Surveyor Colin Rose found that Simpson seemed reluctant to discuss issues with him.

Simpson it was end-Silurian, and was designated F_3. The idea was unacceptable to Helm, for it appeared to him (and Roberts) that it would mean that a single N–S compressive force could generate both an ENE-trending anticline and NW-trending minor structures. Hence he had increased the number of proposed deformations. In fact, Helm and Roberts were generally sceptical about making structural comparisons between rocks in separate localities. They doubted they could be made for Black Combe and the main Skiddaw inlier, and were even more sceptical about Simpson's structural correlations between the Skiddaws and the Manx Slates. On a more specific point, they maintained that Simpson's (1967, fig. 4g) stereogram for F_3 linear structures showed two maxima (in the NW and SE quadrants) and formed a girdle about a NE–SW axis. They interpreted this to mean that the NW–SE trending structures had been refolded by the movements that produced the Lake District anticline.

Simpson (1968*b*) replied tartly. He claimed that Helm's F_3, F_5, and F_6 were very 'weak', not associated with pronounced folding, and were without regional significance. So, in effect, one could only talk about three phases of deformation. Certainly, he thought, Helm's weak Black Combe F_3 should not be correlated with his (Simpson's) strong Manxian F_3. Interestingly, Simpson maintained that Helm and Roberts had misunderstood the significance or proper interpretation of his figure 4g (1967). As he put it: '[t]he fact that the poles of a set of linear structures form a girdle does not necessarily mean that such structures have been refolded'. For 'the linear structures associated with the intersection of an earlier fold or folds by a later crosscutting cleavage will lie in a girdle or partial girdle'.

The fact that such a debate could appear in print between a lecturer and a former student was not very 'edifying' and suggested that relations between Simpson and Helm had deteriorated, at least in relation to scientific matters. It also reveals that the interpretation of stereograms could be contentious, and that they could not be treated in a simple 'positivistic' fashion.

Be this as it may, Helm and Simpson agreed that there was an unconformity below the Borrowdales, a view that the geological community as a whole did not accept. Thus Soper examined what he called three 'critical localities', which, he claimed, showed evidence of a *conformable* relationship between the Borrowdales and the Skiddaws. The selected sites were the nicely accessible exposures by Hollows Farm near Grange, Borrowdale; at the aforementioned 'Greenscoe Vent' on the southwestern margin of the Lakes, where there is a small inlier of Borrowdales in contact with Skiddaws; and Crookley Beck, east of the village of Bootle on the SW side of the Lakes (see Fig. 6.3). Here the beck runs E–W along the line of contact between the Borrowdale Volcanics to the north and the Skiddaws of the Black Combe inlier to the south. It would appear that there was formerly a reasonable exposure in the northern side of the small valley according to Soper (1970, plate 21), probably because the site had been cleared, but the area was considerably overgrown when I visited it in 1996.

On one issue there was agreement: the character of the cleavage was as important as visible conformity or unconformity. For if the cleavage(s) ran through the contacts, i.e. if there was the same cleavage system above and below the contacts, then the Simpson–Helm idea of two distinct orogenies (with a total of three foldings – or six, according to Helm) would be suspect. Unfortunately, however, the cleavage did not behave in a simple manner. The volcanics and the slates were so different in their ductilities that they had evidently responded very differently to tectonic forces. Because of ductility differences, the cleavage could swing round locally so that it was almost parallel to the supposed boundary. There was also the question of whether or not there was interbedding between the uppermost Slates and the lowest Volcanics, as might be the case if there were 'intimations' of volcanic activity during late Skiddaw time, with ash beds beginning to appear in the highest slates.

Soper's paper, which was accompanied by large-scale sketch-maps for his three areas, was read before the Yorkshire Geological Society at Hull in April 1969. It evidently generated animated discussion, though Simpson was unable to be present, the 'Birkbeck view' being argued by his co-workers Helm and Brinley Roberts. Simpson later sent in a written commentary, which was duly published.

Soper was willing to countenance more than one distinguishable structural surface arising from tectonic activity, but he only admitted one substantial phase of deformation. From his perspective, it seemed implausible to suppose, as Simpson had done, that 'F_3' could produce substantial folds in the Borrowdales and the Silurian rocks of southern Lakeland, aligned approximately SW–NE (Caledonian trend), *and* the weak NNW–SSE folds claimed to occur in the Skiddaws. At Hollows Farm, Soper asserted the occurrence of a fresh exposure, with tuff filling a channel in an underlying conglomeratic mudstone, the junction dipping gently to the WSW. A steep SSE-dipping cleavage was said to be found in both rock types, though it was more pronounced in the pelitic Skiddaw rock. In general at this locality, there appeared to be a set of conformable beds – from slates to volcanics with intermediate 'transitions' – with cleavage passing through all of them. Likewise at Crookley Beck there appeared to Soper to be a transitional 'Mixed Group', forming a conformable passage from the Slates to the Volcanics, again with cleavage passing through. A photograph by Soper certainly suggested a conformable sequence, but my own visit to the locality (as said, now rather overgrown) in 1996 left me wondering what was Slate and what was Volcanic. Soper's notion of 'Mixed Group' was, I take it, intended to deal with the problem.

As already mentioned, discussion at Hull was animated. Helm claimed that there was an unconformity 'convincingly displayed' at Crookley Beck, the basal group of the Volcanics resting on various different lithological units of the SS. There was, he asserted, a plunging fold-pair in the Skiddaws which struck directly towards the Volcanics. Brinley Roberts wondered how Soper would account for all the structural complexities in the Skiddaws, compared with the relatively simple structure of the Borrowdales, using only one main tectonic episode. Worse, he accused Soper of failing to distinguish correctly between bedding and cleavage at Crookley Beck, so that the section was misinterpreted. Simpson wrote in, asserting that Soper's whole methodology was wrong: you could not establish the relationship between the Borrowdales and the Skiddaws by looking at 'critical localities' in isolation. Soper's 'piecemeal' approach was 'unsound, and, as past history ha[d] shown, indeterminate'. A much broader view needed to be taken to resolve the problem. Regarding Greenscoe, he said that what Soper had taken to be Skiddaws were in fact 'shaly intercalations near the base of the Borrowdale Volcanic Group', a point he had previously made in Simpson (1968*b*).

However, Soper was not without support. Colin Knipe (who will reappear in Chapter 20) and W. Grieve reported that they had found graptolites in the 'shaly intercalations' at Greenscoe, which Oliver Bulman had identified as belonging to the Skiddaws. Frank Moseley (see p. 133) said that he knew of areas in the Buttermere–Borrowdale area where he had found conformable passages, with ENE cleavage passing through. He thought that Simpson's F_1 and F_2 structures were found also in the Silurians, so that the whole system of folding could be end-Silurian. David Roberts (who was studying the complex folding of the Skiddaws in the Caldew Valley, just to the south of Carrock Fell, for his PhD at Birmingham, under Moseley; see Chapter 12), had found N–S folds in both Skiddaws and Borrowdales. Moseley had found similar folding in the Volcanics near Ullswater, as had Firman and Mitchell in their areas. In brief, Moseley thought that there were analogous structures in both the Skiddaws and the Borrowdales, so all might be due to end-Silurian deformation. But he agreed with Simpson that the deformation was polyphase. David Roberts thought that there was conformable passage from the Skiddaws to the Borrowdales, but the folding should be regarded as 'pre-Bala' (pre-Coniston Limestone Series; hence Ordovician) rather than end-Silurian.

Simpson's views were broadly supported by Helm (1970), in a paper read at Leeds before the Yorkshire Geological Society in October 1969, making known the results of his doctoral work. From the discussion following Helm's paper, it is evident that there had also been a field excursion to the Black Combe area earlier that year, and according to Ronald Firman who participated in the excursion it was a pretty heated affair too, like the one at Hollows Farm, if not more so. However, I have not located any published account of the Black Combe meeting. Soper did not participate, as he informed me (pers. comm., 2000). He was off sledging in Greenland at the time.

Anyway, as mentioned, Helm went further than Simpson, envisaging six deformation phases. His D_1 and D_2 corresponded to Simpson's F_1 and F_2, and Helm's D_4 was ascribed to 'end-Silurian'.[16] The other deformations were acknowledged as being less clear and less significant. It is instructive to compare here Helm's sketch of one of his most important sites (near White Hall Knott: 3157 4857; see Figs 6.3 and 9.6a) with a recent photograph of the same locality (Fig. 9.6b). It will be seen by comparison of these illustrations how Helm construed his observations at this particular site, but his argument rested on the analysis of many determinations of dip and strike, plotted on stereograms as had previous workers such as Glennie and Simpson. D_2 is represented as a more open folding than D_1. Personally, I cannot discern the faint D_3 structures claimed by Helm for Figure 9.6a, but they are not important to the argument, though one may say that it is difficult to conceive a sequence of tectonic forces that might have yielded the asserted sequence of deformations.[17] Be this as it may,

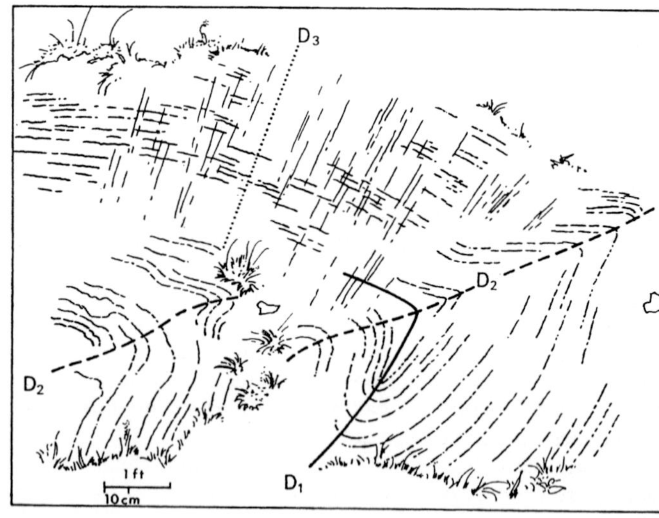

Fig. 9.6a. Sketch showing three phases of deformation of the Skiddaw Slates (Wicham Blue Slates), Black Combe Inlier (Helm 1970, p. 114). Reproduced by courtesy of the Yorkshire Geological Society.

Helm (1970, p. 127) claimed the existence of a major unconformity below the Borrowdale Volcanics, with his first three deformation phases *not* passing through to affect the overlying volcanic rocks. Helm did not acknowledge assistance from Simpson, but only Brinley Roberts and John Temple, who had helped him with fossil identifications.

Fig. 9.6b. Photograph of same site, taken by author (1998).

[16] It may be noted that Helm (1970) did not acknowledge Simpson's assistance, and it would appear from what Simpson's daughter told me (pers. comm., 1999) that relations between her father and Douglas Helm were not the best. It is important to note that Helm stated (Helm & Siddans 1971, p. 529) that he initially finished his PhD in 1965, which would be two years *before* Simpson published his paper on the Skiddaws and 'put the cat among the pigeons', so to speak. I understand that Helm's thesis had to undergo a rewrite (Moseley, pers. comm., 1996), which would explain why it did not appear until 1968. On the foregoing chronology, Helm would have been developing the idea of multiple deformations in the Black Combe area *before* the appearance of Simpson (1967). However, Simpson undoubtedly brought the idea of polyphase deformation with him from the Isle of Man to the Lakes; and in any case it was pretty evident to all interested parties that the SS were multiply-deformed (though the deformations might arise from slumping or tectonic activity). The whole notion of determining multiple deformations by structural analysis was popular at the time. According to Dr Brinley Roberts (pers. comm., 2000), Simpson took Helm's work as confirmation of his idea of polyphase folding in the Lakes; and as far as he can recall, Simpson accepted Helm's observations and conclusions. According to David Roberts (pers. comm., 2000) and Soper (pers. comm., 1998), Helm – from his Black Combe work – first had the idea of tectonic foldings *within* the Ordovician. Simpson used this notion when he was mapping the northern Skiddaws, and published it first.

[17] With hindsight, it may appear that Helm elevated locally developed kink-bands into regional structures.

There was much interesting and complicated discussion following the presentation of Helm's paper. Soper started off by congratulating Helm on his work and saying that he accepted Helm's interpretations of the structure; but then went on to tell a different story about many points! All the participants seemed to follow rather the same tack (or tactic). For Soper, the anticline recognized in Black Combe by Helm (his 'Borrowdales Anticline') might be complementary to the Ulpha Syncline of Mitchell, the whole plunging eastwards. Such an anticline might explain the occurrence of the exposed Black Combe Skiddaws in SW Lakeland. It was, for Soper, a pre-Coniston Limestone – but post-Volcanics – structure rather than end-Silurian. There were also the 'end-Silurian' foldings. On this view, there did not need to be Simpson and Helm's intra-Skiddaw tectonic activity.

Peter Brown (Soper's colleague and former supervisor at Sheffield) reminded the meeting of the debate in the field earlier in the year, led by Helm, where they had been unable to agree at Crookley Beck whether cleavage did or did not pass through from the Skiddaws to the Volcanics. Moseley, like Soper, thought it did. Ian Burgess reported corroborative evidence from the Cross Fell Inlier where he had been surveying with Tony Wadge. Wadge mentioned the evidence coming in from the eastern Lakes where the 'high' *Didymograptus murchisoni* zone had been found in the Skiddaws in the Tarn Moor Tunnel; and the Borrowdales appeared to lie successively across lower zones from east to west, implying unconformity (see p. 136). Michael Nutt (who was to gain his PhD from Queen Mary College, London, in 1970, on the geology of the volcanics of the Haweswater area, with Moseley as external examiner; see Chapter 12) advocated an intra-Borrowdale age for the unconformity in the eastern Lakes (see also Nutt 1968), and doubted that the Skiddaw clasts in the Borrowdales in that area showed an independent, earlier cleavage such as Helm claimed in the Black Combe area.

In his reply, Helm revealed that he and Soper had been in communication and had to some extent used the same data, including information supplied from Firman and Mitchell, but had arrived at rather different conclusions about the relationships between the claimed anticlines and synclines and the directions in which they plunged. This again suggests that the 'positivistic' use of stereograms to reveal 'averaged' structures was not infallible, and could yield dissimilar results from the same database; and that the structural knowledge of the region was still too piecemeal to yield unambiguous results. Complete and systematic resurvey was needed; but that was not to occur until the 1980s.

The situation was, then, by no means resolved in 1970, and it rapidly worsened! There was by this time a deal of knowledge on the point at issue, albeit unsynthesized. Notably, David Roberts had been working on his PhD on the complex foldings of the SS in the Caldew Valley. Some years earlier, Moseley had been busy in the Ullswater area, as will be discussed below. The surveyors Tony Wadge and Ian Burgess had been mapping in the Cross Fell Inlier and also examining rocks in the Tarn Moor Tunnel, which cut through the SS–BVG contact (see p. 136). Moseley's student Peter Jeans, who was writing a PhD on the geology of the SS in the Crummock area (Jeans 1973*a*) (see Chapter 12), had made an extensive examination of the SS–BVG boundary along the whole line of the contacts. Nutt had his PhD on the Haweswater area. The Birkbeck group were continuing to take an interest in the question. Thus we have the 'notorious' excursion to the northern Lake District made by the Geologists' Association in September 1970, when a considerable number of geologists with views on the question were gathered together (Mitchell, Moseley, Firman, Soper, Roberts, Nutt and Wadge; see Mitchell *et al.* 1972). Here I refer chiefly to the events that occurred at the Hollows Farm exposures, where Soper was leading the day's excursion.

Soper's view (at least at the beginning of the day!) was that there was a 'passage' from pelite with minor tuffaceous bands to andesitic tuff, there being slight erosional unconformity but no significant angular unconformity. The SS had a steep ENE-trending slaty cleavage, affected by a gently inclined second cleavage. The former ran through to the volcanics and therefore post-dated that part at least of the Borrowdales.

However, the geologists could not agree where the boundary lay between the Borrowdales and the Skiddaws, with Helm wanting it lower than where Soper allowed. The problem was exacerbated by the transitional nature of the rocks near the boundary, and the fact that the Skiddaws' bedding seemed to disappear as the Slates approached the contact.[18] At one site – the most contentious and most visited one, being conveniently near the lane leading from Grange to Hollows Farm – Soper claimed that a steep cleavage passed from the slates through to the tuff, but was expressed more weakly in the latter. Brinley Roberts claimed to see two cleavages in the pelitic slates, and it was the weaker, less inclined 'fracture cleavage' that passed through to the tuff. Soper would not have this, saying that the 'weak cleavage' correlated with joints in the tuff, not cleavage.

At another spot in the adjacent Scarbrow Wood, Soper's 'flow-banded andesite' was construed by Nutt as 'bedded tuffs'; at which '[t]he Director diverted attention to the possible pre-cleavage age of . . . [a nearby] fault'. Examination of a section below, which had been excavated by Jeans in 1969 and was apparently cleaned up for the purposes of the excursion, revealed 'an apparently perfect transition from pelite to tuff . . . with a single cleavage passing through both rock types'. Understandably, Dr Soper 'hailed this as vindication of his interpretation'! However, according to a recollection of Peter Jeans (pers. comm., 2000), Soper's effort to use a drill rig to extract a core from the junction to see whether a fresh section of the contact might assist resolution of the issue ended in failure. The drill only went in a short distance and then packed up. Darkness fell before success was achieved. Soper has confirmed (pers. comm., 2000) that he brought a core rig up from Sheffield and was assisted by his department's technician, Graham Mulhearn. The idea was to see whether a cleavage common to the Skiddaw Slate and the volcanics could be discerned in thin section, but as said, this effort came to nothing.

A visit to the site today does not cast as much light on the problem as might be hoped. The exposures in Scarbrow Wood (where the drill was applied) are all moss covered and should not be disturbed in the national park. Higher up, in the open ground on the fell, the actual contact is mostly obscured by grass, but Alan Smith of the Cumberland Geological Society has kindly supplied me with a recent photograph (Fig. 9.7a), taken by a windswept holly tree where, according to Shackleton (1966*b*, p. 39), one could see places where 'geologists have broken off pieces of the rock trying to get a specimen of the two rocks side by side on the one piece'. Figure 9.7 shows other exposures in the area of Hollows Farm and the general locality.

On the whole, I suspect that the Birmingham–Sheffield group out-gunned the Birkbeck people, and Soper went home still thinking that the contact was one of conformity. But it was not a satisfactory victory, if indeed it was one at all; and Soper has long since acknowledged that the SS–BVG contact is not conformable overall, though he still thinks it may be so in places, especially to the east, where the time gap decreases as higher zones of the Skiddaw Group appear – perhaps out of sight under the Alston Block (Soper, pers. comm., 2000). To decide the question of unconformity, of course, one really needed to carry out large-scale mapping, as Simpson had said in commenting on Soper (1970), but this had not been done at the time of the controversy. So in 1970, the problem was reduced to a smaller 'crucial experiment' or 'crucial observation': did the cleavage pass through from the Skiddaws to the BVG? This would seem to be a straightforward empirical question. But it was not. Frank Moseley, reminiscing on events many years later, told me that he could not believe that

[18] Perhaps due to being reworked by erosion immediately below the contact.

Fig. 9.7a. View of grass-covered contact zone by holly tree behind Hollows Farm, between Skiddaw Slates and Borrowdale Volcanics. Photographed by Alan Smith, 2000, and reproduced by courtesy of Dr Smith.

experienced geologists could look at the same outcrop and apparently 'see' such entirely different things (pers. comm., 1996). There was, I suggest, a classic case of the theory-ladenness of observations; or perhaps the controversy-ladenness of observations. Personally, I do not find it at all surprising. As has been said by the philosopher of science, N. R. Hanson, 'there is more to seeing than meets the eye-ball'.

In fact, even the most recent published maps and sections (Black Combe sheet, 1998; Keswick sheet, 1999) do not settle the matter unambiguously. In the former map, the appended section through Crookley Beck shows the two units inclined at about the same angle, but there is a thrust fault marked at the contact (following the line of the beck); and in the latter the bedding of the Skiddaws is not marked on the section, so that the nature of the inter-unit relationship appears to be 'agnostic'. However, geologists are now all agreed that the Borrowdales lie across different subdivisions of the Skiddaws at different places, so there cannot be a conformable relationship overall.

Of course, Rome was not built in a day, and neither was the question of the SS–BVG boundary settled when the Geologists' Association visited Hollows Farm on 2 September 1970. In fact, the debate hotted up even more, soon after that excursion. Helm & Brinley Roberts (1971*a*) published a paper in *Nature*, which set out their interpretation of the situation at Hollows Farm and the SS–BVG relationship more generally. They proposed the following tectonic sequence:

- D_1 tight, upright, isoclinal, approximately N–S folding of the Skiddaws;
- D_2 vertical compression superimposing D_2 folds on D_1 and generating S_2;
- D_3 erosion and unconformable deposition of BVG on the erosion surface;
- D_4 folding along E–W axis generating associated S_4 cleavage, which passed through all the rock types.

It was suggested that Soper had confused or conflated the bedding-plane cleavage in the BVG with the S_2 surface in the Skiddaws.

So in Scarbrow Wood (by Hollows Farm), Helm and Roberts

Fig. 9.7b. Contact zone near holly tree between Skiddaw Slates (right) and Borrowdale Volcanics (left). Photographed by author, 1996.

Fig. 9.7c. Contact between Skiddaw Slates (right) and Borrowdale Volcanics (left), Scarbrow Wood. Photographed by author, 1999.

had a disconformity between the Volcanics and 'erosional hollows' in an underlying tuffaceous conglomerate. Then there was an exposure gap where the section was obscure; and below that were Skiddaw Slates with what appeared to be tight, nearly upright folds.[19] At Crookley Beck, the authors claimed, Soper had confused a bedding-plane cleavage in the BVG with an earlier axial-plane cleavage in the folds of the Skiddaws. Unconformity between the BVG and the SS was claimed because the former sat on top of different graptolite zones at different places along the boundary. (Following the work of Jackson, more was now known about graptolite zones in the Skiddaws through the work of, among others, the surveyor Tony Wadge (Wadge *et al.* 1969), and David Skevington (1964, 1970) of University College, Galway.) An accompanying sketch-map suggested that the Skiddaws formed an anticlinal fold, plunging north, with its axis aligned NNW–SSE, and the BVG ran unconformably across the southern side of this anticline.

Needless to say, the zonal boundaries were imprecise because of the scarcity of graptolites. Writing in *Nature*, Wadge (1971), who had found the Upper Llanvirn zone fossil *Didymograptus murchisoni* within the Tarn Moor Tunnel in eastern Lakeland (which suggested that the Skiddaws went up to a higher zone in the east than had previously been supposed), quickly came up with quite a different structure for the Skiddaws. He had them folded along an ENE–WSW axis, plunging ENE. Jeans (1971) also wrote to *Nature*, informing readers that he had mapped the Hollows Farm area in detail in 1969 and (as we know from the 1970 excursion report) had excavated the contacts as much as was feasible. He seemed tacitly to accept the occurrence of a disconformity within the lowest BVG at this locality, but the substantial unconformity claimed by Helm and Roberts could not actually be *seen* – as was evident from their published sketch profiles (Helm & Roberts 1971*a*, p. 182). On the other hand, Jeans was happy with the idea of a 'regional break' between the BVG and the Skiddaws.

Helm & Roberts (1971*b*) were not apparently much disconcerted by the graptolites or Wadge's objections and made

Fig. 9.7d. General view of Hollows Farm and hill rising to Maiden Moor: rugged volcanic country on left; Skiddaw Slates on lower ground to right. Photographed by author, 1996.

[19] Later interpreted as structures due to sedimentary syndepositional slumping.

things fit again by hypothesizing *two* anticlines in the Skiddaws – perpendicular to one another! This schema fitted the Birkbeck notions of polyphase folding, of course. They claimed to be puzzled to as to why Jeans would allow a 'regional break' and yet be so reluctant to accept the idea of unconformity at Hollows Farm.

Soper, with Moseley's PhD student David Roberts, returned to the contest in a paper in the *Geological Magazine* (1971), using rather different arguments. Roberts' PhD (1973) (see Chapter 12) was well advanced, and some details had already been published (D. Roberts 1971). He had made a detailed examination of the complex folds exhibited in the hornfelsed SS exposed in the Caldew Valley close to the small outcrops of the Skiddaw granite, also seen in the river bed a few hundred metres upstream. Now Soper was able to enlist Roberts, with his knowledge of the Skiddaw aureole and the Skiddaw Slate foldings, to tackle the question of the occurrence or otherwise of intra-Skiddaw orogeny (or orogenies) as favoured by the Birkbeck geologists. Interestingly, Soper also put some of his students at Sheffield to re-examining Rastall's old slides, made when he had been examining the Skiddaw aureole back at the beginning of the century (see p. 76).

By the time that Soper and Roberts were at work, the granite itself had been dated radiometrically to about 400 Ma (Miller 1961; Brown *et al.* 1964), or 'end-Silurian'. (Estimates for the date of the Silurian–Devonian boundary have been changed since 1971, but not sufficiently to affect the argument.) Examining thin sections from the aureole in the area of the corrie to the NE of Atkinson's Pike to the north of Blencathra (also called Saddleback) (see Fig. 12.1), they found that andalusite crystals had been formed *between* the two periods of deformation. That is, the microscopic evidence suggested that the andalusite crystals had 'overgrown [Simpson's] S_1 statically'; and then the crystals had themselves been deformed in various ways (in F_2). So Simpson's F_2 was post-granite and post-Ordovician. Yet according to Simpson, F_1 and F_2 were fairly close in time and both intra-Ordovician. More detailed argument was supplied, which will not be rehearsed here. The point was that there could not have been two intra-Ordovician deformations – provided that the deformations in the Blencathra area were indeed the ones that Simpson had been discussing elsewhere in the SS. Simpson himself made no response to the new argument of Soper and Roberts. Soper's approach, aimed at giving a relative dating to the deformations and cleavage by examination of their relationships to datable igneous rocks, was effective and difficult to counter.

Meanwhile Frank Moseley (who is still waiting to be properly introduced; see p. 133) had put his PhD student, Peter Jeans, to work, examining the line of the SS–BVG boundary in areas other than the 'classic' ones such as Hollows Farm; and some of his undergraduate students too (Moseley, 1973–1974 [1975]).[20] Jeans (1972) examined particularly the locality at the upper reaches of Newlands Beck between High Spy and Hindscarth (or between Derwent Water and Buttermere; see Fig. 7.4). In the stream bed, the Slates, though folded in places,[21] generally dipped quite steeply eastwards, striking approximately N–S. Above them, the unfolded Borrowdale Volcanic rocks dipped gently SE. There appeared to be no indications of faulting, and the single cleavage observed (ENE–WSW Caledonoid; Simpson's F_1; Helm's D_4) seemed to pass through from the Slates to the Volcanics. Jean's view, like that of Soper and Moseley, was that the then supposed tectonic Skiddaw foldings were pre-BVG, and the second deformation episode, which had caused the cleavage, was end-Silurian. However, the actual contact between the two units was not visible in the stream bed due to debris. And, I may say, this situation seems to recur almost everywhere one wants to see the contact. It seems to be wilfully elusive!

In fact, Moseley had been putting his students to work examining the SS–BVG contact in as many places as possible, and he now published a summary account of all this work in the *Proceedings of the Cumberland Geological Society* (Moseley 1973–1974 [1975]), one of the two very active local geological societies serving Cumbria.[22] As he said: '[i]t has become increasingly apparent that this junction is an exceedingly difficult one to interpret, even when it is completely exposed' (Moseley 1973–1974 [1975], p. 128) – which is generally not the case. There always seems to be some 'snag' at the many exposures, making matters obscure and difficult to interpret. Were it otherwise, it would not, of course, have been a controversial issue in Lakeland geology. In fact, it was coming to be realized by the early 1970s that the nature of the contact might be different at different places.

For example, at Newlands Beck, even though the contact could not be seen clearly, there appeared to be angular discordance between the slates and the volcanics; but there also appeared to be faulting. Similarly at Warnscale Bottom, in the beck running into Buttermere (see Fig. 12.3). By contrast, at Hollows Farm, there appeared to be a conformable transition from slates to tuffs. Further east, near the lane connecting St John's in the Vale with the valley of Naddle Beck, between Low Rigg and High Rigg (see Fig. 12.1), the contact, though not exposed, seemed to show a conglomerate at the bottom of the volcanics, which Wadge (1972) construed as an unconformity, but which Moseley thought involved faulting.

Yet further east, at Matterdale Beck (see Fig. 9.9), Wadge (1972) had remarked on another conglomerate apparently lying with clear unconformity on the slates of *D. bifidus* zone. The Birkbeck group had thought that some of the cleavage there did not pass through to the volcanics, whereas the Birmingham–Sheffield faction held that the cleavage did pass through. So consensus was not reached on the issue of whether there was an orogeny between the Skiddaws and the Borrowdales. Near Ullswater, Moseley had claimed that he had found evidence of thrust faulting (as will be discussed below).

Further north, near Whitefield Cottage (3238 5348), and at Overwater (see Fig. 12.1), the Eycott Volcanics appeared (at that time) to be conformable to the Skiddaws, for they were seemingly

[20] Moseley mentioned K. Smith, D. T. Aldiss, B. W. Bull, E. B. Daniels and J. M. Chaytor; and professional geologists such as Tony Wadge were also mentioned.

[21] These structures are also now regarded as syndepositional.

[22] The Cumberland Society, which grew from an earlier Workers' Education Association group, was founded in 1961 by Charles Edmonds and Edgar Shackleton (1903–1991), the latter proving to be the major leading light in the Society in its early days (Shipp, 1990–1991 [1992]). (See also Dodd 1990–1991 [1992].) Edmonds was a miners' unionist with a detailed local knowledge of the geology of west Cumbria. Though leaving school at 12, Shackleton had some scientific training at Blackburn Technical College and he moved to the Lakes (Bowness) in the 1920s, doing work as a guide and local lecturer. During World War II he was drafted into explosives manufacture at Drigg in west Cumbria and afterwards did applied geological work in that region. He did much work in support of adult education and the Lakeland National Park, made extensive geological collections, which are now on display in a room dedicated to his memory in the 'museum' working haematite mine (Florence Mine) near Egremont, wrote three attractive guidebooks to Lakeland geology, and published several valuable historical papers on Lakeland geologists in the Cumberland Society's *Proceedings*, as well as excursion reports and various Presidential Addresses. Shackleton was a staunch supporter of amateur Lakeland geology, though in his old age he vexed his fellow amateur geologists by refusing to countenance plate tectonics, and the Survey geologists did not think very highly of him, I have been told. The Cumberland Society still conducts excellent field excursions in the summer and has a winter lecture programme. Papers by both amateur and professional geologists are published in the *Proceedings*. The Westmorland Society was founded in 1973 by Walter Amis and performs functions similar to its Cumberland counterpart, though its *Proceedings* are not so professionally produced. For a lively history of the Westmorland Society, see Parsons (1998).

interbedded. However, Soper and his colleague at Sheffield, the Glaswegian palynologist Charles Downie (1923–1999),[23] had found evidence that there were Llanvirn microfossils (acritarchs) in the slates (Downie & Soper 1972), which suggested that the Eycotts were older than the main exposure of the Borrowdales (which in any case had a different mineralogical constitution). So conformity in the north did not imply general SS–BVG conformity. Moseley accepted the view initiated by Nutt (1966) that the conglomerates might represent intermissions in the vulcanicity, during which valleys could have been excavated and filled with debris from eroded slates and/or volcanics. The Skiddaws were initially folded without cleavage. The cleavage was imposed, with further folding, at a subsequent end-Silurian Caledonian orogeny. In brief, the SS–BVG contact was complex and diachronous, and not uniform along the outcrop.

With the foregoing information in mind, it is now time to introduce Frank Moseley properly (Fig. 9.8). He was really the major figure in Lakeland geology from the 1960s through to the 1980s. I was at one time quite well acquainted with him, as he tutored me for a year when I was a student, and subsequently he has been most helpful in providing information during the preparation of the present book.

Moseley (b. 1922) hailed from Bradford and then moved with his family to Morecambe. He served in the RAF during World War II and then went to Sheffield University, where he got into geology 'by chance' (pers. comm., 1996). Perhaps the move into geology was not wholly accidental. Moseley's flying over Africa and the Middle East, and some climbing ventures while on leave, stimulated his interest in travel, scenery, rocks and rock climbing, and pointed him in the direction of geology. Immediately after the War he went climbing in the Cuillins, and decided to seek a career where one might be paid to do such things. So he took a degree in geology at Sheffield, and went on to do a PhD there on the geology and geomorphology of the Forest of Bowland in North Lancashire east of the Lune Valley (Moseley 1952; and see p. 246) under the supervision of William Wilcockson, the external examiner being W. B. R. King. From Sheffield, he successively obtained positions at Keele, Cambridge and Birmingham Universities, the latter under Professor F. W. Shotton, who had done important work in the remapping of the Cross Fell Inlier in the 1930s (Shotton 1935). Moseley stayed at Birmingham until his retirement. He did work in East Africa, Libya, Yemen, Oman, Spain and Greenland, but his main life work was devoted to the study of the Lake District, on which he has published numerous papers, field-excursion reports, field guides and maps. He edited a major collection of papers on Lakeland geology for the Yorkshire Geological Society, summarizing the knowledge of the area for the 1970s (Moseley 1978*a*), and trained a number of geologists such as David Millward, who later specialized in Lakeland studies. Moseley also was active in the early application of the plate-tectonic theory to the Lakes, as will be discussed in Chapter 10. Stimulated by his experiences as a wartime pilot, he specialized in, among other things, the use of aerial photographs to assist in mapping and structural understanding (Moseley 1981, 1983*a*, 1984–1985 [1986]). Though not always useful in the Lakes (e.g. for the Skiddaws), they could help in appropriate terrain such as the Borrowdales, and with the help of a stereoscope even dips could be estimated, as well as outcrops. In any case, they were of great help in finding one's way around on the fells. Frank's 'scrambler' motorbike, which (I understand) used to get him around effectively, is remembered with nostalgia by some who knew him in his younger days.

On moving into Lakeland work, Moseley consulted with George Mitchell, and was advised that the area round Ullswater would be suitable for study: it had not received much attention since the days of the first Survey and the work of Harker and Marr. Moseley (1960) provided a detailed map of much of the

Fig. 9.8. Frank Moseley in the field (1985). Copy of photograph supplied by Dr Moseley and reproduced by his permission.

ground to the west of Ullswater. It is interesting that there is a stream (Aik Beck; see Fig. 9.9) running into the NW end of the lake where Green (1915, p. 206) had thought there was a conformable relationship between a coarse andesitic tuff at the bottom of the Borrowdale Volcanics and the underlying Skiddaws; but the claimed contact could no longer be seen. None the less, Moseley accepted Green's observation, and then placed a fault a little way upstream from the tuff (i.e. in the Borrowdales). He mapped the fault to the SE, roughly parallel with the lake shore, but over rather unexposed ground, the Skiddaws only making an appearance briefly in Swarth Beck, between Aik Beck and Howtown (where the steamers presently dock). So the fault was shown as running as far as Howtown and then curving round to the south, running up the valley of Fusedale, as the 'old boys' of the first survey had marked it.

To my knowledge, no clear exposure of fault can be seen in Fusedale, but the way that Moseley and the 'old boys' had the curved line of fault implied that it was a low-angle structure or thrust. Moseley dubbed it the Ullswater Thrust, and endeavoured to establish a structure for the region on that basis. This took account of the asserted thrust and also the various high-angle

[23] See Owens & Sarjeant (1999), Sarjeant (1999, 2000).

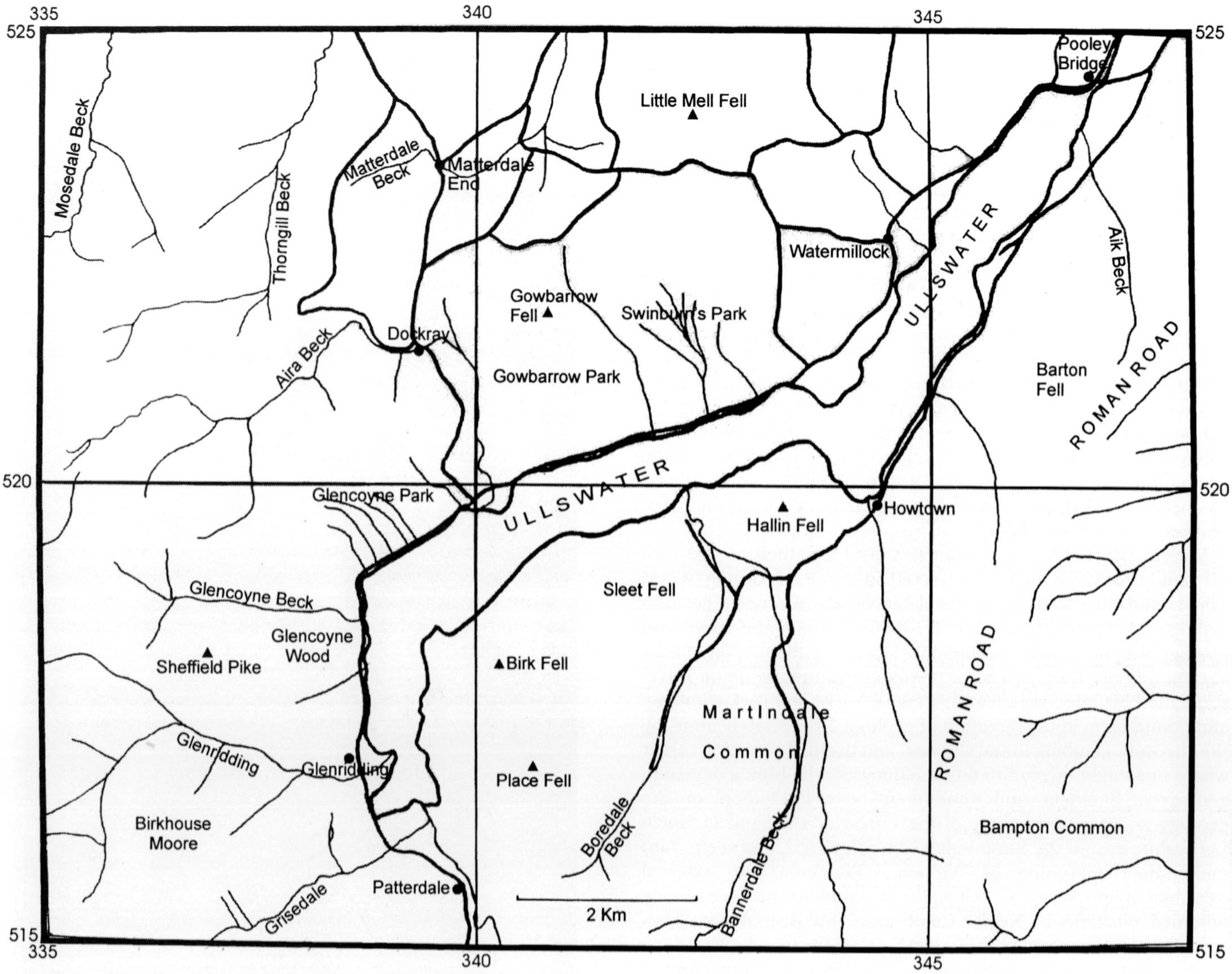

Fig. 9.9. Topography of Ullswater and neighbouring localities.

faults, yielding the rather complicated structure shown in Figure 9.10. It was suggested that thrusting and rather open folding came first; and then the normal faulting.

Moseley also offered a lithologically grounded stratigraphy for the volcanics of the area, but the details need not concern us here beyond saying that it consisted of named andesites, rhyolites, tuffs and 'streaky rocks' – this term being used rather than ignimbrites. (We recall that Mitchell was not initially over-enthusiastic about ignimbrites.) Moseley, it should be remarked, was dealing with the lower part of the Borrowdale Volcanic succession. He suggested how it might be correlated with the work of Mitchell in Kentmere and Shap, and the work of Hartley around Windermere and Helvellyn. However, this was of less importance than the structural innovation. This might, I suppose, be regarded as a development of Marr's notion of there being a major thrust fault to be discovered somewhere in the Lakes – which, however, had never been identified hitherto.[24]

The discussion following Moseley's presentation at the Geological Society was interesting. Tom Eastwood recalled that in the Greenside Mine on the west side of the lake the lode had appeared to follow a junction between the slates and the volcanics, dipping at a low angle to the east – in agreement with Moseley's model. Anthony Millman, who taught mining geology at Imperial College, reported that the lower levels of the mine were now flooded, but one of his students had been examining old mine records and data from drill holes, from which it appeared that the plane of junction of the two units was variable, with a dip up to 45°. Hollingworth (who had worked in the area thirty years previously) had noticed that the rocks on the western side of the lake (at Gowbarrow Fell) were very different from those at Eycott; but if there were thrusting, parts of the sequence might be missing, in which case the old idea of an Ullswater and Eycott Group (Harker and Marr) might still be feasible. Ron Firman thought that there might be several minor thrusts rather than one big one. Colin Rose wondered whether Moseley's 'thrust' might be Marr's 'lag', an idea that Rose seemed to favour. Moseley thought it might be better to remain agnostic on this question for the time being (just having a low-angle fault). He also was

[24] I am not exactly sure when Moseley first had the idea of a thrust in the Ullswater area, but there is mention of one on the western side of the lake in a report of a Yorkshire Geological Society field excursion in July 1958 (Mitchell *et al.* 1958, p. 140). Also, I recall that Moseley was very keen on the idea of thrusts when he was my supervisor in geology for a year at Cambridge (1956–1957), though I do not recall the Lake District being mentioned in this connection. Moseley himself remembers being with Professor W. B. R. King on Hallin Fell (on the east of Ullswater; see Fig. 9.11) one day, when he pointed across the lake to the Knotts and drew attention to the curious pattern of outcrops, suggesting that it would be interesting to map it. Moseley subsequently did just that.

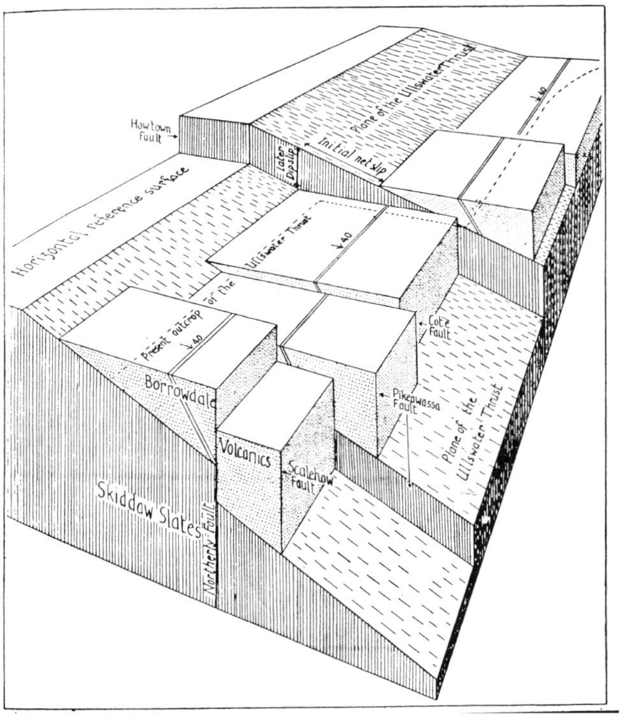

Fig. 9.10. Block diagram representing the structure of the Ullswater region, according to Moseley (1960, p. 77). Reproduced by courtesy of Dr Moseley.

uncertain whether the evidence from the old mine could be deployed in favour of the thrust hypothesis. Other faults, of greater dip, were definitely known on the western side of the lake.

Moseley's mapping then moved to the west of the lake, his results being published in the *Geological Journal* in 1964. On the western side, Skiddaws could be found in the valleys of Swinburne Park and Kirkstile Gill and of Pencilmill Beck, but were poorly exposed, being mostly covered with drift. Borrowdales formed the higher, better exposed, ground of Gowbarrow Park and Birk Rigg to The Knotts. Moseley again invoked thrust faulting, but this time with movement of Borrowdales from the west over the Skiddaws; whereas in 1960 he had had Borrowdales thrusting from the east over the Skiddaws. Both thrusts were called the Ullswater Thrust.

The system of forces that might produce such a result is difficult to understand, and may appear reminiscent of Albert Heim's celebrated theory of the double fold in the Glarus Canton in Switzerland in the nineteenth century. However, Moseley pointed out to me (pers. comm., 2000) that he envisaged a disharmonic junction between rocks of different competencies, rather than a single widespread thrust (or lag). Also (as we see in Fig. 9.11a), he postulated an anticlinal axis running approximately parallel with the length of Ullswater. In the Glencoyne area to the SW of his map, he had a 'slither' of volcanics inserted into the Skiddaw ground by two thrust planes, running parallel to the contours, thus suggesting imbrication (see Fig. 9.11a). His corresponding sections are shown in Figure 9.11b. The 'slither', it may be noted, is not shown on the modern Keswick map (BGS 1999*a*).

These sections may suggest that the thrust faulting was later than the folding (which, incidentally, could only be demonstrated

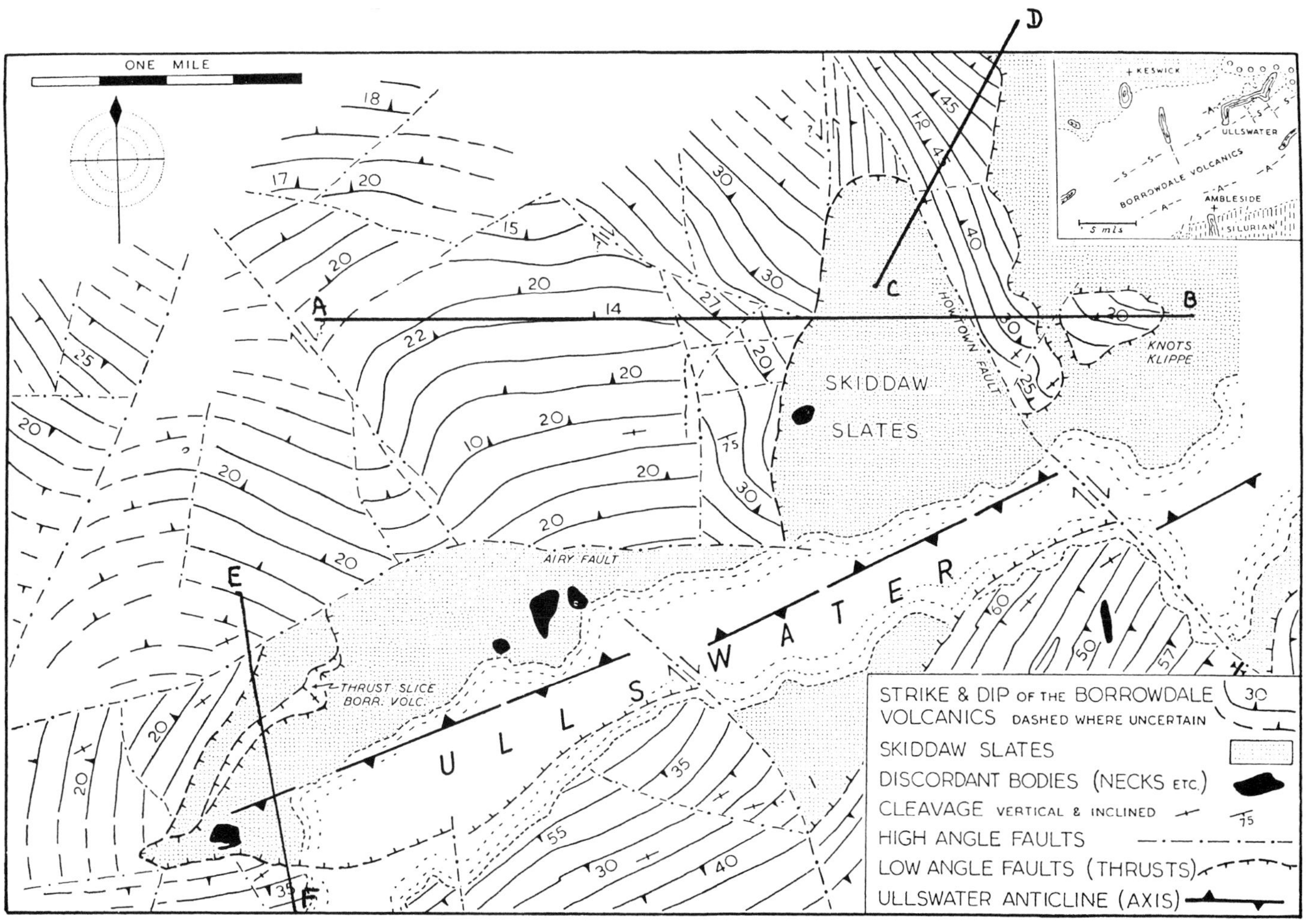

Fig. 9.11a. Structural map of the northwestern side of Ullswater, according to Moseley (1964, p. 133) (lines of section of Fig. 9.11b added). © John Wiley & Sons Ltd. Reproduced by permission.

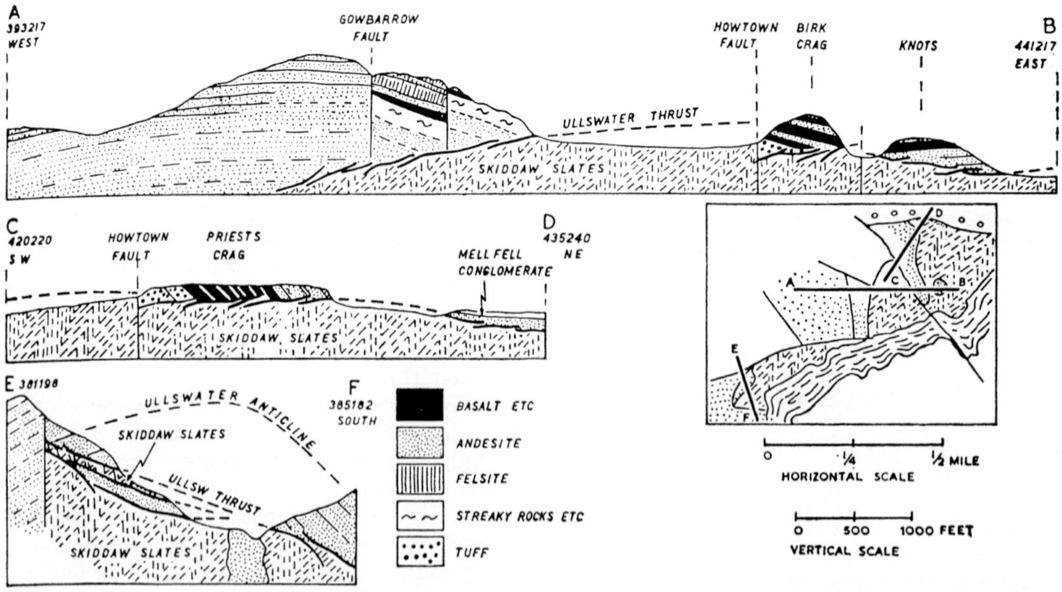

Fig. 9.11b. Figures illustrating the structure of the Ullswater region, according to Moseley (1964, p. 135). © John Wiley & Sons Ltd. Reproduced by permission.

in the volcanics, as the exposure of the slates was and is so poor). It will be remarked that, in the main, younger rocks overlie older ones, contrary to what is usually the case for thrust faults. So, we might say that Moseley's section was compatible with Marr's old idea of a lag fault. However, in the 1960 paper Moseley had Borrowdales involved in local décollement from the south, whereas in 1964 he marked them on his map as thrusting from the north. Moseley reported that the best locality to observe the thrust plane was at the outlier of Knotts (interpreted as a klippe); and slickensiding and shearing could be observed in the top four feet of the slates. Regrettably, I was unable to make this observation during a visit in 1998, due to extensive vegetation cover. Also, as with Marr's lag faults, I have difficulty in envisaging a set of forces that might give rise to a structure such as that represented in Moseley's section.

Others were also sceptical! On 1 September 1970, the day before the Geologists' Association excursion went to Hollows Farm with Jack Soper as leader, they examined the western side of Ullswater with Moseley directing proceedings (Mitchell *et al.* 1972). At quarries by Birk Fell (432 216) the party was invited to examine a faulted junction between tuffs and slates, the latter having two high-angle cleavages, with an orientation rather similar to that of a cleavage visible in the overlying tuff. Drs Helm and Roberts, however, reported perceiving two sets of folds in the slates, the second being of open style and having an analogous structure in the volcanics; only the second (fracture) cleavage seemed to pass through into the tuffs. They were even more unhappy with the putative thrust fault, claiming that the contact occurred at different heights in the exposed face, suggesting that the volcanics 'rest[ed] on an erosional surface cut in the underlying slates'. Soper, of course, was not having any of that. He rejected the pre-volcanic cleavages and thought that the 'small-scale irregularities of the junction did not preclude its tectonic origin' (Mitchell *et al.* 1972, p. 453). Like Moseley, he could perceive a thrust, particularly given that there appeared to be no basal conglomerate at this locality. Moseley refused the objections of Helm and Roberts, saying that the structures in the upper part of the slates were different from those below (because of the thrusting); and the junction cut across volcanics as well as slates.

The thrust idea did not catch on amongst the majority of geologists, though Moseley (1983*b*) advocated it in his field guide, *The Volcanic Rocks of the Lake District* (see Fig. 9.12). However, long before then doubt had been raised by the work of Tony Wadge, who had been surveying in the area in the 1960s, and had made important observations bearing on the question in the Tarn Moor Tunnel (Wadge *et al.* 1972). The 2.6 km tunnel was constructed in 1967–1969 as part of the Lakeland activities of Manchester Corporation Waterworks, and ran from near the NE of Ullswater in a southwesterly direction to Heltondale. It therefore cut right through the SS–BVG plane of contact and allowed it to be examined in a freshly exposed state.[25] The geologists went into the tunnel, and specimens were collected from the northern wall by the contractors every 3 m. However, the inside working conditions were difficult and some of the graptolite specimens had to be collected from spoil heaps. Because the Skiddaw section was rather unstable, the wall was sprayed with quick-setting concrete soon after each tunnel section was driven, so there was little time for detailed examination, and records were not in fact made for the first, Skiddaw, section of the tunnel. Fortunately, however, recording was begun before the Skiddaws ran into the Borrowdales, and the contact was discovered and described. It was a *high-angle normal fault*, which threw a massive Borrowdale tuff against strongly sheared Skiddaw mudstones, with about half a metre of fault gouge (Wadge *et al.* 1972, p. 61).

Another important result was that specimens of *D. murchisoni* were found in rock above that containing *D. bifidus*. Thus a higher

[25] Wadge took the Natural Science Tripos at Pembroke College, Cambridge, mapping the Dent Line and associated structures for his Part II project. With Bulman's introduction, he obtained a post in the Survey and was posted to the Leeds office. He worked near Stoke and then for six years on the 'Pennines Project', which involved him collaborating with Russell Arthurton and Ian Burgess on the mapping of the important Cross Fell Inlier, and writing the associated Memoirs (Burgess & Wadge 1974; Arthurton & Wadge 1981). Later he worked on the Barnsley coalfield, but continued to do work in the Lakes from time to time. In his later work with the Survey he became 'increasingly frustrated' with its 'bureaucratic futilities'. He turned first to doing marketing for the organization, and then resigned to undertake consulting work in the City for insurance companies. At the end of the twentieth century, he was retired and living near Loughborough. Wadge thought that the Survey should be particularly concerned with practical matters. In his early days in the north, he did work on the West Cumberland coalfield, for the Morecambe Bay barrage, and for the construction of the M6 motorway (pers. comm., 1998). Wadge has been a notably eclectic geologist, with work ranging from the study of acritarchs to radiometric dating, and to the aforementioned engineering projects.

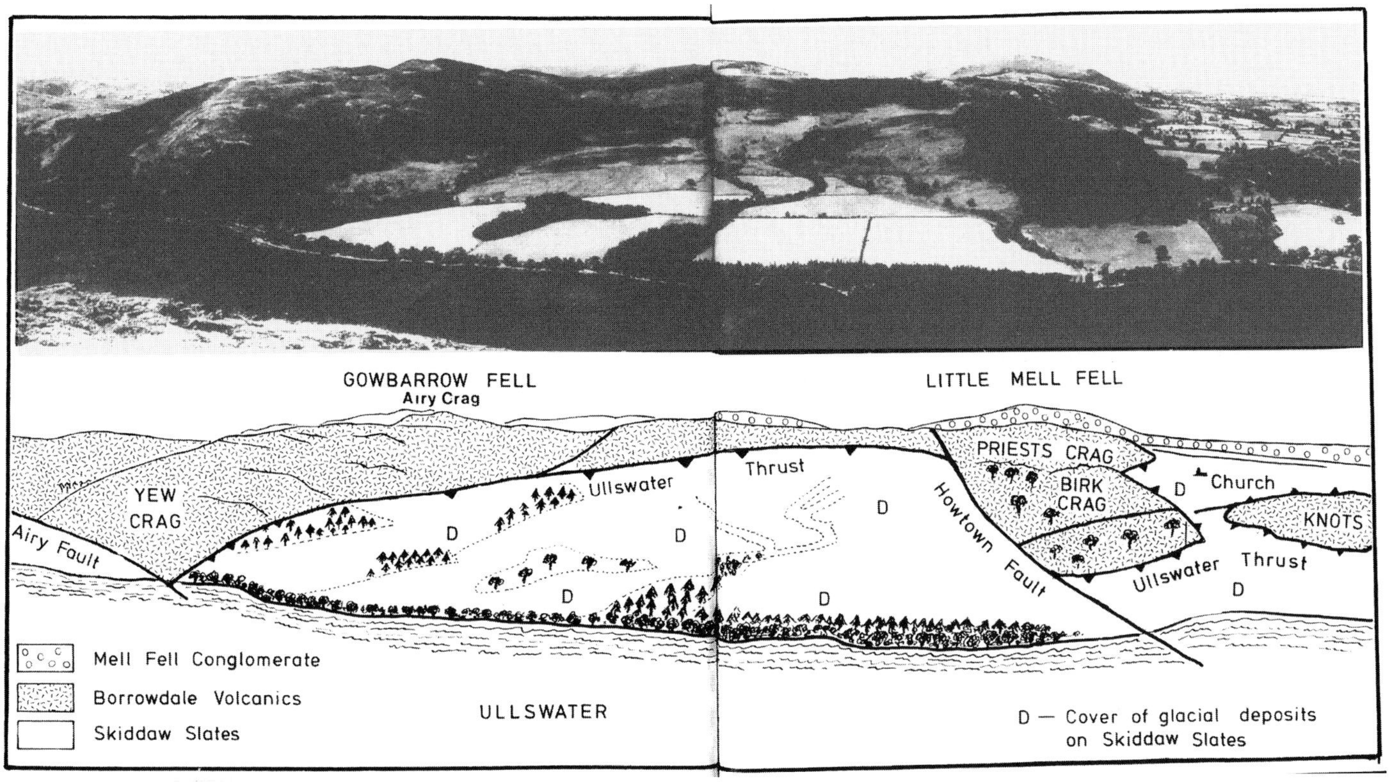

Fig. 9.12. View to northwest across Ullswater to Gowbarrow Fell and Little Mell Fell, showing interpretation of fault structure according to Moseley (1983b, pp. 60–61). Reproduced by courtesy of Dr Moseley.

Llanvirnian zone was discovered in the SS in the Tarn Moor Tunnel than at any other Lakeland locality, before or since. The graptolite expert David Skevington, from Galway, was brought in to provide a detailed description and analysis of the fossils, to support this important finding (Skevington 1970; Wadge *et al.* 1972).

It may be noted that Wadge was visited by George Mitchell and Ron Firman, who went into the tunnel, both being interested in the new evidence about the highly faulted structures of the Borrowdale Volcanics (which need not detain us here, however). Wadge recalls that Mitchell, then in his seventies, scooted into the tunnel so quickly that he (Wadge) could not keep up! (Mitchell was always renowned for his ambulatory prowess.) Unfortunately, Frank Moseley did not visit the tunnel workings. However, he informed me (pers. comm., 1996) that he changed his thinking about the nature of the faulted SS–BVG contact in the Ullswater district after the publication of Wadge's work in 1972. Yet Moseley referred to the Ullswater Thrust in his field guide of 1983. On the other hand, his *Geology of the Lake District* (1978*a*, p. 112) gave the SS–BVG contact as a high-angle fault in the Ullswater area – on the basis of the Tarn Moor Tunnel findings.

A recent BGS Technical Report by Surveyor Richard Hughes (1995*a*) has reviewed the evidence, concluding that the hypothesis of the Ullswater Thrust cannot be sustained. It should be remarked, however, that the outcrop of BVG and Skiddaws at The Knotts, west of Ullswater, though not well exposed, implies a low-angle contact. It follows the contours in a sinuous boundary, which would be compatible with it being a klippe (Moseley, pers. comm., 2000). Also, Moseley has drawn my particular attention to his stereogram (Moseley 1964, p. 138), which depicted the relationships between the BVG and the Skiddaws at Birk Crag (NY 432 217), and which implied a thrust fault at the locality – in an exposure that Hughes (1995*a*, p. 11) reported that he could not find. Moseley acknowledged that it is presently obscured by debris, but, he said, the site could be cleared if desired. Hughes maintained that the contact seemed to be more consistent with a steeply dipping structure than a low-angle thrust plane – though he also said that 'the most recent movement along the contact was indeed at a shallow angle with a north-east to south-west orientation'. It would seem that this issue had a little further to go at the end of the twentieth century.

Wadge (who mapped in the district in the 1960s for the Penrith sheet) also did work on the SS–BVG contact west from Ullswater, near the village of Matterdale. There (at NY 3888 2345) he reported a clear unconformable contact in the bank of Matterdale Beck, with the basal conglomerate of the BVG lying well exposed and with angular unconformity on SS, with the contact traceable continuously for at least 30 m (Wadge 1972, p. 183). I confess, however, that things did not appear so obvious to me when I visited the locality in 1999.

In the discussion following the presentation of Wadge's paper, both Soper and Moseley accepted the evidence for unconformity at Matterdale; but while Moseley, considering the work of his student Jeans in Newlands, was beginning to accept the idea of a more regional unconformity, Soper drew attention to his recent work with Downie (Downie & Soper 1972) to the north, which asserted the existence of a conformable relationship between the Eycott Volcanics and the Skiddaws. Thus he was still looking for evidence of conformity in some localities – though in so doing he was conflating the Eycotts and the Borrowdales – a correlation that was already in doubt as a result of the PhD work of Godfrey Fitton (see Chapter 10), to which in fact Soper referred in his reported remarks. It should be noted also that the Surveyor Michael Nutt thought that the conglomerate of Wadge's asserted unconformity was *intra*-volcanic.

The arguments of the paper of Downie & Soper (1972) command our further attention here. As mentioned, Downie was one of Soper's colleagues at Sheffield, and assumed the chair there in 1972. He was regarded as the leading British authority on acritarchs – those microscopic organisms of doubtful biological affinities but of considerable use in stratigraphy (see Downie 1984; Strothers 1996; Sarjeant 1992). Soper and Downie's material was

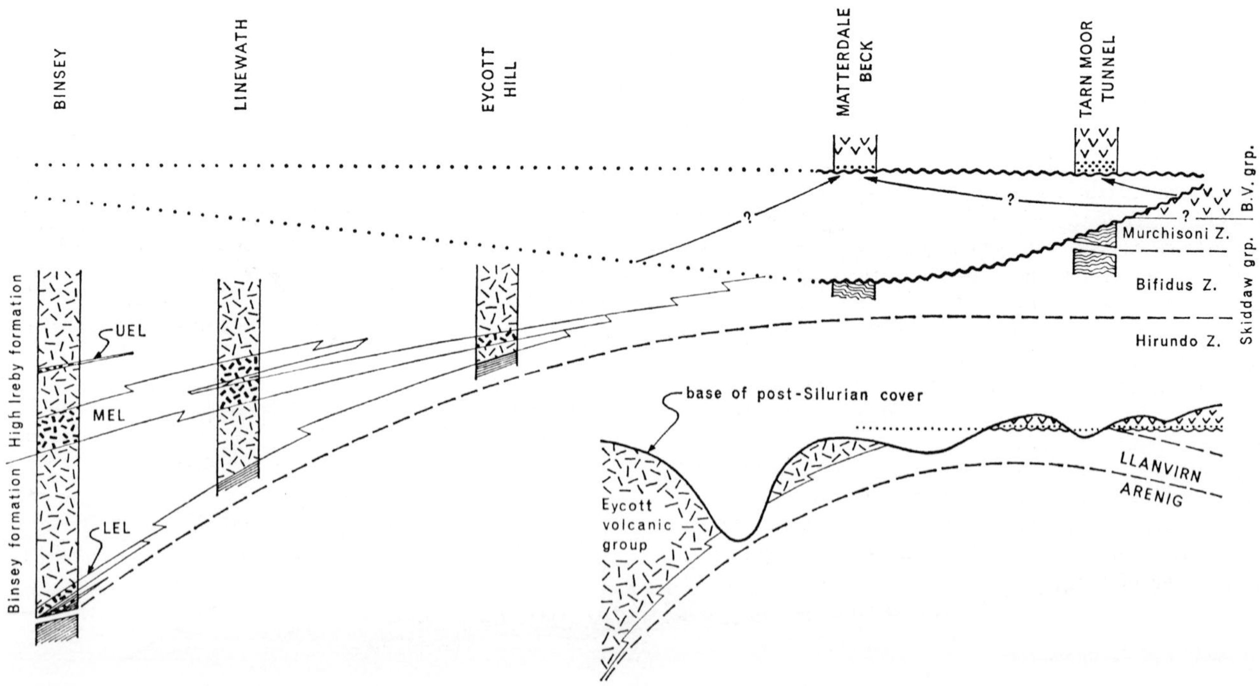

Fig. 9.13. Possible relationships between Skiddaw Slates, Eycott Volcanics, and Borrowdale Volcanics, according to Downie & Soper (1972, p. 266). Reproduced by courtesy of Cambridge University Press.

collected from northern Eycott exposures, which in their lower horizons contained interbedded tuffs and pelites; and near the hill named Binsey (see Fig. 12.1) there appeared to be conformable transition from the SS into the Volcanics, as previously described by Eastwood *et al.* (1968) in the Cockermouth Memoir. The acritarchs from the interbedded pelites seemed to be of Llanvirn age, which made them of the same age as some of the rocks of the main SS inlier. However, Downie and Soper (having regard to the work of Godfrey Fitton & David Hughes (1970), to be discussed in Chapter 10), countenanced a petrological and temporal distinction between the Eycotts and the 'ordinary' Borrowdale Volcanics (the latter being younger), so that while the former might be conformable to the Skiddaws, the latter may be unconformable.[26] Thus they had in mind a possible relationship as shown in Figure 9.13. So the situation was still unclear in 1972, with Downie and Soper pointing out that there were at least four contending hypotheses under consideration (those of Simpson, Soper, Helm and B. Roberts, and Wadge).

There was yet another heated debate about the SS–BVG boundary and related matters, this time at the Geological Society. Moseley (1972) surveyed the structural situation for the Lakes as a whole, as then known, drawing now on mapwork done in the Skiddaws by Colin Rose in the 1930s, but never published because of World War II. Moseley now acknowledged that there *was* unconformity (at least in some places) between the BVG and the SS. He considered the effects of Caledonian, Variscan and Alpine orogenies, but still rejected the Birkbeck idea that there were substantial earth movements within the pre-Borrowdale part of the Ordovician.

In the discussion, the Birkbeck group rejoiced at Moseley's conversion to the occurrence of a SS–BVG unconformity, but were disappointed that he would not accept the pre-Borrowdale movements. Understandably, when one considers the extraordinary complexity of Skiddaw folding, as described by David Roberts (1971) in the Caldew Valley (see Chapter 12), Simpson could not envisage how such things could be ascribed wholly to end-Silurian deformation. Also, he queried why Moseley had not used the available stratigraphic, as well as structural, evidence for the Skiddaws. Helm objected that Moseley had 'not appreciated the severity of deformation in Black Combe' for he had described the character of the folds in that area incorrectly. Brinley Roberts claimed to be surprised that Moseley was now accepting the sub-BVG unconformity, but doing so partly on the basis of Rose's unpublished work, rather than the published evidence of Helm and Simpson. Helm objected that Moseley had Helm's D_1 making an appearance twice: once above and once below the unconformity!

Moseley replied (not in so many words) that he preferred Rose's stratigraphy to that of Simpson. When Moseley started his paper he had not seen Rose's unpublished field maps and had been prepared to accept the reliability of Simpson's; but having seen the older maps he evidently preferred them. Apparently 'several research workers mapping the same areas'[27] had 'commented on the accuracy of the detailed observations'. It seemed to Moseley that one should take account of all the evidence available, not just published material. Besides, unpublished large-scale field-maps could sometimes be preferred to smaller-scale published items.

In sum, the situation was still obscure and contentious in the early 1970s. Perhaps it still is. The use of acritarchs is criticized by some geologists, on the grounds that they may be wide-ranging age-wise; also they may be reworked from older into younger sediments quite easily. Nevertheless, their use has been of great importance for the elucidation of Lakeland geology, as we shall see. And what about all the mysterious contortions of the Skiddaws in the Caldew Valley? Could it be that they were due to some kind of gravity collapses, and not tectonic activity? Moseley (1972, p. 583) did begin to lean in this direction somewhat. Further, should one assume that what was true for Black Combe was applicable more widely in the Skiddaws? They could well have had rather different tectonic histories, the Black Combe rocks being well away from the main Skiddaw Slate exposures.

[26] Fitton, however, did not regard the Eycotts and the Borrowdales as being of different age. Indeed, as he pointed out to me (pers. comm., 1999), his 'theory for the Lakes' (see Chapter 10) would not have worked if he had thought of them as being of different ages.

[27] These were unnamed at the time, but Moseley informed me (pers. comm., 2000) that Peter Jeans was one of them.

So the matter was not settled in the early 1970s. In any case, a geologist whom we shall meet in Chapter 16, Michael Branney, informs me that he believes that the 'base' of the Volcanics, about which people had been arguing in the 1970s, was not in fact the base at all. In his view, the previously supposed oldest BVG rocks were in fact relatively high in the volcanic pile (as Nutt had suggested), as can be shown if one traces the rocks southwards from the contact. Thus the whole issue of the SS–BVG contact would repay further study, though in Branney's view it would probably be too much for a single PhD topic, so complicated is the question (Branney, pers. comm., 1998). Branney also pointed out that in the 1970s, at the time of the controversy, unconformity meant 'uplift and erosion', or orogeny of some kind, but where the eruption of ash from volcanoes is involved, unconformities can readily be formed by other means; so one can – or should – think of the problem in different terms.

In retrospect, the question of the SS–BVG relationship may have been something other than what might at first appear. Geologists are always, of course, interested in unconformities, since their correct identification and interpretation are essential to the understanding of the geological history of any region. Thus when the SS–BVG contact appeared ambiguous it was natural that it would become an issue that had to be sorted out. In the case of the Lake District it might well seem essential to settle that question if one was to have any hope of understanding the geological history of the region. However, there was in fact a more fundamental and broader context to the debate: namely, had southern Britain suffered an Ordovician orogeny? In the 1970s, with the arrival of the idea of Iapetus closure as a major 'explanatory device' to account for many questions in British geology (see Chapter 10), it was obviously important to know whether there was some event or events that had significantly deformed the Lakeland rocks *before* closure (then supposed to be 'end-Silurian'). Simpson's work implied that there had been such an event or events. This ran counter to ideas then being developed about terrane accretion and their application to the Lakeland region. So what appeared on the surface as an essentially empirical issue was actually one of deeper theoretical import. That is, the SS–BVG contact was not perhaps the *big* issue, even if it may seem to have been so, judging by the amount of ink spilled over the question. To the bigger issue we now may turn.

Chapter 10
From micro to macro: plate tectonics in the Lake District – a tangle of theories

One might, at this stage, move to a consideration of the rocks in what came to be called the Windermere Group in southern Lakeland (Moseley 1984) – Otley's third subdivision of the region's rocks – for by 1970 a number of significant developments had occurred 'down south' since we last mentioned these rocks. However, it is probably better at this stage to defer discussion of the Silurian rocks and the upper Ordovicians to later chapters, where ideas about these younger rocks can be given separate attention. Let us carry on from where we finished in the preceding chapter and see how ideas about the relationship between the Eycotts and the Borrowdales were developed, and led to the introduction of plate-tectonic theory to the Lakes. The leading light in this development in the first instance was a Durham research student, Godfrey Fitton, who now holds a chair at Edinburgh University.

As he told me, Fitton (b. 1946) wanted to be a geologist from as early as seven or eight years old, through holidays in Wales and experiencing the early joy of finding fossils. He attended Bury Grammar School in Lancashire and went on to Durham University, where he came in contact with Professor Kingsley Dunham (whom he described as an 'inspiring but rather formidable character'), Dr Henry Emeleus, and Professor Malcolm Brown (see p. 187). Fitton graduated in 1968 and found himself 'nudged into petrology' by Emeleus and Brown. His thesis was supervised by Brown and completed in 1971. Fitton found his supervisor 'rather reserved' (pers. comm., 1999).

Fitton had done some work on Ordovician volcanics (rhyolites and andesites) in Wales during his undergraduate mapping project and Brown suggested that he might look at similar rocks in the Lakes, with a view to understanding how they had been formed. When mapping in Wales, Fitton did not employ the emerging plate-tectonic theory, but it was certainly in his mind when he got to work in the Lakes. The idea was to analyse a sequence of volcanic rocks through the Lake District, to see whether one could establish their petrological succession; and then one might try to fit the results into the emerging plate-tectonic theory. Unusually for the time, Fitton's PhD did not involve a substantial mapping component. The work of earlier investigators such as Oliver, Firman and Moseley pointed him where he needed to go so far as localities were concerned.

The old chestnut about whether the Lakeland garnets were metamorphic or igneous, and whether the garnets were phenocrysts or xenocrysts, was one focus of Fitton's work. He used an electron microprobe, then only fairly recently invented[1] (one of the first instruments of the kind in Britain, being installed at Durham by Dunham when Fitton was an undergraduate) to analyse the garnets. Whereas Oliver and Firman had had to extract the crystals mechanically, and then subject them to chemical analysis, Fitton could get results out within 24 hours. However, the Durham instrument did not work as a well-behaved black box. It had to be 'tweaked' a lot, Fitton recalled. He reasoned that the garnets could not be metamorphic as one cannot grow broken crystals. If they were xenocrysts, there would be no chemical correlation between the garnets and the rocks in which they occurred. But there was such correlation. Therefore the garnets were phenocrysts: QED!

Fitton was influenced in his thinking about the garnets by the ideas of T. H. Green & A. E. Ringwood (1968), working at the Australian National University, Canberra. They had examined rhyodacites and granodiorite porphyrites from Victoria and showed that the production of such garnet-bearing rocks could be simulated experimentally by cooling suitable melts under high pressure, such as might obtain in the lower crust or upper mantle. At the time, their ideas were accounted for in terms of geosyncline theory: basalt could sink in a geosyncline, become eclogite, and get into the mantle, where it might re-melt. This idea provided Fitton's initial theoretical framework, and to an extent it had similar implications as subduction theory, as then understood (pers. comm., 1999). However, during the course of his thesis work, he changed to the deployment of plate-tectonic theory.

Another important issue had to do with the relation of the Eycotts to the main outcrop of the BVG, the problem being tackled to a large extent by chemical and mineralogical analysis. Fitton demonstrated[2] that the Eycotts or 'northern volcanics' were tholeiitic,[3] whereas the Borrowdales were calc-alkaline rocks.[4] But he did *not* suppose that the chemical and mineralogical differences were related to age. That is, at the time of preparing his thesis he did not think the Eycotts were necessarily older than the main mass of Borrowdales.

Fitton, then, was able to make immediate use of plate-tectonic theory in the Lakes, and he and David Hughes (see below) were the first specifically to do this[5] – though, as he told me, he would have not been able to do so satisfactorily if he had thought that the Borrowdales and the Eycotts were of different ages (pers. comm., 1999). He recalls being particularly influenced by the ideas of Bryan Isacks, Jack Oliver and Lynn Sykes in their famous paper on the relationship between seismic evidence and the new tectonic theory (Isacks *et al.* 1968; see also Oliver 1996), by Tuzo Wilson (1966), and by John Dewey (1969*a*, *b*). Thus, there was the possibility that the Lakeland volcanics had something to do with an island arc subduction zone, and one might be able to work out the direction of the subduction process for that area. The idea of an ocean (Iapetus) closing so that England and Scotland became linked (Harland & Gayer 1972) was not published when Fitton completed his thesis.

Fitton did not publish his ideas on plate tectonics as applied to the Lake District alone. His paper was co-authored with an old school-friend from Bury Grammar, David Hughes. He had gone to Leicester University, but the young geologists' paths crossed again when they mapped overlapping areas in Wales.[6] Later, they

[1] Professor Fitton has informed me (pers. comm., 2000) that the invention was patented by J. Hillier in the USA in 1947 and the first working instrument was developed by R. Castaing and R. Guinier in 1949. So it would seem that British universities were a little slow in deploying this important analytical tool. However, it took some time for the instrument to become available commercially.

[2] He points out that, in retrospect, the argument was insufficient, simply being based on the fact that the northern volcanics (Eycott) were more iron-rich than the calc-alkaline rocks of the main BVG outcrop.

[3] The name derives from a place called Tholei in Germany, where such basaltic types occur.

[4] Calc-alkaline magmas become silica-rich during early stages of magmatic evolution through fractional crystallization, whereas the tholeiitic magmas become iron-rich. The former (e.g. andesites, diorites) are characteristic of subduction-zone settings such as island arcs. Tholeiite is characteristic of continental flood basalts and the ocean floor, but is also found in young volcanic island arcs and on the oceanward side of mature arcs.

[5] Previous theorists such as John Dewey (1969*a*, *b*) had obviously taken note of the situation of the Borrowdale Volcanics.

[6] Ironically, the two boys had got a year apart at first because Fitton failed the 11+ examination the first time! But then he got ahead because Hughes 'lost' two years learning French. Hughes did his PhD in Wales under John Wadsworth at Manchester University. At one point in their overlapping careers they both applied for the same position at Edinburgh. Fitton was the lucky one, but Hughes later became Head of the School of Earth and Environmental Sciences at the University of Portsmouth. They did some important collaborative work in Cameroon, having decided to go there on the basis of looking at a map in the back of a diary in a pub. Serendipity!

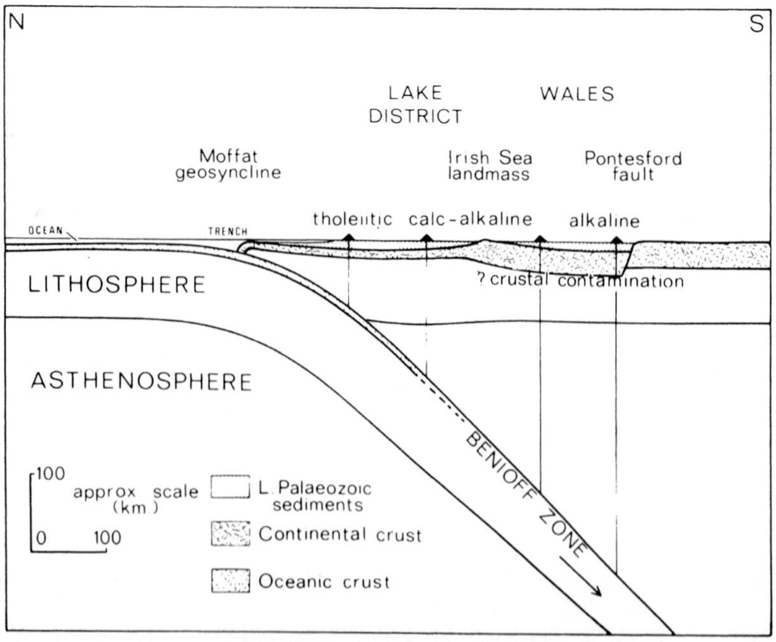

Fig. 10.1. First application of plate-tectonic theory to the Lake District, according to Fitton & Hughes (1970, p. 226). Reproduced by courtesy of Professor Fitton & Elsevier Science.

came together to publish their oft-cited paper on the application of plate-tectonic theory to the Ordovician volcanism of North Wales and the Lakes (Fitton & Hughes 1970), with Hughes providing input from Wales and Fitton from the Lake District.

Their idea was that the garnet-bearing calc-alkaline BVG rocks had crystallized at depth (lower crust or upper mantle) as suggested in the Green and Ringwood paper. In addition, suggested Fitton and Hughes, the geochemical variation between the Eycotts and the main BVG outcrop might be compared to what is seen in modern island arcs (tholeiitic magmas on the oceanic side of island arcs; calc-alkaline on the continental sides) (cf. Kuno 1966), where oceanic lithosphere is being destroyed along Benioff zones.[7] So, if, in the Lakes, there was an oceanic lithospheric slab subducting from the *north* (origin unspecified), it would have reached a greater depth under the site of the present Borrowdales than it would under the more northerly Eycott region, and would supposedly generate volcanics of different chemical composition. The sediments of the Moffat geosyncline of the Southern Uplands might represent the sort of accumulation that could be expected in a trench on the oceanic side of an island arc, as in modern cases. The authors envisaged a closing ocean (cf. Wilson 1966); and with its eventual closure in the late Ordovician the volcanic activity ceased. The young geologists' model is shown in Figure 10.1.

It will be seen, however, that the subducting oceanic slab simply appears as a kind of *deus ex machina* in this figure (though one compatible with early plate-tectonic theory). Clearly the theory needed elaboration and refinement. (Indeed, Professor Fitton himself modestly described the oft-cited paper to me as 'fairly cringeworthy'!) Further development of the theory in relation to the Lake District evidently required consideration of regions further afield, such as Ireland, Scandinavia, Scotland, and even Newfoundland and Greenland, not to mention Wales. The suggestions that were made in the next few years were helpfully summarized by Moseley (1977), on whose work I gratefully lean.

Soon after Tuzo Wilson (1966) proposed his idea of the closure and re-opening of the Atlantic, the notion of an ocean – which later came to be thought of as lying in part between what is now England (as part of 'Europe') and what is now Scotland (as part of 'America') – was proposed by the aforementioned Brian Harland & Rodney Gayer (1972). As is well known, they dubbed this *hypothetical* tract of ocean 'Iapetus' (namely the mythical son of Earth and Heaven, brother of Okeanos and Tethys, and father of Atlantis, who gave his name to the Atlantic). The new name was needed, as what they proposed was not strictly a 'proto-Atlantic' ocean. It was envisaged as covering what are now northern regions, not the main body of the present Atlantic, as between Africa and South America. The argument was based on evidence of initially disparate graptolite and trilobite faunas, which gradually merged as their ages decreased, with the two land masses supposedly coming together during the Ordovician in association with the Caledonian Orogeny. This was a model that supposed the collision of *two* land-masses, not just subduction at an island arc. The concept of a closure of Iapetus in the Ordovician–Silurian was later given further persuasive support by the work of Stuart McKerrow & Robin Cocks (1976), who showed how the faunas of the western Appalachians and western Newfoundland, and NW Ireland and Scotland on the one hand, and coastal New England, southern New Brunswick, Nova Scotia–eastern Newfoundland, and England and northern Europe on the other, gradually converged, with free-floating graptolites in common at about 530 Ma and eventually freshwater fish at about 430 Ma.[8] The faunal evidence was taken to suggest that the closure occurred diachronously: first in what is now the NE, and later in the SW.

Variants of the idea of Iapetus closure in relation to plate-tectonic theory were conveniently collated and figured in the aforementioned paper by Moseley (1977), as shown in Figure 10.2a. The last of these diagrams was Moseley's own suggestion, and his supposed sequence of events that gave rise to this

[7] Hisashi Kuno, of Tokyo University, plotted the distribution of Quaternary volcanoes producing lavas of different chemical composition in the Japanese archipelago (and in other localities round the Pacific Basin). He found that there were three 'bands' of volcanoes in Japan, producing tholeiite, high-alumina basalt, and alkali olivine basalt, with the first band on the Pacific coast of the Japanese archipelago, the second approximately along its axis, and the third on the coast facing Manchuria, in the Sea of Japan, and on the Asiatic mainland. He also found that the different magmas were correlatable with earthquakes at different depths: they became deeper towards the continental mainland. All this was compatible with the idea of a subducting plate moving from the Pacific area northwestwards and down into the mantle, generating volcanoes with magma of different chemical composition from sources at different depths.

[8] These dates would need modification today, but the argument still holds in principle.

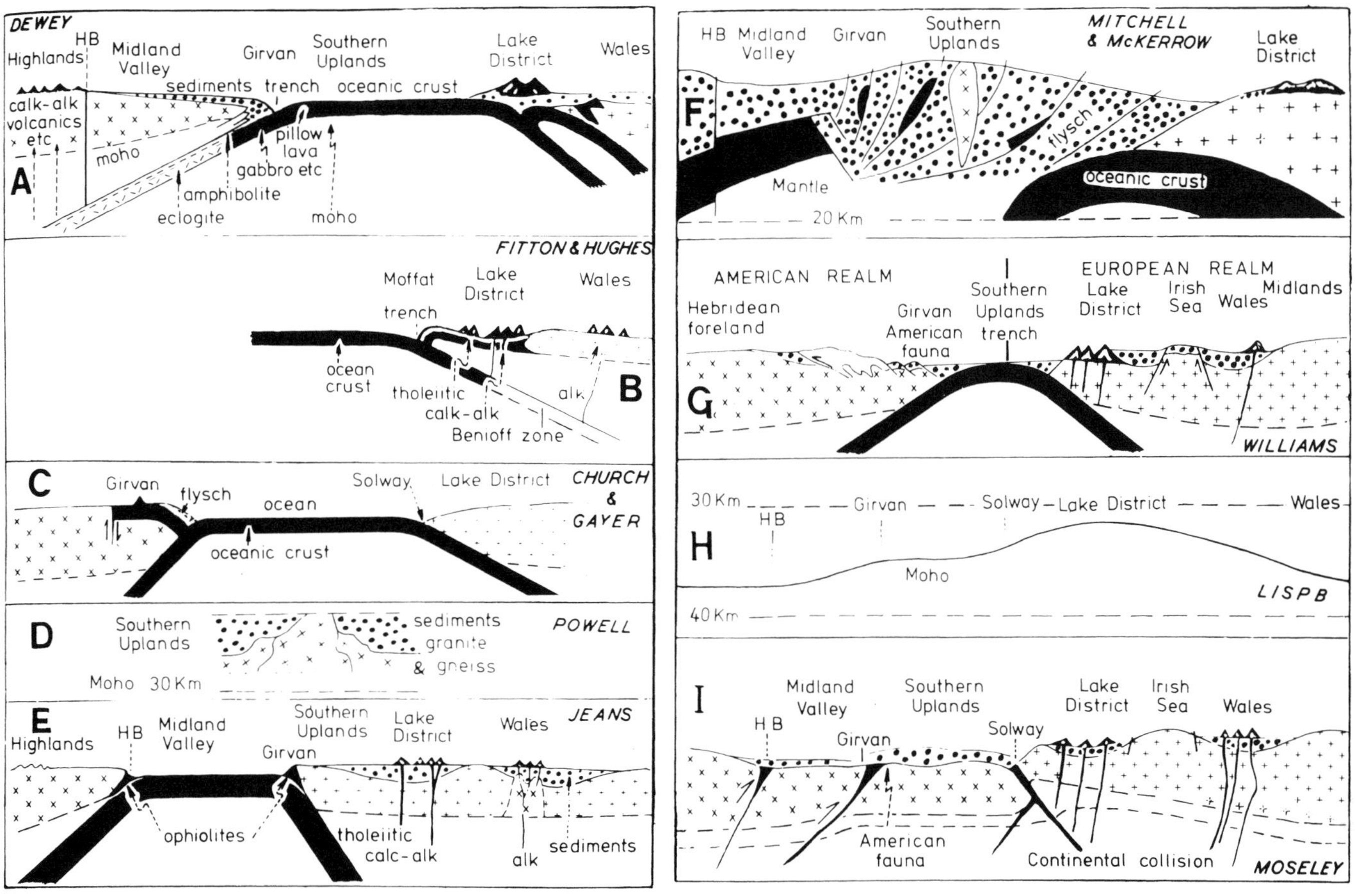

Fig. 10.2a. Figures representing structures of the Lake District, according to various applications of plate-tectonic theory (subduction theory). **A**, Dewey (1969*a*); **B**, Fitton & Hughes (1970); **C**, Church & Gayer (1973); **D**, Powell (1973); **E**, Jeans (1973*b*); **F**, Mitchell & McKerrow (1975); **G**, Williams (1976); **H**, Bamford *et al.* (1976); **I**, Moseley (1977). From: Moseley (1977, p. 765). Reproduced by courtesy of the Geological Society of America.

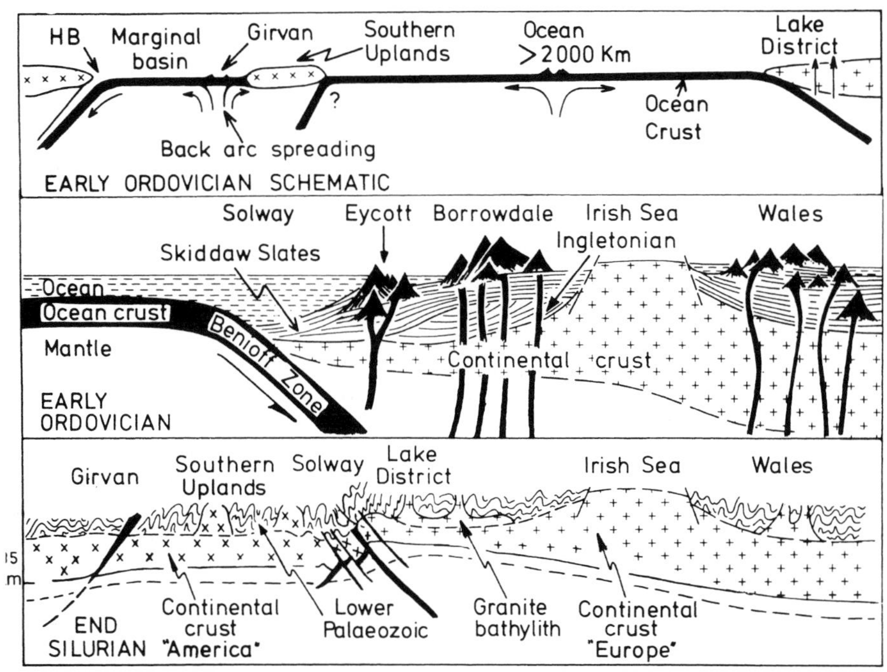

Fig. 10.2b. More detailed representation of Moseley's ideas (cf. Fig. 10.2a [I]: Moseley [1977, p. 767]). Reproduced by courtesy of the Geological Society of America.

situation was given in further diagrams (Fig. 10.2b). D. W. Powell (1971), it may be noted, adduced geophysical evidence to show that there was *continental* crust in the region of the Southern Uplands. William Church & Gayer's (1973) model had the Southern Uplands as a relic of the Iapetus Ocean, with the well-known ophiolites of the Ballantrae complex (on the SW coast of Scotland, just north of the outcrop of the Southern Uplands) as the remains of subducted oceanic crust, descending beneath the Scottish–American former continent. Jeans' (1973*b*) model was somewhat like that of Fitton & Hughes (1970), but added a northern side to the subducting oceanic crust, making it compatible with the 'classic' sea-floor spreading concept. However, his

Iapetus Ocean was situated in the region of the Midland Valley of Scotland, not between England and Scotland. He had the southern subducting slab first generating the Welsh Arenig volcanics; and then supposedly steepening to produce the Lakeland volcanics (Llanvirn–Llandeilo) further north. We note that, by contrast, Moseley had things so that the ophiolites of Girvan–Ballantrae had nothing to do with the hypothesized Iapetus Ocean, but were the residue of some other subduction.

Andrew Mitchell and Stuart McKerrow's model (emanating from Oxford University) was based on analogies with the Burma Orogen (Tertiary), with the British Ordovician and Silurian turbidites being supposedly analogous to the sediments presently being accumulated in the Indian Ocean and forming the fans of the Bay of Bengal. Such turbidites were stacked up into imbricated thrust sheets in the Southern Uplands. The crustal thickening of the region claimed by Powell could perhaps be explained by under-riding of Lakeland rocks from the south. The model of Alwyn Williams of Glasgow (1976) took particular note of his palaeontological evidence for the initial divide as occurring *between* England and Wales and Scotland, *not within* Scotland. It was incompatible with Jeans' model, which had the Girvan area (and fauna) as part of the European rather than the American province.

The LISPB (Lithospheric Seismic Profile in Britain; Bamford *et al.* 1976) figure, which resulted from a large collaborative effort by numerous geophysicists, indicated that the whole of Britain was underlain by *continental* crust. This cast doubt on the Fitton and Hughes model, suggesting as it did that there was no large area of proto-Atlantic remnants in Britain; but the faunal evidence had to be accounted for. Moseley's model (Fig. 10.2b) was intended to solve this dilemma. It will be seen that it was compatible with the idea of Downie & Soper (1972) that the Eycotts were somewhat older than the 'typical' Borrowdales; and it was also compatible with the idea, long held by Soper and Moseley, of a 'big crunch' that gave rise to the major deformations – foldings and cleavages – of the Lakeland rocks at the end of the Silurian. However, the reason for the coming together of the English and Scottish components of Britain was not really explained at that time in terms of 'higher-level' theory. Note that Moseley had the introduction of the Lakeland granites (see Chapter 11) at about the time of the collision of 'old Europe' and 'old America'. The importance of the Lake District for his theorizing needs no emphasis, of course.

All this work received support from the work of geophysicists studying geomagnetism. That is, they found that the different parts of Britain as a whole did not have a unique polar 'wandering path'. Rather, there appeared to be evidence for an oblique convergence of Scotland and England–Wales (Briden & Morris 1973; Morris 1976; Faller[9] *et al.* 1977; Faller & Briden 1978).

There was a degree of symmetry between the northern and southern districts in the early plate-tectonic models such as those of Jeans and Williams (see Fig. 10.2a), but (as these authors were well aware, of course) the situation was by no means the same in Scotland and northern England. The theoretical explanation of the Southern Uplands was perhaps easier and came first, with the hills being construed as an accretionary prism. That is, as the ocean floor of Iapetus moved northwards and subducted under 'Scotland' it formed a trench which filled with turbidite sediments that subsequently became squeezed, folded and thrust faulted in a series of imbricated wedges that eventually came to form the Southern Uplands (McKerrow *et al.* 1977) (see Fig. 10.3). Their

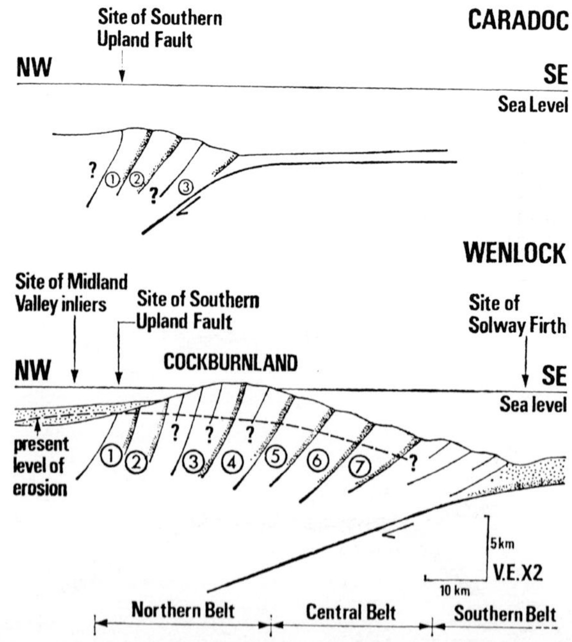

Fig. 10.3. Southern Uplands accretionary prism, according to McKerrow, Leggett & Eales (1977, p. 239). Reproduced by courtesy of Nature Science.

work was greatly assisted by the construction of a gas pipeline through the Southern Uplands, which allowed numerous fresh sections to be examined.

In the Lake District, the Skiddaw Slates might, at a pinch, be thought to be analogous, in rock type at least, to the slates of the Southern Uplands, but their dating was not precisely the same, and they certainly did not have the same structural features. Moreover, there was no equivalent of the Borrowdales in Scotland. So the same model would not do for both north and south of the so-called Iapetus Suture, where the two continental fragments (terranes[10]) supposedly had collided. This suture line was thought to have run across Britain at approximately the line of the Solway Firth (near Carlisle) and into the Northumberland Trough, about 30 km north of the Lake District (see Fig. 10.8). There was no evidence in the field of a 'big crunch' in this locality, but that was where the faunal evidence seemed to suggest that plates had 'coalesced'. Adrian Phillips, Christopher Stillman and Thomas Murphy (1976) suggested that with closure the southeasterly subduction of oceanic crust would have ceased, and this would have terminated the Borrowdale volcanism in the Caradoc. Mitchell & McKerrow (1975) (see Fig. 10.2a), however, had shown that the northwesterly subduction had continued into early Devonian times. Why the difference?

The answer proposed by Phillips *et al.* (1976) – in what Soper (1986, p. 229) called the 'Trinity College synthesis' – was that there had been a large (980 km) dextral strike-slip displacement along the suture zone during the Devonian, the collision being oblique so that contact was first made to the NE. Some 9° of sinistral rotation of the Irish (Waterford)–Lake District volcanic arc was also proposed, to make things fit. So, there was supposed oblique closure first, followed by slip. This idea was intended to

[9] Dr Angela Faller of Leeds University = Mrs Jack Soper.

[10] The notion of terranes became popular in North America in the 1970s, following the work of David Jones, for example. The following elaborate, but partly circular, definition of the term has been suggested by J. D. Keppie (1989, p. 161): '[a] terrane is an area characterized by an internal continuity of geology (including stratigraphy, fauna, structure, metamorphism, igneous petrology, metallogeny, geophysical properties, and palaeomagnetic record) that is bounded by faults, mélanges representing a trench complex, or cryptic suture zones across which neighbouring terranes may have a distinct geologic record not explicable by facies changes (i.e. exotic terranes), or may have a similar geologic record (i.e. proximal terranes) that may only be distinguished by the presence of the terrane boundary representing telescoped oceanic lithosphere.'

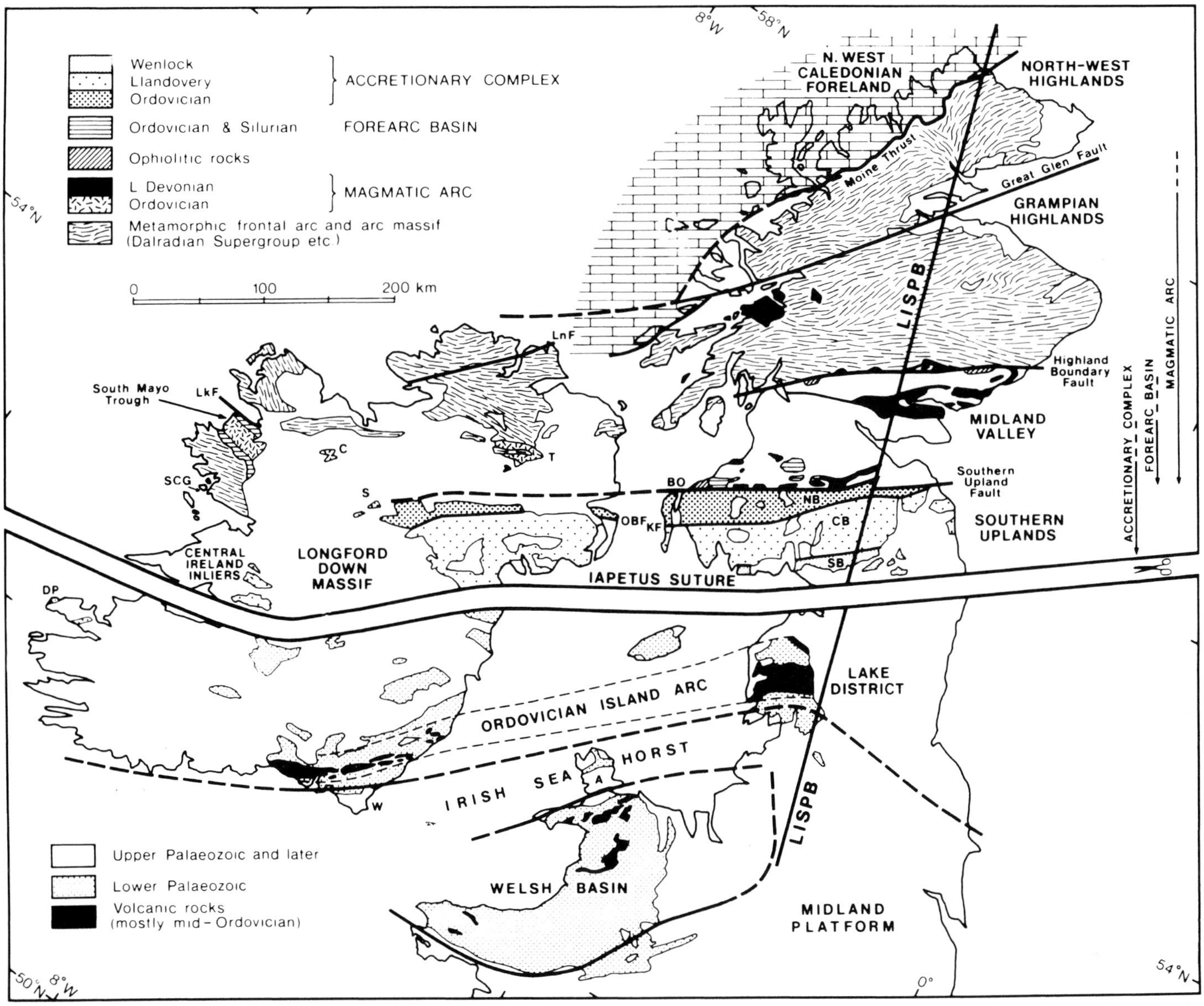

Fig. 10.4. Sketch showing suggested relationship of Lake District to the geology of British Caledonides, and the Iapetus Suture, according to Leggett, McKerrow & Soper (1983, p. 188). (**LISPB**, '**L**ithospheric **S**eismic **P**rofile of **B**ritain' (cf. Bamford *et al.* 1976). Reproduced by courtesy of American Geophysical Union.

account for the present juxtaposition of Southern Uplands and Lakeland rocks, thought to have been involved in collision at different times. It should be remarked that the Irish authors' suggestions were based on general knowledge of the literature of Lakeland geology rather than on detailed field experience there. Field evidence for the existence and location of the suture was adduced from Ireland, not northern England.

Soper, in collaboration with Michael Johnson (Edinburgh) and David Sanderson (Belfast), was not happy with the proposed model (Johnson *et al.* 1979). They objected that it did not adequately account for the deposition of the Silurian sediments of southern Lakeland. They thought that the recognition of an end-Silurian collision orogeny rendered the wrench faulting of Phillips *et al.* unnecessary. Perhaps more persuasively, they maintained that the proposed line of wrench-fault would pass through the Devonian Cheviot lavas in Northumberland, which displayed no evidence of such displacements (though that would all be a matter of timing).

Four years later, Jeremy Leggett, McKerrow & Soper (1983, p. 196) maintained that the Lake District volcanics were too close to the proposed line of the Iapetus Suture (see Fig. 10.4) for them to have been generated by subduction, if the presumed present suture line represented the original trench position – modern arc-trench gaps usually being 150 km or more wide. They therefore proposed that the material of the early Ordovician fore-arc of the southern side of Iapetus might have been subsequently thrust under southern Scotland at the climax of the end-Silurian to early Devonian orogeny. This idea had a number of attractive features: in particular the underthrust continental material might push northwards and provide material for the calc-alkaline volcanics of the Midland Valley of Scotland. Also, the elevation of the Southern Uplands might have provided a source of sediment for the production of the turbidite deposits of the Windermere Group in the late Silurian. The direction from which these deposits originated (according to sedimentological criteria) suggested that the source was located at the place where the Southern Uplands were presently situated (in relative terms) and that there had, in consequence, been no large-scale strike-slip fault along the line of the Iapetus Suture.

An important paper quickly followed (Soper & Hutton 1984). Donald (Donny) Hutton, from Trinity College, Dublin, later at Durham University and presently (2000) Professor at Birmingham, had been working on the Donegal granites of NE Ireland and had concluded that they showed evidence of *sinistral* shearing

(Hutton 1982).[11] Listening to him speaking on this to the Geological Society – where a large 'crack' in the crust was described, filled with granite, which mapped out structurally as manifesting sinistral shear – it occurred to Soper, who was in the audience, that the cleavage in the slate belts of the Lake District and Wales was also sinistral, in the sense of being directed clockwise to the fold-traces in the slates.[12] Earlier, J. Selwyn Turner (1901–1983) from Leeds University had noted that the English–Welsh cleavages turned round to E–SE in northern England (i.e. towards Poland), whereas in Scotland they headed towards Scandinavia. As he put it: '[t]he continuations of the Caledonides of Wales and the Lake District are not to be sought for in Norway, but in the Ardennes and Central Europe' (Turner 1949, p. 285). This had suggested that England and Wales were 'part of Europe' rather than Scandinavia. The Caledonides turning-point – which Turner called the 'North English virgation'[13] – was situated at the apex of the supposed Midland Massif, south of the Lake District, but curvature of the lines of cleavage also occurred further north in the Lakes. The idea was not, however, pursued by Turner. Now Soper thought it could be a relevant consideration, but one would have to think of three plates involved in the closure of Iapetus: Laurentia (Greenland, Maritime Provinces of Canada, part of Newfoundland, etc.); Baltica (Scandinavia); and Cadomia[14] (England, Wales, southern Ireland and parts of Europe). Not only that, but the structures in both Ireland and the Lakes appeared to be *Devonian*. (Hutton dated his granite emplacement and deformation at about 400 Ma or early Devonian.)

Soper's thinking was influenced by the ideas of the Dutch geologist Peter Ziegler (b. 1928), at the time an exploration consultant for Shell International at the Hague, expressed in his palaeogeographical atlas (Ziegler 1982) and in Ziegler (1984). The former publication had the idea of three land-masses colliding, with pieces rifting off from the northern side of Gondwanaland. Ziegler (1982, p. 22) spoke of a 'three-armed orogenic system'. His 'cartoons' (see Fig. 10.5) were evidently noted by Soper, who has acknowledged to me (pers. comm., 2000) that he was influenced by them, as well as by Cocks & Fortey's notion of a Tornquist Sea (see below). As can be seen from the figure, Ziegler's cartoons also involved the notion of sinistral shear.

The significance of the sinistral movement was considerable. In the earlier models, such as that of Stuart McKerrow (1982), England and southern Ireland had supposedly approached Scotland and northern Ireland from the 'southeast'; but to get sinistral movement they would have had to have approached from the 'southwest'. Such a requirement *also* suggested something more complicated than a collision of two land-masses. So Soper and Hutton got together and wrote up their paper in about a week, mostly on the telephone as Soper recalled (pers. comm., 1998).[15] Their model for a supposed three-plate collision is shown in Figure 10.6. The 'waterway' between Baltica and Cadomia was called the Tornquist Sea, previously hypothesized by Cocks & Fortey (1982) on palaeontological grounds.[16] So Laurentia and Baltica supposedly sutured first in mid-Silurian times. Then closure with Cadomia followed, giving the 'big crunch' in the early Devonian so far as the Lake District was concerned. The argument was supported by considerations of cleavage pattern that had been simulated experimentally back in the 1920s with sheared clays. Such artificially produced fractures were claimed to resemble those observed on a large scale in relation to the Iapetus Suture across Britain and Ireland.

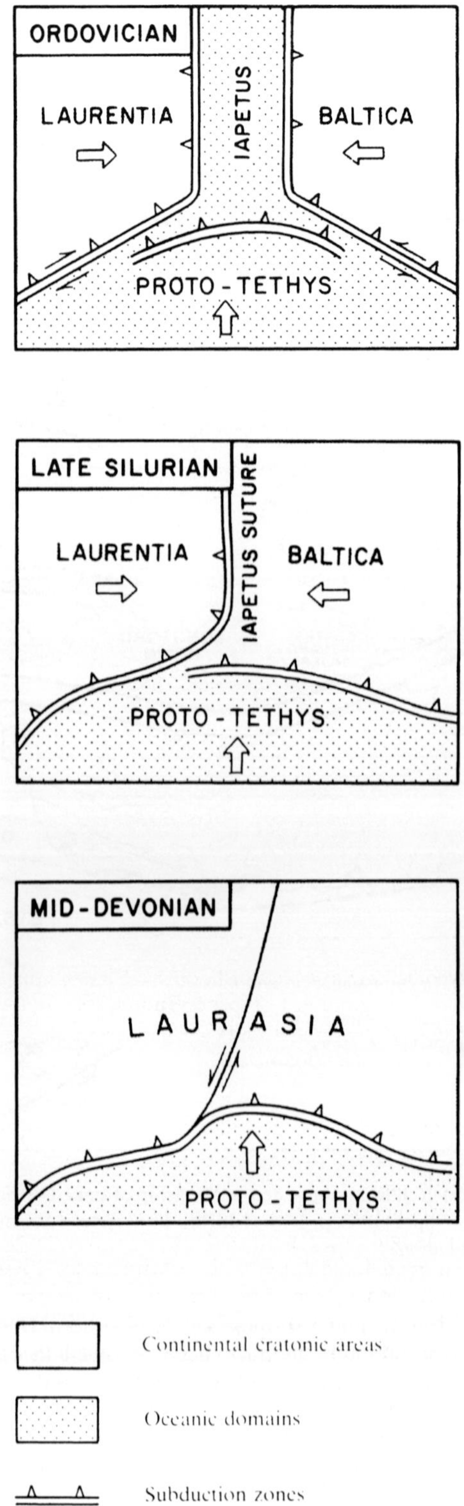

Fig. 10.5. Cartoons showing supposed collision of Laurentia and Baltica and formation of the Iapetus Suture, according to Ziegler (1982, p. 22). Reproduced by courtesy of Professor Ziegler.

[11] Multi-polyphase deformation was then still much in vogue. Hutton had three main phases with a total of nine deformations.

[12] John Dewey was also at the meeting, and according to Soper's recollection was persuaded by the idea of sinistral shear that evening, at Soper's suggestion. Dewey was so excited that he emitted an expletive, which need not enter the historical record here.

[13] Virgate = rod-shaped, or long and narrow. In geology, 'virgation' refers to a system of faults branching like twigs from a bough.

[14] This terrane was so named because it was involved in the 'Cadomian Orogeny' (Cogné & Wright 1980).

[15] Soper recalls how the paper was submitted to the journal *Tectonics*, edited at that time by Dewey. Dewey, it seems, had had much the same idea not long before, but Soper and Hutton had their contribution ready first.

[16] As far back as 1910, the Swedish palaeontologist Sven Törnquist (1840–1920) had noted a 'line' running SE from Denmark towards Prussia, marking a faunal divide (analogous to the modern 'Wallace Line').

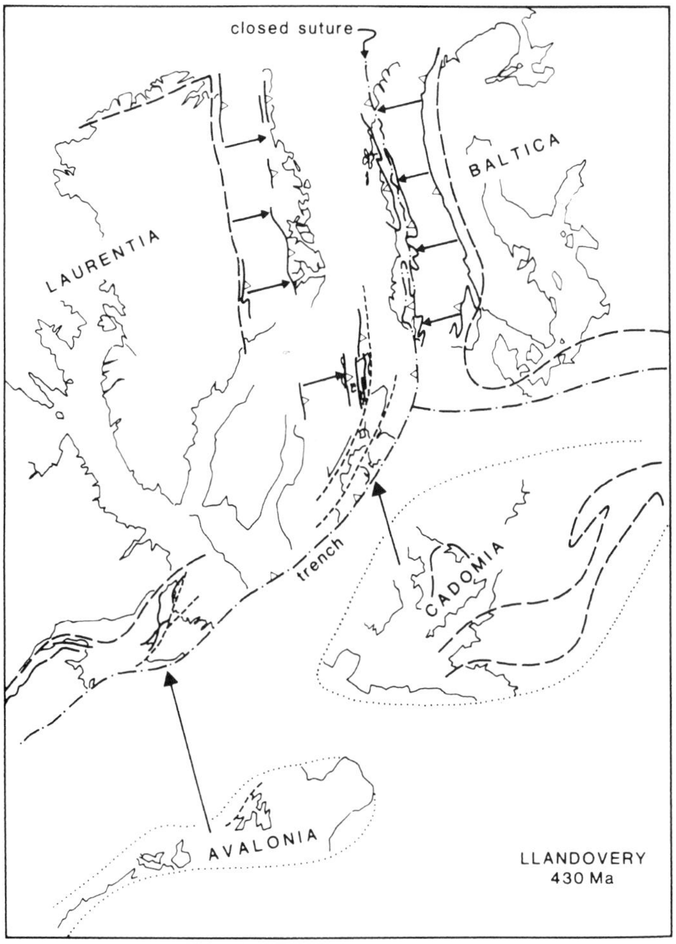

Fig. 10.6. Hypothetical three-plate collision involving suturing in the Solway Firth region, according to Soper & Hutton (1984, p. 788). Reproduced by courtesy of American Geophysical Union.

The ideas outlined above provided the basis for a substantial re-evaluation of the geological history of the British Isles in the so-called Caledonian Orogeny. Soper and Hutton envisaged southward subduction, giving rise to the Borrowdale volcanoes, lasting until Ashgill times, when volcanicity ceased; northerly subduction continued for a time, with continued production of the associated accretionary complex. The timing of the collision could be given with some precision. It affected all the Silurians and consequently post-dated the end of the Silurian (Prídolí Epoch); and it was prior to the emplacement of the (uncleaved) Shap granite in the youngest epoch of the early Devonian (Emsian). Soper has suggested that this was a case where developments in the understanding of Lakeland geology had a substantial impact on regional geotectonic thinking in Britain (pers. comm., 1998).

Soper (1986) developed his ideas further in discussion of what he called the 'Newer Granite Problem'. This was not a revisitation of the old controversy that had agitated the geological community up to the 1950s about 'migmatism' and 'magmatism' (Read 1957). Rather, the question was: how could theory account for the occurrence of 'young granites' (such as that at Shap, but up in Scotland too)? On the basis of plate-tectonic theory, these might be expected to have been the product of subduction-generated magma; but the subduction had supposedly ceased by the time the granites were emplaced.

Leaving aside here Soper's discussions of the Scottish granites, he endeavoured to solve the problem for the Lakeland granites by taking up an idea proposed by Ziegler (1984), which suggested that Cadomia was a composite entity, formed by the coalescence of several smaller 'micro-cratons'. Were this so, it might have had small 'internal' ocean basins, within which subduction processes might have been occurring, some of which could have generated granitic magma. So, it was suggested, the emplacement of the granites in the Lakes might have been *controlled* by the tectonics of Iapetus convergence but the intrusions were *generated* by a separate subduction system further south. If there were a relaxation of stress after the Iapetus closure, with the cessation of movement of the northward-moving 'Cadomia', then there could have been the opportunity for the granite to rise upwards, as at Shap for example.

In Soper's next paper on his three-plate model, he was joined by Barry Webb and Nigel Woodcock as co-authors. We shall introduce Webb properly in Chapter 12, but a few words of introduction for Woodcock are in order here. Born in 1949, he was educated at King Edward VII Grammar School, Sheffield, and decided to take up geology because of good teaching of the subject there and, as with Soper and Webb, an interest in rock climbing. He read geology at Manchester University and did his PhD at Imperial College under the structural geologist John Ramsay, the topic being concerned with the distinctions between slump and tectonic structures in Silurian rocks. Thus he gained expertise in both structural geology and sedimentology – which was recognized by an appointment at Cambridge, where he has been ever since (at Clare College). Later, he also concerned himself with environmental geology and he became involved to some extent in the controversies about the proposed burial of nuclear waste in west Cumbria (see Chapter 20), being interviewed about the matter on television (pers. comm., 1999).

In the 1970s, Woodcock worked on the Mesozoic rocks in the countries of the eastern Mediterranean, but this research was terminated because of the political disturbances in Cyprus, which made it impossible for him to visit localities relevant to his work. So he returned to studies of Lower Palaeozoic sedimentation and structural geology in Wales and also in the Lakes (where his student Louisa King did work that will be considered in Chapter 17). Woodcock was also involved in the Survey–universities collaborations as well as contract work for the Survey, described in Chapter 14.

As part of his teaching duties, Woodcock, in collaboration with Barrie Rickards (see p. 176) regularly took parties of Cambridge students to the Howgill Fells for mapping exercises, and while the students were off doing their individual work he had plenty of opportunity to make his own observations. Over a period of several years he essentially remapped the Howgills on a structural basis. In this way he was able to feed reliable data into Soper *et al.* (1987*a*), along with information from Wales. The paper had essentially the same theoretical basis as that of Soper & Hutton (1984). As can be seen from Figure 10.7, Cadomia had now been divided, so that 'southern Britain' only formed part of 'Eastern Avalonia'. I do not think that Woodcock's data fundamentally altered the way the folding and cleavage had been understood for the Lakes in Soper & Hutton (1984), though obviously more data were made available. Data for the Shap area were also taken from Moseley (1968). It is interesting to remark the contingencies involved in scientific collaborations.

In the Lake District, the granites and granophyres of Eskdale and Ennerdale had been dated radiometrically as Ordovician, and could be associated with the formation of the Borrowdale Volcanics (see p. 157). They could also have exerted a control on the course of the northward movement of 'Cadomia' (or rather Eastern Avalonia) and the deformations associated with its 'docking' with the already united Laurentia and Baltica. The later Lower Devonian granites – Skiddaw and Shap, and perhaps the one recognized by gravimetric investigations under Wensleydale (see Chapter 11) but not exposed at the surface – were supposedly intruded subsequent to the collision-induced deformation and cleavage. But what was causing the cleavage, and when did the 'impact' occur?

Well, according to the arguments of Soper *et al.* (1987*a*), the

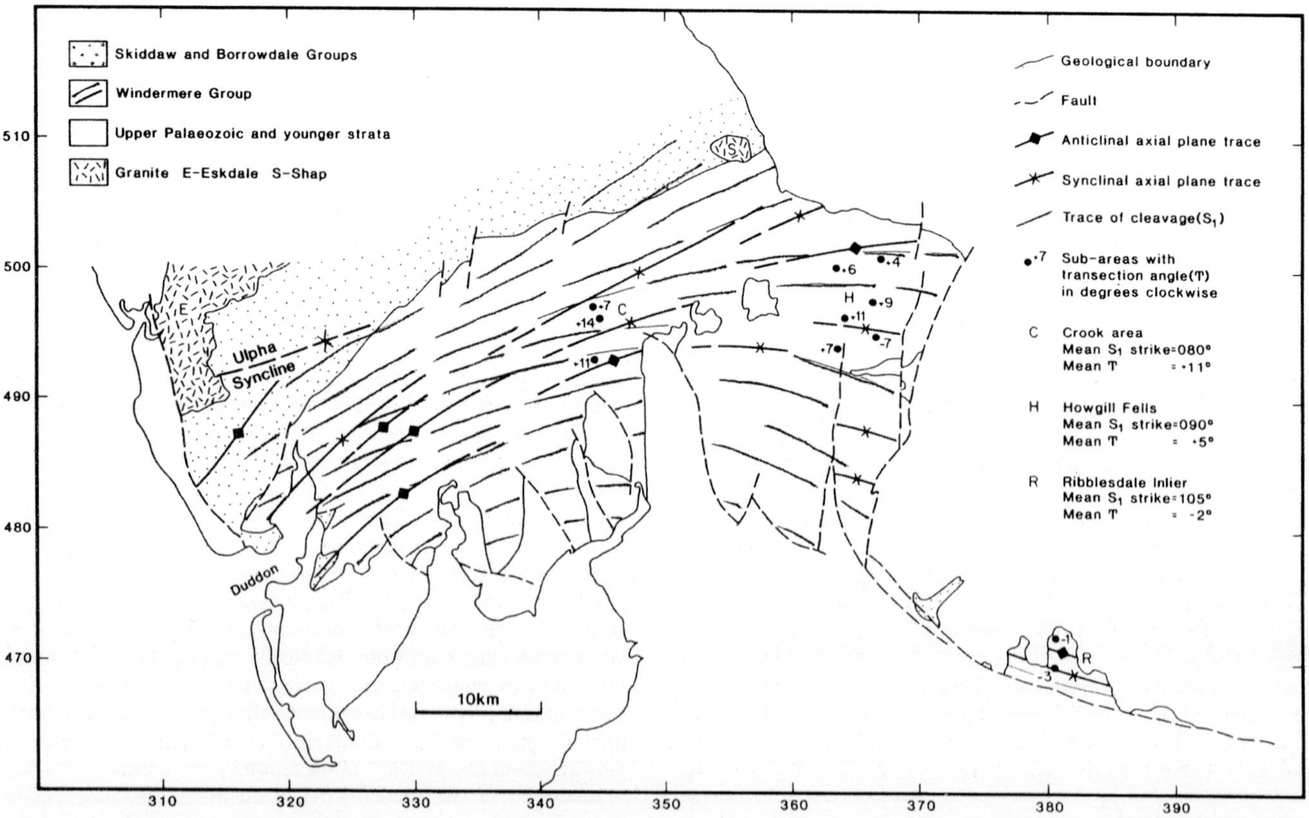

Fig. 10.7. Orientation of cleavages in Silurian rocks in northwest England, according to Soper, Webb & Woodcock (1987, p. 179).[17] Reproduced by courtesy of the Yorkshire Geological Society.

'indenter' had as its main 'body' the Midlands Massif, believed from seismic and borehole evidence to exist in the Midlands, and appearing on the surface in the Precambrian rocks of Charnwood Forest in Leicestershire. As to dating the collision or impact, the authors now suggested that it occurred in early Devonian times, not end-Silurian, which had previously been regarded as the time of the main deformation and cleavage of the Lakeland rocks. However, as Soper *et al.* pointed out, several different events might be taken to mark a 'collision':

(a) appearance on one terrane of material originating from the other with which it is colliding;
(b) cessation of volcanic activity because of cessation of subduction;
(c) deformation of margins (folding, imbricate faulting, cleavage);
(d) erosion of deformed areas and deposition of sediments over the suture.

One might be inclined to say that 'collision' was to be identified with stage (c), but clearly there were other possibilities and in a way it was a semantic question.

Anyway, the theory advanced was that Eastern Avalonia began to approach the accretionary complex of the Southern Uplands with a trench in front. It underthrust the Uplands, raising them to form a rich source of sediment; a basin was formed in front; and movement of the impinging terrane was eventually brought to a halt, causing substantial deformation in the process. The basin gradually filled with sediments during the Silurian (forming the bulk of 'Otley III' or the Windermere [Super]group[18]), with the rate of sedimentation, under these circumstances, *increasing* as the collision proceeded and the basin deepened. (Presumably the 'Windermere' sediments must have been deposited to the north of the Lakes as well as in the south, where they are presently preserved.) The sediments were then deformed (folded and cleaved) as the orogeny culminated in the early Devonian and the advance of Avalonia was finally halted. As Woodcock put it to me (pers. comm., 1999), the situation might be compared to the Ganges Basin to the south of the Himalayas. This basin presumably formed as the subcontinent of India ploughed its way northward and is now filling with sediment. (A further push might convert these Asiatic sediments to slates at some future time.)

The events that occurred in the Southern Uplands attracted much discussion in the 1980s with McKerrow's model coming under challenge (see Murphy & Hutton 1986; Stone *et al.* 1987). The topic continued to attract attention to the end of the twentieth century (and beyond), as evidenced by a symposium devoted to discussion of the geological problems of the area, held by The Royal Society of Edinburgh in 1999, and the Penrose Conference on the Iapetus Suture in 2000; but the issues discussed at these meetings are taken to lie beyond the scope of the present study.

With the idea of shifting terranes, it was likely that some light would be thrown on their past history or histories by consideration of palaeomagnetic evidence. To this end, Soper & Woodcock (1990) gathered together the available palaeomagnetic evidence for the apparent pole-wandering paths of Scotland (part

[17] Soper republished this diagram the following year, but stated that it represented 'End-Silurian' rather than 'Early Devonian' (Soper 1988, p. 488).
[18] The 'Otley III' rocks known as the Windermere Group in the 1970s, were subsequently renamed the Windermere Supergroup (see p. 185).

of Laurentia), Armorica (France and central Europe), and Eastern Avalonia (southern Ireland, southern England and southern Scandinavia), and plotted these on a simple global map, from the Ordovician to the Carboniferous. The data were recognized as being somewhat meagre. Nevertheless, they seemed to indicate a lining up, or 'coalescence' of the pole directions, at some time during the Devonian. Also, the data suggested an anticlockwise rotation of Eastern Avalonia as the terrane moved northwards and docked with Laurentia (Scotland). (Baltica was omitted from the discussion.)

Data on the type and direction of sedimentation were also collated, and the overall scheme for the hypothetical assemblage of the 'parts' of 'Scotland', and then the docking with 'England and Wales' by closure of the Iapetus Ocean, was depicted graphically (see Fig. 10.7). The anticlockwise rotation of 'England and Wales' (part of Eastern Avalonia) should be remarked upon: it implied closure of Iapetus from west to east, not east to west as previously supposed (see p. 144). The closure of the suture was put at Wenlock in Ireland and Ludlow in NW England, when voluminous sandy sediments first came in from Laurentia.

The model was developed further, with modification, in Soper *et al.* (1992*a*) and Soper *et al.* (1992*b*). Iapetus closure supposedly occurred in the Silurian, for sedimentological reasons as mentioned above, but it was suggested that the 'Acadian event' or 'big crunch' came in the Devonian.[19] The theory was now situated in the wider context of the geological history of the North Atlantic. That is, it was suggested that there was a transition from sinistral Caledonian relative displacements to dextral Variscan movement,[20] with the clockwise movement of Eastern Avalonia giving way to anticlockwise movement according to the palaeomagnetic evidence. Thus the 'really big crunch' was not that associated with the Iapetus closure but with a 'mutual twisting' of the terranes. Perhaps this had something to do with the continued force acting from the south as Gondwanaland continued its northward movement (all the preceding events supposedly having taken place in the southern hemisphere according to the palaeomagnetic evidence).

It will be seen from the foregoing that there was a long period during the Silurian when the north of 'England' was supposedly under water and able to receive sediment. What had happened to the Ordovician (Borrowdale) volcanoes? Would they not stand in the way and impede sedimentation? Perhaps not. The theory implied that the volcanoes (with their initially fragmentary ashes) had already been eroded to a great degree. Yet in all probability they were formerly large structures, and today we only see their deep roots – their 'plumbing', as volcanologists like to call it. This is a problem, I think, though the preservation of the tuffs, etc., of the BVG can be accounted for in some measure at least by the hypothesis of caldera collapse, or perhaps extensional rift faulting – ideas that will be discussed in later chapters. On the other hand, as we shall see in Chapter 13, some geologists only a short time previously had been thinking in terms of the 'Otley III' sediments being deposited around some kind of horst structure in the area of Lakeland. However, the two ideas were not necessarily incompatible. The sediments could have been deposited around at first, and then over.

Nevertheless, the theory, if correct, would require the deposition of Silurian sediments all the way to Scotland (so to speak). Why is there no sign of them in the northern Lakes, but only down Windermere way and in the Howgills? The answer would have to be that there were formerly Silurian (or late Ordovician) rocks in the area of the northern fells, but they have subsequently been eroded away. Is there any sign of them there? The only known relevant rocks are the Caradoc (late Ordovician) sediments of Drygill, preserved by downfaulting to the west of the Carrock Fell igneous complex in the northern Lakes; and clasts of chiefly Windermere-type rock found in the poorly sorted, crudely bedded, unfossiliferous, red molasse-type deposit of the Mell Fell Conglomerate, which lies unconformably on the SS and on the BVG to the north and west of Ullswater.[21] Remarkably, the Conglomerate contains few clasts of the Volcanics (and then only at the western end of the outcrop) and no Skiddaws. It is chiefly made up of clasts from the Windermeres – notably the Coniston Grit. Its presence at the north end of Ullswater may betoken the earlier occurrence of Windermere rocks in the northern Lakes, but the pebbles, some well rounded, could have travelled some distance. The pre-Ashgill Drygill Shales are not, of course, Silurian like the Windermeres.[22]

Evidently, we have a problem here, and it will require further discussion when we consider ideas about the Windermere Group and its tectonics (see Chapter 17). For the present, we may simply ask: where are the metamorphic features that might be expected for a suture zone – metamorphics such as occur in the Himalayas or the Urals, for example? The subterranean topography of the supposed Iapetus Suture has now been revealed to a large extent by the seismological work of the Geological Survey (see Fig. 10.8). It does not really look like what one might expect for a collision between two mini-continents. Or does it? Jack Soper (pers. comm., 1999) acknowledged that it must have been the 'most gentle of collisions'. But is it not a remarkable coincidence that Eastern Avalonia came gently to rest at just about the time that it bumped into Laurentia? Perhaps not. There would have been gigantic resistive forces at work, which would have slowed down the motion before bringing it to a halt.

It is true that seismic and geoelectric sounding evidence has suggested the existence of a crustal shear zone dipping NNE at about 25° under the Northumberland trough and the Southern Uplands, penetrating right down to the Moho discontinuity

[19] The term 'Acadian' was initially used to refer to deformation in the Appalachians during the Early–Middle Devonian, but was now being deployed in relation to the tectonic events of the British slate belts. Acadia was a former French colony in Atlantic North America. The idea of a geological 'revolution' (= orogeny) of that name occurring there in the Devonian was suggested by H. S. Williams (1895). The idea of applying the American term to British geology was apparently dreamed up in a pub one evening by Soper and McKerrow!

[20] The term 'Variscan Orogeny' (late Palaeozoic) derives from a mountain system, believed to have extended from southern Ireland and Britain, through central France and Germany, to southern Poland. The name, coined by Eduard Suess in the second volume of his *Antlitz der Erde* (1888), was derived from the name of the land of the Varisci, or Vogtland, being the Latin name of Hof in Bavaria (Suess 1906, pp. 97–122).

[21] The original survey (Appleby sheet) showed the contact as faulted.

[22] The Mell Fell Conglomerate was thought by Richard Oldham (1900) to represent a 'torrential deposit formed on dry land, near the foot of a range of hills, in a generally dry climate, varied by seasonal or periodical bursts of rain', being analogous to ones produced today in Western and Central Asia. He downplayed the occurrence of rounded clasts in the conglomerate, and in fact not all of them are rounded. The ideas of J. F. N. Green on the unit have been discussed previously (see p. 82). A detailed examination of the lithologies of the unit's constituent clasts was made by Joseph Capewell (1955) and Wadge (1978*a*). The latter took the view that the Conglomerate suggested the former occurrence of Silurian rocks in the Ullswater and Haweswater areas in Devonian times, and so had been deposited much further north than might be expected according to the present exposures of Lakeland Silurians.

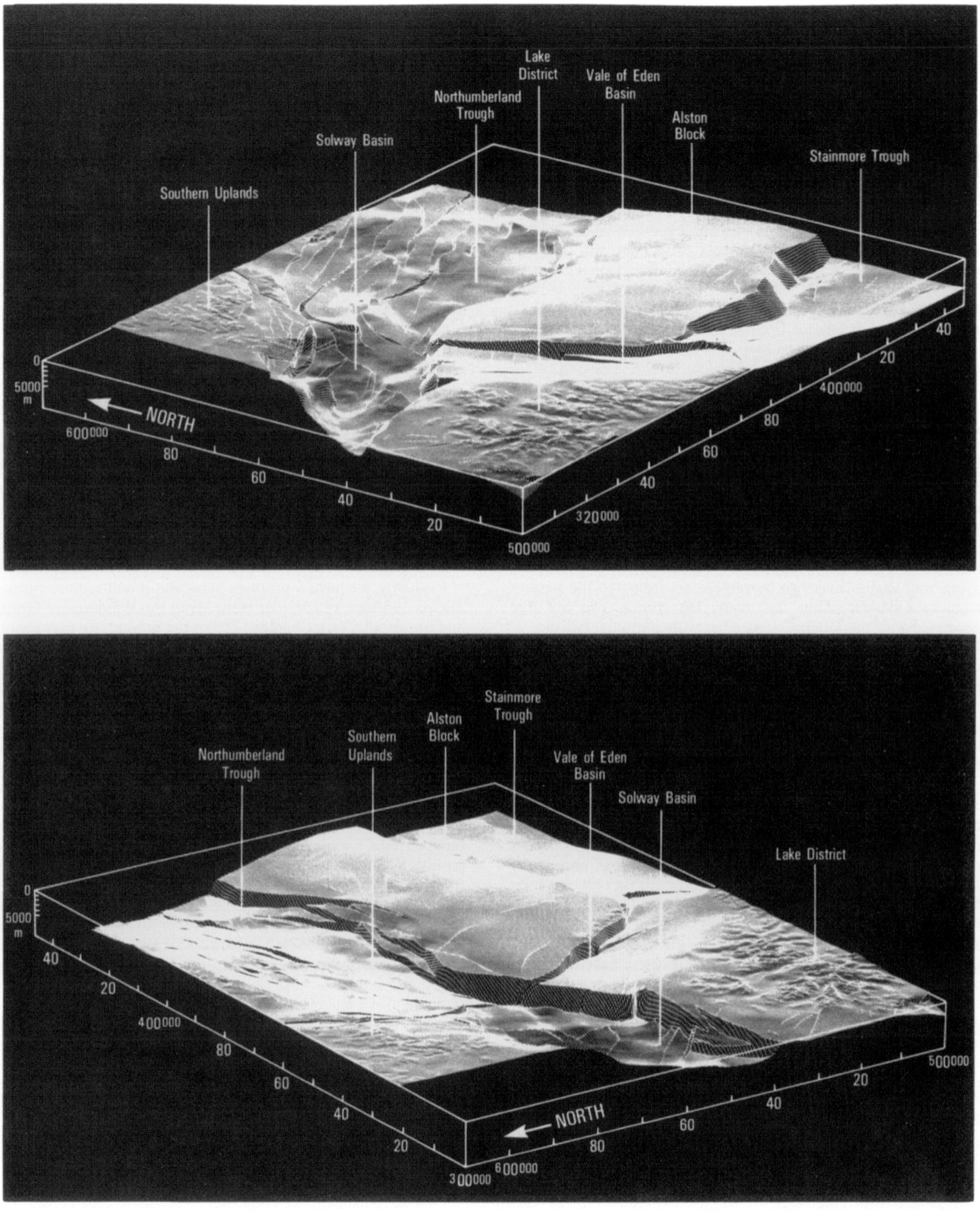

Fig. 10.8. British Geological Survey's model of basement rocks of northern England and southern Scotland, according to Chadwick *et al.* (1995, p. 7).[23] IPR23–7C British Geological Survey.

[23] The surface-mappable faults bounding the Solway Basin, the Northumberland Trough, and the Alston Block are represented as normal in Chadwick *et al.* (1995, p. 2). The Alston Block has chiefly Carboniferous rocks at the surface.

Fig. 10.9. Unconformity between andesite of Eycott group and steeply dipping beds of Skiddaw Group siltstones, at Overwater spillway; from Millward & Molyneux (1992, p. 87) BGS Photograph D4457 (taken by F. McTaggart). IPR/23–7C British Geological Survey. © NERC. All rights reserved.

(Beamish & Smythe 1986), and this may be a manifestation of the suturing.[24] However, the shear zone does not run out to the surface in the Solway Basin, where one might hope to find some evidence for it (though the basin has Carboniferous and some Jurassic sediments at the surface). It may be associated with basement fracture somewhere beneath the SS near the Causey Pike Thrust (see p. 200) – which is not to say that that fracture marks the Iapetus Suture.

Let us now try to refer these large-scale theorizings to the issue that we discussed in Chapter 9 – the question of the relationship between the BVG and the SS, which so exercised geologists' attention in the 1970s. Soper eventually came round to the view that there was indeed a pretty general unconformity between the Borrowdales and the Skiddaws. As we saw in the previous chapter, in his paper with Charles Downie (Downie & Soper 1972), he had thought that there was interbedding of the Eycotts with Llanvirnian *D. bifidus* (later renamed *D. artus*) zone of the Skiddaw Slates. However, the Surveyors David Millward & Stewart Molyneaux (1992) showed that clear evidence for the SS–BVG unconformity could be recognized at the spillway between the small reservoir, Overwater, and the even smaller Chapelhouse reservoir, near Uldale in the northern Lakes (see Fig. 12.1); and what had earlier been supposed to be interbedded Volcanics and Skiddaws were in fact the result of sills being intruded into the Skiddaws. Their mapping showed that the Volcanics were laid down on an undulating Skiddaw surface, and at the spillway they were able to clear a good section and produce an excellent photograph (see Fig. 10.9) of an unconformable contact between andesite and SS.

Soper accepted these results; so palaeontological dating of the Skiddaws did *not* date the Eycotts, which were subsequently regarded as having essentially the same age as the main body of the Borrowdales, themselves by then dated by acritarchs as Caradoc (Molyneux 1988). In fact, the earlier dating of the Eycotts by Downie and Soper was suspect, according to Molyneux, because of possible reworking of acritarchs. Also, the Carrock Fell Complex was known to contain xenoliths of Eycott Volcanics; and taking an age for the former of 468 Ma (cf. Rundle 1979), Millward & Molyneux (1992) concluded that the Eycotts would be *pre*-Caradoc, and thus belonging to the earliest BVG. So if the low Borrowdales were unconformable to the Skiddaws, this must apply *a fortiori* to the upper Borrowdales – or there must be faulting. In any case, as we shall see (p. 221), the upper Borrowdales were undoubtedly produced subaerially, at least in part – implying uplift.

The sharp SS–BVG contact at the Overwater exposure showed no indication of thrust faulting, but every indication of unconformity. On the other hand, there are surely contacts that do display faulting (Moseley 1973–1974 [1975]). What may we make of all this, in the light of the high theory being developed by Soper and others?

Millward & Molyneux (1992)[25] took the view that the angular

[24] On David Smythe, see Chapter 20.

[25] The work of these geologists will be described further in Chapters 15 and 16, where they will be properly introduced.

unconformity displayed at the Overwater spillway implied that there must have been *some* pre-BVG tectonic episode[26] (a partial vindication of Simpson?), and they averred that the new evidence warranted the old arguments about pre-Borrowdale earth movements being revisited; but we shall not attempt to do this here.[27] What we may do is consider how, if at all, the question of the SS–BVG relationship might be related to the theorizings about terrane collisions that we have considered in the present chapter.

I do not think it all fits neatly – or not yet in the present narrative. We have some tectonic activity during or immediately after Skiddavian time, evidenced by the Overwater unconformity, where there is a nice clean, unfaulted contact between the ductile Skiddaw rock and the solid BVG andesite. There is the great pile of Borrowdale Volcanics, displaying some cleavage. There is the strongly and arcuately cleaved slate-belt of the Windermere Group, with its curious structural relationship to the Volcanics, which we have hardly discussed yet. There is the contact between the Skiddaws and the Volcanics that is possibly sometimes conformable, sometimes unconformable, and sometimes (thrust?) faulted. There is the 'mystery' of the Iapetus Suture; and the large Lakeland granites still have to be brought into the picture satisfactorily.

The oddity of it all is that with such massive tectonic activity in the Devonian, with Eastern Avalonia supposedly rotating first anticlockwise and then clockwise, and with Soper's 'indenter', the Midland Massif, driving England under Scotland (!), we might expect the visible effect at the SS–BVG boundary to be greater than it is, given the great disparity in the ductilities of the two units.

However, for the sake of a simple narrative many details of the story of Lakeland research have so far been omitted. So let us retreat somewhat in time, first to tell the story of the work on the Lakeland granites through gravimetric investigation, and then to look at the large-scale effort that was made to remap the Lakes and sort out the geology of the region, starting from 1982. Of course, there is still much to be said about the Volcanics and 'Otley III'. It may be noted, incidentally, that if Soper's 'indenter' theory is correct, with Eastern Avalonia 'drifting' northwards and with Lakeland rocks being driven *under* the Southern Uplands, then Marr's old lag fault theory might have been correct, with the ductile Skiddaw Slates being pushed further north than the solid Borrowdale Volcanics. However, the nature of the contact between the Skiddaws and the Borrowdales is far from what might have been expected according to Marr's idea. Also, if the Lakeland rocks were driven under the Southern Uplands, why are the latter not uplifted and metamorphosed, as one might reasonably expect? Should one be taking a different tack altogether?

As suggested, there could have been some tilting of the SS associated with the incipient Borrowdale volcanism; eruption of the volcanoes so that sometimes the contacts appear conformable and sometimes not; and subsequently (or contemporaneously) a variety of earth movements associated with the various possible terrane movements, with different results in different places according to different ductilities and configurations. Very likely there is more than one kind of SS–BVG contact. If this be so, we can see why geologists of the 1970s, even with their detailed investigation of the contact, found such difficulty in reaching agreement about its nature.

To get a clearer picture of the matter we need more information. What about the granites? What do they tell us? Let us continue!

[26] Other types of deformation that may have occurred in the Skiddaws will be discussed in Chapter 15.

[27] Perhaps the Skiddaw Slates were tilted by doming, arising from the oncoming Borrowdale volcanicity?

Chapter 11
Gravity, geomagnetism and granites

We shall now consider aspects of Lakeland research where geophysical (particularly gravimetric) investigations have played a significant role in the elucidation of the geology of the region. The two main figures have been Martin Bott, Professor at Durham, and his former student Michael Lee, now a geophysicist working for the Survey. Let me first say something about Bott's work.

Bott (see Fig. 11.1) (b. 1926) was awarded a scholarship to Magdalene College, Cambridge, to read mathematics, but decided to do his national service first and served with the Royal Signals in East Africa. On going up to Cambridge in 1948 he started off in mathematics as planned, but halfway through the first term he switched to the Natural Science Tripos and took on mathematics, physics, chemistry and geology. At the end of his first year he participated in an expedition to Spitzbergen led by the lecturer Brian Harland and, enthused by this experience, Bott turned to geology and took the subject for his Part 2. Given that he is now a Fellow of the Royal Society, I was most surprised to learn (pers. comm., 1999) that, as one of the most distinguished scientists whose work is referred to in the present book, he scored a lower second.[1] This was a result of his indifferent performance in palaeontology – a part of geology much emphasized at Cambridge in those days. (Bott had had no previous background in biology or anatomy.)

His overall result notwithstanding, Bott obtained a DSIR grant to do a PhD, and was intending to take up a problem in geomagnetism, but when he got back to Cambridge after a second expedition to Spitzbergen with Harland he found that the slot reserved for that field had been allocated to the Australian Ted Irving (of pole-wandering fame). The topic Bott eventually took up had to do with the question of the possibility of there being granite under the north Pennines, as had been suggested by Kingsley Dunham in his PhD under Arthur Holmes in the 1920s, on the basis of the mineralization in the region. The idea was to investigate the problem by gravity survey, looking for the presence of subterranean granite by finding 'gravity lows' or negative gravity anomalies. Bott was supervised by Brian Harland, and he also collaborated with David Masson-Smith. Bott's attention had been drawn to the possibility that there might be a hidden granite mass beneath the mineralized area of the north Pennines when a second-year undergraduate, by his supervisor Percy Allen.

Bott worked in the Weardale area (Northumbria) and made comparisons with the Cornish granites. Hypothetically, the gravity low in Weardale might be due either to a sedimentary basin or to a subterranean granite, but the mapped shapes would be different in the two cases: a granite sloping outwards and a basin sloping inwards. The gravity profile indicated that it was the former. Bott and Masson-Smith had extensive discussions with Dunham in Durham, and in 1954 Bott obtained a two-year Turner & Newall fellowship at Durham – and he has been there ever since, though he was (nominally) retired when I interviewed him in 1999. The idea was to extend the Pennine work to cover the Lakeland granites and determine whether they were connected underground with the one found under Weardale. Also, in an early paper, Bott (1956) showed that the shape of granite plutons, as determined by examination of gravity anomalies, was more congruent with the magmatic theory than the migmatitic theory, then being promulgated from, for example, the chair of geology at Imperial College. Bott deployed the stoping theory of the American, Reginald Daly (1933), to deal with the 'space problem'.

During Bott's time at Cambridge there was only a rather old German gravimeter available, obtained from war reparations, but at Durham they were able to use a more modern 'Worden' instrument from the United States, which was reasonably portable in a backpack without heavy batteries. Even so, Bott recalls that they cut a hole in the floor of the departmental field van so that the instrument could be lowered directly onto a road to take a reading. In consequence, most of the measurements in the Lakes were taken on roads, but some measurements in critical areas were made off the road. The Lakeland fieldwork was finished in a couple of months in 1954, with about a thousand measurements made overall. It took another two or three months to reduce the results; and then terrain corrections had to be made, which involved complicated calculations such that only about ten could be done per day. Bott recalls that he began to get eye strain at that time. For this reason, he began to work on a computer program that would permit rapid calculation of the corrections. The task was completed in 1958 and published the following year (Bott 1959). A large Pegasus computer at Newcastle was used.

Fig. 11.1. Professor Martin Bott. Undated photograph, Geological Society archives (LDGSL P55/7).

However, before publishing his Lakeland map and results, Bott got interested in other topics such as theoretical problems of stress and strain, and marine gravimetry, and the Lakeland results were not actually published until 1974 (Bott 1974). The task in hand was to devise a model for the subsurface rocks' shape and density that would satisfy the empirical gravity anomalies. The inverse

[1] The Cambridge examiners seemingly did not judge matters aright!

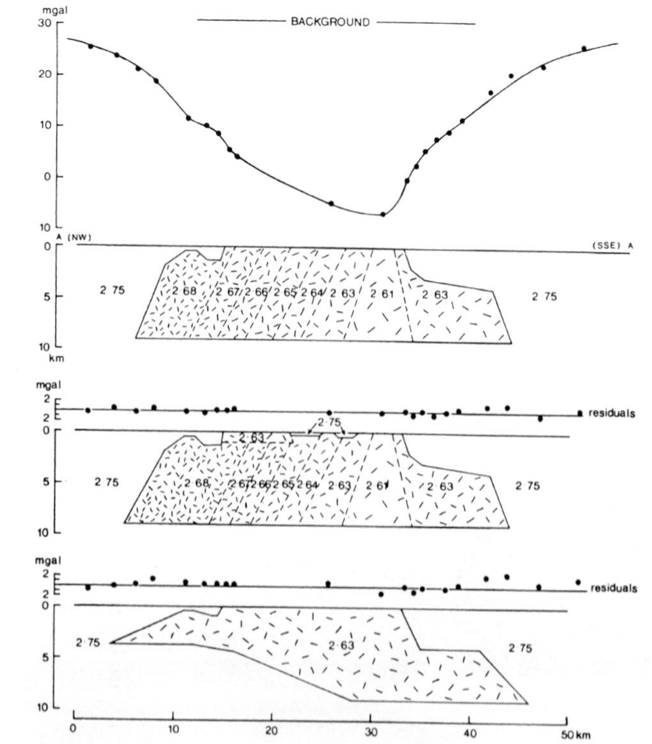

Fig. 11.2. Models for the Eskdale Granite and Ennerdale Granophyre, to match the gravity anomalies, according to Bott (1974, p. 315).

problem (determination of the anomaly from a granite mass of known position, shape and density distribution) could be solved directly, but working the other way there could be many possible solutions, particularly considering the fact that the granite itself need not necessarily be of uniform density throughout, because of material assimilated during its formation. One had to work iteratively, therefore, using known geology at the surface as a guide. Three of Bott's five or six models for the Eskdale Granite and Ennerdale Granophyre are shown in Figure 11.2. The second one was his preferred option, as there was a mismatch between the observed and hypothetical density for the Ennerdale Granophyre for the first model; and the third one was not really plausible, given the known variation in the petrology and density of the pluton at different localities. The horizontal base to the pluton assumed in the first two models was arbitrary, being assumed for the purpose of simplified calculation. Bott had in mind a kind of stock structure, as had been assumed by Trotter *et al.* (1937) in the Gosforth Memoir on the basis of field mapping. (Earlier writers such as Dwerryhouse (1909), Green (1917) and Simpson (1934) had thought in terms of a laccolith for the Eskdale Granite.)

In his 1974 paper, Bott was inclined to think that there might be a subsurface connection between the Lakeland plutons and the one under Weardale, which he had investigated in his PhD. In his retirement, he still thought that such a connection might exist, but the issue was unresolved at the end of the twentieth century (pers. comm., 1999). Bott (1974) further suggested that the general elevation of the Lakeland region (and the north Pennines) might be due to the isostatic effects produced by the granites in the *upper* crust – something not normally considered in the context of general isostasy theory at that time. He also offered an account of the structural history of the Pennine faults. Gravimetric measurements across the Vale of Eden suggested a Permo-Triassic basin over Lower Palaeozoic basement, with the younger rocks thinning to the east. Building on the work of Frederick Shotton of Birmingham University (see p. 172), who had proposed a thrust fault structure for the Cross Fell Inlier (Shotton 1935), Bott envisaged E–W compressive forces acting during the Hercynian Orogeny, causing the formation of the Vale of Eden basin and folding and thrusting up the Palaeozoics of the Cross Fell Inlier at the Inner Pennine Fault system. Following further deposition, and subsequent erosion, the Outer Pennine Fault system was formed during the Tertiary. The whole was supposedly stabilized in the northern Pennines during the Carboniferous by the subterranean granites, *vis-à-vis* the subsiding area to the south. However, while the reported discussion that followed Bott's paper was favourable to his work, the idea of major thrust fault in the Cross Fell Inlier such as he and Shotton had in mind has not been upheld by later investigations of the Survey.

Bott did a considerable amount of work in northern England similar to that described above, and he extended his efforts to the Irish Sea region even before his Lakeland work was published (Bott 1964, 1965; Bott & Young 1971), but his investigations to the east and west of the Lakes need not be discussed here. The ideas given in Bott's chapter in Moseley (1978*a*) were essentially the same as those put forward in his Lakeland paper of 1974. However, as he told me in our conversation in 1999, he had, in his retirement, come back to the Lake District in the sense that he was developing some fundamental ideas about the emplacement of granite, in part on the basis of Lakeland evidence (Bott 1999). Something will be said about this work at the end of this chapter.

Following Bott, the main person to do gravimetric work in the Lakes was Michael Lee (see Fig. 11.3).[2] Lee initially studied applied physics at Brighton Polytechnic as a 'sandwich course', sponsored by the General Electric Co. Ltd. After obtaining a first, he decided to move into geophysics and took an MSc at Durham under the 'grand master of the time', Martin Bott. Lee's degree involved the analysis of the power spectra of magnetic anomalies.[3] In 1972, Lee joined the Applied Geophysics Group at the Survey and worked on a range of projects including resistivity studies for geothermal exploration in St Lucia. He then turned to the geophysical study of northern England. Bott and his Durham students were still working on the Weardale Granite, so the Survey people decided to look at the suspected granite under Wensleydale. Lee devised a computer program for locating the granite and advised on a site for a borehole – and his prediction proved successful. He then became involved in the geothermal programme in Britain, looking to see if there were radiogenically hot granites that might serve as sources of energy.[4] Localities were examined in Cornwall, Weardale, Scotland, the Lakes, etc. Boreholes were drilled on Shap and Skiddaw and the heat flow and radiogenic heat production were measured. However, although the Lakeland granites had above-average heat flow they were not sufficiently hot to provide a commercially viable source of geothermal power.

There was also a considerable amount of geomagnetic data available, as the whole country had been aeromagnetically surveyed in the 1950s and 1960s by the geoprospecting company Huntings, under contract to the Institute of Geological Sciences (precursor to the British Geological Survey). This survey provided

[2] Since April 2000, Lee has been Director of the BGS's Lands and Resources Programmes, which makes him the institution's first geophysicist to be in charge of its mapping programme!

[3] The power spectrum of a magnetic profile is examined, and the gradient of the logarithm of the power spectrum serves as a measure of the depth of the magnetic source. Hence one can use the technique to determine the depth of the 'magnetic basement', which is useful in oil exploration, the study of basic igneous rocks, etc.

[4] The 'Hot Dry Rock' programme was undertaken on Caledonian granites of Britain by the Survey, Imperial College and the Open University, on behalf of the Department of Energy. Other Lakeland investigations were Webb & Brown (1984) and Wheildon *et al.* (1984).

Fig. 11.3. Michael Lee at the British Geological Survey, Keyworth, contemplating a computer-modelled profile of the Lake District. Photographed by author, 1998.

readings every 2 km from a height of 300 m. However, the data later had to be digitized by the Survey (Lee, pers. comm., 1998).

When the minute re-examination of Lakeland geology was initiated in 1982 (see Chapter 14), Lee was an important member of the team, working on a much more detailed gravimetric survey than had been undertaken by Bott in the 1950s. Readings were taken on and off roads, with a coverage (so far as was feasible) of one determination every square kilometre. Additional readings were taken near the assumed granite margins. Remote areas were reached by helicopter. Heights were determined accurately from triangulation points and photogrammetric spot heights. The fieldwork was carried out over three seasons and led to the publication of a Survey Report (Lee 1984) and a Geological Society paper (Lee 1986); also a re-evaluation of the role of the granites in Lakeland geological history was undertaken in collaboration with Ron Firman (Firman & Lee 1986). The techniques for modelling were improving rapidly at that time, with the availability of increasing computer power and improving image-analysis techniques. It is interesting that the structure known as the 'Crummock line', associated with the 'Crummock aureole' overlying a concealed intrusion and also with the important Causey Pike thrust, was first recognized by gravimetry, and the structure was correlated with the geochemistry when the area was investigated geochemically in the Lakeland resurvey (see p. 200). Eventually Lee collected all his Lakeland work together for a massive Open University PhD (Lee 1990).

To make anything of the gravimetric data, it was necessary to have estimates of the densities of the different Lakeland rock types, to separate out the effect of the low-density Permo-Triassic, Carboniferous and Silurian sediments, and also the effect of a regionally eastward-decreasing gravitation field. For this purpose, various models had to be tried and their resultant gravimetric effects calculated and then compared with the empirical data. In this manner, one could approach a solution iteratively. It could not be achieved by direct calculation. Then the gravimetric anomalies due to the plutons could be properly discerned. Lee's chart for the residual gravity anomalies for the Lake District is reproduced in Figure 11.4.

Lee soon came to the view that the idea of a single Lakeland batholith, with a possible extension under the Eden Valley and connection with the Weardale Granite, as envisaged in Bott's early model, was not tenable. However, it appeared that the Eskdale Granite was linked below the surface with the granite at Wasdale. Lee's model for the Eskdale–Ennerdale pluton – the largest in the Lake District, and the most interesting for our purposes – is shown in Figure 11.5. It was obtained using an unpublished computer program devised at the Survey by J. M. C. Tombs, which provided a three-dimensional model of the batholith. As will be seen, it assumed a uniform density for the granitic material. Bott's two-dimensional modelling, by contrast, had sought to construct a profile showing density variations, and showed more detail along his specific profiles. So the two results were not incompatible. Lee (1986, p. 432) pointed out that the base of his model (at 9 km) represented the depth at which the

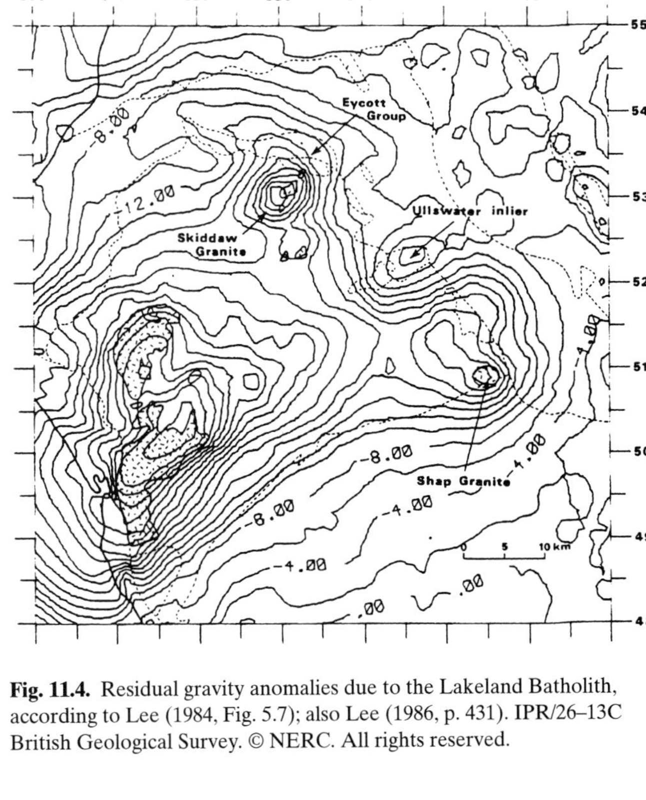

Fig. 11.4. Residual gravity anomalies due to the Lakeland Batholith, according to Lee (1984, Fig. 5.7); also Lee (1986, p. 431). IPR/26–13C British Geological Survey. © NERC. All rights reserved.

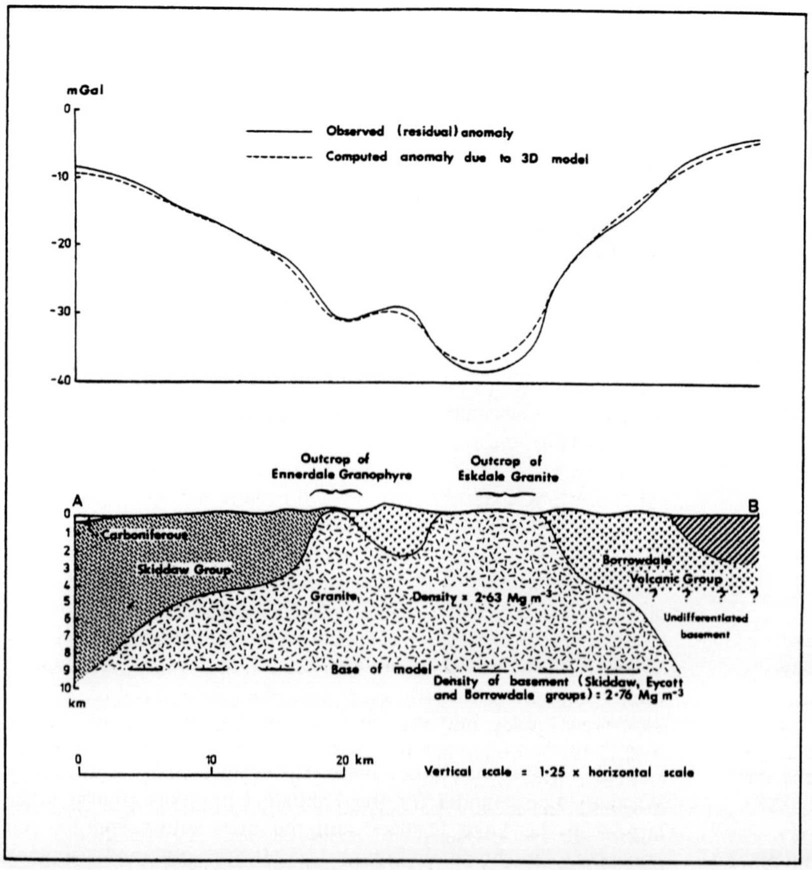

Fig. 11.5. Profile of the Eskdale–Ennerdale pluton obtained by gravimetric analysis, according to Lee (1984, Fig. 5.9). IPR/26–13C British Geological Survey. © NERC. All rights reserved.

batholith was assumed to '[merge] geophysically with the basement for the densities assumed in the interpretation'.[5] It did not necessarily have any structural significance. The central Lake District also appeared to be underlain by a relatively shallow granite, being an extension of the Eskdale Granite, whereas the steeper-sided Skiddaw Granite and Shap Granite were apparently separate bodies (not unexpected, given that they were believed to be younger entities according to the results of radiometric dating determinations). Lee (1986, p. 435) also adumbrated the notion that the 'Lakeland batholith' was a compound body. In his later work, as finally reported in the Memoir for the Ambleside map (see Millward *et al.* 2000), he modelled along a series of 2-D profiles to define the details of the batholith's components and then constructed its overall 3-D form from the 2-D profiles, guided by the earlier simplified model. Also, following the seismic investigations conducted near Sellafield on the western side of the Lakes (see Chapter 20), Lee began to regard the density differences in the different parts of the batholith as being due to different admixtures of country rock with the granitic matter; and that the batholith should be regarded as closely packed granitic sills rather than uniform plutons (Millward *et al.* 2000, p. 29–30).

Much earlier than this, according to Lee's recollections (pers. comm., 1998), he was already, in 1986, beginning to think that there were several 'concentrations' of granite at Ulpha (Duddon Valley), Ennerdale, Eskdale, Wasdale, Buttermere, Dunmail and Rydal; and there was also the linear structure near Crummock Water. There appeared to be some kind of pattern. Each low-density entity (a granite; or was it a thick mass of low-density ignimbrite?) seemed to be associated with a depression or some kind of synclinal structure. Granites seemingly 'sat under' – or were associated with – the Scafell Syncline, the Haweswater Syncline, the Ulpha Syncline, etc.[6] All this could hardly be coincidental. We shall subsequently see how this 'coincidence' began to find an explanation.

Here, however, mention should be made of the interesting collaborative paper by Firman & Lee (1986), published in a collection produced in honour of Kingsley Dunham's 75th birthday. First, how old were the granites? The earlier general supposition had been that they were chiefly end-Silurian, in accordance with the prevailing view of Soper and Moseley at that time – and in agreement with earlier age-determinations such as those of Miller (1961) and Brown *et al.* (1964) – that the main Lakeland deformation was end-Silurian.[7] The Survey radiometrician Christopher Rundle (1979) had, however, shown by Rb–Sr dating that some of the acid igneous outcrops were Ordovician, with the Eskdale Granite having an age of 429 Ma (which he put at Caradoc). In fact, on the basis of his age determinations, he proposed that there had been four phases of intrusion *during* the Ordovician. Yet Rundle did not propose that the batholith as a whole was Ordovician. Rather, he thought it had come up underneath the Eskdale Granite at end-Silurian.

Now, however, Firman and Lee began to contemplate the thought that the whole batholith (leaving aside the Skiddaw and Shap intrusions) might be Ordovician *and* of compound or multiple origin; for with Firman's petrological knowledge of the granites (or granitoids) brought into play, emphasis was now laid on the fact that they were chemically and mineralogically distinct (whereas Lee had earlier downplayed the differences). Further,

[5] Lee acknowledged the assistance of his BGS geophysicist colleagues K. S. Wilson, G. S. Kimbell, R. M. Carruthers, L. F. Smith, B. C. Chacksfield and S. F. Pease, and workers at Imperial and the Open University, in the preparation of his 1986 paper. Though it was a single-author publication, a large number of geologists were actually involved in one way or another. It was in Lee's 1986 paper that he first mentioned the existence of an igneous body under the Crummock Water aureole, 'discovered' gravimetrically, as mentioned above.

[6] Not all of this syncline is underlain by a granitoid; and the Nan Bield *Anticline* in the eastern Lakes seems to be associated with the Haweswater pluton.

[7] Interestingly, J. F. N. Green, so often out on his own, had long before (1917) suggested an Ordovician age for the Eskdale Granite.

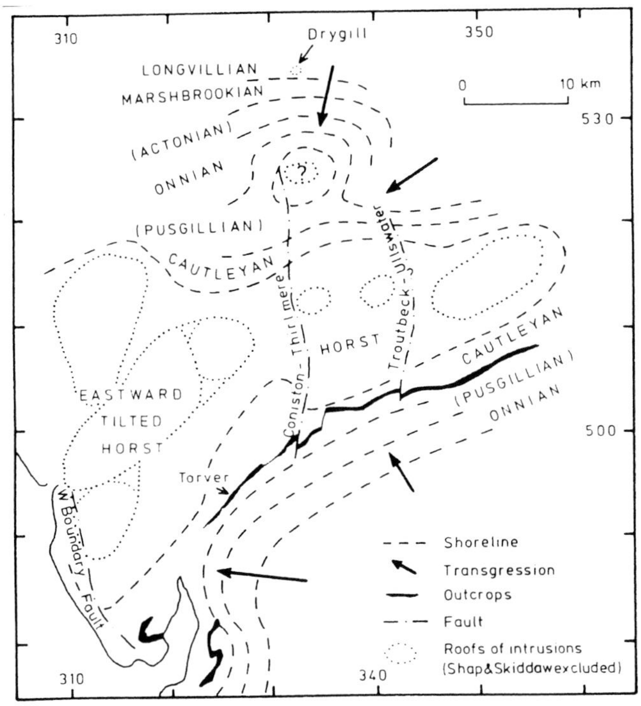

Fig. 11.6. Suggested palaeogeography for the Lake District during Upper Ordovician (Longvillian to Cautleyan), according to Firman & Lee (1986, p. 121). Reproduced by courtesy of Drs Firman and Lee.

the radiometric age determinations might be pointing to dates of metasomatic alteration when the radiometric clocks were 'reset', rather than dates for the original emplacement of the 'granites'. Lee had been doing further gravimetric and geomagnetic work in the western Lakes, which suggested that the Ennerdale and Eskdale intrusions had different geomagnetic signatures and so should be regarded as distinct bodies. Perhaps the whole batholith was composite?

Firman & Lee (1986) now suggested that there could have been several phases of deformation: late Ordovician–pre-Ashgill; end-Silurian; and Devonian and later. The first seemed to be associated with a N–S stress system producing the wrench faults in the Eskdale Granite that Firman had noted in his mapping in the 1950s. He had then thought they were end-Silurian but they were now proposed as pre-Ashgill. Structures such as the Scafell and Ulpha Synclines could either be 'drape structures developed during the intrusion[s]' (p. 121); or they could (somehow) be the result of isostatic uplift by the Eskdale plutons. Further, it was suggested that the granites could have marked an uplifted area, or horst, during Late Ordovician times, so that the several units then deposited could have been formed as it were round an island (see Fig. 11.6). So these sediments were not regarded as having been deposited over the whole of the Lakeland mountain range at that time. Uplift of the granite mass would account for the obvious unconformity below the Windermere Group ('Otley III'). Firman & Lee (1986, p. 121) were apparently sympathetic to the idea of Soper (and by then others too) that there had been subduction in the Ordovician during closure of Iapetus, but their model was not closely linked to plate-tectonic theory.

Then an end-Silurian deformation was proposed, as in the Soper–Moseley theory, being held responsible for the formation of the slate belts to the north and south of the proposed Eskdale–Haweswater granite ridge.[8] It was noted that well-cleaved slates in the Borrowdales seemed to be lacking in the areas over the hypothesized ridge, which suggested that the Ordovician plutons were somehow controlling the end-Silurian structures. On the other hand, good cleavage was known well to the west of the region of the Haweswater pluton, and also over the crest of the Nan Bield anticline, over a supposed low-density mass. The model was not entirely tidy or consistent.

Finally, there was the intrusion of the Shap and Skiddaw Granites, and also the hypothesized Crummock igneous mass, in late Silurian–early Devonian. The two former, it was suggested, produced rising horsts, and perhaps a depression approximately in the locality of the Mell Fell Conglomerate, which could have been deposited with clasts washed in from the northwest, west and south. Thus for Firman and Lee much of the geological history of the Lakes could be accounted for by the emplacement and subsequent influence of the plutons.

We are now well into the period when the Survey–universities collaborations were under way, and not long before a new theory of the structures of the Borrowdales was developed by the Sheffield PhD student Michael Branney, which will be discussed in Chapter 16. In a nutshell, his theory was that as volcanic ash was blown out from the vents of the Ordovician volcanoes, large calderas were formed, which structures then produced 'volcano-tectonic faults', as material collapsed into the cavities below, whence had come the ash. However, granitic material could also have slowly risen into the cavities from below; while structures such as the Scafell Syncline could be sags, rather than the product of lateral forces. In talking to Lee in 1998, I found that he had accepted this idea, supposing that granitic magma had risen up from below at a late stage of the volcanic activity, filling and plugging the cavities. So the association of synclines with granite, mentioned above, was not merely fortuitous.

Later, as we shall see (p. 225), Lee got together with Soper to generate a model that linked the known geology at outcrop with the geophysical data and the likely history of the region in terms of plate-tectonic theory, producing the figure that represent this synthesis in a section accompanying the Ambleside map (BGS 1996). However, discussion of this work at the present juncture would carry us too far ahead for convenience, so the matter will be held over until we consider the Survey–universities collaborations of the 1980s and later (Chapter 16).

I conclude this chapter with mention of some suggestions made by Bernard Leake and John Cobbing (1993) and some remarks about very recent work of Bott (1999), which tells a rather different story about the emplacement of granite than that it simply fills a hole in the crust left by the eruption of volcanic ash – an idea that he thought to be implausible as a 'single' solution to the question, as there was, in his view, just too much granite to be emplaced in such a fashion (Bott, pers. comm., 1999).

For Leake and Cobbing, the question was: why is it that so many granites, in Britain and worldwide, crop out with their upper surfaces approximately coincident with the present erosion levels, such that we can conveniently see the roofs of the granites in contact with the country rock? Numerous examples were cited by Leake and Cobbing from Scotland and Ireland, and the Lakeland batholith was included in the list. Indeed, the top of the Skiddaw Granite is exposed in a small cliff beside Sinen Gill, to the east of Skiddaw (see Fig. 12.1), just a few metres below the moorland erosion surface. That such examples should be quite common, millions of years after the emplacement of the granites, seems to be more than a coincidence and calls for an explanation.[9] Leake and Cobbing suggested that the explanation lay in terms of the theory of isostatic adjustment. During emplacement, the top of the granite might form approximately at the level of the water-table, wherever that might lie. After that, there would be a

[8] Much of the igneous rock near Haweswater is basic, being gabbroic at one locality to the east of the lake.

[9] The generalization is not universal, though. For example, the sacred mountain, Hua Shan, near Xian in China, which I have had the pleasure of ascending, stands high above the plains to the north, and there appears to be no evidence of a cover of country rock at the top of the mountain.

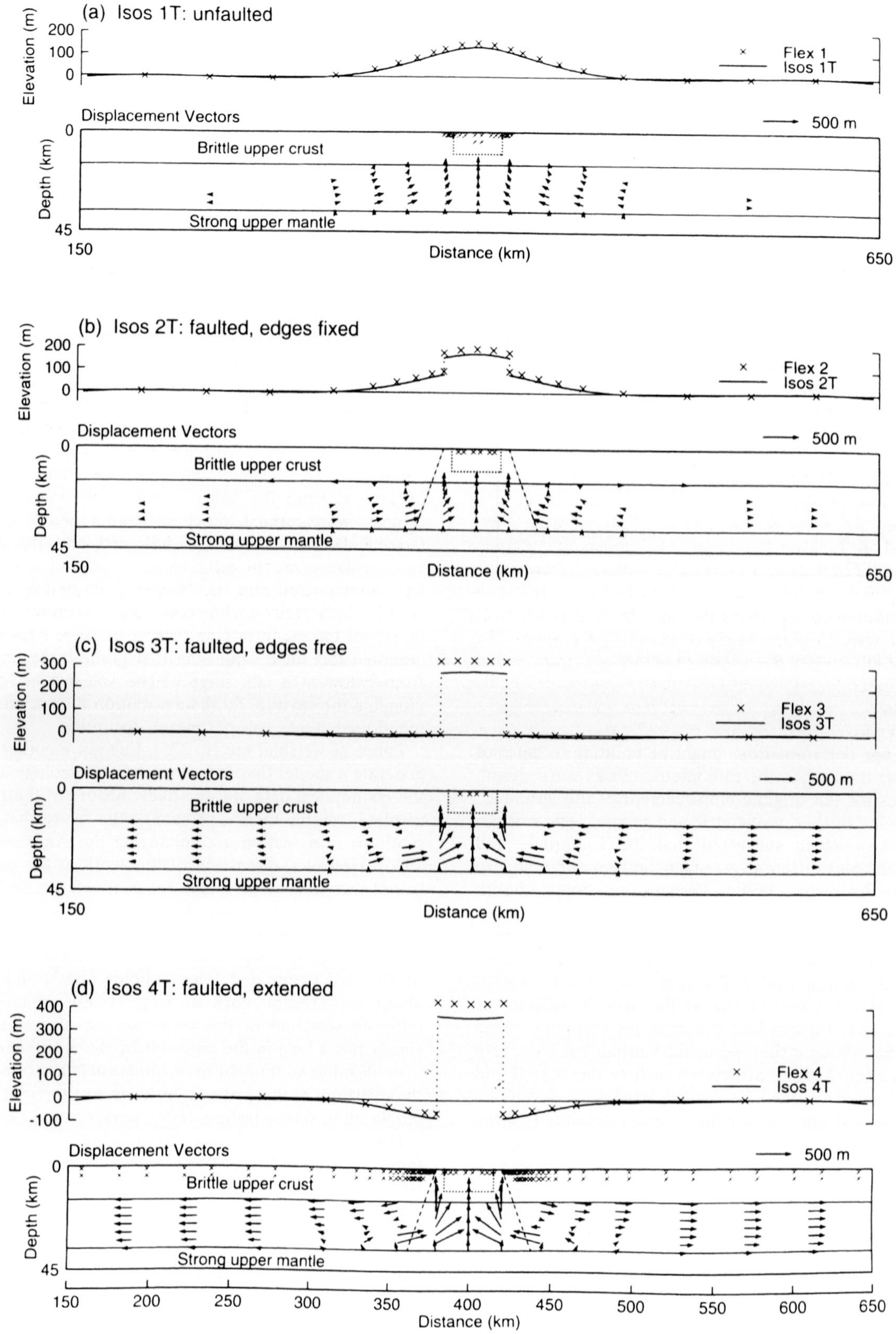

Fig. 11.7. Computer modelling of movement of ductile material in mantle, resulting from upward movement of buoyant mass of granite in faulting crust, according to Bott (1999, p. 20,354, Fig. 4d). Reproduced by courtesy of the American Geophysical Union.

tendency for adjustment to occur in such a manner that the generalization remarked by the authors holds. Thus, as several Lakeland geologists have suggested to me, the key to the whole structure and form of the Lakeland mountains may lie with the granites. This issue will come up again in Chapter 18, when we consider the present general form of the Lakeland mountains and what may have happened to the region between the Devonian earth movements and the events of the Pleistocene glaciation.

Bott's paper of 1999, developed from his early work done in Weardale, the Lakes, Cornwall, etc., was more abstract than his earlier Lakeland work. Regarding the Lakes, Bott estimated that the mass deficiency due to the subterranean granite was approximately equal to the mass of the overlying elevated ground, which suggested a causal relationship. As mentioned above, in his earlier work Bott had been thinking of isostatic compensation in relation to the upper crust, but in his 1999 paper he began to consider the changes that might be operating at a deeper level in the more ductile lower crust. To this end, he deployed a computer model developed at Durham simulating the attainment of isostatic equilibrium assuming the possibility of *lateral* flow of material in supposed ductile lower crust (Bott 1997).

Bott (1999) assumed in his two-dimensional model, depicting a section 800 km wide, the existence of an 'viscoplastic' upper crust 15 km thick; an underlying 'elastoplastic' lower crust of thickness 24 km; and a strong layer of upper mantle extending below that to a depth of 60 km. Also, the presence of a mass of granite 30 km wide and 10 km deep was postulated (like a *deus ex machina*!) in the top–centre of the upper crust. Appropriate physical quantities were allocated to the different parts of the physical model; and the computer was allowed to generate predictions as to what would happen after five hundred, fifty thousand, and five million years. In each case, of course, the granite rose upwards; but different geological results and different subterranean movements of matter occurred according to whether or not one assumed that faulting occurred at the margins of the rising granitic mass. In the simple case, without faulting, the upper crust domed upwards, and the ductile material of the lower crust moved laterally towards the region of the uplift to fill the space. In the most complicated of the four cases considered (but perhaps the one that was geologically most plausible), faulting of the brittle upper crust was assumed to occur some distance to the side of the hypothesized granite mass, with the faults extending down into the plastic lower crust. Under these hypothetical conditions, the granite rose up; material from the lower crust moved laterally inwards to fill the 'space' (so to speak); and beyond the assumed faults material from the *upper* part of the lower crust moved laterally *away* from the central region (see Fig. 11.7). The net result was that one obtained an elevated horst and basins at either side.

Bott suggested that this result was comparable (or compatible) with what one finds in the Weardale area, where there is a known granite mass underlying the elevated Alston Block of the north Pennines. Equally (though Bott did not say as much) one would have a structure approximating that of the Lake District, with basins (those of the Irish Sea and the west Cumbrian Carboniferous to Permo-Triassic sediments, Carlisle, the Vale of Eden, and the Windermere Trough) surrounding the elevated land of the Lakeland mountains with their underlying pluton(s). Also, we have some appropriate structures bounding the Lakeland mountains, such as the Lakeland Boundary Fault at their western margin, or the Windermere monocline – where the Windermere (Super)Group contacts the Borrowdale Volcanics.

All this, it may be noted, was, in a sense, based on abstract considerations, and independent of the theory of plate tectonics *per se*.[10] As said, in Bott's model, the granite just appears as a *deus ex machina*. It says nothing about the possible relationship of the granites to the formation of the Borrowdale Volcanics (or to volcanoes more generally). It does, however, offer much food for thought, and provides another instance where the Lake District has provided nutriment for the development of basic geological theory. What influence the modelling may eventually have on the development of ideas about the geological history of northern England, and how it may be integrated with plate-tectonic theory, remains to be seen by future geologists and future historians. The theory of uplift during the Tertiary favoured by officers of the Survey at the end of the twentieth century (see Chapter 18) does not run along quite the same lines as those laid down in Bott's paper of 1999.

[10] However, Bott (pers. comm., 2001) has pointed out that the theory is based on current understanding of the rheological structure of the crust and uppermost mantle, which itself has an empirical basis in measurements of borehole stresses, the study of earthquake distributions, experiments on the rheological properties of rocks, and so on.

Chapter 12
Some more PhDs, and the extension of Lakeland surveying in the 1970s

In the previous two chapters we have been considering some rather broad and grand themes, but the Lake District has also been a site for detailed fieldwork, and often it has been PhD students who have provided the greatest detail. Indeed, the grander themes have commonly rested on the minuter work of the doctoral students. So let us look at some more of these 'micro-studies'.

In the present chapter, I shall discuss some of the PhD investigations undertaken in the Lakes in the 1960s and 1970s, chiefly Frank Moseley's students at Birmingham and Jack Soper's at Sheffield, concentrating here on the work done on the Skiddaw Slates and the Borrowdale Volcanics. In the next chapter we shall at last begin to look at the studies of the palaeontology, structure and sedimentology of the strata of the Windermere Group.

The first person to be mentioned is Michael Nutt, who worked on the geology of the area around Haweswater in eastern Lakeland while at Queen Mary College, London University, his thesis being completed in 1970 with Moseley as external examiner. Nutt published a few small papers on Lakeland geology in the 1960s (Nutt 1966, 1968), these being written accounts of various field excursions in which he participated, sometimes as leader when they were in 'his' area. Before going to QMC he had been in the merchant navy, and afterwards he obtained a position with the Survey, but he did not do further work for that organization in the Lakes. In 2000 he was retired and residing near Blackpool.

Nutt's thesis, which is notable for the clarity of its writing and its attention to the early literature on Lakeland geology, made extensive use of results obtained from the tunnels dug from Ullswater to Heltondale (Tarn Moor Tunnel, 1968–1969), from Cawdale to Haweswater (Howes [*sic*] Tunnel, late 1950s), and from Haweswater to Longsleddale (Mardale Tunnel), in the establishment of an expanded water supply for Manchester. (There were also subsidiary tunnels – Naddle–Swindale in the late 1950s and Tailbert in 1963 – and aqueducts taking water from Swindale and Wet Sleddale into Haweswater.) The tunnels were examined at the time by Survey staff. The results were not published, but Nutt was able to gain access to the records of the observations and utilize them in his thesis (Nutt 1970, p. 24). He worked in the Tarn Moor Tunnel with the Surveyor, Tony Wadge, but this came after most of his thesis data had been collected.

Nutt's area included an eastern outpost of the SS, called the Bampton Inlier; the unique (for the Lakes) igneous rocks adjacent to Haweswater, where dolerites crop out, with even gabbro at the centre of the intrusion; and various Borrowdale Volcanics, including ignimbrites. Graptolites discovered by him and David Skevington indicated a *D. bifidus* age (Llanvirn) for the Skiddaws of the Bampton Inlier. Nutt found rocks corresponding to Green's (1915*a*) 'Mottled Tuffs', but he rejected Green's suggestion that they passed conformably upwards into the BVG. Rather, from the section in the Tailbert Tunnel, one in Swindale Beck, and elsewhere, the two appeared to be *interbedded*. Nutt construed the Tuffs as occurring not by the intermingling of ash with normal sedimentary material, as Green had envisaged, but as a product of the reworking of volcanic and sedimentary deposits, the material perhaps being derived from the Cross Fell area.

In the Tailbert Tunnel, the 'Mottled Tuffs' appeared to form a wide, open synclinal structure, with the Skiddaw mudstones above and below apparently forming sharp isoclinal folds. With the assistance of stereographic analysis, Nutt found a series of quite open folds for the BVG rocks, striking approximately NE–SW and nothing like Green's (1915*a*) complicated structure. Approximately parallel faults were mapped in Heltondale and running in to the NE end of Haweswater. There appeared also to have been stress exerted NNW–SSE, generating a regional cleavage apparently unrelated to the folding.

Regarding the problem of the SS–BVG relationship, Nutt claimed that isolated evidence for unconformity could be seen near Hungerhill, not far from the road from Bampton to Haweswater (NY 5145 1744 and NY 5133 1738) (even though the Skiddaws' bedding was not evident at that site); and a conglomerate in the area with both BVG and Skiddaw clasts (NY 5140 1824) also suggested unconformity. However, his wider mapping suggested faulting, and probably thrust faulting, with the fault plane having a westerly dip. There was also the evidence of the high-angle fault in the Tarn Moor Tunnel (observed after Nutt's main fieldwork was completed) to be taken into consideration. Nutt ended up with the suggestion that there had been at least some initial unconformity, with thrust faulting occurring subsequently at the approximate SS–BVG horizon, and then the thrust plane being folded afterwards. Evidently, this satisfied his examiner, Frank Moseley. Indeed, it very likely pleased him!

Nutt was not, however, satisfied with Alex Simpson's tectonic history of the Lakes, and was aboard the bandwagon rolling against him. So far as Nutt was concerned, the first Skiddaw folds in his Bampton area were aligned approximately N–S, whereas Simpson's F_1 trended NE. They could not, therefore, be ascribed to the same set of movements. Nutt found a cleavage set, trending NE, that was parallel to Simpson's S_1, but it penetrated both the Skiddaws and the Borrowdales in eastern Lakeland – which was again at odds with Simpson's model. Nutt claimed that the volcanic clasts in the Bampton Conglomerate were similar to actual BVG types (but see p. 208), and so he supposed that though there probably was a local unconformity in his area it was not necessarily at the actual base of the BVG. In consequence, he rejected Simpson's major pre-BVG unconformity – though of course the unconformity could have been diachronous. Nutt did not work on Simpson's Buttermere area; and Simpson had not worked on Nutt's Haweswater area. Nutt did, however, accept Simpson's suggestions about post-BVG tectonic events.

Regarding the complex of gabbros, dolerites, microdiorites and andesites to be found in the Haweswater area, Nutt thought it might represent the remnants of a magma chamber, in which magmatic differentiation had occurred. The several pipe-like bodies of andesitic agglomerate in the area might have been vent breccias. The faults cutting the complex appeared by their alignment to be pre-Caradoc; so the complex as a whole was plausibly Ordovician, forming part of the BVG.

The next student to work on the Skiddaws from a structural perspective was Moseley's previously mentioned PhD candidate, David Roberts.[1] His study was focused on the bewildering structures that can be seen so clearly in the Skiddaw Slates in the bed of the Caldew River, immediately south of the Carrock Fell

[1] Roberts (b. 1944) attended Aberdare Boys' Grammar School, and went on to study geology at Birmingham University. After completing his PhD in 1973, he worked first at the University College of Wales (Aberystwyth), then for the National Coal Board in South Wales, and then at Bristol University. He had a brief spell in the private sector as a consultant with Stefen, Robertson & Kirsten (UK), returning to academia in 1991 to become Head of Geology at Staffordshire University, Stoke. After his Lakeland days, he came to the view that practical and engineering geology are more important than the more abstract problems he tackled during his time as a research student, and he specializes in the evaluation of geological resources, from metals to construction raw materials, with an eye to risk analysis, legislation with regard to the minerals industries, mining finance, etc. (Roberts, pers. comm., 2000). His very recent work (2000) has been in North Africa.

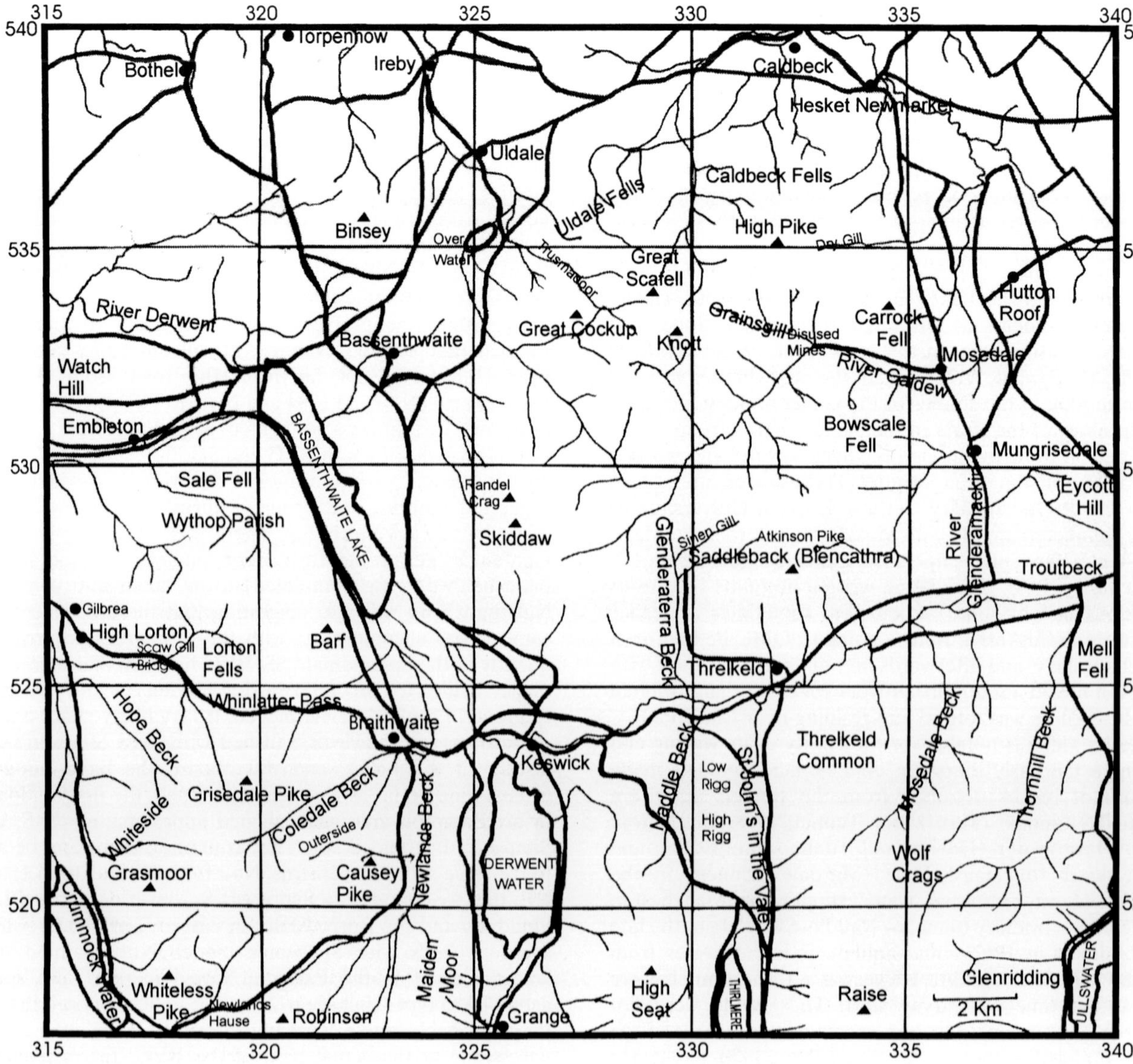

Fig. 12.1. Topography of northern Lakeland, including the Skiddaw massif, Carrock Fell, Eycott Hill, and Mell Fell.

Complex, running from west to east, out of the hills, towards the domain of the Eycott Volcanics, and on into the Vale of Eden (see Fig. 12.1) . From the edge of the northeastern Lakeland hills at the hamlet of Mosedale, the Caldew River runs upstream for about 3 km almost due west, and then turns SW towards the main Skiddaw Slate mountains of Blencathra and Skiddaw itself. Where the bend occurs, a tributary called Grainsgill Beck enters from WNE, and near the confluence is found the previously mentioned outcrop of greisen, where the area was formerly mined for lead, copper, tungsten, etc. In fact, a vast number of minerals, some quite rare, have been found there – more than from any area of comparable size in Britain (for details, see Cooper & Stanley 1990). The area has been actively mined for many years, but anything of economic value has long been removed, and today there are only relics of old workings, and decrepit and dangerous adits.

The source of these interesting minerals and ores is evidently the large mass of the Skiddaw granite, which appears as a few small exposures – apparently the roof of a stock or batholith – in the river bed in the Caldew, just upstream from its confluence with Grainsgill Beck. But Roberts was more interested in the strange structures that could be observed in the Skiddaw Slates, near the confluence of the two streams and down the Caldew for about a kilometre. Due to the proximity of the granite, the slates are hornfelsed and bleached, and their structures have been beautifully etched out by the rivers (see Fig. 12.2). The question was: could the structures of the slates be analysed, and if so, did they support the idea of three deformations such as Simpson had supposed, or six such has Helm had recognized in the Black Combe area, or something quite different?

According to Roberts (pers. comm., 2000), he was 'academically lost for the first 18 months of his research'. He 'measured and recorded' and 'mapped on to a totally inadequate base map but did not know why'. (I suppose many PhD students will know the feeling.) But he recalls that, in writing up his thesis, he hoped it would settle the question of the timing of the Skiddaw deformations. Naturally, though, he did not have the last word on the topic.

According to a statement made in the introduction to Roberts' thesis (Roberts 1973, p. 13), his project was started 'when Simpson's results were made public', and the idea was to test them by examination of the eastern part of the Skiddaw Slates outcrop, around Blencathra and in the Caldew Valley, as described above. It will be recalled that he did some work on the dating of the deformations in collaboration with Soper (see p. 132); and that his supervisor Moseley was of the view that Simpson's F_1 was not the product of an intra-Ordovician deformation but was due to end-Silurian earth movements. However, by the early 1970s, Moseley

Fig. 12.2. Structures in the hornsfelsed Skiddaw Slates, Caldew Valley. Photographed by author, 1999.

was accepting the idea of an unconformity below the BVG, and beginning to contemplate the idea that some of the structures in the SS might be due to deformation of *unconsolidated* material (Moseley, in discussion of Helm 1970, p. 135). The point was that in Helm's Black Combe area the foldings did not appear to have associated cleavages; they were all seemingly cut by a single cleavage (Simpson's S_1). The implications of this hypothesis, as applied to the hornfelsed slates of the Caldew Valley, were examined in detail in Roberts' thesis and in a prior publication (Roberts 1971).

In the bed and banks of the Caldew River, Roberts could discern (especially when the water was low) a wondrous variety of complex structures: concentric folds, similar folds, chevron folds, conjugate folds, irregular folds, composite folds, refolded folds, 'crumples', and sheared folds (see Fig. 12.2). If there had ever been any cleavage associated with this array of structures it had been obliterated by the metamorphism. By analysis similar in principle to that discussed in Chapter 9 for the work of Glennie, Roberts sought to determine axial planes and the plunges of his folds. For the exposures in the river bed and banks, his work was extremely detailed (with mapping at 1 inch to 20 feet), but he also ranged over the hills to the south, southwest and west, examining all the numerous outcrops of slate. A regional northerly dip of 70° – as observed on the overlying BVG rocks – was subtracted from the measured dips for the Skiddaw rocks, leaving a N–S trend discernible for the fold plunges plotted on a stereogram. This, however, Roberts only construed as the first phase of his folding.

Roberts also 'recognized' three subsequent sets of structures: upright E–W folds, with associated cleavage; reclined folds with crenulation, attributed to gravity collapse of incompetent vertical strata; and minor N–S flexures with fracture cleavage. Thus he had four deformations in all. It is interesting to see how his structural determinations panned out. In his paper of 1971 he had *three* phases of folding; in his thesis he had *four*. In the thesis, after he had worked with Soper as well as Moseley, his F_1 corresponded with the gravity-collapse structures envisaged by Moseley in the unconsolidated Skiddaw sediments; and F_2–F_4 corresponded to the D_1–D_3 envisaged by Soper in his discussion of Helm (1970). One might think that Roberts' earlier threefold view of polyphase folding was derived from Simpson, but this does not seem to have been the case. The alignment of his F_1 was different from that of the Scottish Londoner.

Roberts published the results of his thesis in three further papers (Roberts 1976, 1977*a*, *b*). Perhaps the most important suggestion was the idea that the first generation of folds (which he could only discern clearly in the stream bed and the rock at the side of the Caldew) was attributable to the slumping of unconsolidated sediments. This suggestion had been foreshadowed by Moseley, but Roberts has informed me (pers. comm., 2000) that the idea came to him when he saw structures in North Wales 'that were definitely slumps' similar to those he had been studying in the Lakes; and at the time his supervisor 'did not seem too keen on the idea'. Anyway, the notion of syn-sedimentary structures henceforth began to play an increasingly significant theoretical role in the attempted understanding of the Skiddaws, as we shall see.

Roberts' second deformation, which brought on the main cleavage, was taken to be end-Silurian, the supposed main tectonic phase for the Lake District. The third deformation seemed to Roberts to betoken crustal shortening, not gravity collapses as Simpson had suggested. For Roberts, this phase might have been associated with the emplacement of the Lakeland batholith. His fourth deformation was the product of axial shortening, phases 2–4 *all* supposedly being end-Silurian.

Moseley's second student working in the Lakes at that time was Peter Jeans (b. 1948), whom we have already noticed as clearing the exposures at Hollows Farm, examining SS–BVG contacts in the Newlands valley and near Buttermere, and offering an account of the geological history of the Lakes in terms of plate-tectonic theory. His career subsequent to his PhD at Birmingham eventually led him to Royal Dutch Shell. He took up palynology, rather than structural geology such as he did for his doctorate, and became Business Adviser for Indonesia and New Zealand Shell International. In 2000, he was residing in Holland. Jeans found Moseley to be a most supportive and stimulating supervisor (pers. comm., 2000).

Jeans' thesis topic (1973*a*) was an analysis of the structures of the SS in the area to the east of Crummock Water and down to Buttermere village (see Fig. 12.3). It included mountains such as Whiteside, Grasmoor and Whiteless Pike. His method was to examine the area thoroughly, map it, measure all possible dips and strikes for joints, bedding and cleavage, and determine the directions of such lineations as might be discernible. Data were averaged by plotting of appropriate stereograms. Naturally, he looked for faults. He was also interested in the SS–BVG relationship and ideas about the tectonic history of the Lakes.

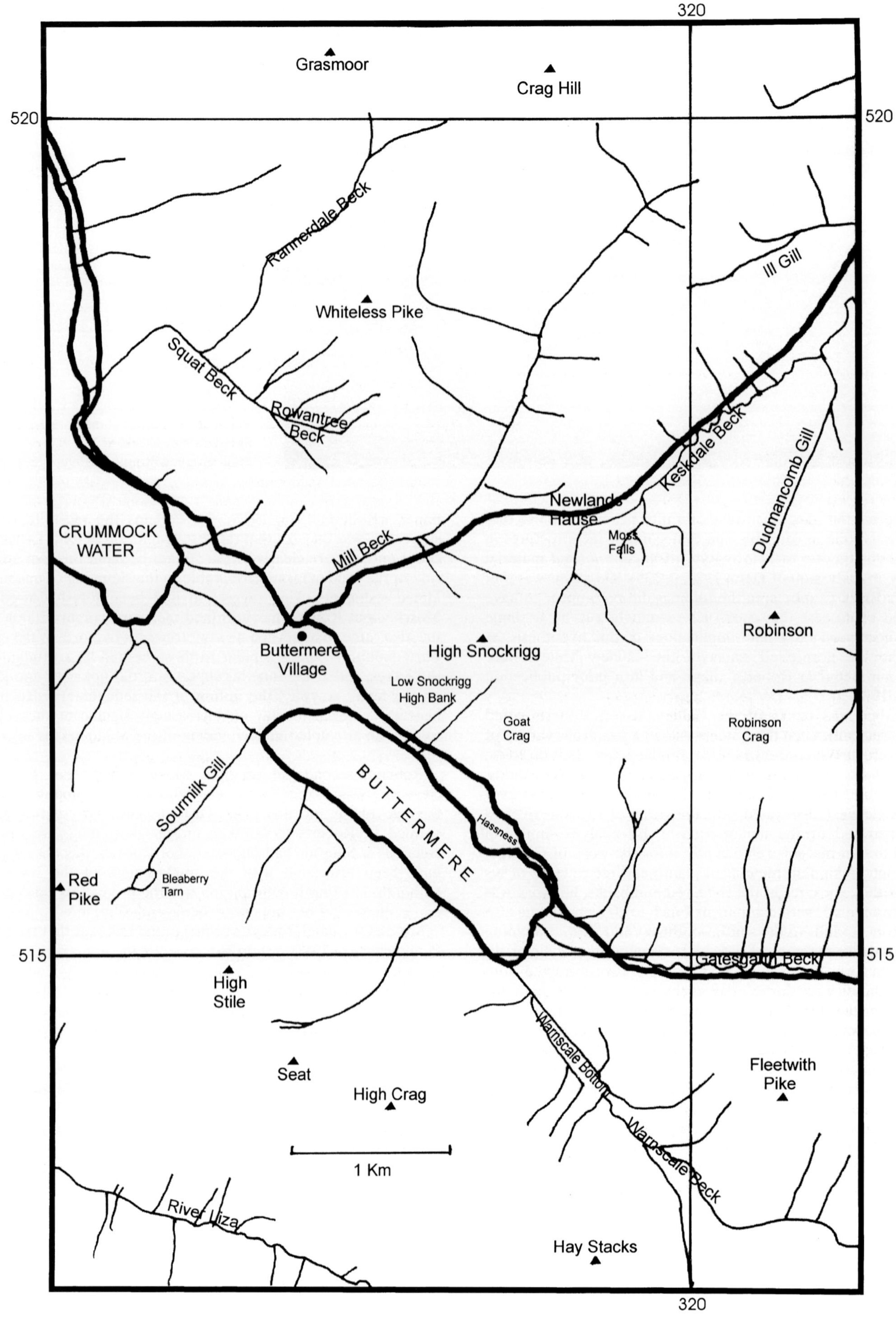

Fig. 12.3. Topography of Buttermere and adjacent areas.

Table 12.1. *Phases of deformation, according to Simpson and Jeans (Jeans 1973, Table 5.1)*

Jeans (1973)	*Simpson (1967)*	*Style/orientation of fold axes*	*Cleavage*	*Age*
F_4	F_3	Variable; generally NW–SE. Steep plunge	NW–SE	
F_3	F_2	Horizontal, recumbent; generally ENE–WSW	Horizontal	End-Silurian
F_2	F_1	Upright, open–tight; ENE–WSW	ENE–WSW Variable	
F_1	–	Overturned N–S	–	Pre-Borrowdale Volcanics
–	–	Generally E–W, recumbent	–	Syn-depositional

From all this, Jeans inferred the occurrence of four phases of deformation, which are summarized in Table 12.1.

It will be remarked that Jeans, like Roberts and like Soper, had multiple deformations. The two students did not have the same number of foldings or styles of deformation, but it will be remarked that they both envisaged syn-sedimentary deformations. The task of analysis was difficult, and the reading of structures from stereograms was becoming something of a 'black art'! Minor crumplings could easily be regarded as representations of significant tectonic events. I would not venture to say what the correct answer should be, particularly given that the 'definitive' Survey Memoir on the Skiddaw Slates remains unpublished at the time of writing (2000). It may be noted, however, that Jeans did not discuss the grand system of thrust faults that are to be found in Gasgale – the valley between Whiteside and Grasmoor – on the southern side of Whiteside. This is a remote part of the Lakes, but it is remarkable, none the less, that Jeans did not discuss this matter, for once the idea is suggested – as in modern guidebooks such as Moseley (1990, p. 72), it seems impossible to overlook such structures. On the other hand, Jeans (following Colin Rose) identified the hornfelsed rock of Whiteside – the bleaching of the SS seemingly being due to some subterranean body that did not crop out at the surface, but which 'burned off' carbonaceous matter from the slate.

Both Roberts and Jeans went on to other things after completing their Lakeland work, and we shall not, therefore, attempt to pursue them here. Meanwhile, up at Sheffield, two interesting and important theses were undertaken in the early 1970s by students of Jack Soper: an Iraqi Kurd, Nazur Numan, and an Englishman, Barry Webb, who later joined the Survey and did much work in the Lakes (see below). Soper was getting restive about the proliferation of phases of deformation. Each tract examined seemed to have been deformed before the next one, but the oldest ones affected the northerly Lakeland tracts and the youngest only the southerly tracts. Polyphase deformation ideas were getting out of hand in the Scottish Highlands too, as well as in the Skiddaws; and Soper was interested to study the supposed 'pre-Bala' deformations in the Borrowdale Volcanics. He was confident that the Ulpha Syncline, mapped by Aveline, was something real, but the situation at Kentmere in the southeastern Lakes was less certain, where, as we have seen, Mitchell had revised Green's notion of isoclinal folds. However, the ground had not been looked at closely for more than a quarter of a century. It required investigation.

An opportunity to do something in the Kentmere area arose when Nazar Mohammed Saleem Numan arrived in Sheffield from Iraq, armed with a scholarship provided by the Barth Party.[2] Numan was supposed to work with Professor Leslie Moore at Sheffield, but was handed over to the tender mercies of Soper, who put him to work in the Kentmere area (see Fig. 3.6), near where another of his students, Andrew Bell, was also soon at work. The idea was to revise Green's Kentmere work, and see what, if anything, could be made of the idea of isoclinal folding. Numan found this task beyond him, which is scarcely surprising. When one considers the unhospitable ground with its complex geology, his lack of familiarity with Britain, the absence of an agreed stratigraphy for the Kentmere volcanics, and in general the considerable number of opposing ideas in the Lakeland marketplace in the early 1970s, his problems are readily understandable. So after an unsuccessful initiation at Kentmere, Numan moved over to the Ulpha area (Dunnerdale), where Mitchell's relatively recent stratigraphy could be, and was, deployed (Soper, pers. comm., 1998). Soper himself went on with the Kentmere work, and found it exceedingly difficult, to the point where he swore he would never again map an area made up primarily of volcanic rock! Numan had access to the results of his supervisor's Kentmere work. Their joint effort appeared as Soper & Numan (1974), and in Numan's thesis (1974).

In the Kentmere area, Numan deployed standard sedimentological 'way-up' criteria – first introduced into Lakeland work by Hartley (1925), but apparently not deployed by Mitchell – and reached the conclusion that the evidence was against Green's old anticlinorial hypothesis. However, when it became evident that a substantial revision of the Kentmere stratigraphy was needed Soper took over there, as mentioned. Numan and Soper argued that the lower BVG was rather more marine in character than previously realized, as it contained interbedded pelites with marine microfossils (at Holehouse Gill in the Duddon Hall Tuffs in the lower part of the BVG succession; see Fig. 5.2, pp. 169 and 211), so there was the possibility of more widespread deposition (and hence correlation) for the lower units than for the upper tuffs, deposited subaerially on the flanks of volcanic cones.[3] Thus a correlation between the units in Kentmere and the Ulpha–Coniston area was proposed, using characteristic accretionary lapilli tuff that appeared in both the Wren Gill Formation (in Kentmere) and the Tilberthwaite Formation (in the Ulpha–Coniston area) as a marker. This suggested that the BVG in Kentmere was generally younger than at Ulpha and Coniston. Also, by knitting together the maps of Mitchell, Oliver and himself (though there were some gaps), Numan suggested that the Ulpha Andesites, the Cockley Beck Andesites of Mitchell (1963) and the Birker Fell Andesites of Oliver (1961) were all correlates, or at least approximately so.

In terms of BVG structure, Numan suggested that there was an E–W pre-Caradoc folding; a NE–SW Caledonian flexural folding; and an associated NE–SW Caledonian cleavage. In Kentmere, the dominant structure was the Nan Bield Anticline, plunging to the east, and prolonged to the west as the Wrynose Anticline. In the upper Duddon Valley area, the dominant structure was the Ulpha

[2] Numan subsequently returned to northern Iraq (with his PhD) and became a lecturer at Mosul University and Director of a Remote Sensing Centre there. He published textbooks and translations into Arabic, but, unsurprisingly, given the political situation in Iraq, he lost touch with his former colleagues in Britain. He shared a room at Sheffield University with Andrew Bell (see p. 166), who remembers that they got on well together.

[3] In the event this proved to be incorrect. Following the work of Branney (see p. 217), it was found possible to trace tuff bands of the upper Volcanics over wide areas, by looking for the weathering patterns (and other characteristics) for different tuffs. This could be done more easily than tracing the andesite flows of the lower BVG by their chemical signatures (though that was also possible).

Syncline, also plunging eastwards, the southern limb of the anticline being complementary to the northern limb of the syncline. Numan suggested that these principal structures were already established by the late Ordovician, and cleavage was imposed subsequently during the late Silurian or early Devonian. So Soper's student was allowing the idea of tectonic activity affecting the BVG during the Ordovician, as would have been satisfactory to Simpson.

Actually, this was also the view of Soper at that time, as evidenced by their joint publication (Soper & Newman 1974), and presumably got into Numan's thesis from that source. The joint publication provided (chiefly) Soper's map of the complex geology of Kentmere and the hills to west and east (over to Longsleddale). Soper found no evidence for Mitchell's (1929) NNE-trending folds. Much attention was given to cleavage, and the many determinations were averaged to reveal a regional trend of azimuth between 050º and 060º (= Caledonoid), with a steep NW dip; and this was attributed to end-Silurian compression, acting on pre-existing (pre-Caradoc) broad folds. Subsequently, as we saw in Chapter 10, Soper came to the view that the 'big-crunch' cleavage was Lower Devonian.

It is interesting that there is an unusual rock type in the area called the 'bird's-eye tuff', displaying a particular kind of accretionary lapilli. It crops out abundantly, for example, in a disused quarry at the mouth of Wren Gill, a tributary running into the upper reaches of Longsleddale; but it is of wider distribution and has served as a useful marker band for the purposes of mapping. Each 'bird's-eye' has a greenish chloritic tuffaceous centre and a white finer-grained, more quartzose rim (see Figs 12.4 and 3.7). It is supposed that the accretionary process occurred when ash from an exploding volcano came in contact with raindrops or hailstones – and this may even be true!

The interest of these lapilli from the point of structure is that the lapilli spheres have evidently been systematically deformed by compessive forces acting during the process of cleavage formation and could thus serve as convenient strain indicators. They were the subject of close investigation by several workers in the 1970s and into the 1980s (Oertel 1970, 1971, 1972; Helm & Siddans 1971; Bell 1975, 1981). Andrew Bell's Sheffield thesis (1975) was specifically devoted to the problem, and his work under Soper also contributed to the mapping of the area. It appeared that the rocks had been compressed perpendicularly to the cleavage by upwards of 50%. Some of the lavas in the Wren Gill area also contain large ellipsoidal vesicles filled with chlorite, similarly deformed into ellipsoids. The results of such detailed empirical and mathematical analysis provided substantial background information to the attempts towards grand theoretical explanations in terms of plate-tectonic theory, developed in the following decade, and discussed in Chapters 10 and 15–17.

Another of Soper's students was Barry Clifford Webb (b. 1942), who came to play an important role in the Survey–university collaborations of the 1980s, discussed in Chapters 14 and 15. Webb (see Fig. 14.3) was a 'late starter' in geology and in academic life. Born in Birmingham and mainly brought up in Sheffield, he got to know Soper during his teens through a common interest in rock climbing. From this, Webb started his career as an instructor in mountaineering, caving, canoeing, etc., but after doing one A-level and some self-taught geology he obtained a place as a mature-age student at Sheffield University, naturally gravitating there through his earlier friendship with Soper. Webb got a first (with a thesis on the structure of the Cross Fell Inlier, for which he was awarded a prize), and went on to do a PhD under Soper on the structures of the Skiddaw Slates (Webb 1975). After getting his doctorate, Webb obtained a position on the staff of the Survey. He worked in North Wales for three years, then in Bolivia, and again in Wales. Then he worked in the Leeds Survey office, from whence he moved to the Newcastle office and did work in the Southern Uplands, where the 'accretionary wedge' hypothesis was coming into favour. Following this, he returned to work in the Lakes, working on the sediments of 'Otley III' and on the Skiddaws. Webb was with the Survey for 15 years and was at one time in charge of its Lakeland Project, but he became disenchanted with the bureaucratic procedures required of him in his more senior position, and so resigned in 1988 to live near the remote defunct mining village of Nenthead, on the North Pennine moors near Alston. At the end of the twentieth century, he was again working as an outdoor instructor and also teaching for the Open University (Webb, pers. comm., 1998, 2000).

Fig. 12.4. Bird's-eye tuff from Wren Gill, upper Longsleddale. Photographed by Janet Latham and reproduced by her permission.

For his Skiddaw work, Webb worked particularly round Buttermere and on the mountain called Robinson above the Newlands valley. As an experienced rock climber, he was willing to tackle treacherous places such as Goat Crag, on the NE side of the lake (see Fig. 12.3). He was allowed to use Colin Rose's old field-slips during his thesis mapping and found them most useful, for Rose had been much interested in structures, though he did not deploy stereographic techniques. Webb's own work had a strong structural bias; and Soper has described him to me as an adept three-dimensional thinker. Webb would walk around with the thumbs and first two fingers of both hands at right angles (as is done by physicists employing Fleming's left- and right-hand rules). His thumb was taken to represent the dip, the first finger the strike, and the second finger the pole of a bedding or cleavage plane. So as he walked around, he used one hand to represent the bedding and the other the cleavage, and his data were plotted directly onto two stereograms. He would say something like, 'The pole of this cleavage is *X*, and you'll find that it plots on a great circle round to *Y*; and the pole of the great circle is the axis of these folds; so these folds are re-folding that cleavage' (Soper, pers. comm., 1998). It could all be done very quickly and simply. Webb seemed to Soper to have an intuitive grasp of the geometry of the deformed rocks.

Webb's thesis area was quite small, but horribly complex. It covered the Skiddaw ground to the NE of Buttermere, and the mountains of Robinson and Hindscarth, and to the NW his ground abutted that of Jeans, working on Grasmoor, with whom he co-operated. Neither Webb nor Jeans accepted Simpson's stratigraphic sequence, nor his structural findings, and thought they could recognize a syncline at Buttermere where Simpson had had an anticline. It is now thought that neither party was correct (see Chapter 15), for the structure at Robinson is today regarded as involving major slumping, whereas Webb sought to interpret his observations in terms of tectonic rather than syn-sedimentary structures, though (Webb, pers. comm., 1998) while he was doing his thesis he did consider – but chose not to adopt – the idea of slumping.

Webb's methods soon bore fruit and he had a paper in *Nature* (Webb 1972) three years before completing his thesis. The essence of his thinking was that of refolded folds. These yielded 'interference' patterns, which he was able to comprehend with the assistance of stereographic analysis. He worked particularly on the fold interferences in a *roche moutonnée* (186 159) at Hassness – an area near the northeastern side of Buttermere (see Fig. 12.3). The '*roche*', about 20 m long, was made up of psammitic and pelitic elements, such that the interference features stood out clearly.

Without going into the details of Webb's stereographic analysis, we may look at his 'cartoons' in his *Nature* paper, showing two types of structure that he described for Hassness (see Figs 12.3 and 12.5a); and in Littledale between Hindscarth and Robinson (see Figs 7.4 and 12.5b). Figure 12.5a represents a fold (F_1) at its region of closure, which has been squeezed and folded in consequence (F_2) – like a sheet of iron folded and then having corrugations put into it. Figure 12.5b shows an analogous situation in which a sheet of iron is folded, and a set of corrugations is again superimposed. The difference (which I found difficult to understand from the figures) was that for Figure 12.5*a* the F_1 fold is refolded 'in and out (horizontally)' whereas for Figure 12.5*b* the folding of F_2 is 'up and down (vertically)' (Webb, pers comm., 2000). The figures represent two different ways in which the rocks responded to the forces producing F_2. Actually, Webb did not find the two types completely distinct. Rather, there was a continuum between them, for which Figures 12.5*a* and 12.5*b* represented the extreme cases.

More important for Webb, in terms of the debates about the tectonic history of the Lakes, his N–S F_1 structures (revealed as fold 'relics' by an E–W aligned girdle showing up on his stereograms) were thought to be *prior* to the folds described by Simpson as F_1. So Simpson's F_1 became Webb's F_2. This possible earlier deformation opened the way to there having been a movement between the accumulation of the Skiddaws and the deposition of the Borrowdales, and hence the possibility of unconformity between the two units, without having to postulate intra-Ordovician deformation of the Skiddaws. Webb spent time clearing a contact at Warnscale Bottom in the valley where water runs off the hills into the head of Buttermere (see Fig. 12.3). Following up Warnscale Beck, one eventually finds the Skiddaws running into the Borrowdales, with a contact that Webb construed as an unconformity, the bottom of the latter displaying a basal breccia (only well seen in thin section). To my eyes, the contact does not call out 'unconformity' loudly and unambiguously and there could well be faulting there too, a proposition that Webb might be willing to accept. However, he and Jeans examined all

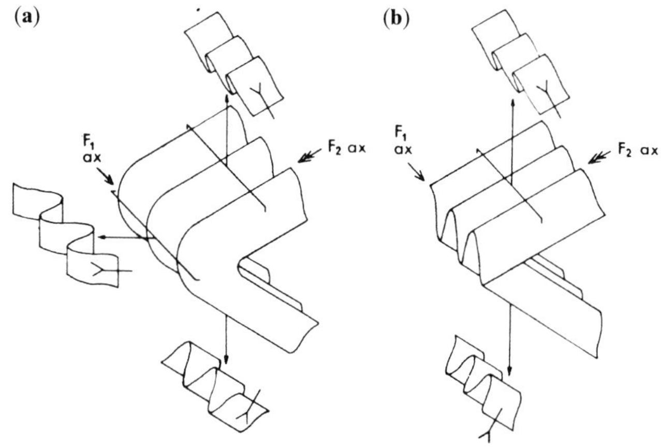

Fig. 12.5. Cartoons of multiply-folded Skiddaw Slates, Hassness and Littledale, according to Webb (1972, p. 139). Reproduced by courtesy of Nature Science.

the possible SS–BVG contacts in the vicinity, and concluded in favour of unconformity (an interpretation by then accepted by their supervisors Soper and Moseley).

Webb's thesis also gave considerable attention to sedimentology. He found examples in the Skiddaws of the divisions proposed by the Dutch sedimentologist A. H. Bouma (1962), who envisaged five different textures according to the distance of a turbidite from its point of origin (fine pelitic material being carried furthest and coarser psammitic material carried the least distance). Close to the point of origin, one might (ideally) expect to find a complete sequence from the former to the latter, as a 'pulse' of sediment was deposited. At a distance there would only be the pelitic material. However, Webb pointed out, the presence or absence of Bouma's five subdivisions could depend on the rate of water-flow as much as the distance from the source.

Be this as it may, Webb sought to determine the directions of original sedimentation, discerned from ripple marks and flute casts, by 'reversing' the deformations that he had revealed in his structural analyses. The result of such geometrical manipulation appeared to show that the currents had originally flowed from the SE quadrant, and predominantly from the SSE. Webb made some conjectures about there having been an initial approximately N–S trough in the Buttermere area, into which sediment had been carried from the eastern side of the trough, and later from the south.

For the Robinson area, Webb simply sought to explain it all by reference to standard tectonic folds and faults. But he came to different views not long after, when in the field with Michael Nutt, who saw the whole structure as a gigantic slump – an upside-down mountain with huge rafts of sandstone. Such ideas were made known to British geologists at the time that Webb was doing his thesis, in a lunchtime discussion at the Survey given by the Surveyor Russell Arthurton, which Webb attended. Arthurton described recent discoveries in Afghanistan and Iran of great 'free-floating' rafts of sandstone and limestone. However, the idea that analogous structures might be found in the Lake District seemed implausible, and Webb made no attempt to apply them in his mapping of Robinson. The idea only came to prominence there in the 1980s, when the Survey began its major revision of the district. Discussion of the idea in its application to the Lakes will therefore be deferred until Chapter 15. Meanwhile Webb's analytical skills impressed people and, as mentioned, he landed a job in the Survey.

Another PhD from this period was that of Frank Moseley's student, David Millward (see Fig. 14.2, who continues working in the Lakes to the present, and was an important contributor to the

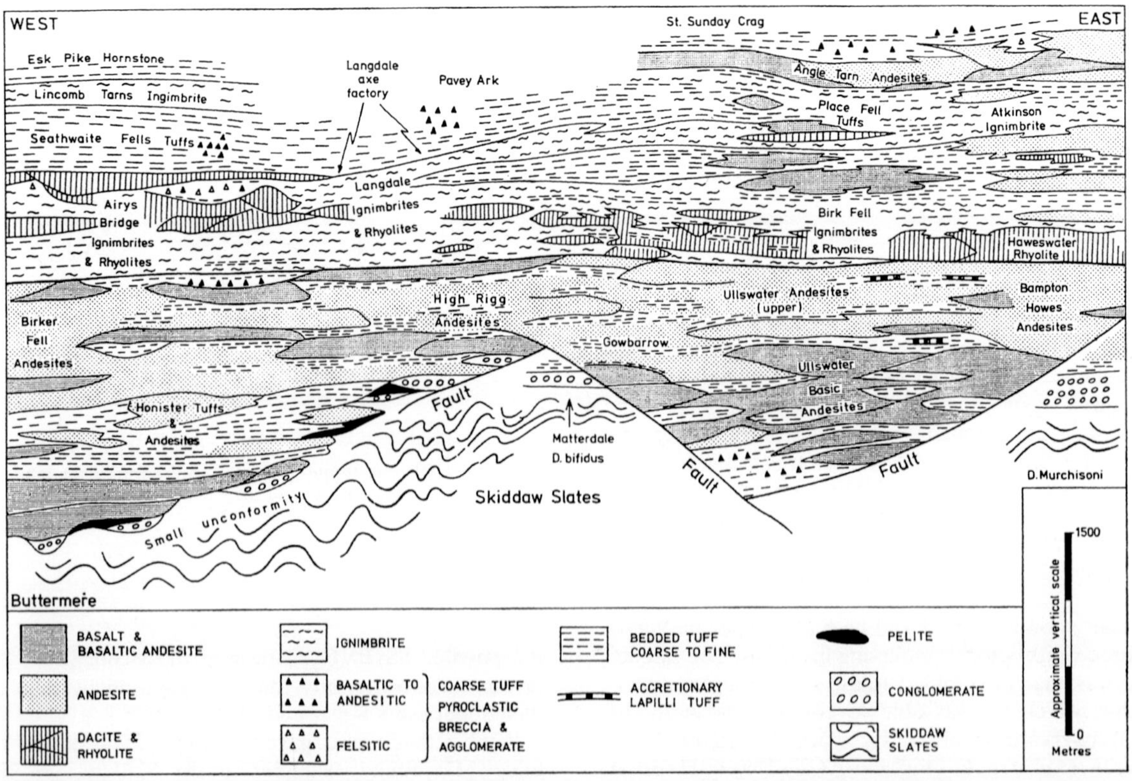

Fig. 12.6a. Lower part of Borrowdale Volcanic succession between the Duddon Valley and Longsleddale, according to Millward, Moseley & Soper (1978, p. 108). Reproduced by courtesy of the Yorkshire Geological Society, Drs Moseley and Soper, and IPR/26–13C British Geological Survey. © NERC. All rights reserved.

work done by the BGS for Nirex (see Chapter 20).[4] Millward was well aware of the great variety of suuggestions that had been proposed for the subdivision of the BVG, going back to the time of Ward, and the great difficulty that had been experienced in establishing correlations from east to west. Green, of course, had been the boldest stratigrapher in this regard, and for a time he had carried people like Hartley and Mitchell with him. But by the 1970s it was becoming clear that the method – based upon comparisons of hand specimens and thin sections – was not working satisfactorily. There was too much muddle and uncertainty. On the other hand, the incubus of the 'streaky rocks' had been exorcized by Oliver back in the 1950s, and Oliver had established a sound, though broad-brush, stratigraphy for the rocks of the Central Fells.

A great deal of Millward's thesis (1976) was devoted to the determination of the chemical constitution of the myriad rock types of the BVG. He concentrated particularly on the ignimbrites, for which, without going to the extent of full chemical analyses for all specimens, he was able to plot proportions of fiamme and pumice, crystals, and lithics on triangular diagrams for different specimens and show that there were determinable geochemical and petrological differences between what had been taken to be different ignimbrites by field examination (Millward 1979). The chemical approach was fraught with difficulty (as Millward himself realized) since the rocks could have suffered changes, post-emplacement, with recrystallization, ion exchange with vapour phases, and groundwater leaching.

I shall not attempt to follow all the details of Millward's thesis here, but we may look at two 'cartoons' from a paper he published with Moseley and Soper (1978), showing the *kinds* of results that were obtained using a combination of field mapping, common sense, and the results of geochemistry and petrology (see Fig. 12.6). The following features are noteworthy: (1) only the Coniston Limestone formed a continuous unit across the Lakes (and it of course did not belong to the BVG); (2) there appeared to be a distinction between lower and upper BVG units, with andesites predominating in the former and ignimbrites in the latter (with tuffs present in both); (3) almost no correlation from the SW to the SE of the Lakes was possible, except for the Wren Gill Andesite; and (4) the boundary between the BVG and the Skiddaws was mostly represented as faulted, but indications of unconformity were recognized in the western Lakes.

Millward *et al.* (1978) emphasized the provisional and uncertain nature of such 'synthetic' sections and pointed out that there were in fact unconformities *within* the Borrowdale succession. Even across a single drift-filled valley, differences could be due to lateral variation of a deposit or faulting so far as one could tell in the field. Broadly, there seemed to be different eruptive centres in different parts of the Lakes, so that time-equivalences of seemingly similar deposits could not be assumed. It was further remarked that the association of acid lavas and ignimbrites was (elsewhere) often associated with the formation of calderas, and it was suggested that there might have been caldera collapses after the blowing out of large amounts of pyroclastics in an eruption,

[4] Millward (b. 1952) did A-level geology at a school in Bath and carried on from there to Birmingham, where he took his PhD under Frank Moseley, having previously been on field trips with him to the Lakes while an undergraduate. Millward's thesis (1976) was on the BVG and leaned heavily on geochemistry – a fact which, I think, had quite an impact on the subsequent development of Lakeland geology. To equip himself for his task, he sat in on an MSc course on mineral chemistry at Birmingham. On completing his doctorate, he was offered a job with the BGS and received his initial Survey training in southern England, working first on the Brighton sheet and then on the Epping sheet. From 1980 to 1982 he had the opportunity to work in Colombia, on deep oceanic fault volcanics (Millward, pers. comm., 1998). On his return, he joined the Lakeland Project in September 1982, based at the re-opened Newcastle office, where we shall follow him in Chapter 14. In 2000 he was based at the BGS office at Edinburgh and was in charge of the editing of the important Ambleside Memoir (Millward *et al.* 2000).

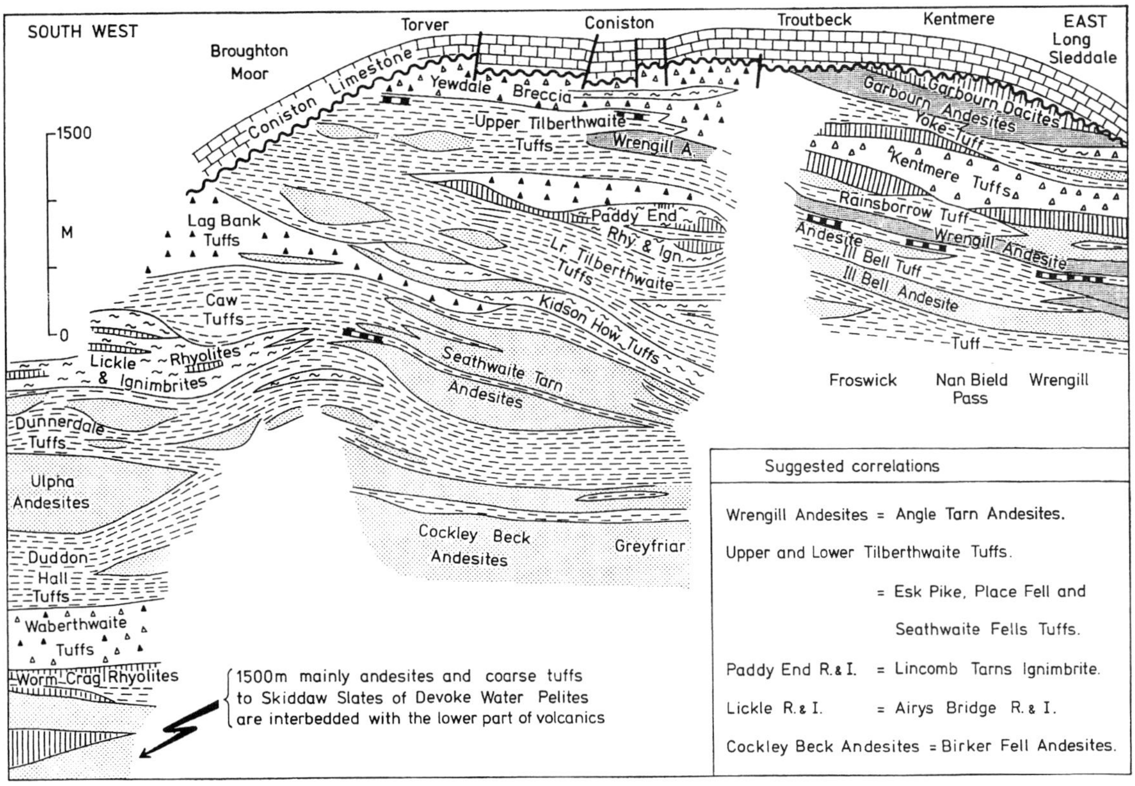

Fig. 12.6b. Lower part of Borrowdale Volcanic succession between Buttermere and Haweswater, according to Millward, Moseley & Soper (1978, p. 109). Reproduced by courtesy of the Yorkshire Geological Society, Drs Moseley and Soper, and IPR/26–13C British Geological Survey. © NERC. All rights reserved.

such as that which gave rise to the Airy's Bridge lavas and ignimbrites (Millward *et al.* 1978, p. 118). So the idea of caldera collapse in the BVG – a concept nowadays chiefly associated with the work of Michael Branney (see Chapter 16) – was foreshadowed here. Mention was also made of Numan's report (1974) of the discovery of marine microfossils (acritarchs) at Holehouse Gill in the Duddon Valley area. These remain the only fossils found in the BVG other than trace indications. Their presence suggested the idea of an eruption of the Ordovician volcanoes in some kind of island arc situation, such as envisaged in the theories discussed in Chapter 10, though the marine status of the Holehouse Gill unit remains somewhat controversial. Some parts of the deposits in the upper BVG succession, showing ripple marks and mud cracks, suggested deposition of ash in shallow pools. However, these could merely be the residues of transient freshwater ponds, and not indicative of a marine environment. Millward (1978, p. 12) thought that there *might* have been just one major Lakeland volcano, with dimensions similar to those of Mount Aetna, but the considerable variety of rock types and the existence of various circular outcrops of agglomerate in the Lakes suggested, rather, the occurrence of a number of different eruption centres. Millward (1978) also mentioned the concept of caldera collapse as being applicable to the Lakeland volcanoes.

The virtues, or otherwise, of a geochemical approach to Lakeland stratigraphy will be canvassed in Chapter 16, where we shall examine some of the results of the grand collaboration between the BGS and the staff and students of a number of north British universities, itself discussed in Chapter 14. However, before we approach this important matter we must now, at long last, give more detailed attention to the rocks of 'Otley III'.

Chapter 13
The stratigraphy of 'Otley III'

So we turn to the work done in the twentieth century on the elucidation of the sequence and structure of the rocks of Jonathan Otley's third major subdivision of the Lakeland rocks – up to the time of the establishment of the 'Lakeland Project' in 1982, and with a preliminary forward glance to the end of the century. The rocks of 'Otley III' consist of limestones, mudstones, shales, slates, grits and conglomerates, with one substantial igneous unit, the Yarlside (or Stockdale) Rhyolite, which runs westwards from the Shap district, parallel with the strike of other rocks low in 'Mr Otley's third great physical subdivision' as Sedgwick (1852*b*, p. 137) called the unit. The 'Otley III' rocks are of generally quite low relief, but, as mentioned previously, many are highly cleaved, the cleavage frequently being more pronounced than the bedding. As we have seen, some of the beds are highly fossiliferous (or were so before they became the subject of close attention by generations of amateur and professional geologists, dealers, or even historians of geology). There is (or once was) a rich fauna of graptolites, molluscs, corals, cystoids and trilobites, and in the past more exotic types like starfish were reported by Sedgwick.

As we have seen, the lowest part of 'Otley III' was formerly called the Coniston Limestone (Series) and was initially likened (by Sedgwick) to the Bala Limestone in Wales. Later it was placed in the Caradoc subdivision of the Ordovician; and still later it was subdivided and the several components variously placed, the lower members being allocated to the Ashgill. The name Coniston Limestone Series is no longer formally extant, though it survives in informal discourse, the unit sometimes being affectionately called the 'Connie Limestone'. The limestone is blue when fresh, and is sometimes mingled with ashy material. It weathers to yield a characteristic yellowish outcrop, with typical limestone weathering features. At Timley Knott (NY 284 971), near Coniston, for example, it lies against Borrowdale Volcanics at a high angle (see Figs 5.2 and 13.1) and would appear to be unconformable thereto at that locality. But as we have seen (p. 83) it was not until the 1920s that geologists fully accepted that there was an unconformity below the Coniston Limestone Series, as it overstepped the various units of the Borrowdale Volcanics during the course of its long outcrop from Shap to the Duddon Estuary.

The rocks of 'Otley III' form outcrops throughout southern Lakeland. They also appear at Drygill in the northern Lakes, in parts of the Cross Fell Inlier, in the Howgill Fells, where there are fossiliferous stream sections near Sedbergh, and down in Yorkshire with the inliers of Horton-in-Ribblesdale, Austwick, Craven and Ingleton. Whereas in the Skiddaw Slates ('Otley I') a major problem was the lack of fossil materials, in some parts of 'Otley III' the problem has been rather the reverse: an excess of materials, now widely scattered in different collections. There was surely more than enough material to enable stratigraphers to use their classical methods to establish firm subdivisions of the strata. What it might all mean in terms of palaeogeography, and the depositional and tectonic history of the area will not be treated fully in this chapter, for all was not made clear before the remapping of the Lakes in the 1980s, and even now there is no final agreement on some important matters. Even some of the basic mapping remained unfinished at the end of the twentieth century. It should be remarked that in the Lake District proper the successions are by no means complete, and stratigraphers have had recourse to the exposures in the Cross Fell Inlier and the Howgills to build a more complete picture.

The task of stratigraphic correlation was not, of course, restricted to the exposures of NW England. It was important to tie those in with standard sections of Ordovician and Silurian rocks developed in other areas, such as Shropshire and the Welsh Borders, the Southern Uplands, Scandinavia, or even further afield. Naturally, the work of the inter-war years built on the investigations of Marr and his co-workers. We look first at what was done in the Cross Fell Inlier.

As previously mentioned, the Inlier was mapped by Frederick Shotton (1935), who later assumed the geology chair at

Fig. 13.1. The 'Coniston Limestone' lying unconformably at a high angle against Borrowdale Volcanic rocks, Timley Knott, near Coniston. Photographed by author, 1996.

Table 13.1. *Summary of zonal subdivision for the Caradoc of the Cross Fell Inlier, according to Bancroft (1933)*

Series (Epoch)	*Stages*	*Zones*[a]	*Lithological units in the Cross Fell Area*
Caradoc	Onnian	*Onnia superba* (T)	Dufton Shales
	Onnian	*Onnia gracilis* (T)	Dark grey mudstones
	Onnian	*Onnia cobboldi* (T)	Not seen
	Actonian	*Onnia grandis* (T)	Black shales, grey shales
	Marshbrookian	*Onniella reuschia* (T)	Dark grey mudstones
	Marshbrookian	*Wattsella unguis* (B)	Tough grey mudstones
	Marshbrookian	*Wattsella wattsi* (B)	Tough grey mudstones
	U. Longvillian	*Kjaerina geniculata* (B)	Corona Beds: grey mudstones
	U. Longvillian	–	Grey and black mudstones
	U. Longvillian	*Raymondellae gigantea* (B)	Grey mudstones
	U. Longvillian	*Kjaerina bipartita* (B)	Band with *Brongniartella bisulcata* (T)
	L. Longvillian	*Raymondella typa* (B)	Corona Beds (part)
	L. Longvillian	*Wattsella indica* (B)	Fossiliferous mudstones (fauna not examined)
	L. Longvillian	–	Ditto
	L. Longvillian	*Watsella lepta* (B)	Ditto
	L. Longvillian	*Wattsella horderleyensis* (B)	Mudstones
	Soudleyan	[9 zones]	Volcanics (Ash, Dufton Pike Rhyolite and Knock Pike Rhyolite)
	Harnagian	[5 zones]	Borrowdale Volcanics
	Costonian	[5 zones]	Borrowdale Volcanics

[a] T, trilobites; B, brachiopods.

Birmingham.[1] Though an authority on Quaternary geology, in the Cross Fell Inlier he was chiefly interested in structures and had the idea of a major thrust fault, dipping gently to the SW, bringing Skiddaw Slates up and over the basal Carboniferous rocks. By means of an unpublished MS section, held at Birmingham, one can discover that this fault corresponded to what the Surveyors in the late 1960s regarded as the Lad Slack Fault (Geological Survey of Great Britain 1972).[2] Shotton was not much interested in the stratigraphy of the Upper Ordovician and Silurian rocks of the Inlier, but its fossiliferous Upper Ordovicians were actively investigated by [J.] B. B. Bancroft.

Following his earlier investigations of Ordovician strata in Shropshire and the Welsh Borders, (John) Bevis Beeston Bancroft (1897–1944)[3] published privately an important, but now rare, work, containing correlation tables for the Caradoc rocks of Wales and England (Bancroft 1933). It appears that he wrote a monograph on the Upper Ordovician (presumably in the 1920s), but could not get it published because of the expense of reproducing photographs of the many trilobites he collected. Then he had a lucky break in 1930 when he was invited by the Department of Scientific and Industrial Research to write a monograph on the Llandovery faunas, and he worked on this project for two years at Cambridge, studying the collections of V. M. Turnbull and O. T. Jones, during which time he took the opportunity to undertake a 'rapid survey' of the Ordovicians of the Cross Fell area. Then, it seems, his funding was discontinued (unsurprisingly in the Depression, though perhaps it was because he was being paid to look at the Llandovery but was spending too much time on the Caradoc! – and seemingly he did not 'fit in' at Cambridge). So Bancroft retired to his family home at Blakeney, Gloucestershire, on the western side of the Severn, from whence he published his tables privately. Having been severely wounded in World War I (and having been recommended for a *Croix de Guerre*), Bancroft died on active service during World War II. His fossil collection was purchased by the British Museum (Natural History) in 1936 (Dean 1959, p. 187).

After World War II, Bancroft's friend, Archibald Lamont, from the Edinburgh University Geology Department, published some account of Bancroft's work on the basis of his manuscripts, with maps showing where Bancroft had collected his specimens, and photographs illustrating a new species of trilobite that he had discovered and named (*Onnia superba* Bancroft) (Lamont 1948). Lamont's work was most useful in revealing where certain fossil types had been collected, and by whom.

We discover from Lamont that Bancroft worked in three principal areas in the Cross Fell Inlier (see Fig. 1.3): (1) in Pus Gill, half a mile NE of Dufton; (2) in Swindale Beck, a mile and a half north of Dufton; and (3) at a site about half a mile NE of Melmerby. Subdivisions were proposed (Table 13.1), and correlation was provided with the standard zones of the Onny River section in Shropshire.

All the stages mentioned in Table 13.1 are still in current use, but the zones are now defined by graptolites and conodonts, and the Upper Longvillian was later renamed the Woolstonian by J. M. Hurst (1979). Also, it has been thought that Bancroft's original subdivisions for the Caradoc were too fine for stages and are now

[1] F. W. Shotton (1906–1990), who came from Coventry, obtained a first in geology at Cambridge in 1927, and the following year, at the suggestion of Marr, he did his work in the Cross Fell Inlier. The following year he was on the staff at Birmingham, and after doing 'military geology' during World War II he held the Sorby Chair at Sheffield for a brief period before taking up the Wills Chair at Birmingham in 1949, where he remained until his retirement in 1975. His principal work was in Triassic and Quaternary geology. He was elected FRS and was President of the Geological Society from 1964 to 1966. See Worsley (1991).

[2] Burgess & Wadge (1974) accepted the idea of reverse faulting in the Inlier, but did not have a major low-angle thrust fault as did Shotton. Shotton seems to have had an *a priori* commitment to the presence of a thrust fault, as his mapping of the lower part of the Skiddaw Slates in the northern part of the Inlier (near Cuns Fell, see Fig. 1.3) showed the outcrop tending to follow the contours, suggesting a low-angle contact. But this did not agree with the results of mapwork conducted by the Survey in that area in the 1960s. See Geological Survey of Great Britain (1972).

[3] Bancroft was born in Northwich, Cheshire, and educated at Warwick School and Trinity Cambridge, where he read English, getting a third in Part I in 1920 and a second in Part II in 1922. Subsequently he took an MSc in geology at Manchester. Apparently he had interests in botany and zoology as well as geology and wrote short stories. Only one obituary is known to me (Lamont 1946), and this dealt chiefly with his military activities. It contained a photograph of Bancroft, which suggests that he may have been a gentleman amateur – which would be consistent with his career so far as it is known. I thank Hugh Torrens for the reference to Lamont's obituary notice, and Jonathan Smith for information about Bancroft's time at Cambridge and his date and place of birth.

treated as sub-stages, with just four stages for the Caradoc as a whole. See Fortey *et al.* (1995, p. 16), which adds some further sub-stages below the units named as stages by Bancroft.

The most recent stratigraphic table for the Lakes that I have seen, unpublished in 2000 (see Fig. 13.10), places the top of the Borrowdale Volcanics in the middle of the Soudleyan, much as Bancroft had it back in the 1930s. However, when William Dean (1962) produced his monograph on the Caradoc trilobites of the Cross Fell Inlier, he did not carry the succession down below the Lower Longvillian. He placed the base of the Dufton Shales at the bottom of the Upper Longvillian, rather than in the Onnian, so Bancroft and Dean diverged concerning the range of that unit.[4]

Bancroft (1945, p. 182) was also responsible for proposing (posthumously) another stage of the Ordovician: the 'Pusgillian'. Based on fossils found in strata in Pus Gill in the Cross Fell Inlier near Dufton (see Fig. 1.3), he had the Pusgillian as a stage in the Caradoc Series, occurring between the Onnian and Ashgill Stages; but he said little about the unit other than that it was characterized by the Trinucleid trilobite, *Tretaspis kjaeri* (or *kiaeri*) (so named by Lamont 1941), and that it had its 'type deposit' in the Dufton Shales of Pus Gill. The Pusgillian survives as a stage name in the modern literature, but is now regarded as belonging to the Ashgill, rather than the Caradoc, this change having been implicit as early as 1948 in a paper by the Cambridge geologists William Bernard Robinson (W. B. R.) King[5] and Alwyn Williams, on the grounds that beds near Cautley (Sally Brow) in the Howgills (see Fig. 5.9) possessed a fauna that was overall Ashgill but which appeared to be similar to strata from Bancroft's Pusgill near Dufton. (In fact, Bancroft himself seems to have allowed for the possibility that the Pusgill Stage was wholly or partly in the Ashgill.)

Following difficulties perceived by the Bristol professor Walter Frederick ('Fred') Whittard (1952) in Bancroft's Shropshire work regarding the presence or otherwise of Ashgill rocks in that county, Dean (1959) was not happy with Bancroft's would-be correlations between his shelly fauna and graptolites, describing them as 'varied and confused'. Further, he complained that Bancroft's analysis of the strata in the Inlier was based on incomplete sampling of the fossils. On the other hand, Dean did not go along with King and Williams's allocation of the Pusgill strata to the Ashgill. The argument focused particularly on the questions of whether the trilobite *Dicalymene marginata*, Shirley (Marr's '*Calymene*') should or should not be regarded as indicative of membership of the Ashgill; and whether Bancroft's shelly fauna at Pus Gill correlated with the Ashgill zone fossil *Dicellograptus complanatus* (as Bancroft thought). Dean (1959, p. 216) thought that the graptolite evidence at Pus Gill was inconclusive, and he therefore preferred to use trilobites and brachiopods – which seemingly kept the Pusgill out of the Ashgill. The arguments were based on analogies with rocks in the Howgills, in Shropshire, in Scotland, and even in Norway. Dean suggested that the Pusgill Stage of the Cross Fell Inlier belonged to the zone of *Pleurograptus linearis* (then regarded as pre-Ashgill, but subsequently thought to span the Caradoc–Ashgill boundary) since it was evidently above the Onnian. As stratigraphers know, but students are not always taught, faunal-based stratigraphy can be anything but a clear-cut enterprise.

Dean (1962) assisted matters considerably by publishing detailed descriptions and photographs of what he regarded as *Caradoc* Cross Fell trilobites. He reaffirmed his belief that the Pusgill Stage belonged to the Caradoc, but acknowledged that its Trinucleid trilobites were members of genera thought to be characteristic of Ashgill strata. Thus he suggested that Ashgill elements might have invaded the Anglo-Welsh area early on (Dean 1962, p. 130). This was a somewhat 'Barrandian' notion, particularly considering that Dean contended that there was a 'marked stratigraphic break above the Pusgillian' and that the overlying limestone belonged to 'a relatively high Ashgill horizon' (Dean 1962, p. 71). However, the Pusgill Stage eventually became firmly situated in the Ashgill (see p. 174).

From the Cross Fell Inlier, Dean moved on to examine the fossils of the Upper Ordovician in the Drygill area and the Stile End [Grassing] beds between Kentmere and Longsleddale, where they lie unconformably on the southern side of the Borrowdale Volcanics. Again his concerns were chiefly palaeontological, but efforts were made to achieve stratigraphic correlations with other areas. Photographs were published of the trilobites and brachiopods, and a new trilobite was formally described (Dean 1963). He used his own collections and material previously collected by W. B. R. King for the Stile End beds and by Elles and Wood for the Drygill strata (see p. 107), all held at Cambridge.

As we have seen, the highly fossiliferous Stile End beds attracted the attention of early workers such as Harkness & Nicholson (1877). Marr (1892*a*) thought that the beds were a low unit of the Coniston Limestone Series, and had them in the Ashgill, but by Dean's analysis they were Actonian, i.e. belonging to the upper part of the Caradoc; and in consequence the adjacent Stockdale or Yarlside Rhyolite could be regarded either as late Caradoc or early Ashgill. (Dean did not mention the possibility of its being intrusive.) His placement of the Stile End beds in the Caradoc held the field for about a decade; but his allocation of the Drygill Shales to the Longvillian (mid-Caradoc) remained accepted in 2000.

Much additional palaeontological and stratigraphic work in the Ordovician inliers of the Howgill Fells was undertaken by Keith Ingham, who had done a PhD under John Neale on the area at Hull University (where W. B. R. King had assisted in the establishment of the geology department), and then obtained an appointment at the Hunterian Museum of Glasgow University. While doing his PhD, Ingham obtained assistance from King, and collaborated with another Hull PhD student who was a year behind him, Barrie Rickards (see p. 176). The extensive collections made by Ingham and Rickards are still held at Hull. Ingham's results were published in a lengthy paper in the *Proceedings of the Yorkshire Geological Society* (1966), in three splendid volumes by the Palaeontographical Society (Ingham 1970–1977), and in a paper discussing the status of the Ashgill Series co-authored with Anthony Wright, then a lecturer at Queen's University of Belfast (and subsequently Professor there) (Ingham & Wright 1970).

Most of the Howgill Fells (see Fig. 5.9) are made of Silurian rocks. However, there are also the seven quite small, but interesting and important, Ordovician inliers, previously mentioned, on the eastern side of the hills, NE of Sedbergh, with good stream sections in the valleys and side-valleys. They are situated between the Howgill Fells proper and the faulted contact to the east, where Carboniferous rocks meet Lakeland types in a great 'smash-up' at the Dent Fault (a structure, it will be recalled, well known to Sedgwick, who hailed from Dent, of course). In his thesis and his 1966 paper, Ingham provided clear and detailed black-and-white

[4] The Dufton Shales were introduced as a lithological unit by Harkness & Nicholson (1877). Their type locality is in Swindale Beck, near Knock, in the Cross Fell Inlier (see Fig. 1.3).

[5] King (1889–1963) (see Fig. 8.2) came from near Aysgarth in the Yorkshire Dales. After studying at Cambridge under Marr, Nicholas, Elles, Harker, etc., he joined the Survey, from where he did geological work for the war effort. From 1920 to 1931 he was Demonstrator and assistant to Marr at Cambridge, from which position he gained the Chair at University College, London, and made further geological contributions during World War II. Then from 1943 to 1955 he held the Woodwardian Chair at Cambridge. (See Shotton 1963.) In view of his Yorkshire ancestry and Cambridge connections – especially with Marr – it is unsurprising that King should have worked on Lower Palaeozoics and Pleistocene geology, but much of his work was actually done in Wales. King did important work in the Yorkshire Palaeozoic inliers, near Horton-in-Ribblesdale, etc.

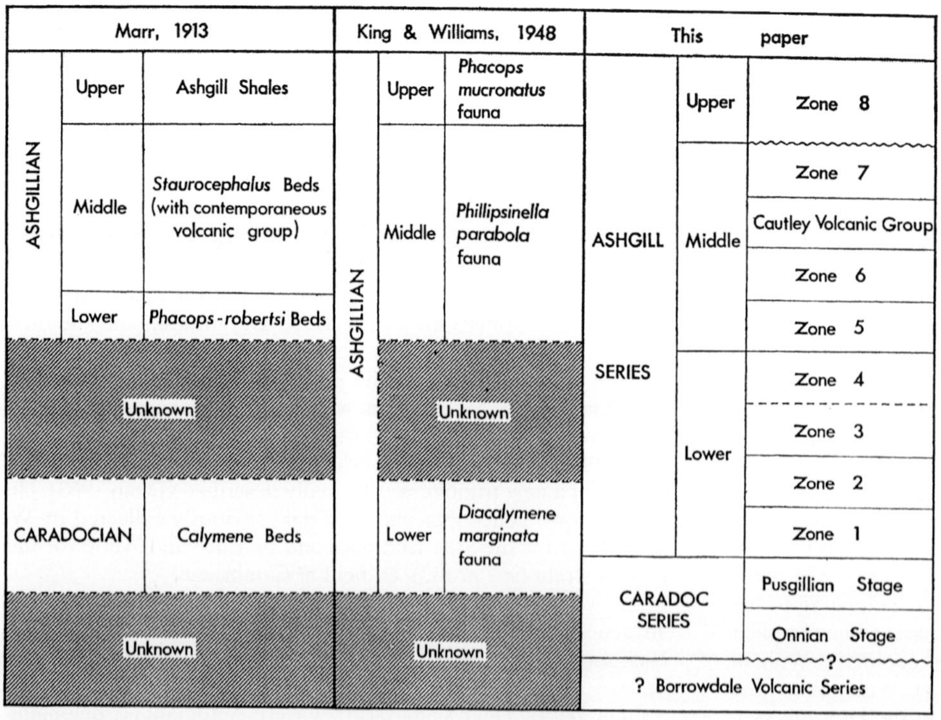

Fig. 13.2. Development of ideas about the stratigraphic succession at Cautley, according to Ingham (1966, p. 460). Reproduced by courtesy of the Yorkshire Geological Society.

maps of the inliers, marking on them his fossil localities and proposed zonal boundaries. The Palaeontographical Society memoir provided chiefly descriptions and photographic illustrations of the trilobites collected; and the paper with Wright discussed the question of the placement and limits of the Ashgill Series.

A major feature of the work was a proposed zonal scheme for the Ashgill. For the inliers overall, Ingham found representation of the Onnian and Pusgill Stages (both regarded as Caradoc), and eight zones for the Ashgill, based on trilobites.[6] His scheme allowed correlation betwixt and between the strata of the seven inliers, and across to the Cross Fell Inlier and beyond. Each zone was represented by characteristic trilobite assemblages rather than particular index fossils. It will be noted that at this stage Ingham followed Dean's work in the Inlier by placing the Pusgill Stage in the Caradoc, though this was queried by Jack Shirley from Newcastle-upon-Tyne (commenting on Ingham 1966), who thought that the Pusgill fauna seemed to be more Ashgill than Caradoc in character. Ingham helpfully showed how his scheme was to be understood in relation to the earlier proposals of Marr, and King and Williams (see Fig. 13.2).

In fact, Ingham did not retain the Pusgill Stage in the Caradoc for long, as can be seen from Figure 13.3. The reasons for the change were explained in Ingham & Wright (1970). As we have seen, Marr (1905) introduced the Ashgill on the basis of outcrops near Coniston, but no type locality was then specified. He subsequently (Marr 1913, p. 13; 1916, p. 189) stated that the Cautley area – as at Backside Beck in the Westerdale Inlier (the north-western of the seven Ordovician Howgill inliers) – would serve this purpose better as it gave a more complete and less 'compressed' succession. Unfortunately, however, Marr erred (1913) by placing his *Calymene* beds (with what he identified as *C. planimarginata*) in the Caradoc. The fossil was later identified as *Dicalymene marginata*, which occurred in Scotland in a zone defined by *Dicellograptus complanatus* as belonging to the Ashgill. Ingham, working at Girvan, found that some rocks in Scotland contained *D. complanatus* in association with fossils that had been found in the Pusgill Stage at Cautley (Ingham & Wright 1970, pp. 234–235), so he moved the Pusgill Stage up to the Ashgill in his tables, as Shirley had suggested might be appropriate when commenting on Ingham's paper of 1966. The three lines of argument were as follows:

1. The Pusgill Stage was absent in the Caradoc type area, where the Onnian was the highest member; and the Pusgill Stage overlay the Onnian in the Cross Fell Inlier.
2. At Cautley, the Pusgill Stage was partly equivalent to Marr's '*Calymene* Beds', which were *not* Caradoc according to King and Williams.
3. The Pusgill Stage contained the horizon of *D. complanatus*, which was accepted as belonging to the Ashgill.

In addition, Ingham & Wright (1970) now renamed the three subdivisions of the Ashgill at Cautley – 'Lower', 'Middle' and 'Upper' – as Cautleyan, Rawtheyan and Hirnantian, with the same zonation as that of Ingham (1966); and they inserted the Pusgill Stage between the Onnian (top of Caradoc) and Zone 1 of the Cautleyan.[7] These stage names are still used in the modern literature (e.g. Harland *et al.* 1989, p. 35; Fortey *et al.* 1995, 2000).

There is a peculiar feature in one of the units in the Cross Fell Inlier: the fossiliferous Keisley Limestone, so named by Nicholson & Marr (1891) (see p. 65), and the rich Ashgill fauna of which was described early on by Cowper Reed (1896, 1897). (Keisley is a hamlet about 4.5 km NE of Appleby; see Fig. 1.3.) Marr revisited the area with several colleagues in the Easter vacation of 1906, and suggested that the Keisley Limestone was equivalent to the neighbouring Swindale Limestone in Swindale Beck, to the NW of Keisley, near Knock. Marr regarded this limestone as the lower part of his *Staurocephalus* Limestone, and this suggestion was adopted in the Survey Memoir on the Cross Fell Inlier (Burgess & Wadge 1974, p. 20), but was not accepted by subsequent workers (for details, see Wright 1985).

Near Keisley Farm there is an old quarry (NY 713 238), from

[6] Zones 1–4, Lower Ashgill; Zones 5–7, Middle Ashgill; Zone 8, Upper Ashgill.

[7] The Rawthey is the principal river of the Cautley area, flowing down by a circuitous route from the moors of Baugh Fell to Sedbergh (see Fig. 13.2). The Hirnant Limestone was regarded by Sedgwick (1844) as a unit with a distinctive fauna at the *top* of the Upper Bala in Merionethshire. The term 'Hirnantian' was, however, introduced by Bancroft (1933), for the highest stage of the Ordovician in the Bala district. The name comes from Cwm Hirnant.

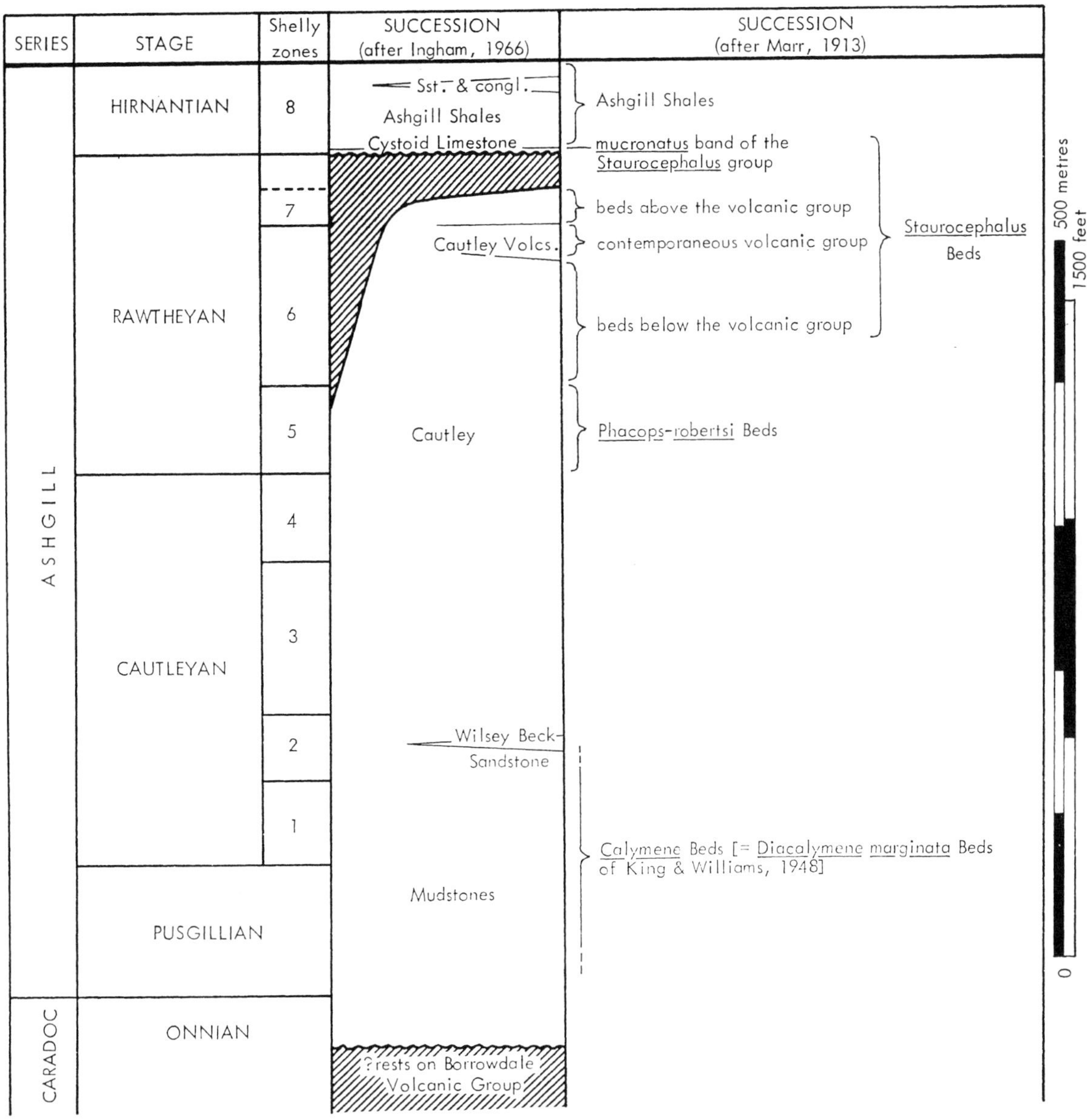

Fig. 13.3. Ordovician succession in the Cautley–Dent inliers, according to Ingham & Wright (1970): Ingham (1970–1977, **1**, p. 2). Reproduced by courtesy of the Palaontographical Society.

which limestone was formerly obtained for a nearby limekiln. The nodular limestone 'structure' at Keisley, which might appear to have been an ancient reef (though not made of coral), was called a 'knoll' by Marr (1906*a*, p. 485), with the suggestion that it had been produced by some kind of process of folding, producing thickening and crystallization of the limestone. Marr thought that the Keisley Limestone was faulted against Skelgill Beds. However, the Surveyor Ian Burgess (in Shotton *et al.* 1968, p. 343), who remapped the Cross Fell area with Tony Wadge in the late 1960s, thought that there was an apparently conformable sequence between the limestone and the overlying graptolite-bearing shales, which – according to John Temple, who published a monograph on the brachiopods from the locality (Temple 1968, p. 3) – were identified by Barrie Rickards as being of *Monograptus atavus* age.

Thus the Keisley locality was apparently close to the Upper Ordovician (Ashgill)–Lower Silurian (Lower Llandovery) boundary, but no exposure of the contact was visible. Such boundaries are, of course, always of interest to geologists, and there is competition between different countries as to which may have the 'honour' of hosting the various stratotype boundaries, suitably ratified by the appropriate commission of the International Union of Geological Sciences. To my knowledge, Keisley was never a candidate for the Ordovician–Silurian type section, but it was evidently worthwhile having as close a look as possible at what was there. The contact seemed to fall under the roadway leading from the farm to the quarry. Accordingly, some years later Anthony Wright organized a mechanical digger to cut a trench across the track leading to the quarry so that he could log the section (Wright 1985, 1988).

The outcome was that a stratum of siltstone containing *Akidograptus acuminatus* and various Hirnantian-type (Upper Ashgill) shells was found *below* the *atavus* zone. Just a little (2 cm) lower, Rickards identified various graptolites including *Glyptograptus persculptus*, indicative of the top of the Ashgill. So the Ordovician–Silurian boundary had been located with considerable precision, but the locality could not serve as a stratotype boundary. The farmer would not have wanted his roadway to have been dug up every time a geologist wanted to inspect the boundary; and meanwhile Lapworth's old site at Dobs Lin in the Southern Uplands had been chosen as the international Ordovician–Silurian boundary stratotype. As for the knoll, it is now interpreted as a carbonate mud-mound (Wright 1985, p. 261).

Turning now to the 'Otley III' strata exposed in Lakeland

Fig. 13.4. Barrie Rickards at the Sedgwick Museum, Cambridge. Photographed by author, 1998.

proper, we find that the lithological subdivisions used in the 1930s were summarized by Robert Blackie (1933), who was working from Imperial College and Liverpool University on a Senior Studentship of the Commissioners of the 1851 Exhibition.[8] His subdivisions, essentially the same as those developed in the nineteenth century, were:

Kirkby Moor Flags	
Bannisdale Slates	
Coniston Grit[s]	
Coldwell Beds	[= Upper Coniston Flags (Lower, Middle, Upper Coldwell Beds) (Marr 1916)]
Brathay Flags	[= Lower Coniston Flags (Marr 1916)]
Stockdale Shales	{ Browgill Shales Skelgill Shales

Not much additional research was done on these rocks until the work of a Cambridge PhD student, Peter Llewellyn (1960), who worked in the area between Longsleddale and Shap.[9] He suggested that these cleaved rocks were the products of turbidity currents in many instances, derived from the NW, and showing evidence of slumping. He acknowledged that he had been unable to establish any Wenlock graptolite zones for his area, but his interests were chiefly structural and sedimentological rather than palaeontological or stratigraphic.

So when serious zonal work was begun in the Upper Silurians of 'Otley III' in the early 1960s by Barrie Rickards (b. 1938), there was not a well-established knowledge-base to build on. Rickards attended school at Goole, and went to Hull University, planning to read chemistry. But like many of us he started geology at university and was soon seduced by the charms of the subject. Initially, he intended to do hard-rock geology from a geochemical perspective, but perhaps fortuitously it did not work out like that. His undergraduate mapping introduced him to Ordovician trilobites in the Lake District, and when it came to the choice of a thesis topic these engaging fossils won out over Scottish migmatites. So work was started in the Berwyn Mountains in Wales under the supervision of John Neale. However, after only three weeks in Wales, Rickards ran into William Dean working there, also on Ordovician trilobites, who kindly suggested that Rickards might like to look for greener pastures elsewhere. This he did. He returned to Lakeland work and teamed up with Keith Ingham, at work on his PhD on Ordovician trilobites in the Howgills. Rickards took on the Silurians therefore. He found that graptolite zonation was needed, so that he could situate the trilobites he found; and before long he was writing a thesis about graptolites in the Silurian parts of the Howgill Fells (Wenlock and Ludlow). It was completed in 1963 with Isles Strachan from Birmingham as external examiner. It was a monumental piece of work, based on the collection of over 22 000 graptolites (now archived in Hull). It is no surprise that the examiners said that they were 'very happy' with the thesis (though it apparently took them four hours to reach that conclusion) (Rickards, pers. comm., 1998).

So, though only young, Rickards was already 'Prince of Siluria' as far as British Silurian graptolites were concerned. Today, I might use Murchison's sobriquet, 'King of Siluria' (at least for Lakeland graptolites)! From Hull, Rickards took a post at the geology museum at University College, London, and librarian at the Garwood Library there. But he did not remain long. Oliver Bulman saw the monumental thesis and, knowing something good when he saw it, requested that Rickards be allowed to move to the Sedgwick Museum as a 'University Assistant in Research' – in practice Bulman's assistant. After Bulman's retirement, however, the new professor Harry Whittington wanted to work on trilobites, so Rickards decided to move to the Natural History Museum in London to take charge of the graptolite work there. He found it congenial from a collegial point of view, but did not care for the bureaucracy of the place; so he took a lectureship at Dublin for two years, before being called to Cambridge again, where he became the Curator of the Sedgwick Museum in 1969, upon the retirement of A. G. Brighton (who had the misfortune to tutor me for a year). Rickards (see Fig. 13.4) has since been at Cambridge, where he is a Fellow of Emmanuel College, and he was promoted professor in 2000. Of course, his specialty was graptolites even during his PhD, but he must have had a great flair for the work (like Dennis Jackson), for as compared with Rickards' 22 000 specimens, Llewellyn (whose work was, as said, primarily structural and sedimentological) recorded only six types in his thesis.

[8] He later became Geological Curator of the Royal Albert Memorial Museum, Exeter.

[9] Llewellyn went on to a post at University College, London, and thence into various positions in the petroleum industry.

When Rickards started in the Howgills, he picked up from the work of two students of Gertrude Elles and John Marr: the Misses G. R. Watney and E. G. Welch, from Girton and Newnham Colleges respectively.[10] They had presented a zonal scheme for the Wenlock and Ludlow strata at Cautley to the British Association meeting at Sheffield in 1910 (Watney & Welch 1911*a*), the same year that Marr and William G. Fearnsides published a paper on the same theme. Then Marr read Watney & Welch's more detailed paper to the Geological Society in January 1911 (Watney & Welch 1911), so we may assume that their fieldwork and mapping were chiefly done in 1910.

By close examination of the exposures in the stream banks around Wandale Hill, the River Rawthey, and Ecker Secker Beck in the Westerdale Inlier (see Fig. 5.9) Watney and Welch obtained the following graptolite zones:

Lower Ludlow	*Monograptus leintwardinensis*
	M. nilssoni
Wenlock	*Cyrtograptus lundgreni*
	C. rigidus
	M. riccartonensis
	C. murchisoni

A map was produced, tracing out the outcrops of these zones. It was, however, as Rickards put it to me (pers. comm., 1998), a 'broad brush' production, and I would say not altogether compatible with the modern map for Kirkby Stephen (BGS 1997*b*). Watney and Welch's work, which was undertaken at Marr's instigation, sought to deploy the graptolite zones suggested for Wales and the Welsh Borders by Gertude Elles & Ethel Wood (Shakespear) (1901–1918), and also the work of Sven Tullberg (1852–1886) in Scandinavia (1883). Marr's (1913) paper on Cautley provided no general map to assist Rickards. There was, however, an unpublished Birmingham University thesis by D. W. R. Wilson (1954) on the Valentian[11] rocks of the area.

When Rickards got to work, he remapped his area (between Ravonstonedale to the north and Sedbergh to the south, with the Lune Valley on the west and the Carboniferous Limestone outcrop of the Pennines on the east), giving more attention to the Brathay Flags (Wenlock) than had his predecessors. He already had a head start from his undergraduate mapping in the area and knew where to look for fossils, and how. Rickards adopted a method (independently of Tressilian Nicholas, who also used it) for getting at the fossils on the bedding planes successfully. Wearing a stout glove, he held a cleavage slab horizontally and gave it a heavy blow with a sharp-edged hammer in the direction of the bedding planes, and as often as not the rock then parted along a bedding plane rather than the cleavage, perhaps revealing graptolites. (Rickards recalls that he got on well with Nicholas, but found that he did not always reveal his techniques to others. Rickards had to learn for himself how to break the Howgill slates as described.)

As a result of his investigations, Rickards gave the following zones in his thesis and in a paper he published in 1967:

Ludlow
- *Monograptus leintwardinensis*
- *M. leintwardinensis incipiens* (a zone not recognized by Watney and Welch)
- *M. nilssoni* and *M. scanicus* (these two zones being found indistinguishable in the Howgills)

Wenlock
- *Cyrtograptus lundgreni*
- *C. ellesi* (or *ellesae*) (a zone not used by Watney and Welch)
- *C. linnarssoni*
- *M. flexilis/M. belorophus* (situated by Watney and Welch in their *riccartonensis* zone, the *flexilis* and *belorophus* zones being found to be indistinguishable in the Howgills)
- *M. anttenularius* (also situated by Watney and Welch in their *riccartonensis* zone)
- *M. riccartonensis*
- *C. murchisoni*
- *C. centrifugus* (a zone not recorded by Watney and Welch)

Subsequently, Rickards (1970*a*) published a Palaeontographical Society monograph on the Llandovery graptolites, with the following zonation:

Upper Llandovery
- *Monoclimacis griestoniensis*
- *M. crispus*
- *M. turriculatus*
- *Rastrites maximus*
- *M. sedgwicki*

Middle Llandovery
- *M. convolutus*
- *Pribylograptus leptotheca* (*M. argenteus*)
- *Diplograptus magnus*

Lower Llandovery
- *M. triangulatus*
- *M. cyphus* (*Coronograptus cyphus*)
- *M. acinaces*
- *M. atavus*
- *Akidograptus* (*Orthograptus*) *acuminatus*

Such work required the assemblage of large numbers of specimens (which Rickards certainly accomplished), keeping tabs on them all, making judgements as to what groups of fossils represented distinct zones, and doing this with an eye to linking in with zonal schemes developed by workers elsewhere. Moreover, the zonation could be used as a basis for mapping the Howgills, as Watney and Welch, Ingham, and others had done. The use of graptolites for this purpose was well established, needless to say, but one also had to try to make the graptolite zones mesh with those based on trilobites (and later with conodont zones). Besides using his zones to correlate exposures in different stream sections

[10] Gwendoline Rose Watney (1885–1976), daughter of a London doctor, was at Girton from 1904 to 1908, when she obtained a lower second in geology; but she held a Harkness Scholarship in 1907–1908. She is reported as doing geological research at Birmingham in 1908–1909, presumably under Lapworth, Elles and Wood, and presumably on graptolites. She married Joseph Barclay in 1915 and they went out to Japan to undertake missionary work. She published a book on her experiences in 1928 (*If I lived in Japan*) and retired to Hoddesden, Hertfordshire. The *Girton College Register* for 1907 records that Gertrude Elles, from Newnham, gave lectures at Girton and that students from that college attended various university and intercollegiate lectures, including ones by Marr, Harker and Elles, and the palaeontologist Henry Woods. Eleanor Gill Welch (1884–1963) was daughter of a Staffordshire doctor. She was at Newnham for the same years as Miss Watney and also read geology. She only obtained a third, but was President of the Sedgwick Club in 1908. Miss Welch taught at Perse Girls' School in Cambridge from 1909 to 1915, and resigned that year to marry Stephen Grove Lee. We can understand why Watney was the senior author of their joint paper as she was employed as a geologist at the time of their Howgills research, and obtained a higher examination mark. It would be interesting to know why the two young ladies obtained indifferent academic results, but were encouraged by Marr to engage in a quite difficult piece of fieldwork. I am indebted to Anne Thompson and Kate Perry, archivists at Newnham and Girton respectively, for providing information concerning the Misses Watney and Welch.

[11] An old subdivision introduced by Lapworth, approximating the modern Llandovery.

in the Howgills, determine thicknesses, and hence build up a composite picture of the area, Rickards also sought to make stratigraphic links into the Cross Fell Inlier, further north into the Lakes proper, into the Southern Uplands and the Welsh Borderland, and as far afield as Czechoslovakia.

It would have been convenient if the Howgill zones, named above, had meshed tidily with the lithological subdivisions; but unfortunately such tidiness could not always be achieved. For example, the boundary between the Coniston Grits and the Bannisdale Slates was placed *within* the *incipiens* zone. However, Rickards (following Wilson 1954) placed the Llandovery–Wenlock boundary close to where greenish Browgill beds gave way to blue-grey Brathay beds, near the bottom of which *C. centrifugus* made its appearance. There were two limestone bands at the base of what was taken to be the Ludlow, with graptolitic mudstone of *nilssoni–scanicus* type in between. The graptolites below the lower limestone were clearly Wenlock. So this left the age of the lower limestone somewhat in doubt.

Rickards (1967) paper was accompanied by a 1: 25 000 map for the eastern Howgills or the Cautley and Ravonstonedale districts. It is similar in broad outline to Watney and Welch's map, but besides having finer stratigraphic divisions it also shows many more faults. Some of these I find peculiar, notably the ones in the northern Ravenstonedale section (where Watney and Welch showed no faulting). Rickards envisaged two substantial faults running E–W and several others cutting across, approximately N–S. None of these faults were represented as producing any offset on the others: they intersected almost like crossing tramlines. This seems hardly plausible, requiring all the faults to be vertical and without strike-slip. None the less, Rickards' mapping for this area was utilized almost directly by the Survey in its Kirkby Stephen one-inch map (Sheet 40) (1972), and he was named as one of the co-authors. I raised the matter in correspondence with Rickards, and he responded that his mapping of the faults had met with the approval of the tectonics aficionado John Ramsay, then of Imperial College, in that the arrangement might be feasible if the faults were without strike-slip (Rickards, pers. comm., 2000). I await with interest to see what the map of the area looks like when eventually republished, as remapped by Jack Soper and Nigel Woodcock on contract to the BGS. In fact, according to Soper (pers. comm., 2000), the remapping did not bear out the pattern of faults indicated in the present Kirkby Stephen sheet, and Woodcock (pers. comm., 2000) informed me that the faults were being reviewed.

Rickards (1964) also attended to the varied facies of the Silurian strata. Several graptolitic mudstones were found in the Llandovery and Wenlock, but they became infrequent in the Ludlow. They contained carbonaceous banding (possibly arising on an annual basis), the dark matter being attributable to algae. The sediments showed minimal graded bedding in the generally fine-grained mudstones, and turbidity currents must have been of low intensity. The conditions of deposition were apparently anaerobic. Ascending the stratigraphic column, the sediments became coarser, lighter, and more calcareous, and trilobites and brachiopods became more evident. In general, then, the currents seemed to have become stronger up into the Ludlow.

Rickards (1969) sought to correlate his Howgill findings with those in the Lake District proper. He now used *Cyrtograptus rigidus* instead of *Monograptus flexilis* as the index fossil for one of his Wenlock zones and *Saetograptus ludensis* was added as a ninth zone above that of *C. lundgreni*. The Brathay Flags were run from *C. centrifugus* up to *C. lundgreni*, and the Middle Coldwell beds were allocated to *Monograptus ludensis*. But the Lower Coldwell beds (not found in the Howgills) also appeared to be *C. lundgreni*, producing a kind of stratigraphic misfit.

The situation was clarified by reference to sections examined in Lakeland proper (Rickards 1970*b*). Back in the nineteenth century, Aveline (in Aveline & Hughes 1872, p. 8) had introduced the lithological unit of the Coldwell Beds,[12] designated as Wenlock, being sandy flags within his Coniston Flags and Grits; and Marr (1878) introduced Upper, Middle and Lower Coldwell Beds, with Coniston Grits above and Brathay Flags below, his Middle Coldwell being equivalent to the Coldwell Beds as named by Aveline. Examining exposures near Torver (near Coniston) and to the west of Kentmere, Rickards described better developed sections containing graptolite bands corresponding to the 'doubtful' limestone band between the *nilssoni* and *lundgreni* zones in the Howgills. These sections suggested that the unzoned Howgills limestone band was equivalent to the *ludensis* zone (top of Wenlock) in Kentmere, which in turn was equivalent to the Middle Coldwell Beds near Torver. This did not, however, solve the problem of where to situate the coarse grey grits of the Lower Coldwell Beds satisfactorily (but Watney & Welch (1911*a*) had had them as *lundgreni*).

Further work was done on Llandovery graptolite zonation by Jana Hutt (Mrs D. J. Batten) for her PhD at Leicester (1975*a*), who subsequently worked as a research assistant to Rickards at Cambridge (Rickards 1999, p. 76). She produced two volumes for the Palaeontographical Society (Hutt 1974, 1975*b*), with minute details and excellent illustrations of the graptolites of the Lakeland Llandovery strata, adding a zone of *Monograptus crenulata* above, and *Glyptograptus persculptus* below Rickards' zones (see p. 177). The zones were also allocated to four stages, but these have not all survived to the present; and Hutt's *Glyptograptus persculptus* came to be regarded as Ashgill. Her work focused particularly on Yewdale and Church Becks near Coniston (see Fig. 5.2), and Pull Beck (which runs into the northwestern shore of Windermere; see Fig. 2.2), where she found stream sections that followed the Llandovery outcrops for some distance most satisfactorily. The total outcrop was traced from Appletreeworth Beck near Broughton Mills in the SW through to Browgill near Longsleddale in the east. For a correlation of Hutt's graptolite zones with the earlier graptolite- and trilobite-defined zones of Marr & Nicholson (1888), see the table in Scott & Kneller (1990, facing p. 19).

Another palaeontologist to work on 'Otley III' in the 1970s was the Cambridge PhD student Kenneth McNamara (see Fig. 13.5),[13] who went down to the Ashgill again, studying its trilobites and seeking to understand the palaeogeography of the Lakeland region in the late Ordovician (McNamara 1975, 1979*a*, *b*; Ingham *et al.* 1978). His PhD was supervised by the celebrated trilobite man, Professor Harry Whittington, and during his studentship McNamara was able to travel to Oslo to examine the trilobite collections of the Paleontologisk Museum. Apart from his systematic work on trilobites, McNamara was interested in reassessing the

[12] 'Cold Well' is a locality near Outgate, by the Ambleside–Coniston road, where there are several slate quarries.

[13] McNamara (b. 1950) went to school in Brighton and acquired an interest in geology through fossil hunting on the South Downs. This led him to take a geology degree at Aberdeen, from whence he might have proceeded to a PhD in metamorphic petrology. However, during a field excursion in his last year as an undergraduate he rekindled his interest in fossils by discovering a previously unknown cache of Cambrian trilobites in a roadside quarry, in the 'Fucoid Beds' of the Northwest Highlands of Scotland (by the road north of Ullapool, near Loch Awe). These had evolutionary implications and led to a paper with the Bristol stratigrapher John Cowie (who is mentioned again in Chapter 18). From this auspicious beginning, McNamara proceeded to his PhD at Cambridge. He told me that he thought his life-work really developed from his discovery of the trilobite-rich quarry in the Highlands. At Cambridge, McNamara came in contact with a 'number of cheerful Australians', including John Jell, who told him about a tutorship at the University of Queensland, which position he obtained. From there he went briefly to Sydney University, and thence across to Perth in 1979, where at the end of the century he was Curator in Palaeontology at the Western Australian Museum (McNamara, pers. comm., 2000). McNamara's Scottish work proved to be important in general ideas about heterochrony (see Fortey 2000, pp. 170–172).

Fig. 13.5. Kenneth McNamara, with a momento of his Cambridge days. Photographed by author, Western Australian Museum, Perth, 2000.

subdivisions of the Ashgill, and the Coniston Limestone, and revising some of Dean's work. His main hunting ground was among the exposures between Longsleddale and Windermere. Being at Cambridge, McNamara was able to utilize Tressilian Nicholas's forever unpublished maps, which he did not need, being no longer active in geology.

To remind ourselves of the situation at Stockdale and Longsleddale (see Fig. 3.6): we have the River Sprint, the valley of which runs northwards into the domain of 'Otley II'. Abutting the mountains we have the long band of the lower part of 'Otley III', running approximately NE–SW. Just south of the entrance to the mountains (see cover illustration), Stockdale Beck runs into the Sprint as a tributary from the east, and its main side-branch to the right (SE) is Brow Gill. Sedgwick passed this way in 1823, going up Stockdale Beck. On his left (to the north) there is a hill consisting of igneous rock, today called the Stockdale (or Yarlside) Rhyolite, but thought by some to be an ignimbrite. On the other, northern, side of this hill is a band of highly fossiliferous calcareous ashes and blue-grey shales or siltstones, the Style End Grassing Beds (Harkness & Nicholson 1877) or Stile End Beds (Marr 1892), which Sedgwick did not see. Ashgill Shales crop out near the confluence of Stockdale Beck and Brow Gill (to the south of the Stockdale Rhyolite), and can be traced up the latter stream, where contacts with Silurian strata are exposed. Marr included within his unit some coarse clastics below the fossiliferous siltstones.

Marr regarded his Stile End Beds as lateral equivalents of his Applethwaite Beds (ashy shales and limestones, exposed at Applethwaite Common, Kentmere, where the type section was situated) and overlying '*Calymene*' Beds, both belonging to the Caradoc. The latter were subsequently transferred to the Ashgill (King & Williams 1948, p. 206), as we have seen, but when McNamara was at work the Stile End and Applethwaite Beds were regarded as Caradoc, being thus enshrined by Dean in the Ordovician volume of the *Lexique Stratigraphique International* (Whittard & Simpson 1960). Dean (1963) put the Stile End Beds in the Actonian Stage of the Caradoc. It would appear, however, that the fossils did not speak unambiguously, for there were facies changes over quite small distances. It was, then, McNamara's task to try to mesh together palaeontology, stratigraphy and palaeoecology, and this involved him in some speculative thinking. He worked his way along the 'Coniston Limestone' outcrop, looking particularly for alterations in lithology from east to west, and correlations between the trilobite anatomies and the facies changes.

To try to sort out the stratigraphy, McNamara first introduced or redefined lithological units (which he now called 'formations'), with the type sections described. These were as follows:

- Ashgill Shale Formation: dark blue-grey shales, with little calcareous matter.
- Troutbeck Formation: fossiliferous grey-green calcareous shales, called *Phacops mucronatus* Beds by Marr.
- White Limestone Formation: hard pale-grey micritic limestone, so named by Marr at Ashgill Beck. Overlain by a rhyolitic tuff in the Torver–Coniston area.
- Torver Formation: dark blue, cleaved mudstones (called the *Phillipsinella* Beds by Marr).
- High Pike Haw Formation: nodular calcareous siltstones at the western end of the Coniston Limestone Group, with fossils similar to those in both the Stile End and Applethwaite Formations. (High Pike Haw is near Torver Common.)
- Applethwaite Formation: limestones, calcareous mudstones and shales (regarded by Marr as equivalent to the *Calymene* Beds of the Coniston and Cautley areas).
- Stile End Formation: fossiliferous, calcareous, bioturbated, ashy siltstones and sandstones.
- Longsleddale Formation: conglomerates, breccias, siltstones, sandstones, lying unconformably on the BVG.

All were placed in the Ashgill. Disconformities were stated to exist above the Torver Formation and the White Limestone, on the basis of comparisons made with more complete sequences in the Cautley district and the Cross Fell Inlier.

At Cambridge, McNamara became friends with the young palaeontologist, Simon Conway Morris, sharing a house with him. It was a time when there was much excitement over the discoveries concerning the strange fossils from the Burgess Shale in Canada, including the renowned *Hallucigenia*, so called by Morris. Stephen Gould (1989, p. 154) has said that the name was coined 'as a memorial to an unlamented age of social experiment', which might, I suppose, refer to the heyday of marijuana smoking. However, McNamara insists, he and Morris merely came up with the name one night on a bus on the way home from the pub. I mention this to make the point that McNamara was a bit irreverent at Cambridge; so we find that his thesis (1975) was curiously entitled 'Upper Ordovician stratigraphy and trilobites of the southern Lake district: a geofantasy [*sic*] in four movements'.

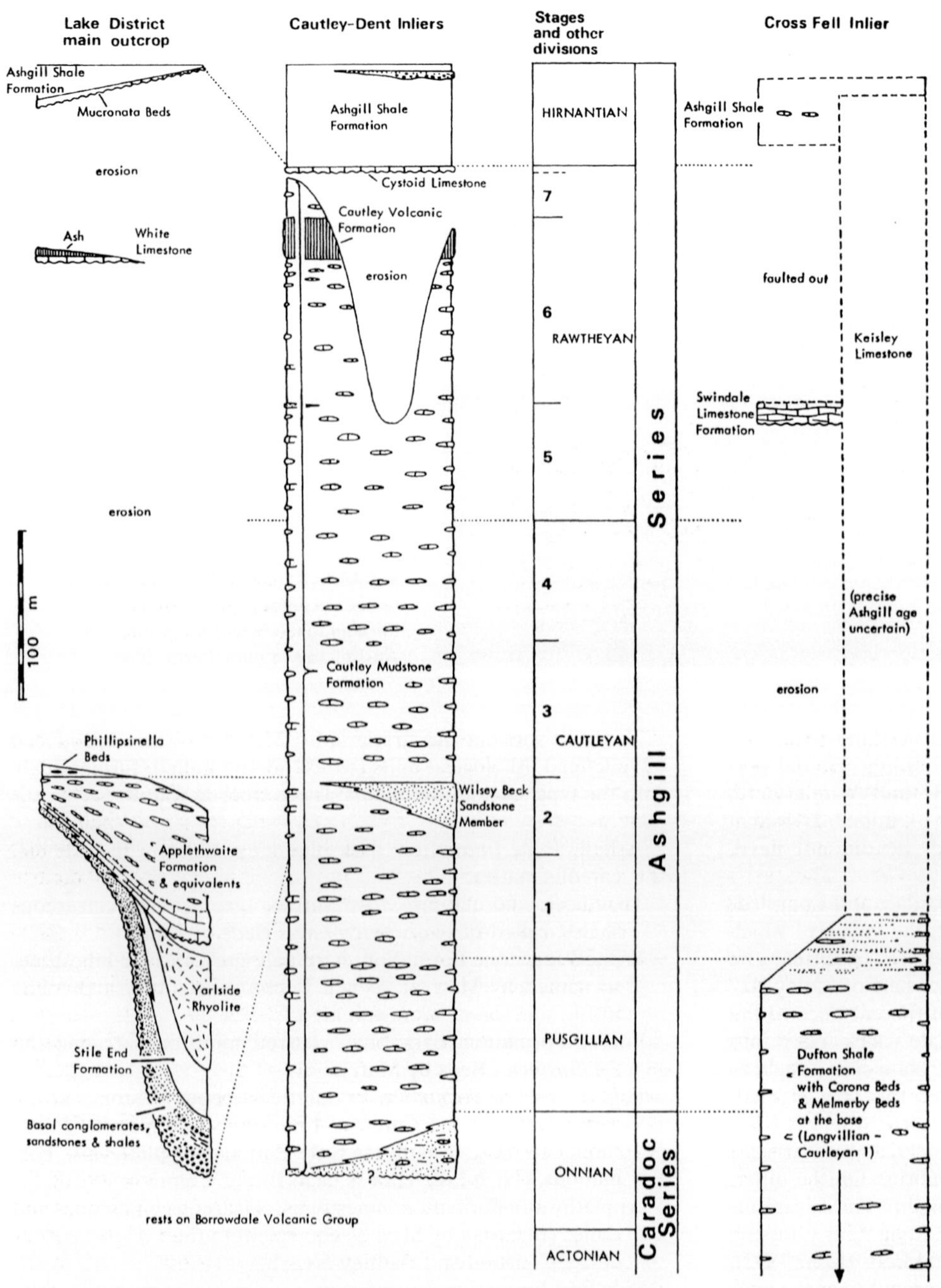

Fig. 13.6a. Correlation of Lakeland Ordovician succession with Howgills, Cross Fell, and standard British succession, according to Ingham, McNamara & Rickards (1978, p. 124). Reproduced by courtesy of the Yorkshire Geological Society.

Really, this meant no more than that it contained some hypotheses, not fully proven. However, according to McNamara's recollection, the examiners (Whittington and Keith Ingham from Glasgow) thought he was flippant, and subjected him to a rather heavy eight-hour grilling, in which questions of taxonomy and grammar were the principal foci of attention, whereas McNamara thought the main issues were his revised stratigraphy and his ideas about the relationship between trilobite anatomy and conditions of deposition. Anyway, he duly got his degree, and he and Ingham subsequently collaborated on Lakeland papers.

The proposed 'geofantasy' was described in the paper by Ingham, McNamara & Rickards (in Moseley 1978*a*) and in McNamara (1979*a*) (see Fig. 13.6). It will be seen that so far as the Lake District *per se* was concerned, the bottom of the old Coniston Limestone Group (Series) was correlated with Zone 2 of the Ashgill Series (Cautleyan) in the Howgills; so all Dean's Caradoc rocks, apart from the Drygill Shales, were banished from the Lake District proper, and they have not been readmitted subsequently.

Perhaps McNamara's boldest 'fantasm' (which others apparently shared) was his attempted palaeogeographic reconstructions, as written up in his thesis and in Moseley (1978*a*), for example, which, however, developed ideas earlier proposed by Ingham & Rickards (1974). The proposal was based on the limited and scattered exposures of Upper Ordovician rocks: at Drygill; in the Cross Fell Inlier; in the Howgills; and along the general outcrop of the Ashgill at the southern margin of the BVG. Ingham, McNamara and Rickards had the idea of there having been a horst[14] in the general area of southern Lakeland, paralleling the outcrop of the Coniston Limestone, onto which, during the late Ordovician and Silurian, there had been, first, a marine transgression from the north and southeast, from Longvillian to Cautleyan time (see Fig. 13.7a); then a retreat of the sea during the Rawtheyan (see Fig. 13.7b); and then a transgression that covered the horst during the Lower Llandovery (*atavus* time), with deposition of the Stockdale Shales. Higher Silurian sedimentation then followed.

Sediment was supposedly supplied (according to McNamara) from rivers running off the horst. The 'fantasised' (hypothesized!) volcano shown in Figure 13.7b was suggested as having been

[14] But the horst idea was actually due to Ingham and Rickards.

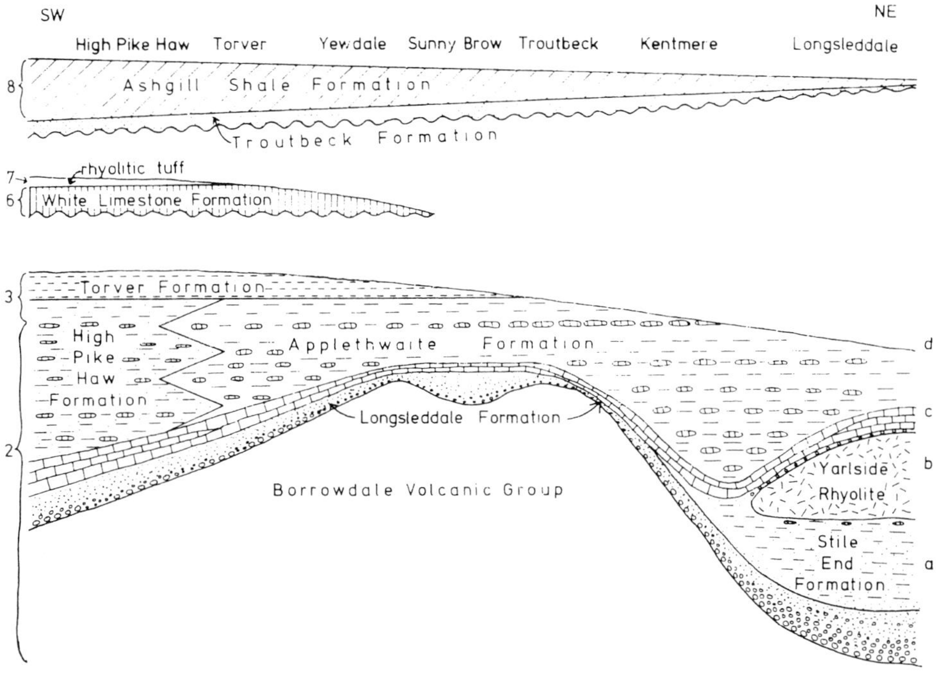

Fig. 13.6b. Suggested overlap of Borrowdale Volcanics by Coniston Limestone Group, according to McNamara (1979*a*, p. 61). © John Wiley & Sons Ltd. Reproduced by permission.

responsible for the production of the Stockdale (Yarlside) Rhyolite, and its production as a large mass was thought to have caused a shift to the SW of the river-mouth through which sediment was transported from the horst into the Upper Ordovician sea (McNamara 1979*a*, p. 62). Thus emplacement of the Rhyolite could have been responsible for some of the facies changes that McNamara observed along the outcrop of the Coniston Limestone.

All this was suggested without recourse to plate-tectonic theory.[15] It does not quite 'fit', so far as I can see, for it is odd that there should have been a horst at the postulated locality, given that we know that there were huge volcanoes active to the north during the Ordovician, the remnants of which might still have been standing high in the areas where Ingham, McNamara and Rickards had their Caradoc and Ashgill deposited (see Fig. 13.6). Also, when Rickards eventually had the horst submerged during *atavus* times (Rhuddian Stage, Lower Llandovery) the sediment deposited was seemingly graptolitic mudstone (Rickards 1978, p. 133) – perhaps not what one might expect on the top of a sea-mount?

Above the Coldwell Beds in 'Otley III' come the Coniston Grits; the Bannisdale Slates; and the Kirkby Moor Flags, which today are regarded as forming the top of the Silurian succession in the Lake District (Ludlow to Prídolí), though Shaw (1971) (see p. 182) suggested a higher unit, the Scout Hill Flags, running into the Devonian. The Coniston Grits and the Bannisdale Slates are rather monotonous and relatively unfossiliferous units compared with the strata below them, and for a long time they attracted little attention, compared with the older units discussed above, though in the last twenty years they have been carefully remapped and have been the focus of important discussions amongst sedimentologists and structural theorists, and those interested in the general tectonic history of the Lakes. Indeed, as we shall see in Chapter 17, they have attracted considerable recent interest and the origin of the basin in which they accumulated was still controversial at the end of the twentieth century.

The extensively cleaved Bannisdale Slates are structurally interesting. They form a major feature for southern Lakeland, the Bannisdale syncline, which runs NE–SW, parallel to and south of the important boundary between 'Otley II' and 'Otley III'. There is downfaulting associated with this syncline so that Kirkby Moor Flags – early on correctly recognized by Sedgwick (1846, p. 118; 1852*b*, p. 141) as being at the top of the Lakeland Silurian succession – crop out to the north of Staveley, between Kendal and Windermere, whereas the main outcrop is further south, to the east of Kendal and near Kirkby Lonsdale (see Fig. 1.1).[16] South of the Bannisdale syncline, and running parallel with it, mapping reveals the complex Crook Synclinorium and then the Underbarrow Anticlinorium, on the SW-dipping southern slope of which the Kirkby Moor Flags crop out.

With a rich benthic fauna, the Kirkby Moor Flags are more fossiliferous than the preceding Coniston Grits and Bannisdale Slates and have characteristic dewatering structures. They attracted the attention of a PhD student at Newcastle University, Richard William Leonard Shaw (1969), and his results were published two years later (Shaw 1971). His work was supervised by John Shirley.[17] Shaw thought that the older geologists such as

[15] Indeed, McNamara recalled (pers. comm., 2000) that he had little interest in applying the model in his doctoral work, and that it received little emphasis when he was at Aberdeen, or indeed among his particular circle at Cambridge.

[16] At the end of the twentieth century there was no type section for the Kirkby Moor Flags and the origin of the term is obscure. A Kirkby Moor is marked on modern maps near Ulverston, but far away from any exposures of Kirkby Moor Flags, which were first recognized in SE Lakeland, either by Otley or Sedgwick. Kneller *et al.* (1994, p. 240) have given a grid reference (NY 4645 9933) for the type locality, but this does not appear to be historically correct. Their spot was near Staveley, where the Kirkby Moor Formation is indeed well exposed, but there is no locality called Kirkby Moor in the vicinity. Soper (pers. comm., 2000) suggested to me that the high ground to the NW of Kirkby Lonsdale might formerly have been called Kirkby Moor; but there is no such indication on the nineteenth-century Ordnance Survey map. It would agree, however, with Sedgwick's (1846, p. 118) statement that the 'well-defined upper group' lay in the hills between Kendal and Kirkby Lonsdale, and with some vague statements he made as to localities in his third letter to Wordsworth (Sedgwick 1842). Soper also stated that he and Woodcock would probably define a new type section for the unit in 2001 when they revised the upper Windermere stratigraphy for the Kendal and Kirkby Lonsdale maps. They were thinking of a quarry site near where the Kendal–Sedbergh road crosses the M6. However, the cancellation of fieldwork in the area in 2001 due to the outbreaks of foot-and-mouth disease in 2001 delayed that determination.

[17] Shaw's first degree was also at Newcastle, but following the closure of the University's Geology Department, little record of him survives there, and I have been unable to make contact with him. I thank Moira Bent of the Robinson Library, Newcastle, for providing me with a copy of parts of Shaw's thesis.

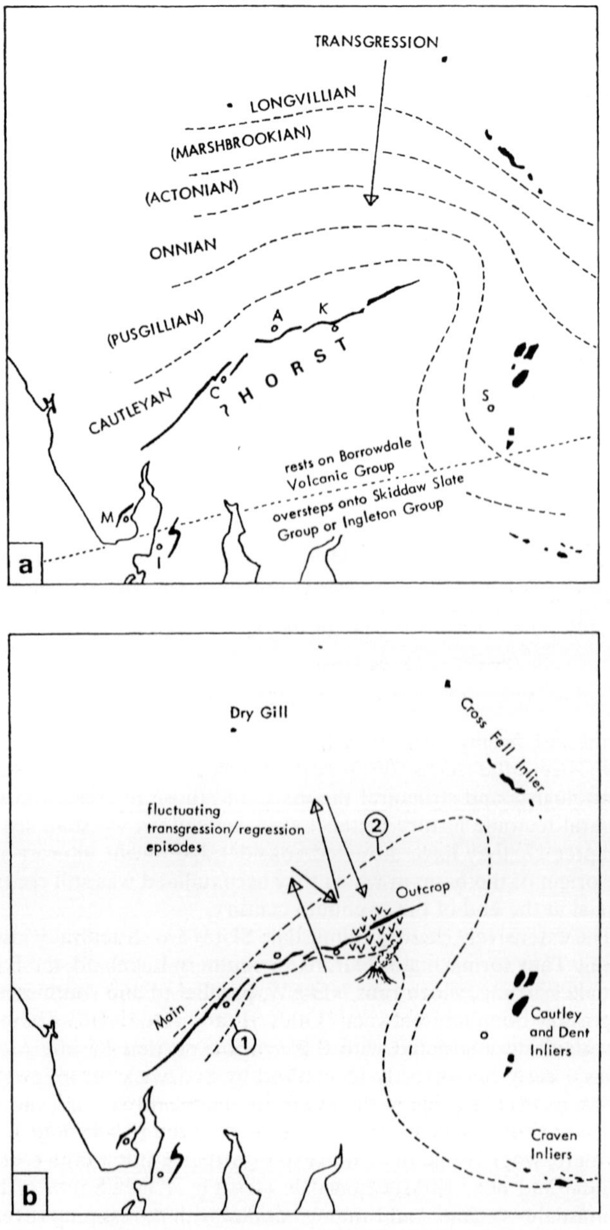

Fig. 13.7. Transgression of hypothetical South Lakeland horst during the Upper Ordovician (Caradoc to Ashgill), according to Ingham, McNamara & Rickards (1978, p. 122). Reproduced by courtesy of the Yorkshire Geological Society.

Murchison (1872, p. 136) had been correct in attempting some subdivision of the upper part of the Ludlow, identifying 'Passage Beds' in the Welsh Border country from the marine Ludlow into the non-marine Old Red Sandstone.[18] The first Survey had envisaged the correlation of the upper part of the Kirkby Moor Flags with the Welsh Border Passage Beds, but the distinction had not been sustained by twentieth-century authors such as Marr (1916) or Mitchell (1956). Now, basing his arguments chiefly on the evidence of ostracods, Shaw supposed that sedimentation continued through into the Downtonian, or the lowest stage of the Devonian, in the Scout Hill Flags above the Kirkby Moor Flags.[19] The Scout Hill Flags had some red beds, perhaps suggestive of early Devonian. Shaw also differentiated so-called Underbarrow Flags below the Kirkby Moor Flags. The sediments appeared to him to be derived from the NW.

Shaw (1971, p. 360) envisaged the following stratigraphic scheme, the four units being distinguished on faunal grounds:

System	*Stage*	*Unit*
Devonian	Downtonian	Scout Hill Flags
	Whitcliffian	Kirkby Moor Flags
		Upper Underbarrow Flags
Silurian	Upper Leintwardinian	Lower Underbarrow Flags

His evidence for carrying 'Otley III' up into the Devonian was not, however, based on much *positive* evidence. Fossil species were mostly disappearing rather than appearing and the argument hung chiefly on the appearance of a single new kind of ostracod: *Frostiella groevalliana.* Rickards (1978) accepted Shaw's suggestions, but the Scout Hill Flags have subsequently disappeared from the literature and from the Devonian (e.g. Lawrence *et al.* 1986). As said, the Kirkby Moor Flags are regarded as belonging to the top of the Silurian, namely the Přídolí.[20]

On the basis of his faunal evidence, Shaw regarded the base of the Kirkby Moor Flags as diachronous, with progradation *from the NW to the SE.* This argument will later be seen to be of considerable and controversial significance (see Chapter 17). It may be remarked here that Shaw's subdivisions do not show up satisfactorily in the field as separate, mappable lithological units, so that one might wish to omit the palaeontologically defined Underbarrow Flags as a distinct mappable unit and simply map the 'visible' boundary between the Bannisdale Slates and the Kirkby Moor Flags. Soper informed me (pers. comm., 2000) that he was contemplating doing just this when he remapped the relevant area of the Kendal sheet with Nigel Woodcock in the twenty-first century.

I have entitled this Chapter 'Otley III' because the designation of the unit and its subdivisions was by no means settled during the period that we have been discussing, from the 1930s through to the 1980s, when the detailed remapping of the Lake District began. This situation was unsatisfactory to geologists and it was Moseley (1984) who took the lead in trying to get some consistency into the terminology, according to the proposals of a Geological Society Report (Holland *et al.* 1978), and following discussions with the Survey staff who were then beginning their work in the Lakes. As Moseley pointed out, what I have been calling 'Otley I', 'II' and 'III' represented very different stratigraphic problems. The Skiddaws are complexly folded sediments, with an indifferent fossil record. The Borrowdale Volcanics represent a great variety of volcanic products, emitted from different vents at different times; or the same volcano might simultaneously pour out lava on different sides during a single eruption, which would not be physically linked in the deposits; or contemporaneous volcanoes might yield quite different products. Ashfalls might be produced at much the same time as lavas from the same volcano. There was naturally a plethora of local sequences and lithological units in 'Otley II'.

In 'Otley III' there are again many lithological units, and there was frequent renaming and subdivision and refinement (see Fig. 13.8). Running in 'parallel' with such changes, there have been modifications and refinements of the fossil zones (as also in the SS). However, inconveniently, 'Otley III' straddled the Ordovician–Silurian boundary – which is unsurprising, since that boundary was not initially established in the Lakes.

With all these complications, there have inevitably been terminological inconsistencies in the earlier literature. Marr (for example) referred to the Borrowdale Volcanic *Series* and Oliver referred to the Airy's Bridge *Group*. But, following the Geological Society's recommendations (Holland *et al.* 1978), the term 'Series'

[18] The evidence for such beds was based on observations made near Ledbury by W. S. Symonds when the Hereford–Worcester railway was under construction.

[19] Scout Hill (NY 3563 4827) is situated to the north of the A65 road between Kendal and Kirkby Lonsdale (see Fig. 1.1).

[20] This stage name derives from one of Barrande's classic sections in Bohemia, where the Silurian–Devonian boundary is firmly established (Bassett 1985).

	SERIES		FORMATIONS	MEMBERS
	DOWNTON	WINDERMERE GROUP	KENDAL FORMATION	SCOUT HILL FLAGS
SILURIAN	LUDLOW			KIRKBY MOOR FLAGS UNDERBARROW FLAGS
			BANNISDALE SLATE FORMATION	
			CONISTON GRIT FORMATION	YEWBANK SANDSTONE SHEERBATE FLAGS POOLSCAR SANDSTONE
			COLDWELL FORMATION	UPPER COLDWELL FLAGS (= HORTON FLAGS)
	WENLOCK			MIDDLE COLDWELL MUDSTONE (= STENNERLY MUDSTONE) GAWTHWAITE SANDSTONE
			BRATHAY FLAGS FORMATION	
	LLANDOVERY		STOCKDALE SHALES FORMATION	BROWGILL SKELGILL
ORDOVICIAN	ASHGILL		CONISTON LIMESTONE FORMATION	ASHGILL SHALE TROUTBECK (*Phacops mucronatus* Beds) WHITE LIMESTONE [INCLUDES RHYOLITIC TUFF] TORVER (*Phillipsinella* Beds) HIGH PIKE HAW (WEST), APPLETHWAITE (CENTRE & EAST) (*Calymene* Beds) STILE END [INCLUDES YARLSIDE RHYOLITE] LONGSLEDDALE

Fig. 13.8. Subdivisions of the Windermere Group, according to Moseley (1984, p. 245). © John Wiley & Sons Ltd. Reproduced by permission.

was to be reserved for chronostratigraphic units, such as Ashgill, while the term 'Group' was to serve for large lithostratigraphic units. Thus 'Otley II' became the Borrowdale Volcanic *Group* (terminology which I have already had occasion to use), made up of *formations* such as the Airy's Bridge *Formation*. Likewise, 'Otley III' should be a group, and Moseley (1984, p. 244) helpfully coined the name 'Windermere *Group*' for this purpose.[21] The lithostratigraphic units below the level of formations were designated *members*. Hence Moseley proposed the taxonomy above for 'Otley III' (Fig. 13.8), and we shall follow his general lead henceforth, though there have been numerous subsequent modifications, as we shall shortly see.

Early on in the resurvey it was decided, at the initiative of one of the university participants in the collaboration, Ben Kneller (see p. 231), to drop lithological terms from Moseley's 'members', which were henceforth designated 'formations'. Thus, for example, the Kirkby Moor Flags were renamed the Kirkby Moor Formation. Subordinate units *were* retained as members. For example, the lowest unit of the 'Stile End *Formation*' became the 'Longsleddale *Member*'. Associated with this change, the main Lakeland groups were designated the 'Skiddaw Group', the 'Borrowdale Volcanic Group' and the 'Windermere Group'. The main changes over the years in the nomenclature of the Windermere Group were conveniently summarized by Scott & Kneller (1990, facing p. 2) (see Fig. 13.9, noting that numerous 'members' were also proposed in the memoir as subdivisions of the formations). So during the course of the course of the revision, a considerable number of new lithological units were introduced, generally with good reason. For example, when Marr divided the Coldwell Beds into Lower, Middle and Upper units he yoked together with one main name entities that were lithologically rather different. So the introduction of the Wray Castle Formation and the Birks Riggs Formation had justification; though not all Kneller's formations within the Coniston Subgroup (for example) were distinguishable right across the outcrop of the Windermere Group.

Only four years after the foregoing arrangements were published, the stratigraphic subdivisions of 'Otley III' were revised, using a 'supergroup' and 'groups' (rather than a 'group' and 'subgroups'), as can be seen in Figure 13.10. We notice that the Windermere Supergroup had now got out of kilter with the Skiddaw Group and the Borrowdale Volcanic Group, upsetting the old simplicities of 'Otley I', 'II' and 'III'. In consequence, as it seems to me, we have an uncomfortable taxonomic scheme for the Lakeland rocks. Kneller's proposals, already published as an internal BGS Report (Scott & Kneller 1990), were discussed at a meeting of Lower Palaeozoic geologists at the Survey headquarters in 1991 and the scheme was discontinued so far as Survey maps and memoirs were concerned at the behest of Survey officer Dr Peter Allen (see Chapter 14 and p. 189), who wanted the same principles to be applied in the Silurian as had been adopted by the Survey in the Jurassic strata of Dorset. As Allen has explained to me (pers. comm., 2000), if the conventions of the North American Code were applied to the cases of the Corallian in Dorset and the lower units of 'Otley III' in the Lakes, one found oneself dealing with formations that, because of their thinness, could not be represented satisfactorily on a 1:50 000 map. Consequently, some had to be 'lumped together' (my wording). So even though, overall, they were quite thin, the 'composite' but mappable Stockdale and Dent units were defined as groups, whereas they would have had to have been subgroups if the Windermeres as a whole formed a

[21] Back in the early days of Lakeland geology, Daniel Sharpe (1842, 1843*a*) referred to the Windermere Rocks as a new name to replace Marshall's (1840) Blawith Slate. This old term was not so very different in its meaning (i.e. from the top of the Coniston Limestone to the top of the Bannisdale Slates) from that proposed by Moseley. However, Sharpe differentiated 'Ludlow Rocks' (approximately equivalent to what were later called the Kirkby Moor Flags) from his underlying Windermere Rocks. Moseley's Windermere Group, by contrast, was essentially the same as Otley III, and ran from the bottom of the 'Coniston Limestone' (as initially understood) to the top of the Kirkby Moor Flags.

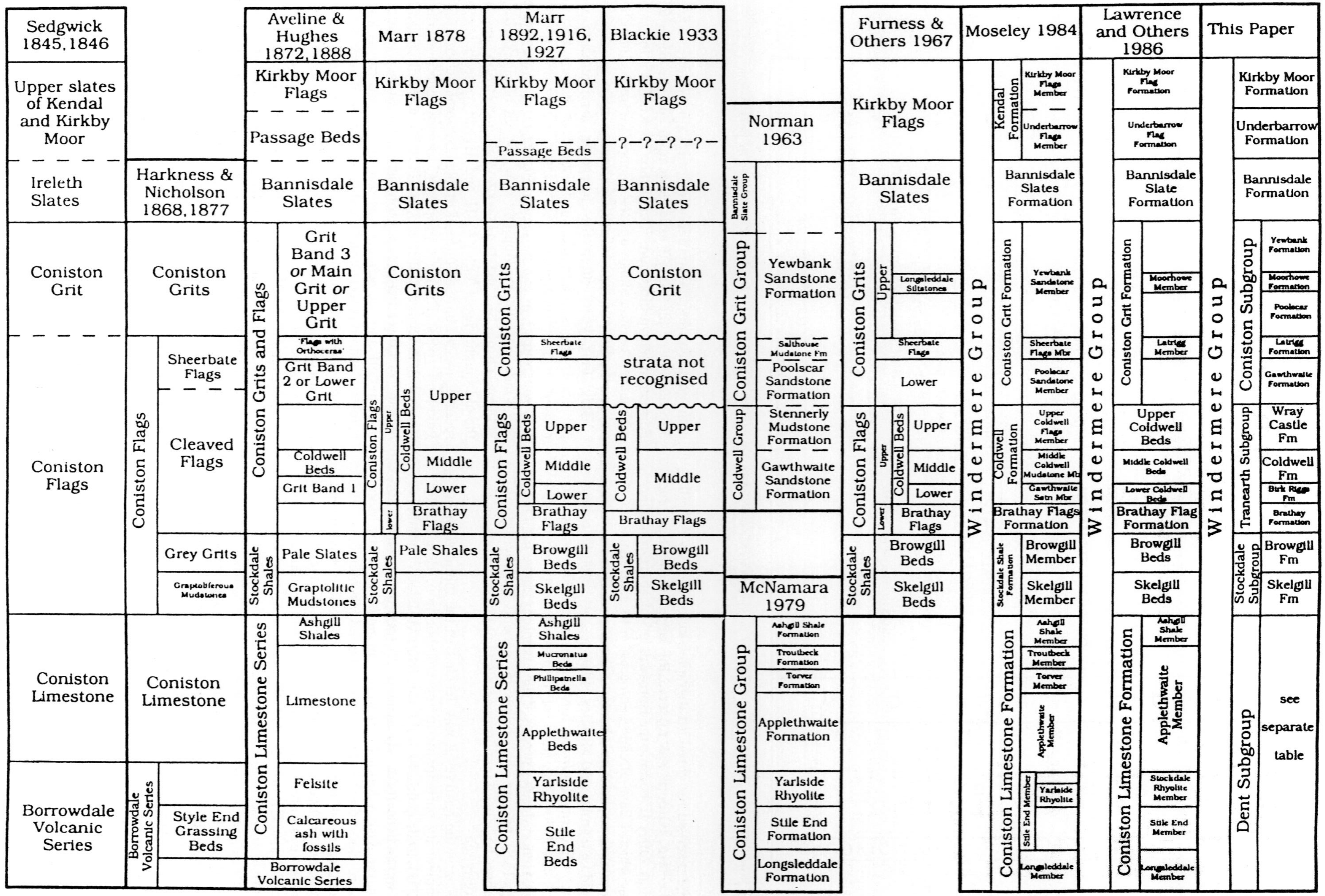

Fig. 13.9. Historical summary of the subdivisions of 'Otley III', according to Scott & Kneller (1990, facing p. 2). IPR/23-7C British Geological Survey.

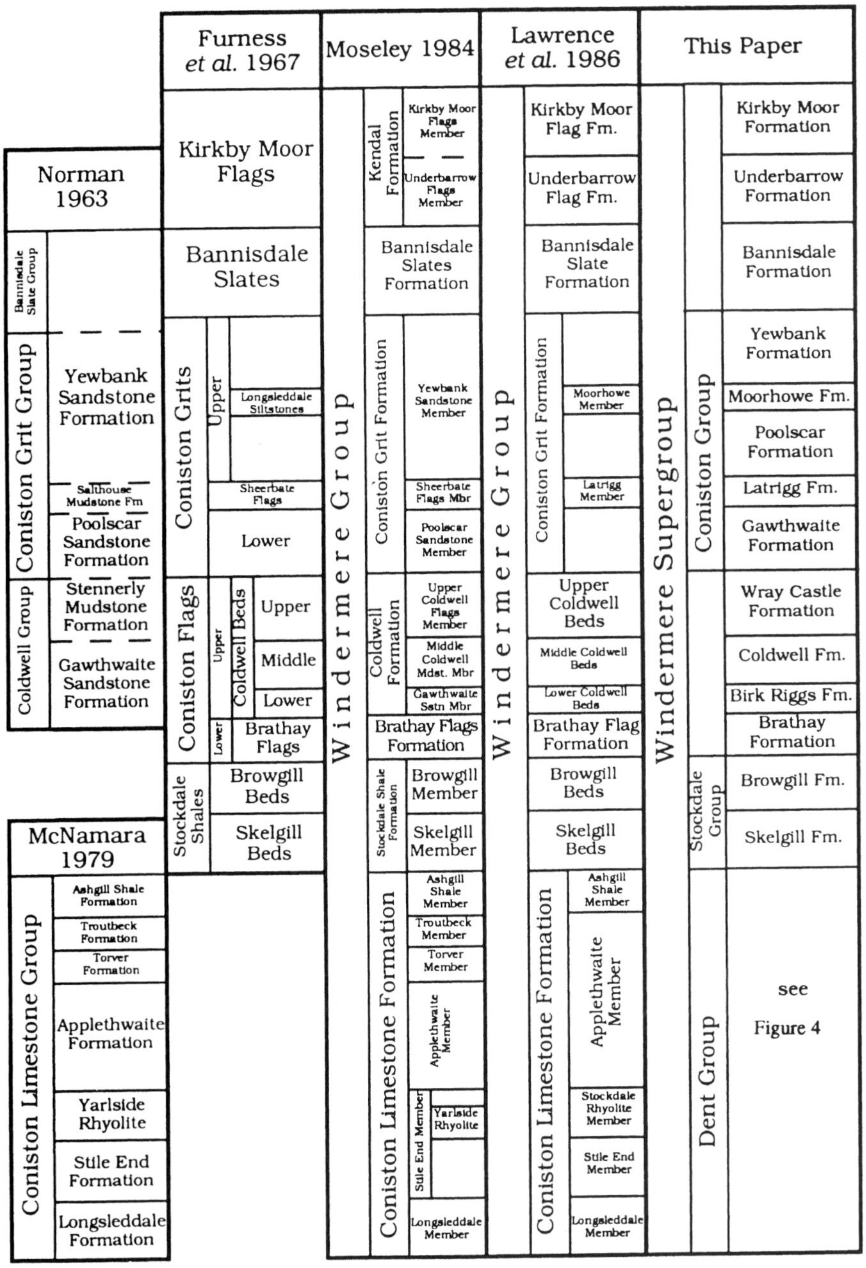

Fig. 13.10. Subdivision of Windermere Supergroup, according to Kneller *et al.* (1994, p. 223). IPR/23-7C British Geological Survey. © NERC. All rights reserved.

'group', as Kneller thought right. The Windermeres were, in fact, much 'smaller' than the supergroups used in North American stratigraphy. And Allen did not want subgroups. The discussions were influenced by Woodcock's (1990) definitions of Dyfyd, Gwynedd, and Powys supergroups in Wales, which corresponded approximately with 'Otley I', 'II' and 'III' in the Lake District.

As I have been informed, the changes in classification between 1990 and 1994 involved rather heated discussions between representatives from the Survey and from universities, who were supposed to be engaged in a collaborative revision of Lakeland geology, as we shall discuss in Chapter 14. Some of Kneller's earlier ideas (such as the introduction of a Tranearth Subgroup;[22] see Figs 5.2 and 13.10) were already incorporated into the Geological Society's *Revised Correlation of the Silurian Rocks in the British Isles* (Cocks *et al.* 1992, p. 13) and thus had precedence in the literature.[23] Nevertheless, they were dropped in 1994. Things seemed to have settled down by the end of the twentieth century, and geologists were growing, or had grown, accustomed to the Windermere *Supergroup*. The issue is interesting, however, in that it illustrates the existence of three alternative, usually collaborating, but occasionally contending, centres of authority in questions of British stratigraphy: the Survey, the Geological Society, and university geologists. Perhaps unsurprisingly, the desires of the Survey prevailed, for after all they had always been the guardians and producers of the official maps – though Allen thought in retrospect (pers. comm., 2000) that 'Windermere Group' might have been better after all. Certainly some of the amateur geologists I met in Cumbria seemed to be a bit puzzled about all the fuss, and wondered why they suddenly had to get used to the idea of a supergroup.

In fact, compromises were made: Soper and his Sheffield team, who were participating in the BGS–university Lakeland Project (see Chapter 14), conceded on the question of subgroups, thereby permitting the birth of the Windermere Supergroup, even if it had a painful delivery. Reciprocally, Allen accepted a number of the new stratigraphic names proposed in Scott & Kneller (1990), and

[22] Tranearth is a farm near Birk Riggs, close to Torver Beck (see Fig. 5.2).

[23] The Society's correlation table did not follow Scott & Kneller (1990) exactly. It put the Wray Castle Formation (named after a castle on the NW side of Lake Windermere) in the Coniston, not the Tranearth, Subgroup.

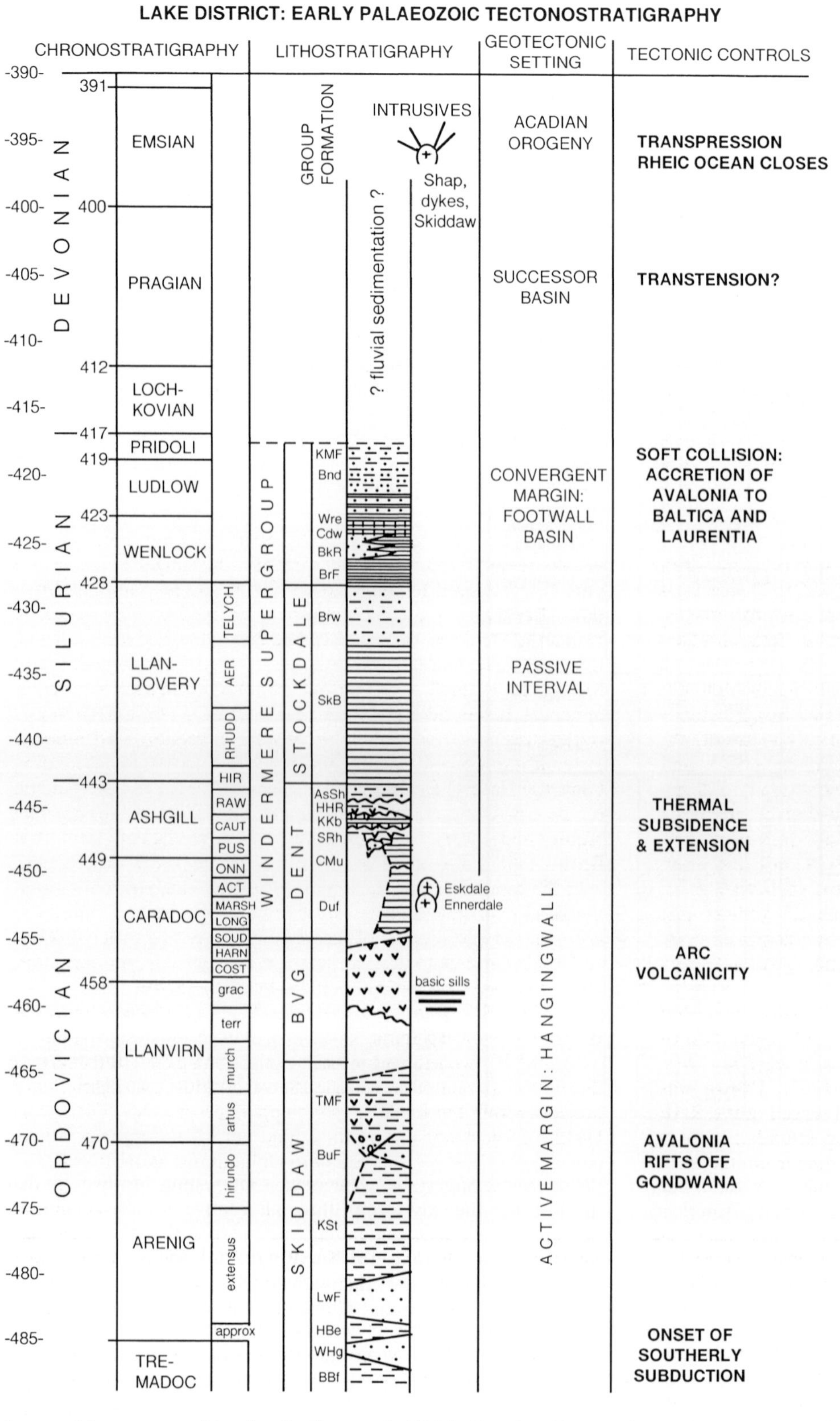

Fig. 13.11. Scheme for Lake District tectonostratigraphy (Soper n.d. [2000]). Reproduced by courtesy of N. J. Soper. (This diagram is a 'forerunner' of one published in the Survey's Ulverston Memoir: Johnson *et al.* 2001, p. 78.)

he put his name to them in Kneller *et al.* (1994). But Kneller's 'Tranearth Subgroup' was refused, as Allen thought its proposed constituent formations were too disparate in their lithologies to warrant a single designation (though in fact they are largely laminated hemipelagites, no more disparate than the constituent formations in the Borrowdale Volcanics Group). Thus we have the rather curious blanks in the table for the Windermere Supergroup in Kneller *et al.* (1994).

The latest scheme for the subdivision of the Lakeland rocks that I saw in my researches for the present book was one compiled by Soper, unpublished at the time of writing (2000), but which he kindly allowed me to reproduce (see Fig. 13.11). It is convenient to place it here, since it provides a useful conspectus of the stratigraphy of the Lake District as a whole, as viewed at the end of the twentieth century, and also relates this stratigraphy to the theoretical processes that may have given rise to the rocks' formation – processes that will be discussed in subsequent chapters.

We note, of course, that Soper's table also has a missing group, but he informed me (pers. comm., 2000) that he would be happy to have it filled by a 'Tranearth Group'. This would be tidy, though it would alter the meaning of the name, as introduced by Scott & Kneller (1990). My feeling, then, is that even at the beginning of the twenty-first century, the taxonomy for 'Otley III', or indeed for Lakeland rocks as a whole, had not fully gelled. But making stratigraphic subdivisions gel is the work of geologists, not historians.

Chapter 14
The great collaboration

At the beginning of the 1980s there were no modern 1: 50 000 or inch-to-the-mile maps available from the British Geological Survey (at that time named the Institute of Geological Sciences) for the central Lake District, and a good many other areas of Britain too. For the main body of the Lakes, the *only* official maps were those of Clifton Ward and his colleagues, which were regarded as outmoded not long after they were issued, and were not widely distributed, for each map was individually hand-coloured. These old maps did not even use Lapworth's Ordovician. For the dates of issue, see p. 40.

Understandably, the paucity of maps was regarded in some quarters as deplorable. For example, Glasgow Professor Bernard Leake, in a submission to the Study Group on UK Geological Surveying dated 5 December 1985, a copy of which he has kindly made available to me, described the state of affairs at that time as 'dismal and a national disgrace', with 'progress derisory in relation to the backlog'.[1] About ninety of the sheet map areas in the UK had last been fully surveyed before 1870, and in some areas of south-central Wales the only 'official' maps went back to 1845! By contrast, France expected to complete its remapping by 1990, having initiated a remapping programme in 1960. The lack of maps was particularly felt by the general public for the Lakes, where so many amateurs wanted to wander the fells and 'see where they were' geologically speaking.[2]

Needless to say, the Lake District is a geologically 'difficult' area, and not one of major importance for mining or agriculture. To remap the Lakes to standards appropriate to late twentieth-century geology would be time-consuming and expensive, and from one point of view such work would not be warranted if one regarded the BGS's remit as being primarily that of providing information for industry and agriculture, and for transport engineering. Such was indeed the view of some BGS staff (e.g. Tony Wadge, pers. comm., 1998) – perhaps rightly so. Other staff, however, were more interested in pure science. Others just wanted to see the whole country mapped as soon as possible, no matter what. Such differences were of long standing in the Survey and will perhaps never be fully resolved.

The Survey's pure–applied pendulum began to swing somewhat towards its pure end when a new Director, Malcolm Brown (1925–1997), was appointed in 1979 (for obituaries see *The Times*, 9 April 1997 and *The Guardian*, 10 April 1997). Brown was another Yorkshireman (from Redcar). He had served in the RAF during World War II, and during his training attended a short cadet course at Durham University, where he came in contact with Lawrence Wager, who interested him in geology. After the War, Brown went back to Durham to read geology and duly acquired a first in 1950. According to Brown's *Times* obituary, Wager regarded him as the best student he ever had. The year he graduated, Brown followed Wager to Oxford to become one of his research students and was put to work on layered igneous rocks (on Rhum). He also collaborated with Wager in his celebrated work on the Skaergaard intrusion in Greenland. Brown was at Princeton in 1954–1955 and then obtained a lectureship back at Oxford. In 1967, he returned to Durham to take the chair vacated by Kingsley Dunham. Brown was highly respected there, becoming Dean of Science and later a Pro-Vice-Chancellor, thus gaining high-level administrative experience. He was elected FRS in 1975.

Brown was thus admirably qualified and appropriate as Survey Director, when he took over the post from Austin Woodland. However, unlike Dunham, who also moved from the Durham chair to the Director's hot seat, Brown had not had prior experience with the Survey, and naturally he brought a number of 'university' ideas with him to his new institution, sought to encourage pure research, and to some extent shifted the institutional directions. Brown's period of office was a strained one. It coincided with the shift of the Survey's headquarters from Exhibition Road, London, to the buildings of a vacated teachers' training college at Keyworth, a small town just south of Nottingham – a great upheaval. It was, additionally, a period when the organization was contracting under increasing financial constraints, with reduced expenditure necessitated by financial cuts to its main funding body, the Natural Environment Research Council (NERC), made by the new Conservative Government.[3] It had to endure a public inquiry and engage in competitive tendering (Hackett 1999); but Brown (1984) was keen to see the basic mapwork accelerated, with the associated publication of sheet memoirs. He reorganized the BGS management structure from a hierarchy to a 'matrix' – not with the full support of the assistant directors (Allen 1997) – and did his utmost to give his personal assistance to younger staff. During Brown's directorship (1979–1985) the Survey changed its name (in 1984) from the Institute of Geological Sciences to the British Geological Survey.

Among Brown's ideas for his organization's new directions was the notion (circulated in 1981) of establishing multidisciplinary projects within the Land Survey mapping programme.[4] The Lake District Project was one of those that got off the ground, others being in North Wales, the Southern Uplands, the Worcester Basin,

[1] According to Leake (pers. comm., 2000), there were many reasons for this, some going back to the loss of staff in World War I, and there was a huge backlog of unpublished Survey material. Sometimes delays occurred when agreement could not be reached between Survey staff members; or the ground rules for a particular map were changed when senior staff positions changed hands. Thus Leake recalls one map that started as a solid and drift edition; then, with the advent of a new senior geologist, it was changed to a solid map; and when a new senior man came in they reverted to a solid and drift map. Both surveyors and drafting staff lost heart! It was said jokingly, but as a wry joke, that new staff would be lucky to see their first mapwork or memoirs published before they retired! The rather frequent movement of senior staff in a period of constant 'restructuring' made things very difficult. (Incidentally, Professor Leake is known affectionately to some of his colleagues as 'the Learned Beak'!)

[2] Even today, delays are producing incongruities. For example, the use of the Llandeilo as a Series was 'officially' discontinued in Fortey *et al.* (1995) and again in Fortey *et al.* (2000). However, it appears in the Ambleside 1:50 000 map (albeit as '? Llandeilo–Caradoc') and this was, in consequence, carried through to the accompanying Memoir (Millward *et al.* 2000), though I understand that it will not be used in any subsequent Lakeland memoirs.

[3] The Government departments were frequently required to engage in competitive tendering for their research contracts, and it was a natural tendency to go for the lowest tenders, which were not necessarily the most technically competent. So, with NERC transferring funds for Earth science from the Survey to universities, the Survey came under pressure and had to reduce its staff from about 1200 to 850. Also, the Survey held a Price Waterhouse review in 1990, which, in accordance with the ethos of the times, recommended that the various groups in the BGS should be, as it were, semi-autonomous businesses. This had the undesirable consequence of generating competition *within* the BGS and the building of 'sub-empires' within the whole. For this and other reasons, the management system was overhauled yet again in the late 1990s, with a return to another 'matrix management' system, rather similar to that which had operated in the 1980s. For a brief account of some of the management travails in the 1990s, see Cook (1998).

[4] There had been earlier multidisciplinary projects in the Survey, but not ones chiefly concerned with basic mapping.

East Anglia, and an engineering geology project in Greece. (Later, another project was established in the Grampians.) These projects were to have secure funding for three years, and were then to be absorbed into the general mapping programme. Eventually, in 1989, the Lake District programme became absorbed into the 15-year Core Programme for mapping, which started in 1990 and was still in place at the end of the century.[5] But it began in 1982 as one of the new multidisciplinary projects.

There was a second element to the Lakeland Project, namely the idea of accelerating the Survey's work with the help of university expertise – or for that matter any expertise or knowledge that might be available in the country. As mentioned, there was dissatisfaction with the Survey's mapping progress, some of which may well have been justified. One of those who best understood the problems was the aforementioned Bernard Leake, who chaired a 'Visiting Group' to the Survey in the 1980s,[6] and encouraged industrialists to write in urging that the completion of mapping for the whole country should be regarded as a top priority. Among various suggestions made to expedite matters, the Group proposed that mapping could be accelerated by collaboration between the Survey and universities that might be interested in collaborating, with joint supervision of PhD students – who would receive expert training in mapping and at the same time find topics that would enable them to make contributions to basic science.[7] Leake also recommended the establishment of a 'monitoring group' to make sure that things were maintained at a good pace, whereas the suggestions of some earlier 'Visiting Groups' had not always been followed through.

The plan was implemented, with a team of PhD students, and some more senior men, assembled by Jack Soper at Sheffield, who was chiefly involved so far as the Lake District was concerned (see Figs 14.4 and 14.5). The plan differed from earlier Survey–university collaborations in that under the new scheme there was to be a Joint BGS–Academic UK Geological Mapping Committee, which would be administrated by NERC, who would let out the contracts (whereas the contracts had previously been made directly between the Survey and universities) and generally keep an eye on progress. This programme continued until 1993. It should be noted that in its recommendations the Visiting Group specifically suggested that those non-BGS personnel engaged in the collaborative projects should be permitted to publish the results emerging from their work in the scientific literature, either during or after the periods of contract. However, they would be required to deposit their notebooks and field maps, or copies thereof, with the BGS (Leake, pers. comm., 2000).

It all seemed a good idea at the time. Students would get worthwhile PhD topics and the Survey would gain access to the theoretical knowledge and practical skills of university staff. The Survey would get additional fieldwork done, in a sense, on the cheap (or it would get its hands indirectly on funding that had been partly withdrawn by NERC, for several PhD students involved in the projects were funded by NERC). Various BGS–university collaborative projects were initiated, for example in North Wales, SW England, and in the Lake District. Some, like the Welsh Project, went well; others, like that in SW England, got into difficulties, while others, as in the Lake District, did succeed but were fraught with problems.

The BGS staff were by no means wholeheartedly enthusiastic about the collaborations. Some thought that the PhD students and their university supervisors would get all the scientifically interesting areas to examine, while the BGS staff would get the mundane work. Perhaps this did happen to some extent (though it was never the intention); but it could cut the other way too. According to the recollection of the former Surveyor Eric Johnson (pers. comm., 2001), the collaboration lacked an ethos of collective understanding and some of the students in some of the projects received insufficient support from their university supervisors. I have heard of a student who was given endless areas of monotonous mudstones in Central Wales to work on, and was unsuccessful in generating a PhD-worthy project there. In the event, the idea of having PhD students on the job was subsequently dropped, and in the Lakeland Project's later phase the work was being done by Survey staff, by university teams under direct contract to the BGS, or by private contractual arrangements. However, so far as the Lake District was concerned the PhD students undoubtedly made significant contributions, and they all got their doctorates.

There were inherent problems involved in the BGS–universitiy collaborations so far as mapping was concerned. A PhD student coming to an area might well wish to erect his or her own stratigraphic arrangement, as Oliver had done in the Central Fells in the 1950s. But this would not do if one were endeavouring to establish a unified and coherent system of classification of strata for the country as a whole, for which all the maps had to 'fit together'. So some real stoushes developed from time to time between university workers and Surveyors on the question of the stratigraphic column to be used as the basis of Lakeland mapping. The problem was exacerbated by the fact that the Surveyors had already got to work in earnest in the Lakes in 1982, and were sometimes displeased when students or university people came in and wanted to make changes to the stratigraphic 'arrangements', no matter how good the arguments for modifications might (or might not) have been. An example of a problem (though not one that caused 'trouble', to my knowledge) was the reconciliation of Soper's BGS–Sheffield group's 1: 50 000 Ulverston map (eventually published in 1997) with the Survey's overlapping 1: 25 000 Dalton-in-Furnace map of 1977, which had been remapped by Surveyors Colin Rose (in his retirement) and Wyndham Evans in the 1970s (largely as a 'desk' revision of the earlier map of Dunham and Rose, without much additional fieldwork) (Johnson, pers. comm., 2001). The stratigraphic subdivisions were substantially different and it is somewhat difficult at first glance to see that the two maps in fact cover the same ground to a considerable extent. This problem was 'seen coming' back in the 1980s.

I have interviewed a number of university and BGS staff who were involved in the Lakeland Project, and among those has been Peter Allen (b. 1940) (see Fig. 14.1), who retired from his position as Head of the Corporate Services and Business Development

[5] Looking back on the period, Jim Briden (pers. comm., 2000) took the view that the greatest influence of 'that period of NERC dirigisme of BGS' was to establish a policy and plan for the completion of UK mapping in 15 years, so that comprehensive rather than selective mapping became the goal. He emphasized that Peter Allen expounded this policy more consistently than anyone else.

[6] Leake (pers. comm., 2000) recalled that his group operated from 1982 to 1984, and it took up about a third of his time for two years. There were nearly forty other specialists who came in to look at particular aspects. Leake interviewed more than half the scientists in the Survey. He found that there was an extraordinary backlog in the writing-up of work, and in some cases items had receded to such an extent that they had got lost sight of entirely. It took someone three months just to compile a list of the items in the backlog, and this excluded those for which there was no one available to complete the work, as a result of retirement or death! Leake described his report (which is not in the public domain) to me as 'hard-hitting but constructive'.

[7] There had been earlier BGS–university collaborations, notably in North Wales, and a Visiting Group in 1978 had recommended that there should be more such collaborative efforts. However, the situation was somewhat different in the 1980s because of the unfavourable financial situation in which the Survey then found itself.

Fig. 14.1. Peter Allen. Photograph courtesy of the British Geological Survey. IPR/23-7C British Geological Survey.

Division of the Survey in 2000.[8] He kindly looked through his old files for me and provided copies of some of the documents that related to the initiation of the collaborative project (Allen, pers. comm., 1998). First came the Survey's own work. In 1981, Brown circulated Survey staff, inviting them to put forward proposals for various collaborative projects, as described above. They were to be *multidisciplinary* projects, so that the full panoply of Survey expertise could be brought to bear on the problem areas: basic mapping, geophysical and geochemical work, palaeontology, sedimentology, tectonic theory, etc. (Of course, all these cannot be neatly separated from one another.) It was Allen who proposed the Lakeland Project in the form that was eventually accepted, though Michael Calver, Assistant Director for Northern England, based at Leeds, had already proposed the extension of mapping into the Lakes in 1980 (memo dated 9 October 1980), stressing the economic significance of work in the transport corridor from England to Scotland, past Kendal and over Shap Fell, occupied by the old London, Midland and Scotland Railway and the present M6 motorway. Calver also mentioned the possibility of gaining information that would be useful in the subsequent search for a site for radioactive waste disposal in the Silurian (*sic*) strata. He had in mind standard routinized mapping that might be undertaken by two new staff who would become available for the work in 1981 – but which would never, I think, have unravelled the complexities of the Lakeland rocks that were eventually revealed.

Two other proposal documents from Allen's files – by Gordon Smart, acting head of the Northern English Field Unit based at Leeds (the first dated 26 March 1981, and the second, undated, apparently a refinement of the first) – seemingly reflected ideas emanating from the Director's office about multidisciplinary approaches. Referring to such topics as the relationship of an accretionary prism (the Skiddaws) to island-arc volcanism (the BVG), the relationship of mineralization to thrusting in the prism and of mineral veins to the large igneous bodies in the Lakes (Firman's old problem), they placed much more stress on theoretical issues than did Calver's proposal. There were to be palaeontological, geophysical, isotopic and petrological studies too, with geothermal investigations and studies relating to nuclear waste disposal sites. Both brief proposals were entitled 'Lake District maps: a major project requiring an interdisciplinary approach'. The word 'interdisciplinary' was surely put there to appeal to Malcolm Brown's ideas. Smart optimistically thought that the project would require 20 man-years. In the event it consumed hundreds of man-years, and was not completed by the end of the century. His proposal made no mention of university participation.

Smart's interdisciplinary proposal as such never got off the ground, but in another memorandum (15 April 1981) he suggested continuation of the work on the Penrith sheet to the Appleby sheet, and this was in fact started, though only in a small way, with the palaeontologist Jack Patterson mapping the Carboniferous Limestone in the Vale of Orton in 1981. The Appleby mapping was completed in 2000 but the 1: 50 000 map remained unpublished at the time of writing.

Peter Allen, however, developed a much more ambitious project, copies of two drafts of which he has kindly given me. The first (dated 22 May 1981) envisaged airborne and terrestrial geophysical survey, ground mapping, palaeontological collection and analysis (particularly of acritarchs), geochemical survey, detailed geophysical, geochemical and petrological investigation of the granites (relevant to their potential as 'hot dry rock' energy sources), and studies of the structure of the Lakeland basement rocks. It mentioned possible university collaboration, with students being trained by Survey staff, who would co-supervise the students, along with university staff. The proposed mapping area was concentrated in the eastern Lakes, covering areas such as the Mell Fell Conglomerate, the BVG of Ullswater and Haweswater, the Bampton Inlier of Skiddaw rocks in eastern Lakeland, the Shap area, and the classic Ordovician and Silurian areas in Kentmere, Longsleddale, etc., and south from there. Some special projects already going on in the Lakes, particularly to do with geochemistry, would be utilized. The work was expected to carry through to 1989. The cost estimate for the geological work for 1982 was £163 000, with the aerial survey costing an additional £350 000.

Allen's proposal was not what one might expect in a number of ways, in that it did not envisage looking at the volcanoes of the Central Fells, omitted the main outcrop of the Skiddaws, and ignored the main granitic areas of Skiddaw, Ennerdale and Eskdale. But I understand that he initially had in mind the idea of running a kind of 'section' through the Lakes to get an idea of the main features of the stratigraphy. His proposal also followed the traditional Survey procedure of clearing off one sheet at a time, and writing a memoir about each, and it would conveniently involve a westward extension of mapping already done or being done in the North Pennines and Eden Valley.

[8] Allen (pers. comm., 2000) has provided me with the following details of his career. Born in North Wales at Connah's Quay (a steel town on the Dee Estuary, near Chester), he attended Hawarden Grammar School and read geology at Manchester University. He worked for two years for the Geological Survey of British Guiana and then went to the Research Institute of African Geology at Leeds University to do a PhD on 'The geology of part of an orogenic belt in Sierra Leone', supervised initially by W. Q. Kennedy and then by R. M. Shackleton. Allen joined the BGS in 1967 and was posted to Hong Kong to map Jurassic volcanics. Returning to Britain in 1970, he was assigned to the BGS team in North Wales, where he worked on the Harlech Dome, acquiring experience on British Lower Palaeozoic volcanics. Thence he moved on to mineral exploration in various parts of Britain, before setting up the Lakeland Project in 1982. Much of the detail of the bureaucratic arrangements associated with Survey and the establishment of the Lakeland Project is based on a lengthy document he kindly wrote for me (Allen, pers. comm., 2000).

Clearly Brown thought Allen's initial proposal a bit unsatisfactory, for, as I understand (Allen, pers. comm., 2000), it did not fit in with what Brown had in mind. As said above, Brown brought his university background to the Survey: he wanted problems solved as well as maps produced. (Very likely he was right, for if there are unresolved 'problems' the maps are likely to be deficient.) However, the 'Brown' policy could lead to a rather piecemeal approach to the mapping process, with emphasis on specific areas of interest, rather than arbitrarily defined rectangular sheets. (Thus it subsequently came about that the earliest maps from the Lakeland Project were 1: 25 000 special sheets for areas deemed to be of special importance.)

So Allen prepared an alternative plan with a different approach, which was submitted on 25 November 1981. He now laid emphasis on the interest of the Lakes as exhibiting an accretionary prism and being part of an associated volcanic arc, and the relationship of these to the Iapetus Suture. He pointed out that most of the research done in the Lakes was 'pre-plate tectonics'. The proposed five-year programme would study controls on sedimentation, magmatism and deformation within a volcano-sedimentary pile in terms of modern orogenic theory. It would look at changes in sedimentary facies throughout the succession, deep structure and metallogenesis. Traditional mapping techniques would be used, supplemented by geochemical and geophysical mapping. There would be a concentration on four main areas: Skiddaw rocks from the Derwent Fells NW to Watch Hill; the BVG rocks for the Langdale Fells and Scafell area; the Ordovician and Silurian rocks from Kentmere to Kendal; and the granitic rocks of Eskdale and Ennerdale. These proposed areas were intended to provide a 'cross-section' of the region, though obviously not a continuous one. Seven or eight field staff were to be committed to the Project each year, and there would be several ancillary staff. It was envisaged that six PhD students should be involved, with a budget for the years 1982–1987 of £828 500.

Allen's second proposal was discussed 'internally' by Tony Wadge and Michael Nutt, and Gordon Smith from the Leeds office, who wrote a memorandum on the proposal (15 January 1982) to the Assistant Director. Smith was evidently opposed to the Project. He thought it too vague and too ambitious, and that the proposed areas for special investigation were unsuitable, particularly in that they omitted the area of the Eycott Volcanics. Allen's proposal anticipated that a 'unified stratigraphy' might be established for the BVG, but Smith doubted this was possible. He questioned the relevance of some of the proposed 'sub-projects', and noted overlap in some cases with work being done in universities or elsewhere. Allen's proposed programme of publication was described as a 'pious hope'. His appendix on an assessment of previous Lakeland work was described as 'inaccurate and inadequate in many respects'.

In the event, Smith's comments were discounted, and Allen's Project was given the go-ahead by the Director (memo, 21 December 1981). A budget allocation of £250 000 was made for 1982–1983, with Allen as Project Co-ordinator under Denys Smith as Project Manager, operating out of the office newly opened in Newcastle in 1981, which had Gordon Smart as head. To get together his team, Allen simply rang around among his colleagues at BGS (something that would not have been possible in the 1990s, though I understand that the subsequently re-introduced 'matrix structure' for the BGS management made it easier to assemble multidisciplinary teams when required). The prospect of working in the Lakes was naturally appealing, and it was not difficult to interest people in the Project. It began in April 1982. Looking back on the work, Allen (pers. comm., 2000) considered that the initial 'skipping about' approach – opposed by Smart, and about which he himself held reservations – was unlikely to yield the desired systematic production of maps in the minimum time. In the event, the solving of problems in key areas facilitated more rapid completion of the mapping later on. Nevertheless, after a couple of years the Assistant Directors, who were concerned about the 'skipping around' approach, persuaded Brown that it had flaws so far as BGS practices were concerned, and the mapping reverted to its traditional 'sheet mode'.

During the first year of the Lakeland Project, the following team from the Survey got to work. One of them (David Millward) we have already met, through his Lakeland researches for his PhD. Some of the others we shall meet from time to time in the pages that follow.

Peter Allen	Co-ordinator and mapping of Borrowdale and Skiddaw Groups
Brian Young	Mapping Eskdale Granite
Eric Johnson	Mapping BVG
David Millward	Mapping BVG
Tony Cooper	Mapping Skiddaw Group
David Lawrence	Mapping Ashgill and Silurian
Adrian Rushton	Ordovician macropalaeontology
David White	Silurian macropalaeontology
Stewart Molyneux	Micropalaeontology
Michael Lee	Geophysics
Richard Carruthers	Geophysical feasibility studies
Adrian Walker	Geophysical field surveys
Derek Cooper	Geochemistry
Neil Fortey	Mineralogy and petrology

Later major participants were Barry Webb (SS), Richard Hughes (SS), Brett Beddoe-Stephens (BVG), Michael Petterson (BVG), Philip Stone (SS), and Tony Reedman (BVG and Project Co-ordinator).

Already there was some university collaboration at this early stage, with the Survey using university equipment for remanent magnetism work (Leicester) and for chemical analysis (Nottingham). By 1983, three PhD students were at work, bringing in their own scholarship funding: Neil Mathieson from Sheffield (known informally as 'Big Neil'), working on the Black Combe area; Michael Branney also from Sheffield, working on the BVG around Langdale; and an Iraqi student A. F. al-Jawardi from Newcastle, working on the minor intrusions in the BVG. The Sheffield men were supervised by Soper, but during the course of his degree Branney began to be informally co-supervised by a volcanologist from Liverpool University, Peter Kokelaar. Subsequently, Soper took on other students to work with the Project: Richard Moore, Neil Davis (known informally as 'Little Neil'), Robin Scott and Bill McCaffrey. In 1989, a contract was established between the Survey and Liverpool University to do mapping on the Keswick and Ambleside sheets, involving the rocks of the BVG in the high Central Fells. This group was headed by Kokelaar (see Fig. 14.2), with Brian McConnell (at the time of writing, with the Irish Geological Survey), Ben Kneller (until recently at Leeds University, but at the University of California at Santa Barbara as of 2000), and Michael Branney (now of Leicester University) as post-doctoral assistants.[9] Separate contract work has also been done by Andrew Bell, who was with the Survey for a time, then went freelancing, then was with the

[9] Kokelaar got into geology through his interest in mountaineering. He read geology at Aberystwyth, where he obtained a first in 1971, and went on to do a PhD on the volcanic complex of southern Snowdonia, supervised by John Ray – but he broke off this work for a year to do a Master's degree in geochemistry at Oxford. Returning to Wales, he completed his doctorate in 1977. His research area was also being mapped at that time by the Survey, who took him on, and provided funding for his work, as one of the first students to receive joint training in BGS mapping procedures (from, among others, Gordon Smith) and doctoral research. Kokelaar also formed close contacts with the BGS geologist, Malcolm Howells, and generally found the collaborative work most satisfactory. In 1977–1978, Kokelaar worked for NERC, examining all the Ordovician volcanic sites in Britain. Then he was with the Ulster Polytechnic (subsequently University of Ulster) for 10 years, but moved to Liverpool University in 1988, where he was a Reader in 1998 (Kokelaar, pers. comm., 1998). He was at that time working for the British Government, advising on the Montserrat Volcano.

Fig. 14.2. Peter Kokelaar in his office at Liverpool University. Photographed by author, 1998.

Open University, and at the time of writing was freelancing again from Penrith. Soper, too, has done freelance work for the Survey in the southern Lakes and Howgills; and Rickards and Woodcock from Cambridge have also been associated with the Project and continued to be so at the time of writing (2001). From time to time during the course of the work, Lakeland geologists such as Frank Moseley would pay a visit, as did volcanologists such as Timothy Druitt, John Wright and Colin Wilson.

Several members of the Survey team appear in Figure 14.3. Jack Soper's group is pictured in Figure 14.4.

Fig. 14.3. Members of the Newcastle office of the British Geological Survey, 1991 (BGS Photograph MNS 5615). Back row (left to right): David Lawrence, Michael Petterson, Richard Hughes, Martin Boland, Ian Jackson, Eric Johnson, Brian Porteus, Ruth Ward. Front row (left to right): Tony Cooper, Jo Dunkley, Brian Young, Douglas Fettes (Regional Geologist), Donald Frost, Sandra Clothier, David Millward. Courtesy of the British Geological Survey. IPR/23-7C British Geological Survey.

Fig. 14.4. Members of the Sheffield team for the Lakeland Project. Left to right: Barry Webb, Michael Branney, Jack Soper, Robin Scott, Andrew Bell, William McCaffrey, Benjamin Kneller, Richard Moore. Copy of undated photograph supplied by Dr Soper and reproduced with his permission.

Table 14.1. *Maps issued as products of the Lakeland Project, with dates of earlier issues*[10]

Sheet		*Scale*	*Date of latest issue*	*Date of previous issue*	*Date of first issue*
Sheet 54°N–04°W (Lake District)		*1:250 000*	*1982*	*1982*	*1982*
Sheet 22	Maryport	1:50 000	1995	1930	1892
Sheet 23	Cockermouth	1:50 000	1997	1975	1885
Sheet 24	*Penrith*	*1:50 000*	*1974*	*1974*	*1893*
Sheet 28	Whitehaven	1:50 000	1979[11]	1979	1895
Sheet 29	Keswick	1:50 000	1999	1875	1875
Sheet 30	*Appleby*	*1:50 000*	*1893*	*1893*	*1893*
Sheet 37	Gosforth	1:50 000	1999	1980	1881
Sheet 38	Ambleside	1:50 000	1996	1882	1882
Sheet 39	*Kendal*	*1:50 000*	*1887*	*1887*	*1871*
Sheet 40	Kirkby Stephen	1:50 000	1997 ('Provisional Series')	1972	*1889*
Sheet 48	Ulverston	1:50 000	1997	1877[12]	1877
Sheet 49	*Kirkby Lonsdale*	*1:50 000*	*1892*	*1892*	1869
Sheet 50	*Hawes*	*1:50 000*	*1997*[13] ('Provisional Series')	*1971*	*1889*
–	*Cross Fell Inlier*	*1:25 000*	*1972 ('special sheet')*		
–	*Dalton-in-Furness*	*1:25 000*	*1979*		
–	Kentmere and Crook	1:25 000	1986 (uncoloured)[14]		
–	Lorton and Loweswater	1:25 000	1990		
–	Devoke Water and Ulpha	1:25 000	1991		
–	Black Combe	1:25 000	1998		
–	Scafell (Central Fells)	1:25 000	Reported in 2000 as 'in press'[15]		

As we shall see in chapters that follow, an enormous amount of work emerged from the Lakeland Project. The students got their PhDs. The mapping got done, though less quickly than had been anticipated. Important theories were developed, and numerous publications flowed from the work. The maps listed in Table 14.1 were issued.

It can be seen from Table 14.1 that several of the intended new Lakeland maps remain unpublished. A new Whitehaven map is needed, but is on its way. At the time this book was written, Jack Soper (aged 66 in 2000; see Fig. 14.5) was working on the Kendal sheet, and had been committed to finishing it by March 2001, but there were delays that year because fieldwork ceased due to the farming crisis associated with foot-and-mouth disease. As of 2000, he said he had no idea when the map would eventually appear (Soper, pers. comm., 2000). Work on the Kirkby Lonsdale sheet by Richard Hughes had been due to start in 2001. The mapping for a 1:25 000 map of the Central Fells (Scafell) by Kokelaar's Liverpool group was completed, but was stuck in the pipeline. The revised Whitehaven map was heralded. The timetable for the issue of Lakeland maps fell within the 15-year time-frame for the completion of British maps, starting from 1990 (Allen, pers. comm., 2000), and the deadline of 2005 might yet be met. However, I understand (Michael Lee, pers. comm., 2001) that budget cuts forced a rescheduling of the completion date for the 1:50 000 maps to 2009.[16] The foot-and-mouth outbreak exacerbated the problem.

It is evident that the time-gap between the initiation of the Lakeland Project and the completion and publication of mapping has been considerable, though it must be said that Allen's initial proposal, as approved by Brown, did not contemplate a total mapping of the Lakes. It may be noted that the special Lorton and Devoke Water maps, which appeared relatively early as an outcome of the Lakeland Project, reflect a fast-track status in that Allen regarded these areas as critically important, even at the time of his submissions in 1981. Though Allen hoped to see a full set of such 1: 25 000 maps for the Lake District, such as has been done for much of Snowdonia, this was precluded by financial constraints. Individual copies of the 1: 10 000 maps may be obtained by special order (at great expense!).

In addition to the maps listed above, there should be associated sheet or topic memoirs, but at the time of writing, only a few have appeared. There have been the important *Geology of the Ambleside District* (Millward *et al.* 2000), an impressive *Regional Geochemistry of the Lake District and Adjacent Areas* (BGS 1992), *The Geology of the Irish Sea* (Jackson *et al.* 1995), *Geology of the West Cumbria District* (Akhurst *et al.* 1997), and the 48-page sheet memoir, *Geology of the Keswick District* (Woodhall 2001). The Ulverston Memoir (Johnson *et al.* 2001) also appeared recently. At the time of writing (2001), the Skiddaw Group Memoir was largely completed, and was to be published in 2002; there was talk of a BVG memoir to accompany the Liverpool group's Central Fells map, but no firm commitment to publication. I have also heard mention of an Eycotts memoir, but am not sanguine about its publication.

The geochemical memoir reflects the BGS's strong geochemical interests throughout the Lakeland Project, about which more will be said in due course. The publication of the West Cumbria Memoir resulted, in part, from funding provided by the atomic

[10] Italics denote those maps not produced as an outcome of the BGS–universities collaboration.

[11] A reprint at 1:50 000 of the one-inch map that was originally issued in 1929. A new edition was at proof stage in 2000.

[12] A map of this area was published at the scale of 1: 25 000 in 1977.

[13] This map depicts only a few outcrops of Lakeland-type rocks, and is essentially concerned with geology of the Yorkshire Dales, where, however, there are some Palaeozoic (or perhaps Precambrian) inliers near Ingleton, Austwick, etc. Some geologists disagree with the stratigraphic allocations for these rocks on the Hawes 'provisional' map.

[14] This map appeared in association with a BGS report: Lawrence *et al.* (1986).

[15] The new Penrith, Whitehaven, Appleby, Kendal and Kirkby Lonsdale sheets will, when published, be products of the Lakeland Project.

[16] The revised programme allowed for the provision of more detailed geological information, particularly with regard to the Quaternary, so important for land-use planning, climate modelling, etc. There was also a focus on understanding three-dimensional geology, both shallow and deep, involving interdisciplinary investigations (Michael Lee, pers. comm., 2001).

Fig. 14.5. Jack Soper in the field, at work on the Kendal Sheet. Photographed by author, 1999.

energy industry via Nirex (see Chapter 20), allowing this publication to 'jump the queue' so to speak. There have also been numerous publications emerging from the Lakeland Project appearing in academic journals and in internal BGS reports. In addition, the Survey stores much unpublished information, such as a lexicon of named rock units, and geophysical, geochemical and palaeontological data, which can be consulted on request.

Clearly not all went well with the Lakeland Project, judging by the information provided above. So in the rest of this chapter we shall try to analyse some of the reasons for the delays. On the other hand, it should be emphasized that the Project *will* eventually reach completion, whereas some of the other collaborative projects failed, and the Project has unquestionably led to substantial increases in understanding of British geology.

Perhaps unsurprisingly, one problem facing the Project was money. Though it started with Brown's assurance of a budget for three years there was also a warning that long-term funding could not be guaranteed; and during the period of the Project the Survey was living in an era – still continuing – when it was constantly being asked to find external funding. Eventually, the Survey began to do this in relation to Lakeland geology, through its work for the nuclear industry, but that was chiefly in the 1990s, and had its own problems, as we shall see in Chapter 20.

A second reason was that the 'goalposts kept moving,' as Brian Young (who was in charge of the Newcastle office subsequent to Gordon Smart and then Denys Smith) put it to me (pers. comm., 1998);[17] Phil Stone (see p. 197) said the same (pers. comm., 1998). When Young got into the Lakeland work it was assumed that he would be suitable to work on the western granite outcrops as he had previously worked in Thailand on mineral exploration in a granite terrain. In fact, his Thailand work had been on sedimentary rocks. Nevertheless he worked on the Eskdale Granite for three seasons. The Department of the Environment then began to give more money to applied geology, and so Young was shifted to work on the Cumbrian coalfield. He recalled that he was quite happy with the change as he was getting into a rut with endless examination of granite exposures. The coal work was relatively well funded, whereas by that time the Lakeland Project was getting short of money. So the coal work began to subsidize the Lakeland Project, as there was overlap of ground, even though the questions were often totally different. (Is this the *D. bifidus* zone? How hard is this clay here, and do we need explosives to dig a hole in it?) Anyway, the coal group and the Lakeland group were able to collaborate to get the Cockermouth sheet done. However, that took them over towards the Eycotts and Carrock Fell, and (according to Young's recollection) surprisingly the Department of Environment got rather interested in that and seemed to lose interest in the applied geology! Because of the association with the coal work it was the small Maryport map that was the first 1: 50 000 sheet to be published for the Lakeland Project's efforts (Young, pers. comm., 1998).

Another reason for the slow progress was the fact that in the first instance there was little direct accumulated expertise in the group in relation to Lakeland geology. Amongst the BGS participants, only Millward and Webb had worked there to a significant extent – for their PhDs. There was little historical collective memory between the older generation of geologists such as Mitchell, Firman or even Moseley, and the newly formed group, though from the university side Soper knew the Lakeland rocks well. Moseley and Firman were invited to participate (Allen, pers. comm., 2000), but they decided not to do so. It was, in the view of one of my informants, unfortunate that BGS staff such as Malcolm Howells or Diarmid Campbell who had done such successful volcanological work in North Wales did not participate extensively in the Lakeland Project. Thus the newly established Group did not perhaps draw sufficiently on the knowledge of some of those in the Survey who had experience with similar types of rocks in other places, notably in ancient dissected volcanoes – though Campbell did work in the Lakes for a year in the 1990s.

One should also emphasize the complexity of Lakeland geology. It was a difficult task to work out a coherent stratigraphy for the SS, with their multiple deformations, of uncertain date and cause. The BVG exhibited a bewildering variety of lithologies, to which rather recent advances in volcanology had to be applied. The Windermere rocks, as we shall see in Chapter 17, presented numerous problems, some of which were still not resolved at the theoretical level by the close of the twentieth century.

Another reason for the apparently slow progress was the very thoroughness with which the work was done. Young told me that *every* accessible outcrop was examined – and there are a great many outcrops in the Lakes. (No wonder he got a bit bored with granite!) Even so, the rate of work was different for the Surveyors and the students. In general, the latter covered the ground more slowly, either because of inexperience or because they were working on problems, not areas; and problems could not be solved at a routinized rate. Interestingly, Allen informed me that some of the younger BGS staff were initially a bit overawed by the university staff. (Knowing the people involved, I find this a little difficult to believe, but I have met them some years after the

[17] Young did his geology degree (1965–1968) at Queen Mary College, London, under John Francis Kirkaldy. Immediately after this he was appointed to the Survey, and was promptly sent to Thailand. On his return, he worked at the Edinburgh office and then at Newcastle, from where he was one of the founding members of the Lakeland Project. His special interest has been mineralization and the mineral veins of northern Britain.

events and doubtless they have grown bolder as well as older.) As to the rate of progress, Young recalled that Malcolm Brown was disappointed with the way things developed, and within two years was saying that the Lake District publication record was appalling! But the Surveyors thought that the new Director did not understand how the Survey worked, i.e. with extreme thoroughness. Perhaps they were right!

One problem was that the initial team lacked a specialist volcanologist, needful for the study of the ancient Borrowdale volcanoes. To be sure David Millward had done his PhD in the Lakes and subsequently worked on Tertiary and Recent volcanoes in South America before returning to the Lakes. However, his Lakeland doctoral work had to a considerable extent been geochemical rather than concerned with volcanic processes. Peter Allen, it should be mentioned, had worked on volcanics in four continents before turning to the Lakes. In any case, subsequently, through the contribution of Peter Kokelaar from Liverpool and latterly from the former Project student Michael Branney, at Leicester University since 1996, expertise in volcanology was increased. There has also been a volcanologist, Sue Loughlin, working at the BGS office in Edinburgh, editing the Central Fells map.

Yet another reason for the delay – one that I have already emphasized – was that the goals of the university geologists were not necessarily the same as those of the Survey staff. To get their doctorates, the PhD students needed scientific *problems*. This meant that they did not necessarily want to be confined to the requirements of working on a particular map sheet, covering it all in detail. Their problems might require them to wander over several sheets, and perhaps onto ground that had already been mapped by the Surveyors. The Surveyors, by contrast, were accustomed to work systematically over a given sheet area; and as the Project proceeded, the original problem-oriented approach envisaged by Allen and Brown began to recede and the sheet-oriented approach began to reassert itself.

Then there were problems of personnel and leadership. In 1986, Allen got promoted to head the Exeter regional office, and so he left the Lakeland Project for a while. Charge of the Project passed to Barry Webb, but he was irked by bureaucratic procedures, to the extent that he decided to resign from the Survey in 1990. In fact, Allen did not remain long in Exeter, as the following year he became head of the UK South Land Survey, and returned to Keyworth as an Assistant Director. Then in 1991 he became head of the whole of Land Survey for England and Wales; the Lake District thereby came under his jurisdiction again and he began to do field inspections, etc., in the Lakes. However, in the mean time, personality problems had been breaking out between a few of the BGS staff and some of the university people, and in 1989 Tony Reedman was brought in as a 'troubleshooter' to take charge of the Project (Reedman, pers. comm., 1999). He had, in fact, worked previously with Branney in Korea, and they had a good relationship. This helped, but the change of leadership came after Branney had completed his PhD in 1988. In any case, Reedman was only there for a year, as he was promoted to take charge of BGS International. (This illustrates the problems that have for so many years afflicted BGS mapping: shifting of personnel before the completion of projects.) The Lakeland Project was then divided into two, with Phil Stone in charge of the Lower Palaeozoics and Brian Young in charge of the Upper Palaeozoics, and this arrangement still stood at the time of writing, with direction for the Project coming from the Edinburgh office, following the closure of the Newcastle office in 1994. This made sense, since Stone, though not a volcanologist, had much experience with the rocks of the Southern Uplands, some of which are similar to (but also different from) the SS, which he also mapped.

Another point worth emphasis was the increasing tendency of the Survey (like universities also) to place greater emphasis on administrative or management attributes, with corresponding rewards going to those willing and able to move around. There have therefore been increasing movements of personnel and tasks, with concomitant disruptions of continuity of work, which is obviously a hindrance in the completion of a complex research programme.

Yet another problem, already touched on, had to do with the *source* of the funding. As Allen explained, during the 1980s the funding for the university parts of the intended collaborations came from NERC, but the actual taxpayers' money was taken from BGS and given to NERC, who put it into a fund for contracts, for which the universities could make bids; yet in some cases, with matters now in the hands of the BGS–Academic UK Geological Mapping Committee under the control of Jim Briden at NERC, the BGS was not itself allowed to make independent bids for what would previously have been its own work.[18] Naturally, this led to ill-feeling. In fact, some of the university contracts, awarded under such an arrangement, got into considerable difficulties, as I understand from Allen. This did not happen in the Lakes, despite all the problems referred to above. But funding was a big issue throughout. The costs proved to be far greater than anticipated. Allen and Reedman have independently estimated that the production of the Ambleside sheet alone eventually cost about £2 000 000.

Necessarily related to money was the question of the time available to write up the scientific results of the mapwork – a matter of considerable concern to the young post-doctoral geologists employed on the NERC mapping contracts, such as that of the Liverpool University group. After they had finished their mapping, they had insufficient time to write up their work for formal publication. Their positions were temporary and they had to spend time writing research proposals in order to obtain their next source of funding, rather than getting out their ideas in the form of refereed articles. This state of affairs (the research-proposal treadmill) has become one of the chief bugbears of modern scientific life. It is not surprising that it should have arisen in a situation where 'soft' rather than 'hard' money was involved.

Finance lay behind yet another important aspect of the Lakeland Project. As will be recounted in Chapter 20, in the 1990s the BGS began to do consultative work for the nuclear industry in relation to the building of a repository for nuclear waste in subterranean BVG rocks near the Cumbrian coast. With the expertise in Lakeland geology achieved by then, the Survey was well placed to undertake, such work, which was both of national importance and lucrative, as the nuclear industry never seemed short of money. So a number of BGS Lakeland staff, such as Millward and Beddoe-Stephens, went off to do contract work for the nuclear industry, thus depleting the resources available to complete the Lakeland mapping (though they were working on BGV rocks). In the event, the nuclear repository work was to have a profound effect on the Lakeland work and indeed on the BGS as a whole. But this story must wait until the denouement of Chapter 20.

Finally, there was the question of personalities, which everyone whom I have interviewed regarded as central to the problems involved with the Lakeland Project, and which caused it to stagger so severely. I do not wish to dwell on such matters, but personalities are always relevant in the history of science. Peter Kokelaar told me (pers. comm., 1998) that he had earlier got on well with the BGS geologist Malcolm Howells when collaborating with him in the study of Snowdonian volcanics. But I gather that he was, by

[18] As Director of Earth Sciences for NERC, Briden had responsibility for all NERC money going into Earth sciences other than that for ordinary university research grants and studentships. BGS had to report to him, and he was responsible to the NERC Council for the BGS budget. He did not, however, sit on the BGS–Academic Mapping Committee (Briden, pers. comm., 2000). Before joining NERC, Briden was Professor of Geophysics at Leeds University and was at Oxford at the time of writing.

contrast, dissatisfied with the way things worked out in the Lakes and that there are several strong letters from him to Douglas Fettes, head of the Edinburgh office, complaining about the scientific ability and conduct of the BGS staff (Stone, pers. comm., 1998). (I have not sought permission to see this correspondence.) In fact, various people I have interviewed from the BGS and the university 'camps' have described members of the other camps in unflattering terms. Very likely things could have been different if the university people and the BGS staff had worked together as a unified team (as had happened in the North Wales collaboration). But that was not the way it was in the Lakes. The Surveyors had started first and the different groups covered different ground, with for example the Liverpool team working largely on the Upper Borrowdale Volcanics, while the BGS people focused on the Lower Borrowdales.

The rows had partly to do with differences as to who was to be responsible for what, within the programme of work, and who was to take credit for what, as the arrangements were not formally agreed to in advance. Then there was the fact that the research students sometimes moved into ground previously mapped by BGS people and reinterpreted it, without informing the BGS staff concerned. This so annoyed some of the BGS people that certain students were barred from consulting the Surveyors' field slips. What sort of collaboration was that?[19] (One might conjecture that students might have felt reluctant to inform the Survey if they felt they might be refused permission to reinterpret a senior person's area.) There were also some lesser tensions amongst university workers as to who was to claim credit for what. And finally, there was the ironic situation that it was one of the students, Michael Branney, who (with, it must be pointed out, the advantage of the previous results of Survey mapping) came up with key evidence for the Borrowdale volcanoes having been subaerial, an idea that (with others) revolutionized the interpretation of the BVG. I do not suppose that, when the collaboration started, it was anticipated that a postgraduate student would suggest ideas that would lead to a substantial revision of thinking about the Lakeland volcanoes on the part of the seasoned BGS staff.

All this sounds a bit depressing and unseemly. It was an expensive exercise and consumed much time and energy. However, it is all part of the history of geological research in the Lakes. And as we shall see in the next three chapters, an enormous amount of successful work has been done since the commencement of the Lakeland Project. So let us try to encapsulate it in discussions relating to 'Otley I', 'II' and 'III', in three separate chapters. Some of the matters have, of course, already been broached in earlier chapters.

[19] In fact, such problems have sometimes occurred within the Survey itself. E. B. Bailey, for example, started reinterpreting the Dalradian structures of areas mapped by his colleagues, and there was such tension that the Director barred Bailey from consulting maps not his own! (Leake, pers. comm., 2000.) Bailey left to take the Glasgow Chair, but returned as Survey Director just before World War II.

Chapter 15
Collaborating on the Skiddaw Group

When the Lakeland Project started in 1982, the first Survey men to go into the field among the outcrops of the Skiddaw Slates were Peter Allen, Tony Cooper, Adrian Rushton and Stewart Molyneux, and they were soon joined by Barry Webb, with his previous Lakeland experience. Later, Philip Stone did a good deal of Skiddaw work as he had experience of somewhat analogous rocks in the Southern Uplands, and he is currently (2001) leader of the Lakeland Project.[1] Richard Hughes (see Fig. 14.3), a former student of Barrie Rickards at Cambridge, did a considerable amount of Skiddaw work after he joined the Survey group at the Newcastle office in 1989, and undertook the scientific editing of the Survey's Skiddaw Memoir, currently in press.[2] The surveyor Eric Johnson (see p. 209 and Fig. 14.3) did work on the Black Combe area, as did Andrew Bell from the Open University on independent contract to the Survey,[3] and Soper's Sheffield PhD student Neil Mathieson, co-supervised by Peter Allen, did work in the area too, but more specifically on the volcanic rocks.[4] Co-supervised by Soper and Cooper, Richard Moore from Leeds University did important work for his PhD (1992), which influenced the interpretations of the Skiddaw Group in northern Lakeland. However, broadly speaking, the input from university geologists to the Lakeland Project has been less important for 'Otley I' than for 'Otley II' and 'III'.

Cooper joined the Survey in 1975.[5] Having worked on Ordovician rocks for his PhD, he expected to be put to work somewhere in the Lower Palaeozoics, but, as was the custom with the BGS, he was required to acquire a general experience of geology and so started his training on Carboniferous and Quaternary strata and then went mapping the Permian and Triassic in the Vale of York, working from the Leeds office (Cooper, pers. comm., 1998). Nevertheless, it was unsurprising that Allen should have invited Cooper, with his 'Ordovician background', to join the Lakeland Project, and in consequence he joined the team and moved to the Newcastle office (see Fig. 14.3).

The now-retired Survey palaeontologist Adrian Rushton (b. 1937) (see Fig. 15.1) was at school at Oundle and read geology at Cambridge, where he had been brought up. Supervised by Bulman, he wrote his PhD on the Cambrian rocks and fossils of the Nuneaton area in the Midlands (where Lapworth had done important work). After appointment in the Survey, Rushton worked in many Lower Palaeozoic areas, including the Southern Uplands, where he had to become well acquainted with their graptolites, making comparisons with the overseas literature (Australasia, Scandinavia, Newfoundland) (Rushton, pers. comm., 1998). His knowledge of graptolites made him a natural choice for an invitation to join the Lakeland Project.

The palynologist Stewart Molyneux (see Fig. 15.2) was also a natural choice.[6] He was recruited to the Survey to study Jurassic microfossils in relation to North Sea oil, but then he returned to the study of Lower Palaeozoic rocks, initially in Snowdonia, before being one of the first on the ground in the Lakeland Project (Molyneux, pers. comm., 1998). Given the dearth of macrofossils in Skiddaw Slates, it was obviously appropriate for Allen to recruit Molyneux to help provide a stratigraphic framework for 'Otley I', and try to link it to the graptolite zonation.

Cooper recalls that his first task, on approaching the problem of the Skiddaws, was to read the various relevant PhDs, such as those of Jackson, David Roberts, Webb and Jeans; and he soon found that they were not all in agreement. Then he went into the field to look at Barry Webb's ground in the Buttermere area, which he found to be 'absolutely horrendous', with the geology being a 'complete and utter mess', though it seemed that Webb's thesis had made some sense of it. After a week, Peter Allen came out to see how Cooper was getting on and they decided to try something easier and moved north to the area of the Lorton Fells, Bassenthwaite, and Whinlatter Pass (see Fig. 12.1), where

---

[1] Stone was born in 1947 at Mountain Ash (between Aberdare and Pontypridd) in Glamorganshire, South Wales, but went to school at Forest Grammar School near Wokingham, between Reading and London. He read geology at Cardiff University (BSc, 1970) and went on to a PhD at Birmingham (1975) under Raymond Adie (then Deputy Director of the British Antarctic Survey) on the geology of South Georgia (pers. comm., 2000). In 2000 Stone was stationed at the Edinburgh office of the Survey, and was also District Officer (Cumbria).

[2] Hughes (b. 1959) attended Grove Park Boys Grammar School and then Yale Sixth Form College at Wrexham. He read geology at the University College of Wales at Cardiff, and then moved to Cambridge to do a PhD under Barrie Rickards and David Price on 'Mid-Ordovician graptolite biostratigraphy in southern Britain'. Hughes joined the Survey in 1985 and was put on to Carboniferous sediments in NW England, but with his PhD experience it was unsurprising that he should have been transferred to the Newcastle office and the Lakeland Project in 1989 to work on the SS. This occurred at the same time that Tony Reedman took over the Project after Barry Webb's resignation. Hughes subsequently did work for the BGS Nirex consultancy (see Chapter 20), and in the Southern Uplands, and he spent much of his time during the last five years of the twentieth century working in the Andes of Ecuador (Hughes, pers. comm., 2000).

[3] Bell (b. 1950) read geology at Sheffield, where he became a PhD student under Soper (1975), studying the deformations of accretionary lapilli (such as the bird's-eye tuff in Kentmere; see Figs 3.7 and 12.4) as a means of gauging the stresses and consequent strains that these rocks had undergone. (He had got interested in the topic as a result of attending MSc lectures on structural geology by John Ramsay at Imperial College.) The big name in the 'stress–strain' field at that time was Gerhard Oertel, who visited Sheffield in 1972 and gave Bell help in the mathematical analysis of the problems, and subsequently acted as his external examiner. On completing his doctorate, Bell got a position with the Survey in Edinburgh, where he shared an office with Phil Stone, so conversations naturally turned to problems in the Lake District and the Southern Uplands. Bell was with the Survey from 1975 to 1979, and worked with the computer unit, but following a long tradition he was also put to work on a topic about which he had little or no previous experience – in this case the mapping of sands and gravels in the Scottish Lowlands. Bell then moved to the Irish Survey, where he did mapwork in Donegal and Leitrim. Next he joined the staff at University College, Galway, following which he became a freelance geologist, prospecting for oil companies in the Midland Valley of Scotland. He worked for the Open University (1991–1994), and in 2000 was again a freelance geologist, based at Penrith (Bell, pers. comm., 2000). I had the pleasure of meeting him briefly in 2001.

[4] Mathieson, N. A., 'Structure and stratigraphy of the volcanic rocks round the Black Combe Inlier, English Lake District', PhD Sheffield University, 1986. Mathieson did not do Lakeland work after finishing his PhD.

[5] Cooper (b. 1951 near Oswestry) moved around with his family in the forces in his early years. They subsequently settled in Rayleigh, Essex, where he attended Swain Grammar School. He studied geology at Sheffield University and went on to do a PhD on 'The stratigraphy and palaeontology of the Ordovician to Devonian rocks of the area north of Dornes (near Figueiroa Dos Vinhos), Central Portugal' under Michael Romano. This was completed in 1980, several years after Cooper joined the Survey (pers. comm., 2000).

[6] Molyneux (b. 1953) was brought up in Salford, Lancashire, where he attended De la Salle College. He read geology and botany at Sheffield and went on to do his PhD under the acritarch man, Charles Downie, on 'The micropalaeontology and age of the Manx Group, Isle of Man'. Molyneux completed his thesis in 1981, two years after joining the Survey, and not long before he became a member of the Lakeland Project (pers. comm., 2000).

Fig. 15.1. Adrian Rushton in the field, 1997. Copy of photograph supplied by Dr Rushton and published by his permission.

previous PhD work by Jackson and others could provide a starting point in a less complex area. There the geology seemed more intelligible, with reasonably well-behaved dipping sequences, though there were indications of thrusting and inversion of strata. Soon Webb joined them from the Southern Uplands, and discussions led them to think that the Lorton and Buttermere areas were essentially different in structure.

Cooper settled on the exposures in Hope Beck and logged a SS section there, metre by metre, such as had never been attempted previously in the 'Skids'. He worked in collaboration with Rushton and Molyneux, and as he put it to me: 'the main tools for the stratigraphy were Adrian and Stewart'. As intended in the multidisciplinary approach, the palaeontologists assisted the structural geologists, establishing a zonal scheme, based on both graptolites and acritarchs, so that the structural men could tell where they were in the sequence, as well as the shapes of the structures. This early work in 1982 provided the foundation for all the subsequent labour in the Lakeland Project on the SS.

Like Dennis Jackson (see p. 116), Rushton was not altogether happy with Gertrude Elles's work. He found that some of her identifications could not be sustained, especially where her idea of 'transient' or intermediate species was involved (which notion, it will be recalled, was invoked in accordance with her evolutionary ideas, probably traceable to Marr). Rushton also thought Elles was mistaken in placing so much reliance on the *proportion* of fossils of a given type present. For her, an abundance of *Tetragraptus* specimens meant that you were in the *Tetragraptus* zone; but this ignored the graptolites' probable planktonic life-style. A 'bloom' of the creatures could produce numerous fossils on a slab, but that was not necessarily a demonstration that one was in the *Tetragraptus* zone (say). To determine that, one needed to ascertain all the associated fossils, which could now, with Molyneux's assistance, include the microfossils. In recognizing a zone, Rushton's usual practice was to rely on the appearance of new taxa rather than the appearance of a dominance of forms of some long-ranging species. As well as questioning Elles's methodology, Rushton also thought that some of Jackson's work was too much influenced by 'population' arguments. So some of his stratigraphy needed revision too. In particular, the Watch Hill Grits were construed by Rushton as being older than Jackson had envisaged (Rushton, pers. comm., 1998).

In fact, the Skiddaw Group was soon found to have rocks significantly older than previously supposed. Molyneux recalled that as early as May 1982, when they were having a reconnaissance look at the SS in the area of the River Calder in the western Lakes (near Calder Hall, of fame for its adjacent nuclear power station; see Chapter 20 and Fig. 20.1), they discovered a *Tremadoc* trilobite; and subsequently a few more trilobites were found in the same locality, in an area previously assumed to be of Arenig age. The first-found key specimen was identified as *Niobina davidis* (Lake), which although lacking its cranidium was thought to be identifiable. Rushton (pers. comm., 2000) vividly remembered Molyneux taking a large sample, to be dissolved in acid and examined for acritarchs, when a flake fell off exposing the trilobite. The party was galvanized by what seemed a sensational find and eagerly searched for further specimens. Other trilobites such as *Peltocare olenoides* were found complete, and the fauna as a whole appeared to be Upper Tremadoc (probably *Angelina sedgwicki* zone [Rushton 1988]), on the basis of Welsh comparisons.[7] The extracted acritarchs were also compatible with known Tremadoc forms. Thus an acritarch assemblage was found that would likely identify Tremadoc rocks in the SS in areas where no trilobites could be found (see Molyneux & Rushton 1984).

Then, at the suggestion of Dennis Jackson, Rushton went north to poke around in the dreary little dry valley called Trusmadoor, forming a small pass on the fells between Cockermouth and Skiddaw (see Fig. 12.1). To his delight and astonishment, he found a huge retiform graptolite, *Dictyonema murrayi* (Hall),[8] about four times the usual size for such fossils, which showed that there were Tremadoc rocks up in the northern area too (Rushton 1985).

[7] SS trilobites were known from the nineteenth century, having been collected by Harkness and local amateur naturalists, notably Kinsey Dover and J. Rushton Harrison. Some specimens were described and figured by Survey Palaeontologist Robert Etheridge in an appendix to Ward's Memoir (1876*a*) and placed in the Arenig. Specimens were deposited in the Keswick Museum and described by Postlethwaite (1884–1885) with figures by Surveyor J. G. Goodchild. During the resurvey, Adrian Rushton looked over the nineteenth-century material in the Museum and found Harrison's material useful in being dated (e.g. 1889) and with letters indicating their localities (e.g. OS = Outerside). The Arenig specimens did not contribute substantially to the establishment of stratigraphic subdivision of the Skiddaws, but confirmed that the Lake District had originally been part of Gondwana, not Laurentia or Baltica (Rushton, pers. comm., 2001).

[8] Later referred to *Araneograptus murrayi* (Hall).

Fig. 15.2. Stewart Molyneux in his Keyworth office. Photographed by author, 1998.

Jackson's Arenig rocks were just over the Trusmadoor col, so Rushton hoped to find fossils of the *Tetragraptus approximatus* zone in the intervening space; but he was unable to do so at that time. Subsequently, more material was found and the Trusmadoor locality has been recognized as the only place in Britain where a transition of graptolitic rocks (also containing diagnostic acritarchs) can be observed from Tremadoc to Arenig. Thus the locality is regarded as a site of international scientific significance (Owen & Rushton 1999, pp. 262 and 264).

With the resurveying, the issue of the age of the Watch Hill Grits (or Formation) necessarily recurred (see Fig. 8.6). Jackson (1978, p. 96) had followed Rose (1954) in placing the unit in the Loweswater Flags, i.e. at about the middle of the SS succession, but Molyneux's acritarch work on the Grits, together with the discovery of *Dictyonema* specimens, suggested that the unit was Upper Tremadoc or Lower Arenig (Molyneux & Rushton 1988), so it had to be brought down in the succession, below Jackson's Hope Beck Slates (his lowest unit). However, the fossils seemed to be somewhat younger than those in the Calder area. The trouble, as always, with the Watch Hill Grits was that they were at the top of a hill (NY 148 318) above what appeared to be younger rocks. The classic solution to such a situation was, of course, to invoke a thrust fault, as had been done by the earlier surveyors such as Hollingworth and Trotter in the Cockermouth Memoir, who suggested thrusting from the south (Eastwood *et al.* 1968, p. 14). However, the resurvey, involving the work of Cooper and Webb, and the palaeontological results of Rushton and Molyneux, led to the suggestion of thrusting from the north – as indicated in a sketch map in Molyneux & Rushton (1988, p. 45); then later in a general account of the Skiddaw rocks (Cooper *et al.* 1995, p. 187); and still later in the section for the revised Cockermouth 1: 50 000 sheet (BGS 1997*a*). However, as already mentioned, I have seen no topographic expression of the fault in the valley to the south of Watch Hill, at the line indicated on the Cockermouth map. Thus considerable structural theory was or is dependent on the diminutive, but stratigraphically powerful, acritarchs! It was Rushton and Molyneux who thought that the palaeontology of the Northern Fells demanded a thrust structure, and on putting the idea to the structural geologist Barry Webb they found that he agreed that this was structurally feasible.

The grid references given for the few specimens of *Dictyonema* found (Molyneux & Rushton 1988) do not appear to refer to areas mapped on the Cockermouth and Maryport sheets as Watch Hill Grits but to the Hope Beck or Bitter Beck Formations. Be this as it may, the interpretation of the Watch Hill Thrust has not been based upon consideration of the outcrops of the Northern Fells belt alone, but is integrated with studies further south (Hughes *et al.* 1993; and see p. 207). Also, despite the structural complexities, with the particular assistance of the acritarch determinations, the stratigraphic sequence for the northern SS outcrops soon seemed clear. It was as follows:

Kirkstile Formation	Arenig
Loweswater Formation	Arenig
Hope Beck Formation	Arenig
Watch Hill Formation	Arenig–Tremadoc
Bitter Beck Formation	Tremadoc

The placement of the Latterbarrow Sandstone, exposed in the western Lakes (see Fig. 20.1), was also an issue, and was examined particularly by Allen and Cooper. They interpreted the unit as an 'estuarine sandstone deposited unconformably on Skiddaw Group rocks in the late Llanvirn or early Llandeilo times' (Allen & Cooper 1986, p. 74). That is, they regarded it as a basal member of the BVG, not a member of the Skiddaw Group at all. (This was, of course, before the demise of the Llandeilo.)

There were, however, more pressing difficulties at the southern end of the Skiddaw outcrops, near Buttermere, Newlands Hause, and on Robinson, where Barry Webb had done his PhD and, as we recall, had interpreted the area in terms of complex polyphase folding. As has already been mentioned, early in the Lakeland Project, Webb came down from the Southern Uplands and, following a spell on the Windermere Group, he and Cooper reviewed the rocks in Webb's old PhD area, with Molyneux providing new palynological information.

Tremadoc acritarchs were found near Newlands Hause (the pass that links the areas of Braithwaite and Buttermere; see Fig. 12.3), and close by there appeared to be Arenig acritarch assemblages. However, contrary to the situation for the northern SS, there was no 'layer-cake' stratigraphy in this area – rather, a tangle of disparate blocks of sediments of different lithologies, some of them large 'rafts' of sandstone, with adjacent rocks containing acritarchs of significantly different ages. To account for this confused stratigraphy, the possibility began to be considered that there was a major slump structure, rather than multiple post-depositional tectonic deformations.

Meanwhile, early on in the Lakeland Project, geochemists and geophysicists became involved, as intended in Allen's original proposal. An early attempt (1982 and 1983) to use variations in resistivity measurements as aids to mapping the SS in the Lorton Fells area did not prove useful and was not pursued (Carruthers 1986). However, a vast amount of effort was put into geochemical measurements all over the Lakes, using a team of university students collecting samples of stream water and stream sediments, and making chemical determinations by emission spectroscopy in field laboratories using procedures designed by B. A. Toms of Birmingham University (British Geological Survey 1992, p. 1). The results provided information about the disposition of ore bodies and mineral veins, but failed to lead to substantial new mining projects. By the 1980s, after centuries of mining, the Lakeland ores were largely exhausted for economic purposes.

Nevertheless, the geochemical investigations assisted advances in structural understanding and ideas about the geological history of the Lakes, including the SS. For example, thirty years earlier, Colin Rose (1954) had mapped a whitish or 'bleached' area of SS between Grisedale Pike and Whiteside and had suggested the presence of a southward-directed thrust fault near the summit of the hill, Causey Pike, which overlooks the village of Braithwaite (see Fig. 12.1). Rose suggested that the whiteness did not represent a separate unit of the Skiddaw Slates (the 'Blakefell Mudstone' of Dixon; Smith 1925, p. 70) but, transgressing as it did different types of Skiddaw rock, it was taken to be an area of metamorphism around a hypothetical igneous intrusion, situated under the Causey Pike Thrust. Rose had drawn a suggestive section, showing the supposed subterranean igneous body, a thrust fault, and a metamorphic aureole affecting both the 'Mosser–Kirkstile Slates' and the 'Loweswater Flags'. Peter Jeans also interpreted the whitened slates as representing a zone of alteration.

This older work attracted the interest of the interdisciplinary group, and the geochemists Derek Cooper and Neil Fortey[9] began to undertake 'geochemical traverses' over the area, while Michael Lee (1989) investigated the gravity and geomagnetic details. Cooper built up a systematic archive of the geochemistry of the Skiddaw Group, doing 'megatraverses' – taking samples every 100–200 metres and getting everything conceivable analysed (Fortey, pers. comm., 1999). These data were not fully published at the time of writing, but they provided evidence for the interpretation of what came to be known as the Crummock Water Aureole.

There is a good deal of tourmalinization visible even to the casual observer near the track westwards from Causey Pike, and the whole area proved interesting when examined geochemically. Two papers were published on the topic (Fortey & Cooper 1986; Cooper *et al.* 1988), the latter being a multidisciplinary production, involving the work of stratigraphers and mappers (Cooper and Allen), a structural geologist (Webb), a geophysicist (Lee) and a radiometric dating man (Rundle). The 'bleaching' was ascribed to the removal of carbon by the agency of a hot pluton, but more importantly the chemical distributions suggested the addition of arsenic, boron, potassium and rubidium; and loss of chlorine, nickel, sulphur, zinc and water. A model was developed according to which an underlying granite had brought up solutions, which moved by convection in the heated zone over the granite, thereby bringing about mineralogical changes. Such a hypothesis was called for, as the area did not display thermal metamorphic zones. Lee's (1989) gravimetry revealed the granite, of course. Christopher Rundle from the NERC Isotope Geology Centre in London obtained a rubidium and strontium radiometric age determination of about 400 Ma, in the Lower Devonian, which linked the intrusion to the Skiddaw and Shap Granites, date-wise, rather than the older (*c.* 420–430 Ma), more neighbourly, Eskdale or Ennerdale granitic bodies.

These results were published a few years after the initiation of the Lakeland Project, but I mention them here as the Lakeland team (especially Tony Cooper) became exceedingly interested in the approximately E–W 'Causey Pike Line', which, besides having the same orientation as other lineations in the Lakes such as the 'Otley II–Otley III' boundary or the linear region of metasomatism of the Crummock Water aureole, also had the Causey Pike Thrust itself, *and* marked out a regional difference within the SS. North of the line, there was some faulting, slumping, even thrusting, but the area could nevertheless be broadly understood as a sample of 'layer-cake' stratigraphy. South of the line, there was the area of great confusion that had driven Cooper and Allen to look for somewhere easier to start their Skiddaw investigations. Along the line, a thrust, *per se*, with apparent movement from north to south, could only be identified between the southern end of Crummock Water and a spot to the NW of Derwent Water (about 2 km south of Braithwaite). The fault has also been depicted on the Keswick map (BGS 1999*a*) as extending well to the east of Keswick, and also to the west of Crummock Water. In these lateral extensions it is there represented as a strike-slip fault with sinistral shear, not thrust. In the area to the east of Keswick, however, much of the supposed line of fault marking the 'Causey Pike Line' is in poorly exposed ground. The Causey Pike Thrust itself is not represented on the 1999 map as having undergone strike-slip movement (see Plate V).

There could be several explanations of the extraordinary structures found to the south of the Causey Pike Line, i.e. south to where the famous fault, unconformity, or ?, occurred between the SS and the BVG; for example, in the valley at the head of Buttermere (Warnscale Bottom), in Honister Pass, across to Hollows Farm, and eventually to the eastern Lakes on the eastern side of Ullswater, in the Tarn Moor Tunnel, and so on. One might perhaps be dealing with a 'melange' of tectonic origin, or a giant slump structure or 'olistostrome'. Or possibly one might be dealing with 'mud diapirs'.[10] Or something else entirely? It was the idea of an olistostrome that began to be favoured. There is some uncertainty as to who initiated this idea to account for the confusion of rocks round Robinson and overlooking Buttermere. Apparently Michael Nutt (by then in the Survey) talked about slumps to Barry Webb (see p. 166), such as Nutt had seen in the Denbigh Grits of North Wales. The Surveyors I consulted were not themselves certain who first had the idea of interpreting the rocks north of Buttermere as an olistostrome, though they remembered it coming up in one of the fairly regular discussion groups held by the members of the Lakeland Project. Webb recalled that he was influenced by a talk by Russell Arthurton at a BGS lunchtime lecture in 1975 on great slumps in Iran and Afghanistan, which had huge 'rafts' of sandstone and limestone embedded in other sediments. Webb also recalled (pers. comm., 2000) that he thought they might be dealing with either a melange or an olistostrome, neither of which, he thought, was adequately defined; and he eventually plumped for an olistostrome. Allen

[9] Fortey (no known relative of the palaeontologist, Richard Fortey) was born at Orrell, near Wigan, Lancashire, in 1949. He went to University College, London, with the intention of reading civil engineering, but a friend interested him in geology and he soon transferred courses, carrying on to do a PhD under Maurice Wells on the Ardnamurchan Complex in Mull, following which he joined the Survey's mineral reconnaissance programme, in which he was instrumental in the discovery of the Aberfeldy baryte deposits in 1977–1978. When this programme 'faded out' he became involved with the Lake District multidisciplinary project, supplying expertise in mineralogy and geochemistry and studying the metamorphism of Lakeland rocks (see p. 237) (Fortey pers. comm., 1999, 2000).

[10] This is an idea that Bell has thought plausible, following discussions he had with Tony (A. J.) Barber and Barry Webb in the field (Bell, pers. comm., 2000). It has not been taken up, though it is mentioned *en passant* in reference to the Bampton Inlier of Skiddaw Slates on the as yet unpublished Appleby Sheet (see Bell 1997, p. 15). Barber had proposed such structures for rocks in Timor, somewhat similar to the SS.

(pers. comm., 2000) credited Molyneux with the idea. Molyneux himself (pers. comm., 2000) was hesitant about making such a claim, though he recalled being puzzled to find Arenig and Tremadoc acritarchs in outcrops only a few metres apart.

The term 'olistostrome' was first proposed by G. Flores, a geologist with Gulf Italia Co., at the Fourth World Petroleum Congress held in Rome in 1955, and following subsequent investigation it was described in detail in the *Proceedings* of the Fifth Congress in 1959 (Flores 1959).[11] Flores' company had been examining central Sicily in search of oil, and found there, amongst Upper Tertiary deposits, distinct areas of what was interpreted as 'chaotic "slump" material, intercalated with "normal" sediments', which could be regarded as stratigraphic units (Flores 1959, p. 259). The 'chaos' did not appear to be purely tectonic in origin, or generated by thrusting. The pelitic materials examined in Sicily could not, Flores supposed, have transmitted the mechanical stresses in such a way as to produce the observed deposits. On the other hand, a good 'Earth shake' might have set off subaqueous slump movements – and in commenting on Flores' paper, W. H. Butcher of Columbia University drew an analogy with known large-scale submarine slumps off Newfoundland that had been responsible for breaking submarine cables back in 1929. Flores explained that 'olistostrome' meant 'accumulation due to sliding', and 'olistolith' (one of the large lumps or blocks in the olistostrome) meant 'slided rock'. He suggested that a number of areas of chaotic stratigraphy previously interpreted as tillites, nappes, ancient mud volcanoes or diapirs, might in fact turn out to be olistostromes. Olistostromes were, he claimed, mappable units, though they did not display bedding. The fragmentary blocks within them could range from pebbles through to blocks several cubic kilometres in volume.

Olistostromes have not been universally accepted in the way Flores proposed. Some Russian geologists (e.g. Leonov 1978) have regarded them as, in effect, melanges of tectonic origin. The distinction between melanges and olistostromes has been discussed by Hsü (1974), who suggested that the former were typically bounded, top and bottom, by shear surfaces, while the latter are normally stratigraphic units intercalated in a normal sedimentary sequence. He mentioned, however, that the upper surface may seem to display shear in some places and depositional contacts at others, if the overlying element is autochthonous. Some geologists I know, such as Celâl Şengör (pers. comm., 1998), maintain that there is no clear empirical distinction between olistostromes and melanges, especially given that there is no agreed type locality or 'type specimen' for the former (though Flores (1959, p. 264) referred to a type locality in Sicily). As Hsü pointed out, a sheared olistostrome would be difficult to distinguish from a tectonic melange. A 'melange' is, by its French derivation, simply a name for a mixture of rock types, but the term is most often used in reference to mixtures having a tectonic origin.[12]

In the case of the exposures of the SS south of Causey Pike (between the Causey Pike Thrust and the SS–BVG contact) the putative olistostrome does not fall neatly into a 'layer-cake' sedimentary sequence. It is bounded by a thrust fault to the north and the mysterious SS–BVG contact to the south – the latter possibly faulted or unconformable, or both. Be this as it may, the Surveyors took up the olistostrome idea with enthusiasm. Certainly 'things' that could be construed as olistoliths could be mapped out, and sometimes seen directly (see Plate V and Fig. 15.3). So Webb's old PhD area was revisited, with further mapping undertaken on Robinson and on the crags above Buttermere, and a paper was published in which the new interpretation was proposed, the unit containing the claimed olistostrome being named the Buttermere Formation (Webb & Cooper 1988).

Fig. 15.3. 'Olistolith' exposed near top of Newlands Hause (Moss Falls). Photographed by author, 1998.

In this publication, Webb modified his earlier idea of 'straightforward' polyphase folding, developed in response to Simpson's ideas. Webb had previously (*c.* 1974–1975) looked at structures in the Denbigh Grits (Wenlock) in North Wales, thought by Michael Nutt and Tony Wadge to be due to slumping, but he did not immediately accept the idea that the Skiddaws offered analogies (Webb, pers. comm., 2000), for he found the Denbigh slumps 'indecipherable' and in any case an order of magnitude smaller than the structures in the Skiddaw Slates around Robinson (Webb, pers. comm., 2000). Webb also recalled (pers. comm., 1998) that after finishing his PhD (1975) he had discussions with Nutt, who, as mentioned, had worked with a Survey team in Denbighshire and was inclined to think that Webb's area north of

[11] However, as Bernard Leake pointed out to me, the essential concepts of subaqueous slumping and gravity sliding had already been established in well-illustrated papers by Robert Baldry (1938) and Charles Barrington Brown (1938) on their studies of Tertiary sediments in Peru and Ecuador – the two having worked in this region for a good many years prior to these publications. One might say that Flores really just added the *name* for the phenomenon.

[12] Despite the protests of some sceptical structural geologists, the idea of olistostromes, which have modern analogies, finds considerable favour today. For an 'animated' representation of the slumping of sediments to form an olistostrome, off the coast of Cadiz, with accompanying map and section, see http://terra.redivis.es/pro/CYTMAR/TASYO/home.html. (I thank Barry Webb for this reference.) There are references in the literature to 'flow melanges', produced during subduction, which would appear to be a hybrid between the 'slump' structures of an olistostrome and something of tectonic origin. See Cloos (1982) and Murphy & Aalto (1983). For a report and discussion of a recent (1998) slump off the coast of New Guinea, apparently triggered by tectonic activity, see Tappin (2001); and for a geologically recent slump, likewise of tectonic origin, see Anon. (2001*b*).

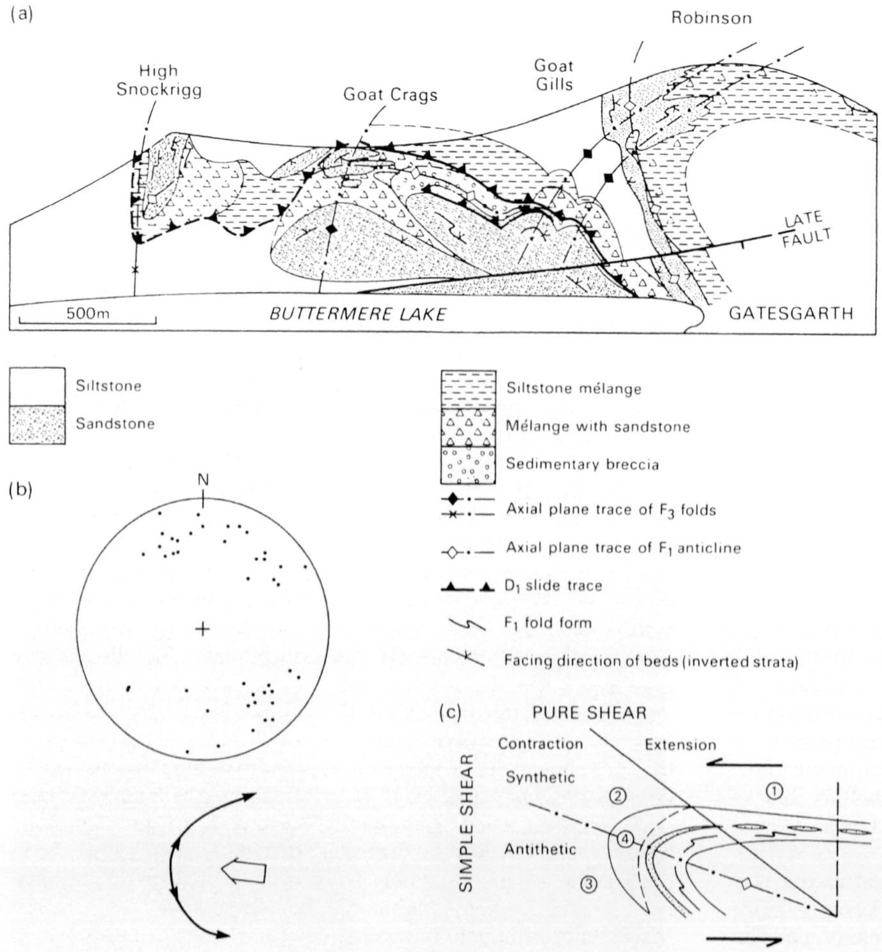

Fig. 15.4. Profile of the Buttermere olistostrome, viewed looking north from Buttermere Lake, according to Webb & Cooper (1988, p. 469). Reproduced by courtesy of Elsevier Science. IPR/26–13C British Geological Survey. © NERC.

Buttermere might be some kind of slump structure. While Webb was initially reluctant to accept analogy with the Welsh rocks because of the difference of scale, the idea that the SS in the Buttermere–Robinson area contained an olistostrome was undoubtedly plausible, or at least feasible, when the mixed ages of the rocks there began to be apparent from Molyneux's palynology, and considering the lithological differences between the sandstone 'rafts' and their slaty matrix.

Stratigraphically, then, the idea was that materials within the Upper Tremadoc or Lower Arenig had been randomly incorporated into the Buttermere Formation, judged to be formed in the early Llanvirn (Cooper 1988–1989 [1990], p. 223) or late Arenig (Cooper *et al.* 1995, p. 202). This would imply an age difference of about 20 Ma. One might suppose that that would be sufficient time to allow a considerable degree of consolidation in the older deposit and hence make submarine slumping difficult – though Flores' Sicilian observations suggested incorporation of Eocene or even Cretaceous material into his Miocene olistostrome; and Cooper has pointed out (pers. comm., 2000) that some of the sandstone blocks within the olistostrome are angular and coherent, whereas others have nebulous margins where they have started to disintegrate, so a 20 Ma age difference may not be implausible. The large 'floating' olistoliths within the olistostrome were collectively called the 'Robinson Member', being mapped as such on the Keswick sheet (BGS 1999*a*). They have lithological similarity to, but not identity with, rocks in the Loweswater Formation.

So Webb remapped the Buttermere–Robinson area, again using the stereographic methods deployed in his thesis, and partly reinterpreted the structures. The task was exceedingly complex, as there were other folds superimposed on those produced by the suggested slumps. Webb & Cooper's (1988) envisaged overall structure above Buttermere is shown in Figure 15.4 and a 'cartoon' of how the slumping might have occurred is reproduced in Figure 15.5. The supposed direction of the slumping on the southern side of the Causey Pike Line was suggested as being somewhat to the north of west, down the steeper slope of the E–W basin in which the Skiddaw sediments were supposedly accumulating. This was not stated expressly in words but is indicated in a diagram showing the general vergence of structures in the Robinson Member (Webb & Cooper 1988, fig. 6b).

In a later publication, Webb (in Moseley 1990, p. 76) stated that the southern slide was directed westwards. The details of the structure were modified again (Webb in Treagus 1992) (see Fig. 15.6), but the essential idea was retained. Here we may note that Webb referred to a 'melange' and to a 'thrust or slide surface'. However, he was not referring to a melange created by tectonic processes, for he also specifically referred to a slump or slide. In fact, Webb (pers. comm., 2000) thought that the melange *versus* olistostrome issue was more semantic than substantive. Besides, as he pointed out, the structures in the area are so complex that there will necessarily be uncertainty.

A NNW direction for the slumping south of the Causey Pike Line was reaffirmed in Cooper *et al.* (1995, p. 202). The Keswick sheet shows the 'fragments' of the Robinson Member mostly disposed in the Buttermere–Robinson area in an alignment compatible with NNW slumping. The unit's character was formally defined in Cooper *et al.* (1995).

Some slumping (towards the SE) was also proposed for the 'northern Skiddaws' (Webb & Cooper 1988, pp. 465–468), but it was apparently less severe than that which occurred south of the Causey Pike Line. Webb mapped the hills on the NE side of Loweswater (Darling Fell and Loweswater Fell). At first he interpreted his observations as showing that there had been a huge overturned F_1 anticline, which preceded the formation of cleavage folds. Later he began to interpret the structure as revealing two

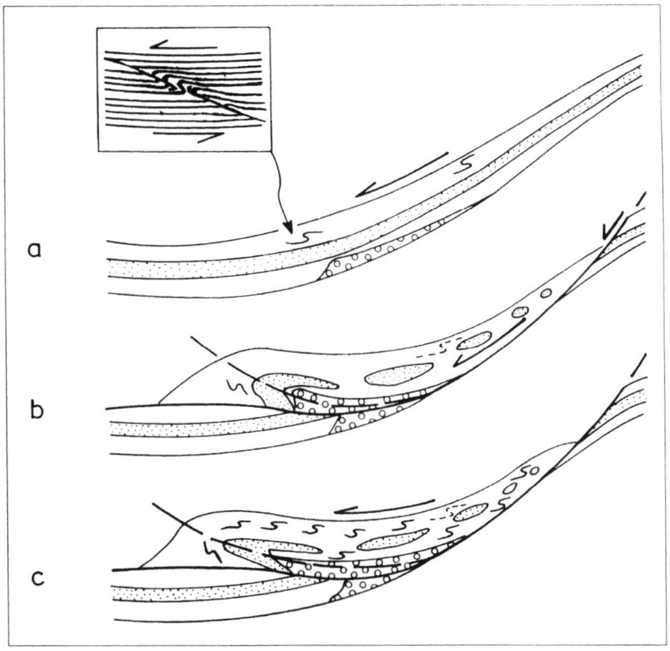

Fig. 15.5. Development of slump structures (olistostrome) in Skiddaw Slates, according to Webb. *In* Moseley (1990, p. 77). Reproduced by courtesy of Drs Webb and Moseley. (See also Webb & Cooper [1988, p. 471].)

stages of what was essentially a single deformation. The published version construed the area as a recumbent anticline with shear, which he interpreted as a slump structure (Webb & Cooper 1988, p. 466). Cooper mapped west of Sale Fell (a locality east of Bassenthwaite Lake; see Fig. 12.1) and thought the anticline there was a tectonic fold, whereas Webb thought it was a slump fold, subsequently tightened tectonically. Cooper was persuaded of this interpretation for their co-authored paper (Cooper, pers. comm., 2000).

The slumping of the Skiddaw sediments on the northern side of the Causey Pike Line was, however, later queried by Hughes as a result of his mapping in the Sale Fell and Loweswater areas. He regarded the Sale Fell rocks as showing folds of tectonic origin and was able to change Cooper's mind on this question again. The PhD student Richard Moore, drawing on sedimentological argument, arrived at the same conclusion independently. At the time of writing, Cooper thought that Sale Fell and Loweswater were too far apart for slump folding to maintain its style over such a distance (about 9 km). Hughes informed me (pers. comm., 2000) that he remained convinced of the purely tectonic origin of the Sale Fell anticline, but he had not visited the Loweswater area with Barry Webb and therefore remained open-minded about the possible slump-fold origin of the Loweswater Anticline. The issue did not, therefore, reach consensus as of 2000; and Webb's hypothesis remained open for further investigation.

The sedimentological work of Moore (1992) on the SS, focusing on the Watch Hill and Loweswater Formations (which show relatively little cleavage), should attract our particular attention, as his was the first study of the Slates to employ modern sedimentological techniques. Moore identified 21 facies classes, separable into six main facies groups, or (alternatively) four principal facies associations. For the fourfold taxonomy, some represented proximal channel and levee facies associations; some distributary channels; some depositional lobe facies; and others inter-lobe facies associations. In general, the palaeocurrents for the Watch Hill area for the proximal channel or levee facies appeared to Moore to have been from east to west, with a thinning westwards. However, further east, near Great Cockup (see Fig. 12.1), the Watch Hill palaeocurrents seemed to have been directed more towards the NE. Moore suggested that there might have been flows along an approximately E–W trough; or there might have been radial flows upon a submarine fan (Cooper *et al.* 2001, p. 38). Moore further thought it possible that the change from the Watch Hill greywacke, siltstone and mudstone to the Bitter Beck mudstone and siltstone might be as much a lateral facies change as a time marker.

For the distributary channel facies in the Loweswater Formation at Darling Fell, sole structures indicated erosive

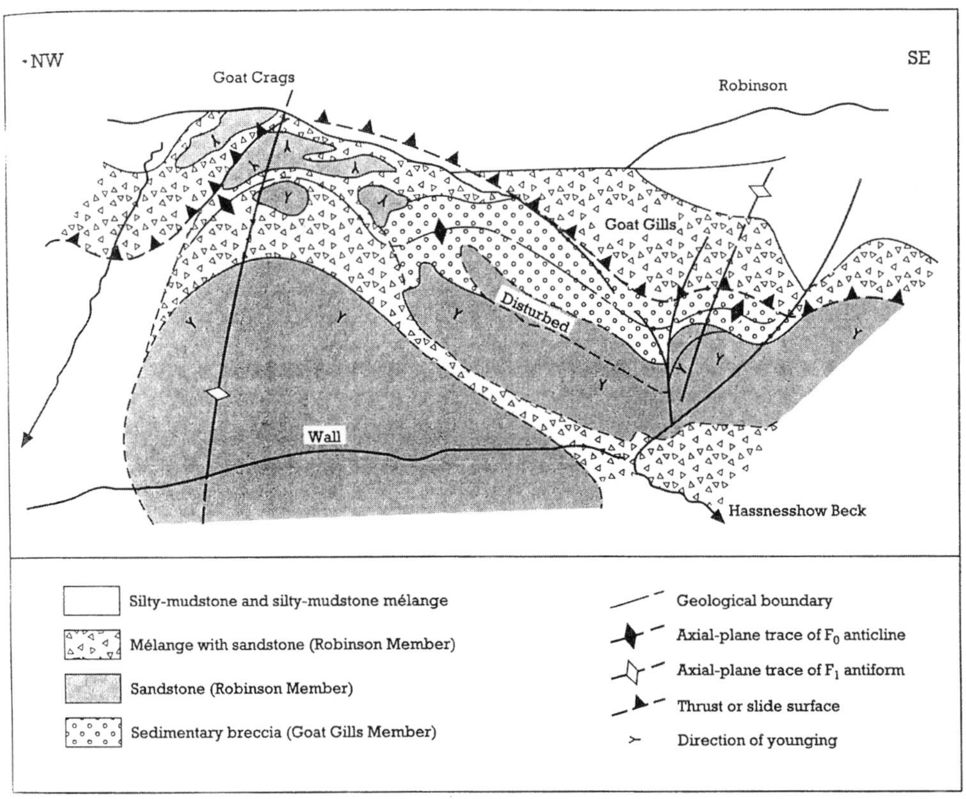

Fig. 15.6. Detail of section shown in Fig. 15.4, as represented by Webb. *In* Treagus (1992, p. 59). Reproduced by courtesy of Dr Webb.

currents flowing to the NW, and a bi-modal variety of ripple-mark orientations, indicated overbank flows or the spreading of flows at channel termination (Cooper *et al.* 2001, p. 39).

For the depositional lobe facies association, Moore found a general northward-directed palaeo-flow; but to the west, such as between Loweswater and Whinlatter Pass, sole marks suggested NW-directed palaeo-flow, whereas ripple marks indicated NNE and WSW water movement. Thus he had to construct a rather complicated model, according to which turbidity currents were controlled by a fault-controlled series of topographic troughs, with the ripples representing reflected waves generated at the trough margins. The inter-lobe facies was taken to represent lower fan or inter-channel circumstances.

The Skiddaw Memoir (Cooper *et al.* 2002), of which I was able to see some proof pages before the end of 2000, sought to relate the variations in the SS to global changes in sea levels, as did Moore in his thesis. Thus the Bitter Beck Formation was taken to represent a Tremadoc sea-level high. The Watch Hill Formation was linked to a lowering of sea level in late Tremadoc or early Arenig. Sea levels supposedly rose again during the deposition of the Hope Beck Formation, fell in the Loweswater Formation, and rose once again as the Kirkstile Formation was deposited. Additionally, there were more local tectonic effects produced by the rifting of Avalonia from Gondwana, and the disturbances produced by the onset of Lakeland volcanism.

Developing Moore's work, the recent studies described in the Skiddaw Memoir also considered the provenances of the Skiddaw rocks. It was thought that the group was made up of deep-water and slope deposits that were formed in extensional basins developed on the Gondwana margin. The preferred view, based on Moore (1992, p. 246) and Cooper *et al.* (1995), was that the sediments derived from an old, inactive and probably Precambrian continental arc terrane, lying to the SE of the Skiddaw depositional basin.

However, full consensus was not reached by the end of the twentieth century as to the form of the Skiddaw basin. As we have seen, Webb & Cooper (1988) envisaged an approximately E–W trough, with slumping towards the SE for sediments north of the Causey Pike Line, and to the NW for rocks south of the Line. But Moore's work suggested that there was a west-facing palaeoslope during Llanvirn times (Kirkstile Formation), when the main slumping and olistostrome emplacement supposedly occurred. The Skiddaw Memoir has broadly supported Moore's interpretation. Moore (1992, p. 243) specifically rejected the older approach of workers such as Simpson and Helm, focused on the study of polyphase deformations. In his view, it was only through detailed observation and mapping of cross-cutting structures that the geometries and relative ages of the different SS components could be determined; and even then he doubted whether their pre-Iapetus closure deformations could be widely correlated. Thus it would be illusory to suppose that one could determine a definite sequence of deformations applicable to all parts of the SS, such as earlier workers had sought.

Moore considered that there was a range of possible geometries for the collision processes. The preferred model was that advocated by his supervisor (Soper), with transpression and sinistral shear (see p. 146), which could have controlled sedimentation orientations. It was suggested that the inversion of the Skiddaw basin could have been due to compression at right angles to the Tornquist Sea convergence zone in mid-Ordovician; or uplift could have been due to doming associated with the emplacement of the BVG rocks (with some tuffaceous material in the Tarn Moor Formation (see note 13) – the youngest Skiddaw unit presently recognized by the Survey – heralding the oncoming volcanism). Alternatively, by analogy with the situation with the modern situation off the Chilean coast of South America, there might be oblique southward subduction of an oceanic plate, which would produce some uplift where submarine ridges or sea-mounts were 'swallowed' (Moore 1992, p. 242). All this involved (it seems to me) drawing a long bow from the sedimentological minutiae of the Skiddaw Slates, which is not to contest the interpretations.

Be this as it may, geologists actively involved in the Skiddaw work have come to the view that there *is* an olistostrome south of the Causey Pike Line. Even so, granted that there is slumping in the Buttermere area, there is naturally a question as to how far it – or the Buttermere Formation – extended or extends laterally. On the Keswick sheet (BGS 1999*a*), the Formation is mapped as extending a significant distance to east and west, but for want of other maps I am uncertain how far the Survey regarded the unit as extending laterally. Webb & Cooper (1988, p. 471) stated that the presumed trailing edges of the slumped mass lay outside the area that they had mapped at that time. On the Keswick Sheet, the Buttermere Formation is shown to the east as being faulted against the youngest member of the Skiddaw Group, the Tarn Moor Formation (Upper Arenig and Llanvirn mudstones, including beds of volcaniclastic sandstones and occasional bentonites), which also crops out further east in the Ullswater and Bampton Inliers.[13] Cooper *et al.* (1995, p. 199) mentioned that the Tremadoc sediments near the River Calder (west Lakeland) display some slumping, but these slumps may be older than the Buttermere Formation. Andrew Bell (pers. comm., 2000) thought that the eastward extension of the Causey Pike Thrust as an extended strike-slip fault, forming the northern boundary of the Buttermere Formation, was questionable, given that the ground is poorly exposed and the rocks are not lithologically distinct.

The Survey officers have been more sanguine on this point. In previous chapters, I have frequently referred to the three principal Lakeland units as 'Otley I', 'II' and 'III'. However, from the geophysical work of Michael Lee (1989)[14] and his co-workers, it began to appear that there might be other ways of 'dividing the cake'. Geophysical survey seemed to reveal a significant lineament, *within* the SS domain, crossing the Lake District approximately from west to east, associated with the elongate Crummock Water Aureole and the Causey Pike Fault. Eastwards, it was claimed that the fault could be extrapolated under the younger sandstones of the Vale of Eden to the Cross Fell Inlier (Cooper & Molyneux 1990),[15] and that it showed itself again in the aligned Ninety Fathom Fault in Northumbria (Kimbell *et al.* 1989). The Line was also traceable by geophysical evidence (Chadwick & Holiday 1991). In the Cross Fell Inlier, Cooper & Webb (1988) thought they could see lithological differences (noticed, but not underscored, by Wadge in his earlier examination of the area), similar to those in the Lakes proper. Cooper published on the matter, with Molyneux providing palynological argument (since Webb went off to Spain at about that time). Cooper (pers. comm., 2000) pointed out that the Causey Pike Line truncates the northern outcrop of the Mell Fell Conglomerate, which, he suggested, could have been deposited in a palaeovalley parallel to the Line.

There was also the lineament running approximately along the

[13] The mudstone and siltstones Tarn Moor Formation, formalized in Cooper *et al.* (1995), was at one time thought to belong to the Eycott unit (Wadge 1978*b*, p. 69), but Millward & Molyneux (1992) followed Jackson (1978) in allocating it to the Skiddaw Group. By virtue of its fossils and pattern of outcrop it is believed to be unconformable to the 'inchoate' Buttermere Formation, but the base of the Tarn Moor Formation is not exposed. Its correlative in the Cross Fell Inlier is called the Kirkland Formation (Cooper & Molyneux 1990). It does not crop out in the area of the Keswick map depicted in Plate V.

[14] Much of the work described in this bulky report was undertaken earlier in the Lakeland Project. See also Cooper *et al.* (1988).

[15] The corresponding extension of the Causey Pike Fault in the northern Cross Fell Inlier was called the Thack Moor Fault, which separated a newly defined unit, the Caterpallot Formation (Tremadoc–Arenig) from the Kirkland Formation (Llanvirn). Acritarchs provided most of the evidence, though the Kirkland Formation had previously been dated by graptolites.

long-recognized line of division between 'Otley II' and 'Otley III'. So Cooper (1988–1989 [1990], p. 223), for example, referred to North, Central and South divisions of the Lakes, with the boundaries at the two lineaments. In Cooper *et al.* (1993), the divisions were referred to as the Northern Fells Belt, the Central Fells Belt and the Southern Fells Belt. The question, then, was whether these lineaments might be related to ideas about the closure of the Iapetus Ocean and such matters, and wider theories of plate tectonics. For example, it may have been the case that, at the back of their minds, some surveyors in the Lakeland Project were under the influence of the idea of 'terranes', associated with plate-tectonic theory, then much in favour. Perhaps the suggested lineaments marked terrane (or micro-terrane?) boundaries (Soper, pers. comm., 2000), though Webb (pers. comm., 2000) reported that they were not thinking of terranes as being entities so small as to fit the Lakeland structures. As of 2000, Tony Cooper certainly believed that the Causey Pike Fault and lineament were significant features – though he too did not accept that his work was 'driven' by adherence to terrane theory (pers. comm., 2000).

The lineaments may indeed mark deep fractures in the basement rocks. Even so, I find it difficult to know how the model is supposed to work overall, especially for the area west of Keswick that we have been discussing. The 1: 50 000 Keswick map shows thrust faulting from the north in the Causey Pike area, with no indication of strike-slip on the map (though Cooper informed me (pers. comm., 2000) that he thought there had been some strike-slip on the Causey Pike Fault, in addition to the southward thrusting specifically indicated on the map). To the west, the thrust is mapped as terminating at a NNW aligned normal fault near Crummock Water; and to the east at an approximately parallel fault with some sinistral shear running towards Derwent Water. (Slightly to the west of this a parallel normal fault, with downthrow to the east, is indicated.) Beyond these faults (i.e. to both east and west of the terminations of the thrust) the Causey Pike Fault is shown as having a sinistral strike-slip movement. The situation is shown in Plate V, which is a part of the 1:50 000 Keswick map (BGS 1999*a*).

The idea of a sinistrally sheared Causey Pike Fault or Line would cohere with Soper's previously developed idea of sinistral shear associated with the 'docking' of Eastern Avalonia and Laurentia (see p. 149), but to my knowledge the shear between the Kirkstile and Buttermere Formations shown on the Keswick map does not rest on strong and straightforward field evidence for sinistral movement. Non-Survey geologists such as Webb (pers. comm., 2000) and Bell (see above) have expressed doubt. Cooper himself has acknowledged that the evidence for sinistral shear is not particularly good, but in so doing he drew attention to the bending of some structures and the apparent refraction of cleavage towards the fault, which might suggest shear (pers. comm., 2000). Thus while the idea of the Causey Pike Line exhibiting sinistral shear along its whole length accords with Soper's theory, it would appear not to be fully demonstrable in the field.

Also, Molyneux (pers. comm., 2000) informed me that there are problems about the interpretation of the eastward extension of the Buttermere Formation as a coherent unit. Near Troutbeck, there are 'foreign' acritarchs of late Tremadoc–early Arenig age in the mix-up of rocks of the 'Buttermere Formation', but they are different from ones found in the Buttermere area itself. In fact, the 'foreign' acritarchs at Troutbeck have affinities to those found in the Bitter Beck, Watch Hill and basal Hope Beck types of the *northern* fells – which could suggest a slump from the *north*. However, one does not know what may lie under the BVG rocks or what may formerly have existed to the south of the present Skiddaw limit of outcrop, so the presence of the acritarchs with northern affinities at Troutbeck does not prove that there was a slide there from the north.

It should be remarked that the outcrops of the olistoliths of the Robinson Member are, with but one exception SE of Portinscale, mapped within the region between the aforementioned NW faults and south of the Causey Pike Thrust (see Plate V). This may suggest that the formation of the olistostrome and the Robinson Member was causally associated with tectonic activity. So what sort of story (scenario) could be proposed to fit with the map?

Given that the eastern of the two NW faults cuts the (extended) Causey Pike Fault (see Plate V), the CPF must, according to the map, be the older fault. Given that (for the sake of discussion) the whole of the Buttermere Formation, as mapped, is indeed a higgledy-piggledy olistostrome, slumped from the SE before the eruption of the Borrowdales (let us here disregard the acritarchs at Troutbeck), it would presumably have required a tectonic trigger, perhaps associated with the onset of BVG activity or with terrane collision. Given that the contact between the Buttermere Formation and the Kirkstile Formation is mapped as a fault (with strike-slip), the two units must have been in existence prior to the faulting. Given the topological relationship (close, but not quite 1: 1, correlation) between the Robinson Member and the Causey Pike Thrust part of the asserted Causey Pike Fault–Line, one may, as said above, suspect a causal relationship between the olistostrome and the thrust fault.

It is, however, difficult to make all this cohere. Let us, for example, suppose that the SS were deposited in sequence in a basin running approximately SW–NE, there being land to the south or the SE. Let there be some tectonic disturbance that produced the olistostrome of the Buttermere Formation, with slippage from the SE (and perhaps some slumping in the northern part of the Skiddaws too?). Then let there be tear faulting with sinistral shear such as to produce the Causey Pike Fault (coincidentally juxtaposing the Buttermere and the Kirkstile Formations). Then let there be (relative) movement from the NE such as to disrupt part of the already existing Causey Pike Fault and generate thrusting in the part of it near the Crummock Water aureole, perhaps creating also the tear fault that runs SE from Bassenthwaite towards the head of Derwent Water. As said, from the geometry of the situation the Causey Pike Thrust may seem to be causally linked to the production of the sandstone rafts of the Robinson Member to the south of the Thrust. However, the olistostrome was, by the scenario above, already in existence before the Causey Pike thrusting occurred, so that the topological relationship between the Thrust and the outcrops of the Robinson Member would have to be coincidental. (Indeed, the thrusting must have come late, as it seemingly cuts out a portion of the southern side of the Crummock Water aureole (depicted by a line of dots and crosses in Plate V) at the line of the Thrust – though there could have been several periods of movement at the line of the Thrust.) And finally let the eruptions of the BVG occur, so that they can overlie the Skiddaw Slates, with unconformity and perhaps with faulting also.

This cannot be correct if the initial slumping was triggered by the emerging BVG materials. So we can construct another scenario. The Skiddaw Slates were deposited in sequence, but were shaken up in their southern parts by the activity of the BVG producing the olistostrome of the Buttermere Formation. Or as Barry Webb put it to me (pers. comm., 2000): 'slumping was probably a result of volcanic doming'. Perhaps some of the chief 'N–S' faults were also initiated by the BVG activity (with its caldera collapse; see Chapter 16). Subsequently, sinistral shear occurred along the Causey Pike Line; and then there was thrusting from the north (relatively speaking), producing the folds and thrusts of the northern Skiddaws and the Causey Pike Thrust. Also, pre-existing 'N–S' faults could have been reactivated to a greater or lesser degree.

This is a more coherent story, but not altogether satisfying. As Richard Hughes reminded me (pers. comm., 2000), it is now generally agreed that the BVG volcanism was subaerial, with the earliest eruptive events depositing material onto an exhumed basement of SS. Further, the foregoing scenario leaves the association of the Robinson Member rocks (in the Buttermere olistostrome) with the Causey Pike Thrust unexplained (though the association may simply be coincidental); and, as we have seen, the occurrence of strike-slip displacement along the Causey Pike Line is debated by the geologists who have worked on the map. Some would have shear run along the line of the Thrust, as well as to the east and west of it, where it is marked. Others seem to doubt whether there is lateral shear at all. All agree on the existence of thrusting at Causey Pike and its association (topological at least, if not directly causal) with the Crummock Water pluton. They also agree that the pattern of faults is uncertain, since exposure is poor in so much of the ground. As Webb emphasized (pers. comm., 2000), the actual outcrop of faults is rarely seen in the field so that some of the ones marked on the map should really be regarded as conjectural. Indeed, he doubted whether the pattern of faults had been thought through on a regional scale. One may add that we do not know, from simple inspection of the Keswick map, the absolute ages of the faults. Some (or all?) of the N–S faults may in fact be Tertiary (see Chapter 18). In fact, this is quite likely (or older faults were reactivated in the Tertiary), in which case the scenarios sketched above may all be imperfect.

So, although the Keswick sheet was published as one of the later products of the Lakeland Project, it may prove not to be definitive (although that is true of all geological maps, it seems). Needless to say, these matters were part of current geological debate at the turn of the century, and are *not* matters for an historian to settle. Nevertheless, they are worth mentioning in the context of attempting to see what was or was not accomplished or agreed upon by the end of the twentieth century.

In the foregoing, I have focused on the details of the geology of the Keswick sheet. However, one may also consider the wider context. The Skiddaw rocks can be seen as having accumulated on the southern side of the Iapetus Ocean, and were presumably folded or cleaved in part as a result of the terrane docking associated with the closure of that Ocean. But where is the actual suture arising from the closure? Most writers (e.g. Phillips *et al.* 1976) have supposed that it is somewhere in the region of the Solway Firth. But this idea was challenged by Peter Allen (1987) who suggested that the Solway Line did *not* mark the suture. Rather, he thought, it might be situated further south. Allen mentioned that Beamish & Smythe (1986) (see p. 151) had, through geophysical work, identified a northward-dipping high-conductivity zone, which seemed to pass under the outcrop of the SS; but he did not make any firm suggestion that this was in fact the Iapetus Suture.

In fact, Allen may not have been fully committed to the idea, even in the 1980s. He informed me (pers. comm., 2000) that he wrote his short communication (1987) in response to a challenge from Soper, and in a sense as an act of devil's advocacy. They were thinking at the time that the leading geological theorists were beginning to lose interest in the Lakes, and that it would be a good idea to stir up the geological community somewhat. The gambit did indeed achieve something in that line, in that it elicited a strong response from Richard Fortey (Fortey *et al.* 1989), who argued that the SS trilobites – which, though rare, had been known from the time of Postlethwaite in the nineteenth century – clearly belonged to the Gondwana Province, having affinities to types known from Bohemia, France, Spain and Morocco. Therefore the SS must have been deposited on the southern side of the Iapetus Ocean, not the north. The Natural History Museum geologists, Fortey and Owens, were drawing on their knowledge of the Arenig trilobites of South Wales. They started their riposte to Allen's hypothesis as a duo, and then turned to Rushton for assistance with Lakeland data. He was able to supply information about Lakeland stratigraphy not known to the Museum men, and thus there was a useful synergy between the Museum and the Survey, even if, in this instance, it undermined a suggestion that had originated within the ranks of the Survey (Rushton, pers. comm., 2000). The issue of what happened, where and when, for Iapetus closure and subsequent events did not take off for further debate in the literature on the basis of Skiddaw work, and the question of the tectonic history of the Lake District and Iapetus closure was only fully reignited (in public) through later work on 'Otley II' and 'III', discussed in the next two chapters.[16]

As the Skiddaw Memoir remained unpublished at the time of writing, the most complete account of the tectonic history of the Skiddaw Slates available to me was Hughes *et al.* (1993). According to their views, there was the syndepositional deformation of the SS, and the formation of the olistostrome in the Buttermere Formation. Then, on the basis of the discovery of clear evidence for an angular unconformity at Overwater (see p. 151), it appeared that there was some substantial uplift and erosion of the SS prior to the eruption of the Borrowdale Volcanics. This uplift could be accounted for by the processes of magma generation and the incipient development of volcanoes; and of course the volcanoes were evidently subaerial to a considerable extent. Hughes (1990–1991 [1992], p. 381) described the Borrowdale Volcanics as having 'punched through' the Skiddaw Slates (which, however, is not incompatible with some BVG rocks lying unconformably on the Slates in some places),[17] but the Survey remapping had not revealed any 'pre-volcanic tectonic deformation event, related to collision or accretion' (Hughes *et al.* 1993, p. 623). So the earlier ideas about pre-volcanic polyphase foldings were now discarded or radically modified by the Survey. Things had swung round in favour of relatively small-scale soft-sediment deformation and/or olistostrome structures. The many deformation phases of Simpson, Helm, or those of David Roberts, producing the wondrous structures in the Caldew Valley, were left behind. Perhaps (following Fyfe & McBirney 1975) the uplift was caused by water being carried down into the mantle by a subducted plate.

Then, for the Skiddaw Slates, we can, following the volcanism, leave the events in the Silurian concerning 'Otley III' aside for the moment and jump to the Acadian event in the Devonian (see p. 148), which produced the presently observable regional cleavage and the broader folds of the Slates. The cleavage appears to have preceded the emplacement of the Crummock Water granitic body (dated about 400 Ma), as both bedding and cleavage appear to have been deformed across the granite's aureole, in such a (sigmoidal) way as to suggest sinistral movement. However, the aureole also appeared to have been affected, post-emplacement, by reverse fault movement on the Causey Pike Thrust. So the movements along the Causey Pike Line, both wrenching and thrusting, were late tectonic events: the olistostrome came first and all the other movements *followed*. Even so, not all the geologists involved agree about the interpretation of the cleavage near the Crummock Water aureole (Webb, pers. comm., 2000).

A model for all this is represented in Hughes *et al.* (1993) (see Fig. 15.7). We shall leave for the moment the question of the emplacement of the large mass of granite, of which the Crummock Water rock is merely represented as a small offshoot. Certainly there would appear to be an association between the Crummock Water pluton and the Causey Pike Thrust. But it is worth mentioning that in the field it is the Gasgale Thrust that appears as much the greatest area of tectonic disturbance. The

[16] Allen (pers. comm., 2000) pointed out that the question of the how and why of the Iapetus Suture was perhaps still not fully resolved as of 2000, though he agreed with Fortey's arguments about the relevance of the trilobites. He thought that the magmatism in the Cheviots to the east of the Lakes might suggest some closure pre-Wenlock.

[17] In retrospect, Hughes thought the word 'punched' was inappropriate. He was, at the time of using it, aware that there was an unconformity at the Overwater spillway, but it did not appear to be such as to suggest a violent tectonic event.

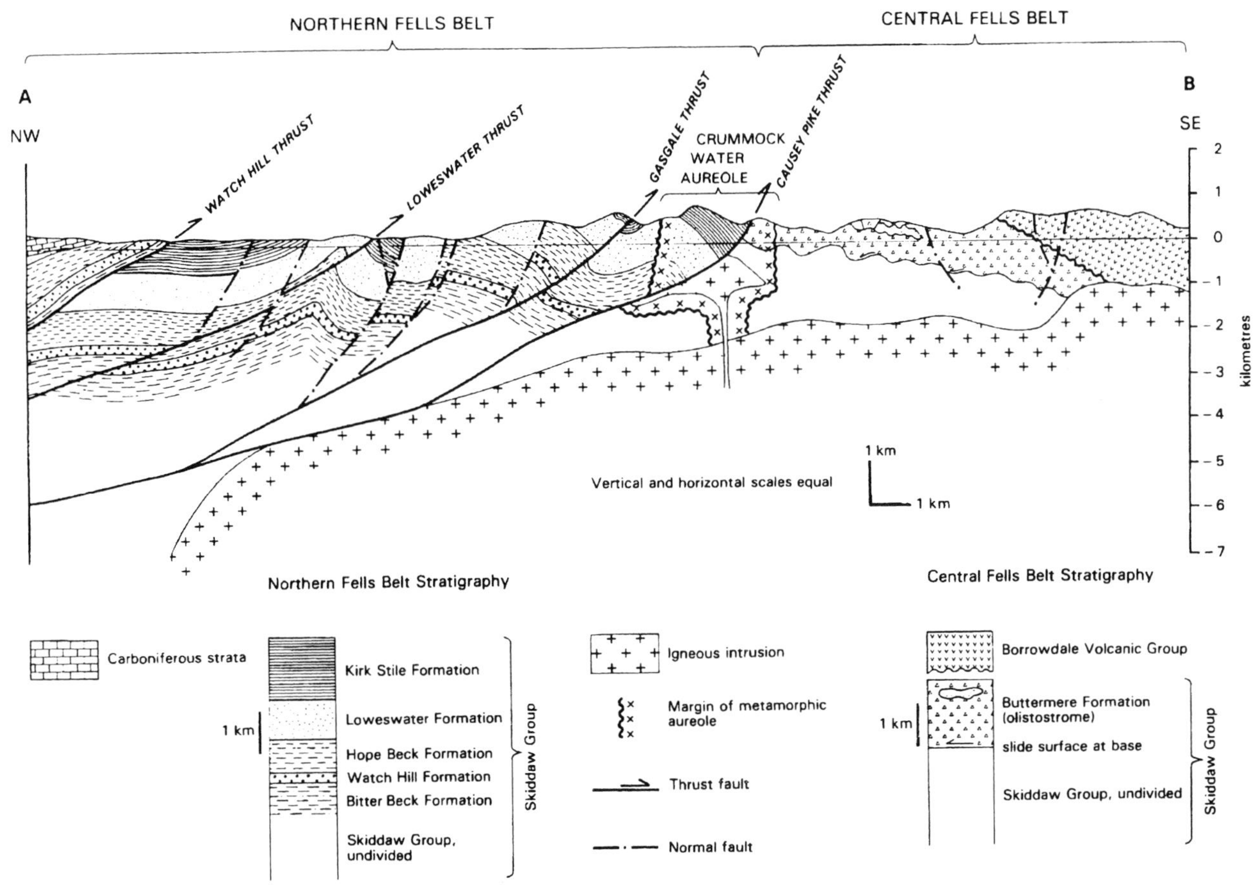

Fig. 15.7. Profile of the Skiddaw Group, *etc.* in the Lakeland northern fells, according to Hughes, Cooper & Stone (1993, p. 626). Reproduced by courtesy of Cambridge University Press. IPR/26–13C British Geological Survey. © NERC. All rights reserved.

Loweswater Thrust is certainly visible as a topographic feature and disturbed rocks are exposed there, but the Watch Hill Thrust is essentially inferred from palaeontological evidence. Still, one would expect the faulting to die out the further one is from the main disturbance; and the presence of a large mass of granite could well form a structure 'against which' the thrust faulting might have acted.

The part of this figure that may seem puzzling is that to do with the Buttermere Formation olistostrome. It seems to me to be remarkably 'neat' the way it is 'sandwiched' between the Causey Pike Thrust to the north and the SS–BVG 'contact' (I use a neutral term) to the south. Could it, after all, represent a tectonic melange? Stone (pers. comm., 2000) informed me that he was comfortable with the section, though he would wish it to be construed as 'the pre-Acadian situation', representing a structure on which cleavage was subsequently imposed, with reactivation of faults. Hughes (pers. comm., 2000) pointed out that the Buttermere Formation (olistostrome) does not have the high-strain features commonly associated with tectonic melanges,[18] and he held that one should *not* assume a causal connection between the thrust and the olistostrome: the proximity of the rafts of the Robinson Member to the Causey Pike Thrust might simply be coincidental. Of course, if the diagram of Figure 15.7 represents a pre-Acadian situation, then we cannot have it that all the movement on the Causey Pike Line was due to a late set of tectonic events.

Again, taking a broader view, it should be remembered that the interpretation of the Lake District was necessarily made in conjunction with the development of ideas about the Southern Uplands, Stone in particular doing work in both areas. It will be recalled (see p. 149) that Soper had in mind an initial 'gentle' collision of terranes and closure of Iapetus in the Llandovery, followed by 'tightening' in the Devonian, with formation of the main cleavages in the slate belts and associated igneous activity as at Shap for example – in the 'Acadian deformation'. Stone's views were rather different (Barnes *et al.* 1989). He envisaged the development of a 'thrust stack' in the Southern Uplands, with an associated 'foreland basin' (i.e. on the southern, seaward side of the thrust stack), which ultimately led to the development of the basin in which were deposited the sediments of 'Otley III' (see Chapter 17). Stone was, then, inclined to think that deformation gradually spread southwards, subsequent to the closure of Iapetus. So the cleavage in the SS could be older (late Silurian) than in the Windermere rocks (early Devonian). Solution of the problem of the dating of the Skiddaw cleavage(s) might, he said, be achieved by determining the ages of the big igneous intrusions, and ascertaining whether they were or were not cleaved along with the SS. Stone (Hughes *et al.* 1993, p. 627) found the evidence in the literature on this issue ambiguous, but it was 'permissive' of a late-Silurian cleavage. Soper's view, by contrast, was that the initial closure of Iapetus must have been the most gentle of collisions, for the simple reason that nowhere in the Lakes does one find the metamorphic rocks that would have to be there if there had been a 'hard' collision. We should note, of course, that the situation in the Southern Uplands and the Lake District is substantially different for their slate belts. Neither the Skiddaws nor the Windermeres form an imbricated thrust stack, such as McKerrow and others envisaged for the Southern Uplands.

[18] On the other hand, Hughes (pers. comm., 2000) considered that the Sale Fell Anticline in the NW Lakes (a supposed continuation of the Loweswater Anticline), which Webb & Cooper (1988) believed was a soft-sediment slump fold, is in fact an ordinary fold of tectonic origin.

Work, based on U–Pb analyses, published not long after the appearance of Stone's paper (Hughes *et al.* 1996), gave ages of 452 Ma and 450 Ma for the Ennerdale and Eskdale plutons respectively (i.e. Caradoc), pushing back earlier determinations of 420 Ma and 429 Ma of Rundle (1979) (i.e. Wenlock), based on Rb–Sr data. The new datings would, of course, still be 'permissive' of a late-Silurian cleavage, but would no longer necessitate such a late cleavage deformation. The new dates did, however, lend support to the Branney–Soper view (Chapter 16) that the granites were emplaced at about the time of the formation of the BVG; and late BGS field mapping on the northern margin of the Ennerdale aureole (Hughes & Fettes 1994) showed that the emplacement of this pluton predated the cleavage in the slates, as Allen (pers. comm., 2000) had earlier determined for the Eskdale pluton. With now just two main dates for the Lakeland plutons likely (Ennerdale and Eskdale, Caradoc; and Skiddaw and Shap, early Devonian) the radiometric datings of the 'granites' no longer served to fix the date of the SS deformations and cleavage. The deformation could, of course, have been linked with subduction and the formation of the granites and the BVG during the Caradoc; or it could have occurred in the Silurian, for all that the dating of the plutons revealed.

For a summation of Survey ideas about the history of the SS, so far as work on them during the twentieth century was concerned, one should consult the Skiddaw Memoir. By courtesy of Philip Stone I was, as mentioned, able to see proofs of some parts of that work while completing the present book. The 'synthesized scenario' for the twentieth century was, then, as follows.

Some syndepositional folds are first proposed. Then came the large-scale deformation associated with the production of the olistostrome of the Buttermere Formation during the late Arenig. This was followed by

> fault-block rotation and tilting, as regional uplift of the Skiddaw Group was initiated during the late Llanvirn in response to subduction and magma genesis prior to the main volcanic episode. The Eycott and Borrowdale volcanic groups, along with the older part of the Carrock Fell Complex, were probably then subjected to further large-scale folding, and preserved by synvolcanic subsidence before the mid-Caradoc. . . .
>
> The Skiddaw Group was affected by pervasive tectonism during the early Devonian Acadian event . . ., which deformed all the Lower Palaeozoic parts of the Lake District. Crenulation . . . indicates that the Acadian tectonic event consisted probably of more than one pulse of activity. . . .
>
> The main deformation event (D_1) produced the regional north-east to east-trending penetrative cleavage and its associated folds in the Skiddaw Group. The intensity of the S_1 cleavage is variable, and domains of contrasting intensity appear to be terminated or offset laterally by north-trending wrench faults (Skiddaw Memoir proofs, p. 45).

However, it is still difficult to relate the foregoing to the pattern of faults, as depicted on the Keswick sheet (1999*a*) (Plate V), discussed above (p. 205). Note that the Skiddaw Memoir postulates faulting of the Skiddaws *before* the emplacement of the BVG. As noted above, some of the mapped faulting could be Tertiary.

Before leaving the Skiddaws for the time being, attention should also be given to the work on Black Combe, as part of the Lakeland Project, and work over in eastern Lakeland on the SS Bampton Inlier. The main Project workers on and around Black Combe were the PhD student Neil Mathieson, Andrew Bell, whose sceptical views have been referred to above, and the Surveyor Eric Johnson, whose work will be considered again in Chapter 16.

As we shall see (Chapter 17), Bell played a part in developing ideas about the structure of the Windermere Supergroup in collaboration with Benjamin Kneller, but here I wish to say something about his work at Black Combe (1992), which he did on contract to the Survey while freelancing, with fieldwork in 1989; and on the Bampton Inlier, which he mapped on contract for the BGS in 1995–1996 (Bell 1997). It will be more convenient if we break the chronological order and consider the Bampton work first.

Bampton is a small village near the northern end of Haweswater, on the road from Penrith to that lake. The ground round there had not been seriously surveyed since the nineteenth century, apart from the PhD work of Michael Nutt (1970), and some unpublished work by Tony Wadge in the late 1960s and early 1970s. The Skiddaw Slates in the generally low-lying district are for the most part poorly exposed. Palaeontological work by Skevington (1964, 1970), Wadge *et al.* (1969), Wadge (1972), and Rushton (1996) indicated that the Slates in this area are younger than those to the north and west, and of Lower Llanvirn age, belonging to the *Didymograptus bifidus* (= *D. artus*) graptolite zone.[19] These eastern rocks might also extend up to the *D. murchisoni* zone, as evidenced by specimens obtained from the Tarn Moor Tunnel, situated about 3 km to the north of the Bampton Inlier.

Though, as mentioned, the rocks in the inlier are poorly exposed, Bell located a useful section by Keld Gill, in the lonely and dreary country of Ralfland Forest, SW of Shap between Swindale and Wet Sleddale (see Fig. 2.2). There, at 35397 51340, he examined a clear exposure of the SS–BVG junction, always important for Lakeland geologists. Bell observed bedded mudrock with dips approaching the vertical, overlain unconformably by half a metre of pale-grey, bioturbated mudstone, followed by tuffs belonging to the BVG.

Wadge *et al.* (1972) had reported a basal BVG 'Bampton Conglomerate' from the Tarn Moor Tunnel, and Nutt (1966, 1970) had referred to exposures of conglomerate in the district, taken to be basal to the BVG; but Bell was unable to locate conglomerate at the sites mentioned by the earlier workers, as they were either sealed up in the tunnel or were perhaps now covered by alluvium. Reinterpreting the previously described observations, he wondered whether the conglomerate did, in fact, represent a basal BVG unit, and suggested instead that it might be 'sub-Carboniferous'. The reasons were that some of the clasts in the conglomerate were microgranite, similar to that exposed at Threlkeld, over towards Keswick; that eroded garnets were to be found; and that there were BVG-like clasts in the conglomerate, which would be unlikely if the formation of the rock occurred at the onset of volcanism. The Threlkeld Microgranite had been dated Ashgill–Llandovery, which would be incompatible with its being a source of clasts for a basal conglomerate of Caradoc age, belonging to the Lower BVG. So the Bampton Conglomerate, which had reddened clasts, might belong to the Devonian (like the Mell Fell Conglomerate) or be otherwise 'sub-Carboniferous' (Bell 1997, p. 21) – and the same might well apply to the conglomerate found years before in the Tarn Moor Tunnel. (Its clasts were dark coloured, but could perhaps be tinted thus through long burial in damp, reducing conditions.)

Whether all this cast doubt on the supposed unconformity below the BVG at Matterdale Beck, on which Tony Wadge had placed much emphasis (see p. 132), I do not know; possibly it could, raising yet again difficulties about the understanding of the SS–BVG relationship. The most recent information I have on this point (Richard Hughes, pers. comm., 2001) is that Bell's suggestion about the Bampton Conglomerate was looked into by the Surveyors David Millward and Brett Beddoe-Stephens, who

[19] The name was changed as a result of difficulties in correlation of British and North American types, and the seeming absence of the American *D. bifidus* in the British rocks at the supposed corresponding stratigraphic horizon. So the former British *D. bifidus* zone required a new name and zone fossil. *D. artus* was chosen (Fortey & Owens 1987, p. 90).

satisfied themselves that the Conglomerate did indeed belong to the BVG. So the older view of Wadge *et al.* was upheld by the Survey at the century's end. However, to my knowledge, the evidence on which Millward and Beddoe-Stephens based their conclusions was not published before the end of 2000.

Bona fide BVG rocks were, however, to be found above the SS of the Bampton Inlier, and apparently belonging to the upper part of the Volcanics, which is the part typically well supplied with garnets. (However, there were also signs of igneous activity within the upper part of the Skiddaws themselves, suggesting the onset of volcanicity occurred before the end of 'Otley I'). The Skiddaw rocks of the Bampton Inlier showed no systematic bedding or cleavage orientation, but appeared to be inclined, thus according with the earlier observation of dipping rocks lying unconformably under BVG rocks at the Overwater spillway (see p. 151). Bell did not accept a full-scale olistostrome hypothesis for the Skiddaws so far as the Bampton Inlier was concerned, but suggested that the data allowed 'either complex refolding, or random fabric orientations as might be generated by soft-sediment slumping, mud diapirism or basement disruption during thermal doming prior to vulcanism' (Bell 1997, p. 30). Here, then, we have a hint of alternatives to the olistostrome hypothesis.

Let us now say something about the work undertaken in the early 1990s in the Skiddaws of the Black Combe area (see Fig. 6.3), which was remapped chiefly by Eric Johnson and Andrew Bell, with palaeontological support from Adrian Rushton and Stewart Molyneux. A glance at the photograph of the Coniston Limestone lying at a high angle against the Borrowdale Volcanics at Timley Knott (see Fig. 13.1) may be sufficient to suggest a monoclinal structure. Curiously, given the early suggestion of Green (see p. 000), this idea was not, so far as I am aware, elaborated greatly until the 1980s (see Chapter 17). However, when Bell and Johnson were at work in the Black Combe area, the idea was widely accepted. It is interesting to consider *why* Skiddaw rocks should be exposed at all in the Black Combe area, and forming substantial hills too. Geophysical work suggested that the Black Combe slates may be underlain by low-density igneous rocks in the northern part of the Inlier, compatible with indications of metasomatism in this area, which could have caused hardening of the slates. The situation would be analogous to the alteration of the rocks around the Crummock Water aureole, except that at Black Combe the alteration was presumably due to the large mass of the Eskdale Granite, exposed at the surface to the north, and plausibly occurring under the northern part of the Slates also. However, there is not a perfect match between the Eskdale aureole and the areas of bleaching and tourmalinization in the Slates. So there may also be a subterranean pluton in the Black Combe area, as at Crummock Water. As to surface occurrence of Skiddaw-type rocks at Black Combe, thinking of the proximity of the 'Westmorland Monocline' (the name given to the structure found where the 'Otley III' rocks meet those of 'Otley II'; see Chapter 17), one could envisage the Black Combe rocks as having been caught in the hinge of this structure and pushed towards the surface, where they are now revealed for the attention of the geologist's hammer.

Back in 1970, Helm divided the slates of Black Combe into three lithostratigraphic subdivisions on the basis of colour, and mapped accordingly, yielding what Bell (1992, p. 4) described as 'an almost unimaginably complex refolded structure', which was now found to be unsubstantiated. Eric Johnson (1992)[20] explained things more simply by proposing that the lighter coloured slates represented metasomatic bleaching, though Bell pointed out that in some places there were evidences of granitic sills around which there was no bleaching. Anyway, the colouration was 'not a primary feature of stratigraphic significance' (Johnson 1992, p. 3). Hence the geometrical pattern of the colouration might have nothing to do with polyphase folding, contrary to what Helm had supposed.

If colour could not be used to provide lithostratigraphic markers, and with much of the rock strongly cleaved so as to obscure the original bedding, the prospects for refined mapping in the Inlier were bleak. The rocks contained but few macrofossils, though assiduous searching by Rushton and Molyneux revealed a handful of poorly preserved, though identifiable, graptolites and a fair number of acritarchs in the southern unaltered 'blue slates'. These were allocated to the late Arenig (*Didymograptus hirundo* zone), but fossils were evidently not going to be sufficient to establish a stratigraphic sequence in the SS of Black Combe (Rushton & Molyneux 1989).

Regarding structure, we recall that Helm proposed six phases of deformation, the first two being (in accord with 'Simpsonian' theory) supposedly imposed prior to the BVG event. Johnson (1992) suggested that Helm's D_1 might be due to soft-sediment deformation, but the independent Bell (1992) looked at the matter differently. Indeed, he did not address the possibility of soft-sediment deformation, which was becoming increasingly popular amongst the Surveyors. Rather, he discerned a 'pervasive, relatively flat-lying fabric' either visible directly, or identifiable if the effects of later foldings were subtracted by stereographic rotation. He called it the 'main fabric' and equated it with Helm's D_2. Bell also identified a later quite upright folding of this 'main fabric', evidently caused by the emplacement of the granitic sills. This was Helm's D_4, which he had regarded as end-Silurian. Helm's D_5 and D_6 were weak, being represented by subordinate cleavages and kink-bands, and were dismissed by Bell as insignificant for understanding the general structure. They might, in fact, be relatively recent. The main fabric appeared to cleave sills too; so the 'main' event and the upright folding were perhaps approximately coeval.

The main fabric was itself folded, then. In addition, it exhibited an intersection lineation arising from folding of an earlier penetrative cleavage, and a stretching lineation (exhibited as grain alterations and alignments). With his earlier experiences on lapilli deformation and studies of strain, Bell was evidently searching for indications of different degrees of strain (and hence of stress) in the slates. Indeed, as he pointed out, the world had changed since Helm's time. In the 1960s and 1970s, geologists wanted to establish sequences of deformations (the more the better, sometimes!), without worrying so much about what might have caused them. By the 1990s, some at least wanted to develop kinematic models to explain the deformation sequences, and these models could perhaps be related to higher-level theory concerning the movements of terranes according to plate-tectonic theory.

According to Bell's analysis, the pale (altered) slates exhibited higher strain than the dark blue slates to the south. In the region of higher strain, the main-phase folds apparently tightened, the main fabric intensified, and the earlier fabric and bedding both rotated towards the main fabric. The high strain also evidently passed through into the BVG. Regions of strike-slip shear were discerned; and thrust faults were mapped, dipping NW, which

[20] Johnson (b. 1948) was brought up in a mining village near Sunderland, where he attended Robert Richardson Grammar School. From his early days he used to collect fossils from the Coal Measures and the local Permian, and minerals from the Pennines and the Lake District. He read geology at Queen Mary College, London University, and went on to write a thesis on amphibolites in south Norway, supervised by Ian Starmer. From 1971 to 1976 he worked on mineral exploration in Norway, and then obtained a post in the BGS, working in the Settle area from the Leeds office. This work involved looking at the Palaeozoic basement in the Craven inliers, and in writing this up he was assisted by Malcolm Howells respecting the thin Ordovician volcanics there. Johnson moved to the Newcastle office in 1982 to participate in the Lakeland Project, in which he was involved until 2000, when he was retrenched as a result of the restructuring of the Survey (Johnson, pers. comm., 2000). This, I think, was partly a result of the funding crisis in the Survey, following the termination of its Nirex work (see Chapter 20).

brought together the SS and the BVG, the latter cropping out between the Slates and the Eskdale Granite to the north.

Bell suggested, then, that the main fabric, the late upright fabric, and the granitic sills might all be associated with the main Acadian deformation of the Lake District in the late Lower Devonian (post-emplacement of the Eskdale Granite), with sinistral transpression, as advocated by authorities such as Donald Hutton (1987) or Bell's former supervisor, Jack Soper. What, then, were the earlier deformations due to? Bell did not state a firm view on this question, but there was the suggestion (among some other possibilities) that they might be related to the emplacement of the Eskdale Granite, perhaps in the Lower Devonian.[21]

So we begin to come around, once again, to the question of the role of the granites in the matter of the tectonic history of the Lakes. But that matter is (arguably) best treated in relation to the events associated with the formation of the BVG – which is the topic of the next chapter, to which, therefore, we now turn.

* * *

Additional note

The present study carries through to the end of 2000, making use of such information as was available to me at that date. By courtesy of Richard Hughes and Phillip Stone I was enabled to see advance copies of some pages of the BGS Skiddaw Memoir, and have referred to them, but I have not made use of the full publication. Possibly changes in my text would be appropriate if I had seen the whole text. But that will have to be dealt with properly by the next person to write on the history of geology in the Lake District.

[21] Johnson's (1992, p. 5) schema was generally simpler than all this. He applied the ideas on the northern Skiddaws from Webb & Cooper (1988) directly to the Black Combe situation as follows:

Time	*Structural event*	*Deformation episode* (equivalent terminology of Helm 1970)
Arenig–Llanvirn	Syndepositional slumping and sliding	D_1–D_3
Late Llanvirn–Caradoc	Uplift, deformation and erosion (folding of BVG to produce pre-Ashgill unconformity)	D_4
Early Devonian	Main Caledonian deformation (Ordovician strata folded and cleaved).	D_5–D_6

Chapter 16
Collaborating on the Borrowdale Volcanics and the granites

If working out a plausible story satisfactory to the majority of geologists was difficult for the Skiddaw Slates, and is probably still incomplete, things proved even more difficult for 'Otley II' and 'Otley III', as we shall shortly see. Getting an agreed mapping for the Borrowdale Volcanics, and an agreed interpretation as to the structure and history of the unit has proved to be the most controversial part of the whole story of Lakeland geology (though debates about 'Otley III' were perhaps more current by 2000); and the work done on this part of the sequence spun off into the debates about the burial of nuclear waste described in Chapter 20. Yet the basic idea that the Ordovician volcanoes gave rise to the BVG as part of a system of arc volcanism, in accordance with orthodox plate-tectonic theory, was never a controversial issue, though not all have agreed as to whether the volcanism was or was not wholly subaerial. Even so, the *method* of dealing with the BVG proved controversial. So too have been the tectonic implications for the structure of the Lake District, the relationship of the ancient volcanoes to the emplacement of the granites, and the role of these in the geological history of the region.

As for the Skiddaw Slates, the Surveyors were again first into the field in the examination of the BVG. The principal workers on the volcanics were Peter Allen, David Millward (see pp. 168 and 191), Eric Johnson, Michael Petterson and Brett Beddoe-Stephens. Tony Cooper also lent a hand at times. Brian Young did much work on mapping the granites, as we have seen. The PhD students involved with the BVG at Sheffield (see Fig. 14.3) were Neil Mathieson (1986), Michael Branney (1988*a*) and Neil Davis (1989). The later Liverpool Project, concerned chiefly with the remapping of the Central Fells, was set up initially by volcanologist Peter Kokelaar (see Fig. 14.2) and Branney in 1989, but in 1992 Branney went off to Korea for a while to work on calderas there with Surveyor Tony Reedman. So Benjamin Kneller (see p. 231) moved over from Leeds, where he had been with Soper, to take his place. Kneller subsequently obtained a position back at Leeds but in 2000 he moved to the University of California at Santa Barbara. Another post-doctorate working on the Liverpool contract was Brian McConnell, presently with the Irish Geological Survey. Kokelaar was leader of the Liverpool group, which Branney rejoined after his return from Korea until he obtained a lectureship at Leicester University in 1996. However, he continues his collaborative work with Kokelaar. Branney and Kneller had been together as undergraduates at Sheffield.

The older view of the Borrowdale Volcanics – derived from workers such as Mitchell, Oliver, Firman, Moseley, Soper, Millward and others – was that they formed chiefly a succession of various andesite lavas in the lower part of the succession, with a predominance of tuffs and ignimbrites in the upper part of the succession, forming particularly the high ground of the Central Fells, e.g. Great Gable, Scafell, Bowfell, the Langdale Pikes and so forth. Thus the frontispiece map to Moseley's edited volume on the Lake District (1978*a*) showed a distinction between Lower and Upper Borrowdale Volcanics. The various igneous rocks could not all be correlated from west to east, though much effort was made to do so earlier in the twentieth century by workers such as Green and Hartley, and Mitchell too in his younger days. Moseley's 1978 volume showed the uneasy state of correlation between the eastern and western BVG (see Fig. 12.6). The BVG rocks as a whole seemed to have been thrown into broad folds, running approximately SW–NE, such as the Scafell Syncline, the Wrynose Anticline, and the Ulpha Syncline in central–west Lakeland, or the Nan Bield Anticline to the east. There might be thrust faulting near Ullswater, according to Moseley.

It was natural that the resurvey of the BVG should try to build on previous work, and thus it was that the Surveyors got to work in SW Lakeland, near to, but not exactly on, the old terrain of Mitchell and Firman, leaving for the time being the more daunting high ground of the Central Fells, where the work of Oliver in the 1950s was still the most important contribution. (This high ground was subsequently mapped to a large extent by the PhD students and the geologists working on the Liverpool contract.) Also, given that David Millward was one of the initial team, and that his PhD had had a strong geochemical slant, it is understandable that the Surveyors sought to use geochemical techniques so that they could 'fingerprint' the bewildering assortment of andesites, ignimbrites, etc., for stratigraphic purposes.

Eric Johnson (pers. comm., 2000) recalled for me how things went with his mapwork with Millward. They began on the southern limb of the Wrynose Anticline in the upper Duddon Valley (see Fig. 16.1), where there had been no systematic surveying since the primary survey in the nineteenth century. This was a region with a complex succession of andesite lavas, but a link was established between this area and Mitchell's 'Grey Friar' and 'Seathwaite Tarn' andesites, the succession seemingly being repeated by folding and faulting. Correlation across to the Central Fells was achieved via the Whorneyside and Airy's Bridge Formations (see pp. 98 and 217), using further pyroclastic markers (the Cockley Beck Group and Grey Friars Tuffs). Firman's 'Lower Andesites' were regarded as equivalent to Mitchell's Cockley Beck unit (Johnson & Millward 1987, reporting a Yorkshire Geological Society excursion of 1984).

In 1985, the group moved south and west to help complete one of the special 1:25 000 sheets, the Devoke Water and Ulpha sheet (BGS 1990), which Allen had from the start regarded as an area of special interest. Thus mapping of the Duddon Valley was undertaken in detail, from which it appeared that it had a succession analogous to that in the Central Fells, but considerably thicker. The apparent area of subsidence was called the Duddon Basin, and the discovery of just one seeming marine unit (named the Holehouse Gill Formation, at 3182 4926, in Dunnerdale; see Fig. 5.2) by the primary survey in the nineteenth century, also referred to in Numan's thesis (1974), was confirmed. The unit was found to contain acritarchs regarded, perhaps with some uncertainty, as Caradoc (Molyneux 1988); but there is the possibility that these microfossils have been reworked, so an age-determination made solely on such evidence is necessarily somewhat doubtful. Nevertheless, this was an important find, providing the only direct palaeontological evidence for the age of the BVG (the Drygill Shales being faulted in).[1]

In the mid-1980s, however, the Surveyors began to run into problems of establishing correlations with the work being done in the Central Fells by the PhD students, of which more anon. Then in 1992 the Survey men moved towards the SW Lakes, onto the Ulverston sheet, the aim being to link the area to the NE of Millom (Millom Park) with the PhD work of Neil Mathieson to the north of the Black Combe area. This was the area where J. F. N. Green had first tried to establish his succession for the

[1] As I understand from Peter Kokelaar (pers. comm., 2000), work on the Holehouse Gill acritarchs, unpublished at the end of the century, suggested that they may extend back to the age of the SS, so the Holehouse Gill rocks may *not* represent a marine incursion, but a reworking of the sedimentary basement, perhaps arising from a debris flow. If this be so, then the BVG rocks could have been formed wholly subaerially, apart from the sediments formed in transient volcanic lakes. In fact, the Holehouse Gill unit is a curiosity. Michael Branney (pers. comm., 2001) drew my attention to the fact that on the Ambleside map it appears to crop out both above and below an andesite sheet. Could it be a slither of marine sedimentary rock emplaced by thrust faulting?

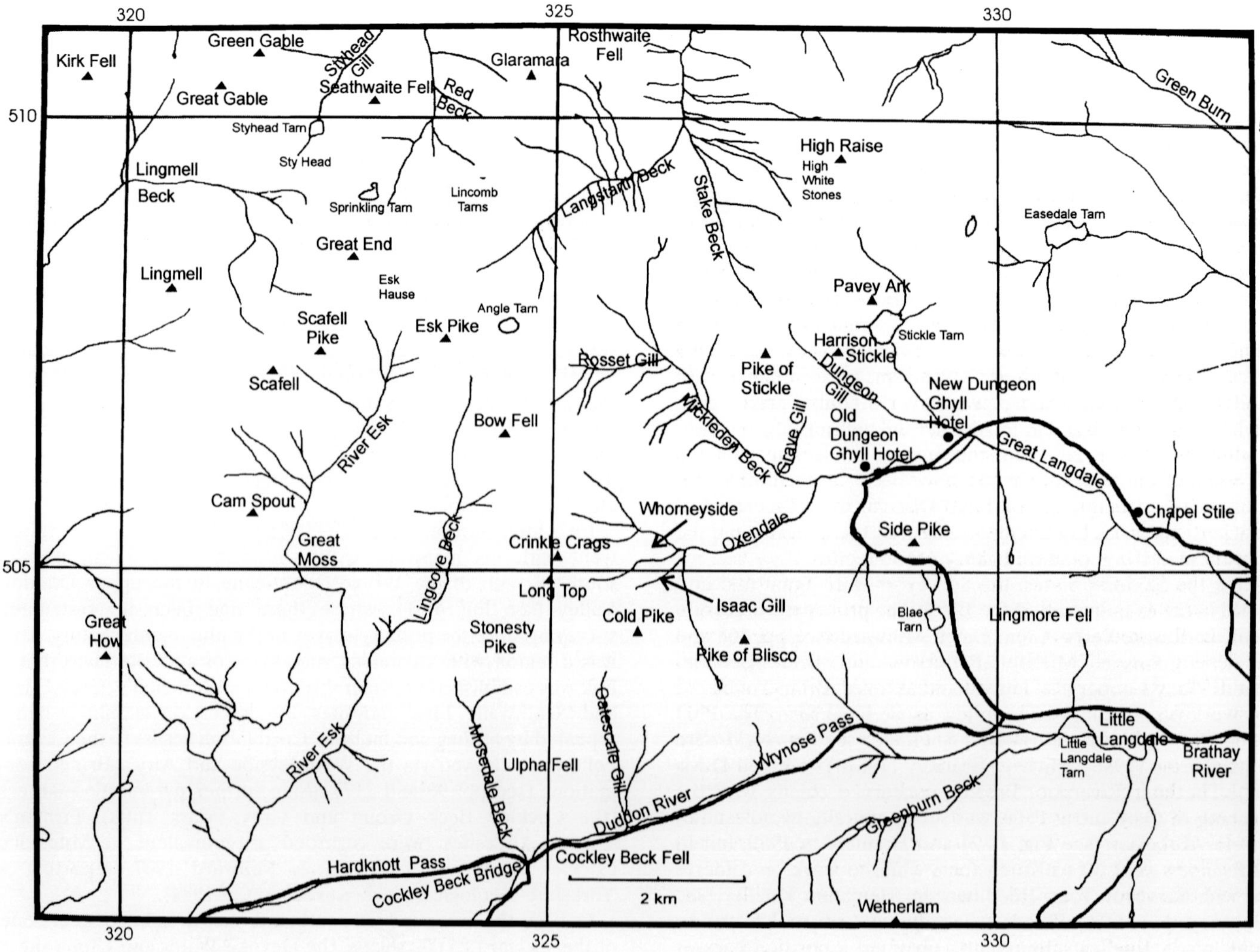

Fig. 16.1. Topography of the area of the upper Duddon Valley, Langdale, and the Central Fells.

Lakeland volcanics. The Surveyors interpreted their observations as indicating that many of the volcanics there were intrusive sills rather than lavas, and that the host rocks (in that relatively low-lying region) were part of the *upper* BVG succession, even though they lay in places directly on Skiddaw Slates. They were not able to account for this in theoretical terms at the time. However, lower BVG rocks were confirmed on the northern side of Black Combe, where thrusting was proposed (and is shown on the recently published 1: 25 000 Black Combe sheet; BGS 1998). This fitted in with ideas being developed regarding the structure of the Lakes and the formation of the Westmorland Monocline (see pp. 226 and 233). From 1995, Johnson did work on the BVG rocks of the eastern Lakes on the Kendal sheet.

A somewhat younger Surveyor working with Johnson in the early days of the Lakeland Project was Michael Petterson (see Fig. 14.3).[2] He mapped initially on the Wasdale area of the Ambleside sheet, working northwards towards Seatallan, Scoat Tarn and Scoat Fell, and the Ennerdale Fells (see Fig. 7.4), attending to the contacts between the Eskdale Granite and the BVG. Then he went over towards Grasmere, Rydal Water and Rydal Valley, and Ambleside (see Fig. 2.2). From 1993 he was involved in the contract work for Nirex (see Chapter 20), but this project collapsed, the details of which we shall see – though not through any fault of Petterson! Petterson also worked in the Solomon Islands and Guiana and in 1998 was leader for the mapping project in the Southern Uplands.

Another Surveyor in the Lakeland Project was Brett Beddoe-Stephens[3] He was not involved in the Ulpha–Duddon area, but starting from Eskdale he worked across the southern side of Wasdale, the Ennerdale area, Dale Head, lower Borrowdale, west of Thirlmere, the Easdale Tarn area, then north of Ambleside, and across the Kirkstone Pass (which runs between Ambleside and Ullswater). Towards the end of his BVG mapping he worked on the rocks of the Haweswater area in eastern Lakeland, where

[2] Petterson went to school in Durham and on to Leicester University, reading geology and graduating in 1981. For his PhD he participated in the Karakoram Project in northern Pakistan, doing reconnaissance mapping near Gilgit. There he found himself looking at 'an island arc, turned on end, with a huge batholith' (pers. comm., 1998). His work relied heavily on isotope geochemistry and he considered the complex problems of tectonics and petrogenesis in that Himalayan region. Following such adventures and back in Britain, Petterson did a Dip. Ed. and taught for a year at Hexham, but perhaps unsurprisingly he did not stick with this and in 1987 he obtained a position with the BGS in the Newcastle office, where he was promptly put onto the Lakeland Project for five years. From 1997 to 2000 he was leader for the Southern Highlands Project in Scotland and in 2000 became Manager of the Onshore Minerals and Energy Programme.

[3] Beddoe-Stephens (b. 1952) did his PhD at Durham on metamorphic rocks in British Columbia. Following this, he obtained a post with the BGS as a microprobe specialist, and in 1983 became involved with the Lakeland Project, doing analyses of minerals from the Eskdale Granite aureole in the London office. He was then posted to the Edinburgh office and started on the Lakeland work with Michael Petterson, David Millward and Brian Young. Thereafter, he continued work on the Lakeland volcanics every year of his time with the Survey except one (Beddoe-Stephens, pers. comm., 1999).

he collaborated with Derek Woodhall. Beddoe-Stephens saw a lot of the BVG; and he too was part of the BGS contingent working on the Nirex contract. He told me (pers. comm., 1999) that he thought there was a considerable need for a BVG memoir, but doubted that one would ever be produced.

With the members of the BGS Lakeland team mentioned above, and with Neil Fortey, Derek Cooper, David Millward and Brian Young also involved, there was a strong interest in geochemistry, and it seemed to these workers that geochemical techniques could be used to provide a stratigraphy for the Lower BVG succession, in which they were chiefly working. However, Petterson recalled (pers. comm., 1998) that he, Beddoe-Stephens and Millward initially had difficulty in persuading the BGS authorities to adopt a geochemical approach. They eventually succeeded and began sampling the andesite pile of the Lower BVG (collectively called the Birker Fell Formation), layer after layer, through a depth of more than two kilometres. (There were also some basalts, dacites and ignimbrites, and some interbedded tuffs. Some of the layers were construed as sills, though the proportion of these became an issue that was contested with the university participants in the Lakeland Project.) Chemical analysis of the layers in the pile provided the Surveyors with a method for sampling the 'plumbing system' in the depths of the volcanoes' magma chambers, and the fractionation processes occurring therein. In this lower part of the BVG sequence the Surveyors were dealing essentially with an effusive volcanism (and the Upper BVG was, as we shall see, interpreted as a product of explosive volcanism). The distinction between upper and lower parts of the volcanic pile was not new to geology with the Lakeland Project. As mentioned earlier, it was represented, for example, in the frontispiece map in Moseley (1978*a*).

An early Survey publication with a geochemical slant on the Lower BVG rocks was one by Allen *et al.* (1987), which discussed the lava flows in the area of the upper Duddon Valley. Fortey examined thin-sections and did the mineralogical identifications and petrography. Cooper undertook the geochemical analyses, while Allen did the field mapping and provided the main interpretations underlying the observations. It was specifically stated (Allen *et al.* 1987, p. 957) that the 'distinctive geochemical signatures' of the lava flows were to be used for the purpose of mapping, which was to become a feature that to some extent differentiated the approach of the Survey workers from that of the university collaborators.

The objective of the paper was to map some of the lavas north of the upper Duddon, analysing their compositions by X-ray fluorescence spectrometry, and attempting to develop a theory as to how the composite lavas might have been generated. The area mapped can be compared with that subsequently published in the 1:50 000 Ambleside sheet (1996) (see Plate VII). There are significant differences in marked outcrops, lines of faults, and rock designations. Notably an area mapped as 'Andesite, undivided' in 1987 becomes 'dacite' a decade later. But disregarding this (since a dacite is essentially a quartz andesite), the earlier map has some internal problems and it is difficult to link it to the generalized section proposed and to the theoretical model, which was presented in diagrammatic form (see Fig. 16.2). There was an unkeyed rock type in the 1987 map that became 'composite andesite flow' in 1996; and similar rock types were mapped with different symbolism in different areas of the 1987 map. The point of this was, I think, to suggest that there was one essential process at work in the generation and emplacement of the lavas, but different parts of a sequence of events were sectioned or sampled in the different areas.

Allen's model (I think it was essentially his, but it was based on ideas then current in the literature – especially Eichelberger 1980) was that there had been simultaneous emplacement of magmas of different composition. Hybridization could take place to some extent, but also there could be interlayering between lavas of different composition (basalt, basaltic andesite and andesite).

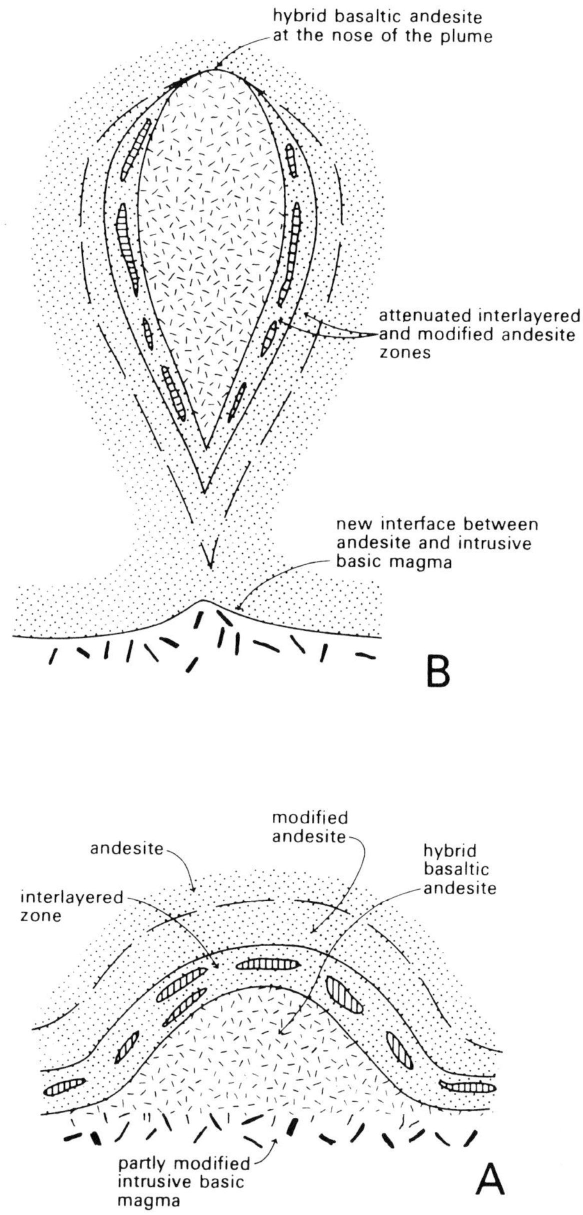

Fig. 16.2. Hypothetical development of a magma plume, suggested by mapping in the upper Duddon Valley, according to Allen, Cooper & Fortey (1987, p. 959). IPR/26–13C British Geological Survey. © NERC. All rights reserved.

Relatively hotter, more basic, matter could have intruded a cooler, more acidic, melt. Crystallization of the more basic magma could have increased the volatile content of the residual liquid thereby reducing its density. So andesite magma could have been emplaced first, followed by basaltic magma, which formed a layer at the bottom of a magma chamber. Crystallization of plagioclase from the more basic magma could produce a less dense residual liquid which could rise as a plume or diapir through the andesite melt, hybridizing therewith, but also carrying up some crystallized basalt on the 'nose' of the plume. The process could be repeated, and repeated again if the magma chamber were recharged. It was supposed that various 'sections' of the hypothetically modelled rocks might be seen by ground traverses of the lavas exposed in the Duddon Valley. However, as will be seen, it does not appear that the later Surveyor theorists sought to develop or apply this particular model further, and with Allen already away in his new post in Exeter the idea seems to have been dropped. The 'phenomenological profile' of the composite lavas in the Wrynose area survived in the Ambleside Memoir (Millward *et al.* 2000,

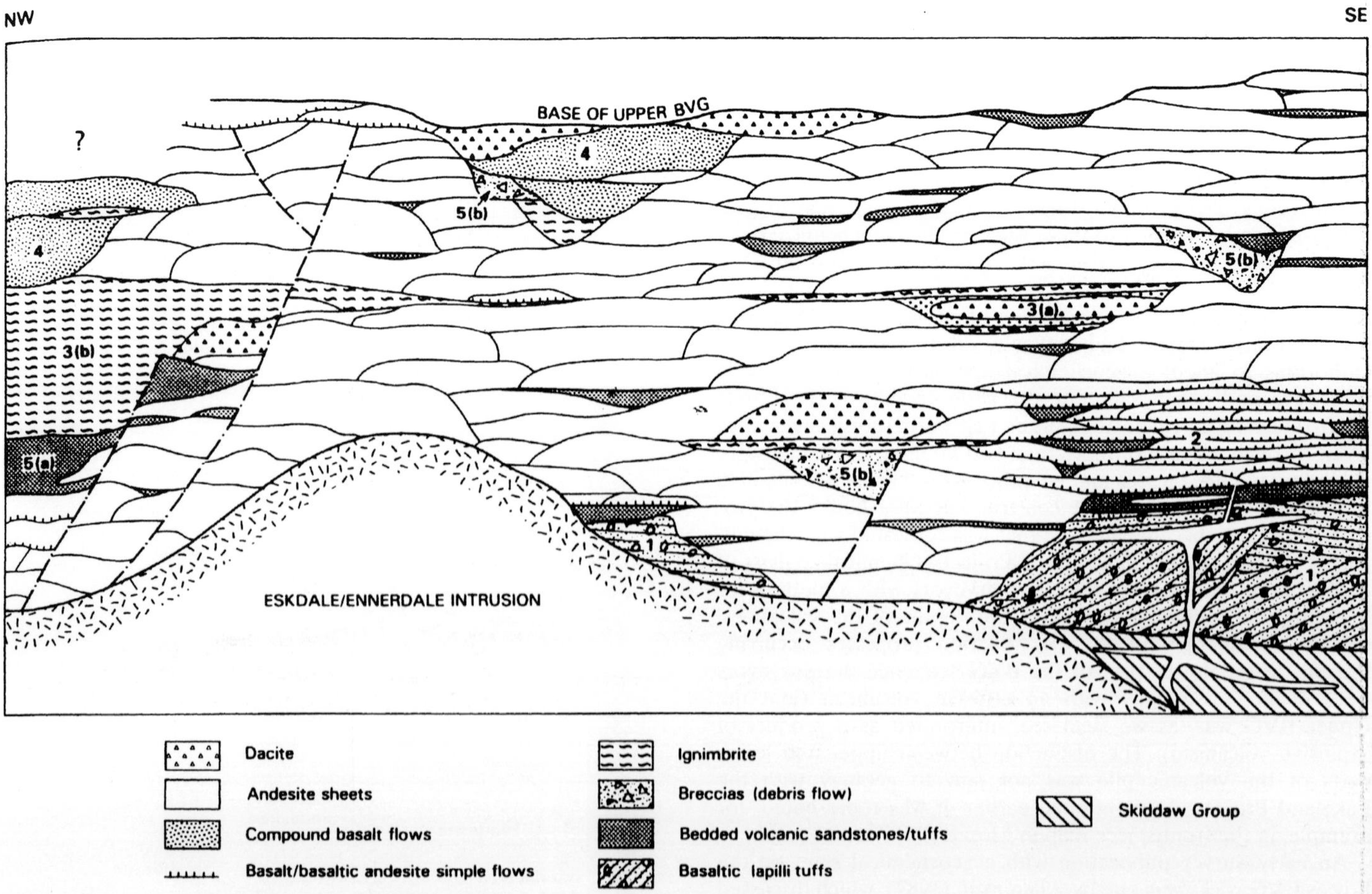

Fig. 16.3. Schematic representation of Lower Borrowdale Volcanic Group, according to Petterson *et al.* (1992, p. 899). IPR/26-13C British Geological Survey. Cf Plate VI.

p. 67), but the theoretical explanation proposed in Allen *et al.* (1987) did not.

In several areas of Lower BVG outcrops, for example St John's in the Vale, south of Keswick, the lavas exhibited a 'stepped' topography such as may be seen in basaltic lava fields (as in the Inner Hebrides or in the Deccan Traps of India). However, the 'steps' in the Lake District were made up of andesites, typically flow-brecciated, not basalts as in the Hebrides; so the region exhibited a somewhat unusual volcanic field. There were not obvious facies changes around central volcanoes, according to the distances from supposed eruption sites (as one sees, for example, around Mount Ruapehu, in New Zealand's North Island). So deploying an earlier idea analogous to that of the Keele PhD student, Roger Suthren (1977), who used analogies with Mexican volcanoes, and works of Branney (1988*a*, 1991), the Surveyors, with Petterson in the lead on this issue (Petterson *et al.* 1992), developed the idea of an andesite-plateau field, with eruptions from fissures or low shield volcanoes, not from strato-volcanoes as previously envisaged (Fitton & Hughes 1970; Moseley & Millward 1982).

However, by the time Petterson *et al.* (1992) was published, fresh ideas about the structure were being incorporated into the Surveyor's thinking, arising from the work of the university collaborators, particularly the student Michael Branney (see p. 216). As we shall shortly see, for the upper part of the Borrowdale pile Branney & Soper (1988) developed the idea of piecemeal caldera collapse with associated 'volcano-tectonic' faulting. That is, there were mappable faults that were formed at the period of the original eruptions of the volcano, not subsequently due to earth movements associated with Acadian deformations, or whatever. The idea of volcano-tectonic faults was also used by the Surveyors in their interpretations of the lower, andesitic part of the pile, as can be seen from a work-in-progress drawing from 1989, kindly made available to me by Michael Petterson (see Plate VI). The ideas in this sketch were subsequently refined and published (Petterson *et al.* 1992) (see Fig. 16.3).

From Figure 16.3 and Plate VI we can see

1. how a distinction was made between the Lower and Upper Borrowdales;
2. how volcano-tectonic faults were named thus in Plate VI and appeared similarly in Figure 16.3, and were represented in such a manner that they must have occurred *within* the Ordovician volcanic episode;
3. how some of the andesites were regarded as unconformable to one another;
4. how ignimbrites were taken to characterize the Upper BVG but occurred occasionally in the Lower BVG;
5. how a mix of rocks was found in the Lower BVG, but with andesites predominating;
6. that Skiddaw Slates and the large granite masses were added to the cartoon in the published version (which is perhaps relevant, as the Surveyors were later to consider the effects produced by the magmatic assimilation of Skiddaw Slate material);
7. that in the published version a vent for the emplacement of basalt was shown, but not for the andesites.

The Surveyor's mapping – around Wastwater, for example – showed a complex tangle of faults in the andesite field. The structure was interpreted as representing a number of discrete fault-blocks, which seemed to have resulted from brittle fracture, with some rotation of blocks, rather than being the product of regional compressive folding. Thus a fault structure for the Lower BVG was suggested that had some analogy to that proposed by Branney & Soper (1988) for the Upper BVG (see pp. 218 and 220).

The chemical approach to the study of the Lower BVG yielded

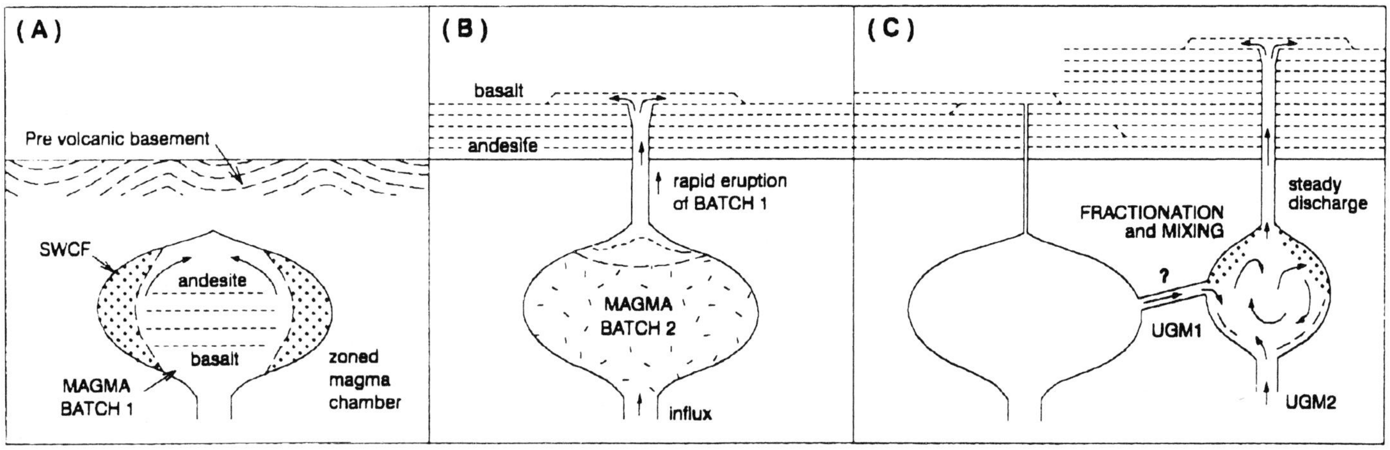

Fig. 16.4a. Hypothetical development of magma chamber for area of Devoke Water (see Fig. 5.2), according to Beddoe-Stephens *et al.* (1995, p. 106) (cf. Fig. 16.2). Reproduced by courtesy of Elsevier Science. IPR/26–13C British Geological Survey.

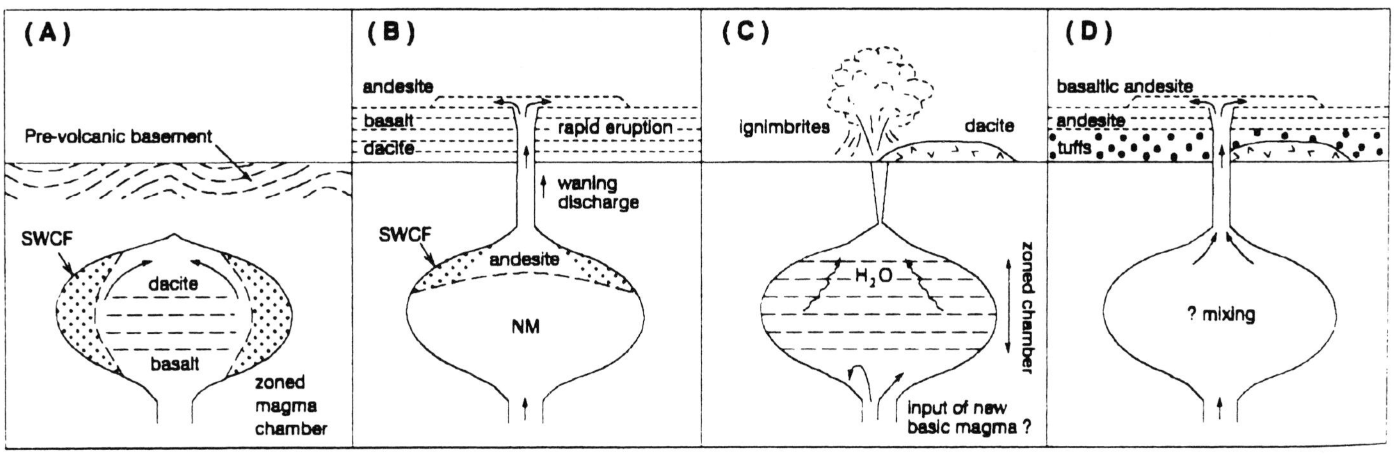

Fig. 16.4b. Hypothetical development of magma chamber for area of Haycock, according to Beddoe-Stephens *et al.* (1995, p. 106) (cf. Fig. 16.2). Reproduced by courtesy of Elsevier Science. IPR/26–13C British Geological Survey.

some perhaps unanticipated results in that analysis revealed an apparent cyclicity in the effusion of the plateau field. For example, near the mountain of Haycock (14 10) (between the valleys of Ennerdale and Wastwater; see Figs 7.4 and 16.2), the number of parts per million of cerium was found to rise and fall in an apparently regular fashion as one ascended the volcanic 'heap'. So a particular layer that had an anomalously 'low' reading by comparison with the others was conjectured to be a sill rather than a lava flow.

The view developed for the Lower BVG was, then, that the volcanics were emplaced with low topographic relief, and where fault scarps or erosional gullies formed depressions these were filled by the ponding of subsequent lava flows, or occasional ash falls, which might sometimes have been reworked. Analogies were proposed with plateau-andesite fields in Chile, Mexico and Central America. But to achieve the postulated low relief there had to be regional subsidence to allow the accumulation of the thick volcanic sequences; and this implied the operation of intra-arc *extensional* forces. Such a state of affairs might, it was suggested (Petterson *et al.* 1992, p. 905), be associated with 'rifting within a destructive plate margin environment'.

The matter was not left like that. Three years later another paper – much more complex and theoretical, examining the chemical cyclicity and proposing explanatory volcanological models – was published by the BVG Surveyors, joined by Giz Marriner from the Royal Holloway College, London University (Beddoe-Stephens *et al.* 1995). Working up the stratigraphic sequence, at Haycock and five other localities in the andesite field, they prepared 'Harker' variation diagrams, plotting the proportions of TiO_2, Al_2O_3, Fe_2O_3, MgO, CaO and K_2O against SiO_2, and likewise for less common elements such as vanadium and cerium; also the cerium/yttrium ratio. The exact chemical variation differed somewhat in the six localities examined, but the overall pattern of variation was essentially the same at all the places sampled, so that averaged values could legitimately be calculated and tabulated – which was done. (For this, the Haycock results were omitted as they seemed to differ significantly from the others.)

One then needed to propose theoretical geochemical and petrological models as to what might have been going on during the formation of the magma, its rise through the crust and its extrusion. It is difficult to summarize the arguments briefly, but in sum the data seemed to be compatible with the hypothesis that the magma originated as a primary melt in the mantle; that as it rose it assimilated crustal material, perhaps from the Eskdale Granite; and that it underwent crystal fractionation before extrusion. The data for zirconium, for example, plotted on a Harker variation diagram and compared with theoretically calculated values assuming assimilation and fractionation, fitted the data better on the assumption that granitic matter was assimilated, rather than material of the bulk composition of Skiddaw Slate.

As with Allen *et al.*'s (1987) findings, the geochemical sequence was not observed to be identical at all the sites investigated. At Devoke Water, a trend from more basic to acidic product was found; while at Haycock the sequence was reversed; and at Throstle Garth (in the upper reaches of Eskdale SE of Scafell) there seemed to be no overall trend with time. So the sequences of events, shown in 'cartoon' form in Figure 16, were postulated.

At Devoke Water, the process seemed to have been as follows. Mantle-derived magma entered a sub-volcanic chamber and partial crystallization occurred as cooling took place, chiefly at the walls of the chamber. Gravity differentiation of melts of different densities occurred, with the more siliceous material accumulating uppermost. The influx of a second batch of magma pushed out the previous material – that of andesitic composition, which was extruded first – so that in the external flows basaltic sheets came to lie atop andesite. This process appeared to have been twice repeated. Then a second magma chamber might have formed, yielding material that was extruded at the top of the Lower BVG sequence, possibly with the admixture of material from the first chamber. This model was not wholly different from that of Allen *et al.* (1987) (Fig. 16.2), but greater emphasis was placed on the assimilation of material from the walls of the hypothetical magma chamber.

At Haycock, the first stage was supposedly similar to that at Devoke Water except that the magma in the upper part of the chamber was this time supposedly dacitic. Thus the arrival of new magma (NM) caused the surface production first of dacite and then of basalt. As the influx of new magma declined, a pause allowed the occurrence of side-wall crystal fractionation and mixing, yielding an andesitic melt. The arrival of an additional magma batch pushed out dacitic matter; and then the arrival of volatiles in the chamber led to the production of a modest explosive event, with the production of tuff or ignimbrite. The subsequent effusion of andesite and basaltic andesite could have come from the lower parts of the previously zoned chamber. The different, more basic material found at Throstle Garth suggested eruption from yet another magma chamber. All this was regarded as compatible with the model previously developed by Petterson for the production of a plateau-like lava-field, with a number of discrete magma sources.

The foregoing gives but a sketch of a complex process of geochemical reasoning, which involved a vast amount of chemical analysis, calculation, comparison with sites elsewhere in the world, and the mapping of a large number of rocks, often similar to one another as they presented in the field, with relatively few marker bands recognized by the Surveyors. It was to be rather different for the university researchers when they got to work with the Upper BVG in the high country where, as we shall see, there were many different tuffs to be mapped, which could often be identified relatively easily by their external appearance in weathered outcrops. Doubtless it was natural for the Surveyors to begin their work in the terrain of lower altitude and to try to attach their work to that of previous investigators such as Mitchell. But in doing so, they may perhaps be said to have drawn the short straw. As Michael Petterson put it to me rather ruefully at a rainy lunchstop in the Rydal Valley, when he was leading a Cumberland Geological Society field excursion in August 1998, the Surveyors drew the short straw while the university geologists picked the cherry off the top of the cake! It should be remarked that the map accompanying Petterson *et al.* (1992, p. 891) differed substantially from that of the Ambleside sheet (BGS 1996), especially for the area of the Hardknott–Wrynose Fault taken to run down the Upper Duddon Valley (see Fig. 16.1).

So what did the university people get up to in the BVG? As mentioned, Soper at Sheffield had three of his students working on the volcanics in the BGS–university Lakeland Project: Neil Mathieson (1986), Michael Branney (1988*a*), and Neil Davis (1989). Of these, the work of Branney (see Fig. 16.5) proved particularly important, and he will enter the account that follows in various ways. In fact, Soper was not really a volcanologist (see p. 165), and during the course of his PhD Branney began to get advice from Kokelaar at Liverpool; [4] and, as already mentioned,

Fig. 16.5. Michael Branney as a student in the Lake District. Copy of photograph supplied by Dr Branney and reproduced by his permission.

after graduating Branney joined the Liverpool team, doing contract work on the Central Fells for the BGS. Soper (pers. comm., 1998) described Branney as the best student he ever had, and he contended that it was Branney who made the breakthrough in understanding the Lakeland volcanics: 'he got further in a year than all the Survey staff had with their years of experience'!

Branney (pers. comm., 1998) has told me how he proceeded initially.[5] He was given a piece of ground to work on and went over it with geologists who already knew the area. He was then let loose in the field by himself. He read the literature on modern volcanoes during the winter: people like George Walker (Imperial College, Auckland, Hawaii, and latterly Bristol), Stephen Sparks (Imperial and subsequently Professor at Bristol), Stephen Self (Imperial, and subsequently Professor of Geology and Geophysics at the University of Hawaii), and John Wright (a student of Walker at Imperial, subsequently working in Puerto Rico and Australia). At his own expense, Branney visited the Italian volcanoes in the Mediterranean: Pantelleria (between Sicily and Tunisia), the Aeolian Islands, Stromboli, and the Lipari Islands. There he could see recently formed tuffs, ignimbrites and lavas, and volcanoes in action. Later, after he had come to think he could see evidence for a former volcanic caldera in the Scafell region, he went to the western United States to look at places in southern Colorado and New Mexico, and found that little was known about the more recent caldera structures at depth.

[4] In addition, Branney was co-supervised by Raymond Satirio at Sheffield.

[5] Branney (b. 1960) was schooled at Sale Grammar School, south Manchester. Like many geologists, he became interested in his subject through an early interest in fossils. However, his attention shifted from palaeontology to volcanology.

Fig. 16.6. The Whorneyside phreatomagmatic tuff at Stonesty Pike. The bedding records successive showers of ash and accretionary lapilli from one of the early voluminous explosive phases of Scafell caldera. The soft-state thrusting is thought to have occurred when the ground tilted as the incipient caldera started to subside. Photographed by author, 1999.

However, an ancient caldera could be examined from a different perspective, and more satisfactorily, in the Lakes, where extended erosion revealed its deep structure. In 1987, Branney participated in a field trip to Santorini, led by Stephen Sparks, Timothy Druitt and Robin Mellors, which he remembers as being 'hugely instructive'. There he saw andesites, ponded within old collapse depressions, which rang a bell in relation to what he had seen in the Lakes (Branney, pers. comm., 2001).

Branney started mapping around Side Pike, one of the lower hills in the Langdale Valley, and to the south on the rough terrain between the Langdale Valley and Wrynose Pass (see Fig. 16.1), fairly near where Peter Allen was working. But the area proved exceedingly perplexing, as it is to all intents and purposes a kind of megabreccia (analogous to one in the San Juan Mountains, Colorado), and lacks continuously mappable elements. Branney worked there for a season and then moved northwards into the area around Crinkle Crags, which form part of the high ground ringing the head of Langdale Valley. This shift of ground did not please Allen, who wanted the collaborating students to follow BGS methods and get a continuous area mapped. Also, it entailed Branney moving into some territory previously covered by the Surveyors and offering reinterpretations thereof (Allen, pers. comm., 1999). But Branney does not recall that his mapwork differed greatly from that of Allen, though he did interpret one of Allen's composite lavas as a peperitic sill.

Branney's work depended on following tuff marker bands, rather than making geochemical analyses such as the Surveyors were emphasizing. In particular, Branney noted a tuff that served as an important marker band, apparently at the bottom of the Upper BVG, which conveniently separated it from the layered andesites of the Lower BVG: the Whorneyside (pronounced 'Wurneyside') Tuff (now Formation). It is an andesitic lapilli tuff, the lower part being an ignimbrite (called the Wet Side Edge Member by the Survey); and the upper part came to be interpreted as the product of a phreatoplinian ash-fall (Branney 1991; Branney & Kokelaar 1993). The type locality is at Whorneyside Force (waterfall), on the south side of Crinkle Crags (see Fig. 16.1), but the rock had distinctive weathering patterns and Branney found that they could trace it, up hill and down dale, all the way from Seathwaite Tarn in Dunnerdale, across the north Coniston Fells, through Borrowdale, and as far east as Thirlmere. The Whorneyside Tuff formed the lower part of Oliver's (1961) Airy's Bridge unit, but was not recognized by him as a separate entity. This was partly because Oliver's mapwork did not extend to the Whorneyside area – though he must have viewed the corresponding rock in the Seathwaite area, by Sour Milk Gill, where he *did* do a lot of work, and where the Tuff is well exposed. A profile of the important Tuff, as exposed at Stonesty Pike, is shown in Fig. 16.6.

The recognition of the Whorneyside Tuff as a distinctive marker band separating the Lower and Upper Borrowdales has seemed important to others (but Branney himself, in interview, thought it over-simplistic). Be that as it may, he was able to recognize other tuffs too, higher in the sequence, each with its own characteristic appearance, especially in weathered outcrops: Stonesty Pike Tuff, Cam Spout Tuff, Glaramara Tuff, and so forth (see Fig. 16.1). This allowed lateral correlation between rocks of different facies. For example, if one found

	Area A	*Area B*	
Stonesty Pike Tuff			Stonesty Pike Tuff
	andesites	lacustrine deposits[6]	
Whorneyside Tuff			Whorneyside Tuff

then there was good reason to believe that the andesites and the lacustrine deposits in Areas A and B were of about the same age, for the tuffs had 'draped' the landscape over large areas during individual eruptions. Different ignimbrites might likewise be correlated.

The method worked well in the BVG where biostratigraphy could not be used, and where geochemical evidence could be ambiguous, but it was demanding in the field, as one had to 'chase' a tuff over long distances, around, up, down and over the mountains. Earlier workers such as Green and Mitchell had, as we know, sought to correlate lavas with lavas, tuffs with tuffs, lacustrine sediments with lacustrine sediments, etc. In Branney's view, this had caused a deal of confusion in the earlier investigations of the BVG (pers. comm., 1998). However, one had to watch out for the fact that the tuff marker bands could also show lateral variation in thickness and grain size, according to their distance from a volcanic vent, and sometimes they might be no more than a metre thick. So the tuffs had to be followed with care, and volcanological 'insight' was needed as well as 'mechanical' mapping. The methodology was new in application to the Lakes, but had standardly been used by volcanologists working on more recent volcanoes. It was gratifying to use it on Ordovician volcanoes also.

The trouble was that the Surveyors had already mapped in Dunnerdale using different methods from Branney's, so that even at the end of the twentieth century there was disagreement between Kokelaar and Branney on the one hand, and the Surveyors on the other, as to the interpretation of the rocks in that area. Unsurprisingly, the Surveyors did not want to go over their work again just because one of the students in the collaborative programme had deployed a different approach to the Lakeland mapping and stratigraphy! In fact, they have maintained that the stratigraphy *is* different in the Dunnerdale area, which is why the Ambleside map looks different there from elsewhere on the sheet (Beddoe-Stephens, pers. comm., 1999) (see Plate VII). Certainly, the rocks of the Scafell and Langdale areas are less cleaved than those of the Dunnerdale and Coniston areas, so the sequence of markers could be established more easily in the universities' domain than in the territory of the Surveyors. The method advocated by Branney would therefore have been less likely to have been initiated in Dunnerdale, where marker tuffs are less prominent. So the differences in approach were in a sense contingent, being dependent in part on where the survey work was conducted – regardless of whether you got a short straw or a cherry. Kokelaar and Branney would, I think, like to see how well their pyroclastic markers could be traced into Dunnerdale and what effect that would have on the mapping. But that could involve a substantial reworking of the Ambleside sheet, which is not in prospect.

Examination of the tuff bands also gave Branney information about the faulting in the upper units. For example, a unit that he named the Bad Step Tuff[7] in Langdale shows great changes of thickness across faults, which might suggest a 'filling of topography', but the underlying stratigraphy on both sides of the faults is the same. From this, Branney inferred that the fault had moved either immediately prior to or during the eruption that produced the tuff, which obviously suggested a causal connection. In fact, according to Branney's interpretation, the faults (of which there are many) were in effect subordinate caldera scarps. They were formed by collapsing material as the tuffaceous material was spewed from the erupting volcano.

The 'scarps' were found all over the place, without much system to the pattern. Certainly they did not form tidy circles round the deposition centres, though they had the effect of providing different centres for the ponding of various tuffs. But, as it appeared to Branney, these deposition centres all formed different collapsing parts of a *single* large caldera. The floor of this caldera kept fracturing. Hot welded ignimbrites could be erupted. Or if water got in and came in contact with the hot magma below, there could be even more violent phreatomagmatic eruptions. The repeated violent eruptions produced, as Branney put it to me (pers. comm., 1998), a 'ticker-tape' sequence of tuff marker bands. Thus, during the course of his PhD work, he formulated his theory of 'piecemeal caldera collapse', as applied to the Central Fells of the Lake District.

Branney's mapping was very detailed, on a scale of 1:7000, and much of it (Branney 1988*a*) survived largely unscathed into the modern Ambleside map, and into the associated Memoir, with regard to the distribution of faults around Scafell (Millward *et al.* 2000, p. 165). Also, as Branney claimed, the observed structures for the Upper BVG could be reinterpreted in the light of his theory of the way the volcano 'worked'. Previous investigators such as Moseley & Millward (1982) had construed the Scafell Syncline as a fold produced by compressional forces, but on Branney's interpretation it was a giant 'sag' structure, resulting from caldera collapse, and was only later affected ('tightened') by the Acadian compression. The difference in interpretation can be seen from a figure that Branney published with his supervisor, Jack Soper (1988) (see Fig. 16.7).

The general theory used to account for the structure of the Scafell Caldera was, then, one of 'piecemeal caldera collapse'. It may conveniently be understood with the help of a schematic representation published by Branney & Kokelaar (1994) in the *Bulletin of the Geological Society of America* (see Fig. 16.8), a paper which served as the effective theoretical culmination of Branney's Lakeland investigations, resulting from his PhD work and his collaborations with Soper, Kokelaar and others. The *BGSA* paper presented also the main results of the collaborative mapping of the Central Fells, which served as the empirical basis for the caldera collapse theory.

Some of the Surveyors in the collaborative team were initially sceptical about the youthful Branney's ideas. They had not all appreciated the usefulness of the Whorneyside Tuff as marker for a key horizon. Having worked for some years on the Lower BVG where the layered andesites were manifestly *not* 'layer-cake' in their stratigraphy, they were reluctant to accept the idea of widespread marker layers in the Upper BVG. Soper (pers. comm., 1998) recalled that there were tremendous rows and some of the Surveyors would not allow Branney to see their field-slips and the lines thereon, but Soper encouraged Branney to get his ideas into print ('otherwise the Survey would be claiming them for themselves'!), and they composed a joint paper in 1987. Branney read this at a meeting of the Volcanic Studies Group of the Geological Society that year and the paper appeared in the Society's *Journal* the following year (Branney & Soper 1988). It was one of the most important papers published on Lakeland geology.

[6] Such materials are regularly to be found in the Upper BVG. For example, there are excellent mud-cracks to be seen by Sour Milk Gill, running down to Seathwaite in upper Borrowdale (see Fig. 7.4), and there is a magnificent slab of ripple-marked tuff near the summit of Bow Fell. But the lakes in which these structures were formed must have been short-lived. They left no organic remains.

[7] So called because there is a place on the path up Crinkle Crags that crosses the unit, and where a rather awkward scramble is needed to ascend (though one can walk a little further and thereby avoid the obstacle).

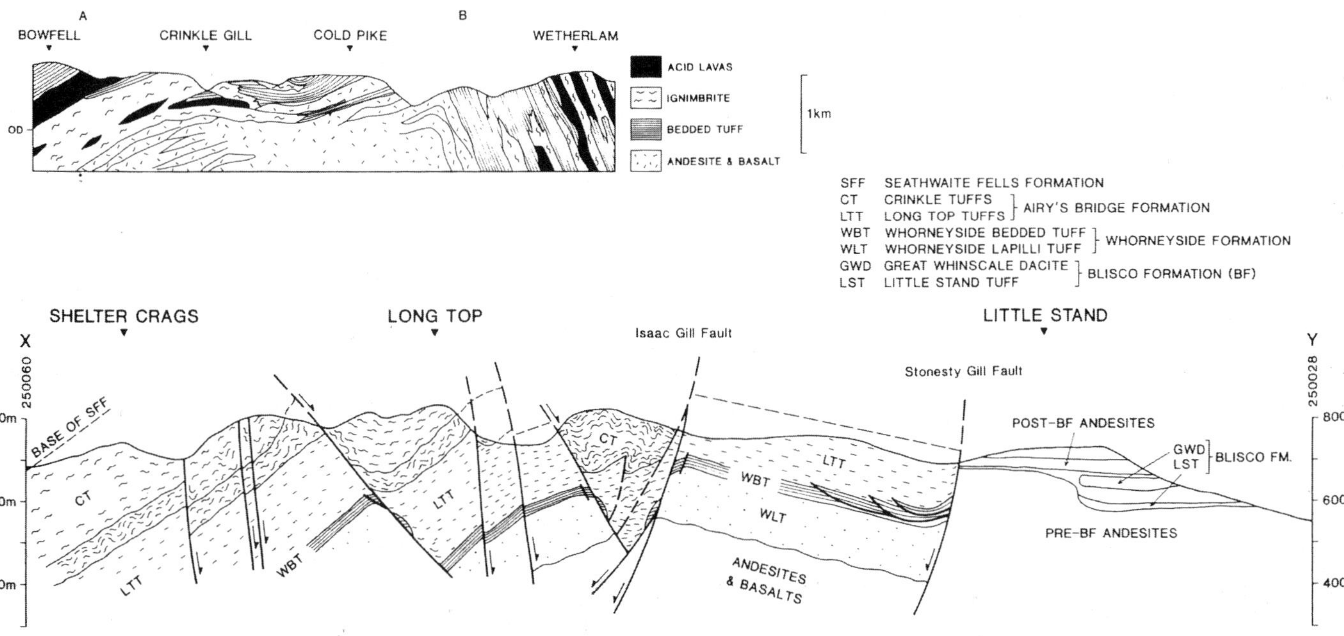

Fig. 16.7. Comparison of interpretation of structure of Upper Borrowdale Volcanics, according to Moseley & Millward (1982) and Branney & Soper (1988). Branney & Soper (1988, p. 370).

In the normal course of events, one might have expected the geologists to go to the localities in the field together and talk through the conflicting evidence and conflicting opinions, but to my knowledge this did not always happen, perhaps in part because Peter Allen left the project fairly soon after it started. The two camps had fallen out with one another. As I understand, amongst those working on the BVG it was Eric Johnson who was most willing to talk things over in the field; and Michael Petterson began to adopt and publicize Branney's ideas, as for example in a paper he published with the Cumberland Geological Society in 1990 (Petterson 1988–1989 [1990]).

One of the main points of difference, which remained unresolved at the end the period covered in this book, had to do with distinguishing between sills and lava flows. Near Clermont Ferrand in the volcanic province of the French Auvergne, there are rocks called *pépérite* (Cordier 1816, pp. 367 and 385) interpreted by French geologists as being produced by intrusion of magma into wet sediments (Giraut 1902). However, their interpretation had long been controversial. Archibald Geikie (1903, vol. 2, p. 1254), for example, thought that such rocks might be produced by volcanic ash falling into lakes and becoming interbedded with lacustrine sediments. There are numerous examples of such rocks in the BVG, which had been investigated by Keele University PhD student Roger Suthren (1977*a*).[8] Branney got together with him to re-examine some of Suthren's exposures and came to favour the view long before proposed by Green (1913, 1915*a*), but contra Marr (1916) and Moseley & Millward (1982), that many of the supposed autobrecciated lavas were in fact peperitic sills (Branney & Suthren 1988). Here they were following a lead from Peter Kokelaar's (1982) work in Scotland and Wales. An example of such a rock, from Brown Knotts, in the hills of the Lower BVG to the east of Derwent Water (see Figure 2.3), is shown in Fig. 16.9. (See also Fig. 16.10.)

The peperites showed enormous variation from one locality to the next, and all possibilities cannot be described here, but a 'typical' section for the rocks at Brown Knotts, was figured by Branney & Suthren (see Fig. 16.10). They pointed out that the intrusion of a sill into a wet sediment need not have produced the usual baking, since the steam produced could have insulated the host rock. The 'peperitization' of the *upper* margins precluded the rock from being an extruded lava. But one needed to be able to distinguish the typical peperite structures from the brecciation produced in a flowing lava that has had its upper layer solidifying as flow continued below. To spot the difference, it is helpful to look for vesicles (former gas bubbles) in the sediment just above the sheet. Their presence indicates a peperitic sill; but unfortunately their absence does not prove the converse.

Needless to say, *if* a good many of the Surveyors' 'lavas' were in fact sills, this threw doubt on their general stratigraphy for the BVG. Branney & Suthren (1988, p. 186) put it as follows:

> Despite over one hundred years of research, the stratigraphy of the BVG remains unresolved. The local successions which have been established largely on the basis of the relative abundance of andesites within parts of the stratigraphy have been only speculatively correlated together, and then with many difficulties (Moseley and Millward 1982). The common occurrence of sills throws considerable doubt on this approach. Establishment of a coherent stratigraphy within the BVG is urgently required to provide a framework for volcanological and petrological research. It is suggested that caution should be applied in designating the genetic terms 'lava' and 'sill' to the many ambiguous brecciated sheets in the BVG, and that correlation of sequences of widespread pyroclastic horizons by high-resolution mapping may prove to be a more effective approach towards the understanding of this complex volcanic pile.

Thus, finding what appeared to be peperites throughout the BVG succession, Branney and Suthren threw down the gauntlet to the authorities on Lakeland geology! However, the Surveyors felt that Branney *et al.* went too far, seeing peperites everywhere, so to speak (while acknowledging that such rocks do indeed occur aplenty in the Lakes). As Michael Petterson put it to me during the wet lunch-stop mentioned above, Branney *et al.* were seeing so many peperites in the Lakes that it would be difficult to squeeze all the lavas into the sediment. There would not be enough sediment! Branney (pers. comm., 1998) countered that there was a great bias in favour of the preservation of the andesites as opposed to the ashes, which are rather easily eroded away soon after deposition. Also, he has pointed out to me (pers. comm., 2001) that the intruding sill can 'excavate' the sediment in a

[8] Presently at Brooks University, Oxford (formerly Oxford Polytechnic).

fluidized state (sediment grains in steam) so that sediment can be 'lost' during the process of the intrusion. There are other arguments, which I leave the geologists to debate. But the issue is one of some importance in the present context, for the application of the peperite concept supported the value of the use of pyroclastic markers in mapping, in that one could know what is part of the stratigraphic sequence, as opposed to intruded material.

Among other aspects of basic volcanological and petrological theory that emerged from Branney's work was his co-authored paper with Kokelaar and Brian McConnell, based on their study of the Bad Step Tuff – a tuff that bore much resemblance to a lava (Branney *et al.* 1992; Branney & Kokelaar 1992). They contended that, as the title of the papers suggested, this rock displayed 'rheomorphism': it had fused to the point where it could exhibit viscous flow. Traditionally, such behaviour was supposed to occur *after* the deposition, coming to rest, and welding of extremely hot ash, but the authors argued that the flow need not necessarily occur after emplacement.

The Bad Step Tuff was found to consist of a basal breccia, a central flow-laminated part lacking unequivocal vitroclastic textures but with folded laminations, and an upper autobreccia. The nub of the argument was that the flow textures were syndepositional, and that extremely high-temperature ignimbrites behaved like lavas and had analogous characteristics. So one could have a continuum from non-welded ignimbrites, to non-rheomorphic welded ignimbrites, to rheomorphic ignimbrites, to lava-like ignimbrites, to high-temperature fountain-fed lava flows. It was suggested that the rheomorphic ignimbrites such as the Bad Step Tuff could have erupted rapidly from fissure vents along volcano-tectonic faults during the process of piecemeal caldera collapse. There need not be a distinction between 'primary' welding and 'secondary' viscous flow. The full details of the argument cannot be given here, but it should be noted that Lakeland work stemming from Branney's thesis had some general theoretical significance (which, no doubt, was the reason why the authors offered their paper to a volcanological journal rather than one concerned chiefly with British geology).

When discussing the application of plate-tectonic theory to the Lakes in Chapter 10, we saw that one of the several models on offer was that the Ordovician volcanoes of the Lake District were

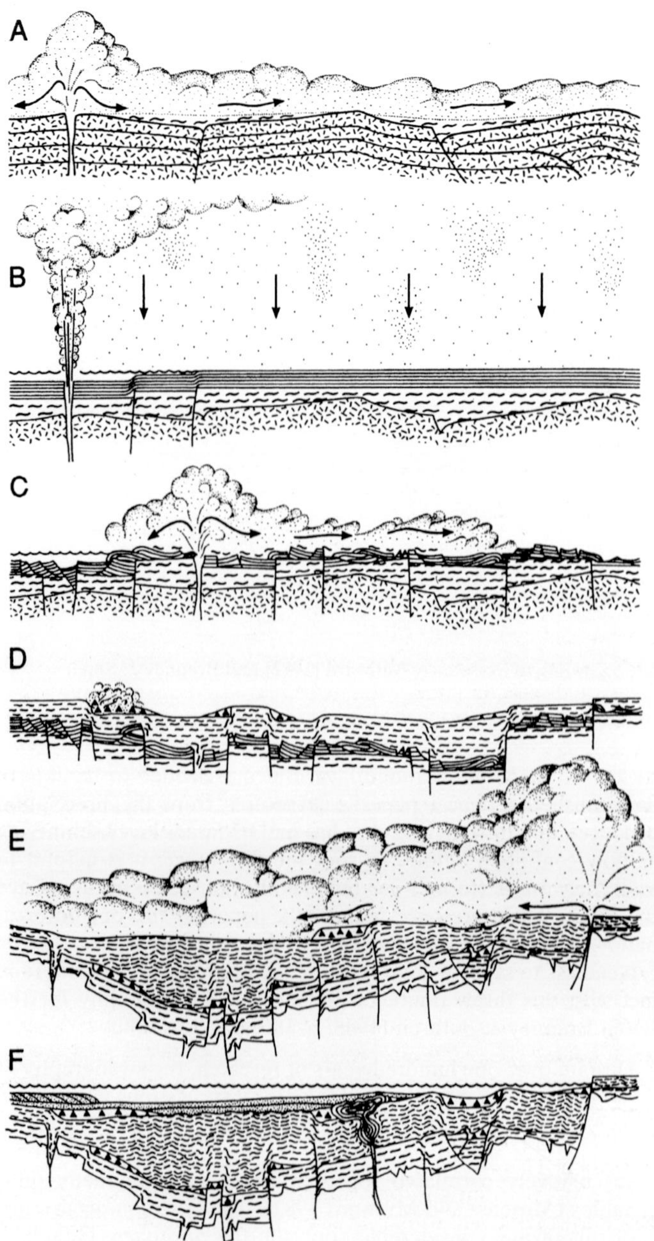

Fig. 16.8. Figures indicating the envisaged (simplified) history of the evolution of the Lakeland Central Fells, utilizing the notion of 'piecemeal caldera collapse', according to Branney & Kokelaar (1994, p. 528). Reproduced by courtesy of the Geological Society of America.

Fig. 16.9. Peperite at Brown Knotts. Photographed by author, 1999.

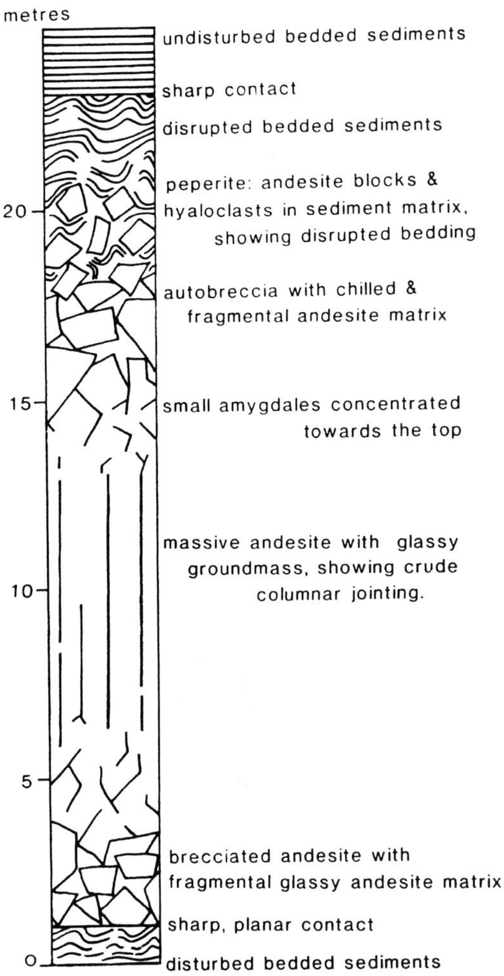

Fig. 16.10. Diagrammatic vertical section of sill at Brown Knotts, Borrowdale, according to Branney & Suthren (1988, p. 173). © John Wiley & Sons Ltd. Reproduced by permission.

generated in the context of an island arc, in which case there was a reasonable supposition that some of the deposits were emplaced under water, though pillow lavas were not known in the Lakes. The occurrence of acritarchs in the Holehouse Gill Formation in Dunnerdale certainly implied a marine environment for the deposition of these rocks, but these they were very much the exception (and see p. 211). On the basis of interbedded mudstones, Mitchell (1956) and Millward *et al.* (1978) had suggested a submarine origin for the Eycott rocks (assumed Lower BVG), but their arguments were queried by Branney (1988*b*) on the ground that the Eycotts and Borrowdales were distinct entities; and (as was by then agreed by all parties) there was an unconformity between the Borrowdale rocks proper and Skiddaw rocks (e.g., Wadge 1978*b*). A rock consisting of volcanic fragments in an acritarch-bearing mudstone matrix was known down in the far SW of Lakeland, north of Millom (at Whinneybank Quarry, 3169 4844; see Fig. 6.3) – the 'Mottled Tuffs' of Green (n.d. [1913]) – but Branney pointed out that much of the mudstone matrix was in fact derived from accretionary lapilli and thus represented explosive reworking of older marine strata. So the acritarchs could be inherited from earlier sediments.

Another important issue arising from Branney's work has been the nature of the volcano-tectonic faults in the Upper BVG, from which extrapolations were also made to some extent to the Lower BVG. Some of the faults merely reveal themselves by surface changes in lithology, but others have been strongly eroded and appear as great steep gashes on the hillsides, generally with small becks running therein. Good examples are provided at Grave Gill (278 064) on the northern side of the Langdale Valley, and in Isaac Gill (257 046 to 263 048), towards the head of that valley, below Crinkle Crags (see Figs 16.1 and 16.11).

This locality exhibits such interesting structures as different dips on opposite sides of the fault (indicating relative rotation of fault blocks during caldera collapse); tuffs and ignimbrites displaying almost vertical fiamme textures, apparently draped into the fault; slumped Whorneyside Tuff with incorporated blocks of Whorneyside ignimbrite; ponded 'Long Top Tuff' with steep fiamme close to the fault; the Whorneyside ignimbrite with stratified basal tuffs overlying the seemingly autobrecciated top of a pre-caldera andesite; apparent soft-state faulting, shearing and partial homogenization of Whorneyside phreatomagmatic tuff within the fracture zone of the tuff. These features had been noticed by Branney back in his PhD days, but are conveniently displayed in Branney & Kokelaar (1994) (see Fig. 16.12). Branney later observed such features as vertical *fiamme* in Korea with Surveyor, Tony Reedman; and they can be seen elsewhere in the Lakes, as at the head of Rydal Valley. Vertical *fiamme* can be ascribed, for example, to welding compaction against a fault scarp, fault movement beneath a hot and still viscous ignimbrite, or possibly to textures produced during the emergence and welding of hot ash from a vent.

An analysis of a possibly analogous sequence of events for Grave Gill was described in the Ambleside Memoir (Millward *et al.* 2000, p. 166), which also stated that 'many of the volcanotectonic faults within the Scafell Caldera complex c[ould] be traced into the underlying Birker Fell Formation [Lower BVG]' (Millward *et al.* 2000, p. 167), thus marking the Survey's acquiescence to the 'university' view of the BVG rocks. Kokelaar & Branney (1999) went further, though, in that they took a 'large-scale' view of the structure of the complex: the Crinkle Crag range was construed as a 'keystone graben', or a fractured block that had collapsed during the events involved in the eruption of the Upper BVG materials. One can see this idea expressed earlier on, in Branney & Soper (1988) (see Fig. 16.7).

The various matters to do with faulting in the BVG may seem to be related solely to 'pure' geology – or perhaps '(semi-)private' debates within the geological community, but as we shall see, they eventually came to be a matter of public debate, and arguably came to exert a significant influence on industrial and social policy in Britain. This claim will be developed in Chapter 20.

One other point of considerable significance emerged from Branney's work. According to plate-tectonic theory, the Ordovician volcanoes *might* have formed part of an island arc, as in the modern Aleutians say; or they might have been generated in a mountain range at the edge of a continent, as in the modern Andes. The sum of the evidence collected in Branney's PhD investigations was that the latter alternative was to be preferred. That is, Branney (1988*b*) argued that the Borrowdale Volcanics had been formed in a 'subaerial setting'. Thus at Side Pike (see Fig. 16.1), for example, he remarked an ignimbrite conformably overlain by a 'co-ignimbrite ash-fall deposit', in turn overlain by a pyroclastic surge deposit, incised by the erosive base of a coarse debris-flow breccia. Such an arrangement indicated a subaerial facies. Then there were fluvial facies found at, for example, Pike of Blisco. There were examples of subaerial lava flows, as in the Lingcove Formation; but no pillow lavas. To be sure, there was evidence of subaqueous deposition as indicated by mud-cracks and ripple marks, as at Sour Milk Gill, Seathwaite (see Fig. 7.4); but there were (apart from the doubtable Holehouse Gill rocks) no indications of marine fossils and sedimentation. The mud-cracks and ripple-marks could reasonably be interpreted as having been formed in short-lived freshwater lakes. In short, there was 'no evidence that the BVG represent[ed the] development of volcanic islands' (Branney 1988*b*, p. 889). So one would expect to find the overlying strata deposited unconformably on the BVG; but this did not necessarily imply tectonic uplift of the BVG rocks before deposition of the Windermere rocks. The BVG rocks were

Fig. 16.11a. View of Isaac Gill, Langdale. Photographed by author, 1999.

Fig. 16.11b. Close-up of same, showing tuffs draped into fault-fissure. Photographed by author, 1999.

already 'up' when first formed. However, they needed to fall or be severely depleted by erosion if they were later to be covered by Windermere sediments, as favoured by various theorists we have previously encountered.

While the foregoing account gives much attention to Branney's work, it must be emphasized that he was part of a team, and that the area which he mapped in such detail in his thesis, and with such interesting theoretical results, was relatively small. The important contributions of Neil Davis (1989) should also be emphasized, who, like Branney, came to be co-supervised by Kokelaar at Liverpool.[9] Developing the stratigraphy of Oliver and Branney, Davis mapped (at 1:7000) all the way from Hard Knott, through Scafell, Lingmell and Great Gable, round to Base Brown and Borrowdale – in about the toughest country that the Lake District could offer. He also covered ground with rocks of similar type that cropped out to the west (towards Wasdale) at Yewbarrow and Ill Gill Head on Kirk Fell (Davis 1989, fig. 3.1; see Fig. 7.4), and correlated the rocks of the Yewbarrow area with those of the Central Fells, from which they were separated by volcano-tectonic faulting (Davis 1989, fig. 1.2).[10] This work was essential to the demonstration of the idea of the Scafell Caldera and its piecemeal collapse (see Fig. 16.8). It was published by Branney & Kokelaar (1994), with due acknowledgment to Davis. By the time of writing up this paper Davis had obtained a position with British Petroleum and was too busy to be formally involved in the publication. Kokelaar (pers. comm., 2000) has regarded him as an 'unsung hero' of the story.[11]

Given Branney's subaerial thesis, one would not expect to find much evidence of fossils in the BVG, the slopes of volcanoes or lava plains not being the sorts of places where life would be abundant or preserved. Arthropod traces have, however, been found at the River Lickle in SW Lakeland near Broughton Mills (Dunnerdale Formation) (see Figs 5.2 and 6.3), and in the evident lacustrine deposits at Sour Milk Gill (see Fig. 7.4) above Seathwaite (Whorneyside Formation), both belonging to the Upper BVG (Johnson *et al.* 1994); [12] and these formed part of the subject of an interesting debate that was conducted 'out of sight' concerning Branney's 'subaerial setting' communication.

---

[9] It was when Branney realized the magnitude of the task facing him in his PhD, mapping a large caldera complex, that he suggested to Soper that another research student should be taken on to help with the work.

[10] Kokelaar continued Davis's work across Rosthwaite Fell to Bell Crags above Thirlmere (Kokelaar, pers. comm., 2001).

[11] As I understand (Kokelaar, pers. comm., 2001), some mapping results of Ben Kneller and Brian McConnell were also incorporated in this important paper at a late stage, after it had been reviewed.

[12] In this work, the Surveyors collaborated with Derek Briggs of Bristol University, who provided the chief palaeontological input. Interestingly, Johnson informed me (pers. comm., 2000) that he never met Briggs, communication being made by email. Thus technology modified publication procedures as well as field and laboratory practices.

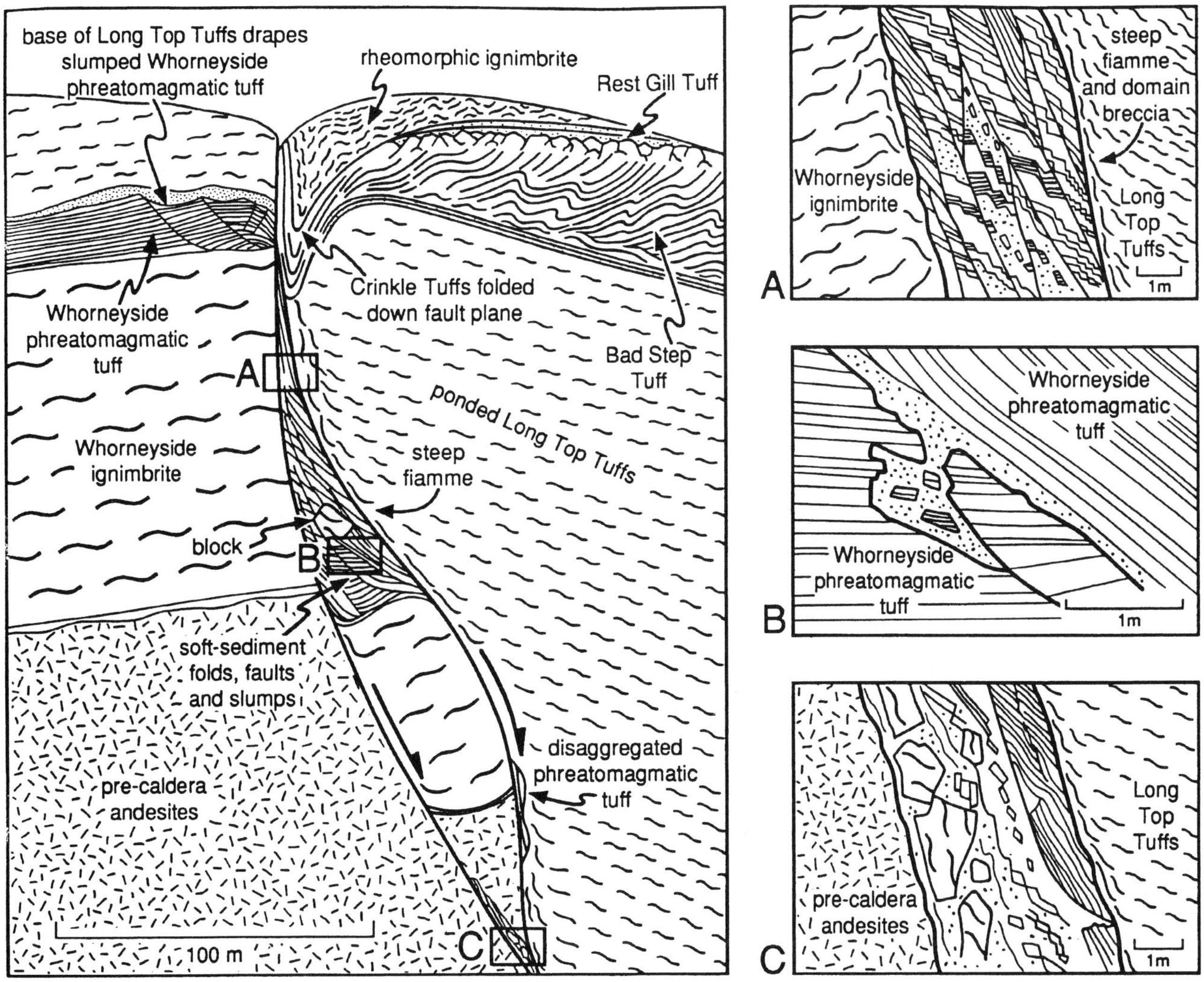

Fig. 16.12. Profile of volcanotectonic fault at Isaac Gill, according to Branney & Kokelaar (1994, p. 515) (cf. 16.10b). Reproduced by courtesy of the Geological Society America.

The Surveyors did not, it seems, like the 'subaerial' thesis initially, and Johnson *et al.* (n.d.) prepared a 'reply' for publication. In this, they argued that the Latterbarrow Formation was estuarine; that recent BGS work in SW Lakeland had suggested that the Ulpha Basin was an area of subsidence – and thus, I take it, not a syncline due to tectonic compression; that the aforementioned arthropod traces in the Whorneyside Formation were of *marine* origin; that there were the acritarch-bearing marine sediments of Holehouse Gill; and that acritarchs had also been found near the base of the Tilberthwaite Tuff (a BVG unit exposed in Lower Langdale and near Coniston). So it was suggested that the volcanic field developed on a land area, but close to sea level. Branney was construed as advocating crustal *extension* for his model of the BVG volcanoes.

Branney replied energetically, pointing out that the Latterbarrow Formation was irrelevant, as it was much older than the rocks under discussion and was separated from them by an unconformity. He was, in any case, only arguing that the BVG was *predominantly* non-marine. He also made the point that the theory of caldera subsidence did not necessarily require that the volcanic field be under extension. The dominantly calc-alkaline chemistry of the BVG did not suggest rifting, and the question of the relationship between the subsidence and rifting remained an open one at that time.[13]

In the event, the discussion described above did not get published; but its existence is of interest, exemplifying ongoing tensions, and perhaps misapprehensions too.

There was also rather animated discussion concerning the question of the batholith supposedly underlying the Lake District, according to the gravimetrical investigations of Martin Bott and Michael Lee (see Chapter 11, where the ideas of Firman and Lee were discussed). Firman & Lee (1986) had suggested that the western granites were emplaced during the late Ordovician and might have been responsible for *uplift*, producing elevated ground against which the rocks of 'Otley III' were subsequently deposited, with overstep and eventual complete inundation. Stimulated by this work, Soper wrote to the *Geological Magazine* early in 1987, suggesting how the Lakeland granites might be related to the Lakeland volcanicity. He had had the pre-cleavage age of the Eskdale Granite demonstrated to him in the field by David Millward, and this entailed some revision of Soper's earlier thinking.

There were several points to consider: the 'post-cleavage' coarse granites of Skiddaw and Shap; the pre-cleavage microgranites of Ennerdale, St John's in the Vale, and Threlkeld; and the coarse pre-cleavage Eskdale Granite, which Soper had earlier mistakenly grouped with the Skiddaw and Shap rocks on account of its coarseness. Soper now accepted, in part at least, the Firman

[13] I am indebted to Dr Branney for giving me copies of the unpublished papers.

and Lee idea that the Eskdale Granite might be causally connected to the formation of the Ulpha Syncline (which they were now regarding as a syn-magmatic structure), though he pointed out (Soper 1987, p. 481) that since the granite cut the syncline's axial trace it must post-date it in part at least. Be this as it may, separating the Eskdale from the Shap and Skiddaw Granites in time allowed the former to be causally linked to the volcanic activity involved in the formation of the BVG and the structures of that unit.

So Soper associated himself with his student Branney's idea that the extrusive BVG volcanism and the emplacement of the granitic magma could be related, the magma rising to fill a space from which the pyroclastic deposits were ejected. Also, the buoyant granite could engender a regional uplift, so that there could have been subaerial volcanism building up a volcanic pile *and* regional uplift due to the buoyancy of the granite. The Ulpha Syncline could be a sag structure, but one that had been tilted eastwards as a result of the emplacement of the granite. Then, suggested Soper, the granite might have cooled and contracted after its emplacement, leading to a reversal of the uplift, thereby facilitating the overstep of the Windermere ('Otley III') sediments onto the BVG rocks.

The Surveyors did not respond favourably to this suggestion and replied in a letter in the same issue of the *Geological Magazine* (Webb *et al.* 1987). They did not accept what they took to be Soper's suggestion, namely that the upper parts of the pluton actually gave rise to the emitted, largely dacitic, pyroclastic sheets. They thought that the chemistries were different for the BVG rhyolites and the Eskdale Granite, so, 'any relationships between the volcanic and plutonic activity must be seen as part of the long term evolution of magmas generated in the sub-crustal and/or crustal sources within the destructive plate margin environment that existed in the Lake District during this time'. They maintained that their mapping did not support the view that the Ulpha Syncline was a drape (or sag) structure developed about the time of the intrusion of the batholith. In their view, while there might well be a late Ordovician (Caradoc) deformation it should not be taken to be causally related to emplacement of the batholith, which, they pointed out, actually 'truncated' the Ulpha Syncline. Their preferred scenario was as follows:

1. A tensional episode in which the Borrowdale Volcanic Group was erupted.
2. A compressional episode to produce the Caradoc folds and uplift.
3. A tensional episode in which the batholith was emplaced and the Silurian basins opened.
4. A compressional episode, being the main, early Devonian, deformational phase [corresponding with the docking of Avalonia with the re-entrant between Laurentia and Baltica] (Webb *et al.* 1987, p. 483).

Stages 1 to 3 might represent transpressional and transtensional episodes as Avalonia sheared sinistrally along the irregular margin of Laurentia.

Soper and his Sheffield students, Branney, Mathieson and Davis, replied rather tartly (Soper *et al.* 1987*b*). It had not been suggested, they said, that the Borrowdales were erupted from the same magma body as that of the Eskdale Granite. Essentially, their view was that the ignimbrites, etc., of the Upper BVG were blown out, with caldera collapse, and the Eskdale Granite had subsequently been emplaced with a more chemically evolved magma than the chemically different pyroclastic deposits. But some of the subterranean matter might be material equivalent to the Airy's Bridge Formation (say), which had not reached the surface. The connection between the BVG rocks and the hidden batholith might have been recognized years earlier from the work of Oliver and Bott 'had we the wit to see it', said the Sheffield group. Soper *et al.* continued:

> [W]e question whether ... [the Caradoc deformation of the Borrowdale Group] took the form of flexural folding related to regional compression, and find it surprising that several years into the mapping programme the B. G. S. group is still firmly wedded to this traditional view. Detailed mapping in Langdale by M. J. B[ranney] has shown the structure to be dominated by volcanotectonic faulting and block tilting, and the folds which have previously been inferred to affect the area [cf. Figs 16.7 and 16.8] do not exist. ... Many previously inferred fold traces in the volcanic tract are illusory, resulting from Devonian compression of fault blocks with opposing dips; perhaps the Ulpha Syncline will prove to be of this type (Soper *et al.* 1987*b*, p. 484).

Here we see an epiphenomenon of a substantial theoretical and personal tussle that was erupting as a result of the BGS–university collaboration. Figuratively, we might compare it to the pyroclastic outbursts of the BVG itself, from which we can infer the existence of the hot magma below, which only appeared occasionally as mappable boundaries on the historical record.

Firman and Lee, whose paper in the Kingsley Dunham memorial volume gave rise to the foregoing exchange, also contributed to the debate (Firman & Lee 1987). They emphasized that the Lakeland batholith should be recognized as composite in both space and time. Dates for the several plutons needed to be considered, bearing in mind that the figures obtained by radiometric dating might not represent dates of original emplacement, but could well be 'reset' dates due to subsequent geological events. If the Sheffield group's scenario were correct, the Eskdale Granite and the Ennerdale Granophyre should be Ordovician (Caradoc) (say 464 to 443 Ma), and compatible or comparable with the radiometric estimate for the age of the BVG. (A date of 457 ± 4 Ma for the latter was available for BVG rocks of eastern Lakeland (Thirlwall and Fitton 1983), but was not referred to by Firman and Lee.) Unfortunately, however, the Eskdale and Ennerdale ages then known were 429 and 420 Ma (Rundle 1979), which did not match the geological requirements. But these dates could be regarded as representing resetting, probably due to 'a post-tectonic metasomatic event' – presumably Firman's suggestion, given his interest in such matters.

So one had to resort again to the 'geology of the situation'. In Firman and Lee's view, Webb *et al.*'s (1987) idea of a Silurian emplacement of the plutons under a tensional regime was unlikely, given that there was no evidence of Silurian Lakeland volcanic activity. Firman's investigation of the jointing of the Eskdale Granite and adjacent rocks years before (see p. 94) did not suggest cooling of the pluton under tension, but rather a N–S compression, probably operating before consolidation was completed. Also, the mineral grains in the granite showed evidence of strain, indicative of compression. So in their original paper they had opted for an Ordovician age for the emplacement of the plutons. The granite cut the Ulpha Syncline, but this did not necessarily mean that it was causally related to the fold. Thus, Firman & Lee (1987) pointed out, diapirs can pierce the very structures that they have produced. It seemed more than coincidental that the Eskdale joint system suggested a stress system that could have also produced the Ulpha Syncline. However, they also emphasized that the batholith was not presumed to be wholly responsible for the Caradoc (and Ashgill) structures, but could have '*aided* and *abetted*' the responsible tectonic forces. Firman and Lee pointed out, *contra* Soper, that while the western plutons might have produced the Ashgill marine transgression as they cooled and contracted, they could hardly have been responsible for the part of the transgression that occurred in eastern Lakeland.

The contribution of the Survey geophysicist Lee to the debate is noteworthy as he was in fact presenting views at odds with those of some of his colleagues. But he was not a member of the 'field geologists surveying in the Lake District for the British Geological Survey', as Webb *et al.* introduced themselves in their letter to the

Geological Magazine (1987). There could be public differences of opinion within the Survey as well as without.

As to the age question, some years later the Survey staff did co-author a paper (Hughes *et al.* 1996) which, by U/Pb dating, gave ages of 452 and 450 Ma respectively for the Ennerdale and Eskdale plutons, comfortably within the Caradoc. So the Sheffield group's suggestion of Ordovician emplacement of the west Lakeland plutons was apparently vindicated, and their ideas became ensconced in the Ambleside Memoir (Millward *et al.* 2000).

Amongst the university staff who collaborated on the Lakeland Project, Jack Soper was probably the most active, and in 2000 he was still doing contract mapping for the Survey in southern Lakeland. We have already seen how Michael Lee became involved with, and undertook geophysical work in, the Lakes. During the 1980s, he and Soper got together to produce a profile for the rocks on the Ambleside sheet that was compatible both with the geophysical data and with the known empirical and theoretical ideas for the Lake District. Lee recalls that they met over his computer for several days, locking the door and turning off the telephone!

Soper (pers. comm., 2000) recalled that his collaboration with Lee began in 1985 or 1986, at one of the Lakeland Project meetings held at the BGS Newcastle office. The Surveyors had mapped out a fault running northeastwards from Seathwaite Tarn (see Fig. 5.2) into Little Langdale, cutting across Greenburn Beck (see Fig. 16.1). This fault is now named the Greenburn Thrust, with thrusting from the NW (see Plate VII), but at the time the Surveyors were interpreting it as a normal fault with downthrow to the north. Soper, by contrast, thought that it might be a syn-cleavage fault, associated with contraction, and he suggested the idea of it being a backthrust, rooting below the Windermere Group and climbing up the subterranean batholith's shoulder. Lee was attracted to this idea, and pointed out that it would entail the batholith margin having an 'overhang', but this would mesh with what he knew of the gravity and magnetics. So he and Soper got together to try to produce a profile from the Solway Firth right across the Lake District. The outcome of their behind-doors collaboration was the profile published along with the Ambleside map (see Fig. 16.14), which also appeared in a somewhat more refined version in the Ambleside Memoir (Millward *et al.* 2000, p. 163).

As Lee showed me when I visited him in 1998, he and Soper used an iterative 'trial and error' method of working. They knew the lines of outcrop of the different rock units on the ground; they knew the magnetic and gravimetric profiles for the chosen line of section; and they knew (approximately) the densities and magnetic properties of the different rocks thought to exist along that line of section. Soper also had ideas as to what might be the shape of the rocks according to his theoretical notions, though thicknesses were uncertain. Then the two of them 'fiddled' (computer-wise) with the 'geological' variables (thickness, density, rock type, fault structure, etc.) until magnetic and gravimetric profiles were produced compatible with those known actually to exist and with what was geologically plausible.

Soper and Lee gave a talk on the 'backthrust model' at Keyworth in 1992, which stimulated Benjamin Kneller (see p. 231) and Andrew Bell (see p. 197) to develop their ideas about the structure of the 'Westmorland Monocline', as will be discussed in Chapter 17 (see p. 233). Soper kindly made available to me the series of transparencies used to illustrate this talk. For lack of space they cannot all be reproduced here. Nevertheless, something can be said about the modelling. There was a figure representing the hypothetical situation at the end of the Silurian. It showed variously shaped plutons underlying a severely faulted caldera collapse structure of BVG rock, with a hypothetical 'north fault' and a 'south fault' marking the northern and southern margins of the Central Fells caldera. The Windermere Group was shown as *covering the whole*, to a total depth (at one time) of some six kilometres.[14] The next section showed the situation at the end of the Devonian, after the whole had suffered severe (Acadian) N–S compression. In the northern Lakes, we find the series of southward-directed, north-dipping, thrust faults (Watch Hill, Gasgale and Causey Pike), as identified in the field and in the palaeontological laboratory by the Survey staff; and to the south, one could see represented a possible south-dipping 'backthrust', developing between adjacent plutons.

The next transparency represented this hypothetical southern structure in greater detail, and is reproduced in Figure 16.13. Here, the Scafell Syncline and parallel Ulpha Syncline (unmarked) were to be regarded as sag structures resulting from the original caldera collapse of the Lakeland volcano (though doubtless 'tightened' during Acadian compression). The folding of the Windermere Group and the series of reverse faults indicated in the Tilberthwaite area resulted from the same compression. The northern Honister and southern Tilberthwaite Slates were also supposedly produced from ash deposits during this phase of compression. The model showed how the granitic plutons were seemingly resistant to the compressive forces, whereas the Skiddaw and Windermere rocks yielded, becoming cleaved, faulted and folded. Shearing across the deep basement rocks was also indicated. Thus a story was developed about the geological history of southern Lakeland, based to a significant degree on the results of geophysical determinations.

The concepts contained in the profile of Figure 16.13 reappeared in the aforementioned section accompanying the Ambleside sheet (BGS 1996), reproduced in Figure 16.14. It was this profile that Michael Lee was kind enough to show me on his computer when I visited him in 1998, and which he could 'manipulate' by putting in different numbers for rock densities, thicknesses, etc., so as to get better or worse agreement with the known gravimetric and geomagnetic data. Naturally, the figure published with the Ambleside map represented the best 'fit' that the collaborators could achieve when they locked themselves up for a week! This figure, we note, chose (for simplicity, I suppose) to omit the fracture within the granite, depicted in the 1992 lecture. In fact, such a reverse fault has not (to my knowledge) actually been mapped out at the surface, but it was possible that if there were initially multiple subterranean plutons there could be movement between them under the action of severe compressive forces, as suggested in the figure.

In any case, when I visited Lee in 1998 he was already working on a more elaborate model, showing the granites as multiple, as he had supposed them to be years earlier when he was doing his gravimetric studies of the Lakes – with a granite perhaps situated under each main syncline. We did not then look at his new modelling in detail, but the outcome of his further work was later published in figures in the Ambleside Memoir (Millward *et al.* 2000, pp. 28–31). One of these figures is reproduced in Figure 16.15.

At depth, the section is naturally conjectural in part, though resting on seismic as well as geomagnetic and gravimetric evidence, so far as the northern Lake District is concerned. In fact, the evidence for deep structure rests to a considerable extent on the interpretation of 'fuzzy' reflective layers revealed by seismic survey (mostly undertaken by profiles constructed from data collected at sea to the west and east of northern England and southern Scotland). While this has revealed evidence compatible with theory for the northern Lakes and the area of the supposed Iapetus Suture (Freeman *et al.* 1988; Soper *et al.* 1992), the situation is less clear for the southern Lakes. The existence of 'ramps' in the basement rocks of southern Lakeland rested, at the

[14] The way in which this depth was estimated will be discussed in Chapter 17.

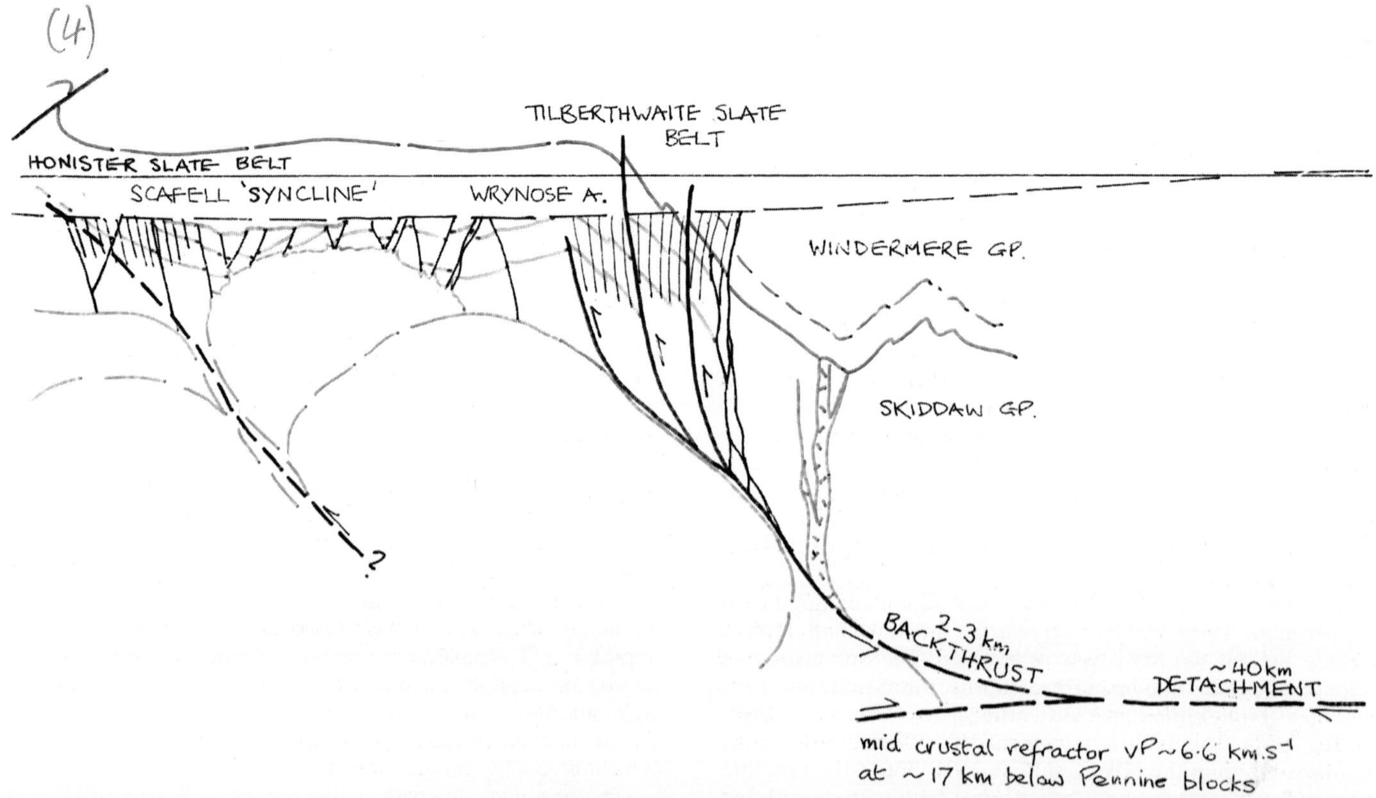

Fig. 16.13. Copy of transparency used in lecture by Soper & Lee, British Geological Survey, Keyworth, 1992 (reproduced by courtesy of N. J. Soper and M. K. Lee).

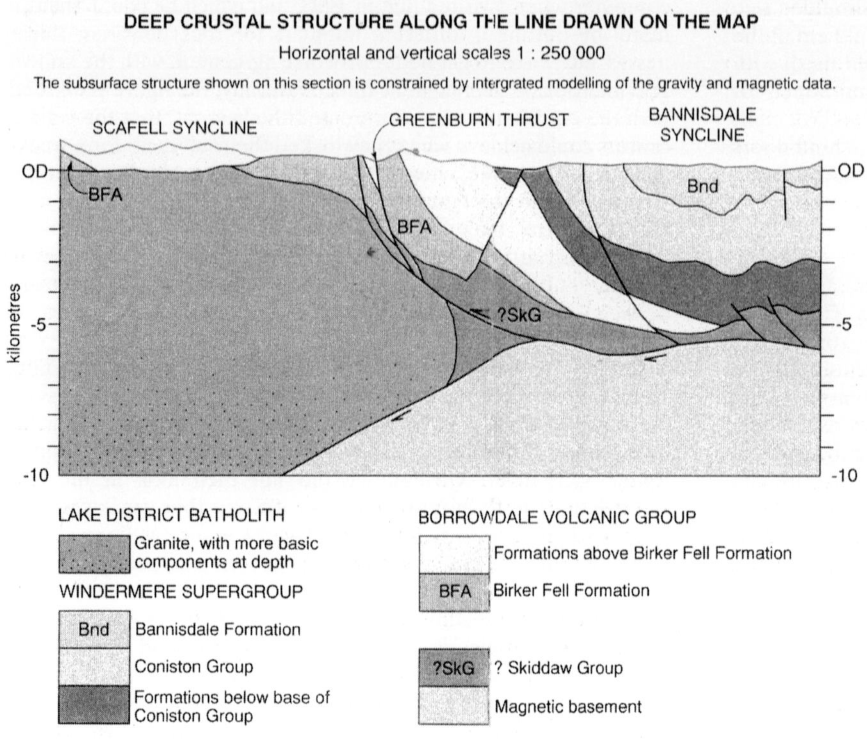

Fig. 16.14. Crustal structure along line of section, NW–SE, across the Ambleside 1: 50,000 Sheet (British Geological Survey 1996), according to Lee & Soper (coloured in original). IPR/23–7C British Geological Survey.

end of the twentieth century, on geomagnetic evidence and seismic observations and interpretations for the eastern Irish Sea (Millward *et al.* 2000, p. 32).

We shall see in Chapter 17 how the ideas about the structure of the Westmorland Monocline and the strata of the southern Lakes were developed by Kneller, Bell and others, and how Soper came to contemplate some alternative views, so the matter will not be taken further here. It should be remarked, however, that Figure 16.15 is obviously schematic. It would seem, on structural evidence, to allow a range of possibilities for the location of the Iapetus Suture, although, as we have seen, palaeontological evidence points to its existence to the north of the Lakes, in the Solway region. The prior southward-directed subduction leading to the eruption of the BVG seems to have left no structural trace in this figure other than the *existence* of the BVG rocks and the plutons; and the states of affairs at the lower edges of the plutons

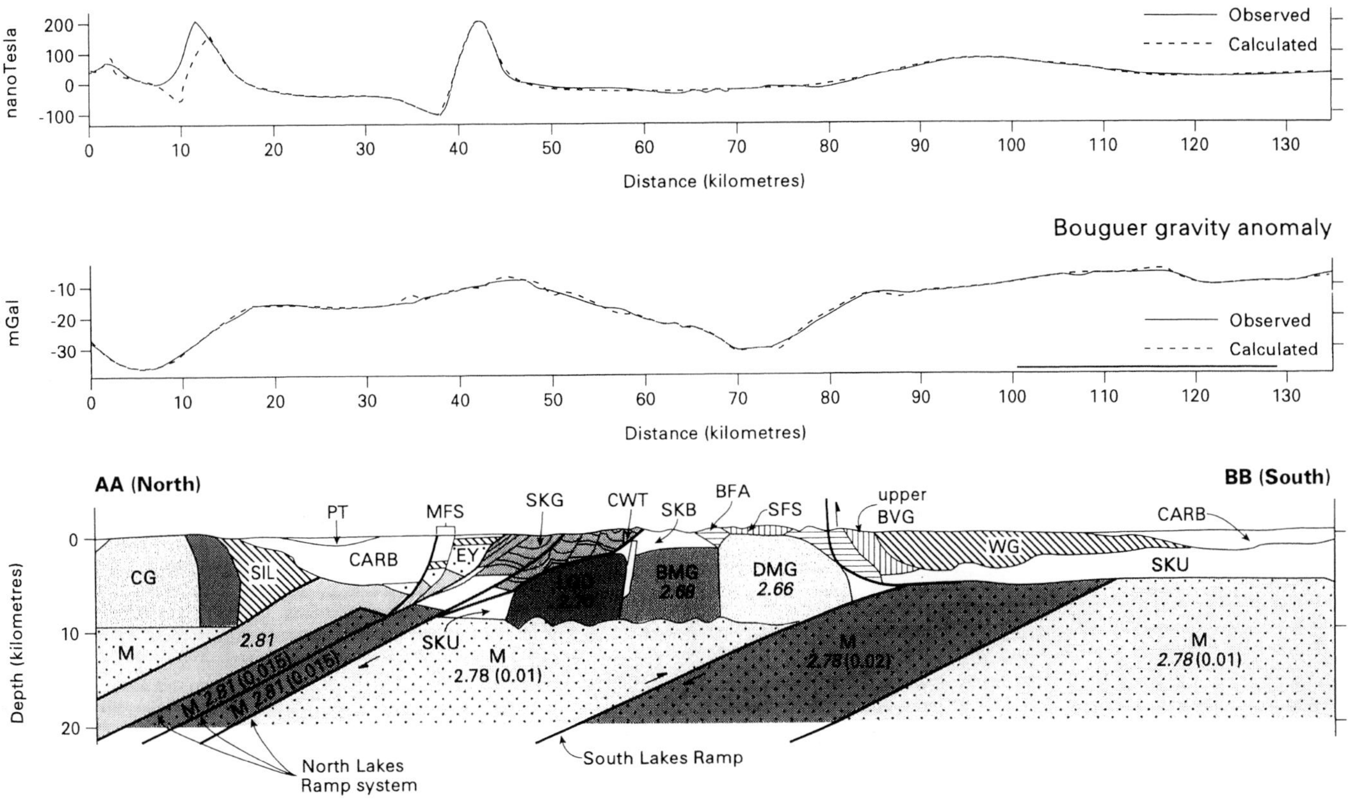

Fig. 16.15. Hypothetical section from southern Scotland (Criffel Granite (CG): see Fig. 1.2), through the Solway Firth region, the northern Lake District, and the Central Fells (passing East end of Wastwater), to southwestern Lake District – based on matched observed and calculated geomagnetic and gravimetric data, and seismic information concerning the supposed 'ramp' systems; according to Lee, *in*: Millward *et al.* (2000, p. 31), developed from Lee (1989) (coloured in original). M, Magnetic basement.[15] IPR/23–7C British Geological Survey. © NERC. All rights reserved.

is necessarily the result of assumptions made for the purpose of the theoretical modelling. Nevertheless, while this figure is likely to be refined during the course of the twenty-first century, it assuredly gives a picture with which one can think about the geological history of the Lake District.

Returning now to the question of the mapping of the BVG rocks, we have seen that this was accomplished by the collaborative efforts of the Survey staff, and geologists from Sheffield, Leeds and Liverpool Universities, with inputs where required for matters such as radiometric or geophysical survey. The Survey people worked chiefly on the Lower BVG, making much use of geochemistry for their stratigraphic work. The university people worked chiefly on the Upper BVG, making greater use of tuff marker bands for mapping purposes, and developing ideas about general processes of volcanic eruption and the formation and evolution of calderas. What had earlier been regarded as synclines formed by compressive forces were now, in some cases, construed as sag structures, albeit 'tightened' by subsequent compression.

I have, however, referred previously to the 'rows' that erupted in the process of arriving at this consensus. Indeed it is amusing to record that some witnesses to the disputes thought that the volcanologists were, collectively, of rather fiery temperament, as befitting their trade, whereas by contrast the geologists working on the Skiddaws, say, were of quieter, more contemplative temperament. (I dare say one does not get so 'worked up' if one sits for many hours on a scree slope in the rain, looking for obscure indications of graptolites)

Be this as it may, the eventual outcome of the BVG collaboration may seem quite magnificent, judged by the published 1:50 000 maps (Gosforth, Cockermouth, Keswick, Ambleside and Ulverston). Focusing here just on the Ambleside map and accompanying Memoir, it may appear that in the end everything worked out well, and that a real *tour de force* in mapping was achieved. Thus, consider the portion of this map that is reproduced in Plate VII. One can hardly fail to be impressed by the intricate detail (and beauty) of the thing. I certainly was when I first saw it, and remain so.

Nevertheless, my enquiries have revealed that tensions continued to simmer beneath the surface of this splendid map, and it had anything but an easy birth. The dark purple, brown, yellow and pale blue-green areas, etc., represent the Upper BVG, with the dark purple and brown denoting the Wet Side Edge Formation (WSE) and Whorneyside Formation (Wny) respectively at its putative base. The lighter dull mauve colour represents the Birker Fell Formation (BFA) of the Lower BVG. The Eskdale Granite is represented by the large patch of red. And various units of the Windermere Supergroup (blue-green, mauve, lilac, etc.) can be seen represented as extending unconformably over the SE portion of the part of the map depicted. To the SW, various igneous units are represented, as for example the Caw Formation (CwT), depicted in green, cropping out in the general area of Dunnerdale. These latter rocks seemingly have a different stratigraphy from that of the rocks of the Upper BVG of the Central Fells.

The production of such a map (costing about £2 million, let it be remembered) involved an immense effort of fieldwork, theoretical discussion, transfer of information from field-slips to a 'standard' 1:10 000 map, checking and revision, reduction to 1:50 000, further checking and revision, and eventual publication. One might suppose, then, that the persons involved in the production of the map would be satisfied and pleased with the product. Sadly, this is not entirely so, and ultimately, I suggest, it all goes back to the

[15] It is suggested that the 'magnetic basement' rock may consist of a thick layer of pre-Skiddaw Group sedimentary rock, extending down to the mid-crust.

different methods that were used in the study of the Lower and the Upper Borrowdale Volcanics, and the fact that the Surveyors built on the work of George Mitchell in the Duddon Valley, whereas the university people developed the mapping of Robin Oliver in the Central Fells.

There were several points at issue. According to Kneller (see p. 231), who worked for the Liverpool group on the BVG rocks around Pillar, SE of Ennerdale (see Fig. 7.4),[16] the Birker Fell Formation was capable of finer division than that which was accepted by the Survey leader in the preparation of the 'standards', and he was annoyed to find that some of his lines were deleted without consultation, in an area where he had put his heart, soul and body into an extremely difficult piece of mapping on treacherous ground, and which he regarded as one of his very best pieces of mapwork (Kneller, pers. comm., 1999). Branney (pers. comm., 1998) has confirmed that the question of the fineness of subdivision was one of the main bones of contention between the Survey and the university geologists. I am not in a position to form a judgement as to whether finer subdivisions could or should have been accommodated on the 1:50 000 map, but I understand they were removed by fiat from the 1:10 000 'standard' well before reduction to 1:50 000 was undertaken (though some were subsequently restored when Kokelaar protested about their removal without consultation). The point, then, is that while on the Ambleside sheet the Birker Fell Formation appears to be more homogeneous and widespread in its outcrop than the various units of the Upper BVG, this *may* be an artefact of the units selected for mapping as much as what is actually 'on or in the ground'.

As a second example, I have to report that, even at the end of the twentieth century, Michael Branney and Peter Kokelaar were not 'happy with' the appearance of the map in its SW corner, down Duddon way (pers. comm., 1998). It can be seen that on the map (Plate VII) the outcrop of the Whorneyside Formation, to the south of the Greenburn Thrust, is terminated at the extension of the Grassguards Fault (where the Whorneyside Tuff 'points towards' the outcrop of the Caw Formation depicted in green), but the university geologists believe that the Upper BVG rocks extend further to the SW. So, I suppose, if Branney and Kokelaar were or are correct, this part of the map carries the imprint of the earlier work of Mitchell and the Survey tradition, and the Survey's particular use, in the remapping of the 1980s, of geochemical techniques for stratigraphic purposes. It would probably look rather different if the area had been mapped according to the Liverpool team's methods. Yet Kokelaar informed me (pers. comm., 2000) that when he went into the field with Michael Petterson at some stage in the Dunnerdale area he was able to show his colleague evidences of 'Central Fell stratigraphy' there, and make successful predictions as to what might be found at the next outcrop (so to speak) from knowledge of the Central Fells tuffs. Yet these Upper BVG outcrops in Lower BVG territory do not appear on the map. Further, Kokelaar informed me (pers. comm., 2000) that he and Branney had observed rocks at Hollin House Tongue, north of Seathwaite in the Duddon Valley – which the Ambleside sheet maps as belonging to the Duddon Hall Formation (DH, brown: chiefly andesitic tuff)[17] – as forming a sequence similar to that found in the Whorneyside Formation in the Central Fells. He had discussed the rocks with BGS staff. The issue was admittedly complicated by the presence of andesites at Hollin House Tongue. It would seem, nevertheless, that the map was published before consensus was reached by those concerned.

However, as Branney pointed out to me in partial explanation of the differences of opinion, some of the Survey's earliest remapping was done on the Eskdale Granite and then extended into its aureole (in the Birker Fell Formation), the margin of which is represented by a dotted line on the map. The alteration tended to obliterate some of the stratigraphic distinctions that might be made. So the subdivision of the Birker Fell Formation (or the lack thereof) depended to some extent simply on the historical contingencies of the mapwork.

Over and above that, Kokelaar has pointed out that the faults marked in the 'Grassguards area' are in poorly exposed ground. He and Branney thought that the structure could involve either a strike or thrust fault generating repetition of units, and there might be peperitic sills too. This takes us back to the previously mentioned debate as to what is a lava flow and what is a peperitic sill. Perhaps a substantial portion of the Birker Fell Formation in fact consists of sills. If this were so, then the sills could be 'stripped out' of the stratigraphic sequence, leaving a considerably reduced overall thickness for the Lower BVG.

In fact, though the university collaborators were not, even in 2000, wholly satisfied with the beautiful Ambleside map, things might have been even worse regarding the representation of different views on the map. In July 1992, Peter Kokelaar wrote to Douglas Fettes, Regional Geologist, expressing his concern that changes had been made by BGS to the 'standard' 1:10 000 maps for which the Liverpool group had been responsible, without giving the field mappers the opportunity to comment on the changes. It appeared that some aspects of Branney's mapping in the Wrynose area had been discounted; Neil Davis's work in the Hardknott area had not been acknowledged by his name appearing on the map; and in particular the subdivisions of Oliver's old 'Birker Fell Formation' (for the hard-to-subdivide Lower BVG andesites) had not been retained, whereas Kokelaar commended subdivision into at least three main packages: the Lingcove Formation, the Blisco Formation, and some sub-Blisco unit, with the Whorneyside Tuff at the bottom of the Upper BVG. Kneller and McConnell had in fact proposed finer subdivisions of the Birker Fell Formation, but Kokelaar did not feel the need to press for these.

The BGS evidently took note of Kokelaar's requests, including for example the Blisco Formation, as requested, and by March 1993 proofs were ready for checking. But again, extraordinarily, no arrangements were made for Kokelaar's group to have the opportunity to examine them. One of the BGS officers at the Newcastle office wrote to his superiors Fettes and Stone, and to the Liverpool group, saying 'I can anticipate that you [Kokelaar] will be denied the opportunity to comment on the compilation before it is returned to the drawing office' (Kokelaar, pers comm., 2000). This proved all too true. It was only in November 1994 that colour copies of the second proofs of the map were made available to Branney and Kokelaar, though Kneller and McConnell were enabled to comment on the portions of the map for which they had been responsible.

Even then things were not altered to the Liverpool group's satisfaction. Thus in February 1995, Kokelaar wrote to the Survey again, complaining that crucial BVG geology had been misrepresented, especially in the line of section. There were cartographic and transcription errors, some removal of geological information, and changes of geological interpretation that conflicted with papers published by Branney and Kokelaar both before and during the compilation of the sheet. To make the points specific, Branney heavily annotated another proof copy supplied, pointing to typographic errors and numerous corrections that he favoured, particularly in the matter of faults in the area round Pike of Blisco, the upper part of the valley of Little Langdale, and the Wrynose Pass (areas where much of his PhD work had been focused). A list of 115 recommended corrections was drawn up by Kokelaar and Branney on 1 March 1995, and submitted to the BGS on 10 March. A meeting was eventually held in Edinburgh on 13 March between

[16] This area is actually at the SW corner of the Keswick sheet, but the point applies, nevertheless, to the discussion here.

[17] The Duddon Hall Formation is represented on the Ambleside sheet's vertical section as occurring *above* the Whorneyside Formation; but it should perhaps not be regarded as a distinct unit – or at least not one that crops out at Hollin House Tongue.

Kokelaar, Branney, Fettes, Millward, Johnson and Stone, and matters were finally thrashed out, with many of the requested changes being agreed to, though not all. Thus whereas the mapping in the Little Langdale area was brought more into line with Branney's original thesis results, the Survey did not accept that there were adequate marker horizons to warrant the establishment of new formations by subdivision of the Birker Fell Formation, as favoured by the Liverpool geologists.

It is interesting to compare the proof section (copy supplied to me by Peter Kokelaar, 2000) with the version eventually published in 1996. The latter shows faults penetrating into the granitic pluton, supposed to underlie the Central Fells. Branney did not want these, as according to his thinking the magma had filled the space created by the volcanic extrusions, so that the volcano-tectonic faults pre-dated the granite. At the meeting, it was agreed (for the most part) that the granite should cut the faults; or that the faults should stop at the granite, but at some stage, even after the meeting, the BGS reinstated some faults (notably the Stonesty Gill Fault) so that they cut the granite – as can be seen in the published version. Other faults previously proposed by the Survey were deleted or modified. Notably, their 'Langdale Thrust', on the Langdale side of Bow Fell and extending some distance to the east as shown on the proof map and section, was taken out – a significant modification. (However, as represented in the finally published section, it would be possible to envisage a thrust at the line of section from the rocks as depicted.)

These are some of the controversies and manoeuvrings. As an historian, it is *not* my task to adjudicate such matters. I think, however, that it is my legitimate role to bring them to readers' attention, and insist that a geological map is not something that should be regarded as an object of unassailable authority. It is something that emerges from the 'agonistic field' of the scientific community. It represents the best compromise that the occupants of that field are willing to agree to, or have the power to impose. We should like to think of a map as being 'consensually persuasive'.[18] In the case of the Ambleside map, however, the participants in its preparation only achieved partial and somewhat reluctant consensus, and after considerable toing and froing, despite the many 'man-years' of work spent on the project and its high cost.

The map did, of course, get published and it is, I think, unlikely that the Lake District will ever be mapped again as a whole, inch by rainy inch. However, techniques of recording information were changing by the end of the century, with the Survey turning to the digitized recording of its cartographic information. In consequence, small areas may now be resurveyed and the maps adjusted accordingly. Maybe the areas near the Grassguards Fault and Dunnerdale will be localities that will one day be revisited. Or maybe not? Either way, one can say with confidence that the production of ever better maps is the stimulus to better research and better geological theorizing. Maps are essential tools for geologists; but they are not, of course, the *ultimate* product of geological theorizing.

[18] I employ a phrase used by historian of geology, Martin Rudwick.

Chapter 17
Collaborating on the Windermeres

We have already seen in Chapter 13 how the stratigraphic divisions of the Windermere Group or Supergroup were elaborated in the latter part of the twentieth century, but there are still major questions to be asked about this large basin of sediments, its structure and its tectonic history. In particular, why is there the strange contact with the Borrowdale Volcanics, with the Windermere sediments evidently unconformable thereto, but also reared up against it, as at Timley Knott (see Fig. 13.1), with other parts of the contact also displaying significant faulting. How did this come about? Also, did the Windermeres[1] formerly go right over the BVG (which they seem, from Timley Knott, to have been trying to do)? Perhaps they did, given the evidence of the Drygill Shales or the occurrence of Coniston Grits in the Mell Fell Conglomerate of the northern Lakes. Or did they 'go round' the mountains only? If they went over the Volcanics, were they folded over, or were they deposited in such a thick sequence that they smothered them entirely? Or were the Volcanics largely eroded away before the deposition of the Windermeres? The 'Otley III' sediments are evidently thick. But how thick? And what were the reasons, in terms of the tectonic history of northern England, for the formation of a basin such as could accommodate the large mass of sediments? Where did these sediments originate? And then what caused the subsequent inversion of the basin so as to bring the sediments into view?

Yet again, one of the main theorists was Jack Soper, who did much mapping in the Windermere rocks (Ambleside, Ulverston, Appleby and Kendal sheets, with work on the Kirkby Lonsdale sheet also contemplated, as of 2000). Soper had two PhD students working on the Windermeres: Bill McCaffrey in 1987–1991 and Robin Scott in 1987–1992 (McCaffrey 1991; Scott 1992). Both these students were co-supervized by Benjamin Kneller (of whom more anon). At Cambridge, Nigel Woodcock and Barrie Rickards also contributed much to the understanding of the Windermeres, in Rickards' case, as we have seen, as far back as his PhD days, long before there was any suggestion of BGS–university collaborations for the Lakes. Woodcock's student, Louisa King, co-supervized by Kneller, also wrote a PhD directly on the kinds of problems mentioned above (King 1992), which will be discussed in what follows, though it was not part of the system of formal BGS–university collaborations. Andrew Bell was another important participant in the work.

The Surveyors David Lawrence and Barry Webb did work quite early on in the history of the Lakeland Project, along with Brian Young and David White. They produced a report for the Kentmere and Crook areas to the north of Kendal (Lawrence *et al.* 1986), which was accompanied by a curiously shaped 1:25 000 uncoloured geological map. It had the form of an inverted L, covering the line of BVG–Windermere contact between Troutbeck and Kentmere and then across a line of country southwards from Kentmere, cutting the Bannisdale Syncline, the Crook Synclinorium and the Underbarrow Anticlinorium, and terminating just south of the villages of Crossthwaite and Underbarrow. It was prepared thus as a result of Peter Allen's initial strategy of mapping a band of country through the Lakes, to sort out the stratigraphy and get a general idea of the structure, and also to look specifically at presumed key areas, knowledge of which might illuminate wider problems.

Benjamin Kneller[2] started work in the Windermeres with Soper in May 1987, and was shown over the ground by Lawrence and Webb, who had fairly recently completed and published their transect map and Memoir. By then, Lawrence had moved on to other things, but Webb had made a strong personal investment in the work and he and Kneller got on well. Peter Allen was 'in the background', as Kneller recalled (pers. comm., 1999). Robin Scott had been recruited from Glasgow University and was responsible for mapping the Dent Group – the newly 'amalgamated' Ashgill units at the top of the Ordovician, between the BVG and the Silurian Lower Llandovery. This was an onerous task for a relatively inexperienced geologist, and, although the job was accomplished successfully, such a role perhaps exhibited the kinds of problems that might emerge in the BGS–university collaborations. Regrettably, Scott's work did not get published as a separate item, but was absorbed into some co-publications.

The fact that there was an unconformity between the Windermeres and the BVG may seem obvious, though as we have seen the matter was not settled until as late as the 1920s, by Bernard Smith (see p. 83). The disposition of the lower rocks of the Windermeres (see again Fig. 13.1) is highly suggestive of the existence of a monocline. Surprisingly, this idea was not given close consideration until the late 1980s when Kneller got to work on 'Otley III', though Green (1920, p. 122) had envisaged such a structure early in the twentieth century.

According to Kneller's recollections, he started thinking about the peculiarities of the Windermeres from the very beginning of his contract work, when he was mapping with Soper in the general area between Windermere and Coniston in 1987–1988 (Kneller & Soper 1990). He wondered why there were unconformities within the lower parts of the unit and why there were such evident disparities in the sedimentation rate through the sequence. Why were there perhaps some 9 km of stratigraphy standing on end? There were several alternative models in the market-place to account for the existence of the sediment basin: crustal loading (Leggett *et al.* 1983); transpression followed by extension in a post-collisional sinistral strike-slip zone (Pickering *et al.* 1988); trans-tension or 'successor basin' theory (Hutton & Murphy 1987); and crustal extension (Webb *et al.* 1987) (Kneller 1991, p. 207).

It is easy to assume that the shoreline of the transgressing Windermeres was formerly more or less directly along the line of the present unconformity, but Robin Scott's palaeontology showed that this was not so. Rather, the transgression was from east to west. That was the first point. The second was to try to 'run time

[1] In the present chapter, I shall generally use the terms 'Windermeres' or 'Otley III', thus side-stepping the question of whether the rocks of 'Otley III' should be regarded as a group or a supergroup.

[2] Kneller (b. 1953) attended High Wycombe Grammar School, where he recalls that he got 'appalling A levels', which meant that he had to take a variety of 'awful jobs' for several years. However, he eventually found his vocation when a mature-age student, reading first environmental science and then transferring to geology at Sheffield, where he was tutored by Soper and obtained a first. This led him to do a part-time PhD at Aberdeen, working also as a research assistant for a BGS collaborative project in Scotland on the Dalradians, but before this was finished he went down to Sheffield again to join a Lakeland contract with Soper, starting in 1987. Kneller continued there until 1990, when he joined the Liverpool group under Kokelaar until 1992, working on the BVG. Kneller then went to Leeds on a variety of 'soft money' contracts, but when we met in 1999 he was a full member of staff, doing consulting work sufficient to buy himself out of administrative duties. At Leeds, he established a centre for the study of turbidites, supported by 'oil money', and I gather that this went well, but Kneller told me that he was finding university life burdensome in Britain in the late 1990s, for indifferent money. In fact, he could be earning more as a school teacher (pers. comm., 1999). It was not surprising to me, therefore, to learn that he moved to the University of California in 2000.

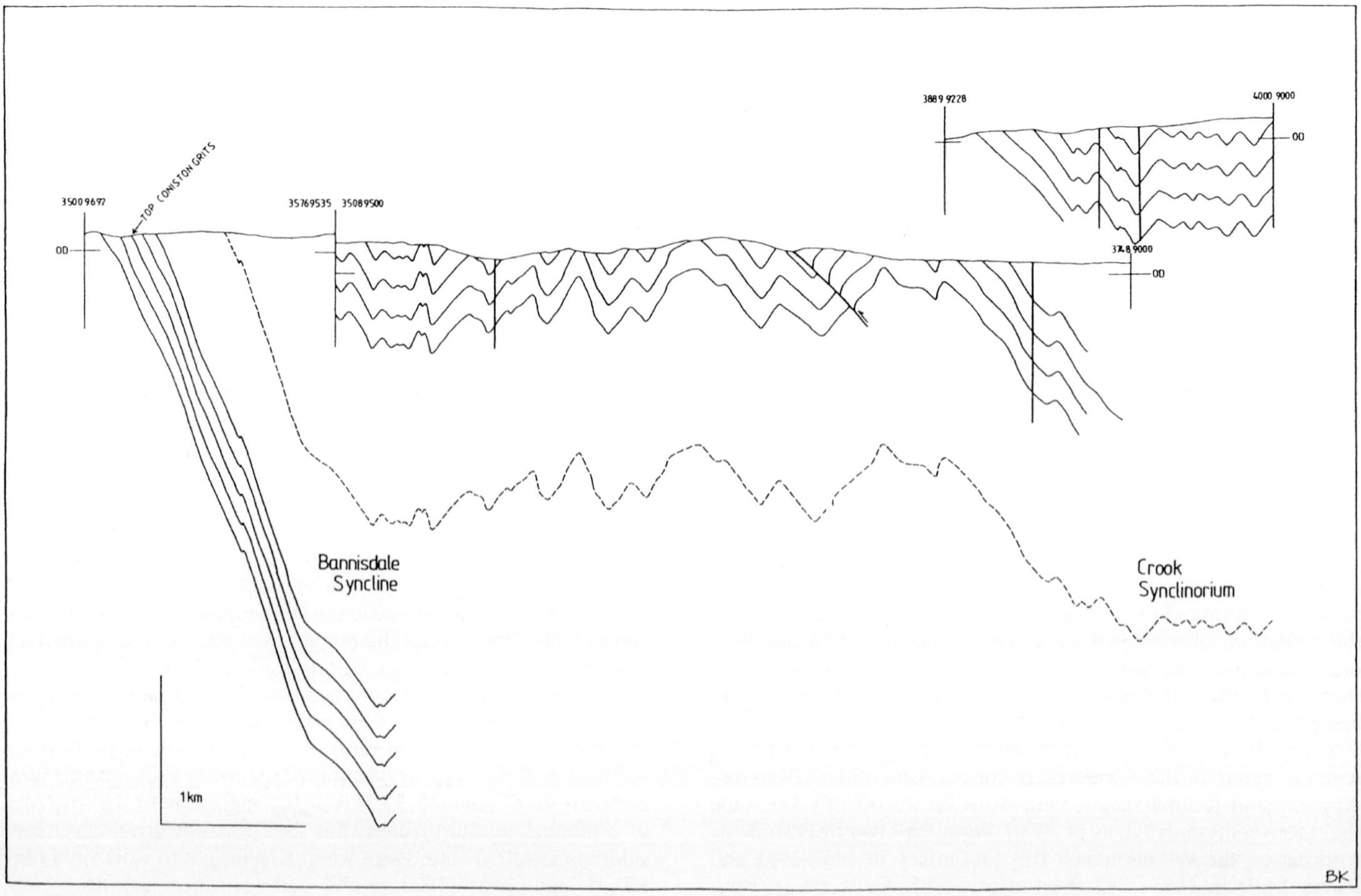

Fig. 17.1. Profile of Windermere rocks, according to Kneller (Kneller & Soper 1990, Fig. 3.3). IPR/23–7C British Geological Survey. © NERC. All rights reserved. See also: Millward *et al.* (2000, p. 163).

backwards' to see what the thickness of the sediments in the basin had been before they suffered deformation. It was essentially a geometrical problem, but also required assumptions about compactions subsequent to deposition, towards the solution of which Kneller used the techniques described in British Petroleum's structural training manual. His internal BGS Report (Kneller & Soper 1990) contained a profile (see Fig. 17.1) of the Windermeres, based upon knowledge of measured thicknesses at outcrop, determination of folds from dips, and faults; along with some information about subsurface structures from BGS geophysical evidence concerning the depth of the magnetic basement, and gravimetric data. This diagram carried through essentially unchanged into the Ambleside Memoir (Millward *et al.* 2000).

Then, 'using tracing paper in the old-fashioned way,' Kneller (pers. comm., 1999) straightened out all the 'kinks' in this section to get an idea of the present thicknesses of the different lithological units. Next, he had to estimate the original thicknesses prior to compaction, making allowances for the types of sediment, the estimated depth of water under which they had been deposited, and the durations of compaction. These were complicated matters to determine, given that the Silurian deposition had been subject to extensive changes due to glacio-eustatic changes in sea level. The reasoning depended in part on considering the oxic or anoxic conditions of sedimentation, which would be an approximate indication of water depth at deposition. Also the faunal evidence provided some indication of depths at time of deposition. Even without allowing for such complications, one can see from Figure 17.1 that the Upper Ludlow strata were generally much thicker than those of the Lower Ludlow, Wenlock and Llandovery.

Now the rocks of 'Otley III' are frequently fossiliferous, and their graptolite zonation was known from the work of Rickards and others, and from analogous investigations elsewhere in the world. From such work, Kneller assumed, as an approximation, that one graptolite zone in the Silurian lasted about 1 Ma.[3] So the sediments could be approximately calibrated as to time. Thus Kneller could plot sediment depth against time (see Fig. 17.2), from which it appeared that the sedimentation rate increased dramatically at about the end of the Wenlock or early Ludlow.

One could then consider what sorts of shape one might expect for such a diagram (Fig. 17.2), according to some of the tectonic models in the theoretical market-place. These were represented as cartoons, as shown in Figure 17.3.

If the basin were formed due to extension or rifting, then one would expect an initial rapid increase in deposition rate, followed by an 'exponential' tailing off of sedimentation. If the basin came into being as the result of strike-slip, one might expect an initial rapid 'pulse' of sedimentation, soon succeeded by a decrease of sedimentation as compression and uplift followed. But if one were dealing with deposition in a 'foreland basin' (see p. 235), the formation of the basin would reflect the deflection of the lithosphere (Schwab 1986), and one could expect a rapid increase of sedimentation as the basin deepened and concomitantly acquired an increased sediment load. This was the model Kneller favoured, since the study of the surface geology of the Windermeres did not provide evidence of either syndepositional faulting or igneous rocks indicative of extension.

In his initial paper, Kneller adumbrated the idea of 'the south-eastward migration of the basin as Eastern Avalonia continued to

[3] Later work by Richard Hughes (1995) would modify this generalization, giving more time to the Prídolí, but this would not have altered Kneller's arguments or conclusions substantially.

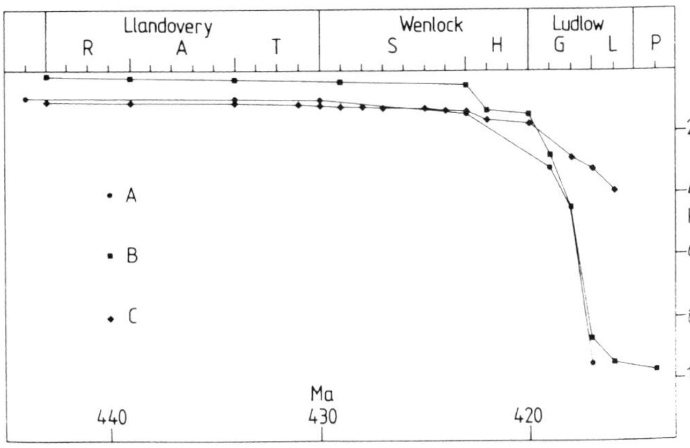

Fig. 17.2. Decompacted sediment thicknesses for rocks of different ages, at different areas of the outcrop of the Windermere Supergroup (A = Western; B = Central; C = Eastern), according to Kneller (1991, p. 208). Reproduced by courtesy of Dr Kneller. IPR/26–13C British Geological Survey.

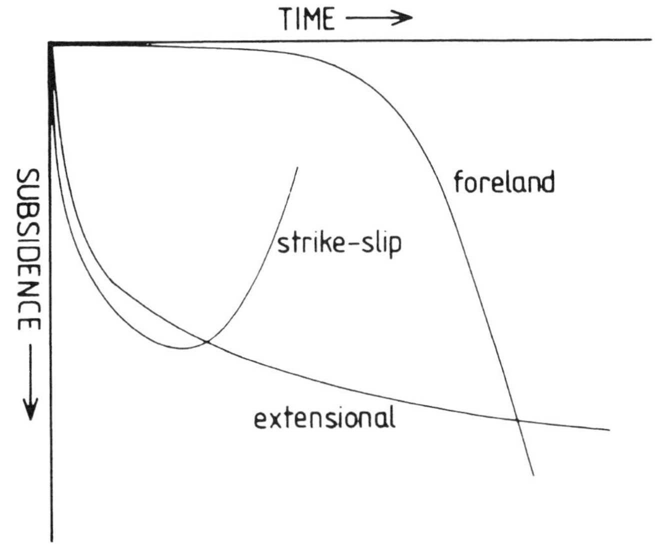

Fig. 17.3. Idealized curves representing changes of depth with time, according to different hypothetical causes of basin formation: Kneller (1991, p. 209). Reproduced by courtesy of Dr Kneller. IPR/26–13C British Geological Survey.

underthrust Laurentia' (Kneller 1991, p. 209). Without giving details, he also calculated a rate of Iapetus closure as being between 8 and 63 mm per year in the Upper Silurian, which was significantly less than the estimated rate of northward drift of Eastern Avalonia of about 100–120 mm per year in the early Ordovician and early Silurian (Soper & Woodcock 1990). This was construed as evidence of a slowing of the rate of closure as Eastern Avalonia approached Laurentia. Two terranes might be expected to decelerate relative to one another before they collided fully.

It will be recalled that Soper envisaged the 'big crunch' as occurring in the Emsian (late Early Devonian), when the slate belts acquired their cleavage. This raised the question of whether sedimentation had continued in the 'Windermere basin' in early Early Devonian (Pragian and Lochkovian) times, with the formation of sediments that were subsequently 'spirited away' by uplift, weathering and erosion. There were some indications that such deposits had indeed been present at one time, indicating a substantial sedimentary overburden, inducing mild metamorphism. One such indication was the state of crystallinity of illite (a form of mica intermediate between montmorillonite and muscovite, so named in 1937 after the American State of Illinois), in the Ludlow sediments. Kneller referred to unpublished information from the BGS mineralogist Neil Fortey in support of this suggestion. We shall come back to the question of 'illite crystallinity' later in the present chapter (p. 237). It is a topic that has attracted a deal of discussion.

Though Kneller was working with Soper, they had some disagreements. Soper thought that Kneller's mapwork was slow (albeit meticulous) and that he was overenthusiastic about devising new lithostratigraphic subdivisions for the Windermeres. There is no need for me to intrude into such matters, but it can be said that Kneller chose to develop his ideas with Andrew Bell rather than Soper; and he moved from the Sheffield group to the Liverpool team under Kokelaar. So it was Kneller & Bell (1993)[4] who sought to provide an explanation of the structure of the Westmorland Monocline (so named for the first time in their paper, though, as mentioned, Green had previously referred to there being a monoclinal structure in southern Lakeland; see p. 82), developing the ideas outlined above and employing Bell's familiarity with the mathematics of strain theory – which he had considered years before in his PhD on the Kentmere area, involving the deformations of the 'bird's-eye lapilli' (see Fig. 12.4).

Kneller and Bell distinguished a 'Northern Flat Belt' (which extended northwestwards from a line 4–6 km north of the 'Otley II'–'Otley III' boundary); a SE-dipping 'Steep Belt' (extending from that line to about 5 km south of the boundary); and a 'Southern Flat Belt' (which ran southeastwards for the remainder of the exposures of the Windermeres). The 'Steep Belt', or the inclined plane of the Monocline, consisted of both BVG and Windermere rocks, with the unconformity between them dipping at about 45°–70°. Examination of the rocks in the monocline showed that the fold was formed prior to the regional cleavage (taken to be Emsian) and prior to the emplacement of the Shap Granite (Soper & Kneller 1990). The actual age of the fold was not pinned down precisely, merely being said to be 'early in the local Acadian deformation sequence'. However, McCaffrey's (1991) Leeds PhD suggested that there were rocks of the Southern Uplands and the Lake District Silurians that had accumulated in the *same* basin, the southern part of which had been floored by the residua of the BVG. Thus the Windermeres could formerly have extended right over the top of the eroded remains of the BVG.

The basin in which the sediments accumulated did not appear to be fault controlled. There were no fan-deltas, slump-sheets, canyon-fills or anything of that kind. Rather, it was suggested, the basin had been there, receiving sediment, and had deepened in late Wenlock and Ludlow times to receive the large thickness of Coniston Grits and Bannisdale Slates. This deepening had supposedly occurred in the southern Lakes as the Iapetus Ocean had closed, and as the Avalonian plate to the south was pushed below the Laurentian plate to the north, as previously proposed by Soper. But the Borrowdales were thought to be underlain by the large solid plutons of the Eskdale and Ennerdale 'granites'. So these had formed a kind of 'wedge', which, as closure occurred, pushed (relatively) southwards, with thrusting over the basement rocks and the Lakeland rocks ('Otley I', 'II' and 'III') being back-thrust up the sloping southern side of the pluton, thus forming the exposures of the inclined Borrowdale and Windermere rocks of the 'steep-belt' of the Windermere Monocline.

Such ideas had already been developed by the geophysical modelling of Lee and Soper (see p. 226), and Kneller & Bell (1993) specifically referred to Lee's massive BGS Technical Report, which contained a N–S profile from (and through) the Carlisle basin and on to the southern Lakes (Lee 1989, fig. 7.37) –

[4] Bell was then working at the Open University.

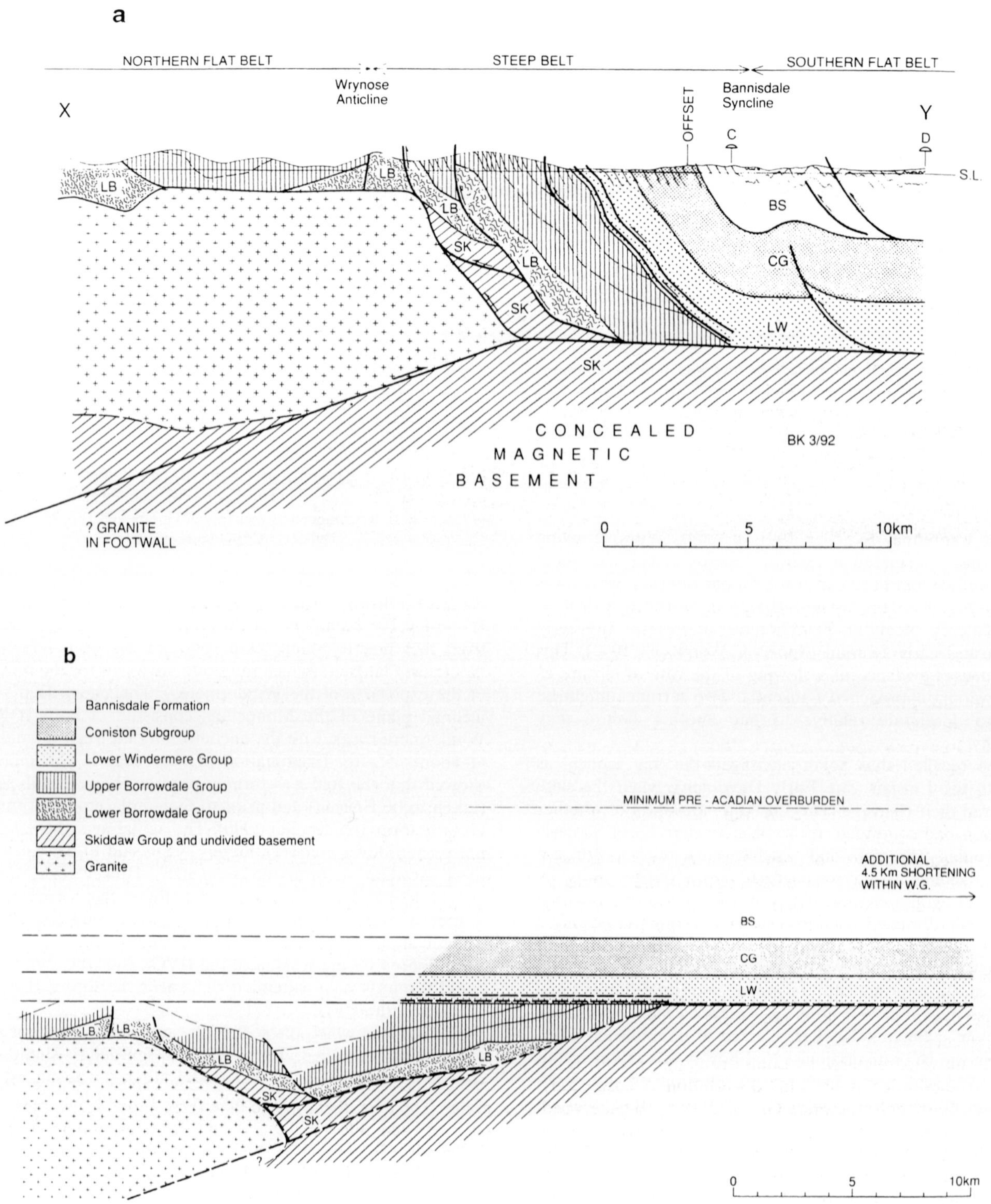

Fig. 17.4. Profiles of strata in southern Lake District. (**a**) Present configurations; (**b**) 'restored' to situation prior to Acadian deformation according to Kneller & Bell (1993, p. 208). Reproduced courtesy of Cambridge University Press. BS, Bannisdale Slates; CG, Coniston Grits; LW, Lower Windermere Group; LB, Lower Borrowdale Volcanics; SK, Skiddaw Slates. Vertical lines represent Upper Borrowdale Volcanics whilst small crosses represent granite.

a precursor of figures that subsequently appeared on the Ambleside 1:50 000 sheet (BGS 1996) and in its accompanying Memoir (Millward *et al.* 2000) and which we discussed in Chapter 16 (see Fig. 16.14).

Kneller and Bell's own diagrams are reproduced in Figure 17.4. It will be seen that they bear some similarity to ones presented by Soper & Lee in their lecture at the BGS in 1992 (see Fig. 16.13). However, the particular thing about the Kneller & Bell (1993) paper was that it also sought to offer a reconstruction of what the arrangement of the rocks might have been when the Windermere sediments had been deposited, but before the 'big shove'. As in Kneller's 1991 paper, there was an 'unravelling' of the present structure; or time was run backwards to the period at the end of the deposition of the Bannisdale Slates, before the 'big shove'. The Silurians were depicted as having been deposited right over the Borrowdales, as most geologists – perhaps all – now accept, on the basis of theory, the occurrence of the rocks of the Drygill Shales and the Mell Fell Conglomerate, and small outcrops of Dufton Shales (Upper Ordovician) and Browgill Beds (Lower Silurian) at the northern tip of the Cross Fell Inlier. The high

Fig. 17.5. Louisa King, sharing space with gum trees, Perth, Western Australia. Photographed by author, 2000.

strain evident in the rocks of the 'steep belt' (such as Bell had investigated in his PhD; Bell 1975) was compatible with the theory. However, while the title of the paper referred to the Westmorland Monocline as a 'mountain front' (cf. Vann *et al.* 1986), Soper did not accept that the structure looked like mountain fronts that he knew personally, such as one at Ellesmere Island to the NW of Greenland (pers. comms, 2000). There were no emergent thrusts and fan-deltas for the Westmorland Monocline such as he knew at Ellesmere Island.[5] Figure 17.4a was reproduced essentially without change in the Ambleside Memoir (Millward *et al.* 2000, p. 163).

Of course, if sediments were deposited in a tectonically formed basin, they must at some stage have been uplifted and 'exhumed' if they are now available for inspection at the surface. How might this have happened? Also, how far did the 'Otley III' basin extend southwards? Its southern edge is obscured by Carboniferous and other sediments, but there is certainly the possibility that the basin of Silurian sediments extended further south than the present limit of outcrop, and might still be found there if the covering of younger rocks could be stripped off.

Such problems were explored by Louisa King (see Fig. 17.5), of Clare College Cambridge, in her PhD dissertation 'A basin study of the early Palaeozoic Windermere Group' (1992), co-supervised by Nigel Woodcock at Cambridge and Ben Kneller, who was by then at Leeds.[6] Kneller, King and Bell wrote several papers in the early 1990s, developing the ideas discussed above (Kneller *et al.* 1993; King 1994*a*, *b*), and produced what was, at the end of the twentieth century, the paradigm for understanding the deposition and structure of the rocks of the Windermere Silurian basin.

In her PhD, King provided a clear discussion of the ideas behind her model, which were in part based on the abstract ideas about basin formation promulgated in Cambridge by Dan McKenzie (1978), which were to the fore at that time. To illustrate (as King did in her thesis): one can think of the case of the Himalayas, the Ganges Basin and the Deccan Peninsula, with collision occurring between the mountains and the Peninsula, the mountains constituting a 'loading plate', and the rocks disappearing beneath them by subduction being the 'foreland plate'. The Ganges area, then, would be an example of a 'foreland basin'. As the collision proceeded one could get imbrication in both the edge of the 'loading plate' and the sedimentary rocks accumulated in the foreland basin; and the formation of a 'forebulge' in the 'foreland plate' to the south of the basin. As the 'loading plate' 'advanced' (relative to the 'foreland plate'), the foreland basin and the 'forebulge' would be shifted southwards. Thus there could be a migration of the basin away from the 'loading plate' – or a 'progradation' of the basin structure.

Similarly, in northern England and southern Scotland, one might have the Laurentian 'loading plate', an imbricated structure in the sediments of the Southern Uplands, and a southward-migrating (prograding) basin in the area of the Lake District, the rocks accumulated therein being now exposed as the Windermeres. The idea, then, was to take these broad ideas, such as one may see exemplified in India or the Pyrenees, and apply them to the small-scale, less obvious case, such as might be provided by the Windermere rocks in northern Britain.

However, as King (1992, p. 205) explained, the situations in India and Britain were not wholly analogous. In India, there appeared to have been episodic collapses of an orogenic wedge, whereas in Britain there was apparently direct impingement of

[5] However, neither are there for the Lapstone Monocline, which forms a front along the eastern side of the Blue Mountains, with which I am familiar near Sydney – and which looks somewhat like the Westmorland Monocline (but there are high-level gravels on the Lapstone Monocline, which presumably derive from rivers formerly present at higher altitude than today).

[6] Louisa King (b. 1968) attended school in Northleach, Gloucestershire, and proceeded to Queen's College, Oxford, where she got the top first in geology in her year. On graduation, she and her partner Christopher Clayton (now husband) obtained NERC scholarships to study under Nigel Woodcock at Cambridge. There was one position to work on the Welsh Silurian sediments and one for the Lakes, which latter position King obtained. We thus find that her thesis shows much interest in the relationships between the Lakeland and Welsh Silurians. During her period as a PhD student she also had the advantage of extended periods of excellent weather for fieldwork on the fells. After obtaining her doctorate in 1992 she worked for a geo-consulting company called 'Intera' (now 'Quanti Sci-Enviros') in Leicestershire, which included reviewing work done by Nirex at Sellafield (see Chapter 20). When her husband obtained a job in Perth, Western Australia, in 1997, she went too, intending to take a bit of a break, but was immediately snapped up by URS Australia Pty as an environmental consultant for the disposal of intractable waste in Western Australia, including the small amounts of radioactive materials accumulated as waste in that state (King, pers. comm., 2000).

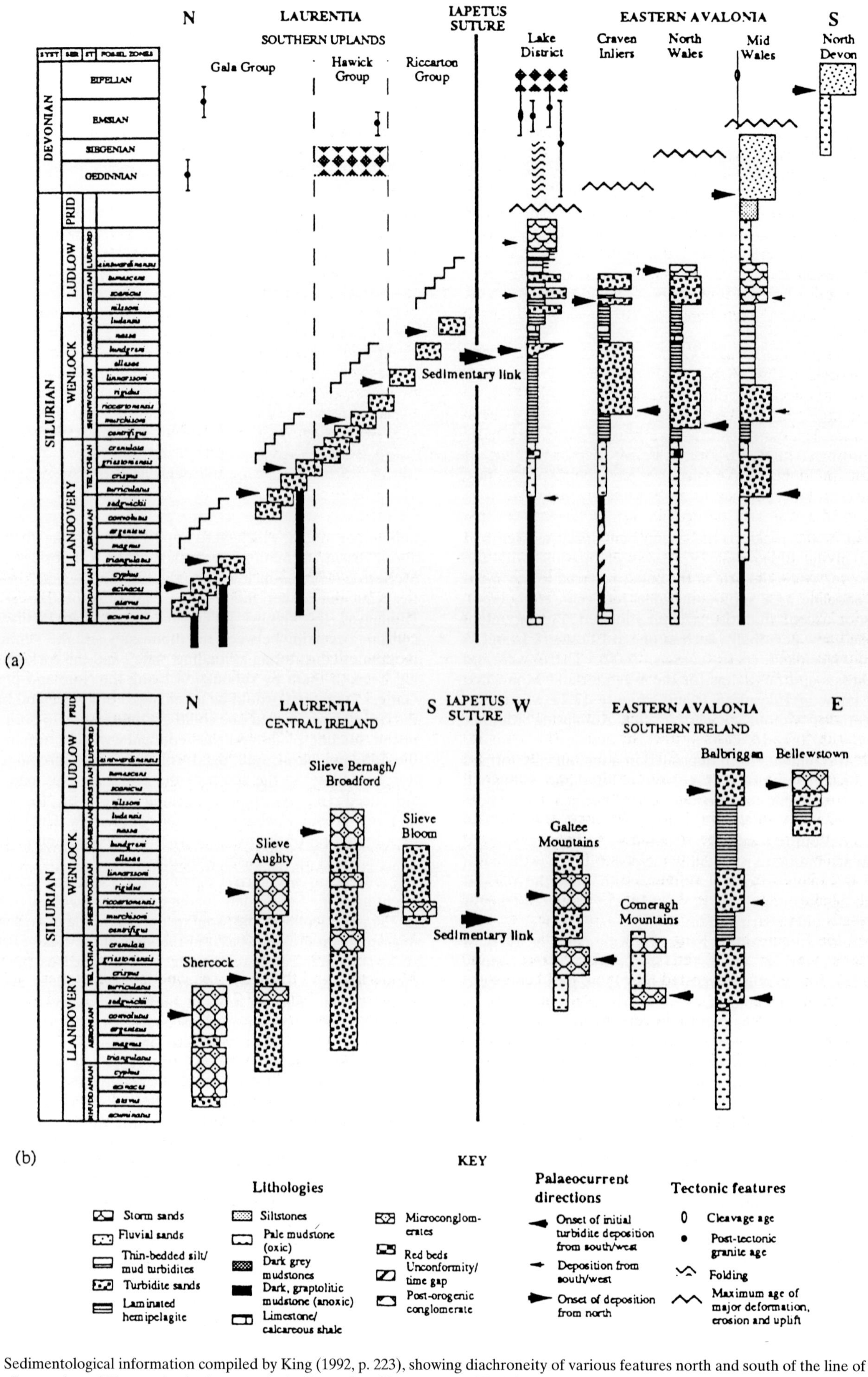

Fig. 17.6. Sedimentological information compiled by King (1992, p. 223), showing diachroneity of various features north and south of the line of the Iapetus Suture, for Laurentia and Eastern Avalonia respectively. Reproduced by courtesy of Dr King.

continental margins and imbrication of the basin fill (in the Southern Uplands). Nevertheless, like Kneller, she drew the idea of a steady southward progradation of the basin from the theory of foreland arcs. Empirically, her idea was based upon a study of sedimentological logs of the Windermeres, especially in the Howgill Fells, and comparison with the situation in the Southern Uplands, Wales and even Devonshire, from published sources. She considered the types, ages and thicknesses of the various strata of the strata, and from her observations she could see when sedimentation appeared to turn from the south to the north. In Scotland, turbidite sands appeared to be deposited further and further south the younger they got (with sediment being derived from the north). This persisted until the closure of Iapetus about mid-Wenlock. Meanwhile, sediments had been accumulating in a basin to the south of the Suture in the Lakeland region, in the Howgills, down in the Craven Inliers in Yorkshire, and in North and Mid-Wales.

King's particular point was that she believed that the sedimentary structures in England and Wales evidenced a shift in the direction from whence the sediment was derived – initially from the south and later from the north – with this change occurring in the Lakes and Wales *after* the Suture closure, with the basin gradually shifting southwards. That is, the shift occurred *later* the further south you looked: Lake District in mid to upper Wenlock; Craven Inliers in late Wenlock to early Ludlow; North Wales in early Ludlow; Mid-Wales in the Prídolí. She even suggested that down in Devonshire northerly sedimentation commenced later still, in the Devonian. The ideas were summarized in her thesis (see Fig. 17.6), utilizing evidence from the work of her partner, Chris Clayton, collecting materials for his thesis in Wales, from Barnes *et al.* (1989) for the Southern Uplands, and from other published sources. Thus she had a southward progradation of the Silurian sediment basin(s) of northern England and Wales. The slaty cleavage also seemed to become progressively younger southwards.

Thus with respect to the Lakes, King did not think that formation of the foreland basin proper started in the southern Lakes in the Ashgill, where the first Windermere sediments appeared ('Coniston Limestone', Dent Group, etc.). Rather, it formed when there was a substantial deepening of water and an increase in sediment supply in Upper Wenlock and through Ludlow times; the process came to a close, so far as the Lake District was concerned, in the Upper Silurian (Prídolí). During the Ashgill and Llandovery, the sediment basin was presumed to be relatively shallow, as evidenced by the Coniston Limestone and its fossils (though there was perhaps glacio-eustatic deepening in the Llandovery). King (1992, p. 206) did not take quite the same view of the situation in Ashgill times as did Kneller (1991). Rather, she suggested that there had been some extensional faulting associated with the emplacement of the BVG, and when that volcanic episode ceased there was a fairly slow subsidence, as the subterranean magma cooled and contracted. Thus the initial subsidence of the basin would, at that time, have been chiefly 'thermally controlled'. The subsequent formation of the foreland basin proper would have begun with the impending closure of the Iapetus Suture and the disruptions caused by the collision of the two terranes of Laurentia and Eastern Avalonia. She regarded the water as having been deep in both Wenlock and Ludlow times and only influenced by eustatic changes to a relatively minor extent.

King's calculations were complicated and subject to assumptions. She needed data on present thicknesses of strata, and estimates of how much they might have been compacted since deposition. This would depend on knowledge of the depth of water in which the sediments were deposited, which would itself depend on knowledge of former sea levels, and the load of overlying sediment. One also had to consider the compression of the sediments by tectonic activity subsequent to their formation (producing cleavage, for example).

Then, following Kneller's example, estimates of the depth of the basin were made by 'back stripping'; that is, having obtained a 'subsidence curve' (subsidence plotted against time), she needed to subtract the subsidence assumed to be due to compaction, etc., from the total subsidence in order to obtain an estimate of the 'tectonically driven' subsidence. The whole process was subject to uncertainties, it seems to me, for, as King herself said, sedimentological 'depth indicators are rare' (King 1994*b*, p. 650). None the less, back-stripped subsidence curves were determined, and it was claimed that their *form* indicated the tectonic 'reason' for the basin formation, which (like Kneller) she found to be 'loading' of the foreland, as discussed in her thesis.

Essential to King's theory (1992) were the notions of migrating shorelines, the existence and geometry of a sediment basin, and a 'forebulge'. These could be tracked by looking for differences in sediment type, palaeocurrents, fossil inclusions, evidences of oxia or anoxia, etc. For data on palaeocurrents, for example, King drew particularly on observations from McCaffrey's PhD (1991), but she did much sedimentological investigation on her own account (King 1994*a*, *b*), providing evidence in favour of a gradual shallowing from the time of the deposition of the Bannisdale and Underbarrow Formations to the more sandy Kirkby Moor Formation.[7]

King did not assume that the shoreline, advancing southwards, was linear. Rather, she regarded it as arcuate – according with the general present shape of the Windermere outcrops (extending down to the Howgill Fells and the Silurian (Craven) inliers of NW Yorkshire) and also the arcuate cleavage pattern of the southern slate belts (see Fig. 10.7). The idea was revealed in a series of diagrams from her paper with Kneller and Bell (Kneller *et al.* 1993) (see Fig. 17.7) and one of her papers from 1994 (see Fig. 17.8). The inversion of the basin would follow from the 'southerly' push of the Lakeland 'massif' and the 'southerly' migration of the basin 'forebulge'. The envisaged migration of the basin in the late Silurian could be traced from the changes in sediments showing

> passage from Wenlock and Lower Ludlow deep water 'flysch' deposits to upper Ludlow and Prídolí shallower water 'molasse' deposits.... An upward progression was preserved from oxygen-poor basin slope turbidite deposits through more oxygenated, bioturbated dilute density flow deposits, to storm and wave-influenced sediments (King 1994*a*, p. 255).

The tectonic hypotheses, showing the relationship between the Southern Uplands, the Iapetus Suture, the thrust belt of the northern Lakes, the Lakeland plutons, the Westmorland Monocline, and the basin receiving the Windermere sediments is displayed in Figure 17.7. The 'sedimentology' of the proposed migrating basin is shown in Figure 17.8.

Besides considering sedimentological information, the Kneller *et al.* (1993) paper also deployed evidence such as that provided by minor folds. Buckle folds in the Windermere sediments suggested shortening, but the oldest Windermeres appeared to be more shortened than the ones that followed. '[T]he lower layers were deforming while the upper layers were being deposited' (p. 693). So it was suggested that the basin became detached from its basement and then inverted, as shown in Figure 17.7b. The authors saw the Windermere sediments as being genetically associated with those of the accretionary prism of the Southern Uplands. Indeed, they proposed that the youngest Southern Uplands sediments and the oldest Windermere sediments were deposited in the same basin, which sediments are now, however, preserved and exposed in two separate 'fragments'. The envisaged foreland basin existed over the site of the former Iapetus Ocean,

[7] However, looking back on her work (pers. comm., 2000), King thinks that the 'forebulge hypothesis' is not proven (and difficult to prove) and that the inclusion of data from Devonshire in her argument was rather a long shot.

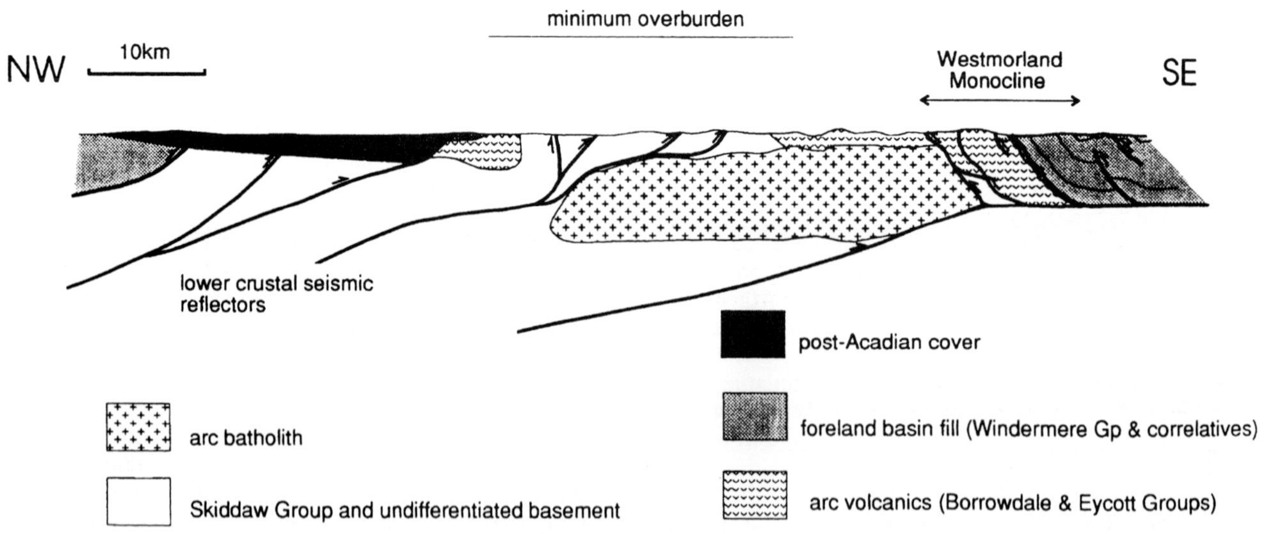

Fig. 17.7a. Profile from Southern Uplands (NW) to 'Windermere basin' (SE), according to Kneller, King & Bell (1993, p. 692). Reproduced by courtesy of Cambridge University Press.

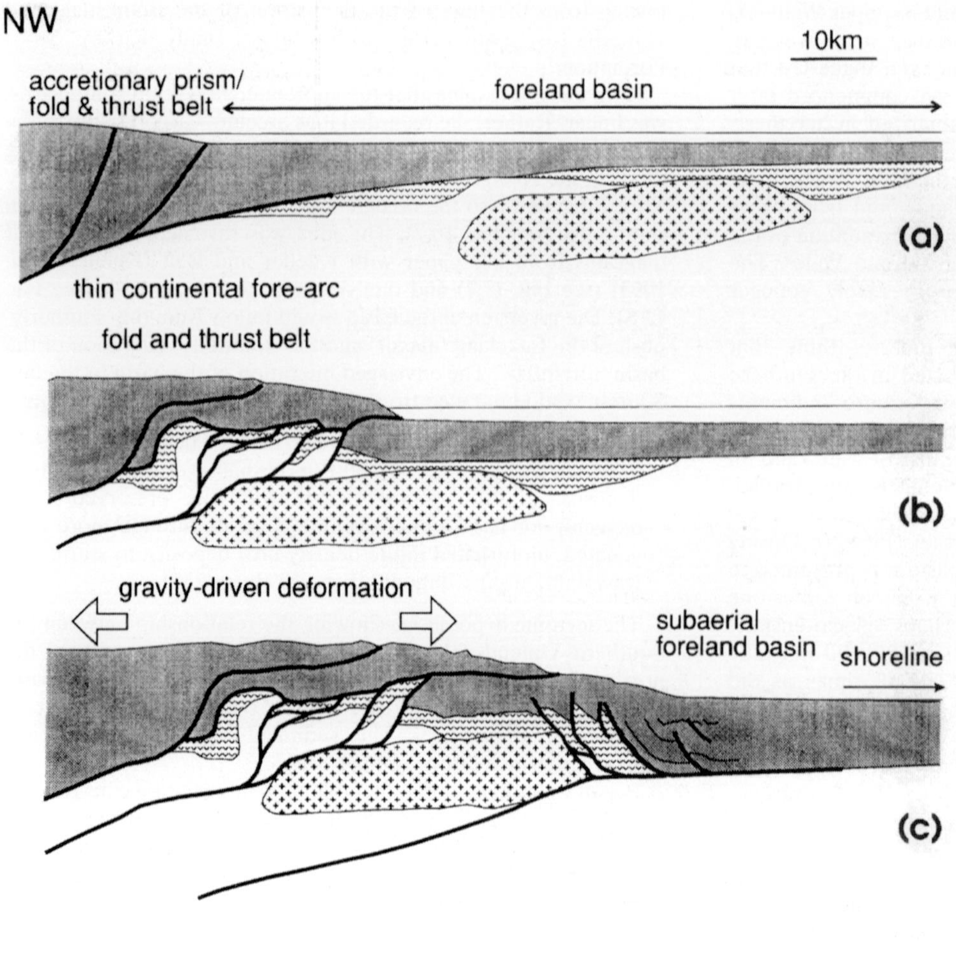

Fig. 17.7b. Evolution of Iapetus Suture zone and Lake District, according to Kneller, King & Bell (1993, p. 695). (**a**) Early Ludlow; (**b**) Mid Ludlow; (**c**) Prídolí. Reproduced by courtesy of Cambridge University Press.

but gradually migrated south as the Windermeres detached from their basement rocks, the whole being associated with one overarching tectonic process. The Devonian cleavage was said to be superimposed on pre-existing folds.

The basin progradation concept largely held the field at the end of the century, or at least it appeared to be the favoured model in the Ambleside Memoir, which stated that, '[t]he Acadian orogeny is here taken to include both the late Silurian foreland thrust-belt development and the Emsian climactic event' (Millward *et al.* 2000, p. 164). This comes close to conflating the processes involved in the Silurian closure of Iapetus and the Devonian cleavage-forming 'big crunch'. Be this as it may, the Memoir gave expression to the Kneller–King–Bell model of a Southern Uplands deformation and a basin prograding through the Lake District, and perhaps beyond, followed by inversion of that basin.

There are, however, several points about it that should be made, for the basin theory has not been universally accepted. King herself informed me (pers. comm., 2000) that she did not actually map her area overall (she had insufficient time, needless to say), and there were no modern maps for much of her ground. There was the uncoloured Crook–Kentmere map (Lawrence *et al.* 1986), which helped, but for a good deal of her area she had to

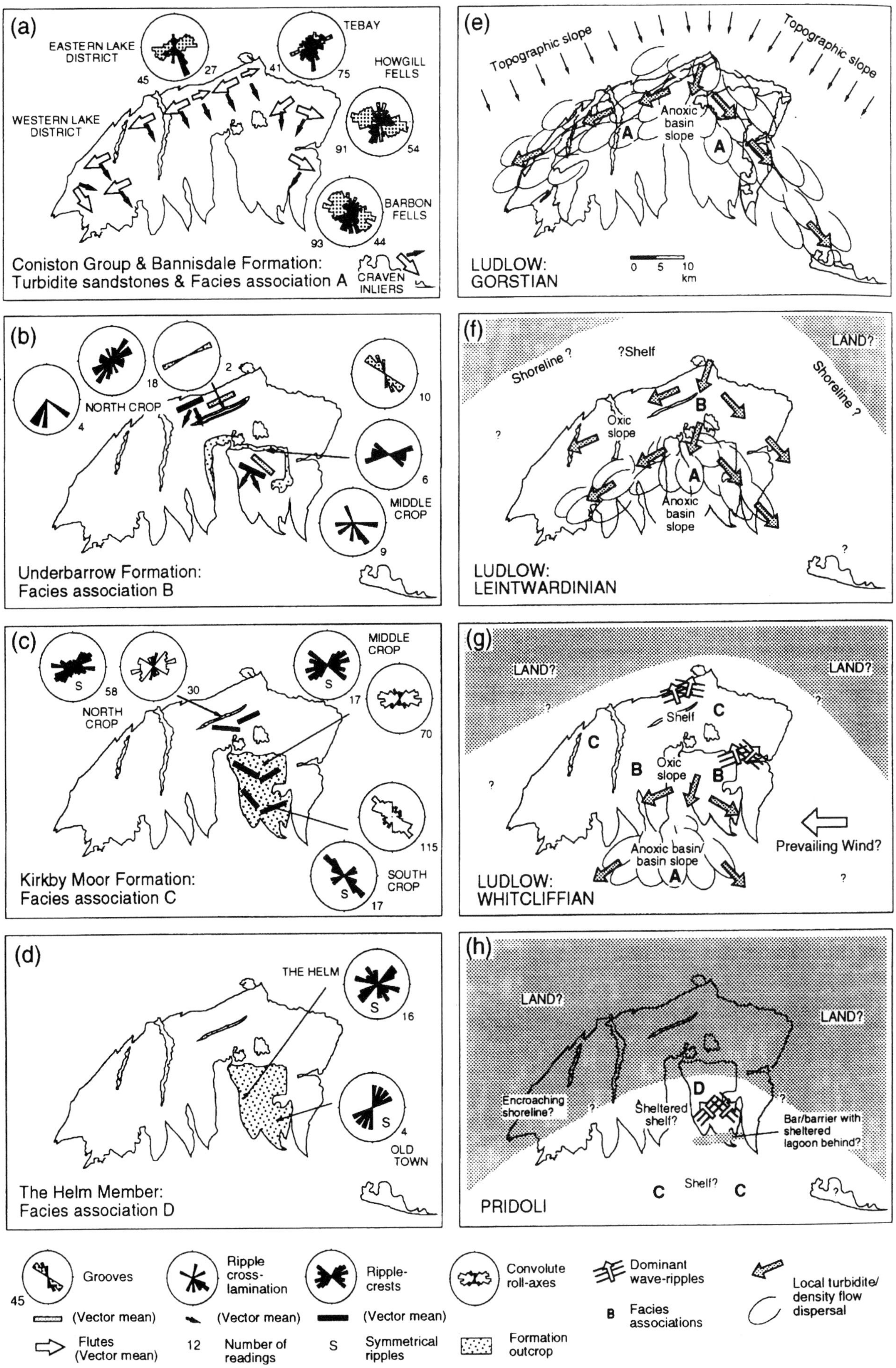

Fig. 17.8. Palaeocurrents and facies changes for the migrating foreland basin of southern Lakeland, Ludlow to Prídolí, according to King (1994*a*, p. 262). Reproduced by courtesy of Cambridge University Press.

rely on rather dated maps such as those of Richard Shaw (1968, 1971), based on palaeontological criteria to a considerable extent. She used Shaw's mapping to estimate some of her thicknesses and for the dates of some of the units.

Soper (pers. comm., 2000) disagreed with the idea that there was an emergent fold and thrust belt in the Ludlow (see Fig. 17.7b) that could have supplied Skiddaw Slate detritus for the formation of the Bannisdale Slates. Instead of fans of sediment from the NW and NE, he thought in terms of turbidites coming from the NE ('Scandian flysch'). If the third cartoon in Figure 17.7b did in fact represent the situation correctly, there should be molasse-type deposits in the Kirkby Moor Flags. But we do not have this. Further, Soper did not agree with the suggestion that there was a basin inversion responsible for shallowing in Kirkby Moor Flags time. He merely supposed that the convergence of Eastern Avalonia and Laurentia was coming to a close and deposition began to exceed subsidence. In fact, he did not accept the idea of basin inversion in the Ludlow for the Lakes: 'sediment could not have accumulated in a basin which had ceased to exist, in the Lakes at least, if it had been inverted in the Ludlow' (Soper, pers. comm., 2000). If there had been a Ludlow inversion he thought there should be evidence of Silurian deformation in the Skiddaw Slates, an idea by then discounted.

Another issue had to do with the so-called 'illite crystallinity' problem, which BGS mineralogist Neil Fortey has kindly explained to me (pers. comm., 1999). As an aluminous mud-rock undergoes diagenesis, or low-grade metamorphism, the clay minerals progressively recrystallize into illite, and then the illite alters into muscovite under the influence of elevated temperature and pressure. As the diagenesis passes into metamorphism and the crystals thicken, the X-ray diffraction peaks for the micas become progressively narrower; and one can measure the widths of the peaks (at their half-height), or their areas, giving a parameter related to the illite's degree of crystallization – the 'crystallinity index' – which serves as an indicator of the degree of diagenesis or metamorphism. This can be 'calibrated' by comparison with rocks for which the metamorphic facies have been identified by the presence of indicator minerals such as prehnite or pumpelleyite. In the Lake District, this mineralogical X-ray diffraction test on illite had, in some cases, been used as an estimate of the pressures, temperatures and overburdens to which the rocks have been subjected (Thomas 1985).

So far as the water-rich Windermere rocks were concerned, they appeared to have acquired an illite component quite rapidly (more so than did the hardened Skiddaw Slates). However, measurements such as those described above, performed on the Windermere rocks, suggested either that they were formerly subjected to a substantial overburden, perhaps of the order of 8 km (Millward *et al.* 2000, p. 186) or 10 km (Merriman *et al.* 1995); or that the rocks were slightly metamorphosed when they acquired their late Early Devonian (Emsian) cleavage. If we accept the first alternative, then it would appear that there was a continuation of deposition in Přídolí, Lochkovian and perhaps Pragian times, presumably with the deposition of a large molasse-type deposit, such as is still seen in the Mell Fell Conglomerate in the northern Lakes, bearing down on the presently observable Windermere sediments. This could have occurred if there were a substantial land-mass to the north, able to supply the overburden indicated by the illite crystallinity. King herself did not undertake illite studies, but she did examine graptolite reflectances, which seemed to confirm the idea of the former existence of a large overburden on the Howgill Silurians. She was willing to contemplate an overburden of well over 10 km of former sediment, now all seemingly lost by erosion.

It must be remembered that the Mell Fell clasts appear to be commonly Coniston Grit, which one might think would have come from the south (though they could also have been in the north if the idea of the Windermeres having formerly extended well to the north is correct).[8] It does, no doubt, seem remarkable that there could have been a large overburden formerly present and then 'spirited away' in a relatively short time (perhaps 20 million years) between the Přídolí and the Emsian.[9] *But*, if there were indeed such a massive overburden at some stage, then the King–Kneller–Bell model for the basin formation, and its south-eastward progradation, might appear to require revision: 10 km of Devonian sediment would be a considerable thickness to 'accommodate' according to the progradation model.

Though 'illite crystallinity' determinations are known to have uncertainties,[10] Soper took the determinations seriously. Thus he thought that the ideas of Kneller, King and Bell were 'driven by their model' (pers. comm., 1999). Whereas King thought that with a widening and southward prograding basin, and with a southward moving forebulge, the Lake District sediments could at one stage have received a molasse overburden from the supposed mountain front (see King 1992, p. 221 for illustrative diagrams). In Soper's view, if the basin were prograding southwards its sediment fill would never acquire the large overburden suggested by the illite crystallinity and graptolite reflectance data. In any case, the model would not account for the approximate coincidence in the date of the cleavage of the Welsh and south Lakeland slates, or indeed for the Acadian Orogeny as a large-scale event, altering rocks all the way from North America in the west to Poland in the east. So, with the Lakeland plutons still 'stuck in the Lakes', so to speak, it would (to me) seem difficult to understand how the forces invoked in the Lakes would account for the extended southward progradation of the basin; at best, they could only explain the formation of a basin immediately to the south of the domain of 'Otley II'. Soper's view was that that there could have been thickening during the Acadian Orogeny, which could have produced some of the pressure responsible for producing the illite crystallinity; and there could have been an additional short-lived Lower Devonian molasse deposit (subsequently removed by erosion) which could account for the rest of the mineralogical change.

However, it may be that the level of metamorphism reached in the Windermeres could have been achieved as a result of the pressure imposed in the 'big crunch' when the cleavage of the slate belt was acquired, without invoking a large overburden. Neil Fortey (pers. comm., 1999) pointed out that the highest level of metamorphism in the Windermeres is found where the rocks are most highly cleaved, though as I understand it the illite determinations indicate temperature and pressure, not pressure only. So perhaps an elevation of temperature, and hence the existence of a former overburden, *is* indicated – not pressure alone.

When I met him in 1999, Kneller was content with his model and Louisa King's idea – an extension of the Kneller–Bell mountain front notion – of a prograding basin (Kneller, pers. comm., 1999). Indeed, he said that he did not understand the relevance of the illite argument. He acknowledged the possible occurrence of a former large overburden, with perhaps 10 km having been stripped away since its deposition. But if King's hypothesis were right, with the basin having been active into the Lower Devonian, then, said Kneller, one would *expect* there to have been a former large overburden. It could have been 'cannibalized' and incorporated into the Lower Devonian of North

[8] On King's model, a northern provenance would be more likely as basin inversion would have occurred first in the north.

[9] Hughes (1995) has re-examined the ranges of the graptolite zones and argued that the Přídolí may be longer than generally assumed (2 Ma), which would allow somewhat more time for deposition and removal of overburden.

[10] For example, if both potassic and sodic micas are present, there will be a 'shoulder' on the side of the observed X-ray diffraction peaks, which can make it difficult to measure their widths accurately.

Wales as a possibility. However, Soper (who incidentally was one of Louisa King's PhD examiners) thought that the evidence from the illite crystallinity determinations presented a significant problem for the basin progradation model.

The doctoral work of McCaffrey (1991) and his subsequent publications (McCaffrey 1994; McCaffrey & Kneller 1996) have also had a bearing on the arguments. Where did all the sediments in the Windermere rocks actually come from? From McCaffrey's work, it appeared that the lower 'shoreline' units of the Windermere Supergroup (Ashgill: Dent Group) were derived directly from the adjacent Upper Borrowdales, which is indeed what might be anticipated. However, the overlying Silurian sediments, often of deep-water character, seemed to have travelled further. McCaffrey determined the Sm/Nd ratios for the grains in samples of Windermere clasts, which could be used to give a measure of the ages of their constituent minerals. The ages of the minerals, thus determined, were regarded as *not* being altered by weathering, transport, sedimentation and diagenesis, and could therefore be taken to be equivalent to the ages of the rock sources, where and when the Sm/Nd-containing minerals were first formed. Thus the geochemist had a 'fingerprint' that might help locate the sources of sediments. McCaffrey found that the 'average ages' of the constituents of the Windermere sediments increased fairly steadily, as the age of these sediments decreased from Ashgill to Ludlow, but then dropped off rather suddenly in the Prídolí. Considering also evidence from microfossils, McCaffrey came to think that the source of the Silurian Windermeres was *not* chiefly the Southern Uplands, or Laurentia or Eastern Avalonia, but the terrane of Baltica to the NE. That is, if the Scandian Orogen was being 'unroofed' during that (Silurian) time, there could have been an increasing component of mineral grains from that ancient source, which would increase the average age of the grains in the Windermere sediments, with the closing Iapetus seaway acting as a conduit for sediment distribution. When that conduit was closed off, the sandy deposits of the Kirkby Moor Formation might have begun to arrive from the Southern Uplands area. As we have already seen, Soper liked the idea of a Scandian origin for Windermere sediments, as envisaged by his former student McCaffrey.

McCaffrey's (1994) findings were not, as it seems to me, altogether consistent with King and Kneller's model of a southerly prograding sediment basin, with sediment derived from the north or northeast, so it is no surprise that we find the matter revisited soon after by McCaffrey & Kneller (1996). They looked in detail at the evidence for the likely source-type for the sediment grains: continental block, magmatic arc and recycled orogen – according to the proportions of quartz (either total or monocrystalline grains), feldspar and lithic fragments. These, when plotted on triangular QFL diagrams, could give an idea of the likely origin of the grains (cf. Dickinson *et al.* 1983). McCaffrey and Kneller also went through the Sm/Nd age determination exercise again, and concluded that there were three main trends in the sourcing of the Windermere sediments. First, there was an *increase* in the age of the grains from early Ashgill to early Llandovery as the BVG components gradually decreased. Then there was a further increase in the age of the sediment grains from early Llandovery to mid-Ludlow, as a greater proportion of material arrived 'from along the Laurentian margin'. And then the grain-age decreased again from mid-to-late Ludlow to Prídolí, after Iapetus closed. The more southerly Welsh basin continued to receive older grains after their supply to the Windermere sediments ceased, which was consistent with King and Kneller's prograding basin hypothesis (though it would be hard to see how Scandian[?] sediments reached Wales if Iapetus was by then closed off).[11]

The interesting point in this proposal was that the Baltica terrane or Scandian Orogen, gradually being unroofed, continued to be seen by McCaffrey and Kneller as an increasing source of the Windermere sediments during early Llandovery to mid-Ludlow; but it was hypothesized that the sediments actually travelled 'along the margins of Laurentia', thus, I take it, preserving the NW–NE supply route envisaged by King, even if the sediment grains actually originated in Baltica.

All this is rather complicated, and, as it appears to me, the geologists who had worked, or were working, in the Lakes did not reach consensus by the end of the twentieth century. It is interesting that this should be so, given that, *prima facie*, the problems of the Windermeres may seem less complicated than those for the Skiddaws or the Borrowdales. However, we must remember that the remapping of the southern Lakes was not completed by the end of 2000. The Kendal and Kirkby Lonsdale sheets were incomplete, and the Ulverston Memoir was not then published.

In a sense, the story all goes back to the biostratigraphically based mapwork of Richard Shaw. Soper, still mapping in the southern Lakes and the Howgills in 1999–2000, was coming to favour chiefly litho- rather than biostratigraphy for his mapwork, mapping boundaries that could be *seen* in the field. Thus he wanted to dispense with the Underbarrow Flags and use as his main datum the diachronous boundary between the sandy Kirkby Moor Flags and the mudstones and flags of the Bannisdale Slates. When this was done, he found that sand deposition apparently began in the *south*, not the north as might be expected according to the southward prograding basin model (Soper, pers. comm., 2000). Soper emphasized, however, that his lithostratigraphic mapwork needed to be reviewed by a palaeontologist, particularly with regard to the diachronous base of the Kirkby Moor Flags, before a strong stand could be taken (pers. comm., 2000). So it is clear that the 'last word' had not been said on this interesting set of problems by the end of the twentieth century, although for Soper's 'last words' we may refer back to Figure 13.11, where we see that he contemplated the notion of *extensional* movement in the Lower Devonian subsequent to the 'soft collision' of Avalonia with Laurentia and Baltica (an elastic rebound?), thereby accounting for the possible development of a basin to receive sediments atop the Kirkby Moor Flags.

As usual, it is not the historian's task to attempt to adjudicate such matters. We must wait to see how it all turns out in the twenty-first century. I may say, however, that I am not comfortable with the idea of the Windermeres having formerly lain across the region of 'Otley II', as implied in Figure 17.4, and as J. F. N. Green had suggested long before. The 'Otley II' rocks were formed in the Llanvirn and Caradoc. Certainly some of their tuffaceous material would have been easily dispersed, but that would hardly have been the case for the andesites and the large ignimbrites. So how could there have been a sufficient reduction in level in a comparatively short time to allow deposition of Silurian sediments (including the graptolitic mudstones usually associated with deep-water conditions, such as the Skelgill Formation) right over the top, especially given the fact that the glacial episodes of the Silurian would presumably have lowered sea levels? Also, the presence of the 'buoyant' or 'upwardly mobile' granitic plutons would not have facilitated lowering of the land surface to form a sediment basin. There are no Ludlow rocks in the Southern Uplands (as King's thesis reported: see Fig. 17.6), so the 'Otley III' basin must presumably have terminated to the south of that area. This was consistent with the 'progradation model'. On the other hand, there is no known northern shoreline near the BVG in Wenlock–Ludlow times and the actual cause of the inversion of the basin still seems to be somewhat mysterious, especially if one is not inclined to accept King's 'forebulge' hypothesis.

[11] Soper (pers. comm., 2000) holds that there was a seaway, or at least several interconnected basins or sub-basins, between NW England and north Wales during Wenlock times, as analogous facies can be identified in the two localities.

The Ambleside Memoir evidently had some unease too. It said:

> During the final stages of closure of the Iapetus Ocean the leading edge of Avalonia collided with the margin of Laurentia and a thin-skinned, south-east-propagating thrust sequence [Causey Pike Thrust, etc.] was one of the results. . . . This linked the Southern Uplands accretionary complex with a foreland basin developed across the suture zone ahead of the thrust belt. . . . These relationships involve diachronous deformation . . . becoming sequentially younger southwards. However, the Early Devonian inversion of the Windermere Supergroup basin and the accompanying imposition of a regional slaty cleavage indicate a climactic orogenic event dated . . . at about 400 Ma (i.e. Early Devonian, Emsian . . .). The relationship between these two phenomena is not fully understood (Millward *et al.* 2000, pp. 163–164).

Here we may also take a final look at the question of the movement of the sedimentary rocks relative to the Lakeland plutons. Late in the century, Soper's thinking (pers. comm., 2000), prompted by a Penrose Conference in 2000 on problems associated with the Iapetus Suture, returned to the questions of the collisions of terranes discussed in Chapter 10. The palaeomagnetic data discussed at the Penrose meeting pointed to a northwards movement of Africa, which could well have been the cause of the Acadian Orogeny, thus supporting Soper's long-held ideas. But a push from the south could mean that the southern Lakes 'back-thrust' (see Figs 16.13 or 17.7b) should be interpreted as a *fore*-thrust, and the south-vergent structures in the Skiddaw Slates such as the Causey Pike or Watch Hill Thrusts should be construed as *back*thrusts! Or more fundamentally, the Acadian movements were not, in Soper's view, related to the closure of Iapetus, or were not a culmination of the Caledonian Orogeny. The two episodes were, in his understanding, separated by a period of extension in the Lower Devonian. One could think of the Acadian movements as 'proto-Variscan' rather than late Caledonian.[12] The Caledonian Orogeny had no business in England in the Upper Silurian and Devonian! This view appears to me to be at odds with arguments earlier expressed by Kneller (see p. 232), whose analysis had seemingly excluded extension as a cause for the formation of the basin in which the Windermere sediments accu-lumulated.

Evidently, the interpretations of the tectonics, sedimentology and palaeogeography of NW England were by no means settled by the end of the twentieth century, despite the efforts of the Lakeland Project, the work of several university investigators, the use of sophisticated analytical techniques, and the application of the latest theoretical ideas on basin formation and sedimentation.

[12] This idea was made explicit to me in a letter from Dr Soper dated 17 January 2002, but he had already been developing such views in the previous decade. It remains to be seen how they will fare in the twenty-first century. Their progress will depend on completion of the mapping programme and further studies of overburden thicknesses, with the help of pressure estimates from the study of fluid inclusions – which would be a distant descendant of work such as Clifton Ward had initiated in the northern Lakes back in the nineteenth century.

Chapter 18
Tertiary uplift

It is one thing to tell the story of the idea about how the rocks of the Lake District were formed, but what happened to them after they came into being? In this chapter, I shall say something about this important issue. One might think that the mountains that we see in the Central Fells have been there ever since the Ordovician, and we are simply observing their eroded remnants. In a sense this is true. The volcanologists do think that we are able to examine, in the field, the eroded interiors of volcanoes. Are we able to do so because erosion has proceeded to such an extent that their very 'guts' are now visible? Presumably that is so, but why, then, are the 'guts' 'up in the air' for our inspection? Uplift can be expected because of the buoyancy of the underlying granites, but is that the whole story? And, presuming there was substantial uplift at some stage, when did that uplift occur, and are there causes other than isostatic forces? These issues will be considered in the present chapter, which covers the geological history of the Lakes subsequent to the 'big crunch', which is thought to have occurred in the Devonian, through to the glacial epoch of the Pleistocene. A review of matters relating to the presumed former sedimentary cover of Lakeland and the techniques for estimating its thickness has been given by Douglas Holliday in his Presidential Address to the Yorkshire Geological Society at York on 5 December 1998 (Holliday 1999).

Attention to the apparent 'domed' structure of the Lakes and its generally radial drainage pattern was given in the nineteenth century by the early investigator, William Hopkins (1842, 1848). He noted that the ('Mountain' or Carboniferous) limestone that frames much of the Lake District appears to dip away from the mountains (this is well seen on the northern end of High Street in the NE of the Lakes; and see Fig. 7.12), but was presumably laid down on an approximately horizontal seabed, and very likely covered the whole region of the present mountains at some stage. So there was evidence of post-Carboniferous uplift of the mountain region. Hopkins related the pattern of waterways and lakes of the mountain area to fractures produced by uplift. He further thought that the New Red Sandstone had also been deposited round the Lakes (though not perhaps covering all the ancient rocks), and it too had been uplifted and then partly stripped away. The sequence of events as envisaged by Hopkins (1848, p. 79) is shown in Figure 18.1, where S is the 'great Pennine fault' near Stainmore. The final elevation of the Stainmore area, might not, Hopkins thought, have occurred until 'after the Tertiary'. Hopkins (1848, p. 93) supposed that the elevation was not necessarily gradual. On the contrary, he regarded the movements as 'paroxysmal and frequent, but not necessarily large'. He sought to relate such movements to his 'catastrophist' theory of the glacial epoch (see p. 256). We remark that even at this early stage of the development of Lakeland geology a late date was envisaged for the final uplift of the region.

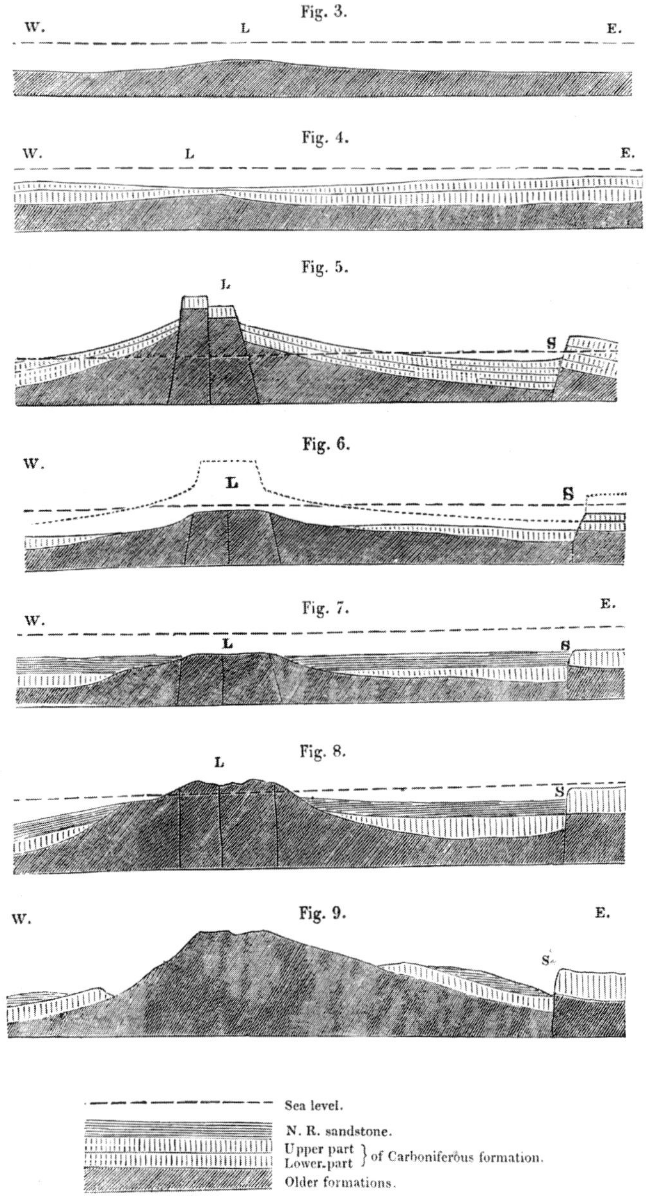

Fig. 18.1. Sequence of events for the developing structure of the Lake District, according to William Hopkins (1848, p. 79).

Subsequently, Ward (1879) supported the idea of the Carboniferous rocks having formerly covered the whole of Lakeland. Goodchild (1885) followed Hopkins in the idea of uplift subsequent to the New Red Sandstone, and in a later paper Goodchild (1888–1889) suggested that the uplift occurred during the Tertiary. He thought that there were formerly 'three plains' in the region: the first being 'fashioned out of the Older Palaeozoic rocks, whereon the Carboniferous strata were spread out'; the second being the somewhat irregular surface that 'received the New Red'; and the third represented the 'denuded surface whereon the Upper Cretaceous rocks were formed'. With remarkable prescience, he suggested that, '[p]ossibly the Upper Greensand, the Chalk, and some of the Eocene strata may have formerly covered the whole of the North of England, and have connected the Later Neozoic rocks of Ireland with those now left in East Yorkshire' (Goodchild 1888–1889, p. 76). The argument, clarified with the aid of a series of four profiles, was made chiefly on the basis of geomorphological information (and particularly with regard to drainage patterns and data from the Howgill Fells).

Considerable attention was subsequently given to such questions by Marr, towards the end of the nineteenth century and into the twentieth century (e.g. Marr 1889, 1896*a*, 1906*b*, 1910, 1916), and he too favoured the idea of Tertiary uplift, having arrived at that conclusion independently of Goodchild. His major statement on the issue was made in his Presidential Address to the Geological Society (Marr 1906*b*), but earlier, in 1889, he put forward the idea that doming of the region might have been due to the action of a laccolitic intrusion.

It is interesting that Marr (1906*b*) sought to link his ideas about uplift to his theories about lag and thrust faults (see p. 68), and he suggested that the Carrock Fell intrusion could have occurred by the faulted point of weakness where the Skiddaw Slates were adjacent to the Drygill Shales. He further suggested that large faults to the east of the Lakes, such as the Dent Fault, occurred during New Red times, and then, after denudation, Jurassic and perhaps Cretaceous rocks could have covered the whole. An outcrop of Liassic rocks near Carlisle was compatible with a possibly wider distribution of Jurassic rocks.

Marr's idea was not so very different from that of Hopkins, and in his Presidential Address he took his point of departure from Hopkins's work. But he added the point that the similar heights of Cross Fell, the Skiddaws and the Borrowdales suggested that there had been insufficient time for erosion to produce substantial differences in elevation. The low elevation of the Windermere rocks in the southern Lakes, compared with their elevation in the Howgill Fells, could be explained by the fact that there had been two separate areas of uplift: in the central and northern Lakes, and to the south in the Howgills. Marr based his arguments chiefly upon considerations of physical geography, discussing possible river captures and such, consequent upon changes in elevation. The details need not be given here, but it may be remarked that he thought the elevations probably occurred in the Miocene or Pliocene. The relatively recent erosive events had, he thought, worked along older lines of fracture, reactivated as shatter-zones.

In a chapter in the Geologists' Association Jubilee volume, Marr (1910, pp. 656–657) stated that he thought that the uplift of the Pennines, and Lakeland and Howgill domings, took place in the mid-Tertiary, and might be related to the famous Tertiary volcanic activity of western Scotland, which generated the vast plateau basalts of the Inner Hebrides (of which such places as Fingal's Cave are relics). In or near Lakeland, the Armathwaite Dyke, which crops out on the hillside to the west of Thirlmere, and the dykes cropping out in the Pennines or the Eden Valley, could, Marr suggested, have been formed in association with the Scottish Tertiary volcanics. Geologically late tectonic activity in the Pennines was subsequently affirmed by Trotter (1929*a*) on the grounds that a peneplanation surface showed evidence of being folded (as, for example, in the area of the Stainmore Gap, which cuts through the northern Pennines from near Brough), with movement in the Pliocene (or even later?).[1]

Marr realized that there was no simple correlation between the Lakeland waterways, valleys and lakes and the underlying geological structures, but the problem was not a major difficulty for him. If Lakeland had been covered by a blanket of Late Palaeozoic and Mesozoic rocks, and then uplifted as some dome-like structure, a drainage pattern would be formed on the slopes of the newly emerged strata that could be independent of the structure of the basement rocks. Then when erosion had cut down to the Lower Palaeozoic rocks the later drainage pattern would be (or could have been) superimposed on the older rocks and structures. But the pattern thus revealed would have been etched on a Devonian surface. The presence of Cretaceous rocks, far from their usual southeastern outcrops, had long been known, as for example the fragment of chalk, about the size of a large house, preserved in a volcanic vent in the middle of the Isle of Arran, Chalk-derived flints in Devonshire, and some Chalk in Ireland. So, the idea that there was formerly a Mesozoic, and perhaps also a Tertiary, cover of the Lakeland mountains, subsequently stripped away following Tertiary uplift, became, as Frank Moseley once put it, 'a treasured belief' (Moseley 1978*a*, p. 13). There must have been a tensional regime in the Mesozoic, with basin structure(s) formed, so as to accommodate the accumulation of Mesozoic sediments such as the Chalk. But with the paucity of Mesozoic sediments in northwestern Britain south of Skye, it is difficult to say much in detail (but see p. 252 for the discussion of Chadwick *et al.* 1994).

In the Lake District, geomorphological investigations were carried forward in the 1930s by Trotter's Survey colleague, Sydney Hollingworth (1936, 1935–1937 [1937], 1938). Geomorphologists of that time were influenced by the ideas of the famous American, William Morris Davis (1850–1934), who had introduced his ideas of landscape planation by subaerial processes, 'base-levelling', the term 'peneplain', and the metaphors of 'youth', 'maturity' and 'old age' of landforms, to British readers towards the end of the nineteenth century (Davis 1895).[2] Davis suggested that there had been a planation of the land area of Britain accompanying the retreat of the widespread Cretaceous sea; and this planed surface was then reactivated during the Tertiary as a result of the collateral effects of the Alpine Orogeny. So, for eastern and southern Britain, two cycles of erosion appeared to be evidenced.

Extrapolating, or extending, Davisian ideas, Hollingworth sought to identify a number of horizontal benches, cut across rock structures, and various planar flats or hilltops, the whole supposedly representing a sequence of rejuvenations of landscape and subsequent planations.

This work, initiated in 1926–1928, was based partly on the examination of topographic maps, and partly on field observations. It was started in the Lakes (Hollingworth 1936) and subsequently extended to Devon and Cornwall, SW Scotland and the Cheviots, and to Wales (Hollingworth 1935–1937 [1937], 1938). Hollingworth pointed out that the concept of erosion platforms had already been successfully applied in Cornwall by George Barrow and J. F. N. Green. In the Lakes, deploying a procedure for extracting information from topographic maps by 'projected profiles', derived from another American, Joseph Barrell (1920), Hollingworth divided the area into approximately radial strips, and investigated each separately. The bench structures were then grouped according to their altitudes, so that a stepped series of erosional levels was identified. The individual levels appeared to extend around the Lake District, suggesting that the area had not suffered significant tilting, though doming might have been accentuated.

A sample of Hollingworth's findings is shown in Figure 18.2, from which we can see that they involved a fair degree of 'imagination'. Thirteen 'steps' were depicted for this part of eastern Lakeland and the coastal plain, but we can see that some were regarded as more definite than others. The higher platforms were taken to be the oldest. It was suggested that both littoral erosion and subaerial base-levelling had been at work, falling sea level being the primary cause. Hollingworth tentatively accepted the Miocene as the time for 'differential movements', and the platforms were presumably produced in the late Miocene and Pliocene. The lower surfaces were thought to be of marine origin, while the higher ones were formed subaerially in a 'Davisian' manner. Those appearing to display warped surfaces were thought to have been affected by the mid-Tertiary Alpine earth movements. Hollingworth suggested that the highest planar surface might be the relic of the domed floor of the old Upper Cretaceous cover, such as Goodchild had contemplated. Alternatively, it could be a Tertiary peneplain produced by a second doming; or some more recent erosion surface.

Such investigations were also pursued by Richard McConnell (1938–1941 [1939], 1938–1941 [1940]), following up many similar types of investigation undertaken by Continental geomorphologists, particularly in the Alps. He recorded the existence of a

[1] Trotter referred to movement post-dating the Red Crag (now regarded as Pleistocene), which he regarded as Pliocene.

[2] Davis visited Britain in 1894 and sought to apply the geomorphological ideas that he had developed in the American West to England. In fact, his paper was largely theoretical, rather than a report of his own empirical studies in Britain. It was furnished with his characteristic 'biological' or 'sociological' metaphors and similes for geomorphology. 'Capturing' rivers were 'victorious', and captured rivers could be 'beheaded'. Rising sea levels could 'betrunk' (as opposed to 'behead') a river. Perhaps it was the charm of Davis's diction that assisted the acceptance of his ideas? Regardless, Davisian geomorphology did well in Britain during the first half of the twentieth century.

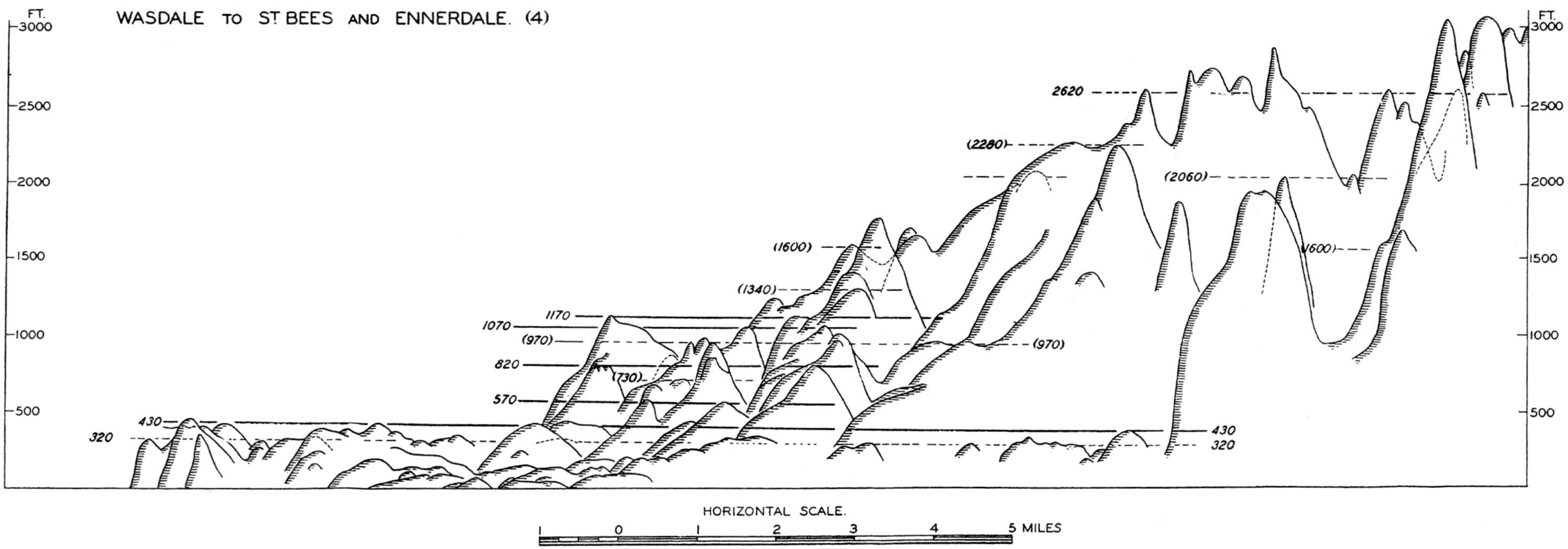

Fig. 18.2. Profile of landforms in eastern Lakeland, showing putative erosional platforms, according to Hollingworth (1935–1937 [1937], Plate VIII). Reproduced by courtesy of the Yorkshire Geological Society.

'staircase' of seven 'residual surfaces' in southern Lakeland and the Howgills, his erosion surfaces being noticeably less rigidly horizontal than those drawn by Hollingworth. Sometimes they appeared to be substantially 'warped' or of varying altitude, which was why McConnell 'recognized' less of them than his predecessor. He found the general physiography to be influenced by rock hardnesses. Nevertheless, there was a general concordance in altitudes of features such as valley basins, bevelled spurs, passes, benches, corries and upland flats, which, in McConnell's view, evidenced the dominant agency of fluvial action.

Further work was undertaken by the McGill geographer J. T. Parry (1960), who mapped ten benches in western and southwestern Lakeland, between 290 and 1300–1500 feet above sea level.[3] Based on their forms, Parry regarded the three upper (older) ones as having been produced by subaerial erosion and the lower ones by marine action. Since the unwarped upper benches appeared to cut across faults of apparently Miocene age, and all showed some effects of glaciation, he inferred that the Lake District had been above sea level in the Miocene and/or, Pliocene, when the upper benches were formed. Parry sought to correlate the benches with a sequence established in southern England by Sidney Wooldridge and D. L. Linton (1939), and in the process redated his lower six as Pleistocene, the one above being a Pliocene–Pleistocene marine level. He regarded the uppermost one as late Miocene and the two below as Pliocene. The idea of marine erosion as the cause of the upper planations was thus well in retreat in the mid-twentieth century.

In his early work in the Forest of Bowland (north Lancashire) our friend Frank Moseley (1961) did not seek to forge links with southern England. Analysing slopes and heights, he identified (or, in the language of geologists, recognized) two main surfaces and a subsidiary one, which were in part the result of differences in lithologies. All were regarded as subaerial, with only the lowest showing indications of marine erosion. Moseley warned against 'flat surfaces closely spaced in height' being interpreted as several distinct surfaces. In other words, he cautioned against a proliferation of surfaces. There was, then, a reaction against the 'staircase' model of the 1930s–1950s. This did not preclude the idea of Tertiary uplift, doming or exhumation of older rocks, but Marr's idea of the Lakeland drainage pattern having been established by imposition on an exposed Mesozoic surface was questioned by Moseley (1972, 1978*a*, p. 13), for a number of the region's waterways in fact followed much older Caledonian structures. Indeed, he regarded many of the Tertiary faults as reactivations of older structures. Stephen Belbin (1985) ascribed the Lakeland uplift to isostatic movement, the granitic plutons being chiefly responsible.[4]

The view of doming of the Lake District as an essentially straightforward process has been questioned by Richard Clark (1986–1987 [1988], 1992–1993 [1994]), a geomorphologist formerly on the staff of The Queen's University, Belfast, who retired early and took up residence in the hamlet of Hartsop, to the south of Ullswater, giving him ample opportunity to devote himself to problems of Lakeland physical geology or geography. When I visited him in 1998, he told me how he had been endlessly pondering the problem of the general form of the Lakeland mountains, especially after the stimulus of the publication of the magnificent coloured poster of a 1: 200 000 satellite image (from *Landsat* 5) of the region (viewed from a height of 700 km) by the British Geological Survey (n.d.).

Clark's principal objection to the old (Marr-type) theory of doming was that the drainage is not truly radial from some central point. Additionally, Clark has noted that there are fault lineaments with rocks of similar types on opposite sides of the fault-dominated valleys, which do not have mountains or hills of the same altitudes on opposite sides. This would not be anticipated if the valleys had simply been initially carved into an undifferentiated surface by subaerial processes. He also referred to watersheds parallel with fault lineaments, various streams running parallel with the lineaments, and streams joining one another at similar angles to lineament junctions (Clark 1986–1987 [1988], p. 26). Further, he remarked that some of the main mountains (e.g. Skiddaw) are undergirded by granites associated with gravity lows, while others of comparable height (e.g. Helvellyn) are not. For such reasons, Clark questioned Marr's idea that the drainage system of the Lakes was established on an Upper Palaeozoic or Mesozoic cover, which system later became superimposed on the underlying Lower Palaeozoics as they were exhumed. Rather, he suggested that the various mountain elements of the Lakes were uplifted as partly separate blocks. But he has not contested the occurrence of Tertiary uplift.

Clark (1992–1993 [1994]) gave special attention to the Skiddaw massif, which Marr (1916) had acknowledged as a discrete centre of uplift. The massif is rather peculiar in that although Skiddaw is one of the largest and highest Lakeland mountains it is made of moderately soft rock, and is surrounded by what Clark dubbed a 'moat' of relatively low-lying ground. It is like an 'island', and probably not an erosional remnant for it lacks the drainage pattern on its flanks that would be expected if it had originated thus (even if fluvial erosion were supplemented by glacial action). Clark took the view that the presence of this discrete mountain mass was due to the isostatic uplift of the area due to the buoyancy of the underlying Skiddaw Granite. So the whole was a relatively recent topographic feature associated with Palaeogene uplift: Skiddaw's 'granite core exercis[ed] a growing independent influence on local relief as removal of thick [Mesozoic] sediment cover progressed'. However, this model would not give a comparable explanation of the existence of the somewhat isolated Helvellyn, which does not show indications of an underlying low-density pluton.

Leaving these points aside, if the general idea of Tertiary uplift for Lakeland is accepted, can it be related to notions of some wider tectonic activity (such as volcanism in the Hebrides) or some wider geological theory (presumably associated with plate tectonics)? It would be attractive if it could.

On a wider front, geologists have 'recognized' a series of rifting events in the area of the continental shelf of Europe, from Carboniferous to Jurassic, associated with the break-up of Laurasia, and sedimentation continuing into the Cretaceous. The resulting basin subsidence is thus well recorded by a series of sedimentary basins (see, e.g. Jackson & Mulholland 1993, fig. 2). In the Lake District, an extensional regime in the Carboniferous is documented by the formation of the Cockermouth lavas (see Chapter 4).

As is well known, with plate tectonics came the idea of 'hot-spots' in the mantle, which gave a pleasing explanation of, for example, the relative ages of oceanic islands (Wilson 1963*a*, *b*).[5]

[3] His work was based on his Liverpool University MSc (Parry 1958).

[4] Belbin, from the Department of Geography, Manchester University, also traced the history of ideas about planation in the Lake District, and I have found his references useful sources in the foregoing account.

[5] In Wilson's (1963*b*, p. 869) theory, a hot-spot was represented as being situated in a part of the mantle away from an oceanic ridge where magma is rising, according to sea-floor spreading theory. Thus an oceanic plate can drift over a hot-spot and if the latter is generating lava as volcanoes a string of oceanic islands of different ages can be successively produced. Wilson showed that for a number of groups of Pacific islands, notably those of Hawaii, there is evidence from their ages and alignments of a general westerly movement for islands west of the East Pacific Rise, and a contrary motion for the islands to the east of the rise. There were several hot-spots generating the several Pacific archipelagos. This idea was later developed by Jason Morgan (1971, 1972, 1981) of Princeton, and many others (see Davies (1999) for a combined historical and theoretical exegesis). Morgan (1972, p. 8) referred to 'plumes of deep mantle rising upward' while 'the rest of the mantle is slowly sinking downward in a pattern analogous to a thunderhead or a coffee percolator'. Material from the hot lower mantle rises in 'pipes' and is added to the asthenosphere.

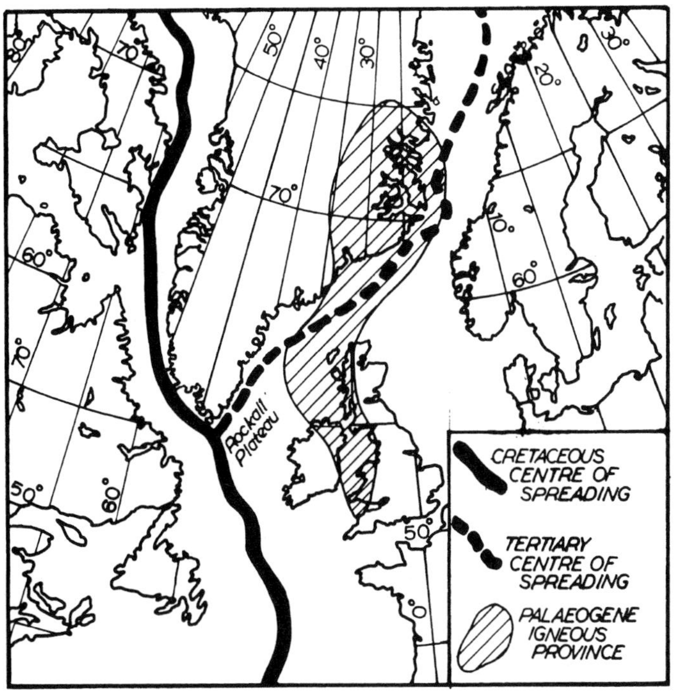

Fig. 18.3. North-Atlantic rifting and the Palaeogene igneous province of Britain and Greenland, according to Brooks (1973, p. 82). © The University of Chicago Press. Reproduced by permission.

At present in the North Atlantic, we evidently have a hot-spot in (or under) Iceland, which sits over the line of the Mid-Atlantic Ridge, from which lineament sea-floor spreading slowly occurs. Plate-tectonic theory, coupled with much empirical evidence, suggests that the North Atlantic began to open around 60 Ma, separating what are now Greenland and Scandinavia. For the eastern side of the ocean, there are traces of the spreading process from the hot-spot, observable in, for example, the Rockall Plateau and the Faeroes. The vast Tertiary lava-fields of the Hebrides, Northern Ireland and eastern Greenland, along with the volcanic centres of Rhum, Skye, etc., and even the igneous rock of Lundy Isle in the Bristol Channel have all been interpreted as part of the same igneous province, for example by Michael Brooks (1973) of Swansea University College (see Fig. 18.3). It should be remarked that Brooks's sketch of the extent of the Tertiary igneous province associated with the Atlantic rifting touches the margin of the Lake District, though we do not in fact have Tertiary lava-flows in west Cumbria – minor dykes at most, the only substantial one being the Armathwaite Dyke. Also, aereomagnetic maps have suggested the presence of Tertiary mafic dykes offshore from the Lake District. There is also the Fleetwood Dyke group to the south and southwest of the Lakes (Arter & Fagin 1993), and the Cleveland Dyke echelon to the north and northeast of the Lakes, running from the Solway Firth region down towards the Yorkshire coast (Kirton & Donato 1985). Dewey and Windley (1988) quoted 56 Ma as the time of maximum tectonic and magmatic activity, associated with the opening of the Norwegian Sea and the final separation of Greenland from Eurasia. The Fleetwood Dyke group, also examined by well drilling in a region of the Irish Sea approximately equidistant between the Isle of Man and the Furness coast, was found to consist of a complex set of dolerite dykes and sills, dated between 61 and 65 Ma. Fission-track analysis (see below) suggested the occurrence of about 2 km of early Tertiary uplift at about the same time as the emplacement of the igneous rocks.

Investigation of the possible former cover of the Lakeland mountains with the help of modern laboratory techniques rather than general topographical or geomorphological studies was undertaken by Paul Green (see Fig. 18.4) of the fission-track laboratory at Melbourne University's Department of Geology in the 1980s (Green 1986).[6] The basis of the technique is as follows. The density of tracks produced by radioactively decaying atoms of ^{238}U in apatite, zircon or sphene crystals depends on the time elapsed since the crystals were first formed; so the rocks can be dated if track density is measured and the decay rate and uranium concentration of the material are known. If the crystals have been heated since their first formation, the tracks become 'annealed' and consequently shortened. So examination of the shortening makes possible the estimation of the temperatures that the minerals may have reached by heating, subsequent to their formation. However, as temperature dominates over time in the kinetics of the shortening process, all tracks are in fact reduced to approximately the same length, irrespective of their different ages, but those tracks subsequently formed in the cooler period after an annealing event will be longer. So the ratio of long to short tracks gives a measure of the time elapsed since cooling. However, with the shortening of the tracks, a lower proportion will intersect any surface that has been cut, etched and examined. So a heating event will make the age of the crystal appear less than its actual age.

Above a certain temperature, which is dependent on the rate of heating and the chlorine content of the material examined, annealing is total and the tracks disappear, thereby resetting the fission-track clock for the uranium decay. Thus there is some uncertainty in the palaeotemperature at which the annealing

[6] The manner of Green's involvement with Lakeland geology from distant Australia is unique for the present story. Born near Birmingham in 1952, he attended Holly Lodge Grammar School, Smethwick, where he became interested in geology and visited the Lakes on a school geology excursion, collecting some nice apatite specimens from the area of the tungsten mines near Carrock Fell – some of which he still possesses in his office in Australia. However, he read physics at Birmingham (but made several further trips to the Lakes while an undergraduate). He then went on to do an MSc in applied radiation physics, studying radiation damage in crystals, including some from lunar rocks. This led to a PhD on the study of etched particle tracks in minerals, and the specimens investigated included the apatite crystals he had previously collected at Carrock Fell, which he found to be about 63 Ma, much younger than expected, presumably due to the uranium 'clock' having been 'reset' by heating at that date. After finishing his PhD in 1978, he moved to University College, London, where he worked with Anthony Hurford, whom he had previously met at Birkbeck College. He too had found an unexpectedly young date for a Lakeland rock – about 60 Ma for the Shap Granite. Green continued collecting in the Lakes, but got a scatter of results. In 1980, he attended the Third International Fission Track Workshop at Pisa, and made contact with people from the Melbourne University Fission Track Research Group, who explained their understanding of the shortening of tracks by thermal annealing. This contact led to his appointment as a Post-Doctoral Fellowship at Melbourne University, to which he moved in 1982, taking with him some Lakeland samples for analysis. In time, the group at Melbourne, doing more and more consulting work (fission-track analysis was proving useful for the study of sedimentary basins, so important to hydrocarbon geologists), became a business entity under the aegis of the University. Eventually (1987), it became a private company (Geotrack International Pty), which some years later acquired its own premises in Melbourne, with Green, Ian Duddy (sedimentary basin research) and Kerry Hegarty (geophysicist) as the principal directors. An increasing demand for work in Europe, while they were still located at Melbourne University, led to the opening of a British branch at Thame in Oxfordshire, run by Richard Bray. Cherry Lewis (see p. 248) was with this group for a while. The upshot of all this was that a Melbourne-based geo-consultancy made a substantial input to the study of Lakeland geology. I am most grateful to Dr Green for supplying me with the foregoing details (pers. comm., 2001), and for providing a preprint of a forthcoming paper in *Tectonophysics*, which gives a succinct statement of methods of fission-track analysis and interpretation (Green 2002).

Fig. 18.4. Paul Green in the field in the 1980s. Copy of photograph supplied by Dr Green and reproduced by his permission.

process has actually occurred: one only knows the minimum value. Given the foregoing considerations, the thermal history of a sample cannot be deduced by means of a direct algorithm. One has to devise a number of thermal history scenarios and use computer modelling to generate the best fit to the figures available for the various parameters.[7]

Possible sources of heating relevant to fission-track analysis could be the emplacement of adjacent igneous bodies, the percolation of hot fluids, or the deposition of an overburden of sediment, which leads to a temperature rise – given that there is a general temperature increase towards the Earth's centre (though the temperature gradient is not necessarily the same everywhere and may conceivably vary over time).

Examining samples from various Caledonian intrusive rocks in the Lakes and elsewhere in northern England and subjecting them to fission-track analysis, Green (1986) suggested that about 60 million years ago rocks in the vicinity of Carrock Fell were heated to about 125 °C,[8] and somewhat lower temperatures (*c.* 70 °C) were suggested for the plutons further south. Assuming a 'fairly normal geothermal gradient' of 25–30 °C km^{-1}, a possible explanation could be that at one time the Lakeland rocks were buried to a depth of 3–4 km, which overburden had subsequently been stripped away, following uplift. Or there could have been a 'short-lived heat pulse, associated with some discrete event' (Green 1986, p. 504). The possibility of the infusion of hot fluids from some centre in the Irish Sea area, previously suggested by Rose and Dunham (1977) (see Chapter 4), was also canvassed. Green argued that the observed distribution of former temperature elevations was incompatible with their having being produced by the same agency as that which might have produced the Lakeland uplift. He did not offer a strong opinion as to which model should be preferred for the heat source.

As a result of Green's contacts with British Petroleum, the Melbourne laboratory was sent samples from the English Midlands to analyse by fission-track methods, and results were obtained rather similar to those in the Lakes: there appeared to have been a temperature maximum at about 60 Ma, and a lost overburden of more than 2 km (Green 1989). Green recalls (pers. comm., 2001) that British Petroleum were most interested in the results, whereas the BGS people were rather sceptical of the existence of such a large thickness of missing overburden. British Petroleum funded Geotrack for further investigations under and around the Irish Sea.

Research in the Lakes was also continued, with the publication of a collaborative paper by Cherry Lewis, Andrew Carter and Anthony Hurford from the Fission Track Research Group at University College, London, with Green contributing from his base in Melbourne (Lewis *et al.* 1992). Lewis had been working on granites in Tibet, and was interested in the estimation of their rates of uplift with the help of fission-track analysis. She visited the Melbourne group in 1990, and was inducted into the techniques of fission-track analysis by Hurford, then working in Bern. Later she joined his fission-track laboratory when this was set up at UCL, funded by British Petroleum to work on the post-Cretaceous thermal history of the United Kingdom continental shelf (Lewis, pers. comm., 2001).

The Lewis *et al.* paper involved the study of a considerable range of samples: from the Lake District and the Vale of Eden, Lancashire and Cheshire, the Isle of Man, and from boreholes in the eastern Irish Sea. British Petroleum provided some well-data. Lewis herself collected samples and undertook the fission-track analyses, using modelling programs devised by Green. The group's results suggested a palaeotemperature of >90 °C, with cooling beginning at 65 ± 5 Ma, consistent with Green's earlier results. Heating due to the few Tertiary dykes in the region seemed to be localized, so that the idea of the phenomena being produced by a 'pulse' of heat from some larger igneous source was discounted. Mineralization around the Cheshire Basin did indeed suggest deposition from warm, saline fluids, but the isotopic signatures appeared to indicate a sedimentary rather than an igneous source. So it was concluded that there had formerly been a substantial Mesozoic cover to the region, responsible for the heat, which cover had begun to be stripped away as a result of regional uplift in the Paleocene. The calculations assumed that the geothermal gradient in late Cretaceous–early Tertiary times was about the same as that found in the region at present.

The investigation suggested, then, that there had been a Cretaceous cover of some 3 km or more, stripped away in the Paleocene as a result of tectonic uplift, aided and abetted by isostatic uplift resulting from the loss of sedimentary cover. Present topographic differences (such as that between the Pennines and the Cheshire Plain) could be due to differential erosion operating on rocks of different hardnesses. The authors noted that the onset of cooling at about 65 (± 5) Ma coincided approximately with the current value for a major phase of rifting in the North Atlantic – at about 60 Ma. The tectonic uplift, *per se*, need only be about 300–400 m. It might be due to the agency of a mantle hot-spot located beneath the European–Greenland plate; or it could be a side-effect of the Alpine Orogeny 'going on' to the south; or both. According to Lewis (pers. comm., 2001), when the group's results were published people used to ask where all the

[7] For additional information, see Green (1986) or Green *et al.* (1989*a*, *b*).

[8] This exceeds the annealing temperature of 110 °C, but fission tracks with a narrow length distribution may represent a distinct heating event followed by rapid cooling from above *c.* 125 °C.

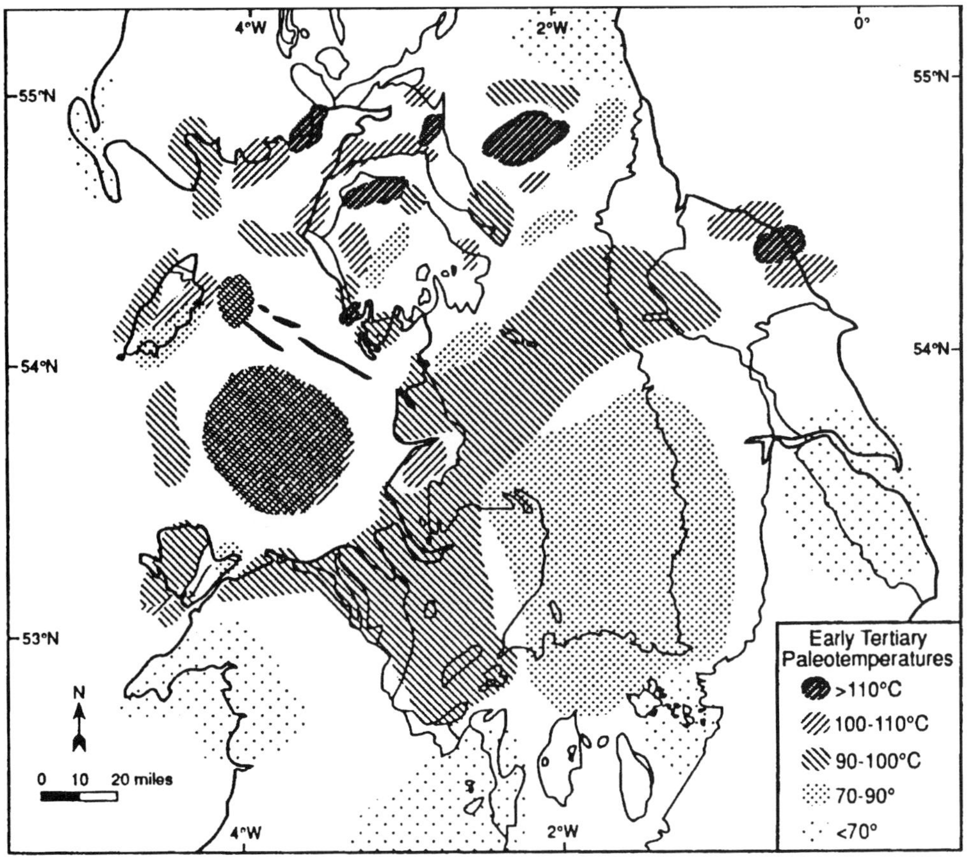

Fig. 18.5. Summary of early Tertiary palaeotemperatures over northern England, *etc.*, derived from fission-track analysis, according to Green, Duddy & Bray (1993, p. 120). Reproduced by courtesy of Paul Green and John Parnell.

huge volume of eroded sediments might have gone to. The authors had in mind the Viking Graben, off the NE coast of Scotland, known to contain thick post-Palaeogene sediments, but that issue was never fully resolved.

Work of a similar nature – a product of the current interest in the possibility of there being oil or gas deposits under the Irish Sea – was also published by Green *et al.* (1993*a*), the whole group now working for Geotrack International, with Green and Duddy in Melbourne and Bray and Lewis in England. Information from studies of vitrinite reflectance[9] was deployed, as well as fission-track analysis. Estimates of palaeotemperatures prior to an early Tertiary cooling were plotted on a map of northern England for a considerable number of sites, both onshore and offshore, the highest figures (>110 °C)[10] being calculated for northern and western Cumbria, the north Pennines, and the NE Yorkshire coast; also the Irish Sea (see Fig. 18.5). The view favoured in this paper was that the early Tertiary heating was due to both burial and the migration of hot fluids from the Irish Sea area.

However, in another paper that year (Green *et al.* 1993*b*),[11] heating due to burial was regarded as the main factor. At the same time, the authors considered the possibility of Atlantic rifting and a mantle plume as agencies. They wrote:

> One process which might be invoked as a cause of such effects is the development of 'Mantle Plumes', thought to be often associated with rifting and to cause kilometre scale uplift over a broad region (*e.g.*, White 1988). While such processes have obvious attractions as likely candidates, the offset of the area in which regional thermal effects have been detected in the UK region from the site of the eventual rift seems to pose a practical problem. In addition, while Mantle Plumes could readily bring about uplift (which might reasonably be accompanied by erosion) it is by no means obvious that they could also cause the observed heating with geothermal gradients close to present values required prior to the onset of cooling (Green *et al.* 1993*b*, p. 1072).

The possible connection between elevated palaeotemperatures in Northern England and plume activity was not developed. Northern England is, of course, 'offset' from Iceland, so any association between the two regions might be tenuous. We find, then, that the authors were inclined to associate the uplift with compression due to the Alpine Orogeny proceeding to the south, following elevated temperatures in northern Britain due to burial. The White (1988) paper sought to relate North Atlantic rifting and plume activity to Hebridean volcanism, but did not discuss Paleogene uplift in northern England.

Aside from any tectonic problems, the Lewis *et al.* (1992) paper attracted attention for other reasons. There were some uncertainties associated with the interpretation of fission-track data, as indicated above. Moreover, the paper differed from the Geological Society's recently published palaeogeographic atlas (Cope *et al.* 1992), and some other previous palaeo-maps, which showed the northern Pennines and Lake District as being above water during the Mesozoic – contrary to the opinion of Surveyors from the Trotter era, who had, like Marr, envisaged a 'Mesozoic

[9] Vitrinite is a component of coals, derived from humus. Its reflectance properties are related to the temperature elevation that a coal may have undergone, and hence to its 'rank'.

[10] The annealing temperature for fission tracks in apatite.

[11] This paper also contains a map showing the distribution of Tertiary palaeotemperatures, which soon superseded the information represented in Figure 18.5. However, the area covered was less than that depicted in this figure, and in particular it did not indicate the presumed Irish Sea 'hot-spot'. The most up-to-date version for the twentieth century (which did cover the Irish Sea) was Green *et al.* (1997, fig. 10). For the present purposes, I think the earlier paper is somewhat more instructive.

blanket' of sediments over northern Britain. Other authors had envisaged a Lakeland Mesozoic island. Either way, Lewis *et al.*'s idea of the topographic differences between the Pennines and their surrounding country being a relatively recent (Tertiary) phenomenon was unexpected. The paper was 'noticed' in less technical journals than *Earth and Planetary Science Letters* (where Lewis *et al.* was published), such as *Geology Today* (Anon. 1993). 'Discuss,' the anonymous reporter commanded.

Discussion was not long in coming. The *Geology Today* summary commented that Lewis and her co-authors assumed an approximately constant geothermal gradient from K–T times to the present; or at least no reason had been found to make this assumption doubtful. The Surveyor Douglas Holliday (1993) accepted the idea of a Mesozoic cover for the Lakes and Pennines, but on the evidence of thicknesses of preserved sediments in basins away from the highs he estimated that the thickness of Permian to Chalk cover-rocks might have been between 700 and 1750 m – significantly less than the figure arrived at by Lewis *et al.*, who assumed the modern average surface-temperature of 10 °C and a temperature gradient of 30 °C km^{-1}. Holliday made three main points: (1) an estimated overburden over the mountains of about 3000 m would imply a greater thickness on the structural highs than in the adjacent basins, where sediment was preserved and thicknesses could be measured; (2) the average subtropical annual temperatures would have been about 20 °C in the Palaeogene, rather than 10 °C as assumed by Lewis *et al.*; and (3) heat flow would vary with depth and according to rock type, so that a 'constant gradient over any significant type of rocks of varied lithology is highly unlikely' (Holliday, pers. comm., 2001). Making the appropriate numerical adjustments (the arguments for which need not be detailed here), he arrived at a figure of 1.7 km for the thickness of lost cover.

So the idea of a Mesozoic cover for the area was sustained, but it was 'cut back to size'. The empirical fission-track data were still valid, but when used in conjunction with different assumptions they yielded lower figures for the supposed lost overburden. Holliday accepted the idea that the track-annealing could be due to elevated temperatures generated by the former presence of a sedimentary cover over the Lakes and elsewhere; but he also thought it possible that igneous activity, or the movement of hot groundwaters, could be relevant causal agents.

Holliday's contribution was followed by one from Andrew McCulloch (1994*a*), writing from a private address in Larkhall, Lanarckshire.[12] Though ostensibly a comment on Holliday's paper, it was in fact (as Holliday (1994) noted in his response) a critique of the way in which Lewis *et al.* (1992) had drawn inferences from their data. McCulloch objected that the sample with the longest track-length (in the range 14.0–15.5 μm), taken to mark the onset of cooling from 110 °C at some time in the range 65 ± 5 Ma, was according to the authority of Gleadow *et al.* (1986), compatible with a cooling event from >110 °C to 50 °C, which put the beginning of the cooling event somewhere in the range 68.0 ± 4.6 Ma and 54.6 ± 2.4 Ma. He further affirmed that the sample used for discussion was atypical and that there were other statistical problems. In his view, the track-length measurements implied that cooling was not necessarily rapid, and that the time of cooling could be approximated by the 'youngest single grain age population present in the most annealed samples', which led to a figure of about 30 Ma. Not only that, fission-track analysis had yielded an age range between 68 and 234 Ma for samples from the same rock (Eskdale Granite) at the same locality! McCulloch was evidently unhappy with Lewis *et al.* His preferred theory was that a maximum temperature at about 60 Ma was due to the direct action of magma and to heat distributed by gravity-driven fluid flow, acting as an agent for the transport of heat through the crust.

Holliday was thus placed in the incongruous position of defending Lewis *et al.* against criticism from McCulloch, even though he had himself criticized their paper. He pointed out that Lewis *et al.* had in fact already considered the possible causal agents that McCulloch favoured, and that if one opted for a lesser Mesozoic cover, and assumed a different palaeo-geothermal gradient (due to changes in thermal conductivities with depth), then satisfactory reconciliation with the fission-track data could be achieved. In other words, he reaffirmed the position adopted in his discussion paper of 1993.

Green and his co-workers in fact responded to McCulloch directly (Green *et al.* 1995*a*, *b*). Details need not be given here, but it may be mentioned that they denied: (1) that there was a cooling event as recently as 30 Ma (an idea that McCulloch seemingly had arrived at from his work along the Irish coast); (2) that they had simply used variation in mean track-length with apatite age to estimate palaeotemperatures but rather had used computer modelling along the lines indicated above (p. 248); (3) that the time of onset of cooling was based on results from totally annealed samples; and (4) that their results were 'model-driven' (in the sense of some *a priori* general geological model, as opposed to a model that fitted all the available fission-track data).

The problem of Mesozoic overburden was also investigated from a different perspective by the stratigrapher and palaeontologist John Cope, of the Department of Earth Sciences at the University of Wales (Cardiff). He too felt that the Lewis *et al.* figure of 3+ km was too high, for there did not appear to be such a thickness of sediment evident anywhere in the British Mesozoic succession. The emended figure of the order of 2 km, suggested by Holliday, was, however, feasible so far as Cope was concerned. As he put it to me: 'I could explain 2 km of erosion for the Cheshire Basin quite readily and the hypothesis gained ground from there' (Cope, pers. comm., 2001).

So Cope (1994) suggested that if there had been a hot-spot located in the general area of the present Irish Sea[13] in the Late Cretaceous (Maastrichtian) the resultant uplift would have led to rapid removal of the soft Chalk sediments, and also much of the underlying Jurassic rocks. The removal of this overburden would have 'encouraged' further doming and erosion would have continued, until the heat source had dissipated itself. The effect of this to the east would have been the exhumation of the old landscape in Wales and NW England, and the well-known curved line of outcrop of the Jurassic and Cretaceous strata, from Yorkshire down to Dorset. Cope drew contours for the estimated amounts of net erosion, which yielded a diagram like a 'target', with the 'bull's-eye' of maximum overburden removal located somewhere to the north of Anglesey. Granites, including the Lakeland granites, mostly cropped out in the 'annular ring' representing an estimated 1–2 km of stripped-away overburden. The supposed erosion would provide an explanation of the exhumation of the ancient Lakeland rocks, but would not in itself account for a general rise of the Ordovician rocks of the Lake District to form upstanding mountains, or for that matter the rise of the Alston Block of the northern Pennines. The Lakeland 'mountains' would simply have to have been there all the time, being exposed when their Mesozoic cover was stripped away and the hard ancient rocks were revealed. But why was there a hot-spot somewhere under the Irish Sea?[14]

Cope's arguments were countered by Kenneth Thomson, then with the Earth Sciences Department at Oxford. He queried Cope's numbers for his contouring, and his presumption that the

[12] See also McCulloch (1994*b*). McCulloch was at the time a PhD student at University College, London.

[13] The BGS Irish Sea Memoir figures an igneous body, '? Granite', in the middle of a horizontal section of the Irish Sea, but does not provide evidence for the existence of such a body (Jackson *et al.* 1995, p. 3).

[14] In fact, as mentioned, various scenarios for the thermal history of the Irish Sea area had already been developed. For a review, see Green *et al.* (1997). The details of these alternatives need not be examined here.

line at which no erosion occurred (at the rim of the supposed Irish Sea dome) coincided with the Chiltern scarp for the area of SE England. This line, Thomson objected, would not be a fixed datum since the scarp would have retreated as a result of erosion. Thomson contended that if there had been doming due to a hot-spot centred below the present Irish Sea, the land would have risen, and then been eroded; and then, after the dissipation of the heat, there would have been subsidence with the result that much of Britain, as well as the Irish Sea, would now be under water. Thomson also queried the suggestion that the present drainage pattern for Ireland and England was what might be expected, given a former doming of the Irish Sea area. So he opted for the idea that the uplift might be explained by the hypothesis of igneous underplating,[15] suggested by Brodie and White (1994) (see below).

Cope (1995) responded with arguments about the extent of erosion in England, which need not be detailed here; but his statement that the Irish rivers tend to rise in the east and drain westwards, whereas the converse holds for the rivers of England and Scotland, would indeed seem to accord with the idea of former uplift and later subsidence of the area of the Irish Sea. He accepted the possibility of igneous underplating as a contributory cause of uplift, but retained the idea of a temporary mantle plume or hot-spot being chiefly responsible for the changes in the Irish Sea area. It was, he maintained, compatible with the production of gases in what is now the Irish Sea gas-field; and with the positive gravity anomaly for the Irish Sea area, which could be explained by crustal thinning (presumably during some period of extension) or perhaps the presence of a buried basic intrusion under the Irish Sea (Cope 1997, p. 49). Faulting near the rim of the Irish Sea (we could think of the Lakeland Boundary Fault as an example relevant to us; see Chapter 19) would explain the stance of NW England above sea level. In Cope's words: '[p]ost-Oligocene downfaulting around the margins of the Irish Sea, as the uplifted dome deflated, caused incursion of the sea and production of the modern Irish Sea by early Neogene times' (Cope 1997, p. 58). However, if there were underplating as a cause of uplift, the thermal deflation might not be sufficient to produce the Irish Sea Basin?

Regarding underplating, the 1994 paper of James Brodie and Nicky White (both from the Bullard Laboratories, Cambridge) accepted the idea of Palaeogene uplift for NW England, doing so in the light of the fact that the region appears to be in a state of isostatic equilibrium, according to the results of gravity and geodetic survey. The authors supposed that there was thermally induced uplift (as well as the massive Hebridean volcanism) at the time of the Atlantic rifting. But why would this uplift have remained to the present, as in the Lake District, if, as might be supposed, there was cooling subsequent to the volcanic outbursts? Their proposed explanation deployed the idea of 'underplating', the basis of which (though not the name) was due to Keith Cox (1980) of Oxford and Dan McKenzie (1984) of Cambridge.[16] By this theory, large quantities of mafic magma produced at the beginning of the Tertiary could have become trapped within the lithosphere, lodged under the plate to the east of the spreading North Atlantic (White & McKenzie 1989) – a hypothesis seemingly supported by seismic survey. More recently, Nicky White and Bryan Lovell (1997) (who worked for British Petroleum Exploration) have claimed that one can correlate pulses of plume activity with the formation of sediment fans in the North Sea (though this is not altogether relevant to what may have happened in the area of the Irish Sea or Cumbria). White informed me (pers. comm., 2001) that his group had seismic evidence, unpublished at that time, which imaged the underplated material.

Along with the ideas of uplift due to some (plume-generated?) subterranean igneous body under the Irish Sea and/or underplating, suggestions have, as mentioned, also been made about the migration of hot fluids into the area of the Irish Sea from a northerly direction (i.e. from the general direction of the main Tertiary igneous activity) (e.g. Green *et al.* 1993; Duddy *et al.* 1994). This would be consistent with the fact that in the Lake District the known palaeotemperatures for outcropping rocks are highest for the areas of Skiddaw, and mineral-rich Carrock Fell in the northern Lakes, and decrease southwards (Green *et al.* 1997, p. 87).[17] These authors were most attracted to the theory that palaeotemperatures in NW England and the Irish Sea area were the combined result of burial and hot fluid flow, but they emphasized that the model required further testing. To my knowledge, the actual *cause* of Palaeogene heating and uplift for NW England, and the apparent doming of the Lake District, was not settled in geologists' minds by the end of the twentieth century (but see p. 254).

The idea of associating the Icelandic plume with the Lakeland elevation found favour, however, with Survey scientists who contributed to the study of west Cumbrian geology in connection with the Nirex project (see Chapter 20):

> at the start of the Palaeogene, north-west England suffered major regional uplift, probably as a peripheral effect of the development of the development of the Icelandic plume (e.g. Brodie & White 1994). This event was associated with emplacement of the Scottish Tertiary igneous province, . . . (Chadwick *et al.* 1994, p. 98).

The idea was echoed in the West Cumbria Memoir (Akhurst *et al.* 1997, p. 90) and the Ambleside Memoir (Millward *et al.* 2000, p. 9), both with citations to Chadwick *et al.* (1994).[18]

The paper of Andrew Chadwick and his co-workers Gary Kirby and Heather Baily (1994) presented some interesting advances on previous work, made possible by the extensive collaborative

---

[15] Here meaning the addition of gabbroic plutons beneath lavas and dykes in a constructive margin.

[16] Cox suggested that, with the upwelling of ultrabasic magma from the mantle, differentiation might occur below the crust into overlying gabbroic materials and 'ultramafic cumulates' forming sills at or close to the base of the continental crust. Some of the gabbroic material might also penetrate the crust as feeder dykes for flood basalts at the surface (or sills emplaced in sediments near the surface). Thus the formation of flood basalts at the surface would be accompanied by the production of large quantities of basic or ultrabasic material emplaced below or at the base of the crust. (For a diagram, see Cox 1980, p. 647.) McKenzie asked himself about the fate of magma generated beneath oceanic and continental crust respectively. The former reaches the surface and spreads over the ocean floors. Some continental equivalents are known such as the Deccan Traps of India, but the volume of these seems insufficient by comparison with that produced by rifting of the oceanic crust. Hence McKenzie suggested that much magma would need to be injected as sills into the lower part of the continental crust, thereby generating uplift. A part of a plate passing over a region where there is hot upwelling mantle could be expected to undergo consequent local uplift, which would cease as the plate moves away from the rising region. A combination of the Cox and McKenzie hypotheses yields that of 'underplating'.

[17] However, the granite is a source of radiogenic heat.

[18] Andrew Chadwick (b. 1956) was reared in Darwen, a town in Lancashire between Blackburn and Bolton. He attended Bolton School, before going up to Oxford, where he completed his geology degree in 1977. The following year he was appointed to the Survey, where he has been to the present (2001), working chiefly on seismic reflection profiling and extension and inversion tectonics. In the north of England, he was involved in the Nirex investigations (see Chapter 20), and did work for the BGS Northumberland–Solway and Craven Basin subsurface Memoirs, as well as the West Cumbria Memoir referenced above. He also has done work related to the Irish Sea and the Isle of Man. From 1977 to 1978 he studied for an MSc in geophysics at Durham, by which university he was also awarded a DSc in 2000 for his published work on the subsurface structure and basin evolution of southern Britain. Like several other geoscientists mentioned in this study, he has had an interest in mountaineering (Chadwick, pers. comm., 2001).

BGS–Nirex investigations, where Chadwick's main contacts were Uisdean Michie and Andrew Boden (see Chapter 20). (It also drew on some Geotrack International results from the Irish Sea.) Chadwick (pers. comm., 2001) has described the paper as 'a simplified version of work carried out in 1993 for Nirex'. It was, in effect, a follow-up to the critique of the Lewis *et al.* (1992) paper by Douglas Holliday (1993).

First, the authors referred to a Jurassic and early Cretaceous extensional regime in the North Atlantic, which allowed the accumulation of Mesozoic sediments in, among other places, NW England and the Irish Sea area. The extensional structures were inferred from the offshore sediments and faults therein revealed by the Nirex and other recent studies (though fault throws could not be determined precisely offshore). As to the former thickness of Mesozoic cover, progress was made by the study of the compaction of sediments obtained from boreholes, allied with the fission-track results discussed earlier. The porosity of a rock naturally decreases with depth of burial, and compaction is retained even after the removal of overburden. So the study of densities and sonic velocities in rocks at different depths in boreholes allows estimation of the former thickness of their overburden. For this purpose, comparison was made with a standard density–depth relationship for sandstones established by the study of boreholes in the Wessex Basin of southern Britain, and also laboratory studies of compaction, conducted at Keyworth. From this, the investigators could estimate the expected degree of compaction at a certain depth, and compare this with the empirically determined value in order to estimate the thickness of lost overburden.

Fission-track results were also used, with updated modelling and data for heat flows, thermal conductivities, surface temperatures and sediment thicknesses, to estimate overburdens. The modelling was accomplished using a program dubbed 'Hotpot', written by Chadwick in conjunction with his colleague John Rowley. Comparisons were made with data from boreholes in Denmark, passing through analogous sediments, which offered information for a nearly complete sequence from Triassic to Neogene, i.e. with little or no lost overburden. Overall uplift was calculated as

$$U = \begin{bmatrix}\text{Thickness}\\ \text{of eroded}\\ \text{overburden}\end{bmatrix} + \begin{bmatrix}\text{Present}\\ \text{elevation}\\ \text{of site}\end{bmatrix} + \begin{bmatrix}\text{Water}\\ \text{depth at}\\ \text{end-Cretaceous}\end{bmatrix} - \begin{bmatrix}\text{Sea level}\\ \text{at}\\ \text{end-Cretaceous}\end{bmatrix} - \begin{bmatrix}\text{Sea level}\\ \text{at}\\ \text{present}\end{bmatrix}$$

Hence uplift since the Cretaceous was contoured over the area of the Lake District. In central Lakeland the uplift appeared to be of the order of 1750 m (in line with Holliday's estimate); but it was greater round the margins, especially to the north and south (up to 3000 m). It was suggested, therefore, that two factors had been at work: (1) a regional uplift of some 1750 m, perhaps due to underplating, the Icelandic plume or whatever; and (2) tectonic events generating crustal shortening and basin inversion. Two interesting sections were provided, showing the topographic situation as it might have been at the end of the Cretaceous; and also the situation as at present, with the supposed removed overburden represented (see Fig. 18.6). The linkage of granites with present structural highs is noteworthy. So too is the fact that such results were possible in large measure as a result of the BGS–Nirex collaborations, to be discussed in Chapter 20. The diagrams were also published in Akhurst *et al.* (1997, p. 95).

These diagrams warrant close examination. They indicate how the Mesozoic overburden was seemingly least over the body of the Lake District and the Alston Block, standing high with their granite plutons, while greater thicknesses of sediment accumulated in the 'sagging' areas of the Irish Sea and the Vale of Eden; and then how, in the Tertiary, there was a general uplift (perhaps due to underplating?), with exhumation revealing the Lake District and Alston blocks, still supported by their plutons. The Lakeland mountains, on this view, did not undergo a localized doming; they are simply erosional remnants produced subsequent to some larger-scale, regional uplift. The softer rocks surrounding the Lakes were uplifted more, but suffered greater erosion. Chadwick (pers. comm., 2001) has suggested that the idea of Lakeland doming has been 'overstated': the dips of the fringing Carboniferous and Permo-Triassic sediments are consistent with their being tilted footwall blocks – towards the Solway Basin, for example. He also thought that the various tiltings developed during extensional basin evolution during the Carboniferous and the Permo-Mesozoic. So the Lakeland block is not primarily a product of Tertiary tectonics *per se*. The rigid Lakeland granites would have been resistent to faulting during the extensional regime. Thus when basins were generated they would tend to have formed around the granites, leaving a relatively elevated mountain block. With inversion during the Tertiary (Alpine Orogeny?), the basins would have been elevated by reversal of their fault margins, but with their sediments being of softer rock these would have been susceptible to preferential erosion. In Chadwick's view, the buoyancy effect of the granites would chiefly have operated during their emplacement, and would no longer be a tectonic factor, once isostatic equilibrium was established – probably in the early Carboniferous, and certainly by the end of the Carboniferous basin developments.

Before we leave these matters, we may like to return to the question of the palaeo-geothermal gradient that Green had assumed in his original investigation of the thermal history of Lakeland, based on his fission-track analyses (Green 1986; Lewis *et al.* 1992), which had been criticized by Holliday. Geotrack International had, as mentioned, done work for British Petroleum (and other oil companies too) in the Irish Sea and the Solway Basin, during the course of which they came across a curious fact, namely that, for wells to the north of the area of investigation, analysis of vitrinite reflectance and fission-track data showed a matching increase of palaeotemperatures with depth, and an exceptionally high gradient of about 50 °C km^{-1}; whereas to the south the two types of determinations gave diverging data: the vitrinite reflectance method seemed to reveal an older (end-Carboniferous?) thermal event, associated with a higher palaeotemperature gradient, while the two together also revealed a more recent (early Tertiary) event that generated a lower gradient of about 10 °C km^{-1} (Green *et al.* 1997). It was while puzzling over these findings, and considering particularly the exceptionally high palaeothermal gradient identified in a well at Westnewton, SW of Carlisle, that Green had the idea[19] that even with mountains of moderate size such as those of the Lake District one could compare palaeotemperatures as determined from samples collected at the top and the bottom of the mountains and hence estimate the approximate palaeogradient (i.e. the mountain could be treated as a kind of inverted well!).

Thus in 1997, Green collected samples from the top and the bottom of Scafell. With an elevation of about 1000 m, some measurable difference might be anticipated if the palaeotemperature gradient there was much the same as in the Westnewton well. Samples from elevations between 122 and 966 m on Scafell were examined, all being from the BVG except one, which was taken from the Eskdale Granite. In 1998, Green collected further samples from Langdale, Helvellyn, High Street and Coniston Old Man, but these had not been analysed at the time of my writing (Green, pers. comm., 2001).

The results from Scafell showed the higher-level samples (above about 500 m) to have an apparent age of about 300 Ma.[20]

[19] It was an old idea really, as comparisons had been made in the early days of fission-track work in high mountain ranges like the Alps, to try to determine rates of uplift.

[20] The stratigraphic age of the Eskdale Granite by U–Pb determination is about 450 Ma. Green had to unravel complications arising from the presence of chlorine in his apatite samples.

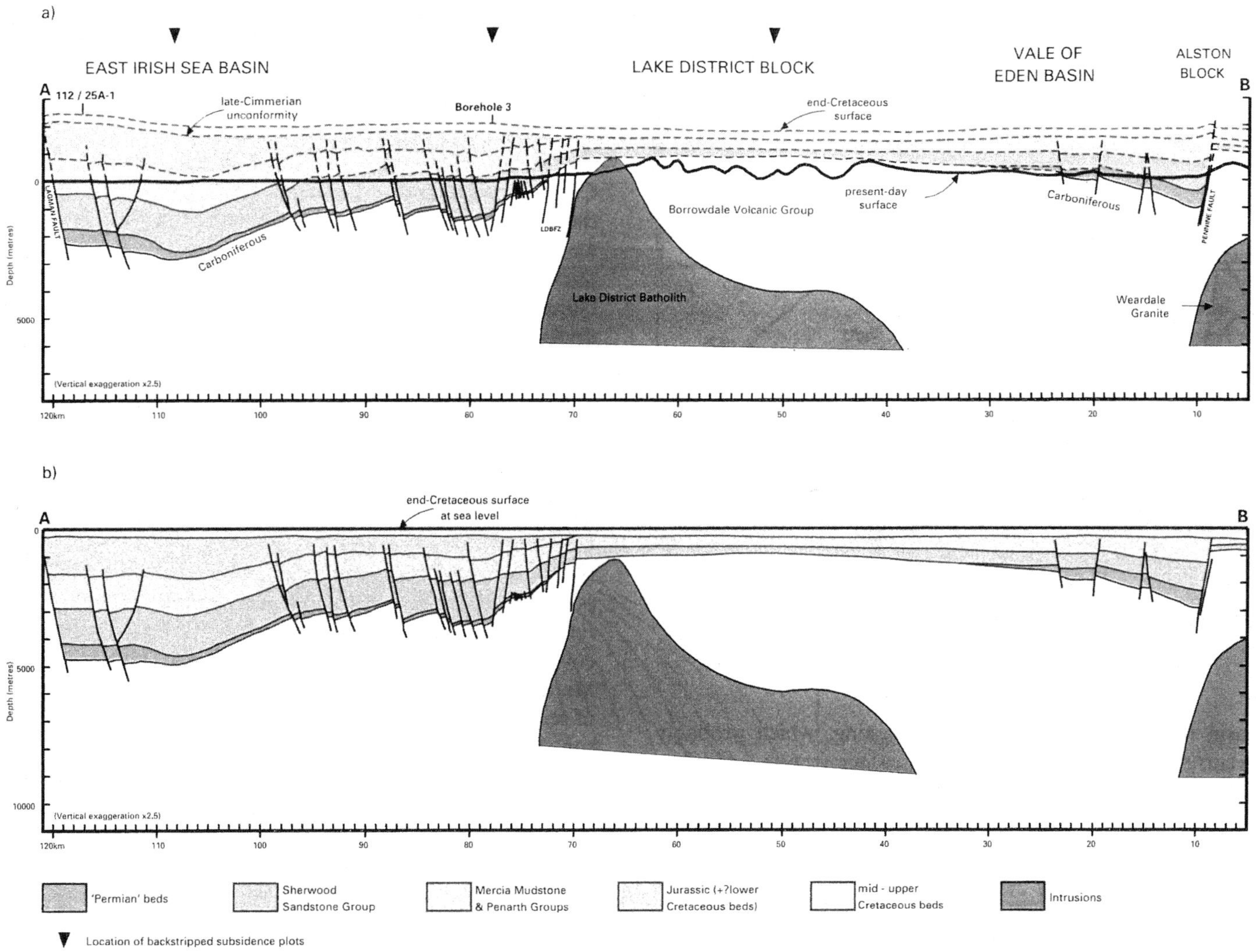

Fig. 18.6. East–west profiles for Irish Sea and northwest England, for end-Cretaceous and present, according to Chadwick, Kirby & Baily (1994, p. 100) (coloured in original). Reproduced by courtesy of the Yorkshire Geological Society. IPR/26-13C British Geological Survey. © NERC. All rights reserved. (See also Akhurst *et al.* 1997, p. 95.)

The ones of lower altitude gave an apparent age of about 60 Ma, but they had been hotter – reaching the temperature of total annealing (110 °C). All samples indicated an onset of cooling in the early Tertiary (about 65 to 50 Ma). Plotting the estimated temperatures at that time against elevation allowed estimation of the palaeotemperature gradient for the early Tertiary. This came out at about 60 °C km^{-1}, i.e. twice the value assumed in the earlier investigations. Using this revised figure, and a different assumed average annual surface temperature for 'Britain' in the early Tertiary of 20 °C[21] (not 10 °C, as in Lewis *et al.* 1992), the estimated thickness for the lost overburden appeared to be about 700 m at the level of the mountain tops, or 1.5–2 km for the coastal plains and Lakeland valleys – figures that appeared satisfactory geologically speaking, and compatible with the results of Chadwick's work.[22] However, Chadwick *et al.*'s paper, published in 1994, did not have the advantage of knowledge of Green's latest figure for the Lakeland palaeotemperature gradient, and it may be that one would find a somewhat different contouring of the Palaeogene uplift (Akhurst *et al.* 1997, p. 94) if the new figure were used in the estimates (and with data from other mountains besides Scafell).

This leaves the question of the cause of it all. The thermal gradient of 60 °C km^{-1} in the early Tertiary implied that there was at that time an episode of elevated heat flow. But how or why was that generated? Was it due to an Irish Sea hot-spot, to the general volcanic activity associated with North Atlantic rifting, linked to underplating, the passage of hot fluids, or what? Why was it that Green *et al.* (1999, p. 355) seemed to find no direct evidence for a high heat flow around the North Atlantic margin of Britain, despite the extensive igneous activity of the region, while there was evidence for such a flow further south and inland, as in the Lake District?[23] At the time of writing (2001), geologists do not

[21] The figure of 20 °C was, Green informed me, taken from Curry (in Duff & Smith 1992, p. 407, fig. 13.7). This temperature estimate was based chiefly on old palaeontological evidence, and newer evidence from the study of oxygen isotopes.

[22] Green's synthesis was completed in 2000, though it was not by then formally published, other than a preliminary statement of his Scafell results (Green *et al.* 1999, p. 355). His more complete results were, however, presented to the Geological Society in a meeting on 'Exhumation of circum-Atlantic continental margins: timing, mechanisms and implications for hydrocarbon exploration, held on 13 and 14 June 2000, and before that at a meeting in Lorne (Victoria, Australia) on 'Fission track dating and thermochronology' in February 2000 (Green, pers. comm., 2001). The paper is in press with *Tectonophysics* at the time of writing (2001), and I am grateful to Dr Green for making a pre-print (dated 2000) available to me. See Green (2002).

[23] A suggested reason was that the igneous activity allowed the escape of heat, whereas further south heat accumulated at the base of the crust, leading to a high heat flow. In the Lakes, one also has to consider radiogenic heat from the granite plutons.

agree on these matters[24] and it seems certain that there is more to be said about them in the twenty-first century. In their West Cumbria Memoir of 1997, which represented the BGS view of the matter at the end of the twentieth century, Akhurst *et al.* hinted that understanding is incomplete, and this is confirmed by Chadwick (pers. comm., 2001): '[w]e don't know the cause but it is likely to have been connected with development of the Tertiary igneous province'.

Gravimetric, geomagnetic and sea-floor topographic maps show the existence of 'bands' parallel with the present Mid-Atlantic Ridge consistent with the sea-floor spreading theory and the migration of traces of hot-spot activity extending southeastwards from the present locus of activity under Iceland. However, this 'hot-spot trace' is directed towards the far north of Scotland, and appears to have little to do with the Lake District (see Braun & Marquart 2001). Did it extend further south, into the region of the present Irish Sea, 'touching' the Lakes, and extending even to Lundy in the Bristol Channel?

In this connection, it is interesting to remark ideas recently presented in summary form by the Caltech geophysicist Michael Gurnis, in *Scientific American* (Gurnis 2001). He discusses large-scale *vertical* movements of the mantle, which may drive whole continents upwards or downwards. Ideas about such movements – the history of which over the last thirty years or so are briefly discussed by Gurnis – do not fall within the scope of 'standard' plate-tectonic theory, nor indeed the present book. However, it is possible that they give some indication of future theoretical developments, already under active consideration (Davies 1999), but not specifically in relation to the Lake District. Plumes – 'blobs' of hot buoyant rock originating from the outer surface of the Earth's liquid core – may rise up through the mantle and cause elevation of whole continents (e.g. Africa). Such activity is something that may be essentially independent of magma rising on mid-oceanic ridges. And perhaps it could link Iceland with Europe? Could there have been a small 'blob' in the mantle, rising under the area of the Irish Sea at some stage?

We should remark, in this context, that the paper of Braun and Marquart, from the University of Frankfurt, mentioned above, sees North Atlantic rifting as proceeding in two main stages: first there was rifting between Greenland and Canada; then a huge mass of hot rock welled up from the mantle between what is now Greenland and Europe at about 60 Ma, at the position of what is now Iceland, producing extension at the old line of the Iapetus suture and a huge rift running to the west of Britain between Greenland and Scandinavia. This rifting and separation of the continents was complete at about 38 Ma, but strong volcanic activity continued to the present at Iceland and sea-floor spreading also continues. This picture is supported by the results of satellite imaging survey and large-scale seismic tomography, which, the authors claim, help unveil events occurring down in the mantle.

This picture is not necessarily at odds with what British geologists have been saying, but, as yet, they do not seem to have related the events of Lakeland to this larger, more global, view. Perhaps the events of the Alpine Orogeny, the North Atlantic rifting, and the massive associated Tertiary volcanism, may have stirred things so as to give localized elevation of the Lakeland massif, as well as the inversion of the adjacent basins? Or are the mountains nothing much more than an erosional remnant? Is the 'doming' that attracted the attention of the old geologists not a Tertiary feature at all, but connected with the Variscan Orogeny at the end of the Carboniferous? Or is the doming 'overstated', as was the end-century BGS view? Like Richard Clark, one may remain puzzled. This part of the story of Lakeland geology is not yet wholly clear and will require further attention in the twenty-first century. Unfortunately, however, at the time of writing it is not obvious that the Lake District will provide the essential key, necessary to unlock the mysteries concerning what happened in NW 'Europe' during Tertiary times. We may hope that it will help.

[24] One of them, Dr X, whom I do not choose to name here, has written to me (2001) saying that he thinks that Dr Y does not understand the distinction between underplating and plume-related dynamic support!

Chapter 19
The glaciation of the Lake District[1]

Assuming that there *was* a Lakeland massif by the end of the Tertiary, let us turn now to discuss the history of investigations of Quaternary geology in the Lakes. For this purpose we must again return to the nineteenth century.

To modern eyes, the Lake District shows numerous manifestations of glacial activity, and many interesting studies of glaciation have been undertaken there over the last 150 years or so. However, it was not the area of Britain where the most important early ideas about glaciation and glacial phenomena were developed. The Scots were the leaders in such investigations, and Andrew Ramsay's work in Wales came before the development of detailed glacial work in the Lakes. Nevertheless, it was inevitable that geologists should turn their attention to glaciation in the Lakes, and in this chapter I examine some of the work done on the Pleistocene geology of the region, considering how it related to the development of theories in other regions and to the general development of Pleistocene stratigraphy.

When Sedgwick did his early work in the Lakes in 1822–1824, he saw plenty of evidence for what, following William Buckland, he called 'alluvium' and 'diluvium'. The latter referred, in the words of Buckland (1819, p. 532), to 'the superficial gravel beds produced by the last universal deluge', and was regarded as a product of marine deposition, even though the material was generally unsorted and often contained angular clasts. Such material was to be found all over the place in the Lakes, and not necessarily associated with rivers or becks; hence it was distinguished from alluvium, which was represented as 'post-diluvian'. That the diluvium had been emplaced by some great inundation was confirmed in the Lakes for Sedgwick by the fact that he found rounded, apparently water-worn, boulders of the granite of Red Pike (2474 feet) on Starling Dodd (2077 feet), high on the chain of hills between the valleys of Ennerdale and Buttermere–Crummock (Sedgwick 1825*a*, p. 31; see Fig. 7.4). Having no idea of glaciation in the district, Sedgwick ascribed the movement of the stones – which he figured in a letter of 7 March 1830 to Henry De la Beche (see Sharpe & McCartney 1998, p. 250) – to some catastrophic inundation. Given that the material was loose, so that the movement had presumably occurred in geologically recent times, such an assumption, in keeping with the commonly held 'catastrophist' ideas of the 1820s, was reasonable, though it may be noted that Sedgwick stated in his notebook, before he climbed the hill, that he 'was persuaded' from its form that Starling Dodd had been shaped by water. Further 'syenite' (granite) boulders were found on the top of Mellbreak (1680 feet) to the west of Crummock Water. Sedgwick found additional evidence, from the lie of the beds, that The Coombe, a valley north of Glaramara, was not structural (i.e. a syncline) but had been produced by some erosive process. Hence he called it a 'valley of denudation'; and the loose boulders on top of the crags were termed 'diluvial blocks' (Field Notebook No. 13, 1824, Sedgwick Museum, Cambridge, p. 109).

As we have seen (Chapter 2), Sedgwick reached the end of his pioneering mapwork in the Lakes with the conclusion that the whole region had been forced up by huge subterranean forces. This 'catastrophic' movement was evidenced for Sedgwick by, for example, the considerable throw of some of the faults of the Coniston Limestone. It is not obvious that Sedgwick associated *this* kind of catastrophe, which he linked to Élie de Beaumont's tectonic theory, with the 'diluvialist' catastrophe that might have led to the movement of erratics or caused the Lakeland valleys to be plastered with diluvium and alluvium, but the link was soon made by Sedgwick's colleague, William Hopkins (1842) (see Chapters 2 and 18).

As is well known (Davies 1969, Ch. 8), the Swiss naturalist Louis Agassiz brought his land-ice theory to the meeting of the British Association in Glasgow in 1840, and promptly acquired Buckland as one of his converts, and also Charles Lyell (albeit temporarily). Buckland (1840) was soon checking out northern Britain for evidences of glaciation, finding moraines and such like in many places. The diluvium now became 'deep deposits of glacier origin' and the 'conical hills laid down on [Joseph] Fryer's large [printed topographical] map of Cumberland [1818]' were now construed as moraines. The 'conical hills' were presumably drumlins. Glacial striations were found in 'Dr Arnold's garden at Fox Howe [Hollow] near Ambleside', i.e. at a house by the banks of the River Rothay, halfway between Ambleside and Rydal Water, which belonged to Arnold of Rugby.[2] Boulders of Criffel Granite from north of the Solway Firth were found near Cockermouth and were thought to have been rafted there by floating ice. Also, ice could account for the transport of boulders of Shap Granite over the Stainmore Gap into east Yorkshire, which had previously been a puzzle to diluvial theorists (including Buckland); for why would water go over a Pennine pass, at an altitude approaching 1500 feet, rather than taking the lower route through the Eden and Ure valleys?

Hopkins (1848) published a geological map of the Lake District, based on his observations, those of Sedgwick, and possibly others. Given that he presented a brief view of his conclusions to the Geological Society in 1842, Hopkins must have started work in the Lakes shortly after the British Association meeting of 1840. He considered three agencies that might have excavated the valleys and transported the erratic boulders: glaciers, marine currents and icebergs.

Considering the glacial (land-ice) hypothesis (invoking a recent Greenland-type of glaciation such as Agassiz (1840, 1840–1841) envisaged in his *Eiszeit* theory[3]), Hopkins thought it inadequate to explain how or why boulders of Shap Granite passed over the Stainmore Gap. They might have been expected to end up as (or in) a moraine somewhere in the Eden Valley below the granite hill near Shap, at a lower elevation.

The floating iceberg hypothesis was rejected because it did not explain why the transported boulders were generally rounded and why the largest of them were found nearest to the source of the boulders (e.g. the Shap Granite outcrop). By the time they had reached the Yorkshire coast, the Lakeland and Shap rocks were generally little more than pebbles or gravel. Yet by the iceberg rafting hypothesis there would seem to be no reason why they should not be both large and irregularly shaped.

Then there was the 'diluvial' theory. Hopkins pointed out that it did not require the blocks to have been moved recently. The Shap debris found in east Yorkshire was assuredly moved *after* the Oolite on which it was deposited, but the movement could, Hopkins claimed, have been a long time ago – possibly *before* the Stainmore barrier was formed by uplift of the Pennines.

---

[1] Material in this chapter has previously been published in Oldroyd (1999*b*) and is reproduced here by courtesy of Taylor & Francis Ltd. (http://www.tandf.co.uk).

[2] Patrick Boylan (1984 & n.d. [1999]) has made a careful examination of all the localities visited by Buckland, helpfully providing grid references for the specific sites.

[3] Strictly, the idea of an ice age was first proposed by the botanist Karl Schimper, a co-worker with Agassiz in Switzerland.

Hopkins' preferred hypothesis to produce a domed form for the Lake District such as we discussed in Chapter 18, was that of 'waves of translation': a variant of catastrophist and diluvial theory. If, for example, the Lakeland mountains had been raised by a series of sudden upheavals (as was compatible with Sedgwick's fieldwork and Élie de Beaumont's theory), they could have produced a series of what we might call 'mega-tsunamis', and in time boulders could have been transported considerable distances – perhaps even from Shap to the Yorkshire Wolds. (Hopkins estimated that a wave of velocity 20 m.p.h. could shift a block weighing 320 tonnes.) Even so, many such disturbances would be needed to carry large boulders such distances, and also excavate the valleys along the lines of weakness created by faulting.

For the next twenty years or so, there was surprisingly little published on Lakeland glacial theory. Expert geologists visiting the district concentrated on trying to sort out the stratigraphic column. However, James Bryce (1806–1877), FGS, a teacher from Glasgow High School, visited the district in 1850 with several amateur geological friends, and they examined striations on rocks near Staveley, between Kendal and Windermere, recently discovered by Edward Wakefield of Kendal; other neighbouring localities were also visited. It appeared that the striations were aligned with the local valleys, such as Kentmere. Bryce (1850) simply deferred to the Cambridge expert Hopkins for theoretical explanation.

During the 1850s and 1860s in Britain, the preferred theory to account for the evidences of glaciation was that of glacial submergence, advocated by Murchison (1851) and Lyell (1863)[4] and indeed the majority of British geologists who interested themselves in such matters. It was to the effect that if sea levels rose, or land levels sank, icebergs could calve off from the ice-sheet and drift across the sea, eventually depositing any boulders that might have fallen onto them; or material scraped from valley floors. The theory might also help explain the interbedding of gravels with boulder clay or till. In retrospect, one can see that the theory was a first attempt to account for the phenomena now explained by the hypothesis of a sequence of glacial and interglacial episodes and deposits, but at the time in Britain it was a response to observations made in Wales. Using the theory, the Scottish agriculturalist Thomas Jamieson (1829–1913), for example, sought to develop quite a complex sequence for the glacial phenomena of Scotland, which can be seen in this light (Jamieson 1860).

In the Lakes, another Scotsman, Daniel Mackintosh (1815–1891), resident of Chester and then Birkenhead, science lecturer, and President of the Liverpool Geological Society, gave close attention to the problem, his first paper on the topic being published in the *Geological Magazine* (Mackintosh 1865).[5] Mackintosh's geomorphology offers an interesting example of the theory-ladenness of observations, for as an advocate of the glacial-submergence theory, and more generally an exponent of the idea that the sea had been, and is, the major agent producing landforms, he assiduously sought evidence of marine erosion in the Lakeland mountains. Many examples of marine cliffs high in the mountains were claimed, such as Pavey Ark above the Langdale Valley, at Pillar in Ennerdale, or on the eastern side of Helvellyn. Mackintosh supposed that some erratics might be accounted for by cliffs being undercut by marine erosion, causing blocks to fall off and roll down into 'erratic' locations. What were later regarded as ice-carved corries – such as Red Tarn below the eastern face of Helvellyn – Mackintosh thought were 'voes' or sea-carved bays. He was referred to by one commentator (Anon. 1869, p. 470) as a 'marinist', as opposed to an 'aërialist' such as the Surveyors Joseph Jukes (1811–1869) or James Geikie (1839–1915), though Mackintosh did of course accept the idea of large-scale glaciation and land-ice, as well as glacial submergence.

Following fieldwork in 1868 and 1869, Mackintosh (1869*a*) described his observations in Lancashire, especially in the Blackpool area, which were then extended northwards into the Furness district. In Lancashire, following the ideas of the Surveyor, Edward Hull (1829–1917) (1865), based on observations also made in that county,[6] Mackintosh claimed evidence for a tripartite division (boulder clay; middle sands and gravels; boulder clay), and mentioned a good vertical section of the same in a river-cliff at Redscar, three miles northeast of Preston. Such strata were then mapped for Furness. In the Lakes, as at Black Combe, Mackintosh found boulder clay and sands and gravels at a higher altitude and, around Coniston Old Man, up to about 1000 feet. But he thought that the boulder clay was produced in part by marine action on material loosened and transported by floating ice. He referred to (early) work by the Scottish Surveyor James Geikie that supported this interpretation.[7]

Mackintosh's work in the Lakes was part of a much wider programme, the results of which were set forth in his *Scenery of England and Wales* (1869*b*), in which he described in detail the results of marine erosion around the coast of Britain, and then sought to apply the idea of marine action to explain the appearances of inland topography. He was contemptuous of the proponents of fluvial erosion, parodying some lines of Robert Southey thus, in reference to the Lodore Falls, Borrowdale (Mackintosh 1869*b*, p. 383):

---

[4] Such theories had their origin in the discovery by Joshua Trimmer (1759–1857) of shells in sand and gravel about twenty feet below the surface, at an altitude of about 1400 feet on Moel Tryfan (the hill of that name SE of Caernarvon, not the better-known mountain in the Snowdonian ranges) in Wales (Trimmer 1834); and Andrew Ramsay thought that drift similar to that in which the shells were found carried elsewhere as high as 2300 feet, suggesting an equivalent subsidence of the land, with the highest of today's Welsh mountains being mere islands (Ramsay 1852).

[5] On Mackintosh, see G. H. M. (1891).

[6] The tripartite model had earlier been developed in the Vosges by Edouard Collomb (1847); in Wales by Ramsay (1852); and in Switzerland by Charles Adolphe Morlot (1820–1867) (1855). Morlot reported and figured a section near Vevey on the NE side of Lake Geneva. He thought the evidence suggested that glaciations could be regarded as *periodic* phenomena. Morlot also stated that his ideas were partly based on suggestions from Robert Chambers in Scotland, with whom he had been corresponding. However, Chambers' publications do not appear to state the notion of periodicity clearly, and Morlot is commonly credited with the idea.

[7] Unfortunately, the reference given by Mackintosh was incomplete and apparently incorrect, and the specific publication of Geikie referred to has not been identified. There is a passage in Archibald Geikie's *Scenery and Geology of Scotland* (1863, pp. 84–85) that might possibly be interpreted as being in support of Mackintosh's ideas.

The cataract weak,
With aspect meek,
First creeps along
The stones among,
Through which it peeps,
Save where it sleeps;
Then sputtering, fluttering, and muttering,
And dallying, sallying, and rallying,
With no downpour,
Or any uproar,
The water steals o'er
The slope at Lowdore.[8]

To reiterate, Mackintosh did not deny the idea of a glacial epoch. It was simply that he associated it with the prevailing marine-submergence theory, attributing drift to ice-borne material acted on by the sea. His knowledge of the geomorphology and topography of Britain, evidenced by his book, was both extensive and impressive. It should be remarked that his interpretations *within* the Lake District drew on ideas developed from his observations *outside* the Lakes in west Lancashire, and very likely from the ideas of more influential geologists such as Ramsay and Hull.

As we have seen, the Surveyors moved into the Lakes in 1866, and continued there until 1888. Geomorphology was not their prime concern, but the drift was mapped and some of them, notably Dakyns, de Rance, Ward and Goodchild, wrote extensively on glacial matters, glacial phenomena by then being easily recognizable to the professional eye. De Rance (1869, p. 490) noted that from the summit of Helvellyn one sees quite a flat surface (the kind of thing that Hollingworth was later to incorporate into his study of the altitudes of Lakeland landforms), with the valleys appearing as 'mere gashes in the upland plain'. (He was right: aeroplanes have landed on the summit of Helvellyn; the top of High Street has served as a race course and Roman soldiers could have held parades there, and very likely did.) De Rance suggested that the Lake District as a whole was like a truncated cone, and that the flat surface perhaps marked a 'pause in the subsidence during which the sea denuded the rest of the country into a gently inclined plane'.[9] He rejected Mackintosh's ideas of corries as voes. Red Tarn (high on the eastern side of Helvellyn) was empty when he visited the locality during a hot August (1869) and examined its bottom. He found numerous angular fragments from the mountain above, showing no sign of marine origin (or morainic character, for that matter).

Like others in the 1860s, however, de Rance accepted the glacial-submergence hypothesis. He thought that the glacial sea rose to about the present 1700 feet contour (below the level of Red Tarn) and that glacial indications extended down to about 400 feet above sea level. Then in the low-lying Furness region he thought (like Mackintosh) that he could discern evidence for an upper and a lower boulder clay, with sands and gravels sandwiched between. The boulder material was supposedly produced by drifting icebergs, but it is difficult to know from de Rance's text precisely what he had in mind for he also seems to have been saying that land-ice gave rise to the indications of glaciation between 400 and 1700 feet.

By 1870, Lakeland glacial history was being actively debated, some eight papers being published on the topic that year. Mackintosh's 'marinist' ideas were regarded as extreme, but the Surveyors subscribed to the marine-submergence and floating-ice theory to account for the diluvium – the view favoured by their chief, Roderick Murchison. However, they also accepted the idea of Ramsay (1862, 1864) that glaciers could 'scoop' out rock-basins. Thus de Rance (1871*a*) suggested the following sequence for the Lakes:

1. Subsidence accompanying the onset of cold.
2. Glaciation of the land remaining above the sea, the ice-sheet scooping out rock-basins and eventually reaching sea level, icebergs calving off and transporting boulders southwards to produce the Lower Boulder Clay in Lancashire, at the same time as the formation of the Lower Moraine Drift in the Lake District.
3. Amelioration of climate, with subsidence still continuing, giving deposits of sand and shingle as the laminated and current-bedded Middle Drift Sand, formed by marine erosion and fluvial transport and deposition, of the lowland regions around the Lakes.
4. Further subsidence and lowering of temperature, resulting in Upper Boulder Clay or Till being deposited on the Middle Drift Sand.
5. Elevation, with the glaciers remaining for a time, excavating some of the Marine Drift and also depositing morainic material.
6. Amelioration to present conditions.

This model meshed with de Rance's (1870) earlier work in Cheshire and Lancashire, and thus, like Mackintosh, he imported ideas from lower-lying regions to the south into the Lakes. The model sustained the idea of two tills and an intervening set of gravels, etc.; glacial submergence; and the formation of rock-basins by ice action. It did not give much weight to marine erosion, and involved uncorrelated changes in temperature–climate and elevation.

The debate continued, with Mackintosh holding to his 'marinist' views. De Rance (1871*b*) emphasized that the real issue was whether phenomena such as roches moutonnées were or were not to be explained by land-ice theory, such as had been favoured by Agassiz, and which was by now being given support by the Scottish theorist James Croll (1821–1890); or by the 'drift' hypothesis of earlier geologists. Croll, an autodidact, joined the Survey's Scottish Branch in 1867 and was soon impressing his colleagues – who, perhaps correctly, regarded him as a genius – with his ideas. He suggested how glaciation might be explained in terms of variation of the heat incident on different parts of the Earth's surface at different eras, arising from secular changes in the ellipticity of its orbit and concomitant changes in the positions of the

[8] Southey's poem (1820), 'The Cataract of Lodore', written for children, contained lines such as:
. . . [G]leaming and streaming and steaming and beaming,
And rushing and flushing and brushing and gushing,
And flapping and rapping and clapping and slapping,
And curling and whirling and purling,
And thumping plumping and bumping and jumping,
And dashing and flashing and splashing and clashing;
And so never ending but always descending,
Sounds and motions for ever and ever are blending,
All at once and all o'er, with a mighty uproar,
And this way the Water comes down at Lodore (Southey 1909, pp. 348–349).

[9] Note that in the nineteenth century there was emphasis on Lakeland subsidence, whereas in the twentieth century the emphasis was on uplift, as discussed in Chapter 18.

equinoctial points due to precession of the equinoxes (Croll 1864, 1867). Such a theory would lead one to anticipate the occurrence of occasional glacial epochs (when the ellipticity was high), with a series of glacials and interglacials during such epochs, arising from cyclic shifts in the equinoctial points. The theory meshed with the land-ice theory; and, unlike the glacial-submergence theory, did not need to invoke large changes of the relative levels of land and sea in quite short periods of geological time. Further, it did not require the unlikely association of cold periods with *lowering* of land level. However, the shells such as had been found on Moel Tryfan by Joshua Trimmer (see Note 4) would have had to have been emplaced by land-ice moving uphill to a considerable (and perhaps implausible) degree.

There were no such grand theorists amongst the Lakeland Surveyors, but given that the Scottish Surveyors such as James Geikie were reacting favourably to Croll's ideas it was unsurprising that the Scottish work began to impinge on the Lakeland debates. Also, with alternative theories to account for the occurrence of land-ice glaciation, action on topography, and distribution of erratics, there was ample work to do of an empirical kind, tracing the distribution and apparent lines of movement of erratics, to try to ascertain whether they had been moved by flood waters, drifting icebergs or slow-moving glaciers. Both the local amateurs and the Surveyors looked into these matters.

Not surprisingly, Croll (1971) thought his theory was applicable to the Lakes. The 'catastrophist' diluvial theory was now outmoded, but geologists such as Harkness (1870) believed that the distribution of Shap Granite boulders over Stainmore was *not* due to glaciers, for the pathway of boulders did not follow the valleys as one might expect glaciers to do; and was *not* due to icebergs, as these could not have got over the Pennine pass, which presumably had shallow water at Stainmore – even if there had been glacial submergence. So Harkness suggested that the transport might have been due to 'pack-ice'.

For Croll – who, so far as I am aware, and as far as his paper indicates, did not visit the Lake District and the Pennines – the explanation was simpler. It was land-ice that carried the boulders across the Pennines at Stainmore (as likewise it had carried shells up Moel Tryfan). The observable glacial striations did not necessarily reveal the movement of the ice carrying the boulders. They could be due to a late phase of glaciation, when the glaciers were confined to the valleys. But at an earlier stage the glaciers could have smothered everything and carried boulders over Stainmore, or anywhere else, marking the routes of glacier movement. Croll pointed out that geologists had been assuming an ice-free North Sea, but that was incompatible with the idea of a glacial epoch.

Still interested in Lakeland glaciation, Mackintosh (1871*a*) was quick to reply. He objected that the area of Wasdale Crag (the hill on the southern side of the Shap Granite, above Wasdale Head Farm – now derelict) was not high enough to generate the massive glacier that Croll hypothesized. And if the ice came from the north, why was it diverted to the east over Stainmore? Such high-level glacial striations as had been found in northern England were not aligned in the direction that Croll required for the easterly movement of Shap boulders. Further, an ice-stream low enough to have Shap boulders roll onto it could not, Mackintosh maintained, have ridden uphill and across Stainmore. Most importantly, Shap boulders seemed to 'radiate' from the Wasdale source. This, said Mackintosh, was incompatible with a land-ice theory, whereas it was reasonable if the boulders were distributed by floating ice, which could be subject to a variety of currents over time. (Of course, if there had been more than one era of land-ice glaciation, as Croll's theory proposed, then the boulders might have been distributed in several different directions at different times. The ice need not always have taken the Stainmore route.)

As said, with the Surveyors active in the field, much empirical information was assembled. In what Kendall & Wroot (1924, p. 83) called a classic paper, Tiddeman (1872) collated information by plotting the directions of glacial striations on a protractor, so to speak, so that the range of motions for north-Lancashire ice could be seen at a glance. They ran predominantly towards the south or SSE and did not align with the present general drainage direction. Hence, Tiddeman suggested, there must have been ice-barriers diverting glaciers from their most obvious or 'natural' courses, as indicated by the present topography. He advocated the notion of a general ice-sheet covering the area of his investigations in northern England.

Clifton Ward also attended to ice work in the Lakes, publishing a detailed account of the glacial phenomena of the northern Lakes for the area round Keswick where he was then residing (Ward 1873). He found glacial striae up to 1500 feet or more, but not on the highest peaks, suggesting that there had been rocks exposed above the general level of the glaciers, like islands in a sea of ice (as Ramsay had envisaged in Wales). Ward deployed the glacial-submergence theory in a fairly uncomplicated form, similar to that of de Rance. There was supposedly a period of intense cold, with the formation of a Lakeland ice-sheet and moraines. Then came a mild period, followed by subsidence of about 800 feet, which allowed remodelling of the morainic material and formation of banks of sand and gravel; then renewal of cold and some further subsidence, so that glacial material could be distributed by icebergs; then subsequent re-elevation, glaciation and formation of the presently visible moraines, followed by climatic amelioration. This theory was also propounded in his memoir on the northern Lake District (Ward 1876*a*). He did not believe that the Lakes had been overwhelmed by a great mass of ice, pushing southward from Scotland, for he contended that no foreign boulders of northern rocks could be found among the Lakeland mountains (Ward 1876*a*, p. 99). His theory was simple, eclectic, empirically based, and avoided the correlation of greatest cold with lowest land elevation. In effect, he had two glacials and an interglacial. Ward claimed that his theory was congruent with James Geikie's, but developed independently.

James Geikie was at this time working on a major volume on glacial theory, *The Great Ice Age* (1874), in which he supported Croll's astronomical theory. He agreed with Tiddeman that the glaciers around Shap might have been diverted from their 'natural' course by other glaciers; but that the ultimate reason for an easterly divergence was the huge mass of ice moving southwards from Scotland down the channel of the Irish Sea. His stratigraphic succession was, as Ward stated, essentially tripartite. However, Geikie (1874, p. 376) associated the submergence of land with a temperate period, as was compatible with the idea of climate changes being due to either astronomical or tectonic causes. The views of Ward and Geikie were thus somewhat different.

However, with the publication of the considerably revised second edition of *The Great Ice Age*, which followed two years after Croll's (1875) *magnum opus*, *Climate and Time*, and was evidently considerably influenced by it, Geikie abandoned the sea-ice theory and largely went over to Croll's theory. Thus Geikie (1877, p. 489) envisaged four glacials and three interglacials in Britain, in accordance with what might be expected from the astronomical explanation of glacial epochs. Geikie gave similar sequences for other parts of the world, including Scandinavia, Central Europe and North America. Croll (1875, p. 254), however, thought that the evidence of boreholes in the Midland Valley of Scotland indicated the occurrence of up to five distinct boulder clays.

Regarding NW England, Geikie (1877, p. 339), in his second edition, envisaged:

4. Upper boulder clay
3. Middle sand and gravel
2. Lower boulder clay, resting on a denuded surface of
1. Unfossiliferous till or boulder clay

He adopted Tiddeman's (1872) suggestion that the Lower boulder clay was formed (like the Caithness till) by an ice-sheet working over an old seabed and pushing marine sediments and organic remains onto the land; likewise with the Upper boulder clay. The Middle sands and gravel represented an amelioration in climate. So Geikie's (1877, p. 340) proposed sequence was as follows:

[7. (youngest) Re-elevation and climatic amelioration.]
6. Reappearance of arctic conditions and formation of an ice-sheet (yielding Upper boulder clay).
5. Depression of land to about 1300 ft and amelioration of climate (involving the reworking of materials, yielding the Middle sands and gravel, some subsequently found at quite high altitude).
4. Disappearance of arctic conditions and denudation of old bottom moraines.
3. Advance of ice-sheet and scooping of former sea-bottom and incorporation marine sediments and shells (forming Lower boulder clay).
2. Recession of ice and submergence of area of present Irish Sea under water.
1. (oldest) Glaciation with land standing higher than at present, and with ice covering Wales and northern England, there being perhaps no Irish Sea (yielding Unfossiliferous till or boulder clay).

It will be noted, of course, that this model dispensed with floating icebergs. Nevertheless, it relied on a significant period of submergence – though not one associated with cooling. Geikie's Lakeland ideas appear to have been based on information supplied by his English colleagues surveying in the NW, not on his own observations. It post-dated Goodchild's work, which is now to be discussed.

The English Surveyors, not in day-to-day contact with Croll, were not so quick to accept multiple glacials and interglacials as their Scottish colleagues, but the land-ice theory *per se* began to attract support and Goodchild (see Fig. 19.1), working with McKenny Hughes, published an important paper on the region's glacial phenomena (Goodchild 1875*a*).[10] The area had been thoroughly searched for evidences of glaciation, particularly glacial erratics. Goodchild recognized that some of the Eden Valley erratics had come from Scotland, and then crossed the Stainmore pass along with pieces of Shap Granite. However, the erratics had not got into Swaledale or the Lune Valley further south, where one might reasonably have expected to find them if they had been moved by water or floating icebergs, there being only small rises in the ground that might block passage into these valleys. Therefore, land-ice movement seemed to be called for. An ice-barrier could have barred the movement of erratics into Swaledale or the Lune Valley.[11]

Goodchild gave close attention to the structures of the drift deposits, many of which could be examined in the new railway cuttings in the area. The observed complex mixtures of slumped and unsorted materials did not, he thought, indicate marine deposition. Moreover, when the distribution of erratics was mapped, it appeared that there had been two sources of ice: one from Scotland carrying pieces of Criffel Granite; the other from the Lakes, carrying Shap and Lakeland rocks over Stainmore.

Fig. 19.1. John George Goodchild (1844–1906). British Geological Survey archives (GSM 1/639, No. 95). Photograph reproduced by courtesy of British Geological Survey.

However, it was difficult to say whether the two flows occurred at the same time. In fact, developing a suggestion of Ramsay, Goodchild entertained the possibility of different 'levels' of ice in a multi-layered glacier (formed by two or more conflated ice-sheets), such that there could be boulders in the ice at the same locality and at the same time, but from different sources. Thus the observed tangle of deposited materials might be explained.

Goodchild (1875*a*) published a map (Fig. 19.2) indicating the distribution of erratics and providing a conspectus of the situation. It suggested how and why ice might have moved through the Stainmore Gap. Goodchild described the deposits at the lower Eden Valley, with their contorted beds, false bedding, and pockets of sand. He regarded such materials as having been formed as the ice-sheet melted away, rather than being formed beneath an active glacier. Goodchild also wrote on the processes of glacial erosion and the origin of cwms, which he regarded as gullies enlarged and hollowed by 'tributaries' of ice (Goodchild 1875*b*, *c*); and he spoke up for Croll's theory.

Mackintosh was unwilling to relinquish his 'marinism' (now somewhat emasculated as a mere floating-ice theory) without a struggle, and he published a paper in which he summarized (and

[10] It may be noted that there are outlets from the Eden Valley to the south at lower altitude than the Stainmore Gap, such as the gap south of Tebay to the Lune Valley or the passage past Ravenstondale to Sedbergh, to the east of the Howgill Fells. However, the way to the Tebay route is partly blocked by the Crosby Fells north of Orton; and the Ravenstondale route is circuitous. Though somewhat higher, the Stainmore Gap involves only a slight divergence from the line of the Eden Valley. Also, as may be admirably seen from the top of Wasdale Crag (Shap) on a clear day, the Gap lies directly to the east. Indications of ice smoothing are evident on the exposed rock at the top of the granite.

[11] It may be remarked that since the work of Jamieson on the enigmatic Parallel Roads of Glen Roy, where the hypothesis of ice-barriers was successfully invoked to account for observations that had puzzled Charles Darwin and other observers, such a hypothesis provided a useful new weapon to assist explanation of observations of glacial phenomena. See Jamieson (1863) and Rudwick (1974). (From 1862, Jamieson was a lecturer in agriculture at Aberdeen University.)

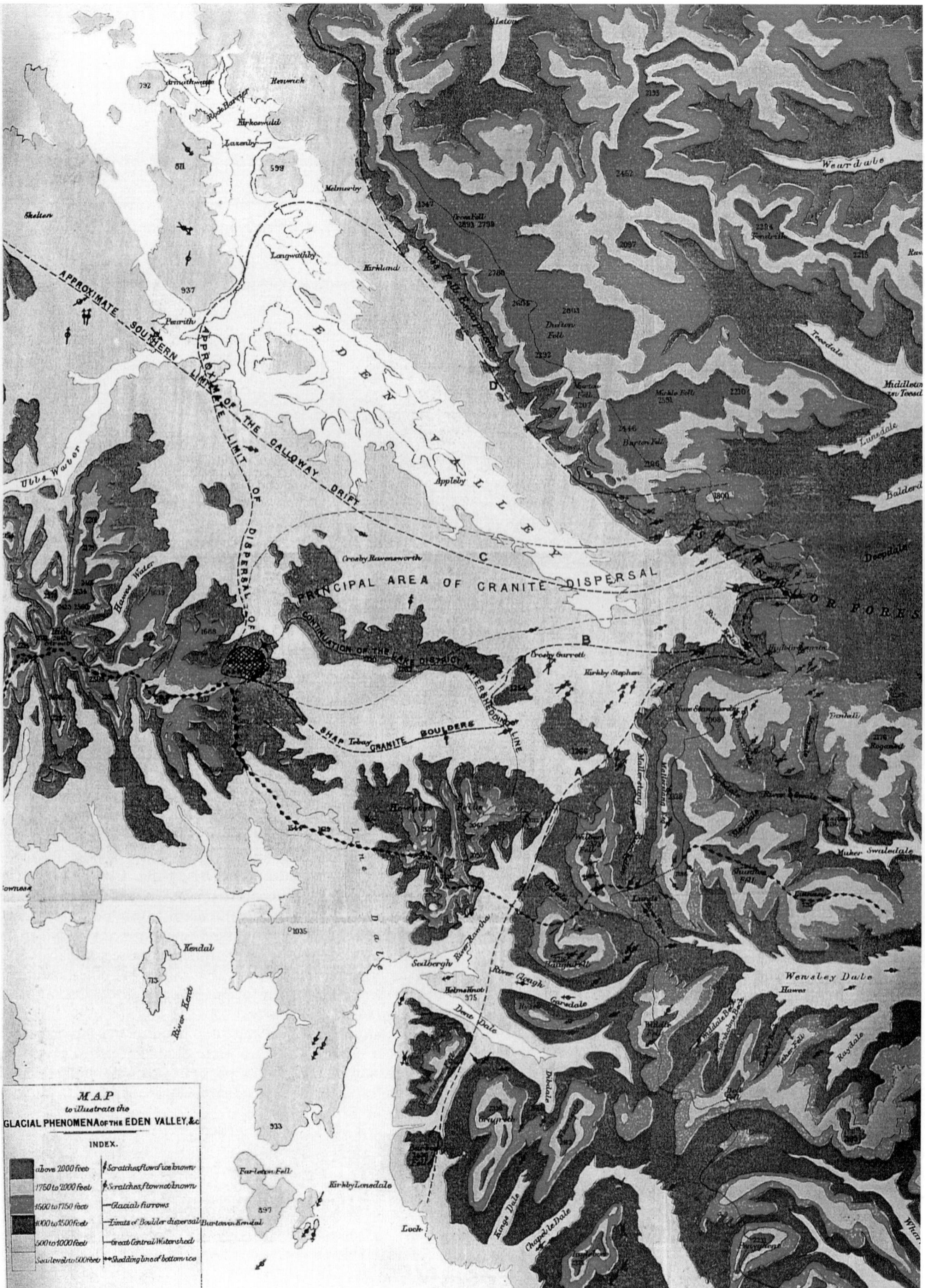

Fig. 19.2. Map showing distribution of erratic boulders in the Eden Valley, near Shap, according to Goodchild (1875*a*, Plate II).

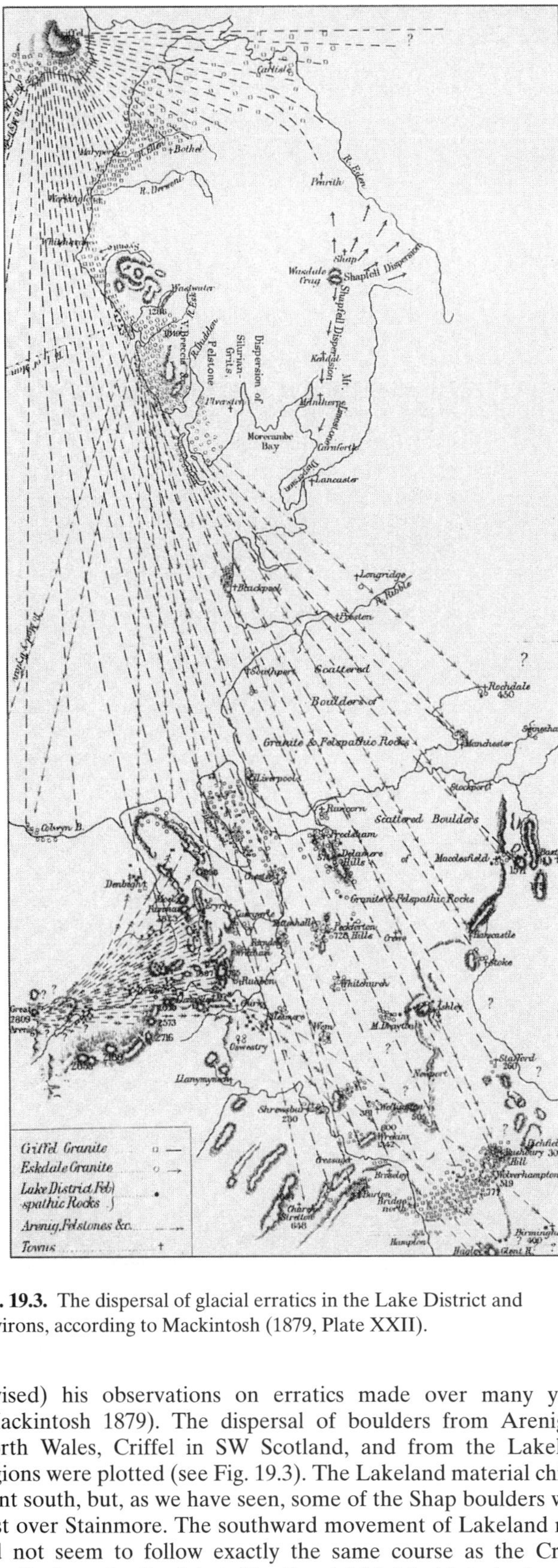

Fig. 19.3. The dispersal of glacial erratics in the Lake District and environs, according to Mackintosh (1879, Plate XXII).

revised) his observations on erratics made over many years (Mackintosh 1879). The dispersal of boulders from Arenig in North Wales, Criffel in SW Scotland, and from the Lakeland regions were plotted (see Fig. 19.3). The Lakeland material chiefly went south, but, as we have seen, some of the Shap boulders went east over Stainmore. The southward movement of Lakeland rock did not seem to follow exactly the same course as the Criffel Granite from Scotland, but, like Goodchild, Mackintosh plausibly suggested that all the movements need not have occurred at the same time, and need not have followed the same routes: the currents could have changed at different periods. He did not claim that his observations proved the marine theory, but he apparently still accepted it. In commenting on the paper, Ramsay suggested that *both* land-ice and floating ice might have been responsible for distribution of the erratics. Mackintosh did not attempt a full chronology of the events of the glacial epoch. It may be remarked that his map was much more schematic than that of Goodchild, who had ready access to the Survey's accumulated database.

The difficulties about the inter-crossings of glacial erratics were examined in an important paper by James Geikie (1882). As mentioned, he had by then embraced Croll's land-ice theory, and was prepared to 'take on' Mackintosh's arguments. Like Goodchild, Geikie contended that two ice-masses might sometimes 'collide' and mingle their contents, as for example in the Midland Valley of Scotland, into which came glaciers from both the Southern Uplands and the Highlands, producing a confusion of materials and striations in the 'debatable ground'. Similar areas, Geikie pointed out, had been examined by Continental geologists such as Hermann Credner, Albrecht Penck and Otto Torell. However, Geikie did not deploy Goodchild's idea of multiple movements of a single glacier, with different 'levels' containing material from different sources simultaneously.

Geikie's explanation of the distribution of the Criffel and Lakeland erratics to the south and west of the Lakes was that in the early stages of glaciation, erratics and other matter had fanned out from the Lake District into the area of the Irish Sea; and then as the climate deteriorated ice came down from Galloway, mingling with the Cumbrian ice and pushing it southwards. Eventually, the Irish Sea area became choked with ice; and then the northerly ice sought an exit through the Eden Valley, escaping over the Pennines via Stainmore. For the ice moving through the Irish Sea area, Lancashire and Cheshire, there could be a further admixture of material, as seen in the Midlands at such places as Wolverhampton, where additional materials were found, derived from the North Sea and Scandinavia.

Thus Geikie (1882) accepted Croll's theory, was an exponent of the land-ice hypothesis, yet also accepted the idea of glacial submergence. He was scathing of Mackintosh, while acknowledging his efforts in searching for erratics. It was a question of seeing the same evidence through the lens of two different theories. The opponents could invoke currents or glacier movements at will to suit the observations; but there was detailed reasoning in Geikie's argument. For example, if western England had been under the sea at the time of deposition of the boulder clay, one would expect to find evidence in the form of appropriate shells in layered deposits, or raised beaches. In fact, the shells were a mixture of deep- and shallow-water forms; and cold- and warm-water types were mingled. What Mackintosh thought were 'lines of bedding' in the till were, for Geikie, often irregular laminations analogous to those of crumpled schists, produced by the shearing action of ice. Where clear evidence of aqueous deposition was seen, such as graded bedding (not so called by Geikie), this was attributed to the action of waters running below retreating glaciers.

Goodchild (1889–1890) likewise extended the land-ice model, proposing that ice in the Eden Valley first escaped to the north through the Solway Firth, and also to the south past Kendal. But then, with a deteriorating climate, the higher ground to the south supposedly became blocked by ice, and the huge mass of ice pressing south from Scotland into the Eden Valley made a northwestern exit impossible; so some ice went east into the valley system of the Tyne, and more went over Stainmore, carrying its Shap debris as previously described. Goodchild envisaged one interglacial, but the final glaciation was not, he thought, severe. However, the land-ice theory did not, even then, carry all before it in the Lakes. The tectonic theorist Thomas Mellard Reade (1832–1909) of the Liverpool Geological Society (where, it will be remembered, Mackintosh had been President) uncharitably described Percy Kendall's description of the theory (1892) as one of the 'fairy tales of science' (Reade 1893, p. 37).

So, following their Scottish colleagues, the Lakeland Surveyors embraced the land-ice theory in the 1880s. However, they did not have a suitable area to recognize a whole sequence of glacials and interglacials. For this, one needed the wide Alpine valleys and the debouchments of glacial materials out onto open ground at the periphery of a mountain region. After studying river terraces in the valleys running from the north of the Alps near Munich, Albrecht Penck (1858–1945) of Vienna University arrived at the idea of three glaciations, based on three sets of outwash gravels (*Deckenschotter* or plateau gravels, *Hochterrassenschotter* or upper terrace gravels, and *Niedterrassenschotter* or lower terrace gravels), which interdigitated with end-moraines (Penck 1882, table II; Bowen 1978). He recognized that fluctuations might be linked to secular climatic variations, and also recognized glacial remains in the Pyrenees. Soon thereafter, Edouard Brückner (1862–1927) of Berne and then Vienna Universities recognized analogous terraces further east (Brückner 1887–1888). Subsequently, the *Deckenschotter* was subdivided, so that four glacials were envisaged (1. Güntz; 2. Mindel; 3. Riss; 4. Würm) and three interglacials. The evidence and the 'model' were jointly published by Penck & Brückner (1901–1909, vol. 3, p. 1169), with estimates of the periods of the glacials made by considering the thicknesses of the outwash gravels formed at a presumed rate; and the interglacials by the amount of dissection of the gravels, at a rate estimated by the observed rate of building of a delta into Lake Lucerne. Following earlier work by the Swiss palaeobotanist Oswald Heer (1809–1883) (1858), who found peats apparently below the glacial debris (later interpreted as interglacial remains), evidence for materials formed during the interglacials, containing plant remains, was described too (such as the Hötting Breccia near Innsbrück). This culminating effort was preceded by much Continental work in the last quarter of the nineteenth century. It should be noted that in the Penck and Brückner model, interglacials were essentially represented by erosional episodes, not actual deposits.

James Geikie made these Continental ideas known to British geologists in his Presidential Address to the British Association in 1889 (J. Geikie 1890), and in papers published in the 1890s, but he claimed that ideas being developed on the Continent were ones that he had long held (Geikie 1892, 1893, 1895), a series of glacials and interglacials having been predicted by Croll's theory. Geikie pointed out that German geologists had been working assiduously on the north German plain and elsewhere, and recognized that if it had formerly been covered by a massive ice-sheet, as that melted a confused variety of deposits would be formed under the ice, some of it sorted or bedded by flowing subglacial waters. Such materials might be mistaken for deposits accumulated in sea water beyond the margin of an ice-sheet. Thus there might be the appearance, but not the reality, of marine action. Moreover, it might be difficult to be sure whether such sediments did or did not form during an interglacial episode. Geikie also suggested that the popularity of the marine hypothesis in Britain might have had to do with the country's 'maritime' status.

So far as Britain was concerned, the claimed two boulder clays with intervening gravel deposits had been taken to represent two glacials and an interglacial at least since the work of Ramsay (1852); but they were compatible with a land-ice theory, a glacial-submergence theory, or the drifting-iceberg hypothesis. Geikie proposed that in Scotland one could see evidence for more than two glaciations, in that ice had apparently ploughed out older high-level terraces of boulder clay (the kind of phenomenon that Penck and others were reporting around the Alps). Geikie was broadly supported by the work of his colleagues Benjamin Peach and John Horne in the Orkneys, which advocated the passage of ice across the North Sea and on to the islands and mainland of Scotland, but with a 'double system of glaciation' (Peach & Horne 1880). The last 'scooping' was held responsible for the formation of rock-basins and lakes, but this glaciation was apparently not as intense as that which preceded it. Interglacial deposits could not be discerned in the Scottish valleys, but Geikie suggested that sediments outside the mountain regions, which had been construed as post-glacial, might represent the products of the last interglacial. In Sweden in 1885, Alfred Nathorst (1850–1921) had indications of a glaciation prior to that marked by the first main boulder clay (*fide* Geikie 1893, p. 296). Likewise, three, or perhaps more, distinct but overlapping boulder clays had been reported in Prussia by Henry Schröder.

Evidently influenced by Continental theory, Geikie (1893, pp. 319–321) argued for four glacials, his proposed sequence being as follows:

1. (oldest)	Weybourne Crag: ground-moraine of great Baltic glacier underlying lower diluvium	Glacial[12]
2.	Cromer Forest-Bed: equivalent of Hötting Breccia near Innsbrück	Interglacial
3.	Lower Boulder Clay	Maximum glaciation
4.	Freshwater alluvia, peat, lignite, etc.	Interglacial
5.	Upper Boulder Clay (Lower Diluvium of Scandinavia)	Glacial
6.	Freshwater alluvia, peat. lignite, etc.	Interglacial
7.	Ground moraines and terminal moraines (Upper Diluvium of Scandinavia; the terminal moraines of the large Alpine valleys of Penck)	Glacial
8.	Freshwater alluvia. Carse-clays and raised beaches formed in Scotland	Interglacial
9. (youngest)	Local moraines in Britain; 'post-glacial' moraines in upper Alpine valleys. Raised beaches	Glacial

Geikie noted that each glacial episode seemed to be associated with a period of submergence, but he could not explain why this was so. While a general supporter of Croll's ideas of climate change, Geikie seemed to hesitate in accepting his further contention that there would (or could) be submergence associated with glaciation, since the weight of ice would depress the Earth's crust – an idea suggested by Croll (1875, pp. 396–397) and earlier adumbrated by Jamieson (1865). The overall situation was necessarily complex, for during a glacial epoch water would be locked up in glaciers, so there would also have been a general lowering of sea level – a point made much earlier by Charles MacLaren (1841), who estimated that an 'Agassiz' ice age might lower the general sea-level by some 800 ft. By 1895, in an article in the *Journal of Geology*, Geikie envisaged six glacials and five interglacials for Europe as a whole. Glacial stratigraphy was suffering 'complexification'.

Such complexities notwithstanding, Geikie's proposed sequence corresponded, more or less, with that later associated with the names of Penck and Brückner. However, it was not developed, or even deployed, by Lakeland geologists. They did not associate themselves with the newly developing theories about the stratigraphic sequence of the Pleistocene, or theories as to the causes of glaciation. Why not?

The following points may be considered. First, the principal local geologists such as James P. Morris, William G. Collingwood and John Postlethwaite were amateurs and probably did not have ready access to the wider geological literature (though in his *Fragments of Earth Lore*, Geikie did an excellent job in explicating Continental ideas).

A second reason might have been that the nearest big city where there was an active geological society and university college (Liverpool) interested in Lakeland geology was in the hands of the marinists Daniel Mackintosh and Mellard Reade. More generally, the evidence in Lancashire, approved by the Survey when it was working there, appeared to support a threefold division of the Pleistocene; and Lakeland geology was much studied by Lancastrians – including Marr.

[12] Later work has revealed remains of deer and other animals in this Crag; but also cold-water shells.

Third, the Surveyors just passed through the Lakes and moved on to other areas; and their job was to survey and produce maps, not to theorize on a grand scale. Of the three who showed greatest interest in glacial matters, Ward resigned from the Survey and died shortly afterwards. Goodchild began to suffer from heart disease and around 1883 was transferred to indoor administrative and library work. De Rance turned his attention to water supplies, but (I have heard in a public lecture) he became an alcoholic and was subsequently dismissed, though only as late as 1898. Ramsay was taken up with his Welsh work and Survey administration and was never active in the Lakes, and Croll was not a significant field worker.

Fourth, the Lakeland region, though the most obviously glaciated part of England, was difficult to decipher so far as its Pleistocene stratigraphic sequence was concerned, lacking the grand system of river terraces of the Alps. In the mountains themselves, later glaciations tended to destroy the evidence that might have been left by earlier ones. Also, because the Lakeland mountains are relatively isolated their glacial debris got mingled with glacial matter from other regions, so that in an area such as the Carlisle Plain material could arrive from (say) both the Lakes and Scotland, producing a daunting stratigraphic record. Some ice might be pushing north and some south; and at any given locality the movement might have been in opposed directions at different times.

Lastly, the Surveyors' main focus of attention at that time in the Lakes was elucidation of the stratigraphic sequence amongst the Palaeozoic strata, not geomorphology.

In fact, as will be mentioned below, though the astronomical theory of glacial epochs, and the idea of a complex sequence of climate changes in the Pleistocene, attracts much support today, modern geologists think there is little stratigraphic evidence surviving in the Lakes for a whole sequence of glacials and interglacials such as the late nineteenth–early twentieth-century theorists envisaged and sometimes tried to justify.[13]

So the work done in Lakeland geomorphology in the decades around 1900 after the first Surveyors had departed was chiefly concerned with examining and recording glacial phenomena, particularly the lakes and tarns, which were intensively studied by Marr. One of his first geomorphological investigations in the Lakes had to do with the forms of the lakes themselves.

The local geologist William Collingwood (1884–1885) thought that a lake such as Windermere did not follow a fault line with vertical movement, but a line of fracture produced by folding, with glaciers then excavating the fractured rock. However, other lakes, such as Coniston, did seem to follow fault lines. Marr examined the question of the origin of the lakes and tarns of the Lake District in an extended series of papers. He was assisted in the first instance by the work of the geographer Hugh Mill, who studied the bathymetry of the principal lakes (Mill 1895). Windermere showed some surprising indications. There was evidence of previous river courses on the bottom of the present lake, but also three furrows running down the lake, which Mill and Marr interpreted as subaqueous valleys between parallel eskers (Marr 1896b). Buttermere, Crummock and Wastwater, on the other hand, had largely flat bottoms, though the latter two, as well as Ennerdale, also showed underwater cliffs.

Mill's bathymetry suggested that some of the lake bottoms showed roches moutonnée structures, but in some cases (e.g. Ullswater) there were great variations in depth, hardly compatible with straight-forward ice-scooping. Marr therefore began to explore the idea that the lakes were in part produced by the damming of river-excavated valleys, which had material deposited both before and after their conversion into lakes. Thus if the deposition of moraine material were an important cause for the formation of the large lakes, the process might not be so different from that in which tarns were formed. Marr gave close attention to tarn formation, regarding such waters such as Rydal, Grasmere and Devoke Water as intermediate between tarns and lakes proper.

In brief, Marr thought that the Lakeland lakes were not 'scooped' by ice, as Ramsay had supposed, but were chiefly dammed by drift. This drift could have been emplaced by land-ice or drifting icebergs, an issue that was still attracting attention at the end of the century. Marr said that he leaned towards the land-ice theory, but it was really immaterial so far as the question of whether the lakes were formed by drift barriers or 'ice scooping'. The lakes did not, on the whole, appear to be of tectonic origin, though the underwater cliffs might indicate faulting.

In 1912, Bernard Smith (see Chapter 4) published a paper on the glacial morphology of the southwestern area of the Lake District, around Black Combe (Smith 1912). This was the region where the Lakeland and Irish Sea ice-sheets might have been pressing against one another (see Fig. 19.3) and, as the ice receded, flow channels might have opened up between the two masses of ice. Smith was able to identify such channels, and map them along the western coastal margin of the Lakeland mountains, but, unlike James Geikie, he did not envisage several glacials and interglacials. Smith's sequence was much like that of the earliest English geomorphologists, the Lancashire Surveyors and the 'marinist' Mackintosh, namely a glacial, an interglacial and a second glacial – there being therefore two types of boulder clay, with intermediate water-deposited sediments.

Marr's *Geology of the Lake District* (1916, p. 147) also adhered to this scheme, and even questioned whether there had been a sequence of glacials and interglacials in the Lakes. Marr pointed out that the sands and gravels supposedly representing interglacials in Lancashire were not much in evidence in the mountain valleys, since, if they ever existed, they had been swept away by subsequent spreading(s) of the ice. Marr (1916, p. 194) did record a section, near Elterwater Bridge, Langdale, where red boulder clay underlay a grey variety, with stratified gravels separating them. However, this only evidenced the simple tripartite sequence of early theory, not James Geikie's more complex 'Continental' sequence. The distinction between the red and grey boulder clays had been noticed by the earliest observers, but often the red material was found uppermost. Marr, then, apparently only accepted the idea of (at most) two glaciations and one interglacial period. Whatever Marr saw at Elterwater Bridge is no longer available for inspection. There are large quantities of waste slate on the right bank and houses and gardens on the left bank.

In his later years, Marr began to lose his sight and concentrated most of his remaining energies on the Pleistocene deposits near Cambridge and on Palaeolithic instruments. In one of his last Lakeland papers (Marr 1924) he did, however, offer interesting geomorphological arguments for the occurrence of more than one glaciation, taking up ideas suggested by the Cornell University professor and student of glacial geomorphology, Ralph Stockman

[13] However, the Furness amateur geologist and fossil collector John Bolton (1862) reported 'black muck' with insect and plant remains from beneath a till in a borehole made at the Lindal Cote Iron-ore Company, SW of Ulverston, in mining operations to obtain haematite from the Carboniferous Limestone. In his *Geological Fragments*, a borehole section at High Cross Gates near Dalton-in-Furness was figured, indicating three distinct layers of 'black muck' (Bolton 1869, facing p. 152). Bolton confessed he was nonplussed as to how they came to be there. The deposits may have represented an interglacial period prior to the glaciation marked by the lower till – a suggestion offered by the mining engineer and Survey critic, J. D. Kendall (1881). He examined a number of nearby localities where deposits similar to those described by Bolton were to be found. However, Bolton's contemporary, Miss E. Hodgson (1863), disputed his observations, suggesting that the plant remains were recent and had been washed into swallow holes. The material was not described rigorously in the nineteenth century and the disused mines are inaccessible today. Even so, the deposit has been tentatively regarded as belonging to the Ipswichian Interglacial. An attempt by the Geological Survey in the 1970s to relocate the site by boring proved unsuccessful.

Table 19.1. *Sequence of glacials and interglacials in Northern England, according to Trotter & Hollingworth (1932*b*)*

	WEST		EAST	
	Southern Part Irish Sea Basin	*Lake District and the Solway Firth*	*Northumberland and Durham*	*Yorkshire*
Fifth Glacial Episode		Retreat phenomena: lakes, channels, sands and gravels, and laminated clays	Not represented	Not represented
		Scottish readvance: Boulder Clay		
		Retreat phenomena: lakes, channels, sands and gravels, and laminated clays (= Middle Sands of Carlisle)	Retreat phenomena: lakes, channels, sands, and gravels	Retreat phenomena: lakes, channels, sands, and gravels
Fourth Glacial Episode	? Upper Boulder Clay of Liverpool district	Boulder Clay of Ice of Lake District–Edenside Maximum and North Pennines[14]	Prismatic Boulder Clay, Cheviot and Scottish Ice with Western Ice in the west[14]	Hessle Clay [Humberside] and its inland equivalents[14]
	Middle sands and gravels	Gravels and laminated clays	Gravels and laminated clays	Gravels, etc.
Third Glacial Episode	? Lower Boulder Clay of Liverpool district	Boulder Clay of 'Early Scottish Glaciation' (including Lake District Ice)[15]	Boulder Clay of Western Ice[15]	Upper Purple Clay[15]
			Gravels	Gravels
Second Glacial Episode		? Weathered Boulder Clay of Upper Caldew Valley	Boulder Clay of Scottish and Western Ice[16]	Lower Purple Clay[16]
	Represented farther south		Loess	
First Glacial Episode			Scandinavian Clay[17]	Basement (Scandinavian) Clay[17]

Tarr (1864–1912) (1906*a*, *b*) – work previously mentioned in *The Geology of the Lake District*. Marr pointed out that Church Beck, running from Coniston Old Man towards Coniston Water, showed 'steps', as it passed through several hanging valleys. At the well-known 'Miner's Bridge' it ran into a narrow gorge, obviously cut by river erosion into the downward end of a hanging valley. Yet there were also evidences of glacial markings within the gorge. The inference was that there had been a glacial retreat, when the river gorge was cut; and then, during a subsequent advance, ice had gouged the glacial striations in the gorge. This did not, however, give Marr a series of glacials and interglacials and was compatible with the idea of just two glaciations, or only one main glaciation.

Thus the Lake District – the main English centre of ice – was by no means the best locality in England to establish the Pleistocene succession. On the contrary, the mappable flat Pleistocene deposits of the Midlands and East Anglia offered better opportunities to unravel the sequence by standard stratigraphic methods, though even there the task was difficult. During the second survey, evidences of an interglacial peat deposit at St Bees on the Cumberland coast were reported by Eastwood and Smith (Smith 1922, pp. 67–68). In the same report (pp. 46–47), Frederick Trotter recorded an upper and a lower boulder clay from just north of Carlisle for the Carlisle sheet, and in the accompanying memoir (Dixon *et al.* 1926, pp. 51–52, 56) and in a subsequent paper Trotter (1929*b*, p. 600) propounded the idea of there having been a late minor glaciation, evidenced by the upper clay, which on stratigraphic evidence from a site near Gretna Green was a separate unit, not merely formed by the earlier ice as it receded. Thus arose the idea of a 'Scottish Readvance' of ice in a late phase of glaciation, occurring in a fairly brief resurgence of glaciation towards the close of the last ice age. This came to be known as the 'Loch Lomond Stadial'.

Subsequently related to this concept was the 'Gosforth Oscillation', proposed as another late minor glaciation inferred for the Gosforth region in the Memoir for that district (Trotter *et al.* 1937, pp. 99–100). But the boulder clay that this generated was thought to be distinct from the similar one further north produced by the Scottish Readvance, the deposits being separated from one another by the supposed St Bees interglacial. Such ideas were fitted into a picture of a continuously receding ice front and rather little attention was given to subglacial or englacial processes. Glacial drainage channels were mapped as lake overflows, and several glacial lakes were suggested on little evidence (Huddart *et al.* 1977, p. 121).

[14] Supposedly equivalent to the Würm Glaciation. This was known as the 'Main Glaciation' in the Lakes, so called by Trotter (1929*b*, p. 560). Much of the work done by Trotter in this paper made its way into the synthesis proposed in Trotter & Hollingworth (1932*b*).

[15] Supposedly equivalent to the Riss Glaciation.

[16] Supposedly equivalent to the Mindel Glaciation.

[17] Supposedly equivalent to the Güntz Glaciation.

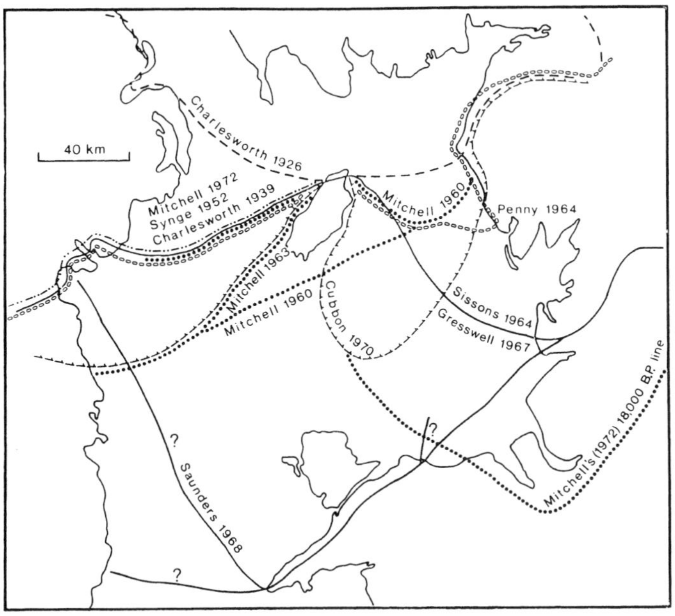

Fig. 19.4. Different views as to the extent of the 'Scottish Readvance', according to Huddart *et al.* (1977, p. 122). Reproduced by courtesy of Professor Huddart.

Trotter & Hollingworth (1932*b*) proffered a more elaborate sequence for the north of England than that of Trotter (1929*b*), involving a sequence of glacials and interglacials. They built on work done in Northumberland and Durham by David Woolacott (1921), and that of several other investigators who had been endeavouring to work out the Pleistocene stratigraphy in their local areas in the Northeast (C. T. Trechmann, E. Merrick, T. Herdman, J. A. Smythe and A. R. Dwerryhouse were mentioned). Woolacott convinced himself that there had been four glaciations on the eastern side of the Pennines, as expected according to the Penck–Brückner theory. He was, however, cautious about the notion of 'interglacials' and suggested that the term 'interval-in-the-glaciation' was more judicious, giving rise to 'interval deposits' during 'interval denudations'. Like earlier northern geologists, Woolacott, it seems, was inclined to contemplate one continuous glaciation, with 'pauses' therein, rather than an alternation of glacial and temperate climates. Be this as it may, it was Woolacott's synthesis in the northeast that provided a kind of 'template' that assisted interpretation of the glacial changes in the Lake District.

Thus Trotter and Hollingworth sought to find a Lakeland sequence analogous to that for Northumberland and Durham, also taking in evidence from the Carlisle area (Solway Firth). Their correlations were as shown in Table 19.1. This scheme was evidently proposed under the continuing influence of the Continental idea of four glaciations, which had become paradigmatic in Europe following the work of Penck and Brückner, though there was no assurance that the Alpine classification of glaciations was applicable all over the Continent, and even less so in the British Isles. The 'Scottish Readvance' idea did not sit well with classifications elsewhere and later became controversial (see below). The Alpine subdivisions of Penck and Brückner have, in fact, gone the way of all flesh, either being substantially modified or altogether replaced (Bowen 1978).[18]

It is noteworthy, then, that later publications for the north of England began to swing back to the idea of a single principal glaciation for the Lakes, so far as the stratigraphic evidence could show. Thus, for example, the Survey's Regional Guide *Northern England* (Eastwood 1946, p. 62) stated that some geologists thought that there was only one major glaciation for northern England (cf. what Trotter had called the 'Main Glaciation' for the Lake District). However, Eastwood favoured an 'Early Scottish Glaciation', which brought down boulders of Criffel Granite; the 'Main Glaciation', when the Lake District itself was a major source of ice; and the 'Scottish Readvance', which only modified the effects of the 'Main Glaciation' in a relatively minor way. This model was thus essentially similar to that envisaged by Trotter (1929*b*) or Trotter & Hollingworth (1932), though making no mention of possibly earlier deposits in NE Lakeland (Caldew area).

A later edition of the Northern England Memoir (Taylor *et al.* 1971, p. 84), correlated the British, Northwest Europe and Alpine successions as follows:

Northwest Europe	*Alpine Region*	*Britain*	
Flandrian		Recent	[Interglacial]
Weichselian	Würm	Devensian[19]	Glacial
Eemian Interglacial	Riss–Würm	Ipswichian	
Saalian	Riss	Wolstonian[20]	Glacial
Holsteinian	Mindel–Riss	Hoxnian[21]	Interglacial
Elsterian	Mindel	Anglian	Glacial
Cromerian	Güntz–Mindel	Cromerian	Interglacial
	Güntz		Glacial

This put three glacials in Britain. It was stated that little was known about the 'Pre-Weichselian' Pleistocene geology for the north of England, the reason being that the last glaciation largely cleared away the evidences that might once have been there for earlier glaciations. Therefore, the synthesis proposed in the foregoing table depended largely on correlations made with deposits in the south of England or the Midlands, as the names indicate.

The question of whether there was or was not a Scottish Readvance in the late Devensian, subsequent to the main glaciation, attracted a deal of attention, with different workers (making use of supposed correlations in the Isle of Man, North Wales and Ireland, as well as NW England) having different ideas as to how far the readvance advanced (see Fig. 19.4). Some authors such as Pennington (1970) and Sissons (1974) did not accept that there was an advance at all. To solve the problem, it was necessary to know whether a particular deposit represented an advance or a retreat, which depended on examination of landforms as well as deposits, and lithological criteria as to which glaciation a particular deposit represented.

In Cumbria itself, at that time, indications of only three glaciations were envisaged by, for example, David Huddart (1970–1971), then at the Geology Department, Newcastle-upon-Tyne, but now (2000) a professor at John Moore's University, Liverpool, summarizing the results of his Reading PhD (Huddart 1970). He deployed the basic three-fold classification of Trotter & Hollingworth (1932) in the Lake District, omitting some doubtful deposits from the Caldew Valley on the northeastern margin of the Lakeland hills.

[18] Modern work on determination of temperature changes in the Pleistocene by determination of oxygen isotope ratios in cores from polar ice or from deep marine sediments (Emiliani 1955, 1957; Shackleton 1969; Shackleton & Opdyke 1973) show patterned regularities, which suggest temperature fluctuations related to regular changes in the Earth's orbit, such as Croll had first envisaged, and which were later calculated more accurately by the Serbian mathematician Milutin Milankovich (1920, 1938) – the so-called Milankovich cycles. These are now routinely correlated with glacials and interglacials as determined by stratigraphic evidence (Bowen 1999).

[19] Named after a tribe, the Devenses, formerly occupying the area now forming North Staffordshire and Cheshire. 'Deva' was the Roman name for Chester.

[20] Named after a site at Wolston in Warwickshire.

[21] Named after Hoxne in Suffolk.

British Stage	*Cumbrian Stage*
Oscillation of the northern Irish Sea Basin ice-sheet, producing a weak readvance. 13 000–14 000 BP	Scottish Readvance
Minor interstadial?	
Late-Devensian ice advance	Main Glaciation: local ice first, followed by Irish Sea Basin ice-sheet
Mid-Devensian interstadial [interglacial], 20 000–40 000 BP	
Early Devensian (Weichsel), *c.* 40 000 BP	Early Scottish?

Huddart's work represented progress in that it made extensive use of thin sections of tillite to reveal which glaciation a particular deposit belonged to. Thus, by studying average grain sizes, the sand-to-matrix ratio and the orientations of the fragments, he distinguished the different deposits of the Main Glaciation and the Scottish Readvance. The orientations enabled him to gauge the directions from which the tills had been deposited, and hence the patterns of ice movement; and to some extent it appeared possible to discern what had happened when the tills of the Main Glaciation were reworked by the (still debated) Scottish Readvance. Red and grey tills were distinguished, along with interbedded sands and gravels. It is not clear whether Huddart was studying the same sections as those where the nineteenth-century workers had made their observations, but he was, up to a point, continuing the trichotomy of two glacials and an interglacial. Lacking palaeontological evidence, correlations with distant areas of glaciation or with classificatory schemes devised overseas were not attempted. In a later paper, Huddart *et al.* (1977, pp. 130–131) suggested that there were two distinct tills in the Furness region for the late Devensian, the 'Furness Readvance' preceding the Scottish Readvance.

However, in the first edition of the Geological Society's Special Report on the Quaternary (Mitchell *et al.* 1973), the Scottish Readvance was referred to as 'so-called' by the Surveyors Wyndham Evans and Russell Arthurton in their section on NW England. They explained (p. 29) how the characteristic lithostratigraphy had usually been taken to be as follows:

Upper Sands
Upper Boulder Clay
Middle Sands
Lower Boulder Clay

but, they pointed out, stratotypes had not been designated, and lenses of till in the Middle Sands might be misinterpreted as Upper Boulder Clay. So we are getting back here to the earlier view of Marr that there was stratigraphic evidence for but one glaciation in the Lakeland mountains; and all the deposits were regarded as belonging to the latter part of the last ice age: late Devensian. Some fossiliferous sediments below a till at Scandal Beck (NY 743 024), down Howgill way, were possibly from the preceding interglacial; and some boulder clay from near Gilsland (NY 622 661), up near Hadrian's Wall, might have belonged to the earlier Wollstonian glaciation. However, both these localities were outside the Lake District proper.

Advances in the study of Lakeland Quaternary stratigraphy were made in the 1970s and 1980s by the geographer John Boardman (at the time of writing with the Environmental Change Unit, Oxford University, but working at Brighton Polytechnic at the time of his investigations), through his studies of the drift deposits of the dreary moorland (Threlkeld Common) in the valley that runs between Keswick and Penrith, and especially in Mosedale and Thornsgill Becks, which run northwards towards the E–W Glenderamackin River by the Keswick–Penrith highway. In the small narrow valleys cut in the drift by the becks, Boardman found a highly weathered paleosol *below* the main till. This weathering he ascribed to the interglacial environment preceding the main Devensian glaciation. That is, it could be the residue of a Wolstonian till, weathered during the Ipswichian interglacial.

Boardman (1980) pointed out that the low relief of the Vale of Threlkeld suggested that the erosive action of the main Devensian glaciation might not have been great in that part of the Lakes; so the earlier pre-Devensian materials survived. He did not, however, find a neat 'sandwich' in the area (glacial–interglacial–glacial) such as the early Lakeland geologists had expected to find following on work in Lancashire. Boardman suggested that if the erosive effects of the Devensian glaciation had been muted in that part of the Lakes then it could well be that the 'classic glacial features' such as the truncated spurs of Blencathra (Saddleback) mountain, to the north of the Keswick–Penrith highway might have been produced pre-Devensian. In fact, he thinks that the main valley system of the Lakes is *pre*-glacial, being Tertiary or earlier (Boardman, pers. comm., 1998), a view that would have been congenial to Marr. Boardman (1985) defined his units in the area of the Vale of Threlkeld and laid out a detailed stratigraphy, with illustrative sections.

It is interesting that in the 1980s and early 1990s there was a revival (in a notably modified form) of the old nineteenth-century debate about 'marinism' (or 'glaciomarinism' as it was now called) in relation to the question of the Scottish Readvance. Given that land-ice can depress a land-mass, and that melting of large-scale ice-sheets can cause a rise in sea level, there is the possibility that the sea stood higher in the relatively recent geological past, compared with the land, than it does today, if the isostatic rebound of the land is slower than the eustatic rise in sea level. Hence, certain phenomena that might be thought to be explicable in terms of the action of melting land-ice and flow of meltwater might also find an explanation in terms of marine agency. That is, there might be phenomena that involved marine action when the overall sea level was on the rise, but which are now exposed inland from the coast because of isostatic rebound following unloading of the land-ice after the ice age. This approach was taken in by workers such as Nicholas Eyles (Toronto University) and Marshall McCabe (University of Ulster).

Thus in a lengthy paper dealing with the Irish Sea area as whole, and only incidentally with west Cumbria, Eyles & McCabe (1989) interpreted some fine red clays, found inland in western Cumbria, as glacial debris deposited in the sea. Certain inland sand and gravel deposits with flat upper surfaces were construed as deltas where meltwater channels from melting ice-fronts formerly met the sea. Banks of drift, which others thought were moraines from land-ice, were interpreted as debris from glacio-marine ice margins. In the Lakes, the highest claimed glacio-marine delta was one at Wasdale (152 m), taken to mark the maximum marine incursion. Since the red clays were found to drape over drumlins it would follow that, for the glacio-marine hypothesis, the marine incursion would have had to have occurred after the formation of these mounds, generally thought to be formed under the ice. Hence the incursion would have had to have followed the ice retreat.

The glacio-marine hypothesis was contested by Huddart (1991) and Huddart & Clark (1992–1993 [1994]).[22] We shall not follow all the arguments here, but will mention one example. The sediments at St Bees were taken by Eyles and McCabe – in common with other deposits round the Irish Sea – to represent materials derived from the glaciers to the east, deposited on quite a steep submarine slope, with cut-and-fill channels, and evidence of turbidity currents and slippage of piles of sediment. But Huddart disagreed. The deposits were, in his eyes, materials of a 'sandur' (a term from

[22] Richard Clark's ideas on Lakeland landforms were mentioned earlier in Chapter 18 (see p. 246).

Table 19.2. *Summary of Quaternary history for Cumbria, based on the BGS West Cumbria Memoir (Akhurst* et al. *1997, p. 99)*

Stage	*Climato-stratigraphy*	*West Cumbria*[24]	*Central Cumbria*[24]
Flandrian (Holocene) 10 Ka BP to Present			Wolf Crags (1–2)
Late Devensian	Loch Lomond Stadial (= Scottish readvance) (= Younger dryas[25]) 11–10 Ka BP		
	Windermere Interstadial 13–11 Ka BP		
	Dimlington Stadial 26–13 Ka BP (= Main Late Devensian glaciation and deglaciation)	Gosforth Glacigenic (2)	
		Aikbank Farm Glacigenic (2)	Blengdale Glacigenic (2–4)
		Seascale Glacigenic (2–3)	
Mid to Early Devensian	Interstadial	Glannoventia (?3)	
Base 122 Ka BP	Stadial	Carlton Silt (?4)	
Ipswichian Palaeosol 128–122 Ka BP	Interglacial	None	Troutbeck
Wolstonian 195–128 BP	Stadial	Drigg Till (?6) (proved in borehole)	Thornsgill (6)

Iceland, meaning outwash sands and gravels) involving pro-glacial lacustrine sediments. The whole had become contorted and otherwise disturbed by the southerly or southeasterly push of ice acting on poorly consolidated sediments during the Scottish Readvance; and the ice from that mini-glaciation laid the uppermost fine-grained till observed at St Bees. (There, in the low coastal cliffs by the shore, thrust till may conveniently be observed in profile, at a conserved Site of Special Scientific Interest.) In Huddart's eyes, Eyles and McCabe had things back to front. The main body of ice was then to the west (in the Irish Sea area), not to the east (in the Lake District). Huddart's earlier PhD experience helped him reach this conclusion, for he could ascertain the provenance and direction from which the sediments had come, and the directions of meltwater discharge. His ideas carried the day, as we shall shortly see.

In recent years, geochemical techniques have assisted Quaternary geologists substantially. By radiocarbon dating, numbers have been put on the main Quaternary events, and some renaming has been done for the Cumbrian deposits, or new names introduced.[23] Much has been achieved through the exceedingly detailed investigations conducted by the company Nirex (see Chapter 20) in west Cumbria, with additional data gleaned from trial pits, boreholes and seismic survey in the region of the Irish Sea, where a more complete record of Quaternary deposits is available than onshore; Nirex's onshore boreholes also provided information that was not so readily or unambiguously obtained by standard surveying methods.

Thus we have the details shown in Table 19.2, drawn from the West Cumbria Memoir of 1997.[24]

The Quaternary section of this new West Cumbria Memoir, developed from Nirex's (see Chapter 20) detailed Quaternary work (Nirex 1997*a*), also provides an excellent synoptic overview of the sequence of climatic events and their results in terms of ice flows, formation of glacial lakes, outwash fans, etc. (see Fig. 19.5).

From Figure 19.5, we can see the general picture, as envisaged at the end of the twentieth century, emerging in large measure from the century-long work of the BGS and the final intensive work of the Surveyors in collaboration with, and funded by, Nirex, linked with contributions from academic Quaternary specialists such as Huddart and Boardman. Examining Figure 19.5, we see that during the Main Glaciation, ice supposedly pushed westwards from the valleys of Ennerdale and Wast Water towards St Bees and Gosforth (Figure 19.5b). At the height of the glaciation (Figure 19.5c), the Lakeland ice merged with that coming down from Scotland in the Irish Sea ice-stream, bringing with it such items as pieces of Criffel Granite. With the relaxation of cold (Figure 19.5d), ice began to melt, outflow channels were cut, and meltwater was ponded in glacial lakes between the remaining land-ice and the continuing mass of ice in the Irish Sea, coming southward from still frigid Scotland. Figure 19.5e shows the resurgence of land-ice in the 'Gosforth Oscillation' (with some exposed hills or monadnocks), in which the Irish Sea ice encroached landwards towards Gosforth. The figure shows places where the new ice acted on the underlying till, producing thrust

[23] For Lakeland Quaternary or Pleistocene stratotypes, as designated at the end of the twentieth century, see Bowen (1999, pp. 95–96); for the stratigraphic column, correlated with the results of oxygen isotope stratigraphy, see Bowen (1999, p. 93).

[24] The numbers in parentheses represent oxygen isotope stages according to Shackleton (1969).

[25] 'Dryas' is the name of an Arctic plant.

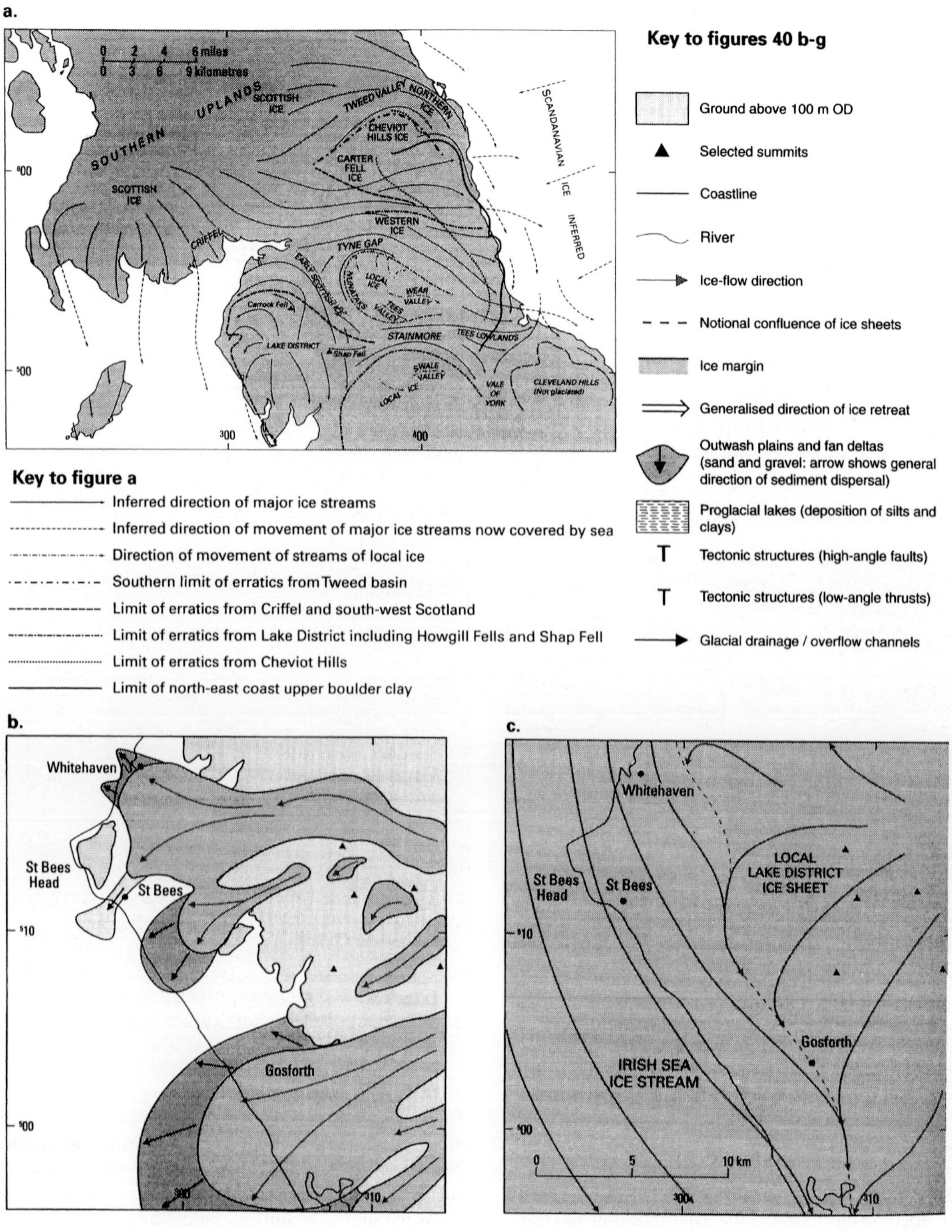

Fig. 19.5. Sequence of events in the Quaternary of west Cumbria, according to Akhurst *et al.* (1997, pp. 100–101). IPR/23-7C British Geological Survey.

faulting.[26] Figure 19.5f shows a subsequent retreat of ice and emergence of land, and the formation of a larger glacial lake as the Irish Sea ice declined more than on the previous occasion. Finally, Figure 19.5g shows the last gasp of the ice age, as the Scottish Readvance occurred and ice once again encroached on the area of the present land. It is envisaged that thrusting of the underlying glacial sediments was associated with both the Gosforth Oscillation and the Scottish Readvance. Figure 19.5a helpfully shows the broad outline of ice movements in northern England and southern Scotland, showing particularly the relationship of Lakeland ice to the movement through the Stainmore Gap and the complexity of the situation immediately to the east of Shap Fell, which formed a kind of Clapham Junction for ice movements, such as had puzzled nineteenth-century observers such as Goodchild.

It will be remarked that most of what has been said in the present chapter has referred to the latest phases of glaciation, and much of it to areas that lie outside the main mountain region of the Lakes. We have, then, the ironic situation that glacial sequences and Quaternary stratigraphy in England may be

[26] Very recent work (Williams *et al.* 2001) shows that seismic techniques have been employed to examine the structure of the thrust tills at St Bees, which shows the detail of current investigations in that area. Presumably such work was completed before the end of the second millennium, but it really lies beyond the scope of the present account.

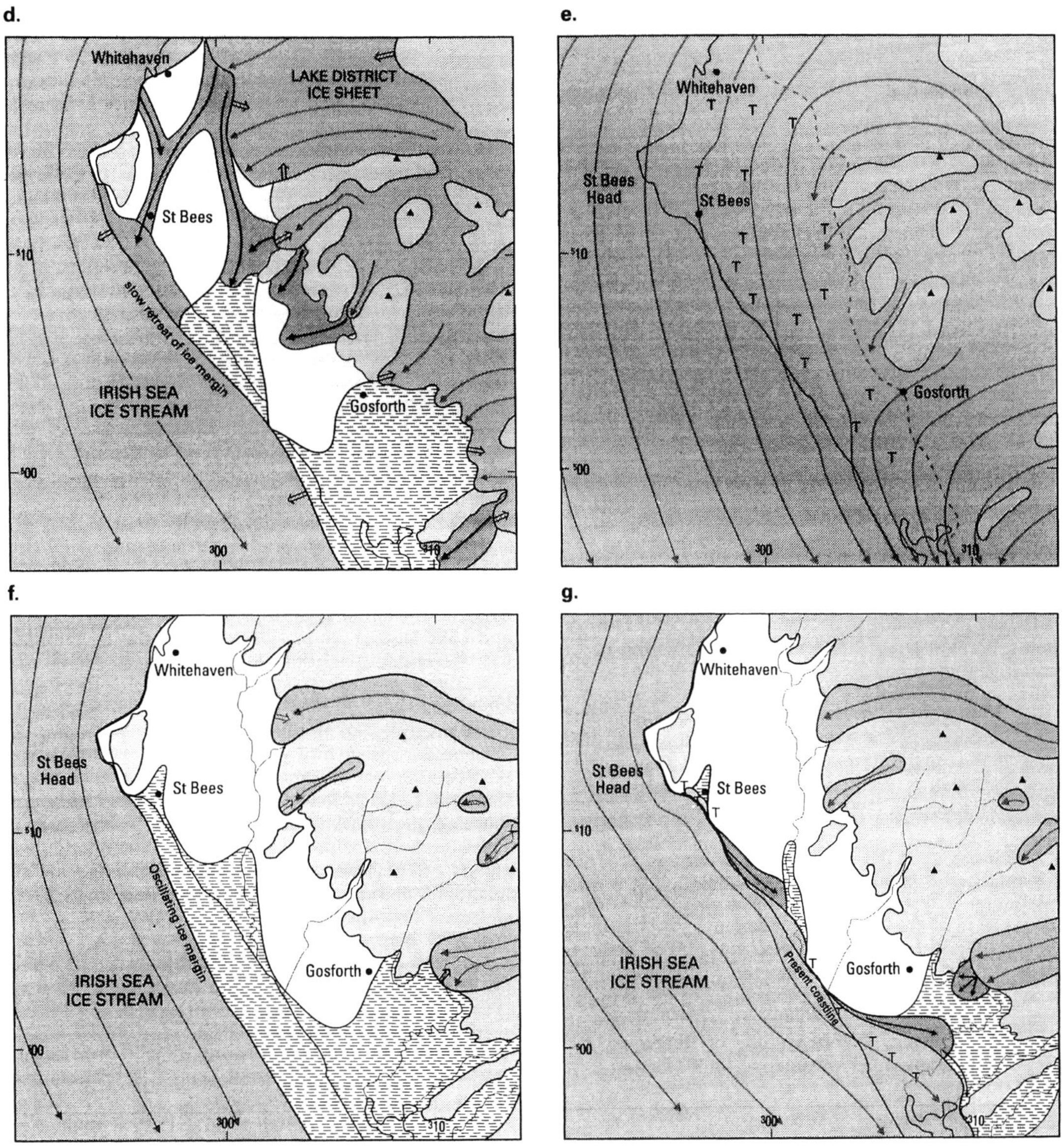

Fig. 19.5. *continued*

studied most satisfactorily in regions such as East Anglia outside the mountainous localities where the ice formation chiefly occurred. Thus it is unsurprising that the nineteenth-century geologists did not sort things out at the first attempt in the Lakeland region. In the early days, the supposed glacial sequence in the Lakes was to a marked degree determined by considering deposits in localities often quite distant from the mountains themselves. In the lower-lying country, the glacial debris could be studied using usual stratigraphic methods, though they were exceptionally difficult to apply, given that the deposits are often unfossiliferous (or may contain mingled fossils derived from different stratigraphic horizons), may contain reworked material, and often form isolated patches rather than continuous strata such as the Chalk or whatever. Further, the early British geologists' experience of actual glaciers was chiefly derived from localities such as the Alps, rather than Greenland-type ice-sheets. Yet it was the latter kind that we now think formerly affected northern Britain in the Pleistocene. What one sees today in the northern hills are chiefly the relics of deposits left by the retreating ice, following on the Main Glaciation; and, within the mountains, the traces of earlier glaciations were obliterated by the work of the later.

So the inclination towards mono-glacialism of one such as Marr – not driven by theoretical considerations like those of Croll, and not using the evidence of vertical boreholes (again like Croll) – is not surprising. In fact, John Boardman (1996, p. 11) offers a model broadly compatible with that envisaged by the nineteenth-century geologists, and early twentieth-century geologists may, I think, have been seduced by the Penck and Brückner model. The whole problem, from a stratigraphic point of view, was much more complicated than was realized in the early days.

Chapter 20
Nirex and the great denouement

We saw in Chapter 14 how the British Geological Survey came under pressure in various ways in the 1980s – the 'Thatcher years'. Some of its basic funding was diverted to the universities, and one of the reasons for the collaboration between the Survey and the universities was that it provided a means whereby it could recoup its 'own' money, so to speak, via the NERC-funded PhD studentships. However, that in itself was insufficient, and increasingly the Survey had to go into the market to help fund its activities through contract or consulting activities. With its large staff and accumulated experience and databases, the BGS (or Institute of Geological Sciences before the beginning of 1984) could readily aspire to be the largest and best geological consulting agency in Britain, although taking on a consultative role entailed a change in its ethos. We can see this manifested in 1991 when three new divisions were established: International, Marketing, and Corporate Coordination and Information. The posts of Chief Geologist, Chief Geochemist and Chief Geophysicist were abolished. Four new programme divisions were established: Thematic Maps and Onshore Surveys; Petroleum Geology, Geophysics and Offshore Surveys; Minerals and Geochemical Surveys; and Groundwater and Geotechnical Surveys (Hackett 1999, p. 7).

A considerable amount of the BGS's consulting work had to do with the nuclear industry and radioactive waste disposal.[1] In the early 1970s, the BGS (then IGS) and the Natural Environment Research Council began to focus on the pending major problem of nuclear waste disposal, and in the mid-1970s the BGS began a research programme into the problem of high-level waste for the UK Atomic Energy Authority, leading to reports such as Chapman (1979). However, this work ceased in 1981, following a change in Government policy. Then in 1984 the BGS started advising a company or organization called United Kingdom Nirex (NIREX = 'Nuclear Industry Radioactive Waste Executive') on shallow disposal options, but this work stopped in 1987, when the Government again rethought its strategy and began to develop a national geological review of options and selection criteria (David Holmes, BGS, pers. comm., 2000). A Government advisory body, the Radioactive Waste Management Advisory Committee (RWMAC), had earlier been established in 1978 to keep an eye on the disposal of nuclear waste in Britain and other matters to do with the nuclear industry.

An opportunity favourable to the BGS emerged in the late 1980s, with the award of a series of contracts to assist in the study of the rocks in the area of the nuclear reprocessing plant at Sellafield, near the Cumbrian coast, to the west of the Lake District National Park,[2] and the provision of hydrogeological information necessary to determine whether this site might provide a suitable repository for the permanent disposal of nuclear waste (see Fig. 20.1). The intended rocks to receive the waste were of BVG type, under the sedimentary cover of the west Cumbrian coastal plain. There was, of course, considerable BGS expertise available with such rocks, given that the Lakeland Project had been active for several years. Thus it was that several Surveyors, including David Millward, Eric Johnson, Brett Beddoe-Stephens, Andrew Chadwick and David Evans (a geophysicist), who had all worked on Lakeland rocks in one way or another, undertook work in relation to the proposed intermediate- and low-level nuclear waste repository near Sellafield. There were also geochemists and hydrogeologists. The team leader was Douglas Holliday, whom we met in Chapter 18. He had not been a part of the Lakeland Project, his specialty being Permo-Triassic, rather than Ordovician, rocks.

The use of the Sellafield site would minimize problems of transporting waste before burial; and it was thought that there would be acceptance of more 'nuclear activity' in the west Cumbrian community, already accustomed to and dependent on the nuclear industry. It was known from the 1930s surveying that BVG rocks existed under the west Cumbrian plain, below a relatively thin sedimentary cover (see Fig. 4.9), but more up-to-date and detailed geological knowledge was needed to accomplish the Nirex plan, even though a new section had been provided with the revised Gosforth Sheet (1980).[3]

Let us now recount some of the events relating to the geological work done by Nirex,[4] the BGS, and other consulting companies at Sellafield. This is a legitimate extension of the present study: first, because it has to do with Lakeland-type rocks; second, because it had to do with the activities and well-being of the BGS; third, because staff from both the Survey and university contributors to the Lakeland Project became deeply involved; and fourth because it raises important social and political questions about the relationship between science, government and citizens in modern democracies.

Nirex's former Manager for Site Characterization at Sellafield, Dr Robert Chaplow,[5] has kindly provided me with information about the early days of the company's Sellafield project (pers. comm., 1999).[6] Nirex was founded in 1982 (then as NIREX) as an executive body, with a remit to find a safe and permanent site for the disposal of nuclear waste in Britain. The intention was to find surface or near-surface sites for low-level waste and a deep underground site for intermediate-level waste. Initially, the organization was owned by the privatized nuclear industry; but in 1985 it became a limited liability company, the shareholders, however, being the various British nuclear industries, the main producers of the waste.[7] The Government, through the Secretary of State for

---

[1] Such work was not altogether new. Indeed, some of it went back to the 1950s.

[2] There had long been a large nuclear industry in west Cumbria, albeit one with a chequered history, notably the serious fire in 1957 at the Windscale works. Following the decision of Parliament in 1978, the construction of a nuclear reprocessing plant was begun at Sellafield close by, and it was the permanent disposal of radioactive waste from this plant that was the main problem for Nirex so far as the present study is concerned. The nuclear plant has long been a major source of employment in the district, of great importance since the near exhaustion of the haematite deposits and the decline of the ship-building industry at Barrow-in-Furness, particularly after the close of the Cold War. It is also of great importance to Britain as a whole, for its nuclear industry.

[3] The compilation of this revised map did not involve a comprehensive resurvey and preceded the Lakeland Project.

[4] For something of the history of the socio-political and scientific developments, see Kelling & Knill (1997); Haszeldine & Smythe (1997); Couples *et al.* (1998).

[5] Bob Chaplow (b. 1947) read geology at Imperial College, London and then went on to do a London external PhD in engineering geology while he was working for the geotechnical and civil engineering consultant company, Sir Alexander Gibb & Partners. He later lectured in engineering geology at Imperial College for four years, but subsequently returned to industry, working for Gibb, during which period his responsibilities included overseeing the geotechnical work at Dounreay. In 1993, he became a member of the Nirex staff, as Manager for Site Characterization at Sellafield (pers. comm., 1999).

[6] For further information on Nirex's research history and strategy, see Nirex (1997*b*).

[7] In 1999–2000 the shareholders were British Nuclear Fuels, Magnox Electric, United Kingdom Atomic Energy Authority, British Energy Generation Ltd (formerly Nuclear Electric Ltd), British Energy Generation (UK) Ltd (formerly Scottish Nuclear Ltd), and the Secretary of State for Trade and Industry (1 share out of a total of 10 001 shares) (Nirex 2000, p. 9).

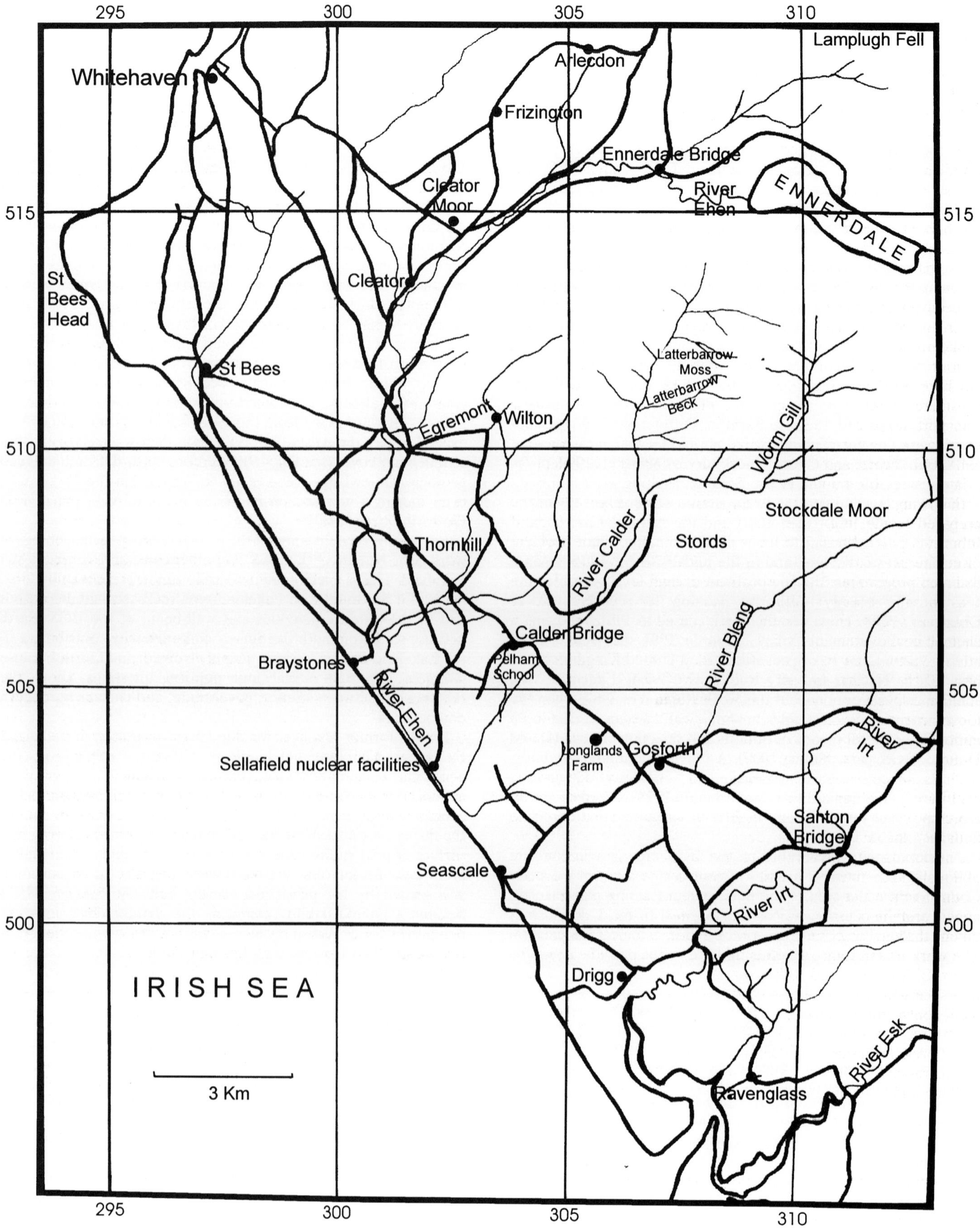

Fig. 20.1. Topography of west Cumbria in the area around the Sellafield nuclear facilities.

Trade and Industry, retained a small amount of ownership, with that Department holding a 'golden share' securing safeguard powers. The Government exercised ultimate control through its ownership of the United Kingdom Atomic Energy Authority, British Nuclear Fuels Ltd, and the Central Electricity Generating Board. Nirex Board minutes are transmitted to the Department of Trade and Industry. The company headquarters are near the nuclear facilities at Harwell.

The main product that Nirex had to sell was expertise, and advice to the Government and the nuclear industry, but it outsourced much of the work to consultants, amongst which the BGS is of special interest to us. Nirex was also responsible for practical matters to do with nuclear waste burial. Its income derives from the nuclear industry, which in turn is derived from electricity purchasers.[8]

A disposal site in an old anhydrite mine at Billingham, near Middlesborough, NE England, was initially considered, but proved unfeasible because of local opposition. A site at Elstow, Bedfordshire, in Oxford Clay, was also considered but got nowhere (Chapman 1986). In the 1980s, then, Nirex undertook a 'desk-top' study of some 500 sites that might, *prima facie*, be suitable for waste burial. Many of them were in areas of low population in western Scotland (Chapman 1986, p. 129). The sites were successively narrowed down to 200, to 50, to 12 and then to 10, setting aside two that were offshore.[9] The 10 remaining sites were confidential and, except for the two where exploratory investigations were eventually started (Sellafield and Dounreay in northern Scotland), they remain so. Numerous factors had to be considered: the geological situation; population densities; transport distances; and (importantly) local acceptance.[10] Thus the sites were subjected to 'multi-attribute decision analysis' (MADA), according to which all relevant factors, weighted appropriately, were taken into systematic consideration.

Following the several culls of the options, three short-lists were prepared (containing three, four and five sites respectively) and submitted to the Nirex Board for consideration. In the event, the three-site list was selected (in 1989). It had the Basement Under Sedimentary Cover (BUSC) site in the Sellafield area (Cumbrian coast, near Gosforth); a site in Palaeozoic to Precambrian rocks at Dounreay, adjacent to a nuclear power station on the north Scottish coast; and a 'Site Six' (one of the 10 then under consideration), said by Dr Chaplow to be a BUSC locality also. On the basis of the information before it, the Board decided that Sellafield and Dounreay should be recommended to the Government for practical investigation. Non-geological factors were clearly important considerations; but that is not to say that Sellafield or Dounreay were not the 'best' sites, all things considered.[11]

Planning permission from local government was required for exploratory drilling, and this was obtained for both Sellafield and Dounreay, 'Site 6' being discarded from further consideration. At Sellafield, the initial plan was to investigate the area directly under the nuclear facility, using a drilling technique that involved boring obliquely and 'steering round a corner'. The Cumbria County Council refused permission, but the decision was overturned on appeal and the design work for the deviated borehole was undertaken, through a complex web of subcontractual arrangements. The work was managed by British Nuclear Fuels Ltd as agents to Nirex. BNFL appointed Robertson Research Ltd of North Wales, in association with another company Hydrotechnica, as management contractor, who awarded the contract for the actual physical work to a Shrewsbury hydrogeological consultancy called Entec. The borehole design was commissioned by BNFL, who in turn commissioned British Coal to do the job. BNFL appointed British Drilling and Freezing Ltd to undertake the actual drilling.[12]

Possibly the web of responsibilities did not help, but whatever the reason, the first borehole did not reach its design depth and was, as described to me by Dr Chaplow, 'a disaster'! The drill got round the corner and down to 1100 m, but then stuck. The drill-string parted and the core-barrel could not be recovered. A second attempt was made, but that was no more successful and the hole was sealed off and abandoned with relatively little data acquired. However, according to Michie (pers. comm., 2001) the geological data obtained were later useful for interpreting the seismic data in the area, and the borehole also provided the first indication of saline water at depth in the Sellafield area.

BNFL then appointed Gibb Deep Geology Group for later boreholes. This group comprised Sir Alexander Gibb & Partners Ltd in association with GeoScience Ltd (a company from Falmouth that specialized in deep drilling investigations, especially 'hot, dry rock'); a geophysical consultancy J. Arthur & Associates; the BGS; Golder Associates (UK) Ltd; and Entec Hydrotechnica Ltd. A joint venture called KSW was to be responsible for drilling the boreholes, comprising Kenting Drilling Co. Ltd, Soil Mechanics Ltd, and Gewerkschaft Walther (a German manufacturer of the wireline drilling string), and the subsequent drilling was successful. The same group undertook drilling at both Dounreay and Sellafield. Geophysical survey was also undertaken, in addition to direct drilling. The Gibb Deep Geology Group was responsible for preparing reports on the results of the investigations, in accordance with the requirements of British Standard 5882. On instructions from Nirex, all the factual reports were classified as 'Commercial in Confidence' (Chaplow, pers. comm., 2001). The various reports were thus published 'in-house' by Nirex, but were formally reviewed. Later, the Royal Society Study Group (1994, p. 90) commented on the strength of the technical reviewing.

Borehole 2, about 2.5 km inland from Sellafield, was begun in August 1990 under the new management arrangements of the Gibb Group and KSW, reaching a depth of 1610 m by April 1991. It was intended to examine and characterize the basement BVG rocks, which were relatively shallow at the locality of the borehole. A fully cored borehole was successfully achieved, and hydrogeological testing, hydrochemical sampling and geophysical logging were undertaken; and an initial assessment of the rock mass was made. Emphasis was given to obtaining a high-quality core for logging. Similar techniques were used for subsequent boreholes. During 1989–1991, regional geophysical work was also carried out in the Sellafield area, both onshore and offshore, involving seismic, gravity and electromagnetic surveys, some airborne. Boreholes 3 and 4 were drilled between December 1990 and July 1991, the former being fairly close to Borehole 1, within the area of the Sellafield works, and the latter not far from Borehole 2 (Alan Hooper, pers. comm., 2001).

By 1991, it was decided that both Sellafield and Dounreay offered potentially feasible sites, the plan at Sellafield being to store the waste in BVG rock, under the overlying Permo-Triassic sediments. In the event, the Sellafield option was preferred and detailed investigation of the site begun the same year. A suitable Potential Repository Zone (PRZ) was considered to be under Longlands Farm, owned by BNFL, about 3 km inland from the

---

[8] I use the past tense here as befits a historical study, but Nirex's work still continues at the time of writing.

[9] Nirex did not rule out the offshore options, and possibly they may be used in the future, if permitted by international agreement.

[10] As one Surveyor put it to me with all seriousness, the site could not be located in Scotland or Wales because of nationalist sentiments; Northern Ireland was not a good idea because of political instability; rural Tory electorates were never thought appropriate by the Tory Government; and high population-density Labour electorates would not do either. So the Cumberland coast was probably always going to be the 'winner', as it had a previous familiarity with nuclear installations; it was a Labour electorate with a considerable unemployment problem; and most of the nuclear waste was at hand and would not have to be shunted round the country. Of course, the procedures used in the selection process were infinitely more complex than this, but perhaps the Surveyor's comment summarizes the public perceptions of the matter.

[11] For additional information on the choice of sites, see Nirex Report 71 (1989); and Cumbria County Council Appeal by United Kingdom Nirex Limited, APP/H0900/A/94/247019, pp. 147–169.

[12] Clarification of the interconnections supplied by Chaplow (pers. comm., 2001); also, according to Uisdean Michie (pers. comm., 2001), Robertson Research was taken over to become in part Hydrotechnica and then Entec.

Sellafield plant (see Fig. 20.1).[13] More boreholes were put down, until no less than 29 were made by 1997, some inclined but mostly vertical, with the highest density of holes being near Longlands Farm (Bowden *et al.* 1998, p. 126). A vast amount of stratigraphic, geochemical and geophysical information was collected.

By 1992, Nirex thought it had sufficient information to suggest that the Longlands Farm site might indeed be feasible, but did not have enough for planning permission to be assured. So they proposed to gather more data, by excavating into the general area where future storage was envisaged, for direct examination of the conditions. Thus the idea of constructing a 'Rock Characterization Facility' (RCF) – an 'underground laboratory' in BVG rock – was proposed,[14] where determinations could be made of the degree of rock fracture and the rate of water flow through the rock at that depth.[15] Such information could be calculated by hydrogeological modelling, but there were many possible models and relevant factors. The RCF would make possible the application of the hypothetico-deductive model for science: make assumptions; build a mathematical model according with those assumptions; insert numbers for the variables in the models; deduce empirically testable information from the model; and then test the models empirically, to see which is the most satisfactory. The RCF was required to assist this last step of the process. However, the very construction of the RCF, with the intended two 5-m shafts to a depth of about a kilometre and galleries opening out into the BVG rocks, might alter the hydrological situation in the proposed site.

In 1992, application was made to the Cumbria County Council to drill 11 boreholes around the proposed RCF site and one in the centre, to give a network for monitoring water-flows. Permission was granted, and some £45 million were spent on the drilling work in 1993 (Chaplow, pers. comm., 1999). The results were published as they came out, and on 9 October 1993, a meeting of the Yorkshire Geological Society was held at the BGS Keyworth headquarters, where the results of the investigations were discussed, the papers subsequently being published in the Society's *Proceedings* in 1994. They included a number of coloured plates with the help of a subvention from Nirex. This donation was acknowledged in the journal and Nirex was flagged as the 'patron' of the Society (founded 1837). The geological investigations near Sellafield became the largest ever carried out in Britain (Holliday & Rees 1994, p. 1).

Perhaps Nirex had concerns, or was feeling pressure from public opinion, for in 1993 they invited the Royal Society to conduct an independent review. This it did, consulting widely and taking submissions from various 'green' groups as well as the nuclear industry. The report (Royal Society Study Group 1994) commented favourably on the quality of Nirex's data, but was critical on some points, particularly to do with the short timetable laid down for the work, and some lack of openness and external peer-review. Their report stated,

> We were forcibly struck during this study by the extent to which some scientific reports are protected from wider scrutiny by being classified 'commercial in confidence'. This applies particularly to reports dealing with PCPAs [post-closure performance assessments] for Sellafield. We were informed about the methodologies used and had models described to us, but we were given few details of specific assumptions, did not have access to the databases of parameter values for models, and were given no detailed results of PCPA calculations. In particular, we were not supplied with all the assumptions and parameter values underlying the calculations in Nirex report 525 . . . [Nirex 1993*a*], nor did we receive any results other than those given in graphical form in that report (Royal Society Study Group 1994, p. 6).

However, Chaplow (pers. comm., 2001) has pointed out that the withheld data referred to radiological, not geological or hydrogeological, matters. Further, Report 525 (which was one in which Nirex summarized its model for the flow of water through the proposed repository site and the time it might take to reach the surface after its closure) was an 'executive summary' of the earlier four-volume Nirex Report 524 (December 1993), which was available to all, and was utilized by the Royal Society's Group. Hooper (pers. comm., 2001), Nirex's Manager for Science, has taken the view that 'the Study Group was given a full account of the development of the models for groundwater flow'. He also has informed me (pers. comm., 2001) that the only documents to which access was not afforded were 'preliminary repository safety assessments'.

In any case, whatever reservations it may have had, the Study Group recommended the construction of the RCF, opining (p. 110) that Nirex had 'provided a relatively robust description of the general groundwater flow regime'. However, its report continued, 'an important unresolved issue has been encountered: the most detailed geological and hydrogeological data yet available have failed to yield a simple relationship between the inferred pattern of fracturing in the repository host rock, the Borrowdale Volcanic, and the organization of groundwater flow patterns'.

By mid-1994 Nirex felt in a position to apply to Cumbria County Council to construct the desired RCF under the Longlands Farm area; but opposition was growing. Despite the fact that there was a degree of local support for the nuclear industry, which brought employment and an element of prosperity to an area with greatly diminished haematite mining, steel and shipbuilding industries, there were memories of radioactive contamination of the Irish Sea in the early days when nuclear waste was discharged directly into the sea; of the Chernobyl disaster; and of the Windscale fire in the 1950s. Moreover, there was a public perception (right or wrong) of above-average radiogenic diseases in west Cumbria. Besides, the Sellafield works was close to the Lakeland National Park. Groups such as Greenpeace, Friends of the Earth, and Cumbrians Opposed to Radioactive Environment rallied against the proposal.

Thus planning permission for the construction of the RCF was refused in December 1994 (*Whitehaven News*, 22 December 1994). Nirex immediately appealed, and this set in train a sequence of events that led to the holding of a major public inquiry into Nirex's RCF proposal. The Inquiry, which ran from 5 September 1995 to 1 February 1996, was not held in London, Carlisle or Whitehaven, but in the town hall of impoverished Cleator Moor, an 'ex'-haematite town, near Sellafield (see Fig. 20.1). A team of five lawyers led by Lionel Reade, QC, and 44 other staff was assembled by Nirex, and 18 expert witnesses were called. Opposing them, also with expert witnesses, were representatives of Cumbria County Council, the Gosforth and Copeland Borough Councils, Greenpeace, Friends of the Earth, local conservation groups, and concerned individuals. The Irish Government also sent representatives. The local Labour MP, Jack Cunningham supported the proposal, as did the unions at the nuclear plant. The Inquiry's chairman (or 'Inspector') was a

[13] Initially a site, 'Sellafield A', under the nuclear facilities, was considered. Later 'Sellafield B' (at the Pelham School Estate near Calder Bridge) came under discussion, and was subjected to 'MADA' in the site-selection process. Eventually the site at Longlands Farm, 2.4 km to the SE of 'Sellafield B' was fixed on, the land there being purchased in 1989, having been offered for sale to BNFL in 1987. Longlands Farm had the advantage of avoiding the Carboniferous Limestone (which could have weathered out underground in all manner of ways) under 'Sellafield B', but the locality was regarded as otherwise similar so far as MADA was concerned.

[14] Such an idea was not new. The general idea of an RCF (not specific to Sellafield) was mentioned by Chapman (1986, p. 132).

[15] The estimated cost was £192 million (*Daily Telegraph*, 18 March 1997).

solicitor, Mr Christopher McDonald, assisted by Mr Christopher Jarvis, solicitor, and by the mining and engineering consultant geologist, Mr Colin Knipe ('Assessor'), who evaluated the technical information.[16] The cost to Cumbria Council was estimated at about half a million pounds, whereas the bill for Nirex came in at about ten million (*Whitehaven News*, 1 February, 1996).

One of the observers of the events, Chris McKeown, of whom more anon (see p. 281), has described them to me rather graphically (pers. comm., 2000). The Greenpeace lawyers and scientists were all women. They were extraordinarily dedicated, McKeown recalls – chopping food for dinner at the same time as discussing, for example, the redox properties of pyrite and whether its presence was or was not a definitive indicator of a reducing environment, relevant to the stability and mobility of certain uranium compounds (which are more mobile in an oxidizing environment). Another of the witnesses who presented evidence for Friends of the Earth, Professor David Smythe (also of whom more anon: see p. 278), recalled that Mr McDonald and his assistants sat on the platform of the Cleator Moor Town Hall; the Nirex people were at tables down the hall to their right, while the 'objectors' were at their tables on the left, each with rows of boxes containing numerous documents (Smythe, pers. comm., 2000). The Inquiry had the status of a court of law. That is, McDonald had to call, hear and assess all available evidence and then form an opinion of the merits of the cases; and he could require witnesses to reveal all they knew on any issue. In his report, Knipe had the daunting task of providing intelligible answers to the questions posed to him by the Inspector on the vast amount of technical evidence tendered. McDonald had to write a report to the Minister of the Environment (Mr John Gummer), who then had to decide whether to accept or reject the Inspector's recommendations, either in whole or in part. It took McDonald until the beginning of 1997 to complete his report.[17]

After deliberating, the Minister accepted the Inspector's recommendations on 17 March 1997, and Nirex's appeal was dismissed, in what *The Independent* (18 March 1997) described as a 'surprise decision'. So the work on the Longlands Farm site was run down and was closed before the end of the century. There was no further search for a British repository site in the twentieth century. Mr Gummer's decision was reached in the context of an upcoming general election, and widespread opposition to the scheme.

It is said that Nirex was shocked by the result, for – in public – they had expected to win the appeal, and certainly wished to do so, having already spent about £400 million on research in west Cumbria and on administrative and legal costs (*The Independent*, 17 March 1997) – a figure confirmed in round numbers to me by Bob Chaplow, with, however, only about half the sum going into actual technical work (pers. comm., 1999, 2001). It was, as it appears to me, something of a David and Goliath contest, in that the Nirex team was defeated by a coalition which, though well briefed, lacked the funding available to Nirex.[18] Having lost the case, the Nirex Board voted its Managing Director, Michael Folger, a pay bonus greater than the annual salary of Dr Rachel Western of Friends of the Earth (*Whitehaven News*, 6 November, 1997), who assisted the Friends' solicitor, John Popham, and did much to lead the campaign.

Before getting into more strictly geological issues, it may be mentioned that those protesting against the Nirex case represented the RCF proposal as a kind of 'Trojan Horse': it was claimed that if the research facility were constructed so much money would be spent on it that there would be an irresistible pressure to continue and turn the 'research hole' into an actual repository. Thus it was that there was a 'demo' outside the Inquiry at the time of the Inquiry, with a large model horse paraded through the street, hauled by hooded figures with death masks, one carrying a 'grim reaper' scythe. An illustrated report in *New Scientist* (Edwards 1996) stated that the estimated cost of the completed repository would be £1.8 billion. Passions ran high.

According to the impression and recollection of the technical assessor, Colin Knipe, the RCF was indeed intended as the first stage in the establishment of a repository near Sellafield. Nirex presented it as a 'stand-alone project' – the RCF was not to be thought of as connected to an eventual repository. However, during the course of the Inquiry Mr McDonald came to regard the two projects as essentially linked, and in his final conclusions he wrote:

> [T]he work on the repository project is much too advanced for Nirex to be able to claim that the potential repository is merely hypothetical, and that it should be ignored for the purposes of the present appeal apart from reviewing the choice of location. Nirex has been working on the Sellafield repository project for several years: . . . and parts of the RCF could well be used for repository construction. The connection between the RCF and the repository is direct and obvious, and so cannot be set aside in the rest of the appeal determination process (Annex 1 to Inspector's Report, APP/H0900/A/94/247019, p. 266).

This, one might say, was the 'Trojan Horse' aspect. However, the Inspector did not ask Nirex to make a case for the whole project.

Nirex *was* required to demonstrate the relative merits of Sellafield, as opposed to other possibilities, and demonstrate how they had arrived at the Cumbrian site as their first choice. In Colin Knipe's view (pers. comm., 2001), the selection process that led to Sellafield emerging as the preferred site was 'by no means transparent', but was 'highly contrived'. Sellafield, he thought, should never have got through the sieving process: it did not have a low hydraulic gradient; it was not a geologically simple locality; it was not a geodynamically stable locality, being close to the Lakeland Boundary Fault; it did not have readily predictable groundwater flow paths; and the flow paths were undesirably short.

Leaving aside now these interesting matters, let us return to the scientific issues, with particular reference to the work of the BGS. The plan for the RCF was to excavate a cavity in BVG rocks below Longlands Farm – in the Fleming Hall Formation (see p. 276), at a depth of about 650 m. There, if results proved favourable, the nuclear waste might be buried to all eternity. But would such a burial be safe – such that the deaths resulting would

[16] Colin Knipe (b. 1943) was reared in Barrow-in-Furness and acquired an interest in Lakeland geology from an early age. He studied geology at King's College, London, and after a short stint of teaching in Barrow he went into the area of local government strategic planning, where he became familiar with planning procedures for town and country planning. He also assisted in the work of a national minerals assessment unit for the BGS. He was offered a position there, but preferred to take one (in 1968) in the geotechnical, land and mineral resources company Johnson, Poole and Bloomer, at Stourbridge, founded in 1844, the oldest extant company of its kind in Britain (see Rayska 1994), of which in time he became the senior partner. He was invited to take on the job of technical assessor at the Cleator Moor Inquiry following his involvement in two other large technical inquiries: a coal-mining site in Wales (Pull Du) and a clay-pit near Wrexham that had been outgassing methane (pers. comm., 2001). For previous mention of Knipe, see p. 127.

[17] Apart from numerous newspaper reports on the case, one can conveniently consult the Inspector's final report to the Government, a copy of which was made available to me *gratis* by the Cumbria County Council ('Appeal by United Kingdom Nirex Limited': File APP/H0900/A/94/247019); and all the evidence given by the Cumbria County Council, Greenpeace, and Friends of the Earth was collected in the form of a book edited by two of the scientists who opposed the Nirex appeal (Haszeldine & Smythe 1996).

[18] Alan Hooper (pers. comm., 2001) has questioned this. He described Greenpeace as 'a well-funded organisation both nationally and internationally and [it] appeared to have more funding for the aspects of the technical work than [did] Nirex'.

be no more than 1 person million^{-1} year^{-1}, as safety requirements stipulated?[19] Obviously, lying near the foot of the mountains, the area was subject to a large hydrostatic pressure, and if the rocks were significantly fractured or faulted, water might be expected to seep through them, so that if the chemical containment of the radioactive materials (by concrete and steel, etc.) proved inadequate then radioactivity might be transmitted to the water, eventually reaching the surface or contaminating drinking supplies. To know whether this might eventuate, hydrogeological modelling was needed; and that required detailed knowledge of the geology of the area, and particularly the pattern of faults and the physical and chemical properties of the rocks intended to form the site for the repository. It will be recalled that, according to the thinking of Branney & Soper (1988), the BVG rocks of central Lakeland were broken into a complex pattern of faults arising from caldera collapse in the area of the Central Fells. But what was the situation over to the west and at depth?

Earlier surveying for the Gosforth Sheet had revealed a substantial fault zone running approximately parallel to the coast, and bringing BVG and New Red Sandstone rocks into contact at the surface. This 'Lakeland Boundary Fault' (a northern extension of which was later named the Thistleton Fault) ran more or less along what is now the western boundary of the National Park. Other faults in the New Red Sandstone were also recognized by the early twentieth-century Surveyors (see Fig. 4.9). The area was now remapped in detail for Nirex by BGS staff, working in collaboration with Nirex geologists such as Uisdean Michie,[20] Andrew Bowden, Catherine Bumpus, Nick Davies, Anna Littleboy and David Mellor. The earlier survey of Trotter, Hollingworth, Eastwood and Rose undertaken in 1932 (Trotter *et al.* 1937) represented the area between Wastwater and Gosforth on the Gosforth Sheet (37) as having outcrops of rhyolites and dacites, with some andesites to the east, beyond a fault. In the reissue of the map (1980), revised by L. C. Jones in 1979, the rocks were indicated as 'Undivided (mainly lavas)', and a modification of the southern part of their boundary line was made, as compared with 1937.

There was an extended description of these rocks in the Gosforth Memoir (Trotter *et al.* 1937), where they were referred to as andesitic lavas, and rhyolites and associated rocks. A photograph showed good columnar jointing in the 'andesitic lava' (Trotter *et al.* 1937, plate IIB). The locality was not specified precisely, but it appears to have been in the area of 'Table Rock' (NY 088 036) near Bolton Head Farm ENE of Gosforth, where they were referred to as follows:

> An interesting feature of these rocks . . . is the frequent occurrence in them of fragments of distinct type, notably microlithic andesite . . . and fragments in varying stages of felsitic devitrification. In some cases the phenocrysts of the enclosing rock are arranged tangentially to the margins of the inclusions. Some, at least, of the fragments appear to be of foreign material caught up in the flow rather than of flow-breccia origin, . . .
>
> In the essentially non-porphyritic types a rough parallelism of the felspar laths induced during flow . . . gives rise to a platy fracture. . . . Narrow veins parallel to the direction of the flow, and consisting of epidote, chlorite and chalcedony . . . or epidote, chlorite and calcite . . . appear to have been formed at a late stage that corresponds with the infilling of the vesicles in other rocks (Trotter *et al.* 1937, p. 32).

This passage was written well before Oliver introduced the category of ignimbrite into Lakeland petrological vocabulary, but the description is, I think, compatible with the rock being of that type. So, as Uisdean Michie recounted to me (pers. comm., 1999), when the resurvey was done in the 1990s there was an attempt to link the evidence from the Borrowdale Volcanics exposed at the western edge of the Lakes with analogous rocks found under Sellafield and Longlands Farm in the boreholes. At the surface, attention was focused on the area east of Gosforth, where, at Table Rock and elsewhere, in Blengdale, or the valley of the River Bleng, a downfaulted area of ignimbrites, etc., was mapped, different from the Birker Fell Andesites to the east, from which they were separated by the Thistleton Fault.

These (re-)examined ignimbrites were not petrologically identical to those of the Scafell Caldera further east, so carefully examined by the Liverpool–Sheffield team. Consequently, a new stratigraphic column was erected for the western area, by examination of rocks exposed at the surface to the east of Gosforth, and in boreholes sunk there and also to the west, under the Permo-Triassic cover around Longlands Farm and Sellafield.

From the surface exposures and more particularly the boreholes, a new stratigraphy for the area below Longlands Farm began to emerge, and the various formations were named after farms in the area above the intended repository between Sellafield and Gosforth (Millward *et al.* 1994): [21]

Unit Name	*Lithology*
Fleming Hall Formation	Massive, homogeneous, andesitic, mainly densely welded medium- to fine-grained tuff and lapilli tuff (up to 520 m)
Brown Bank Formation	Tuffs with dacitic sheets, welded and locally rheomorphic (45–263 m)
Bleawath Formation	Eutaxitic *fiamme*-rich lapilli tuff with pumice and lithic clasts (up to 416 m)
Broom Farm Formation	Variable volcaniclastic sandstone (13.5 m)
Moorside Farm Formation	Massive, unbedded, poorly sorted, lithic-rich lapilli tuff, with coarse breccias (over 114 m)

These units could be identified in the drill cores and their characteristic chemical and radiometric signatures determined. When the geochemical signatures were compared, it became evident that the rocks near Table Rock, previously located in the Craghouse Member of the Birker Fell Andesites (Petterson *et al.* 1992), belonged to the Fleming Hall Formation at a higher level in the stratigraphic column. There was, then, the possibility of correlating the subterranean BVGs of the intended RCF area with surface exposures further east (on the other side of the Lakeland Boundary Fault Zone). That is, the subterranean BVGs might be linked into Lakeland geology (Millward *et al.* 1994). However, as it turned out, the western ignimbrites were not one and the same as those of the Central Fells.

A more detailed stratigraphy for the western rocks was subsequently provided in the west Cumbria Memoir (Akhurst *et al.* 1997), and eventually in the re-issued Gosforth map also (BGS 1999*b*). The Akhurst stratigraphy is reproduced in Figure 20.2.

It is instructive to look at the revised geological map for the

---

[19] This is a paraphrasing of the long-term safety standard, as set by the UK regulator.

[20] Michie (b. 1945), Principal Geologist and Earth Sciences Adviser at Nirex, hailed from Lewis and graduated in geology and mineralogy at Aberdeen. He joined the Survey in 1968 and became involved in several UK Atomic Energy Authority and British Nuclear Fuels research projects, including the study of the geology of the Dounreay area as a potential repository. In 1980 he moved to the Central Electricity Generating Board and developed a programme for the procurement of uranium resources. He joined the Nirex staff in 1990 and was intimately involved with its Sellafield programme. See Michie (1998, p. 257).

[21] This paper gave somewhat different profiles for areas to the north of the PRZ and beneath the Sellafield and Drigg Works.

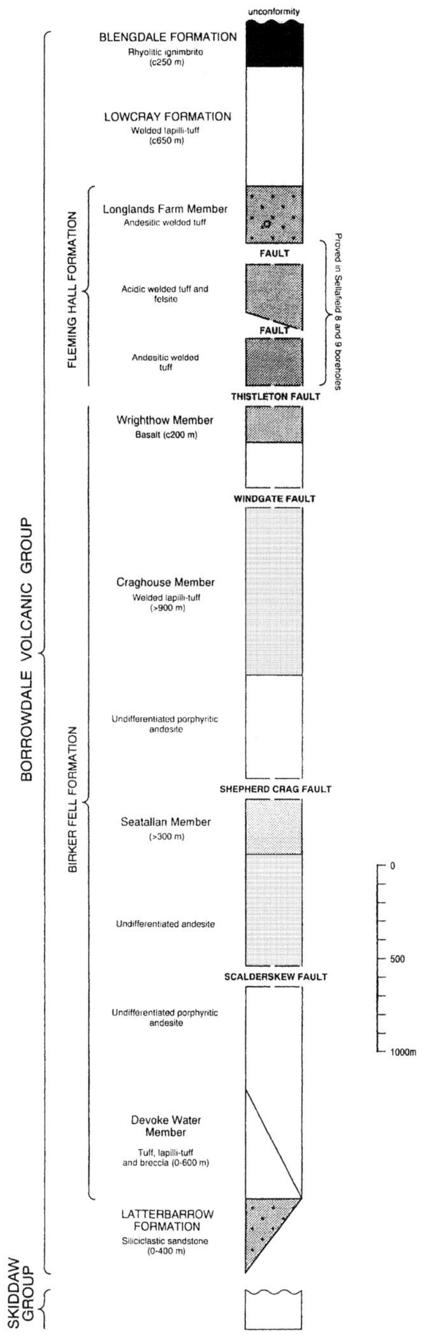

Fig. 20.2. Stratigraphy of the Borrowdale Volcanic Group in the Gosforth area, according to Akhurst *et al.* (1997, p. 44). IPR/23–7C British Geological Survey. © NERC. All rights reserved.

Gosforth area, part of which is reproduced in Plate VIII. The Thistleton Fault forms the eastern boundary of the BVG ignimbrites. The interesting thing to remark in this context is the general form of the ignimbrite complex, which can be construed as the eastern side of a caldera collapse structure, with the western part obscured by Permo-Triassic cover. This was essentially the theory advanced in Millward *et al.* (1994, p. 34) and subsequently developed in Beddoe-Stephens & Millward (2000). That is, the Surveyors had the idea that the Branney & Soper (1988) model for the Central Fells might be applied to the rocks of the western margin of the Lakes, with a separate caldera complex involving ignimbrites ponded within a caldera-type depression. It should be noted that the part of the Gosforth (1999) map reproduced in Plate VIII was also published in Akhurst *et al.* (1997, p. 24), which stated (p. 112) that the area was surveyed in the years 1990–1993 (by Surveyors Petterson, Barnes, Crofts and Auton).

At the end of the twentieth century, a new stratigraphy was published for the western ignimbrites by Beddoe-Stephens & Millward (2000), confirming that the textures and chemical compositions of the ignimbrites of the western caldera differ significantly from those in the Central Fells, being densely welded and of rather low silica content (63%). Hence a separate caldera sequence was proposed for these rocks, in what the authors termed the 'Gosforth succession'. In fact, this became one of four caldera or volcanic basin systems envisaged for the Lakes area: Scafell, Helvellyn, Duddon and Gosforth. The last of these is down-faulted (to the west of the Thistleton–Lakeland Boundary Fault Zone) and supposedly represents a young succession, preserved by the down-faulting. No evidence was found for this former volcano having been underpinned by a granite pluton.

All this had implications for the physical structure of the rocks in the intended RCF zone, for by analogy with the main Lakeland exposures one might also expect to find a tangle of volcano-tectonic faults within the concealed BVG. Also, as we might anticipate, the observations were in line with what might be expected from the theory of caldera collapse. Thus, for example, the Nirex Report (SA/97/032) detailed the results of the investigations in early 1996 for Boreholes 8 and 9, the first of which pierced the Permo-Triassic cover north of Gosforth, penetrating to the BVG rocks below, while the second went directly into the exposed Longlands Farm Member of the Fleming Hall Formation to the ENE of Gosforth. These boreholes were sunk with the intention of comparing the exposed and unexposed western BVG rocks, but the mapping, borehole cores, wireline logs and seismic survey revealed a bewildering complex of faults.

Such results (and others obtained earlier nearer Sellafield) raised the question of whether a sufficiently large block of BVG – of low permeability and good geotechnical properties, such as to allow excavation of underground cavities – existed in the proposed repository area, capable of accommodating radioactive waste safely. The nature of the overlying sedimentary cover was also important, as this influenced the flow paths, the path lengths, and the discharge zones for any material that might escape from the repository. To have a sound opinion on these matters, it was necessary to have a correct understanding of the pattern of faults, and also whether faults did in fact act as conduits for water. It might seem likely that they would; but there was the possibility that they had become sealed, subsequent to their formation, by the deposition of calcite, pyrite or whatever. The pattern of faults in the BVG below Sellafield could not be mapped out from the surface, because of the cover of Permo-Triassic sediments.[22] These sediments did contain some identifiable faults, which might represent subsequent movements along old, underlying fractures, but the likelihood was, considering the mode of formation of the BVG and all the movements related to Iapetus closure, the subsequent 'big crunch' in the Emsian and Tertiary unlift, that there were many more subsurface faults than could be recognized at the surface. These faults had to be located and mapped satisfactorily.

From the beginning of the Nirex work at Sellafield, a considerable amount of geophysical survey – gravimetric, geomagnetic, radiometric, electromagnetic and seismic – was undertaken, both onshore and offshore and with airborne as well as land measurements (Michie & Bowden 1994, p. 7). All this was done in addition to and separately from the study of the general surface survey of the Sellafield region, undertaken by the BGS under contract to Nirex. (This fitted into the Lakeland Project and Land Survey programme rather well, though it drew Surveyors away from areas

[22] The Triassic component of these was called the St Bees Sandstone in the 1930s survey, but this unit was renamed the Sherwood Sandstone with three constituent formations: Ormskirk Sandstone, Calder Sandstone and St Bees Sandstone (Barnes *et al.* 1994).

such as the southeastern Lakes, where they might otherwise have been working to complete the Lakeland mapping.)

So the geophysical data were analysed, with the help of information from the boreholes, yielding mostly two-dimensional sections of the BVG rocks underlying the Sellafield area. These revealed a complex pattern of subsurface faults – but much play was made by Nirex's opponents at the Inquiry about the rather rapidly changing interpretations of the fault systems, with a succession of reports showing interpretations that were not merely becoming more refined but were in some respects inconsistent with one another (for a simplified summary, see Edwards 1996; Haszeldine & Smythe 1997, p. 253). The early interpretations were made using vibroseismic techniques with the tests performed on roads, with 'infill' survey based on reflections from signals generated by small explosions.[23] However, the results were only analysed two-dimensionally, and such profiles did not readily reveal whether there was a mass of rock under Longlands Farm that might serve as a repository. For this purpose, one either had to go down and look, by constructing an RCF, or one needed to use an improved vibroseismic technique, and advanced computing facilities, which could reveal the three-dimensional structures.

As mentioned, Nirex was anxious to move ahead with its programme, and thus it was that it proceeded to the Inquiry largely on the basis of information derived from its two-dimensional profiles, rather than three-dimensional information. But this was, perhaps, a case of where it would have been better to have hastened slowly, for the question of which kind of technique or analysis was necessary and sufficient became a central issue for the Inquiry. Here the evidence and arguments of Professor David Smythe, then of Glasgow University, are particularly important for the present account (Smythe, pers. comm., 2000). We have already met Smythe's name in connection with his paper with Beamish about possible fault structures under the Southern Uplands (see p. 151).

Smythe was employed in the Marine Division of the BGS from 1973 to 1987 doing geological and geophysical work in the western seaboard of northern Britain and the North Atlantic. The work had to do with oil exploration and the search for nuclear waste sites. In 1988, he was appointed to a chair in geophysics at Glasgow University, funded by 'Britoil' (the privatized British National Oil Corporation), but this company was soon taken over by British Petroleum. Smythe found it difficult to get industrial funding from other companies, as they assumed he was funded by Britoil (BP), although BP had apparently lost interest in its Glasgow operations and put little money into the Glasgow department. So in 1989 Smythe joined a group that was investigating the famous deep borehole on the Kola Peninsula near Murmansk, and collected deep crustal seismic reflection data[24] – in a window of scientific opportunity between the end of the communist era and the collapse of the Russian economy and the emergence of the Mafia, which soon made it ill-advised to engage in financial contracts in Russia.

In 1990, Smythe was asked to join a panel set up by British Nuclear Fuels Ltd to review the work being done at Sellafield by Nirex and its contractors, consisting of himself (geophysics), John Lloyd (Birmingham, hydrogeology), Michael Coward (Imperial College, structural geology), and Brian Williams (Aberdeen, sedimentology). Fairly soon afterwards, he decided to resign from this group, partly because of pressure of work at Glasgow, and partly because, as it appeared to him (Smythe, pers. comm., 2000), the review process was being driven by a pre-arranged timetable, which was intended to yield a positive outcome by the end of 1994, going through all the geological assessments, planning permission applications, and reviews by then.[25] As funding from NERC had not been obtained for the Kola project, Smythe felt under pressure to apply for contract research grants to keep the Glasgow coffers supplied. He had the equipment appropriate for a 3-D survey at Sellafield, derived from the Kola project.

Thus Smythe approached Nirex to see whether they were interested in using his facilities and expertise. Alexander Gibb & Co. (or the Gibb Deep Geology Group) had previously prepared a report (GIBB/92127A/RS/TR/046, not seen), which argued that a 3-D survey of the area of the potential repository zone (PRZ) should be undertaken, using dynamite explosions as the source. Smythe was asked to comment on this. This he did in a report submitted in January 1994, in which he recommended that a high-resolution 3-D vibroseis trial survey be conducted, such as he (or Glasgow University) would be in a position to undertake, for a price expected to be a lower figure than what Gibb would charge, though the commercial company could probably deliver more rapidly. In particular, it was necessary to determine whether the vibroseis equipment could yield satisfactory results when used 'off-road'.

Smythe's offer led to much internal discussion at Nirex. There was doubt as to whether the BVG's deep structure could be imaged satisfactorily using a vibroseis source. There might be adverse effects of the work on the local fauna. There was also the large cost of a full-scale survey to consider – perhaps some £4 million. But something was needed, as there had been some inconsistencies between the borehole evidence and that from the earlier 2-D surveys and accurate knowledge of the geological structures was needed if satisfactory hydrogeological modelling were to be achieved. Michie took the view that a full 3-D survey was needed if a planning application for a repository was to be supported by appropriate evidence, but this would need to be preceded by a trial survey such as the Glasgow group might be able to undertake. In the event, Nirex opted for a pilot study, to be undertaken by Smythe and his team, rather than a large-scale investigation such as the Gibb proposal envisaged.

Thus Smythe and Glasgow University obtained a contract (for £270 000) to undertake a pilot study, which he carried out in 1994, in conjunction with his students and in collaboration with an ex-Coal Board company called International Mining Consultants Ltd (later IMC Geophysics Ltd). He thought that the work was scientifically and technically interesting and of national importance. He was not in principle opposed to the idea of the underground disposal of nuclear waste. While in the field, some of Smythe's apparatus was sabotaged by protest groups, but the damage was not too serious and a preliminary report was presented to a Nirex meeting at Harwell on 16 December 1994. According to Smythe's recollection, his work revealed the existence of a hitherto unrecognized N–S fault running through the proposed area of the repository, but according to Michie (pers. comm., 2001), the 3-D vibroseis technique produced structural information that helped resolve the various strands of a fault that had previously been recognized in the area ('Fault 2'). As it happened, four days later the Cumbria County Council refused the application for the construction of the RCF, but the timing was presumably coincidental. An acquisition report, which made some use of material provided by Smythe, was compiled by

[23] 'Vibroseis' is the trade-mark of a device that uses a vibrator plate acting on the ground for a few seconds to generate a wave-train comprising a sweep of frequencies. Several such instruments are sometimes used over an area in conjunction with one another. From the recorded data, a computer can produce a conventional-looking seismic section, such as can be obtained from the use of explosions. The vibroseismic technique is amenable to the determination of geological structures in two or three dimensions.

[24] This important work allowed direct examination of mid-crustal rocks and comparison with their seismic signatures.

[25] This group was disbanded after Smythe left and another one was subsequently formed by Nirex itself. This included Lloyd and Coward, and included the volcanologist professors Stephen Sparks (Bristol) and G. D. Price (London) for geomechanics. It was chaired by Professor Keith O'Nions from Oxford (Uisdean Michie, pers. comm., 2001).

Gibb and submitted to Nirex (Report 622) on 28 December 1994 (Chaplow, pers. comm., 2001).

Subsequently, Smythe submitted his draft report on 31 March 1995 (Nirex Report 760), and a two-page abstract was published in the proceedings of the annual meeting of the European Association of Exploration Geophysicists, which happened to meet in Glasgow that year (Smythe *et al.* 1995). Nothing more than this was published as a fully refereed academic paper.

Given that there was a time-window created by the initial refusal of planning permission, Nirex might have proceeded to a full 3-D survey, which might have been ready by 1996–1997, but Chaplow was concerned that the costs would be high if the whole area of the PRZ were investigated at depth, and he recommended that a full evaluation of the results achieved to date should be made before proceeding to more data collection. Adding complexity, Smythe changed his mind about some details presented at the December meeting, and then reverted to his original position; and he also reported an additional correction (Chaplow, pers. comm., 2001). So Smythe's report was sent off to review by a company called IKODA, who came up with 13 pages of comment, in consequence of which Smythe was requested in August 1995 to revise his report, taking the comments into consideration. His revised report was eventually received by Nirex on 18 September 1996 (Chaplow, pers. comm., 2001). Thus we find that each party viewed the other as being slow in delivery.

In the meantime, Smythe had accepted an invitation from Rachel Western to appear as a witness on behalf of Friends of the Earth for the Cleator Moor Inquiry, but his situation was becoming difficult at Glasgow because of the delay of Nirex's response to his draft report. As he recalled, he felt under pressure for failure to produce publications – the latter being absent partly because of the time spent on the Nirex work (Smythe might have been better off publishing the results of his Kola Peninsula investigations). Also, because of the delay in the response to his draft report he had no publications to show (other than the abstract mentioned above) that could be ready for the 1996 university review process.

Perhaps unsurprisingly, given the sequence of events recounted above, Nirex did not utilize the results of Smythe's work in evidence at the Inquiry. In late 1995 and early 1996, in response to Nirex's criticisms of some details of his preliminary report, Smythe was trying to re-process data, using an expensive computer program, borrowed 'on approval' with a 'sunset clause'. The program was found to have a bug in it, which Smythe had to fix, and the work was not completed before the 'sun set' at the end of February 1996. It was only with difficulty that he was later able to obtain funds from the Robertson Trust to enable the University to purchase the program's licence. By the time Smythe's report was completed, the Inquiry was over. In fact, Smythe had been struggling with his computer while it was in progress. But he made it clear to the Inquiry that, in his opinion, a 3-D seismic survey was needed, and that it would be premature to proceed to the construction of the RCF before such a programme was completed and its results analysed (Haszeldine & Smythe 1996, p. 269).

The question of what I would call the 'prematurity' of the appeal and the Inquiry is thus raised. So too are issues of commercial confidentiality and the crossing of the roles of academic and industrial work. Problems can occur when the activities of an academic become tangled with industrial money, not to mention the law (as at the Cleator Moor Inquiry). We are a world away from the days of the gentleman amateur, J. F. N. Green, or the independent Cambridge professor, J. E. Marr. Smythe's personal position was evidently difficult; but it was industry that supplied the expensive instruments required to conduct his 3-D survey and analyse the results.

Smythe was not, of course, the only person who appeared as witness against Nirex at the Cleator Moor Inquiry. Leaving aside here the 'minor groups', there were also:

For Cumbria County Council

Professor John Mather (Royal Holloway College, University of London) on 'Nirex Site Characterization at Sellafield'[26]

Dr John Starmer (Environmental Resources Management, Virginia, USA) on 'Nirex Performance Assessment for Post-Closure Safety'

For Greenpeace Ltd

Dr Andrew Stirling (Science Policy Research Unit, Sussex University) on 'The Nirex MADA'

Dr Rae Mackay (University of Newcastle) on 'The Effect of Extent of Site Investigation on the Estimation of Radiological Performance and its Relevance to the Proposed Rock Characterization Facility at Sellafield'[27]

Dr Stuart Haszeldine (University of Glasgow) on 'Subsurface Geology, Geochemistry, and Water Flow at a Rock Characterization Facility (RCF) at Longlands Farm'

Mr Philip Richardson (Geosciences for Development and the Environment, Ashby de la Zouche, Leicestershire) on 'International Practice Regarding In-situ Geological Research Facilities'

Dr Helen Wallace (Greenpeace, London) on 'Model Validation and the Role of the Proposed Rock Characterization Facility at Sellafield'

For Friends of the Earth

Dr Patrick Green and Dr Rachel Western (Friends of the Earth, London) on 'The RCF and Government Policy'

Dr Peter Kokelaar (University of Liverpool) on 'The Borrowdale Volcanic Group'

Professor David Smythe (University of Glasgow) on 'The 3-D Structural Geology of the PRZ'

Mr George Reeves (University of Newcastle) on 'Hydrogeological Investigation Programmes: Best Practice'

Dr Shaun Salmon (Aspinwall & Co., Leeds) on 'The Hydrogeology of Sellafield'

Dr Stephen Hencher (University of Leeds) on 'Fracture Flow Modelling'

Dr John Allison (Bullen & Partners, Croydon) on 'The RCF: Engineering Issues'

Dr Roy Wogelius (University of Manchester) on 'The RCF: Geochemical Issues'

The following gave evidence on behalf of Nirex:

Mr Michael Folger (Managing Director, Nirex)
Dr L. D. Phillips (London School of Economics)
Dr John Holmes (Director of Science, Nirex)
Dr Robert Chaplow (Manager for Site Characterization, Sellafield, Nirex)
Dr Alan Hooper (Manager for Science, Nirex)
Dr D. W. Mellor (RCF Science Manager, Nirex)
Professor Keith O'Nions (Royal Society Professor in Earth Sciences, Cambridge, and Professor of Physics and Chemistry of Minerals, Oxford)

together with 11 further expert witnesses from various geotechnical consulting firms. Introductory and closing statements were also made for each group.

The 'proofs of evidence' presented at the Inquiry by the Cumbria County Council, Greenpeace, and Friends of the Earth groups were published by Haszeldine & Smythe (1996) under the imprimatur of the Department of Geology and Applied Geology, Glasgow University. The intention was to put the objectors' side

[26] Mather had previously worked for the BGS, and had been involved in the search for suitable places to bury high-level waste.

[27] I am informed by Uisdean Michie (pers. comm., 2001) that Dr Mackay did not in fact appear against Nirex, but informed the Inspector that he was satisfied by his review of the company's proposals.

of the story in the form of an accessible permanent document. Nirex initially put their evidence on the World Wide Web, but it was subsequently withdrawn, and now can only be accessed from the company by request. Nirex did not like the Glasgow book, and the company's Chief Executive Officer contacted the University, pointing out (as I understand from Chaplow, pers. comm., 2001) that Professor Smythe was required to clear materials relevant to the company with Nirex before publication. In fact, many things had been said on public record at the Inquiry (some at the Inspector's behest) and reported in the media, and the Glasgow book *per se* merely made these matters of public record readily accessible. It would, I submit, be preferable if the Nirex evidence were still on the Web, and likewise readily accessible.

It should be noted that the BGS did not formally present evidence at the Cleator Moor Inquiry, though they did prepare a report (SA/95/002) with maps and drawings that summarized the information that they had acquired. This was submitted as part of Chaplow's evidence. According to Douglas Holliday (pers. comm., 1999), the BGS did not want to become involved in the legal forum and Nirex preferred to rely on its lawyers to a considerable extent, though it also called expert witnesses (see above).

The foregoing account has taken us some distance from Lakeland geology as such. So let us return to our geological theme, and what was done at Sellafield, yet also saying something further about the evidence tendered at the Inquiry. In doing this we re-introduce the volcanologist Peter Kokelaar, leader of the Liverpool team doing the contract mapping in the Central Fells.

Kokelaar took the Inquiry through the theory of caldera collapse developed by Branney (and Soper) for the Central Fells, and pointed to the recent paper by Millward *et al.* (1994), which, as we have seen, sought to apply such concepts to the BVG rocks under the Longlands Farm area. Kokelaar emphasized the general complexity of BVG faulting, the lateral variability of rock types in the unit, and the fact that even the massive ignimbrites frequently had cooling joints of unpredictable extent. Nirex had found difficulty in correlating faults appearing in boreholes only 10 m apart; and, on Kokelaar's view, the company's assumption that the extent of throw of a fault could be correlated with the thickness of fault-rock was not justifiable for the BVG. In Kokelaar's opinion, even if an RCF were constructed, and the particular spot where the excavations were made proved suitable, it would *not* mean that rock close by would be appropriate. This could not be certain, any more than one could form a clear idea of the details of structure from isolated boreholes (Haszeldine & Smythe 1996, pp. 230–231). The BGS workers had not, in Kokelaar's view (pers. comm., 2000), sufficiently emphasized the complexity of the BVG faulting to Nirex in their various reports. Nirex did not seek to make contact with the Liverpool group, which was perhaps regrettable, given its competence on Lakeland volcanics. Nirex did, of course, utilize the BGS expertise. And, as we have seen, the BGS geologists had found that the geology of the western 'Gosforth' caldera differed somewhat from that of the Central Fells, though both were thought to display volcano-tectonic faulting.

It fell to Chaplow to rebut the points made by Kokelaar and Smythe, and we have a full record of their rejoinders, along with summaries of Chaplow's arguments (Haszeldine & Smythe 1996).[28] Chaplow argued that Kokelaar had exaggerated the complexities of the BVG and had made generalizations about it over too wide an area. He had not shown that the proposed repository site was near a caldera margin. Kokelaar countered that he was using information from Nirex's own report (S/95/005), which suggested that the proposed RCF or PRZ was located at the site of a volcano-tectonic fault of a piecemeal caldera. Also, 'geological complexity do[es] not occur only at caldera margins' (Haszeldine & Smythe 1996, p. 234).[29] This was, to Kokelaar, an important point: by the theory of caldera collapse, faults could occur anywhere within a collapsed caldera (see Fig. 16.8).

Chaplow further stated (Haszeldine & Smythe 1996, p. 234) that Kokelaar was treating the BVG cooling joints 'deterministically' rather than 'stochastically', which would be what was needed for dealing with rocks having many small-scale fractures. Kokelaar countered that he did not know of any Nirex report that had attempted a 'stochastic' treatment of the cooling joints. 'Nirex', he said, did not 'comprehend the full significance of their own results' (Haszeldine & Smythe 1996, p. 235).

This exchange raised an interesting feature of Nirex's approach. In 1994 they had assembled an 18-member Expert Group (including four Nirex staff, seven people from the Atomic Energy Authority, and two from the BGS), working with a professional 'facilitator' (from 'Facilitations Ltd') to try to formulate a model for the fracturing of the rocks at Sellafield, and hence to estimate the risks that the fractures might entail. Clearly, every fault, fracture or crack, large or small, could not be examined separately, and the overall permeability thereby calculated. So instead of direct measurement or experimental test, they sought to consider various parameters and 'elicit' the role of each separately. The participants were given training by the facilitator in the art of eliciting – or in 'elicitation methodology' (Nirex 1995, vol. 1, A3). Four types of rock channel were considered: (0) small channels (examined by permeability measurements); (I) joints 10 cm to 50 m long (as revealed by boreholes); (II) fractures 50 m to 5 km long (as revealed by boreholes); and (III) major faults more than 5 km long (revealed by geological and geophysical survey). With this grouping, '[a]ppropriate distribution functions, together with PDF's [probability distribution functions] for the uncertain parameters that characterize the distributions were then constructed for the parameters of the fractures in each category, based on the available data and on *expert judgment*' (Nirex 1995, vol. 1, p. 5.1, emphasis added).

Thus, I take it, there was a kind of 'brainstorming' process in which the experts sought to gauge the nature and contributions of the various factors, and develop formulae to model the flows and estimate the consequent risks. In fact, they spoke of 'effective parameters', since the parameters themselves were rarely amenable to direct determination. The preface to the third volume of the report acknowledged that the study was 'not a comprehensive safety assessment, which would require the systematic evaluation of a wider range of repository features, external events and underlying processes and pathways'. So the report gave the combined 'elicited' views of the expert group concerning a highly intractable problem. I think one can reasonably say that the problem of the fractures was approached on the assumption that it was stochastic in nature. Whether it was thereby solved I leave it to others to judge. The RCF programme was, it should be emphasized, intended to enable Nirex to *test* its model(s).

The arguments of Smythe were rather more complicated than Kokelaar's, though he also introduced some straightforward matters, such as the view of the Royal Society Study Group that the choice of the Longlands Farm site was based on ownership of land and mineral rights and easy access to Sellafield rather than geological suitability. Thus, in Smythe's view, the cart was in a sense before the horse. The site was first selected; then the task was to prove that it was technically suitable. Smythe also claimed that the initial choice of the Longlands Farm site was 'largely based on pre-war geological knowledge' – the work of Trotter *et al.* (1937).[30] He showed that sections of the proposed site, as they

[28] I take it that it worked conversely, but I have no transcript of the cross-examination of the Nirex witnesses or lawyers.

[29] One can see how this would be likely to be so from Branney & Kokelaar (1994); see Figure 16.8.

[30] However, at the Inquiry, and as subsequently reiterated to me in correspondence, Chaplow pointed out that the BVG was not too deep under Longlands Farm, and there was no complicating Carboniferous Limestone there. If the repository were placed too deep there would be problems with additional rock stresses and elevated temperatures.

appeared in Nirex reports of 1990, 1991, 1993, 1993 and 1995, differed considerably. Even two reports for December 1993 showed substantially different profiles. Clearly, knowledge of the structure of the site was still evolving, and one could not be certain that the present understanding was 'robust' or reliable. It was this uncertainty that was picked up by *New Scientist* (Edwards 1996) and presumably did not help to inspire confidence in the wisdom of digging under Longlands Farm.

However, Smythe's main complaint, as might be expected from what has been said earlier, was that there were uncertainties in the knowledge of the structure of the BVG under Longlands Farm because the seismic analysis commissioned by Nirex was 2-D rather than 3-D; or the evidence from Smythe's provisional report on his 3-D work undertaken in 1994 had not been (sufficiently) taken into account. The faults, as determined from borehole evidence and Nirex's 2-D methods in their evolving understanding of the subsurface geology of the Longlands Farm area, were all shown as planar, which was not necessarily correct.

In fact, Smythe reported to the Inquiry that the BVG rocks under Longlands Farm dipped at about 20°–30° to the SE, which, he maintained, was in accordance with borehole evidence and interpretations of gravity and aeromagnetic data previously published by Kimbell (1994) from aeromagnetic survey. Yet Nirex had published a report (S/95/005) based on borehole evidence and 2-D seismic survey that showed the dip of BVG units as varying between south and SW (Haszeldine & Smythe 1996, p. 248). This inconsistency indicated that more work was needed before constructing an RCF.[31] Hydrogeological modelling would be premature if the geological structures in the BVG were not known accurately.

In responding, Chaplow stated that interpretation of a 3-D trial – Smythe's 1994 work, in fact – was proceeding, and that there was a provisional programme for more such work to be undertaken in 1996 before proceeding to construction of the RCF. But on Smythe's estimate, such results would not be ready before 1999; so the hydrogeological modelling could not be done satisfactorily before then.

Another expert witness at Cleator Moor, appearing on behalf of Greenpeace, was the previously mentioned Stuart Haszeldine, at the time a colleague of Smythe at Glasgow, though he later moved to Edinburgh. Haszeldine (an environmental geologist with special interests in diagenesis, the genesis of ore deposits, and sedimentology, along with petroleum geology and the hydrogeology and geochemistry of subsurface waters), together with a PhD student Christopher McKeown (previously mentioned), undertook computer modelling of groundwater flows at Sellafield, using data made available by Nirex (which at that time had a library in Cumbria).[32] Whether such a modelling exercise was warranted, in view of the arguments advanced about uncertainties by Smythe and Kokelaar, is perhaps arguable. The point was that Nirex and the Glasgow group used the same data but obtained different results for the likely flow of water and movement of radioactive compounds through the proposed locality of the RCF.

The way Haszeldine became involved in the arguments is somewhat curious (though he had had earlier contacts with Greenpeace). In 1991, he was taking a party of undergraduate students up to NW Scotland to look at the famous thrust-belt there, and heard that drilling was going on at Dounreay. So they decided to pay a visit on the way back and see what 'geology in action' was like. They saw the rigs, and the people they met were helpful, but it appeared to McKeown that the men working for different contractors and doing the different tasks did not know what the others were doing, and there seemed to be a disconnected work pattern. McKeown and another student became curious and wanted to look into the matter further.

After finishing his honours degree at Glasgow, McKeown worked for a short time on a small hydrogeological modelling project and then Haszeldine asked him to do something similar where radioactive materials were involved. This was completed and McKeown then went off to Australia for a brief period. On returning to Glasgow, he wrote up his report in provisional form and sent it to Greenpeace, thinking they might find it interesting. They did, and invited Haszeldine to put in a small research proposal on geochemical modelling, which gave McKeown six months' research time. Greenpeace liked this too, and invited a proposal for further work, which ultimately funded McKeown for a PhD under Haszeldine (McKeown 1997). Thus the Cleator Moor Inquiry came in the middle of McKeown's PhD work, and he was able to supply information to his supervisor in his evidence to the Inquiry, though McKeown has informed me that he improved and developed his ideas between the time of the Inquiry and the completion of his degree (pers. comm., 2000). The Inquiry was premature for him too!

Working from Nirex's Report 524 (vol. 3) (1993), Haszeldine argued at the Inquiry that the water movement was *upwards* in the area of the proposed RCF, due to the flow of water down from the Lakeland mountains, and the presence of a large mass of underground saline water (known from Nirex's Borehole 3), to the west of the site, adjacent to the coast, which deflected the movement of fresh water.[33] So, '[T]he PRZ is positioned directly on the main subsurface of flow potential of meteoric water moving from the Lake District towards the coast. *This PRZ is in the worst position on this "Site" to allow containment of any waste by present-day hydrogeology*' (Haszeldine & Smythe 1996, pp. 129–130, emphasis in original).

In fact, Haszeldine claimed, the Glasgow group's computer modelling (essentially McKeown's) showed that water might reach the surface in the time-range 10 000–300 000 years, although if there were significant unsuspected fractures the time could be less than the lower figure. Also, the RCF alone could not ascertain whether or not there was such a fracture in the vicinity. Haszeldine opined that the deeper Bleawath Formation was less fractured and would be more suitable as a repository than the Fleming Hall Formation (but then, we should remark, it would cost more to construct a repository, with the greater amount of rock to be excavated, the greater pressures, and the higher temperatures). The Glasgow group's numbers had already been published at the time of the Inquiry (Haszeldine & McKeown

[31] Uisdean Michie has commented on this issue to me (pers. comm., 2001), pointing out that the dips in the BVG are variable, with ignimbrites being draped over different erosion surfaces, with changes of dip across fault zones. However, recent work (Beddoe-Stephens & Millward 2000, fig. 2) shows the general dip in the Fleming Hall Formation in the Gosforth area to be to the south. After the Inquiry was over, Nirex re-processed and reinterpreted Smythe's 3-D trial survey and produced a contoured surface for the base of the Fleming Hall Formation with fairly uniform dips of SSW to SW (Nirex 1997*c*, fig. 16). Michie also construed Kimbell (1994, p. 109) as supportive of a SSW to SW dip; and he stated that the study of Boreholes 8 and 9 supported a S to SW general structural dip. But these arguments were not all available at the time of the Inquiry.

[32] Haszeldine (pers. comm., 1999) recalls that Nirex was generally willing to provide what was requested, but, as he put it, some items had to be 'winkled out of them'.

[33] The saline water was initially discovered during the incomplete drilling of Borehole 1 (see p. 273). Michie (pers. comm., 2001) has asserted that 'Haszeldine and McKeown's modelling was not able to deal with variable density groundwater' and used single values for rock properties. His own generalized model (Michie 1998, fig. 14) gave pathways such that water reaching the surface inland would have travelled only through the rocks of the sedimentary cover. Such water as might pass through the region of the potential repository was represented as remaining in the lower saline waters, exiting below the Irish Sea rather than on land. Michie noted to me that the BVG rocks were shown by the Nirex investigations to be less deep inland from Sellafield than earlier supposed, and hence more conveniently placed for a repository.

1995) and were known to Nirex, but the Company had doubts about the computer programs used by the Glasgow group (see p. 286 and Note 33).

Haszeldine's interpretations were in stark contrast to results previously obtained by Nirex (1993*a*). In this report, Nirex concurred with Haszeldine and McKeown that water would move upwards from the repository area in the BVG, but when it reached the overlying (Sherwood) Sandstone they thought it would move horizontally for about a kilometre, then pass downward to a depth below that of the repository, proceed west for another kilometre, then rise again, before travelling horizontally at a depth of about 450 m, with water eventually reaching the seabed about a kilometre offshore. This was the report about which the Royal Society Study Group had been unable to obtain responses to its enquiries concerning the related database, because of 'commercial confidentiality'; and it was later criticized by the Radioactive Waste Management Advisory Committee (Kelling & Knill 1997, p. 12).

Of course, the Nirex plan was not just to drop radioactive materials into a 'bare' hole in the ground. There was to be engineered chemical containment. The fill material had to be porous, to allow escape of gas, and not too hard, so that material could be retrieved in case of dire necessity, if required. Thus the chemical situation within the country rock itself was relevant to the likely success or failure of containment in the repository; and this was another issue on which Haszeldine focused attention.

It was known that in oxidizing conditions the uranium radionuclides would form mobile compounds and the converse would be true in reducing conditions. Nirex and BGS took the view that there was presently a reducing environment in the neighbourhood of the proposed repository, evidenced by the presence of pyrite. This suggestion might seem remarkable when we consider what has been said in Chapter 4 about the widespread occurrence of haematite in west Cumbria. Kingsley Dunham, we recall, developed a theory about iron-bearing brines moving into the rocks of Cumbria from the Irish Sea Basin and depositing haematite therein (an idea congruent with some of the suggestions described in Chapter 18). Haszeldine (Haszeldine & Smythe 1996, p. 138) now developed this hypothesis. He suggested that the iron was derived from what are now the red beds of the Permo-Triassic sediments of the Irish Sea hydrocarbon basin. Formerly the iron would have existed as ferrous ions in the reducing conditions accompanying the production of hydrocarbons. With a rise of pressure due to hydrocarbon production, the ferruginous (and saline and acidic) solutions could have been driven eastwards towards the sandstones of west Cumbria. The acid would have assisted in developing cavities (sops) in the Carboniferous rocks, and the ferruginous solutions would have penetrated wherever they could, in faults and joints. Then, mixing with oxidizing waters from west Cumbria, conversion to ferric iron occurred with precipitation of haematite, though chiefly in the upper strata, so far as BVG rocks were concerned. This was consistent with Nirex's claim that they had found small quantities of pyrite in some boreholes. However, the very existence of the west Cumbrian ore-field, with abundant ferric iron (albeit in insoluble form), was, for Haszeldine, at odds with the idea that the prevailing environment for the Fleming Hall Formation, at the top of the BVG, was reducing in character.

Haszeldine (Haszeldine & Smythe 1996, p. 173) sketched a diagram ('compiled from Nirex data') showing the variation of oxidizing and reducing conditions at different depths in the Sellafield area, based on the presence in boreholes of pyrite or haematite. According to his understanding, the depth for the proposed RCF or eventual repository put it well within the zone of oxidation, where higher-valency uranium compounds would be mobile. Pyrite might be 'ubiquitous' as the senior Nirex scientist Alan Hooper (at the time of writing, Chief Scientific Adviser) argued in his evidence to the Inquiry, but it was by no means the dominant mineral. However, in the view of Douglas Holliday, 'There's loads of pyrite there; there's been reducing conditions since the Triassic at the repository level'. The BGS kept telling Haszeldine this, said Holliday, but he did not want to take it on board (pers. comm., 1999). Michie (pers. comm., 2001) took the view that once precipitated, the highly insoluble haematite would not be relevant to the redox condition of the circulating groundwaters. And if pyrite were present at all, the solutions could not be oxidizing, as this mineral is unstable under such circumstances. Colin Knipe (pers. comm., 2001) reckoned that the evidence was not conclusive ('case not proved') as to whether the presence of pyrite or oxide indicated truly oxidizing or reducing conditions.

In fact, the question of the redox situation below Longlands Farm was one on which the geologists of the mid-1990s could not agree, and there were several letters on the topic in *Geoscientist* in 1997. Jenny Huggett (April, p. 6) pointed out that pyrite was uncommon and appeared 'corroded', which suggested that it was suffering oxidation. By contrast, haematite was relatively abundant at the depth of the proposed repository. Chaplow (June, p. 6) drew attention to a preliminary paper by Bath *et al.*, which emphasized the uncertainties of the matter. David Savage (June, pp. 6–7) of the BGS pointed out that at the pH of the groundwaters for the BVG at Sellafield, haematite could be stable in both oxidizing and reducing circumstances. Moreover, he took the view that pyrite was 'extremely common in the BVG beneath Sellafield', and he regretted that the abandonment of the RCF project meant that a more careful examination of the problem would not henceforth be possible. Haszeldine & Smythe (July, p. 19) contended that Nirex's borehole tests had revealed oxidizing conditions, but Nirex had '"corrected" these measurements to reducing values by assuming the existence of iron pyrites lining all the fractures within the BVG'. Huggett (September, p. 5) denied Savage's claim that pyrite was abundant, and pointed out that its maximum abundance was at a depth much greater than that of the proposed RCF. She added that Section 6C.164 (p. 203) of the Inspector's (Christopher McDonald's) Report had stated that 'there were considerable uncertainties about groundwater Eh, pH & ionic strength, especially at the preferred repository horizon'. BGS Director Peter Cook (October, p. 7), on the other hand, supported the view that at a high pH haematite might even 'form "reducing" fluids', and in any case, since the haematite had probably been there since the Mesozoic, this product of an ancient mineralization might not be relevant to a discussion of modern redox conditions. Finally, in December 1997 (pp. 4–5), some of the authors of 'Nirex 1995' (Report S/95/012, vol. 3), with others, wrote to say that the redox potential of the water within the rock pores would, following the corrosion of the container giving rise to the formation of ferrous ions in the repository, be determined not by that of the groundwater flowing into the repository but by the solid phases of the materials therein with which the groundwater might interact. It might, they averred, take more than a million years for dissolved oxygen and other possible oxidants brought in by the groundwaters to produce an oxidizing environment overall. Thus were some of the arguments within the scientific community.

There is insufficient space here to continue in this vein and discuss the myriad arguments that were presented at the Inquiry, or were generated by it, and I have only dealt with a few of the protagonists' contributions. Suffice it to say that, in responding to the questions posed to him by the 'Inspector', Colin Knipe broadly agreed with the arguments brought against Nirex, on the basis of the mass of evidence presented. He did, however, say that 'the Sellafield investigation ha[d] involved one of the most comprehensive and technically sophisticated pieces of geological characterization ever carried out in the UK and that the individual items of research and investigation ... [were] of high quality' (Annex 1 to Inspector's Report, APP/H0900/A/94/247019, p. 43).

The arguments about the original choice of Sellafield as a site were also rehearsed. As we have seen, Knipe objected to the manner in which the choice had been made. Various geological options were canvassed, among which the aforementioned 'basement under sedimentary cover' was one of the geological environments earlier recommended as potentially favourable by John Bredehoeft and Tidu Maini (1981).[34] This option is worth some consideration here. In their paper, the authors had in mind a large continental area of low relief, with hard, stable basement rock of low permeability below a blanket of sedimentary rocks whose groundwater flow characteristics were well understood and capable of being securely modelled; or a locality adjacent to a sea or ocean, with dipping basement rocks with a sedimentary cover that would serve as an aquifer to conduct solutions to depth below the sea or ocean, could be suitable.

Such circumstances occurred in some states in America, such as Colorado and Kansas, and Maryland. In Colorado, there are mountains to the west of Denver and Precambrian basement rocks extending eastwards towards Kansas. These have a quite thick topping of shale of low permeability, below which are permeable sediments that could act as an aquifer, but do not reach the surface until well to the east in Kansas. Thus any water that might pass through a repository in the basement rocks in Colorado would be widely dispersed before reaching the surface in Kansas. In Maryland, there are eastward-dipping basement rocks, above which are dipping sediments that could act as an aquifer, leading waters down below the Chesapeake Bay estuary and under the Atlantic (for review, see Michie 1998). The latter case is more similar to that in Cumbria, but whether or not the waters would exit before reaching the sea seems uncertain in both the Maryland and Cumbrian cases. Both envisaged inland masses of salt water that might divert fresh groundwater towards the surface, but for the Maryland case the Appalachians would be more distant from an envisaged repository than are the Lakeland mountains from Sellafield. Nirex contended that their modelling suggested that the exit would be somewhere under the Irish Sea. The description of the Longlands Farm site as a 'variant' of the 'basement under sedimentary cover' situation was, it seems to me, 'hopeful'.[35]

On the balance of the technical arguments, then, Knipe advised against the appellant's case. McDonald accepted Knipe's advice, and John Gummer accepted McDonald's recommendation that 'the appeal be DISMISSED'. Thus for the first time, the tide turned against the nuclear industry in Britain, in a David and Goliath battle. Chris McKeown has told me (pers. comm., 1999) that they never really expected to win. The 'establishment' usually does.

In talking to Bob Chaplow (pers. comm., 1999), I raised with him some of the issues mentioned in the preceding narrative. Regarding the faults in the BVG in the Sellafield area, the pattern of which seemed to change as research progressed, thus creating public unease, Chaplow explained that there was confusion at the Inquiry about the true significance of the faults (regardless of their precise pattern).[36] The flow of fluid did not necessarily follow the faults (or joints), which might have been 'sealed' by deposits of various minerals subsequent to their formation. Study of this problem was still in hand at the time of the Inquiry and evidence about it was tendered, but the results were (unfortunately for Nirex) not fully published until 1997 (Nirex 1996, 1997*d*, *e*, *f*), though a relevant paper by Sutton (1996) appeared earlier and the work was described in the 'proofs of evidence' (Alan Hooper, pers. comm., 2001). What the engineers did in so-called 'pump-tests' was reduce the pressure in different parts of a borehole (isolating different parts with water-inflated rubber packers) and monitor the groundwater pressure responses, either in the same borehole or in one adjacent to it. (The cross-hole pump-tests were started in March 1994 and continued for seven months.) Surprisingly, water was *not* found to follow the known lines of fault, which must, therefore, have effectively been sealed. However, that did not mean that the rock was impermeable. Rather, it showed that it could (at least on occasions) follow a maze-like network of unmappable narrow channels. So the discussions about faults and their alignments at the Inquiry *may* not have been wholly relevant to the envisaged storage area's reliability.

What Nirex found was that water flow tended to follow particular zones, with a characteristic kind of calcite mineralization, rather than clearly recognizable faults. However, even where the calcite was present, it was not always accompanied by flow. In such cases, its presence presumably indicated some earlier flow, and the calcified zone might be a 'potentially flowing feature'. Thinking along such lines, Nirex had been able to achieve results that meshed satisfactorily with their modelling. The technology used involved a high degree of expertise, the cross-hole testing for Boreholes 2 and 4 (shown schematically in Nirex 1997*d*, fig. 17), for example, being the most complex undertaken in the world at that time. Evidence about the borehole pump-tests was, as mentioned, tendered at the Inquiry, but the detailed analysis of 'potentially flowing features' and the conceptualization of faults was incomplete. Whether it would have made any difference to the outcome is a moot point. At the time of my interview with him, Chaplow rather doubted that it would. I am myself, however, inclined to the view that it would have assisted Nirex's case, while noting Colin Knipe's point that open channels might in fact assist subterranean dispersal of fluid (see p. 286).

In retrospect, the whole Nirex enterprise appears to have been in too much of a hurry. (The company had already in 1996 invited three companies to tender for the construction of the RCF; *Daily Telegraph*, 1 August 1996.) In 1996 and 1997 there was an outpouring of publications on the geological structure of the Sellafield site, on 3-D seismic surveys, hydrogeology, etc., with special emphasis on the permeability of rock in fault zones and mineralization in such zones, utilizing the results in the 'pump-tests' used in the boreholes.[37] A meeting was held at the BGS Headquarters at Keyworth in October 1997, co-sponsored by Nirex and the BGS, with sessions chaired by Drs Holliday, Moseley and Chaplow; and the latest results and computer programs were presented and discussed. The papers were subsequently published in December 1998, in a special issue of the *Proceedings of the Yorkshire Geological Society* (the Society again receiving a subvention from Nirex for the colour plates in the publication). Nirex had additionally published its work in a special edition of the Geological Society's *Quarterly Journal of Engineering Geology* (vol. 29, Supplement 1, 1996).

Thus Nirex was moving more into the public domain, recognizing that its previous *modus operandi* had not helped its case at Cleator Moor. It also began to send out its reports for external peer review, whereas earlier they had been scrutinized primarily by the special Nirex Panel of Review. However, one of the new referees, Ben Kneller, told me (pers. comm., 1999) that he found

[34] Bredehoeft was a senior hydrogeologist with the US Geological Survey. Maini was a Senior Research Fellow in Rock Mechanics at Imperial College, London. He had earlier acted as a consultant to the UK Atomic Energy Authority (Northern Division) on the question of the choice of sites for the disposal of high-level waste – a search for which was subsequently abandoned in Britain. I am informed by Uisdean Michie (pers. comm., 2001) that, in fact, Maini developed some of his ideas about 'basement under sedimentary cover' sites from his work at Sellafield.

[35] It should be noted, however, that Haszeldine & McKeown (1995, p. 87) described the Sellafield site as 'a variant of a long-recognized proposal for underground disposal (Bredehoeft and Maini, 1981)'; and their paper was reviewed by Bredehoeft, who presumably accepted the diction. (I thank Uisdean Michie for drawing my attention to this point.)

[36] It was unfortunate, though, that at the Inquiry the Nirex lawyers did not seem to have very clear ideas about faults. According to Haszeldine (pers. comm., 1999), he was asked during his cross-examination to comment on the extent of the fault between two boreholes on a diagram, but the particular diagram only showed the fault on one borehole. He thought at first it was a 'trick question', but then realized it was a 'misguided question', asked by a lawyer, not a geologist.

[37] For a list of the reports see www.nirex.co.uk.

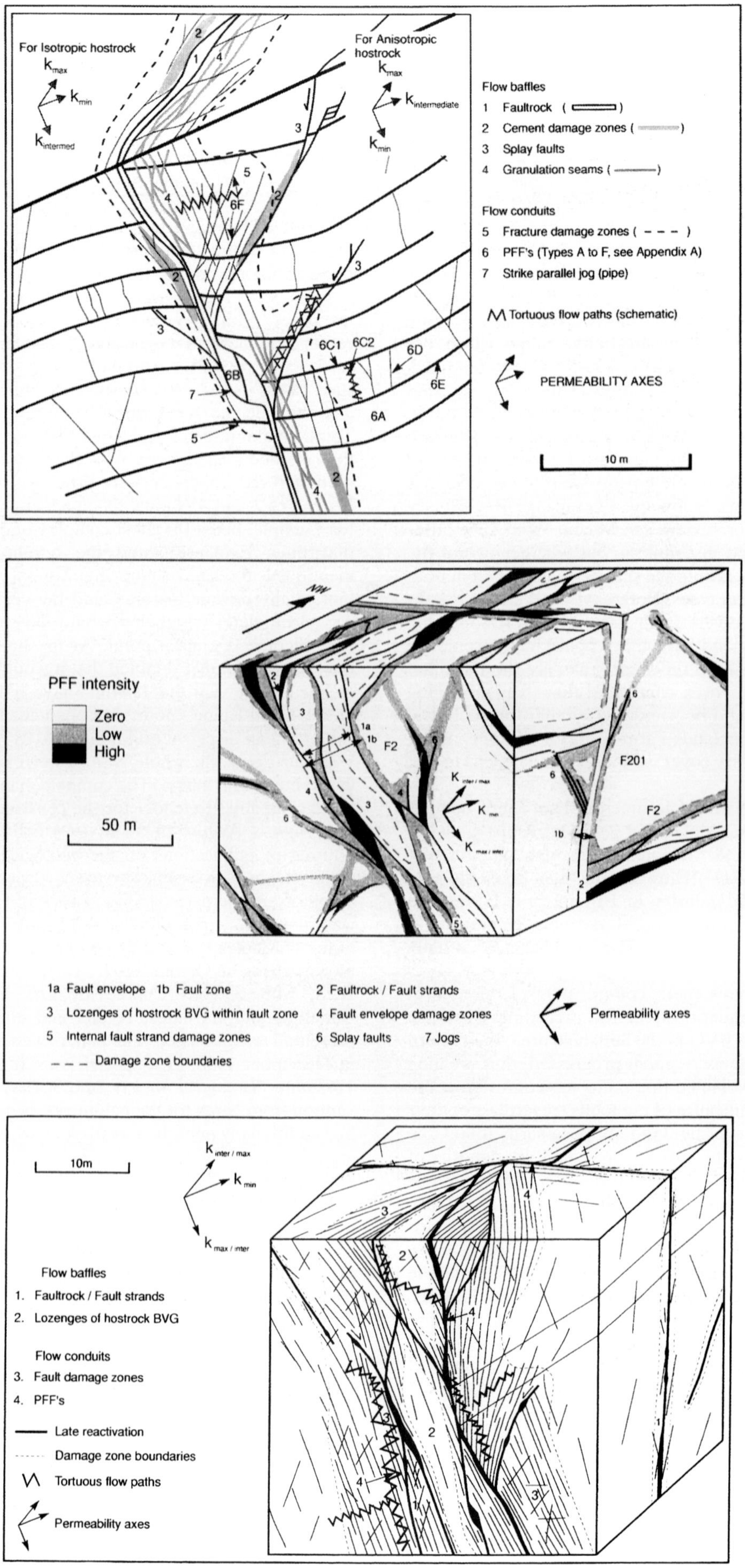

Fig. 20.3. Detailed models of fault structures in BVG rock in Sellafield area, near proposed nuclear waste disposal site, according to Gutmanis *et al.* (1998, pp. 166 & 167). Reproduced by courtesy of the Yorkshire Geological Society.

Fig. 20.4. Rock 'quality' and faults at locality of proposed 'rock characterization facility' (RCF), according to Davies & Chaplow (1998, p. 196 [black-and-white]); and Bowden *et al.* (1998, p. 135 [colour]). Reproduced by courtesy of the Yorkshire Geological Society.

the process a little odd. He was asked to look at work that had been done a couple of years earlier, and found it difficult to form a fair judgement, given that the science had moved on in the interim.

As mentioned, attention was now being focused on the faults within, and the 3-D structure of, the BVG for the proposed repository area. Thus Gutmanis *et al.* (1998), from Geoscience Limited, Falmouth, produced diagrams such as are reproduced in Figure 20.3, which show a fault zone in great detail, with 'potential flowing features' indicated. However, these figures were modestly called 'conceptualizations' and, as I understand from Chaplow, the block diagrams shown here do not in fact refer to an actual locality in the site intended for the RCF, or any eventual repository.

Chaplow with Nick Davies from Nirex (Davies & Chaplow 1998) also described their procedures for determining the 'geotechnical characteristics' of rock, from the huge database by then assembled by Nirex. The boreholes were imaged acoustically and by resistivity determinations, and the physical 'strength' and elastic properties of the different segments of the cores were determined; likewise the stress field in the Sellafield area. Measurements were made from core samples of compressive strength, Young's modulus, Poisson's ratio, 'effective' porosity, and saturated density. Measurements of Young's modulus, Poisson's ratio, porosity, bulk density, compressional sonic velocity, and shear sonic velocity were likewise made by wireline techniques. Geomechanical factors called 'tunnelling quality' (Q) and 'rock mass rating' (RMR) were calculated from empirical (engineering-style) formulae, and the whole was represented in a three-dimensional block diagram with different colours representing different values of RMR (Davies & Chaplow 1998, p. 196; Bowden *et al.* 1998, p. 135; Brereton *et al.* 1998, p. 211). Information derived from the trial 3-D seismic survey to identify the distribution of rock mass properties throughout the block was also included. So one could gauge which volumes of rock were most 'solid' – or, rather, of high RMR, with low apparent permeability and hydraulic conductivity – and thus likely to offer a good site for waste burial. This was technological progress that might be used some day, some place, but by 1998 no tunnelling or waste disposal was to take place in the rocks thus examined so intensively (see Fig. 20.4).

A team from the BGS also looked in detail at the mineralization in the fractures in the BVG under the Longlands Farm area, concluding (from the evidence of what mineralized, or penetrated, what) that there had been nine distinct episodes of mineralization (Akhurst *et al.* 1997, pp. 37 and 87; Milodowski *et al.* 1998). A summary of these is shown below, being related to supposedly known geological events as indicated:

1. (oldest)	Silicate mineralization	Pre-Acadian high-temperature hydrothermal mineralization
2.	Silicate and carbonate	Main Caledonian epithermal mineralization
3.	Sulphide (pyrite) and ?silicate	Fe–Cu–As–Zn–Pb mineralization (associated with mid-Devonian intrusions)
4.	Sulphate	Early post-Carboniferous anhydrite mineralization
5.*	Silicate	Kaolinite–illite–silicate mineralization
6.*	Carbonate ± sulphate	Carbonate-dominated mineralization
7.*	Silicate and oxide	Illite and haematite mineralization
8.	Oxide (iron and manganese oxides)	Oxidative groundwater mineralization
9. (youngest)	Carbonate ± sulphate ± sulphide (pyrite)	Calcite–sulphide–sulphate mineralization from Quaternary and recent groundwaters

Episodes marked with asterisks were taken to be associated with the expulsion of brines from the Irish Sea Basin during the progressive burial of the Carboniferous and Permo-Triassic sequences. Episodes 1–3 were found only in the Ordovician rocks, but the other episodes affected both Ordovician and younger rocks. Episode 1 was taken to be related to silicate re-mobilization shortly after the deposition of the BVG and appeared in all such rocks. Episode 2 appeared to be linked to the copper mineralization in the Lakes. Episode 3 was likewise associated, but was poorly preserved. No Lakeland equivalent was known for Episode 4. Episodes 5 and 6 produced the dominant mineralization in the fractures. It was suggested that the mineralizing fluids (brines) might have originated in organic-rich Carboniferous rocks in the Irish Sea Basin. Episode 7 might be linked with deep burial in the Irish Sea Basin during the Jurassic. Episode 8 was perhaps associated with invasion of meteoric groundwater during Tertiary (Miocene) uplift, which perhaps produced similar phenomena in Lakeland mines. Episode 9, involving fracture infilling at depth, represented recent changes, but not ones previously remarked in Lakeland proper, since pyrite and calcite would have been generally leached from the surface rocks under the conditions of high rainfall; or pyrite would have been oxidized.

It will be seen, then, that the Survey was making substantial advances in basic understanding of Lakeland and neighbouring geology through the Nirex work. It is interesting that the scheme outlined above was not dissimilar to that hypothesized by Haszeldine at the Inquiry, though Milodowski *et al.* (1998) had late pyrite (Stage 9) deposited over the older haematite mineralization (Stage 7), and their oxidative Stage 8 was thought to be restricted to the sediments near the surface (weathered sandstones and Quaternary sediments). However, the interesting results then emerging were not going to get the RCF constructed. It may also be remarked that, in the opinion of Colin Knipe (pers. comm., 2001), the information about the sealing of fractures was in a sense a 'two-edged sword'. In his view, for minimization of risk of nuclear materials, one needed dilution and dispersion. So one *could* argue that permeable faults and joints favoured this, giving longer flow paths. The less permeable the rock, the more vertical would be the flow paths. However, the actual flow of a faulted rock system would depend on the orientation of the faults, which therefore needed to be well understood.

Another issue I raised in discussion with Bob Chaplow in my conversation with him in 1999 was the fact that Haszeldine and McKeown had come up with different estimates from those of Nirex for the time that might elapse between water passing through the repository and its emergence at the surface or getting into drinking water (10 000 years[38] as opposed to a range of 25 000–50 000 years[39]), given that the Glasgow investigators had used Nirex data in their modelling. Had Nirex scrutinized McKeown's thesis and the Haszeldine–McKeown (1995) paper, I asked.

Well, naturally Nirex did look at these documents. They examined 'McKeown's paper' and 'thought it was an interesting study' (Chaplow, pers. comm., 1999), but Nirex objected that McKeown was using an average permeability from measurements from small-scale tests, inadmissibly using these to characterize large-scale features. Also, I was told, McKeown selected the lowest permeability for the overlying sandstones and the highest for the BVG; and he assumed that faults acted as conduits – which as we have seen above was a questionable assumption, though the issue of the sealing of the faults had not been thoroughly investigated and published at the time of the Inquiry. In addition, McKeown's computer modelling programs[40] might not necessarily have been adequate, suggested Chaplow. Nirex had in fact undertaken a study of McKeown and Haszeldine's groundwater modelling prior to the Inquiry, and concluded that the OILGEN program they used had 'severe limitations' in that the assumptions used in its modelling were such as to tend to give low transit times (Chaplow, pers. comm., 2001). Nirex used progams such as NAMMU (Release 6.2) for groundwater modelling, and NAPSAC (Release 3.0) for fracture network modelling (Hooper, pers. comm., 2001). (See further: Heathcote *et al.* 1996.) In any case, Chaplow pointed out, McKeown had only spent three man-years on the task, whereas the company had spent some 50 man-years with 'state-of-the-art' modelling and iterative procedures. Leaving that aside, there is evidently scope for disagreement as to which is the 'best' progam for investigations of this kind. (Of course, the RCF would, I take it, have facilitated the *testing* of the various computer programs!)

While Nirex wound down its investigations at Sellafield, Haszeldine and McKeown subsequently published the essential features of McKeown's thesis (McKeown *et al.* 1999). The paper arrived at a figure of 15 000 years for water to move from the region of the repository to the surface. Comparisons were made between modelled and empirically determined hydraulic heads in Borehole 2, according to three possible values for hydraulic conductivity in the BVG, and the best fit was obtained using a mean value of 1.2 m year^{-1}, from which value the figure of 15 000 years for water to reach the surface was calculated. No reference was made to Nirex's pump-testing results. McKeown's work was done *before* the pump-test information was fully published by Nirex (though it was discussed at the meeting on Sellafield geology held at the

[38] In his thesis McKeown (1997, p. 254) gave a figure of 15 000 years.

[39] Nirex (1995, vol. 3, table 2.3). Nirex also provided estimates of overall risk over time, suggesting that there would be peaks of risk (of about 1 in a million) at 20 000 years, and subsequently at over 1 million years after repository closure (Nirex 1995, vol. 3, pp. 9.1–9.2). The first peak would be due to the mobile radionuclide ^{36}Cl, which was thought to be likely to move more rapidly than water.

[40] OILGEN, 1989 and Geochemist's Workbench™, 1994.

Geological Society in May 1994, at which McKeown was present).[41] In any case, the mean value for flow in the BVG was thought to be the controlling quantity for the overall calculation, and would apply regardless of whether water flowed through solid rock, open joints, or ones sealed by mineralization.

After McKeown, Haszeldine had another PhD candidate working on the modelling task – this time a Chinese student, Kejian Wu, arrived from remote NW China, with a Master's degree in hydraulic engineering geology from Lanzhou University. He came to Scotland to work with Haszeldine in 1995 and successfully obtained his PhD at Edinburgh (Wu 1999), co-supervised by Gary Couples of Heriot Watt University.

Like McKeown, Wu used Nirex data, but sought to do the modelling on a much larger scale (*c*. 120 km long and 7 km deep) than that of McKeown's study. Needless to say, detailed knowledge of the structure of the whole area to the east of Sellafield was not available at depth through boreholes; it could only be inferred from the available maps and geophysical evidence. Despite such limitations, Wu (pers. comm., 1999) endeavoured to develop a general structural picture for the Lakes and link that with the detailed knowledge available for the Sellafield area. He took into consideration the low permeability of the rocks at depth and arrived at a figure of about 50 000 years for the possible emergence of water at the surface from the proposed repository site, which is not dissimilar to the upper end of the range earlier proposed by Nirex (1995).

Haszeldine (pers. comm., 1999) also continued his work on the hydrogeological problems at Sellafield, investigating the possible influence of the radiogenic heat from a repository in producing convection currents in the surrounding groundwater. Such a complication had been considered previously by Runchal & Maini (1980)[42] for high-level wastes and is standardly considered in textbooks. To my knowledge, it did not get into the debates at Cleator Moor, though *prima facie* it is an important consideration. So the end of the story was surely not reached by the end of the twentieth century, but my account has been carried far enough for present purposes.

The Nirex work at Sellafield was the most expensive 'single' geological investigation carried out in Britain, other than the North Sea oil projects. It generated a vast amount of geological knowledge about a small region, with collaboration between Nirex and a plethora of consultancies, of which the BGS has been chiefly of interest for our present purposes. Nirex's money assisted the publication of the BGS West Cumbria Memoir, and the research on the Quaternary of west Cumbria provided a considerable contribution to basic science (see Chapter 19). The same might be said of the contributions to knowledge of the BVG and of various mineralizations in faults and joints. While the flow of money from Nirex to the BGS lasted, some BGS staff that I have spoken to were only too happy simply to have it coming, to fund interesting research. On the other hand, one of them described the company as 'a somewhat maligned organization that deserves more credit than it sometimes gets'. The data collected by Nirex for their investigations at Sellafield were later (2001) transferred to the BGS at Keyworth, where they are likely to provide a valuable resource for future research. Indeed this is already happening. For example, Helen Reeves, a PhD candidate at Durham University, is, at the time of writing, working on a study of the relationship between the local stress field in the BVG and the flow of fluids in the rock fractures, making use of Nirex data (Reeves, pers. comm., 2001).

Even so, Nirex was, I think, in too much of a hurry. It hastened into its planning application and subsequent appeal before it had all the necessary weapons in hand. The procedure for the choice of site was, I suggest, inappropriate; but the truth is, it seems to me, that no site in Britain is wholly suitable. Perhaps the best site, geologically speaking, would be the Mesozoic clays running down under the sea from East Anglia, or possibly the underlying basement rocks.[43] But that would be politically unpopular, and there would be problems in transporting nuclear waste across country from Cumbria or wherever to East Anglia. Nirex were not necessarily wrong in searching for the solution that seemed most acceptable politically, and involved the least transport costs.

In my opinion, another source of problems was Nirex's status as a semi-private company. For reasons of 'commercial confidentiality', it did not at first subject its publications to peer review in the normal scientific manner; and then, when it did open itself to closer public scrutiny, the damage was already done. Smythe's involvement with Nirex led to substantial personal problems. And after the Inquiry, Nirex had to shed jobs (*Daily Telegraph*, 25 March 1997). Though initially granted a pay bonus, the Nirex Managing Director, Michael Folger, was replaced in April 1998 (Nirex news release, 9 April 1998). The BGS lost one of its major contracts, such that whereas Nirex provided £3.2 million in 1995–1996, this had declined to zero by 1999–2000, and there were staff cuts that financial year (Hackett 1999, p. 8).[44] Two geologists whom we have encountered in the present narrative, Eric Johnson and Brett Beddoe-Stephens, left the BGS at about that time.

On a broader front, Britain continues to accumulate waste and has nowhere to dispose of it permanently. Indeed, at the end of the twentieth century Britain had no ongoing search for an alternative repository site; and a company, Pangea, was unsuccessfully sounding out whether Australia would offer its desert spaces. Britain has, however, an immense amount of information (much of great interest to those concerned with Lakeland geology) about a potential hole in BVG rocks that will never become a reality.[45]

So, in this chapter we have entered into some detail about matters that may appear extraneous to the history of geological research in the Lakes. However, it is important to consider both the social and economic context of research and the social consequences that may flow from scientific work, or the manner in which it is applied. It is of interest for the present study that what started, in a sense, as an academic debate about the nature of faulting in the BVG and the structures of the ancient Lakeland volcanoes ended up as a matter of concern in a court of inquiry, of relevance to the safety or otherwise of a proposed nuclear waste repository. There were major long-term implications for Britain's energy policies, not to mention the economic and environmental well-being of the old industrial area of west Cumbria.

---

[41] The Nirex (1996) report on pump-tests was written in December 1996, and some pre-prints were distributed in February 1997, after the Inquiry had finished. See also Nirex (1997*d*).

[42] I thank Uisdean Michie for this reference.

[43] Little detailed knowledge of these rocks was available at the end of the twentieth century, but interestingly they appear to be of 'Lakeland' type: mid-Ordovician volcanics and late Ordovician–Silurian greywackes and slates.

[44] However, Douglas Holliday (pers. comm., 1999) told me that that the effect was not so serious as might be imagined. The Sellafield work tapered off rather than coming to an abrupt halt, and the Survey was able to take up other contracts that it had not been able to accommodate previously, while it was putting so much effort into the Nirex investigations.

[45] Or will it? A newspaper report in 2001 (Lean 2001) stated that the British Government was 'urgently drawing up plans to prevent opposition through public inquiries to the building of nuclear dumps and power stations, motorways, airports and other controversial developments. The plans, which could allow ministers to give the green light to hotly contested projects virtually by decree, pose the greatest threat to democracy in planning since the system was set up by a Labour government half a century ago.' The plans were said to have been 'sparked by frustration at the length of some fiercely fought public inquiries'. A proposed new Terminal 5 for Heathrow Airport was particularly mentioned, but finding a site for nuclear waste was evidently on the agenda again. The 'Davids' may never get another chance to use their sling-shots.

Chapter 21
Some concluding thoughts

What general conclusions, if any, can be drawn from this long story – which may well seem to get more complex and less secure as it approaches the end of the twentieth century?

First, who were the players in the drama? It is noteworthy, I think, that men, and north-countrymen at that, dominate the story to such a great extent. (Few women have been involved, Gertrude Elles, Louisa King and Cherry Lewis being the only ones who really mattered so far as my narrative has been concerned.) If we draw lines across England at about the latitude of Birmingham, Leicester and Nottingham, and from Newcastle to Carlisle, we find that just about all the participants in the story of Lakeland geology have come from within these boundaries, have resided there for extended periods, or have been geologists who attended or worked at universities therein. Think of, for example, Allen, Bell, Bott, Branney, Chadwick, Dunham, Firman, Fitton, Goodchild, Paul Green, Harker, Harkness, Hartley, Holliday, Jackson, Jeans, Johnson, Kneller, Kokelaar, Lapworth, Mackintosh, Marr, Millward, Mitchell, Molyneux, Moseley, Nicholson, Otley, Petterson, Rickards, Rushton, Sedgwick, Soper, Ward, Webb, Woodcock and Young. All these men had, have had, or have, northern associations. The relevant universities have been Birmingham, Durham, Hull, Leeds, Leicester, Liverpool, Newcastle, Nottingham and Sheffield. Two exceptions prove the rule. The first is Cambridge, where several staff and students have had Lakeland connections: Elles, Harker, Holliday, Hollingworth, Hughes, King, Marr, McNamara, Moseley, Nicholas, Oliver, Rickards, Sedgwick, Wadge, Woodcock and so on, although several of these are northerners. The other exception is Manchester, which has evinced surprisingly little interest in Lakeland geology. (Geology as such is not taught at Lancaster University or Bradford.)

The case of Cambridge may well have arisen from what evolutionary biologists call the 'founder effect'. Sedgwick started the Cambridge tradition of interest in Lakeland geology and began collections. Matters snowballed thereafter. Hughes and Harker had experience in northern geology and used to take student excursions there. The chief area of interest of Marr (whose northern speech was such that T. C. Nicholas recalled that he could scarce understand it when first he heard it) was the Lakes. Oliver, arriving from overseas, was put to a Lakeland topic. Researchers from the north, such as Rickards, obtained posts at Cambridge and continued their northern work. Louisa King hailed from the Cotswolds and was an undergraduate at Oxford, but she transferred to Cambridge for her Lakeland PhD. The seeming lack of interest in Lakeland matters at Manchester University seems anomalous. However, its Lower Palaeozoic geologists concentrated on Wales. In its early years, the *Transactions of the Manchester Geological and Mining Society* focused on mining matters.

By contrast with Cambridge, we find rather little direct interest in Lakeland geology at Oxford and virtually all the southern universities, though some Oxford 'high theorists' such as Dewey and McKerrow took Lakeland geology into account in their thinking, and Andrew Chadwick and Louisa King were undergraduates there. Michael Nutt did his PhD at Queen Mary College, London, and there were a few others; but it seems surprising that Hollingworth, who did much Lakeland work in his younger days, did not keep it up after obtaining his chair at UCL, and only had one student, Jacques Konig, working in the area. Hollingworth, hailing from Northampton, came from just south of the imaginary line that I mentioned above. The chief London input to Lakeland geology came from the Scotsman Simpson and his students at Birkbeck.

Simpson got into Lakeland geology from Glasgow via the Isle of Man, but it seems that he did not fraternize much with other Lakeland geologists, and was disinclined to go into the field with them to argue things out on the ground. His theories, stimulating though they were, did not attract wide assent. It may be that, because he was somewhat isolated from the northern community of geologists, he developed ideas different from those of his contemporaries and failed to persuade them to accept his views. Was this because of northern 'tribal solidarity', or because Simpson, being geographically removed from the scene of action, was working somewhat in intellectual isolation so far as the Lakes were concerned? Whatever the case, Simpson was seemingly a somewhat lonely figure in Lakeland geology.[1]

More remarkable still than Simpson's case was the response to a paper by Peter Banham and Frederick Hooper of Bedford College, London University, co-authored with J. B. Jackson (1981) from the Survey. Seeking to determine the 'way-upness' of beds by examining sedimentary structures in quarries in the Skiddaw Slates near Cockermouth, they claimed evidence for inversion and thrust faulting, and proposed the existence of what they called the 'Gillbrea Nappe' in the Skiddaw Slates, named after High Gillbrea Farm, on the eastern side of the Lorton Valley, SW of Cockermouth (see Fig. 12.1). This suggestion, from an 'outside' group, was not thought to be strongly argued or empirically well supported and hardly worth a published refutation by the northern geologists, so I have not chosen to mention it hitherto. However, the manner in which the paper was 'excluded' from Lakeland geology is perhaps significant, illustrating the generalization that northern geologists have been largely responsible for the development of Lakeland geology. This is perhaps surprising, considering the small size of Britain. Yet it meshes with the persistent local divisions in England. Indeed, the Birmingham–Leicester–Nottingham line roughly coincides with a line of speech difference (and, who knows, even an ancient tribal boundary?).[2]

'Tribalism', of course, also meshes with the proprietoriness amongst geologists in some places. The Coniston Limestone was deemed to be T. C. Nicholas's territory, and G. H. Mitchell was made to feel that he should be looking elsewhere for things to do. In Switzerland, I am told, the geologist should stick to his own canton, or 'his' valley, if he wants to avoid trouble. So the Lakeland story occasionally illustrates the notion of geological trespass. J. F. N. Green was an 'outsider' too, and found his first paper rejected, and his later work, though influential, was brought into question, and to some extent into disrepute. The same might be said of Elles's stratigraphy. She perhaps had a double problem, being both a woman and a southerner, though she was well situated in the Cambridge citadel and was highly regarded as an authority on graptolites. Marr had structural ideas no less idiosyncratic than Green's, but they came in for little criticism (except from Green!). Marr's idea of lag faults was ignored in the Lakes, rather than pilloried. Likewise, Moseley retains great respect, though his notion of thrusting near Ullswater no longer attracts general support.

[1] My efforts to contact his former students have only met with limited response.

[2] I hope I can say all this without fear or favour! My paternal grandparents came from West Yorkshire. I was reared in Hertfordshire and Bedfordshire, and have spent most of my working life in New Zealand and Australia. Of course, Banham, Hooper and Jackson's work was not rejected because the authors came from southern institutions. However, their publication was not embedded in the tradition of the 'northern' research, and perhaps for this reason it did not cohere with the ongoing Lakeland work.

However, the 'parochialism' of Lakeland geology may have a less esoteric explanation. Though much studied and loved, the Lake District is a 'postage-stamp' area so far as regional geology is concerned. The issues there were not sufficiently significant to attract a worldwide audience – in contrast with, say, the Auvergne volcanoes in the past, or the Chicxulub Crater in Mexico at present. So perhaps the 'southern' geologists were content to leave their 'northern' colleagues to sort things out.

Thinking about personnel from a different perspective, I would emphasize the importance of research students in the history of modern geology (or indeed in research schools in science more generally). Besides the Survey staff, they are the people with the time and energy to go over the ground in detail with relatively few interruptions. Skimming the titles in my bibliography when converting it at the end of 2000 from a date-based to a name-based order, I noticed that I could readily see from the titles whether I was sorting a PhD thesis or a journal article. The former tended to refer to the mapping of an area, the determination of its stratigraphy and structure, with focus on some particular aspect of geology – sedimentology, graptolites, geochemistry, structural geology, or whatever.

By contrast, the university staff tended to publish on more compact subjects, though they naturally supplied much of the theoretical and conceptual background for their students' work, as did Soper for a string of PhDs (although Branney went off on his own to some extent and came up with novel ideas that shook up the whole of Lakeland geology). The PhD students, then, have traditionally worked on a mapping area as their 'apprenticeship', just as the physics or chemistry student may work with a special piece of apparatus, funded by his or her supervisor, and according to the supervisor's instructions. The apprentice students, whether in geology, physics or chemistry, are commonly the principal data-gatherers and calculators, and have to 'tweak' the apparatus, computers or programs. The traditional approach to a geology PhD, based on mapping, seems to me to have much merit. Many uncertainties are involved in drawing a geological map and both skill and knowledge are required. PhD candidates learn the practical 'craft' and social skills of geology through their mapping exercises. There is risk of theoretical straying if the fieldwork is not firmly map- or data-based, or if published maps are unavailable.

Once a student has completed a PhD there may be a tendency to stay in that general area (either geographical, technical or conceptual), building on the capital generated by the doctorate. Several of the PhD students, such as Millward, Webb or Branney, have remained intimately involved with Lakeland geology and with the types of problems they initially encountered there, long after they completed their degrees. Geologists sometimes retain their interest and expertise in the areas of study in which they started in their twenties throughout their careers. Thus Robin Oliver, when I met him in his seventies in Adelaide, was still concerning himself with semiprecious or precious accessory minerals – rather as he was interested in Lakeland garnets in the 1950s.

What about the Survey? Its staff mostly have their PhDs before they enter the service, but the BGS traditionally has not taken so much pleasure in specialization as have the universities. So, for many years it was the custom to put recruits into areas different from those in which they served their doctoral apprenticeships. In a case mentioned, Kingsley Dunham, with his interests in mineralogy and ore bodies, was initially put to work in the gentle pastures of the Old Red Sandstone. Tony Wadge did his main northern mapwork in the Cross Fell Inlier, but also published on such matters as acritarchs, economic geology and radiometric dating. Such eclecticism may seem ill-advised, but the practice had virtues. The nineteenth-century Surveyors such as Peach and Horne could put their hands to most things and their versatility was invaluable to their institution. In the twentieth century, however, specialization inevitably increased. Michael Lee does geophysics and Stewart Molyneux does palynology; there is no possibility of their swapping roles. The Survey now makes appointments in narrower fields than formerly, such as volcanology or metamorphic petrology. Specialization is inherent in the development of science (Menard 1971) and more of it might have been beneficial in the Lakeland Project. However, the Project was, in fact, specifically designed to bring together various specialists, thus having the advantage of the particular areas of expertise and a 'collective general knowledge'. This was a highly positive feature of the Lakeland work, post-1982.

What is a Survey? What should it be doing, and how? The BGS is, of course, a Government institution and is run according to departmental regulations. Today it is located somewhat awkwardly under the aegis of the Natural Environment Research Council. Its remit is to produce geological maps of Britain, and appropriate memoirs that explain the regional geology to other geologists and interested readers.[3] Its skills are available to advise governments on matters geological: mineral and agricultural resources, highway engineering, earthquake risks, etc. Some of the staff have been, could be, or became university staff. Some relish administration and committee work. Others dislike bureaucratic activities and just want to solve geological problems, get into the field, and make maps. Its libraries are open to the public and constitute a major national resource and geological archive.

As we have seen, the Survey did not make the progress in mapping that might have been expected in the 1960s and 1970s, when France, for example, was pushing ahead with its programme of updating its geological maps. So the British Survey came in for criticism, both from industrialists who sought geological information, and from university men such as Bernard Leake who thought it deplorable that in the 1980s the most recent Survey maps were in some cases hand-coloured productions, over a hundred years old. The Survey endured endless 'restructurings' and, with the staff constantly being shifted, work did not get completed on time. Much of the geology of northern Britain was not known in detail empirically, or understood theoretically, at the beginning of the 1980s, as is evident from the immense amount of information that rapidly emerged after the initiation of the Lake District Project.

So, as we saw in Chapter 14, there came first Malcolm Brown's multidisciplinary projects, and then the Survey–universities collaborations, intended to expedite the land survey programme by pooling national expertise. The scheme did work, up to a point; but there was also trouble. Some of the problems were personal rather than structural; but there were also structural problems. The goals and interests of the universities and the Survey were not always one and the same. The Survey expected that the university collaborators should supply information to it, but would not have control over how that information was used by the Survey (though under the terms of the initial agreement the university workers had intellectual property rights and could publish their work how and where they wished). This arrangement had the potential to ruffle both sides of the collaboration. The Survey staff thought, doubtless correctly, that some of the university workers' ideas depended on knowledge obtained from Surveyors' work. Not surprisingly, the university workers wanted to have a say as to how their data were utilized in the Survey maps.

All this could irk. Historically, the Survey did not always like to receive suggestions as to how maps and sections should be drawn or how the stratigraphic column should be subdivided. 'It' (as an institution) had for long been the arbiter about such matters. 'It' produced the *official* maps. Thus there was a history of 'it' sometimes not wanting to receive or take advice from outside parties – as in the cases of Lapworth in the Southern Uplands, or Kendall on the question of haematite deposits and the Cockermouth Sheet. So suggestions from Liverpool for changes to the

[3] The (expensive) Memoirs are now being phased out, as I understand.

Ambleside map were not, it seems, invited. However, the university men did not like to see alterations made to their lines without consultation. Again there were differences in aims and constraints. The university geologists wanted the finest, or maximum, subdivision of the BVG possible, in order to further their understanding of volcanic processes. The Survey had to look at the matter from a regional perspective, with an eye to extending the BVG stratigraphy onto adjacent maps, and in accordance with international conventions for stratigraphy. (I am not saying that the university people overlooked such considerations.)

There was also the question of joint supervision of the PhD programmes. I do not know how those worked out in specific cases, but it seems there was the possibility of tension, especially if the supervisors from within and without the Survey had disagreements, though ideally different ideas or approaches should complement one another. PhD students, though relatively inexperienced, are expected to do their work as seems appropriate to them, within the limitations of time and money available. They have 'problems' to solve. The map is a means to an end, not an end in itself. And the map for a PhD dissertation is certainly not expected to be a neat rectangle! In the case of the Lakeland Project, the ultimate goals or products were institutionally different: for the PhD candidates a thesis with an argument, and data offering new knowledge; for the Survey a map embodying the most accurate and up-to-date knowledge possible, plus an explanatory memoir synthesizing the geological knowledge of the area concerned. Working towards different objectives, difficulties could arise.

Despite the foregoing, some collaborations worked much better than others. All people I have spoken to expressed satisfaction with the outcome of the Snowdonia programme (attributing much of its success to the efforts of the Surveyor, Malcolm Howells), which in fact involved collaborations going back to the 1960s when Kokelaar was doing some work there for his first degree, and Peter Allen, recently back from Hong Kong, was working on the same (Harlech) Sheet. Kokelaar recalls (pers. comm., 2000) the excellent rapport in the Snowdonian work, with ample frank discussions and exchanges of views and information.[4]

In the Lakeland effort, the work of Jack Soper was also what was wanted. He was able to work with his doctoral students, ensuring that they got their degrees, while collaborating successfully with Survey staff, as for example when he came together with Michael Lee to devise a model for the structure of the Lakes, compatible with field evidence, theory, and geophysical data. Moreover, Soper had the range of interests and abilities appropriate to collaborative commitments. He has seemed to me to be equally at home with quarrymen, Lakeland farmers, Oxford professors, Surveyors, graduate students – even Australasian historians. He produces maps that are both things of beauty (e.g. the Ulverston map) and productions technically acceptable to the BGS. He also ranges into high theory, and it was probably through his application of terrane theory that the Lakes provided its most important contribution to the understanding of the geological structure and history of Britain.

Despite such successes, and the maps and numerous publications generated by the Lakeland Project, the Nirex inquiry throws a long shadow over my narrative. It raises the question of sources of Survey and university funding, and governance, in a world of 'economic rationalism', and the ethos of science that such 'rationalism' may engender. It also raises the question of the social responsibilities of scientists, the social practices of science, and the interactions of science and society. There is even the old question, reputedly first raised by the Roman satirist Juvenal: 'Who shall guard the guardians?'

In 'traditional' sociology of science, as understood according to the scientific 'norms' enunciated by Robert K. Merton (1942, republished 1973), the scientific community should be one of open inquiry, in which 'organized skepticism' prevails, and in which the knowledge acquired is made available to all. Ideas are proposed, and scrutinized by scientists' peers: in conversations, in meetings, and by refereeing. After publication, ideas continue to be tested, especially if thought suspect. (If regarded as wholly suspect, they may simply be ignored, as happened on rare occasions in the story of Lakeland geology.) The community should be one of openness. Most sociologists of science think that, for late twentieth-century science with its intertwinings with industry, the military, civil engineering, etc., Merton's norm of organized scepticism was more of an ideal than a reality, but it continues as an ideal – or perhaps a rhetorical weapon – amongst scientists. It is incompatible with 'secret knowledge'.[5] Such an open approach to science is deemed appropriate to an 'open' and democratic political culture. It is noteworthy that in a (semi-) 'closed' society such as China, even today, the public cannot get easy access to geological or topographic maps. (Even the historian of geology in China may have difficulty in getting access to geological maps.) Also, as I discovered at a seminar with the 'Science and Society' group at Peking University in 1999, interested Chinese academics did not know where nuclear waste is buried in China (or perhaps they chose not to *admit* to knowing).

As said, even in 'democratic' countries, the idealized 'Mertonian' situation is evidently in decline. The intimate connection of science with the military, and the concomitant secrecy, evidently has been in part a product of the Cold War, though of course it goes back long before that. The connection with business is also of long standing, though the patent system has traditionally served as a means of maintaining openness and technological progress. In democratic Britain I can at least purchase geological maps of the Lakes as and when they are published, and I can talk to people. Everyone in Britain who cared knew that Sellafield was the likely site for nuclear waste disposal. The proceedings of the Cleator Moor Inquiry were held in open session, and were reported extensively in newspapers. The forthcoming general election doubtless concentrated Mr Gummer's mind wonderfully when it came to his consideration of the Cleator Moor Inquiry's report.

Direct funding for institutions such as universities and the BGS has been cut. Why? Because of the rise of power of business, and the public's dislike of paying large taxes for things perceived to be of remote benefit. The British public wants safety, health, convenient transport, and entertainment. It will pay, therefore, with relatively little objection, for military spending, health services, roads, sport, etc., but it looks for taxation-restricted options as much as possible. The unhappy privatization of the railways was partly ideological, but was driven by a public desire for good railways to be had on the cheap. 'Industrial' agriculture has also offered the possibility of plentiful and cheap food – but had disastrous concomitants with the serious outbreak of foot-and-mouth disease in 2000. In such a context, the geological mapping of Britain has not stood high in the scheme of things, so it is unsurprising that interested parties looked for expedient remedies. Under financial pressure it was natural that the universities and the BGS might seek to form links, both with each other and with industry.

The situation of Nirex in all this is interesting. It evidently had

[4] As I understand, Kokelaar's collaboration with Howells was not, however, a formal Survey–university collaboration. Kokelaar had his own scholarship from the University of Wales.

[5] Merton recognized that the social situation was different for scientists in commerce and academia, but thought that the fact that some scientists felt their work was in some sense compromised in the world of commerce was an indication that his norms *did* exist in the appropriate social (academic) setting. One may suspect that this situation no longer holds: scientists feel perfectly 'comfortable' doing 'classified' work in the military or for business corporations – given their remuneration perhaps?

large funds at its disposal, in the long run arising from the long-standing nexus between the nuclear energy industry and the military, which developed during the Cold War. At that time, military concerns were paramount, and led to the extensive development of the nuclear energy industry, without too much thought given to the problem of the eventual disposal of the nuclear waste. When this problem could no longer be ignored, a peculiar solution was devised: the establishment of a *commercial* organization to find a means for the disposal of waste, rather than handing the problem directly to the BGS, the Government's chief agency pertaining to the earth sciences. As we have seen, this policy led to the emergence of a complex tangle of contractual, subcontractual, and even sub-subcontractual arrangements. The Thatcher era was one that fostered private business, supposedly because it encouraged efficiency. Maybe? In the case we have been examining, it certainly encouraged complexity.

As a business organization, Nirex was able to shelter to some extent behind the cloak of commercial confidentiality. Thus, as of 2000, the BGS Report WA/95/47C remained confidential so far as BGS was concerned.[6] So the scientific questions relating to the burial of nuclear waste were certainly not dealt with in the routine 'Mertonian' manner. Nirex reports were prepared internally, and a plethora of consultants was engaged, with the BGS being, as we have seen, one of the organizations getting sustenance from this rich resource. As to availability of information, the Nirex website announces that copies of its reports will be sent on request, but I have found that requests have not been answered directly – only when special enquiry was made via Bob Chaplow. However, some of the reports have been put out in the form of CDs, one of which Chaplow kindly gave me when I visited him in 1999, along with some other reports.

We have remarked that it was only some time after the Nirex reports were prepared that they were submitted to external refereeing, and then for fees.[7] As we saw, Ben Kneller found it difficult to undertake such a task, as the reports he looked at were about two years old, by which time it was difficult to assess them, as the science had moved on. Louisa King told me that the consulting firm she worked for in Leicestershire after graduation employed her to assess Nirex work. Many other people must have been employed similarly. It is a kind of work that is foreign to the normal peer-group assessment of academic scientific work (though publishers may now reluctantly pay for books to be refereed). Nirex was doing scientific work for commercial purposes, and the usual interplay of scientific assessment procedures was clouded. The 'guardians' were *paid* to 'guard'.

Such considerations notwithstanding, the debate was conducted in a democratic and tolerably open society, so that when the challenge came to the Nirex planning application, and in the subsequent appeal before an official inquiry, the issues were debated in open court, with the opponents able to present any technical arguments they pleased. Even so, matters were not handled in the usual scientific manner, the circumstances being adversarial. Witnesses accused each other of overlooking or misinterpreting evidence, or of misrepresenting their opponents' views. Such things may be said these days in private emails, or on occasion in discussions at conferences (or more likely at the bar after a meeting), but we do not find them in scientific papers, or if they are there, the language is 'encoded': 'On the evidence presented to date, X's results would appear to be artifactual'!

The philosopher of science, Larry Laudan (1982), has expressed concern about such states of affairs so far as philosophers are concerned, pointing out that philosophers acting as expert witness (as in the case of the 'Arkansas trial' relating to the teaching of evolution) are obliged to operate differently from that which is customary in the philosophical community. Similarly, in a court where scientific or technical issues are argued, expert witnesses are engaged to produce evidence that favours one side or another of a legal battle. Scientific expert witnesses doubtless act with integrity, using their expertise to help oppose what they regard as socially or technically ill-conceived proposals (or vice versa). They will (or should) offer objective opinions. But their expertise is engaged to favour one side of a case. A court-room constitutes a non-Mertonian situation.

In the case discussed in this book, Haszeldine, Smythe and Kokelaar were key witnesses for the anti-rock characterization facility position. For different reasons, all three were scientifically concerned with what they had seen and knew of the Nirex plans. Kokelaar doubted Nirex's understanding of complexity of volcano-tectonic faulting in relation to the BVG, and the details of the fault structure in the proposed repository area were, he thought, insufficiently known. Moreover, in the context of a caldera-collapse structure, detailed knowledge of some particular volume of rock might not be applicable to another one a few hundred metres distant. Smythe thought that Nirex had been too reliant on 2-D seismic survey. Haszeldine and his student McKeown used Nirex data that led them to different conclusions from those that the company had reached and presented to the Inquiry. They also differed from Nirex (and BGS) on the geochemistry of the site (the redox conditions).

Here, then, were three different scientific reasons why witnesses, with first-hand knowledge of the geology of the Lakeland volcanoes or the Cumbrian coastal region near Sellafield, took on the case. Had they not done so, the outcome of the Inquiry might have been different; in which case, the energy policies or 'arrangements' for the United Kingdom might have been substantially different. The course of history might have been different!

The irony of the situation was, of course, that had Nirex been less precipitate it might have won its case, as the arguments about the sealing of fractures by mineralization were incomplete at the time of the Inquiry, some of the relevant reports only being published in 1997 (e.g. SA/97/028 and SA/97/036). Nirex lawyers might have successfully persuaded the Inspector that the company's arguments were sound if everything had been refereed and published at the time of the inquiry. On the other hand, despite the sealing of some fractures, the witnesses opposed to Nirex at Cleator Moor were able to show that there were uncertainties in Nirex's understanding of the pattern of faults in the Sellafield area, and thus its hydrogeological modelling was open to doubt. The circumstances there were such that one could not be sure that the sealing of fractures at one locality would apply over a wider area. Or, as Colin Knipe told me, the question of 'sealing' might be a two-edged sword. Possibly a lack of open fractures could cause water to rise towards the surface?

So, for better or worse, the Nirex case was lost, though whether permanently one cannot say at present. However, without trying to peer into the future in this regard, we can say that many of the arguments tendered at Cleator Moor could only have been made in the light of the information assembled as a result of the Lakeland Project and the Nirex–BGS investigations in the Sellafield area. To that extent, the study of Lakeland geology was a matter of national significance. As a concomitant effect – though one that most of the world would scarcely notice – the BGS lost important contractual work, and partly for this reason it had to make some retrenchments.[8]

[6] C = 'Commercial'. I am informed, however, by David Holmes of the BGS (pers. comm., 2000) that information from this report (on the three-dimensional geological structure of the proposed repository site) was made available in Nirex Report SA/97/026.

[7] There were, as we have seen, earlier specialist advisory panels, presumably suitably remunerated.

[8] One can also report that the Surveyor who urged that the Liverpool group be allowed to comment on the proofs of the Ambleside map and chose to participate in the international volcanological workshops run in the Lakes (and elsewhere in Britain) by Kokelaar, Branney and others, was one of the unfortunates to be retrenched.

Setting all this aside, what can we say about what is now known of the geology of the Lake District, after two hundred years of effort? First, we should remind ourselves of the smallness of the Lake District on the global scale. Taken in isolation – as admittedly I have tended to do in the present account – one cannot deduce the history of the region, or understand all matters geological therein, purely on the basis of work within the geographical region of the Lakes. Further, I have concentrated my account on the problems of 'Otley I', 'II' and 'III', with rather little mention of the bounding strata of New Red Sandstone and Carboniferous rocks, over which we skipped rather hurriedly to a consideration of Tertiary and Quaternary matters.

Even with these limitations, though, there has been much to consider. The Lake District has been subjected to extraordinarily detailed scrutiny. We recall that, according to Surveyor Brian Young, virtually every outcrop was scrutinized during the remapping from 1982. Millions of individual field observations must have been made; millions of instrument readings taken in the processes of geochemical and geophysical analyses. Thousands of maps have been drawn. Millions of words have been written; more millions spoken at meetings. The area around Sellafield has probably received as much or more geological attention per square (or cubic) metre than anywhere in the world. Hundreds of millions of pounds have been spent, with some £20 million going to the BGS for Nirex work since 1984 (David Holmes, pers. comm., 2000).[9]

What do we know as a result of this extraordinary effort over two hundred years? What new theory has emerged specifically from the Lakeland work? To answer the second question first, one may suspect that the answer is 'not very much'. The distinction between bedding and cleavage was probably first made clear in the Lakes, by Sedgwick, or Sedgwick and Otley, but it was very likely known to quarrymen long before. To my knowledge, the concept of lag faults was initiated by Marr for the Lakes, but it is not an idea that has wide application. Graptolite stratigraphy was imported from Scotland. The use of sedimentary structures to determine way-upness came from Wisconsin. The concept of olistostromes came from Italy. The idea of terranes came from California. The concept of subduction came from the work of Hugo Benioff and others. The category of 'ignimbrites' came from New Zealand. Fission-track dating, as used in the Lakes, was largely developed in Australia. The idea of land-ice glaciation came from Switzerland, and the notion of four glacial epochs in the Pleistocene, developed in the northern Alps and Pyrenees, was applied to the Lakes, albeit with only limited success. The ideas of caldera collapse and volcano-tectonic faulting came from a number of locations, including Pantelleria. Gravimetric, geomagnetic and seismic techniques were developed in various places, and with various instruments in the 1930s and earlier, well before they were applied to the Lakes by Martin Bott and others. Andesites were associated with island arc volcanism by Tuzo Wilson. Basin formation theory was developed in abstract form by McKenzie at Cambridge. The idea of mountain fronts was developed and summarized variously by such authors as Vann *et al.* – not Lakeland workers.

To be sure, Martin Bott developed computing techniques to help solve computational problems for gravimetric work, and abstract ideas about the behaviour of granitic masses in the Earth's crust, from long years of experience with the gravimetric anomalies to be found in various parts of Britain, and particularly northern England including the Lakes, but by and large this region has not been one that has led to the development of major new theoretical concepts in geology. The exceptionally well-preserved and exposed silicic caldera of the Central Fells of Lakeland is important, however, as it provides a kind of outdoor laboratory for the study of what goes on within volcanos, and their structure and development. Thus in the last few years there have been several international field workshops where volcanologists have come together in the Lakes to improve their understanding of such matters. The participants have been shown the criteria developed in the Lakes for the recognition of volcano-tectonic faults, and the unusual volcano-sedimentary environments to be seen in the Central Fells with previously unrecorded lithofacies associations. They have been able to study the complex geometry of a collapsed caldera with arcuate graben structures; and have seen how the analysis of pyroclastic facies undertaken on modern volcanoes can also work satisfactorily for the study of ancient successions. They can see a whole range of pyroclastic rocks, from 'ordinary' tuffs through to materials little different from rhyolites. All of this is important, though not such as to change the face of theoretical geology.

For the geo-historian, we can use the area as a place where the development of different 'styles' in geology can readily be discerned. First there were the miners and quarrymen, whose ideas were linked together by the remarkable Jonathan Otley. Then came Sedgwick with his hammer, compass, notebook and hand-coloured maps: the 'amateur' pioneer, chiefly looking at rocks and their structures and trying to make sense of the 'bloomin, buzzin confusion' of them all. In his later Lakeland work, he attended more to palaeontology, began the establishment of a stratigraphic column for 'Otley III', and established that the Skiddaw Slates contained fossils, which helped his Cambrian to be characterized palaeontologically. The first Surveyors walked the ground and coloured in maps chiefly by lithologies and according to Murchisonian stratigraphy for 'Otley III'. Unfortunately, Ward's petrological theory for 'Otley II' was not sustained so that the hand-coloured maps of the primary survey were soon obsolete, as well as being few in number and expensive. However, the outline of the different outcrops was established and the main faults were delineated (though gross uncertainty about the nature of the SS–BVG contact remained).

Harkness, Nicholson and Marr continued Sedgwick's work, until a Sedgwickian–Lapworthian stratigraphy was established for 'Otley III'. But Marr's structural ideas were unorthodox, and as he confessed to his notebook he was uncomfortable working on igneous rocks. Harker found the presence of acidic and basic plutonic rocks adjacent to one another at Carrock Fell interesting, but he pursued such matters chiefly in Skye and elsewhere.

There is little doubt that subsequent progress in Lakeland geology was severely disrupted by the two World Wars and the Great Depression. Geologists were put onto work of economic rather than theoretical or abstract significance, and some geologists, such as B. B. Bancroft, were killed in action. Teaching and learning were disrupted, so that there were staff shortages, even at Cambridge. T. C. Nicholas could not get back to complete his early Lakeland work in the inter-war years. Looking through Survey archives at Keyworth, we find that much emphasis was given to contributions to the War effort in the 1920s job applications. That generation of geologists was not necessarily well trained to undertake the difficult task of remapping and reinterpreting the geology of the Lakeland mountains, and in fact the task was hardly attempted. So far as the geology of the Lake District was concerned it was only the Cumbrian coastal region, with its coal mines and haematite deposits, that received serious attention, though parts of the northern Lakes were also re-examined. The Survey had to enlist the assistance of Gertrude Elles to get its palaeontological work done on the Skiddaw Slates.

So during World War I and the inter-war years, such work as was done was chiefly in the hands of amateurs such as Green; a few students without much supervision, such as Hartley and Mitchell; and a Surveyor such as Mitchell doing Lakeland work in his spare time. The result was that for years there was no systematic attack upon the problems of Lakeland geology, and different investigators tried to develop their own stratigraphies for the BVG in restricted areas, and apply them to the volcanics as a

[9] Holmes is Director for Environment and Hazards, BGS–NERC.

whole, from the Duddon Valley across to the Haweswater region and the Shap district. So we find that, through even to the 1980s, the mapped areas of the Lakes were like a patchwork quilt, with different workers taking on different pieces of the whole, such as seemed to them to be interesting. Higher degrees were obtained, however, and with the work of Oliver (1953) a significant breakthrough was accomplished with the establishment of a general stratigraphy for the Central Fells and the introduction of the category of ignimbrite to Lakeland geology. It is remarkable that it took two young men from New Zealand (Oliver and Battey) to bring forward this petrographic innovation, given that the fells had been walked over by geologists so many times – but the area had not been remapped in a systematic fashion. Up in the north, Dennis Jackson seems to have been an (impressive) 'autodidact', so far as working out a stratigraphy for the Skiddaw Slates was concerned, and he too, as a PhD student, made significant advances. Moseley from Birmingham, through his own researches and those of his students, did achieve a general knowledge of Lakeland geology, through which he was able to begin theoretical synthesis, especially with the advent of plate-tectonic theory.

The new methods of structural geology of Coles Phillips and others began to be introduced to the Lakes following up the work of Glennie in the 1950s and this led to strenuous efforts to work out a sequence of deformations, which might be supposed to connect to a whole sequence of tectonic events. These methods were brought down from Scotland (from Glennie's supervisor, Donald McIntyre), and also up from London (and across from the Isle of Man) in the work of Simpson and his co-workers, Helm and Roberts. The methods have continued to be used to the present, but in the early days not much thought was given to the possibilities of syndepositional deformation, and so the Skiddaw structures were construed in terms of a perplexing sequence of supposed tectonic events. Different workers produced their different sequences: D_1, D_2, D_3, . . .; F_1, F_2, F_3, . . .; S_1, S_2, S_3, . . .; and it proved difficult to relate the different schemes to one another – just as different fossil names or stratigraphic names are often coined by different workers, with resulting problems of synonymy. Quite apart from that, Simpson's idea of two or more pre-BVG tectonic events, apparently demonstrated by structural analysis, caused a rumpus. To check Simpson's results and claims would have required a thorough repetition of his work, but this was not done there and then. Rather, attention was focused on what appeared to be a relatively easy 'test': the issue of the nature of the SS–BVG boundary, and whether or not cleavage 'went through' this boundary. But settling *that* problem did not prove as easy as hoped, and the issue rumbled on for several years, perhaps diverting attention from more fruitful lines of enquiry.

The application of plate-tectonic theory to the Lakes was also initiated by PhD students – the (northern) friends Godfrey Fitton and David Hughes – following which there was a rapid succession of explanatory models in the 1970s and 1980s, as discussed in Chapter 10. Terrane theory was invoked by Jack Soper, with his important idea of an 'indenter' from the south in the form of the Midland Massif or Platform (as part of Eastern Avalonia), producing an arcuate pattern to the cleavage of the slate belts. In this large-scale thinking, Lakeland geology was extended from the local to the global (or at least connecting it with Scotland, eastern Europe and North America). This larger-scale work and thinking was made possible as a result of the new results from seismic survey done in the 1970s (by, among others, Soper's wife Angela Faller) and from the extensive results from the BGS's Lakeland Project, beginning in 1982, not to mention the university contributions from Sheffield and Liverpool in particular.

Understanding of Lakeland geology would surely have long remained in a backward state if Peter Allen, with Malcolm Brown's encouragement, had not initiated the Lakeland Project; and the 'national disgrace' of the official Lakeland maps being more than a hundred years old would have continued much longer. With a team of geologists having different specialities and perspectives, progress was quite rapid, though the publication of maps and memoirs has been regrettably slow and was still incomplete at the end of 2000, with basic mapping unfinished for parts of the southern maps. It is also true to say, I think, that the cross-fertilization of ideas between Surveyors and university geologists proved fruitful and beneficial, albeit sometimes acrimonious. The Skiddaw Slate stratigraphy was sorted out. The concept of 'olistostrome' was productively applied for the southern part of the Slates, partly eliminating the need for all the previous hypotheses about polyphase deformations.

New ideas about the structures and origin of the ancient volcanoes of the Central Fells were also developed, with the folding previously envisaged in the polyphase deformation(s) models being replaced by the idea of caldera collapse and concomitant volcano-tectonic faulting. However, for the BVG there were ongoing problems in getting agreement about basic stratigraphy, arising from the different approaches adopted by the Surveyors and the university men: geochemical and volcanological. Ideas about the Lakeland structures were related to the presence of the granitic plutons, but there seems to be a remaining uncertainty as to how and why the granites got to where we now see them, and what their essential role has been in the history of the Lakeland region since their emplacement.

Down in the gentler pastures of 'Otley III', the basic stratigraphy was sorted out reasonably quickly, after the Lakeland Project had been running for a few years, though there were disagreements about terminology – especially about groups and supergroups, with contention between stratigraphic 'lumpers' and 'splitters'. However, as it seems to me, the issue of the cause(s) of the basin formation for the rocks of 'Otley III', their original extent, their provenance, and the direction(s) by which the sediments filled the basin remains uncertain; as does the interpretation of the Westmorland Monocline. The relationship (if any) between Iapetus closure and the Acadian Orogeny seems to be not yet settled to the satisfaction of all concerned either. Perhaps this is to be expected, for the simple reason that there is mapping still to be done; and until this work is completed we cannot expect to see 'closure'[10] on some fundamental questions of Lakeland geology.

For the interested geo-historian, the Lake District tells us much about the way it has become necessary in the modern world to work as research teams with division of labour, rather than in the old ways of the Victorian individualists. There were controversies in both kinds of geology, which took the form of great personal contests in the old days (Sedgwick and Murchison) but now manifest as contending groups with different agendas. The study has shown contrasting styles of university researchers, the Survey, and industrial geologists (though the case of Nirex is hardly typical for the latter). The difference between science in the culture of academic debate and publication and in the court-room situation has been highlighted. The paramount issue of funding recurs. Francis Bacon held that 'knowledge is power' (or '[h]uman knowledge and human power meet in one'). Today, one might say 'knowledge is funding': give us the money and we will give you scientific knowledge. Maps tell us a lot. They do not reveal 'the truth'. Beneath their surfaces lie the controversies. The maps and stratigraphic subdivisions represent the best consensus that can be achieved or imposed. Modern controversies have to do with approaches or methods, just as much as theories – as was the case in earlier geology.

So what, in brief and in sum, can we say about the geological history of the Lake District on the basis of all the words that precede the present chapter?

[10] Broad consensus about basic issues in scientific debates, with agreement reached (or sometimes forced!) between supporters of opposed opinions in such debates, so that controversy is terminated (or goes into abeyance) for a time.

'In the beginning' (late Cambrian or early Ordovician) came the deep-water turbidites that later became the Skiddaw Slates – into a basin to the north of an ancient northward-moving terrane called (Eastern) Avalonia. Where the sediments came from is still an open question, though Survey geologists have taken the view that they were derived from a land-mass to the south (Gondwana) (Cooper *et al.* 1995), and it appears that the same may be said for the similar rocks of the Isle of Man (Woodcock *et al.* 1999). There are evidently two main structural components to the Skiddaw Slates – north and south of the Causey Pike Line – with a faulted contact. Whether there was or was not strike-slip along this fault, and if so how much, is uncertain, as is the time of the faulting. The sediments to the south of this line have evidently been 'shuffled' at some stage while they were still not fully compacted, so that we have the olistostrome deposits of the Buttermere Formation. Exactly why, how and when (and perhaps even where!) this 'shuffling' occurred appears uncertain, as is the relationship between it and the emplacement of the Borrowdale Volcanics and the formation of the Causey Pike Thrust and associated structures.

The formation of the Borrowdale Volcanics during the Llanvirn–Caradoc was associated with an already established southward-directed subduction, with the formation of chiefly subaerial volcanoes in a volcanic belt (with the central Mexican volcanic belt as a possible modern analogue: Kokelaar, pers. comm., 2000), as can be understood according to orthodox plate-tectonic theory. This process produced, or continued, uplift of the Skiddaw Slates and their erosion in some places, so that the basic relationship between the two is one of unconformity, though there is also faulting in places. The period of Borrowdale volcanicity was intense but relatively brief. Why it was so brief is somewhat obscure, when it is considered that subduction had supposedly been going on for some considerable time. (Perhaps a 'spreading zone' found itself subducted, leading to the abrupt volcanic activity? But this would not seem to fit with the geochemical evidence.) The Eskdale and Ennerdale 'granites' seem to be Caradoc in age, by zircon dating; so their emplacement appears to have been contemporaneous with the occurrence of the volcanoes, or at least approximately so. It is thus reasonable to suppose that the granitic magma welled up and occupied the space(s) vacated by the extruded ashes, ignimbrites and andesitic materials, the rise being helped by the buoyancy of the relatively low-density granite.

With the great eruptions and emission of volcanic materials by the Ordovician volcanoes there would also be the possibility of caldera collapse(s), leading to volcano-tectonic faulting. These could have led to the formation of sag structures, which would produce synforms (that might otherwise be attributed to compression, as the earlier workers envisaged). It is likely that there are several distinct areas of caldera collapse in the BVG, consistent with there having been more than one volcano, as might be expected.

The prevailing view is that during the Silurian sedimentation subsequently occurred right over the top of the area now covered by the Borrowdale Volcanics. This is somewhat hard to explain, given that the area is manifestly underlain by granites, which might be expected to be buoyant and upwardly mobile; but as discussed in Chapter 18, much of the Lakeland uplift may have occurred much later in the Palaeogene. Back in the Ordovician, large amounts of volcanic material would have been blown out of the volcanoes and their ashes would have dispersed relatively easily, or would have weathered easily and been eroded away. Nevertheless, if the deep-water Windermere sediments were deposited over the tops of the Borrowdales there must have been considerable subsidence – perhaps facilitated by thermal contraction of the cooling plutons. The lower members of the Windermere Supergroup (fossiliferous ashy limestones) show indications of having been formed in shallow waters, but ones that followed, such as the Stockdale Shales, were of deep-water character, lacking bioturbation, and there is no evidence of any shoreline at that time. So the eventual subsidence must have been considerable. To my knowledge, it has not been fully accounted for. The 'big ignimbrites' would certainly not have been easy to dispose of by weathering and erosion. Soper has envisaged extension at this time (early Silurian; see Fig. 13.11).

Meanwhile, the Iapetus Ocean to the north was still closing, presumably due to subduction directed towards the area of the Southern Uplands. Eventually, terrane closure occurred, supposedly in the Llandovery, with the creation of the Iapetus Suture, formed diachronously and with sinistral shear. The Suture itself is not associated with intense metamorphism either in the Lakes or in the terranes to the immediate north. It must have been the gentlest of collisions, very different from those that gave rise to, say, the Urals, the Pyrenees, or the Himalayas. Yet the fossils '*require*' that there was suturing in the mid-Silurian, for the life-forms 'come together' at about that time in southern Scotland and northern England. The suturing was seemingly submarine, as a continuity of fossils, N–S, is found post-suturing.

Although there was a suturing during the Silurian, the (relative) northward movement of Avalonia did not cease. One school of thought (Kneller–Bell–King) has linked this to a southward progradation of a sedimentary basin and the formation of an associated mountain front (manifest in the Windermere Monocline), beginning in Bannisdale Slate times. So, as the progradation proceeded it made possible the provision of sediment from one unit, uplifted and eroded, for the formation of the succeeding ones. Thus, for example, the Coniston Grits could have served as a source of sediment for the Kirkby Moor Flags. However, Soper has not accepted this. In his view, the Windermere Monocline does not present the features of a 'classic' mountain front – which usually has massive conglomerates in front of it, such as he saw at Ellesmere Island in his Greenland days, or which the less adventurous, like me, can conveniently see in Switzerland at the Rigi on the northern side of Lake Lucerne, to the north of the Alps. Rather, Soper supposed that the northern edge of Avalonia dipped under Laurentia, subsequent to the formation of the Iapetus Suture. The movement formed a non-prograding basin, which was then filled with sediment. There could in fact have been extension and deposition of a considerable overburden of Lower Devonian sediment. The pressure exerted by this sediment could have generated the illite crystallinity in the underlying Windermere rocks. Then, with 'tightening' due to the further northward 'shove' of Avalonia, there would have been uplift and removal of the overburden and the sediments of the slate belts acquired their presently observable cleavage in the Emsian (Lower Devonian).[11]

This was what I have called the 'big crunch', but others prefer to call the Acadian Orogeny. Besides generating the cleavage, this orogeny was associated with further emplacements of granite (at Shap and under Skiddaw). On Soper's model, it formed the Westmorland monocline with associated back-thrusting, and the 'ramps' of northern and southern Lakeland, and under the Southern Uplands (see Fig. 16.2). However, if Soper's late speculations were correct, one might think of the faults of the northern Lakes, and the ones supposedly underlying the Lakeland plutons, as *back-thrusts* and the faults associated with the Windermere Monocline as *thrusts.* More fundamentally, he envisaged a period of extension in 'northern England' between the Caledonian and Acadian orogenies, which events he would keep separated. This extension would have allowed the formation of the basin in which the Windermere sediments accumulated.

Following the earth movements just referred to, there was, during the late Palaeozoic, and through the Mesozoic, the

[11] However, we may note that in Soper's thinking at the end of the twentieth century (see Fig. 13.11) he had begun to contemplate the collision of Avalonia and Baltica and Laurentia in the late Silurian, with two separate periods of extension: in the Ashgill–Llandovery and Lochkovian–Pragian.

formation of various sediment basins around the present Lake District, and in the area of the present Irish Sea. These movements involved the formation of normal faults around the Lakeland block, stabilized by their underlying granites. Eventually, during the Cretaceous, all had been eroded to a low level, and a wide swathe of sediments (chiefly chalk?) was laid over the Lakes, and large areas of northern England, Scotland, the Irish Sea area, and Ireland, as well as the obvious areas of eastern and southern England where chalk deposits are currently so well exposed. The evidence for a blanket of chalk(?), or some other sediment(s), over the Lake District is circumstantial but persuasive, being based upon a considerable amount of evidence relating to the analysis of the annealing of fission-tracks in apatite crystals, as well as sediment compaction studies and some other procedures. The thickness of a former overburden of Mesozoic sediments has been somewhat controversial, but not its existence.

Then at about the beginning of the Tertiary there was an upward movement over northern England and western Scotland, and the ancient Lakeland Palaeozoics were gradually 'exhumed', so that we are today looking at a 'relic landscape'. The Upper Palaeozoic and Mesozoic sediments in the basins were elevated, along fault lines approximating those that had accompanied earlier basin subsidences, and then suffered 'preferential' erosion. The cause of the upward motion remains obscure. It has been suggested that it related to the rifting of the North Atlantic, and the production of the Icelandic mantle-plume, but the details were not clarified by the end of the twentieth century. In particular, the existence of a dome-like structure for Lakeland itself was not sufficiently explained in a manner that attracted general consensus, though the idea of doming or relative elevation along fault lines was important for understanding the geomorphological features of the mountains and doming had been proposed even in the mid-nineteenth century. By the end of the century the idea that there was a real, as opposed to an 'apparent', doming in the Tertiary was challenged by BGS staff; and the existence of the Lakeland mountains was ascribed to differential erosion as much as preferential uplift or doming. The whole question of the cause of the relative elevation of the Lake District had not been consensually explained by the end of 2000, in terms of developing ideas about movements in the Earth's interior (at the core–mantle boundary?) or due to mantle convection, plumes or 'blobs'. Such issues awaited further attention in the twenty-first century.

With the advent of Lakeland glaciation in the Quaternary, the topography of the Lakeland mountains was substantially modified, with erosion in the hills and deposition of drift in the valleys and surrounding lowlands. Early theorizing (nineteenth century) envisaged a succession of glaciations, along lines of thought pioneered on the Continent, and theorized in Britain by James Croll, but Marr, who knew the Lakeland mountains so well, doubted that there was clear evidence for a succession of glaciations, suggesting there had been perhaps only one. If there had in fact been more than one episode, the last glaciation largely cleared away the deposits that might have been formed during earlier episodes of cold. Nevertheless, from evidence outside the Lakes there is no reason to doubt that there was a succession of glacial episodes, and the Nirex investigations revealed the details for the most recent of these by examination of deposits on the west Cumbrian coast, and under the Irish Sea. With the departure of the last glaciers about 10 000 years ago, we enter the final stages of Lakeland's past: prehistory, and then history. But these are matters for other books and other kinds of historian.

It would be a grand thing, then, if someone would re-write this book in, say, 2100, recording what people say and think in the next hundred years. At present, there remain many mysteries, even after two hundred years of research. The acquisition of geological knowledge is full of contingencies, proceeding as it does on many fronts and in a somewhat unstructured, 'messy' way, not always advancing by the most direct route. Indeed, I am reminded of the metaphor for science suggested by the sociologists of science Harry Collins and Trevor Pinch in their most interesting book, *The Golem: What Everyone Should Know about Science.* They have written:

> The golem [pronounced 'goilem'] is a creature of Jewish mythology. It is a humanoid made from clay and water with incantations and spells. It is powerful. It grows a little more every day. It will follow orders, do your work, and protect you from the ever threatening enemy. But it is clumsy and dangerous. Without control, a golem may destroy its masters with its flailing vigour. . . .
>
> [I]n the mediaeval tradition the creature of clay was animated by the Hebrew 'EMETH', meaning truth, inscribed on its forehead – it is truth that drives it on. But this does not mean that it understands the truth – far from it. . . .
>
> [The golem] is not an evil creature but it is a little daft. Golem Science is not to be blamed for its mistakes; but they are our mistakes. A golem cannot be blamed if it is doing its best. But we must not expect too much. A golem, powerful though it is, is the creature of our art and our craft (Collins & Pinch 1993, pp. 1–2).

It seems to me that this metaphor for science is rather apt when applied to the history of Lakeland geology. The scientific investigators have sought the truth, and to some extent have, I am sure, found it, albeit by circuitous routes, and in a generally 'messy' way. People have surely done their best, but on occasions getting hurt in the process. Maybe some of their suggestions have occasionally been a 'little daft' (others may judge); but the investigators themselves were surely not, nor is the picture that has gradually been revealed. The scientific knowledge has indeed grown a little almost every day. It has been 'crafted' by the geologists carrying out the investigations.

There are, then, almost infinite details that might be related in the historical story, so it is scarcely surprising that it too is 'messy'. As time passes, some of the details, which seemed important at the time, have become obscured and are likely to become more so in the future. So there is benefit in trying to record the modern, as well as the 'ancient', story, before it all gets forgotten. Thus I have attempted to display an image of Lakeland's 'geo-golem'. My aim has been to recount, in as much detail as is reasonable here, what occurred over the first two hundred years of Lakeland geology, in the effort to understand the geological structures and history of just one small part of the globe. No less complex stories could be given for many other regions. But for present purposes, one region is surely more than enough! A larger golem would be even more difficult to comprehend.

Very likely, the future will *not* look like the past; but perhaps one can say that the past is the key to understanding the present. I hope, therefore, that the present book will assist that understanding, even if it does not provide a key that fits the lock perfectly. Infinite complexity can never be described.

References

ABBATE, E., BORTOLOTTI, V. & PASSERINI, P. 1970. Olistostromes and olistoliths. *Sedimentary Geology*, **4**, 521–557.

ADAMS, J. 1995. *Mines of the Lake District Fells*. Dalesman, Skipton (1st edn, 1988).

AGASSIZ, L. 1840. *Etudes sur les Glaciers*. Jent & Gassmann, Neuchâtel.

AGASSIZ, L. 1840–1841. On glaciers, and the evidence of their once having existed in Scotland, Ireland and England. *Proceedings of the Geological Society*, **3**(2), 327–332.

AKHURST, M. C., BARNES, R. P., CHADWICK, R. A., MILLWARD, D., NORTON, M. G., MADDOCK, R. H., KIMBELL, G. S. & MILODOWSKI, A. E. 1998. Structural evolution of the Lake District Boundary Fault Zone in West Cumbria, UK. *Proceedings of the Yorkshire Geological Society*, **52**, 139–158.

AKHURST, M. C., CHADWICK, R. A., HOLLIDAY, D. W., MCCORMACK, M., MCMILLAN, A. A., MILLWARD, D. & YOUNG, B. 1997. *Geology of the West Cumbria District: Memoir for the 1:50 000 Geological Sheets 28 Whitehaven, 37 Gosforth and 47 Bootle (England and Wales)*. The British Geological Survey, Keyworth.

ALLEN, P. M. 1987. The Solway Line is not the Iapetus Suture. *Geological Magazine*, **124**, 485–486.

ALLEN, P. M. 1997. Dr P. M. Allen writes. *The Guardian*, 10 April.

ALLEN, P. M. & COOPER, D. C. 1986. The stratigraphy and composition of the Latterbarrow and Redmain sandstones, Lake District, England. *Geological Journal*, **21**, 59–76.

ALLEN, P. M., COOPER, D. C. & FORTEY, N. J. 1987. Composite lava flows of Ordovician age in the English Lake District. *Journal of the Geological Society*, **144**, 945–960.

ANDERSON, T. B. & OLIVER, G. H. J. 1986. Comment on MURPHY, F. C. & HUTTON, D. W. H., 'Is the Southern Uplands of Scotland really an accretionary prism?' *G*, **14**, 1043–1044.

ANON 1816. A geological sketch of a part of Cumberland and Westmorland. By a correspondent. *PM*, **47**, 41–45.

ANON 1869. Review of: D. MACKINTOSH, *The Scenery of England and Wales: Its Character and Origin* ... *Geological Magazine*, **6**(Decade 1), 465–470.

ANON 1878. Professor Robert Harkness, FRS, FGS. Born 28th July, 1816. Died 4th October, 1878. *Geological Magazine*, **5**(Decade 2), 574–576 and plate.

ANON 1878–1879. In memoriam [Robert Harkness]. *Transactions of the Cumberland and Westmorland Association for the Advancement of Literature and Science*, **4**, 250–251.

ANON 1878–1880. Obituary of Robert Harkness. *Proceedings of the Edinburgh Geological Society*, **10**, 31–33.

ANON 1898–1899. ?, Obituary of Henry Alleyne Nicholson. *Proceedings of the Linnean Society of London*, **?**, 55–56.

ANON 1903*a*. Memorial to Henry Alleyne Nicholson, M.D., D.Sc., F.R.S. *Geological Magazine*, **10**(Decade 4), 451 and plate.

ANON 1903*b*. In memoriam E. Walker. *The Cambridge Review*, 12 March 1903, 245.

ANON 1906. Eminent living geologists. Thomas McKenny Hughes ... *Geological Magazine*, **5**(Decade 5), 1–13 and plate.

ANON 1916. Eminent living geologists. John Edward Marr ... *Geological Magazine*, **3**(Decade 6), 289–295 and plate.

ANON 1932–1935. John Edward Marr 1857–1933. *Obituary Notices of Fellows of the Royal Society*, **1**, 251–257 and plate.

ANON 1960. Obituary of Jerome Hartley. *Irish Naturalists' Journal*, **13**, 133–134.

ANON 1993. Were the Pennines and the Lake District once 3 km under? *Geology Today*, Jan.–Feb., 9–10.

ANON 1997*a*. Professor Sir Malcolm Brown. *The Times*, 9 April.

ANON 1997*b*. Sir Malcolm Brown: from Earth to Moon. *The Guardian*, 10 April.

ANON 2001*a*. Professor Sir Kingsley Dunham. *The Times*, 18 April.

ANON 2001*b*. Giant slump discovered. *Geoscientist*, **11**(7), 9.

ANSARI, S. M. 1983. *Petrology and petrochemistry of the Eskdale and adjacent intrusions (Cumbria) with special reference to mineralization*. PhD dissertation, Nottingham University.

ARCHER, J. B. 1980. Patrick Ganly: geologist. *Irish Naturalists' Journal*, **20**, 142–148.

ARTER, G. & FAGIN, S. W. 1993. The Fleetwood Dyke and the Tynwald fault zone, Block 113/27, East Irish Sea Basin. *In*: PARKER, J. R. (ed.), *Petroleum Geology of Northwest Europe: Proceedings of the 4th Conference held at the Barbican Centre, London 29 March–1 April 1992*. Geological Society, London, **2**, 835–843.

ARTHURTON, R. S. & WADGE A. J. 1981. *Geology of the Country Around Penrith: Memoir for 1:50 000 Geological Sheet 24*. Institute of Geological Sciences and HMSO, London.

AVELINE, W. T. 1869. On the relation of the Porphyry Series to the Skiddaw Slates in the Lake District. *Geological Magazine*, **6**(Decade 1), 382.

AVELINE, W. T. 1872. On the continuity and breaks between the various divisions of the Silurian strata in the Lake District. *Geological Magazine*, **9**(Decade 1), 441–442.

AVELINE, W. T. 1873. *Explanation of Quarter-Sheet 91 N.W., Illustrating the Geology of the Southern Part of the Furness District in North Lancashire*. Longmans, Green & Co and Edward Stanford for HMSO, London.

AVELINE, W. T. 1876*a*. Absence of the Llandovery rocks in the Lake District. *Geological Magazine*, **3**, 282.

AVELINE, W. T. 1876*b*. The graptolitic mudstones of the Lake District. *Geological Magazine*, **3**(Decade 2), 527.

AVELINE, W. T. 1876*c*. The Silurian rocks of the Lake District. *Geological Magazine*, **3**(Decade 2), 376.

AVELINE, W. T. & HUGHES, T. McK. 1872. *The Geology of the Country Around Kendal, Sedbergh, Bowness, and Tebay*. Longmans, Green & Co. & Edward Stanford for HMSO, London.

AVELINE, W. T. & HUGHES, T. McK. (revised STRAHAN, A., DAKYNS, J. R. & TIDDEMAN, R. H.) 1888. *The Geology of the Country Around Kendal, Sedbergh, Bowness, and Tebay*, 2nd edn. Eyre & Spottiswoode, London, Adam & Charles Black, Edinburgh & Hodges, Figgis, Dublin, for HMSO, London.

AVELINE, W. T., HUGHES, T. McK. & TIDDEMAN, R. H. 1872. *Explanation of Quarter Sheet 98 S.E.; Illustrating the Geology of the Neighbourhood of Kirkby by Lonsdale and Kendal*. Longmans, Green & Co. and Edward Stanford for HMSO, London.

BAKEWELL, R. 1813. *An Introduction to Geology*. J. Harding, London.

BAKEWELL, R. 1815. Observations on the geology of Northumberland and Durham; and remarks on Mr. Westgarth Forster's section of the strata, with a sketch of the physical structure of that part of England, from the German Ocean to the Irish Channel. *PM*, **45**, 81–96.

BALDRY, R. A. 1938. Slip-planes and breccia zones in the Tertiary rocks of Peru. *Quarterly Journal of the Geological Society of London*, **94**, 347–358 and plates.

BALK, R. 1937. *Structural Behavior of Igneous Rocks*. Geological Society of America, Memoir **5**, New York.

BAMFORD, D. 1979. Seismic constraints in the deep geology of the Caledonides of northern Britain. *In*: HARRIS, A. J., HOLLAND, C. H. & LEAKE, B. E. (eds), *Caledonides of the British Isles – Reviewed*. Scottish Academic Press for the Geological Society, Edinburgh, 93–96.

BAMFORD, D., FABER, S. *et al.* 1976. A lithospheric seismic profile in Britain – I. Preliminary results. *Geophysical Journal of the Royal Astronomical Society*, **44**, 145–160.

BAMFORD, D., NUNN, K., PRODEHL, C. & JACOB, B. 1973. LI[thospheric]S[eismic]P[rofile] [in] B[ritain] – IV. Crustal structure of northern Britain. *Geophysical Journal of the Royal Astronomical Society*, **54**, 43–60 and charts.

BARRELL, J. 1920. The piedmont terraces of the northern Appalachians. *American Journal of Science*, **49**, 237–258, 327–362, 407–428.

BATH, A. H., MCCARTNEY, R. A., RICHARDS, H. G., METCALFE, R. & CRAWFORD, M. B. 1996. Groundwater chemistry in the Sellafield area: a preliminary interpretation. *Quarterly Journal of Engineering Geology*, **29**, S39–S57.

BANCROFT, B. B. 1933. *Correlation Tables of the Stages Costonian–Onnian in England and Wales*. Published privately, Blakeney, Gloucestershire.

BANCROFT, B. B. 1945. The brachiopod zonal indices of the stages Costonian to Onnian in Britain. *JP*, **19**, 181–252 and plates.

BANHAM, P. H., HOOPER, F. M. W. & JACKSON, J. B. 1981. The Gillbrea Nappe in the Skiddaw Group, Cockermouth, Cumbria, England. *Geological Magazine*, **118**, 509–516.

BARNES, R. P., AMBROSE, K., HOLLIDAY, D. W. & JONES, N. S. 1994. Lithostratigraphical subdivision of the Triassic Sherwood Sandstone Group in west Cumbria. *Proceedings of the Yorkshire Geological Society*, **50**, 51–60.

BARNES, R. P., LINTERN, B. C. & STONE, P. 1989. Short paper: timing and regional implications of deformation in the Southern Uplands of Scotland. *Journal of the Geological Society*, **146**, 905–908.

BARRANDE, J. J. 1850 *Graptolites de Bohème*: *Extrait du Système Silurien du centre de la Bohême*. Joachim Barrande, Prague.

BARROW, G. 1893. On an intrusion of muscovite–biotite gneiss in the south-eastern Highlands of Scotland, and its accompanying metamorphism. *Quarterly Journal of the Geological Society of London*, **49**, 330–358 and plates.

BASSETT, M. G. 1985. Transition and grauwacke – the Silurian and Cambrian systems through 150 years. *Episodes*, **8**, 231–235.

BEAMISH, D. & SMYTHE, D. K. 1986. Geophysical images of the deep crust: the Iapetus Suture. *Journal of the Geological Society*, **143**, 489–497.

BEAUMONT, P. 1968 A history of glacial research in northern England from 1860 to the present day. *University of Durham, Department of Geography, Occasional Papers*, Series 9.

BEDDOE-STEPHENS, B. & MILLWARD, D. 2000. Very densely welded, rheomorphic ignimbrites of homogeneous intermediate calc-alkaline composition from the English Lake District. *Geological Magazine*, **137**, 153–173.

BEDDOE-STEPHENS, B., PETTERSON, M. G., MILLWARD, D. & MARRINER, G. F. 1995. Geochemical variation and magmatic cyclicity within an Ordovician continental-arc volcanic field: the Lower Borrowdale Volcanic Group, English Lake District. *Journal of Volcanology and Geothermal Research*, **65**, 81–110.

BELBIN, S. 1985. Long-term landform development in north-west England: the application of the planation concept. *In*: JOHNSON, R. H. (ed.), *The Geomorphology of North-west England*. Manchester University Press, Manchester & Dover (New Hampshire), 37–58.

BELL, A. M. 1975. *A finite strain study of accretionary lapilli tuff in the Borrowdale Volcanic Group, English Lake District*. PhD dissertation, Sheffield University.

BELL, A. M. 1981. Strain factorizations from lapilli tuff, English Lake District. *Journal of the Geological Society*, **138**, 463–474.

BELL, A. M. 1992. *The Stratigraphy and Structure of the Skiddaw Group in the Black Combe Inlier: Geological Notes to Accompany 1:25 000 Sheet SD 18*. British Geological Survey Technical Report **WA/92/70**.

BELL, A. M. 1997. *The Skiddaw Group and its Contact with Surrounding Rocks in the Bampton Inlier, Cumbria: Parts of the 1:10 000 Sheets NY 52 SW, NY 51 NW, NY 51 SW, and NY 51 SE Part of 1:50 000 Sheet 30 (Appleby)*. British Geological Survey Technical Report **WA/97/9**.

BERGSTROM, S. M. 1979. Conodonts as paleotemperature tools in Ordovician rocks of the Caledonides and adjacent areas in Scandinavia and the British Isles. *Geologiska Föreningens i Stockholm Förhandlingar*, **102**, 377–392.

BICKNELL, P. 1990. *The Picturesque Scenery of the Lake District 1752–1855: A Bibliographical Study*. St Paul's Bibliographies, Winchester; Omnigraphics, Detroit.

BINNEY, E. W. 1848. A glance at the geology of Low Furness, Lancashire. *Memoirs and Proceedings of the Manchester Literary and Philosophical Society*, **8**(2nd Series), 423–445.

BINNEY, E. W. 1855. On the origin of ironstones, and more particularly the newly discovered red stone at Ipstones, near Cheadle, Staffordshire; with some account of the ironstones of South Lancashire. *Memoirs and Proceedings of the Manchester Literary and Philosophical Society*, **17**(2nd Series), 31–45.

BINNEY, E. W. 1868. On the age of the haematite iron ore deposits of Furness. *Memoirs and Proceedings of the Manchester Literary and Philosophical Society*, **7**, 55–61.

BLACK, A. & BLACK, C. 1849. *Black's Picturesque Guide to the English Lakes Including an Essay on the Geology of the District by John Phillips, FFS, GL*, 3rd edn. Adam & Charles Black, Edinburgh.

BLACK, J. H. & BRIGHTMAN, J. H. 1996. The geology and hydrogeology of the Sellafield area: conceptual model of the hydrogeology of Sellafield. *Quarterly Journal of Engineering Geology*, **29** (supplement), S83–S93.

BLACK, J. H., ALEXANDER, J. & NOY, D. J. 1984. Groundwater modelling in selected UK basins and its importance in repository site selection. *In*: British Nuclear Energy Society, *Radioactive Waste Management*. London, 161–165.

BLACKIE, R. C. 1933. The Silurian rocks of the Kentmere district, Westmorland. *Proceedings of the Liverpool Geological Society*, **16**, 88–105.

BOARDMAN, J. 1980. Evidence for pre-Devensian glaciation in the north-eastern Lake District. *Nature*, **286**, 599–600.

BOARDMAN, J. 1985. The Troutbeck paleosol, Cumbria, England. *In*: BOARDMAN, J. (ed.), *Soils and Quaternary Landscape Evolution*. John Wiley, Chichester, 231–260.

BOARDMAN, J. 1990–1991 (1992). Quaternary landscape evolution in the Lake District – a discussion. *Proceedings of the Cumberland Geological Society*, **5**, 285–315.

BOARDMAN, J. 1996. *Classic Landforms of the Lake District*. The Geographical Association, Sheffield.

BOLTON, J. 1862. On a deposit with insects, leaves, etc., near Ulverston. *Proceedings of the Geological Society*, **18**, 274–277.

BOLTON, J. 1869. *Geological Fragments Collected Principally from Rambles Among the Rocks of Furness and Cartmel*. D. Atkinson, Ulverston; Whittaker, London. (A few copies contain a facsimile copy of a letter from Sedgwick, dated 18 January 1873. See example at Barrow-in-Furness Public Library.)

BOLTON, J. 1979. *Geological Fragments: Collected Principally from Rambles Among the Rocks of Furness and Cartmel*. Michael Moon, Beckermet (reprint).

BOTT, M. H. P. 1956. A geophysical study of the granite problem. *Quarterly Journal of the Geological Society of London*, **112**, 45–? and plates.

BOTT, M. H. P. 1959. The use of electronic digital computers for the evaluation of gravimetric terrain corrections. *Geophysical Prospecting*, **7**, 45–54.

BOTT, M. H. P. 1964. Gravity measurements in the north-eastern part of the Irish Sea. *Quarterly Journal of the Geological Society of London*, **120**, 369–396 and plate.

BOTT, M. H. P. 1965. The deep structure of the northern Irish Sea – a problem of crustal dynamics. *In*: WHITTARD, W. F. & BRADSHAW, R. (eds), *Submarine Geology and Geophysics: Proceedings of the Seventeenth Symposium of the Colston Research Society held in the University of Bristol, April 5th–9th, 1965*. Butterworths, London, 179–201.

BOTT, M. H. P. 1967. Geophysical investigations of the northern Pennine basement rocks. *Proceedings of the Yorkshire Geological Society*, **36**, 139–168.

BOTT, M. H. P. 1974. The geological interpretation of a gravity survey of the English Lake District and the Vale of Eden. *Journal of the Geological Society*, **130**, 309–331.

BOTT, M. H. P. 1997. Modeling the formation of a half graben using realistic upper crustal rheology. *Journal of Geophysical Research*, **102**, 24 605–24 617.

BOTT, M. H. P. 1999. Modeling local crustal isostasy caused by ductile flow in the lower crust. *Journal of Geophysical Research*, **104**(B9), 20 349–20 359.

BOTT, M. H. P. & YOUNG, D. G. G. 1971. Gravity measurements in the north Irish Sea. *Quarterly Journal of the Geological Society of London*, **126**, 413–434.

BOTT, M. H. P., JOHNSON, G. A. L. & HODGE, B. L. 1961. Granite beneath the northern Pennines. *Nature*, **190**, 899–900.

BOTT, M. H. P., LONG, R. E., GREEN, A. S. P., LEWIS, A. H. J., SIRKA, M. C. & STEVENSON, D. C. 1985. Crustal structure south of the Iapetus Suture beneath northern England. *Nature*, **314**, 724–727.

BOUMA, A. H. 1962. *Sedimentology of some Flysch Deposits: A Graphic Approach to Facies Interpretation*. Elsevier, Amsterdam and New York.

BOWDEN, R. A., BUMPUS, C. & LITTLEJOHN, A. K. 1998. An overview and update of the site characterization studies at Sellafield. *Proceedings of the Yorkshire Geological Society*, **52**, 125–137.

BOWEN, D. Q. 1978. *Quaternary Geology: A Stratigraphic Framework for Multidisciplinary Work*. Pergamon Press, Oxford.

BOWEN, D. Q. 1989. The last interglacial–glacial cycle in the British Isles. *Quaternary International*, **3/4**, 41–47.

BOWEN, D. Q. (ed.) 1999. *A Revised Correlation of Quaternary Deposits in the British Isles*. Geological Society of London, London, Special Report 23.

BOWIE, S. 1971. Tom Eastwood. *Proceedings of the Geologists' Association*, **82**, 153–155.

BOYLAN, P. J. 1984. *William Buckland (1784–1856): scientific institutions, vertebrate palaeontology and Quaternary geology*. PhD dissertation, University of Leicester.

BOYLAN, P. J. n. d. [1999]. English and Scottish glacial localities of Agassiz, Buckland and Lyell, 1840 (supplement to BOYLAN, P. J., Lyell and the dilemma of Quaternary glaciation. *In:* BLUNDELL, D. J. & SCOTT, A. C. (eds), *Lyell: The Past is the Key to the Present*. The Geological Society, London, 1998, 145–159.) World Wide Web Address: http://www.city.ac.uk/artspol/glaclocs.html.

BRANNEY, M. J. 1986. The Stockdale (Yarlside) Rhyolite – a rheomorphic ignimbrite? *Proceedings of the Yorkshire Geological Society*, **46**, 80–82.

BRANNEY, M. J. 1988*a*. *Subaerial explosive volcanism, sedimentation, intrusion, and collapse in the Borrowdale Volcanic Group, SW Langdale, English Lake District*. PhD dissertation, Sheffield University.

BRANNEY, M. J. 1988*b*. The subaerial setting of the Ordovician Borrowdale Volcanic Group, English Lake District. *Journal of the Geological Society*, **145**, 887–890.

BRANNEY, M. J. 1991. Eruption and depositional facies of the Whorneyside Tuff Formation, English Lake District: an exceptionally large-magnitude phreatoplinian eruption. *Bulletin of the Geological Society of America*, **103**, 886–897.

BRANNEY, M. J. & KOKELAAR, B. P. 1992. A reappraisal of ignimbrite emplacement: progressive aggradation and changes from particulate to non-particulate flow during emplacement of high-grade ignimbrite. *Bulletin of Volcanology*, **54**, 504–520.

BRANNEY, M. J. & KOKELAAR, B. P. 1993. *The Whorneyside Tuff Formation in the Central Fells*. British Geological Survey Technical Report **WA/93/40**.

BRANNEY, M. J. & KOKELAAR, B. P. 1994. Volcanotectonic faulting, soft-state deformation, and rheomorphism of tuffs during development of piecemeal caldera. English Lake District. *Bulletin of the Geological Society of America*, **106**, 507–530.

BRANNEY, M. J. & SOPER, N. J. 1988. Ordovician volcanic-tectonics in the English Lake District. *Journal of the Geological Society*, **145**, 367–376.

BRANNEY, M. J. & SUTHREN, R. J. 1988. High-level peperitic sills in the English Lake District: distinction from block lavas, and implications for Borrowdale Volcanic Group stratigraphy. *Geological Journal*, **23**, 171–187.

BRANNEY, M. J., KOKELAAR, B. P. & MCCONNELL, B. J. 1992. The Bad Step Tuff: a lava-like rheomorphic ignimbrite in a calc-alkaline piecemeal caldera, English Lake District. *Bulletin of Volcanology*, **54**, 187–199.

BRAUN, A. & MARQUART, G. 2001. Die bewegte Geschichte des Nordatlantiks. *Spektrum der Wissenschaft*, June 2001, 50–59.

BREDEHOEFT, J. D. & MAINI, T. 1981. Strategy for radioactive waste disposal in crystalline rocks. *Science*, **213**, 293–296.

BRERETON, N. R., EVANS, C. J. & KINGDON, A. 1998. Hydrogeological and geotechnical property characterization from geophysics. *Proceedings of the Yorkshire Geological Society*, **52**, 199–213.

BREWER, J. A., MATTHEWS, D. H., WARNER, M. R., HALL, J., SMYTHE, D. K. & WHITTINGTON, R. J. 1983. BIRPS deep seismic reflection studies of the British Caledonides. *Nature*, **305**, 206–210.

BRIDEN, J. C. & MORRIS, W. A. 1973. Palaeomagnetic studies in the British Caledonides – III. igneous rocks of the northern Lake District, England. *Geophysical Journal of the Royal Astronomical Society*, **34**, 27–46.

BRITISH GEOLOGICAL SURVEY 1977. *Dalton-in-Furness. England and Wales Sheet SD 27 (with parts of SD 17 and SD 37)*, 1:25 000. Geological Survey of Great Britain, Southampton.

BRITISH GEOLOGICAL SURVEY 1979. *Whitehaven. England and Wales Sheet 28. Solid Edition*. 1:50 000 Series. Geological Survey of Great Britain, Southampton.

BRITISH GEOLOGICAL SURVEY 1990. *Devoke Water and Ulpha. Solid and Drift Geology*. 1:25 000 Series Sheet SD 19. British Geological Survey, Keyworth.

BRITISH GEOLOGICAL SURVEY 1992. *Regional Geochemistry of the Lake District and Adjacent Areas*. British Geological Survey, Keyworth.

BRITISH GEOLOGICAL SURVEY 1995. *Maryport. England and Wales Sheet 22. Solid Geology and Solid and Drift Geology*. British Geological Survey, Keyworth.

BRITISH GEOLOGICAL SURVEY 1996. *Ambleside. England and Wales Sheet 38. Solid Geology.* 1:50 000 Series. British Geological Survey, Keyworth.

BRITISH GEOLOGICAL SURVEY 1997*a*. *Cockermouth. England and Wales Sheet 23. Solid Geology.* 1:50 000 Series. British Geological Survey, Keyworth.

BRITISH GEOLOGICAL SURVEY 1997*b*. *Kirkby Stephen. England and Wales Sheet 40. Solid and Drift Geology.* 1:50 000 Provisional Series. British Geological Survey, Keyworth.

BRITISH GEOLOGICAL SURVEY 1997*c*. *Ulverston. England and Wales Sheet 48. Solid Geology.* 1:50 000 Series. British Geological Survey, Keyworth.

BRITISH GEOLOGICAL SURVEY 1998. *Black Combe. England and Wales Sheet SD 18 and Part of SD 28. Solid and Drift Geology.* 1:25 000 Series. British Geological Survey, Keyworth.

BRITISH GEOLOGICAL SURVEY 1999*a*. *Keswick. England and Wales Sheet 29. Solid Geology.* 1:50 000 Series. British Geological Survey, Keyworth.

BRITISH GEOLOGICAL SURVEY 1999*b*. *Gosforth. England and Wales Sheet 37. Solid Geology.* 1:50 000 Series. British Geological Survey, Keyworth.

BRITISH GEOLOGICAL SURVEY n.d. *The Lake District and Surrounds*. British Geological Survey, Keyworth.

BRODIE, J. & WHITE, N. 1994. Sedimentary basin inversion caused by igneous underplating: northwest European continental shelf. *Geology*, **22**, 147–150.

BRØGGER, W. C. 1890. Die Mineralien der Synit–pegmatit–gänge der südnorwegischen Augit- und Nephlinsyenite. *Zeitschrift für Krystallographie und Mineralogie*, **16**, 1–663.

BRONGNIART, A. 1828. *Histoire des végétaux fossiles, ou recherches botaniques et géologiques sur les végétaux renfermés dans les diverses couches du globe*, **1**, G. Dufour, Paris.

BRONN, H. G. 1859. On the laws of evolution of the organic world during the formation of the crust of the Earth. *Annals and Magazine of Natural History*, **4**(3rd Series), 81–90, 175–184.

BROOKS, M. 1973. Some aspects of the Paleogene evolution of western Britain in the context of an underlying mantle hot-spot. *Journal of Geology*, **81**, 81–88.

BROWN, G. B. 1838. On a theory of gravitational sliding applied to the Tertiary of Ancon, Equador. *Quarterly Journal of the Geological Society of London*, **94**, 359–370 and plates.

BROWN, G. M. 1984. Wither British Geological Survey? A *British Geologist* interview with Malcolm Brown Director of the British Geological Survey. *British Geologist*, **10**, 80–83.

BROWN, P. E., MILLER, J. A. & SOPER, N. J. 1964. Age of the principal intrusions of the Lake District. *Proceedings of the Yorkshire Geological Society*, **34**, 331–342.

BRÜCKNER, E. 1887–1888. Die Eiszeit in den Alpen. *Mitteilungen der Geographischen Gesellschaft (Hamburg)*, **1**, 10–23.

BRYANT, G. E. & BAKER, G. P. (eds) 1934. *A Quaker Journal 1834 of William Lucas of Hitchin*. Hutchinson, London.

BRYCE, J. 1850. On striated and polished rocks and "roches moutonnées" in the Lake District of Westmoreland. *PM*, **37**(3rd Series), 486–492 and plate.

BUCKLAND, W. 1815. Description of an insulated group of rocks of slate and greenstone in Cumberland and Westmoreland, situated on the east side of Appleby, between Melmerby and Murton, in Cumberland [abstract]. *Annals of Philosophy*, **5**, 456–457.

BUCKLAND, W. 1817. Description of an insulated group of rocks of slate and greenstone in Cumberland and Westmoreland, on the east side of Appleby, between Melmerby and Murton. *Transactions of the Geological Society of London*, **4**, 105–116.

BUCKLAND, W. 1819. Description of the quartz rock of the Lickey Hill in Worcestershire, and of the strata immediately surrounding; with considerations of the evidence of a recent deluge afforded by the gravel beds of Warwickshire and Oxfordshire, and the valley of the Thames from Oxford downwards to London; ... *Transactions of the Geological Society of London*, **5**, 506–544 and plates.

BUCKLAND, W. 1840. Memoir on the evidences of glaciers in Scotland and the north of England. *Proceedings of the Geological Society*, **3**, 332–337, 345–348.

BUCKMAN, S. S. 1895. The Bajocian of the mid-Cotteswolds. *Quarterly Journal of the Geological Society of London*, **51**, 388–402 and plate.

BULMAN, O. M. B. 1941. Some dichograptids of the Tremadocian and Lower Ordovician. *Annals and Magazine of Natural History*, **7**(11th Series), 100–121 and plate.

BULMAN, O. M. B. 1949. Robert Heron Rastall. *Quarterly Journal of the Geological Society of London*, **105**, lxxvi–lxxvii.

BULMAN, O. M. B. 1960. Dr. Gertrude L. Elles, MBE. *Nature*, **188**, 1152.

BULMAN, O. M. B. 1961. Gertrude Lilian Elles. *PGSL*, **1592**, 143–145.

BULMAN, O. M. B. 1963. The evolution and classification of the Graptoloidea. *Quarterly Journal of the Geological Society of London*, **119**, 401–418.

BURGESS, I. C. & WADGE, A. J. 1974. *The Geology of the Cross Fell Area (Explanation of 1:25 000 Geological Special Sheet Comprising Parts of Sheets NY 53, 62, 63, 64, 71, 72, 73)*. HMSO, London.

C., A. E. 1977. Obituary W. H. Wilcockson. *Proceedings of the Yorkshire Geological Society*, **41**, 302–303.

CAPEWELL, J. G. 1955. The post-Silurian, pre-marine Carboniferous sedimentary rocks of the eastern side of the English Lake District. *Quarterly Journal of the Geological Society of London*, **111**, 23–46.

CARRUTHERS, R. M. 1986. *An Appraisal of Geophysical Techniques for Mapping in the Skiddaw Group*. British Geological Survey, Report **PDA2 86/1**.

CARY, J. 1811. *Cary's New English Atlas; Being a Complete Set of County Maps, from Actual Surveys ....* J. Cary, London.

CHADWICK, R. A. & HOLLIDAY, D. W. 1991. Deep-crustal structure and Carboniferous basement development within the Iapetus convergence zone, northern England. *Journal of the Geological Society*, **148**, 41–53.

CHADWICK, R. A., EVANS, D. J. & HOLLIDAY, D. W. 1993. The Maryport Fault: the post-Caledonian tectonic history of southern Britain in microcosm. *Journal of the Geological Society*, **150**, 247–250.

CHADWICK, R. A., HOLLIDAY, D. W., HOLLOWAY, S. & HULBERT, A. G. 1995. *The Structure and Evolution of the Northumberland–Solway Basin and Adjacent Areas: Subsurface Memoir*. HMSO, London.

CHADWICK, R. A., KIRBY, G. A. & BAILY, H. E. 1994. The post-Triassic structural evolution of north-west England and adjacent parts of the Irish Sea. *Proceedings of the Yorkshire Geological Society*, **50**, 91–102.

CHALLINOR, J. 1948. Jonathan Otley's geology of the Lake District. *North Western Naturalist*, **23**, 113–126.

CHAPLOW, R. (ed.) 1994. *The Geology and Hydrogeology of the Sellafield Area: Proceedings of the Nirex Seminar, 11 May 1994*. The Geological Society, London (papers reprinted from the *Quarterly Journal of Engineering Geology*, **29**, Supplement 1).

CHAPMAN, N. A. 1979. *Geochemical Considerations in the Choice of a Host Rock for the Disposal of High-Level Radioactive Wastes*. British Geological Survey Technical Report **WA/79/14**.

CHAPMAN, N. A. 1986. Nuclear waste disposal. *British Geologist*, **12**, 128–133.

CHARLESWORTH, J. K. 1926. The readvance marginal kame moraine of the south of Scotland and some later stages of the retreat. *TRSE*, **55**, 25–50.

CHARLESWORTH, J. K. 1939. Some observations on the glaciation of north-east Ireland, *Proceedings of the Royal Irish Academy, Series B*, **45**, 255–295, with plates and maps.

CHARLESWORTH, J. K. 1960. John Jerome Hartley. *Proceedings of the Geologists' Association*, **1582**, 137–138.

CHUBB, L. J. C. 1949. Edmund Johnstone Garwood. *Quarterly Journal of the Geological Society of London*, **105**, lxii–lxiv.

CHUBB, L. J. C. 1950. Edward Johnstone Garwood, F. R. S. *Proceedings of the Geologists' Association*, **61**, 109–112.

CHURCH, W. R. & GAYER, R. A. 1973. The Ballantrae ophiolite. *Geological Magazine*, **110**, 497–510.

CLARK, J. W. & HUGHES, T. MCK. 1890. *The Life and Letters of the Reverend Adam Sedgwick*, 2 vols. Cambridge University Press, Cambridge.

CLARK, L. 1963. *The geology and petrology of the Ennerdale Granophyre, its metamorphic aureole and associated mineralization*. PhD dissertation, Leeds University.

CLARK, L. 1964. The Borrowdale Volcanic Series between Buttermere and Wasdale, Cumberland. *Proceedings of the Yorkshire Geological Society*, **34**, 343–356 and plate.

CLARK, R. 1986–1987 [1988]. Pattern and order in the Lake District Landscape. *Proceedings of the Cumberland Geological Society*, **5**(1), 17–34.

CLARK, R. 1992–1993 [1994]. The Skiddaw massif problem. *Proceedings of the Cumberland Geological Society*, **5**(4), 437–448.

CLOOS, M. 1982. Flow melanges: numerical modeling and geologic constraints on their origin in the Franciscan subduction complex, California. *Bulletin of the Geological Society of America*, **93**, 330–345.

COCKS, L. R. M. 1985. The Ordovician–Silurian boundary. *Episodes*, **8**, 98–100.

COCKS, L. R. M. 1988. The Ordovician–Silurian boundary and its Working Group. *Bulletin of the British Museum (Natural History) Geology*, **43**, 9–15.

COCKS, L. R. M. & FORTEY, R. A. 1982. Faunal evidence for oceanic separations in the Palaeozoic of Britain. *Journal of the Geological Society*, **139**, 467–480.

COCKS, L. R. M., HOLLAND, C. H. & RICKARDS, R. B. 1992. *A Revised Correlation of Silurian Rocks in the British Isles*. The Geological Society, London.

COGNÉ, J. & WRIGHT, A. E. 1980. L'orogène cadomien: vers un essai d'interpretation paléogéodynamique unitaire des phénomènes orogéniques fini-précambriens d'Europe moyenne et occidentale, et leur signification à l'origine de la croûte et du mobilisme varisque puis alpin. *In*: COGNÉ, J. & SLANSKY, M. (eds), *Géologie de l'Europe*. Mémoire du Bureau de Recherches Géologiques et Minières, **108**, 29–55.

COLLINGWOOD, W. G. 1884–1885. The lake-basins of the neighbourhood of Windermere. *Transactions of the Cumberland and Westmorland Association for the Advancement of Literature and Science*, **10**, 1–10.

COLLINS, H. & PINCH, T. 1993. *The Golem: What Everyone Should Know About Science*. Cambridge University Press, Cambridge.

COLLOMB, E. 1847. *Preuves de l'existence d'anciens glaciers dans les vallées des Vosges, du terrain érratique de cette contrée*. V. Masson, Paris.

COOK, P. J. 1998. *A History of the British Geological Survey 1990–1997*. British Geological Survey Technical Report **WQ/98/1**.

COOPER, A. H. 1988–1989 (1990). The Skiddaw Group of Causey Pike. *Proceedings of the Cumberland Geological Society*, **5**, 222–230.

COOPER, A. H., & MOLYNEUX, S. G. 1990. The age and correlation of Skiddaw Group (early Ordovician) sediments in the Cross Fell Inlier (northern England). *Geological Magazine*, **127**, 147–157.

COOPER, A. H., FORTEY, N. J. *et al.* 2001. *The Skiddaw Group of the English Lake District*. British Geological Survey, Keyworth.

COOPER, A. H., MILLWARD, D., JOHNSON, E. W. & SOPER, N. J. 1993. The early Palaeozoic evolution of northwest England. *Geological Magazine*, **130**, 711–724.

COOPER, A. H., RUSHTON, A. W. A., MOLYNEUX, S. G., HUGHES, R. A., MOORE, R. M. & WEBB, B. C. 1995. The stratigraphy, correlation, provenance and palaeogeography of the Skiddaw Group (Ordovician) in the English Lake District. *Geological Magazine*, **132**, 185–211.

COOPER, A. H. *et al.* In press. *The Skiddaw Group of the English Lake District*. British Geological Survey, Keyworth.

COOPER, D. C., LEE, M. K., FORTEY, N. J., RUNDLE, C. C., WEBB, B. C. & ALLEN, P. M. 1988. The Crummock Water aureole: a zone of metasomatism and source of ore metals in the English Lake District. *Journal of the Geological Society*, **145**, 523–540.

COOPER, J. A. 2001. Robin Langford Oliver: geologist with the spirit of adventure. *The Adelaide Advertiser*, 17 February 2001.

COOPER, M. P. & STANLEY, C. J. 1990. *Minerals of the English Lake District: Caldbeck Fells*. British Museum (Natural History), London.

COPE, J. C. W. 1994. A latest Cretaceous hotspot and the southerly tilt of Britain. *Journal of the Geological Society*, **151**, 905–908.

COPE, J. C. W. 1995. Discussion on a latest Cretaceous hotspot and the southeasterly tilt of Britain. *Journal of the Geological Society*, **152**, 731.

COPE, J. C. W. 1998. The Mesozoic and Tertiary history of the Irish Sea. *In*: MEADOWS, N. S., TRUEBLOOD, S., HARDMAN, N. & COWAN, G.

(eds) *The Petroleum Geology of the Irish Sea and Adjacent Areas*. Geological Society, London, Special Publications **124**, 48–59.

COPE, J. C. W., INGHAM, J. K. & RAWSON, P. F. (eds) 1992. *Atlas of Palaeogeography and Lithofacies*. Geological Society, London, Memoir **13**.

CORDIER, L. 1816. Mémoire sur les substances minérales dites "en masse" qui entrent, dans la composition des roches volcaniques de tous les âges. *Journal de Physique*, **82**, 261–268; **83**, 185–161, 284–307, 352–386.

COUPLES, G. D., MCKEOWN, C., HASZELDINE, R. S. & SMYTHE, D. K. 1998. The role of academic environmental geoscientists in radioactive waste disposal assessment. *In*: BENNETT, M. R. & DOYLE, P. (eds) *Issues in Environmental Geology: A British Perspective*. Geological Society, London, 243–276.

COX, K. G. 1980. A model for flood basalt vulcanism. *Journal of Petrology*, **21**, 629–650.

CREESE, M. R. S. 1998. *Ladies in the Laboratory? American and British Women in Science, 1800–1900*. The Scarecrow Press, Lanham, MD and London.

CREESE, M. R. S. & CREESE, T. M. 1994. British women who contributed to research in the geological sciences in the nineteenth century. *British Journal for the History of Science*, **27**, 23–54.

CROLL, J. 1864. On the physical cause of the change of climate during glacial epochs. *PM*, **28**(4th Series), 121–137.

CROLL, J. 1867. On the eccentricity of the Earth's orbit, and its physical relations to the glacial epoch. *PM*, **33**(4th Series), 119–131.

CROLL, J. 1871. On the transport of the Wasdale Crag blocks. *Geological Magazine*, **8**(Decade 1), 15–20.

CROLL, J. 1875. *Climate and Time in their Geological Relations: A Theory of Secular Changes of the Earth's Climate*. Daldy, Isbister & Co., London.

CURRY, D. 1992. Tertiary. *In*: DUFF, P. McL. D. & SMITH, A. J. (eds) *Geology of England and Wales*. Geological Society, London, 389–411.

CURTIS, C. D. *et al.* 1996. The geology and hydrogeology of the Sellafield area (11 May 1994): Chairman's remarks and discussion. *Quarterly Journal of Engineering Geology*, **29** (Supplement), S105–S107.

D'ORBIGNY, A. 1849–1852. *Cours élémentaire de paléontologie et de géologie stratigraphique*, 3 vols. V. Masson, Paris.

DAKYNS, J. R. 1869*a*. Notes on the geology of the Lake District. *Geological Magazine*, **6**(Decade 1), 56–58.

DAKYNS, J. R. 1869*b*. Additional notes on the unconformity between the Green and Skiddaw Slates. *Geological Magazine*, **6**(Decade 1), 116–117.

DAKYNS, J. R. & WARD, J. C. 1875. Volcanic rocks of the Lake-Country. *Geological Magazine*, **2**(Decade 2), 95.

DAKYNS, J. R., TIDDEMAN, R. H. & GOODCHILD, J. G. 1897. *The Geology of the Country between Appleby, Ullswater and Haweswater*. Eyre & Spottiswoode, London; John Menzies, Edinburgh and Glasgow; Hodges, Figgis & Co., Dublin, for HMSO, London.

DAKYNS, J. R., TIDDEMAN, R. H., GUNN, W. & STRAHAN, A. 1890. *The Geology of the Country round Ingleborough, with Parts of Wensleydale and Wharfedale. (Explanation of Quarter-Sheet 97 S.W. New Series, Sheet 50*. HMSO, London.

DAKYNS, J. R., TIDDEMAN, R. H., RUSSELL, R., CLOUGH, C. T. & STRAHAN, A. 1891. *The Geology of the Country Around Mallerstang, with Parts of Wensleydale, Swaledale, and Arkendale. (Explanation of Quarter-Sheet 97 N.W., New Series, Sheet 40)*. Eyre & Spottiswoode, London; John Menzies, Edinburgh; Hodges, Figgis & Co., Dublin, for HMSO, London.

DALY, R. A. 1933 *Igneous Rocks and the Depths of the Earth Containing some Recent Revised Chapters of "Igneous Rocks and Their Origin" (1914)*. McGraw-Hill, New York and London.

DAUBRÉE, G. A. 1860. Études et expériences synthétiques sur le metamorphisme et sur la formation des roches cristallines. *MADS*, **17**, 468–499.

DAVIES, G. F. 1999. *Dynamic Earth: Plates, Plumes and Mantle Convection*. Cambridge University Press, Cambridge.

DAVIES, G. L. 1969. *The Earth in Decay: A History of British Geomorphology 1578–1878*. Macdonald Technical & Scientific, London.

DAVIES, N. & CHAPLOW, R. 1998. Geotechnical characteristics of the Borrowdale Volcanic Group. *Proceedings of the Yorkshire Geological Society*, **52**, 189–197.

DAVIS, N. C. 1989. *The relationship between ignimbrite eruption and caldera collapse in the Borrowdale Volcanic Group of the Central Fells, English Lake District*. PhD dissertation, Sheffield University.

DAVIS, W. M. 1895. The development of certain English rivers. *The Geographical Journal*, **5**, 127–146.

DAY, F. H. 1964. The Rev. J. Clifton Ward. *Proceedings of the Cumberland Geological Society*, **3**, 5–7.

DE RANCE, C. E. 1869. On the surface geology of the Lake-District. *Geological Magazine*, **6**(Decade 1), 489–494.

DE RANCE, C. E. 1870. On the glaciation phenomena of western Lancashire and Cheshire. *Quarterly Journal of the Geological Society of London*, **26**, 641–655.

DE RANCE, C. E. 1871*a*. On the two glaciations of the Lake District. *Geological Magazine*, **8**(Decade 1), 107–117.

DE RANCE, C. E. 1871*b*. Geology of the Cumberland Lakes. *Geological Magazine*, **8**(Decade 1), 238–239.

DE RANCE, C. E. 1871*c*. On the glaciation of the north-west of England. *Geological Magazine*, **8**(Decade 1), 412–418.

DEAN, W. T. 1959. The stratigraphy of the Caradoc Series in the Cross Fell Inlier. *Proceedings of the Yorkshire Geological Society*, **32**, 185–227 and plate.

DEAN, W. T. 1962. The trilobites of the Caradoc Series in the Cross Fell Inlier of northern England. *Bulletin of the British Museum (Natural History) Geology*, **7**, 65–134.

DEAN, W. T. 1963. The Stile End Beds and Drygill Shales (Ordovician) in the east and north of the English Lake District. *Bulletin of the British Museum (Natural History) Geology*, **9**, 49–65 and plates.

DEARDON, J. S. 1968. [John Bolton] The Father of Furness Geology. *Cumbria*, **17**, 542–544.

DEWEY, J. F. 1969*a*. Evolution of the Appalachian/Caledonian orogen. *Nature*, **222**, 124–129.

DEWEY, J. F. 1969*b*. Structure and sequence in paratectonic British Caledonides. *In*: M. KAY (ed.) *North Atlantic Geology and Continental Drift: A Symposium*. American Association of Petroleum Geologists, Tulsa, 309–335.

DEWEY, J. F. 1982. Plate tectonics and the evolution of the British Isles. *Journal of the Geological Society*, **139**, 317–412.

DEWEY, J. F. & WINDLEY, B. F. 1988. Palaeocene–Oligocene tectonics of NW Europe. *In*: MORTON, A. C. & PARSON, L. M. (eds) *Early Tertiary Volcanism and the Opening of the NE Atlantic*. Geological Society, Special Publications, **39**, Geological Society & Blackwell Scientific, Oxford, 25–31.

DICKINS, D. 1998–1999 (2000). Henry Alleyne Nicholson (1844–1899): a renowned Cumbrian polymath. *Proceedings of the Cumberland Geological Society*, **6**, 349–352.

DICKINSON, W. R., INGERSOLL, R. V., COWAN, D. S., HELMOLD, K. P. & SUZEK, C. A. 1983. Provenance of Franciscan greywacke in coastal California. *Bulletin of the Geological Society of America*, **93**, 95–107.

DIXON, E. E. L. 1922 (for 1921). The retreat of the Lake District ice-cap in the Ennerdale area, West Cumberland. *Summary of Progress of the Geological Survey of Great Britain*, Appendix 1, 118–128.

DIXON, E. E. L. 1928 (for 1927). The origin of Cumberland haematite (1). *Summary of Progress of the Geological Survey of Great Britain* (Part 2), 23–32.

DIXON, E. E. L., MADEN, J., TROTTER, F. M., HOLLINGWORTH, S. E. & TONKS, L. H. 1926. *The Geology of the Carlisle, Longtown and Silloth District, MGS, Sheets 11, 16, and 17*. HMSO, London.

DOBSON, M. R. 1977. The geological structure of the Irish Sea. *In*: KIDSON, C. & TOOLEY, M. J. (eds) *The Quaternary History of the Irish Sea*. Seel House Press, Liverpool, 13–26.

DODD, M. 1990–1991 (1992). The Edgar Shackleton Collection. *Proceedings of the Cumberland Geological Society*, **5**, 395–397.

DONALD, T. 1802 *The County of Cumberland Surveyed Anno MDCCLXX and Anno MDCCLXXI by Thos. Donald*, 2nd edn, W. Fadden, London.

DONALD, T. 1810. *The County of Cumberland drawn from the Large Map in Six Sheets Surveyed by Thomas Donald to which have now been added Portions of the Adjoining Counties of Lancaster and Westmoreland thereby Comprehending all the Lakes in that Neighbourhood (2 miles to the inch)*. W. Fadden, London.

DONALD, T. (revised FRYER, J. H.) 1818. *The County of Cumberland Surveyed Anno MDCCLXX and MDCCLXXI by Thos. Donald with Corrections & Additions Made in MDCCCXVIII by J. H. Fryer*. W. Fadden, London.

DOUBLE, I. S. 1932–1933. President's address. *Proceedings of the Liverpool Geological Society*, **16**, 117–118.

DOWNIE, C. 1984 Acritarchs in British stratigraphy. *Special Report of the Geological Society, London*, **17**. Geological Society, London.

DOWNIE, C. & SOPER, N. J. 1972. Age of the Eycott Volcanic Group and its conformable relationship to the Skiddaw Slates in the English Lake District. *Geological Magazine*, **109**, 259–268.

DUDDY, I. R., GREEN, P. F., BRAY, R. J. & HEGARTY, K. A. 1994. Recognition of the thermal effects of fluid flow in sedimentary basins. *In*: PARNELL, J. (ed.) *Geofluids: Origins, Migration and Evolution of Fluids in Sedmentary Basins*. Geological Society, London, Special Publications, **78**, 325–345.

DUFF, P. McL. D. & SMITH, A. J. (eds) 1992. *Geology of England and Wales*. Geological Society, London.

DUNHAM, K. C. 1934. The genesis of the north Pennine ore deposits. *Quarterly Journal of the Geological Society of London*, **90**, 689–720 and plates.

DUNHAM, K. C. 1991. William Colin Rose C. B. E. (1908–1991). *Annual Report of the Geological Society*, 28.

DUNHAM, K. C. & ROSE, W. C. C. 1941. Geology of the iron-ore field of south Cumberland and Furness. *Wartime Pamphlet: Geological Survey of Great Britain*, **16**. Department of Scientific & Industrial Research.

DUNHAM, K. C. & ROSE, W. C. C. 1949. Permo-Triassic geology of south Cumberland and Furness. *Proceedings of the Geologists' Association*, **60**, 11–40.

DWERRYHOUSE, A. R. 1909. On some intrusive rocks in the neighbourhood of Eskdale (Cumberland). *Quarterly Journal of the Geological Society of London*, **65**, 55–80 and plate.

EASTWOOD, T. 1928 (for 1927). The Cockermouth lavas, Cumberland – a Carboniferous volcanic episode. *Summary of Progress of the Geological Survey of Great Britain* (Part 2), 15–22.

EASTWOOD, T. 1930. *The Geology of the Maryport District*. HMSO, London.

EASTWOOD, T. 1931 (for 1930). Cumbrian district [Cockermouth Sheet 23]. *Summary of Progress of the Geological Survey of Great Britain*, 52–55.

EASTWOOD, T. 1932 (for 1931). Cumbria district [Cockermouth, Sheet 23]. *Summary of Progress of the Geological Survey of Great Britain*, 52–54.

EASTWOOD, T. 1933 (for 1932). Cumbria district [Cockermouth Sheet 23 and Gosforth Sheet 37]. *Summary of Progress of the Geological Survey of Great Britain*, 59–64.

EASTWOOD, T. 1938 (for 1937). South Cumberland and Furness: Bootle, Ulverston and Barrow (Sheets 47, 48, 58). *Summary of Progress of the Geological Survey of Great Britain*, 48–50.

EASTWOOD, T. 1939 (for 1938). South Cumberland, Furness and Cartmel area. *Summary of Progress of the Geological Survey of Great Britain*, 43–47.

EASTWOOD, T. 1946 *British Regional Geology: Northern England*, 2nd edn. HMSO, London.

EASTWOOD, T. 1963. Ernest Edward Leslie Dixon. *PGSL*, **1611**, 139–140.

EASTWOOD, T., DIXON, E. E. L., HOLLINGWORTH, S. E. & SMITH, B. 1931. *The Geology of Whitehaven and Workington District*. HMSO, London.

EASTWOOD, T.,, HOLLINGWORTH, S. E., ROSE, W. C. C. & TROTTER, F. M. 1968. *Geology of the Country around Cockermouth and Caldbeck (Explanation of One-Inch Geological Sheet 23, New Series)*. HMSO, London.

EDMONDS, J. M. 1970–1971. Tom Eastwood. *Proceedings of the Cumberland Geological Society*, **3**(1), 48–51.

EDWARDS, A. B. 1936. On the occurrence of almandine garnets in some Devonian igneous rocks of Victoria. *Proceedings of the Royal Society of Victoria*, **49**, New Series, 40–50.

EDWARDS, R. 1996. Sellafield's Trojan horse. *New Scientist*, 6 January, 11–12.

EICHELBERGER, J. C. 1980. Vesiculation of mafic magma during replenishment of silicic magma reserves. *Nature*, **288**, 446–450.

ÉLIE DE BEAUMONT, L. 1828–1830. Recherches sur quelques-unes des révolutions de la surface du globe . . . *Annales des Sciences Naturelles*, **18**, 284–416; **19**, 177–240.

ÉLIE DE BEAUMONT, L. 1831. Researches on some of the revolutions of the globe . . . *Annals of Philosophy*, **10**, New Series, 241–264.

ELLENBERGER, F. 1978. The first International Geological Congress Paris, 1878. *Episodes*, **1**, 20–24.

ELLES, G. L. 1898. The graptolite-fauna of the Skiddaw Slates. *Quarterly Journal of the Geological Society of London*, **54**, 463–539 and plate.

ELLES, G. L. 1922. The graptolite faunas of the British Isles, *Proceedings of the Geologists' Association*, **33**, 168–200 and table.

ELLES, G. L. 1925. Sedgwick Museum notes. The characteristic assemblages of the graptolite zones of the British Isles. *Geological Magazine*, **62**, 337–347.

ELLES, G. L. 1933*a* (for 1932). The Lower Ordovician graptolite faunas with special reference to the Skiddaw Slates. *Summary of Progress of the Geological Survey of Great Britain*, 94–111.

ELLES, G. L. 1933*b*. Graptolite assemblages and the doctrine of trends. *Geological Magazine*, **70**, 351–354.

ELLES, G. L. & WOOD, E. M. R. 1895. Supplementary notes on the Drygill Shales. *Geological Magazine*, **2**(Decade 4), 246–249.

ELLES, G. L. & WOOD, E. M. R. 1901–1918. *Monograph of British Graptolites*. Palaeontographical Society, London.

EMILIANI, C. 1955. Pleistocene temperatures. *Journal of Geology*, **63**, 538–578.

EMILIANI, C. 1957. Temperature and age analysis of deep-sea cores. *Science*, **125**, 383–387.

EVANS, W. B. & ARTHURTON, R. S. 1973 Northwest England. *In*: MITCHELL, G. F., PENNY, L. F., SHOTTON, F. W. & WEST, R. G. (eds) *A Correlation of Quaternary Deposits in the British Isles*. Geological Society, London, 28–36.

EYLES, N. & MCCABE, A. M. 1989. The Late Devensian (< 22,000 BP) Irish Sea basin: the sedimentary record of a collapsed ice sheet margin. *Quaternary Science Reviews*, **8**, 307–351.

EYLES, V. A. & EYLES, J. M. 1938. On the different issues of the first geological map of England and Wales. *Annals of Science*, **3**, 190–212 and plates.

FALLER, A. M. & BRIDEN, J. C. 1978. Palaeomagnetism of Lake District rocks. *In*: MOSELEY, F. (ed.) *The Geology of the Lake District*. Yorkshire Geological Society, Leeds, 17–24.

FALLER, A. M. & BRIDEN, J. C. & MORRIS, W. A. 1977. Palaeomagnetic results from the Borrowdale Volcanic Group, English Lake District. *Geophysical Journal of the Royal Astronomical Society*, **48**, 111–121.

FAREY, J. 1811. *General View of the Agriculture and Minerals of Derbyshire, Vol. 1*. MacMillan, London (republished in facsimile, Peak District Mines Historical Society, Sheffield, 1989).

FENTON, C. & FENTON, M. 1945. *Giants of Geology*. Doubleday, New York.

FIRMAN, R. J. 1953. *Metamorphism and metasomatism around the Shap and Eskdale Granites*. PhD dissertation, Durham University.

FIRMAN, R. J. 1954. Note on metasomatic changes in the rocks adjacent to the Shap Granite. *Proceedings of the Geologists' Association*, **65**, 412–414.

FIRMAN, R. J. 1956. Garnets in the Borrowdale Volcanic Series. *Geological Magazine*, **93**, 435–436.

FIRMAN, R. J. 1957*a*. Fissure metasomatism in volcanic rocks adjacent to the Shap Granite, Westmorland. *Quarterly Journal of the Geological Society of London*, **113**, 205–222 and plates.

FIRMAN, R. J. 1957*b*. The Borrowdale Volcanic Series between Wastwater and Duddon Valley, Cumberland. *Proceedings of the Yorkshire Geological Society*, **31**, 39–64 and plate.

FIRMAN, R. J. 1959. The relationship between the joints, faults and mineral veins in the Eskdale Granite (Cumberland) and the adjacent Borrowdale Volcanic Series. *Proceedings of the Geological Society*, **1568**, 64–67.

FIRMAN, R. J. 1960. The relationship between joints and fault patterns in the Eskdale Granite (Cumberland) and the adjacent Borrowdale Volcanic Series. *Quarterly Journal of the Geological Society of London*, **116**, 317–347.

FIRMAN, R. J. & LEE, M. K. 1986. Age and structure of the concealed English Lake District batholith and its probable influence on subsequent sedimentation, tectonics and mineralisation. *In*: NESBITT, R. W. & NICHOL, J. (eds) *Geology in the Real World – The Kingsley Dunham Volume*. Institute of Mining & Metallurgy, Basildon, 117–127.

FIRMAN, R. J. & LEE, M. K. 1987. The English Lake District batholith – Ordovician, Silurian, Devonian or . . . ? *Geological Magazine*, **124**, 585–587.

FITTON, J. G. 1971*a*. *The petrogenesis of the calc-alkaline Borrowdale Volcanic Group, northern England*. PhD dissertation, Durham University.

FITTON, J. G. 1971*b*. The generation of magmas in island arcs. *Earth and Planetary Science Letters*, **11**, 63–67.

FITTON, J. G. & HUGHES D. J. 1970. Volcanism and plate tectonics in the British Ordovician. *Earth and Planetary Science Letters*, **8**, 223–228.

FITTON, J. G., THIRLWALL, M. F. & HUGHES, D. J. 1982. Volcanism in the Caledonian orogenic belt of Britain. *In*: THORPE, R. S. (ed.) *Andesites: Orogenic Andesites and Related Rocks*. John Wiley, Chichester, 611–636.

FITTON, W. H. 1834. Presidential address [for 1829]. *Proceedings of the Geological Society*, **1**, 112–134.

FLETT, J. S. 1921 (for 1920). Cumberland, Westmorland and Furness. *Summary of Progress of the Geological Survey of Great Britain*, 28–29.

FLETT, J. S. 1937. *The First Hundred Years of the Geological Survey of Great Britain*. HMSO, London.

FLORES, G. 1959. Evidences of slump phenomena (olistostromes) in areas of hydrocarbon exploration in Sicily. *Proceedings of the Fifth World Petroleum Congress. Geology and Geophysics. Section 1, Paper 13*. New York, 259–275.

FORBES, E. 1854. Presidential address. *Quarterly Journal of the Geological Society of London*, **10**, xxviii–lxxxi.

FORD, T. D. & TORRENS, H. S. 1989. John Farey (1766–1826) an unrecognised polymath. *Introduction to*: FAREY (1989), no page numbers.

FORSTER, W. 1809. *A Treatise on a Section of the Strata, Commencing near Newcastle upon Tyne, and Concluding on the West Side of the Mountain of Cross-Fell. With Remarks on Mineral Veins in General, and Engraved Figures of some of the Different Species of these Productions, to which are Added Tables of the Strata in Yorkshire and Derbyshire*. Preston & Heaton, Newcastle.

FORTEY, N. J. 1989. Low grade metamorphism in the Lower Ordovician Skiddaw Group of the Lake District, England. *Proceedings of the Yorkshire Geological Society*, **47**, 325–337.

FORTEY, N. J. & COOPER, D. C. 1986. Tourmalinisation in the Skiddaw Group around Crummock Water, English Lake District. *Mineralogical Magazine*, **50**, 17–26.

FORTEY, R. A. 2000. *Trilobite! Eyewitness to Evolution*. HarperCollins, London.

FORTEY, R. A. & OWENS, R. M. 1987. The Arenig Series in South Wales. *Bulletin of the British Museum (Natural History) Geology*, **41**, 69–307.

FORTEY, R. A., HARPER, D. A. T., INGHAM, J. K., OWEN, A. W., PARKES, M. A., RUSHTON, A. W. A. & WOODCOCK, N. H. 2000. *A Revised Correlation of Ordovician Rocks in the British Isles*. Geological Society, London.

FORTEY, R. A., HARPER, D. A. T., INGHAM, J. K., OWEN, A. W., & RUSHTON, A. W. A. 1995. A revision of Ordovician series and stages from the historical type area. *Geological Magazine*, **132**, 15–30.

FORTEY, R. A. & OWENS, R. M. & RUSHTON, A. W. A. 1989. The palaeographic position of the Lake District in the early Ordovician. *Geological Magazine*, **126**, 9–17.

FOX, M. 1998–1999 (2000). Robert Harkness (1816–1878): a geological pioneer in northern England and southern Scotland. *Proceedings of the Cumberland Geological Society*, **6**, 333–340.

FREEMAN, B., KLEMPERER, S. L. & HOBBS, R. W. 1988. The deep structure of northern England and Iapetus Suture zone from BIRPS deep seismic reflection profiles. *Journal of the Geological Society*, **145**, 727–740.

FRYER, J.[?] 1814[?]. Hand-coloured geological map of the Lake District (Geological Society Archives).

FYFE, W. S. & MCBIRNEY, A. R. 1975. Subduction and the structure of andesitic volcanic belts. *American Journal of Science*, **275A**, 285–297.

G. H. M. 1891. Daniel MacKintosh, FGS. Born 1815. Died 19 July 1891. *Geological Magazine*, **8**(Decade 3), 432.

GALE, N. H., BECKINSALE, R. D. & WADGE, A. J. 1979. A Rb–Sr whole-rock isochron for the Stockdale Rhyolite of the English Lake District and a revised mid-Palaeozoic time-scale. *Journal of the Geological Society*, **136**, 235–242.

GARWOOD, E. J. & THOMAS, H. H. 1934. John Edward Marr. *Quarterly Journal of the Geological Society of London*, **90**, lx–lxiv.

GEIKIE, A. 1863. *The Scenery of Scotland Viewed in Connexion with its Physical Geology*. Macmillan, London and Cambridge.

GEIKIE, A. 1884. The crystalline rocks of the Scottish Highlands. *Nature*, **31**, 29–31.

GEIKIE, A. 1891. Anniversary address of the President. *Proceedings of the Geological Society*, **47**, 48–162.

GEIKIE, A. 1895. *Memoir of Sir Andrew Crombie Ramsay*. Macmillan, London and New York.

GEIKIE, A. 1897. *The Ancient Volcanoes of Great Britain*, 2 vols. Macmillan, London and New York.

GEIKIE, A. 1903. *Text-Book of Geology*, 4th edn, 2 vols. Macmillan, London and New York.

GEIKIE, A. 1904. Vice-President's anniversary address. *Quarterly Journal of the Geological Society of London*, **60**, lxxx–civ.

GEIKIE, J. 1871–1872. Changes of climate during the glacial period. *Geological Magazine*, **8**(Decade 1), 545–553; **9**, 23–31, 61–69, 105–111, 164–170, 215–222, 254–265.

GEIKIE, J. 1874. *The Great Ice Age and its Relation to the Antiquity of Man*. W. Isbister, London.

GEIKIE, J. 1877. *The Great Ice Age and its Relation to the Antiquity of Man*, 2nd edn. W. Isbister, London.

GEIKIE, J. 1882. The intercrossing of erratics in glacial deposits. *The Scottish Naturalist*, **6**, 193–200, 241–254.

GEIKIE, J. 1890 (for 1889). Presidential address: Section C. – Geology [Recent researches in the glacial geology of the Continent]. *Report of the British Association for the Advancement of Science*, 551–564.

GEIKIE, J. 1892. On the glacial succession in Europe. *TRSE*, **37**, 127–149.

GEIKIE, J. 1893. The glacial period and earth movement hypothesis. *In*: *Fragments of Earth Lore: Sketches & Addresses Geological and Geographical*. John Bartholomew, London & Edinburgh.

GEIKIE, J. 1895. The classification of European glacial deposits. *Journal of Geology*, **3**, 241–269.

GEOLOGICAL SURVEY OF ENGLAND AND WALES 1873. Six inches to the mile, geological map, *Sheet LXIV. Area of Borrowdale in Sheets 64, 69, 70, 74, 75. Geologically Surveyed by J. C. Ward under the Supervision of W. T. Aveline*. Ordnance Survey Office, Southampton.

GEOLOGICAL SURVEY OF ENGLAND AND WALES 1875. One inch to the mile, geological map, *Keswick, Sheet 101 S.E. Geologically Surveyed by J. C. Ward under the Supervision of W. T. Aveline*. Ordnance Survey Office, Southampton.

GEOLOGICAL SURVEY OF ENGLAND AND WALES 1877. *Section from the Carboniferous Limestone, 2 1/2 Miles S.W. of Caldbeck, across the Volcanic Rocks North of the Roughton Gill Mines, the Granite and Metamorphic Rocks of Skiddaw Forest, the Skiddaw Slates of Blencathra, the Quartz Felsite of the Vale of St. John, and the Volcanic Series of the Helvellyn Range of Grasmere and Rydal, and of Loughrigg, to the Coniston Limestone and Coniston Flags and Grits West of Windermere Lake. By J. C. Ward, FGS, C. E. De Rance, FGS, and E. J. Hebert MA. Horizontal Section No. 117*. Ordnance Survey Office, Southampton.

GEOLOGICAL SURVEY OF ENGLAND AND WALES 1893. One inch to the mile, geological map, *Appleby, Sheet 102 S.W. Geologically Surveyed by W. T. Aveline, T. McK. Hughes, J. R. Dakyns, J. C. Ward, R. Russell, W. H. Dalton, J. G. Goodchild, G. H. Lightfoot & E. J. Hebert*. Ordnance Survey Office, Southampton.

GEOLOGICAL SURVEY OF ENGLAND AND WALES 1937. One inch to the mile, geological map, *Gosforth Solid*. Ordnance Survey Office, Southampton.

GILLISPIE, C. C. 1951. *Genesis and Geology*. Harvard University Press, Cambridge.

GIRAUD, J. 1902 Hypothèse sur la formation de pépérites. *In*: *Etudes géologiques sur la Limagne (Auvergne), Bulletin des Services de la Carte Géologique de la France et des Topographies Souterraines*, **30** (Bulletin No. **87**), 360–367.

GLEADOW, A. J. W., DUDDY, I. R., GREEN, P. F. & LOVERING, J. F. 1986. Confined fission track lengths in apatite: a diagnostic tool for thermal history analysis. *Contributions to Mineralogy and Petrology*, **96**, 405–415.

GLENNIE, K. W. 1955. *The structure of the Skiddaw Slates in the north-west Lake District*. MSc dissertation, University of Edinburgh.

GOODCHILD, J. G. 1875*a*. The glacial phenomena of the Eden Valley and the western part of the Yorkshire-Dale district. *Quarterly Journal of the Geological Society of London*, **31**, 55–99 and plate (communicated by H. W. Bristow).

GOODCHILD, J. G. 1875*b*. Glacial erosion. *Geological Magazine*, **2**(Decade 2), 323–328, 356–362.

GOODCHILD, J. G. 1875c. On the origin of coums. *Geological Magazine*, **2**(Decade 2), 486–497.

GOODCHILD, J. G. 1882–1883. Professor Robert Harkness, FRS, FGS. *Transactions of the Cumberland and Westmorland Association for the Advancement of Literature and Science*, **8**, 145–188.

GOODCHILD, J. G. 1885–1886. Observations upon the stratigraphical relations of the Skiddaw Slates. *Proceedings of the Geologists' Association*, **9**, 469–481.

GOODCHILD, J. G. 1888–1889. The history of the Eden and some rivers adjacent. *Transactions of the Cumberland and Westmorland Association for the Advancement of Literature and Science*, **14**, 73–93.

GOODCHILD, J. G. 1889–1890. Some observations upon the mode of occurrence and the genesis of metalliferous deposits. *Proceedings of the Geologists' Association*, **11**, 63.

GOULD, S. J. 1989. *Wonderful Life: The Burgess Shale and the Nature of History*. W. W. Norton, New York and London.

GRANTHAM, D. R. 1928. The petrology of the Shap Granite. *Proceedings of the Geologists' Association*, **39**, 299–331 and plates.

GREEN, J. F. N. 1908. The geological structure of the St David's area (Pembrokeshire). *Quarterly Journal of the Geological Society of London*, **64**, 363–383.

GREEN, J. F. N. 1911*a*. The geology of the district around St David's, Pembrokeshire. *Proceedings of the Geologists' Association*, **22**, 121–137.

GREEN, J. F. N. 1911*b*. The older Palaeozoic succession of the Duddon Estuary. *Proceedings of the Geological Society* (Abstracts), **913**, 71–74.

GREEN, J. F. N. n.d. [1913]. *The Older Palaeozoic Succession of the Duddon Estuary*. Hayman, Christy & Lilly, London (privately published).

GREEN, J. F. N. 1915*a*. The structure of the eastern part of the Lake District. *Proceedings of the Geologists' Association*, **26**, 195–223 and plates.

GREEN, J. F. N. 1915*b*. The garnets and streaky rocks of the English Lake District. *Mineralogical Magazine*, **17**, 207–217.

GREEN, J. F. N. 1917. The age of the chief intrusions of the Lake District. *Proceedings of the Geologists' Association*, **28**, 1–30.

GREEN, J. F. N. 1918*a*. The Mell Fell Conglomerate. *Proceedings of the Geologists' Association*, **29**, 117–125.

GREEN, J. F. N. 1918*b*. The Skiddaw Granite: a structural study. *Proceedings of the Geologists' Association*, **29**, 126–136.

GREEN, J. F. N. 1919. The vulcanicity of the Lake District. *Proceedings of the Geologists' Association*, **30**, 153–182 and plates.

GREEN, J. F. N. 1920. The geological structure of the Lake District. *Proceedings of the Geologists' Association*, **31**, 109–126 and plate.

GREEN, J. F. N. 1921. Long excursion to the Lake District. August 7th to 17th, 1920. *Proceedings of the Geologists' Association*, **32**, 123–138 and plate.

GREEN, J. F. N. 1924. The structure of the Bowmore–Portaskaig District of Islay. *Quarterly Journal of the Geological Society of London*, **80**, 72–105 and plates.

GREEN, J. F. N. 1925. Reply on occasion of presentation of Lyell Medal. *Quarterly Journal of the Geological Society of London*, **81**, xliv–xlv.

GREEN, P. F. 1986. On the thermo-tectonic evolution of northern England: evidence from fission-track analysis. *Geological Magazine*, **123**, 493–506.

GREEN, P. F. 1989. Thermal and tectonic history of the east Midlands shelf (onshore UK) and surrounding regions assessed by apatite fission analysis. *Journal of the Geological Society*, **146**, 755–773.

GREEN, P. F. 2002. Early Tertiary paleo-thermal effects in northern England: reconciling results from fission track analysis with geological evidence. *Tectonophysics*, **349**, 131–144.

GREEN, P. F., DUDDY, I. R. & BRAY, R. J. 1993*a*. Early Tertiary heating in northwest England: fluids or burial (or both?). *In*: PARNELL, J., RUSSELL, A. H. & MOLES, N. R. (eds) *Geofluids '93: Contributions to an International Conference on Fluid Evolution, Migration and Interaction in Rocks*. 119–123.

GREEN, P. F., DUDDY, I. R., & BRAY, R. J. 1995*a*. Further discussion on Mesozoic cover over northern England: interpretation of apatite fission data. *Journal of the Geological Society*, **152**, 416.

GREEN, P. F., DUDDY, I. R., & BRAY, R. J. 1995*b*. Discussion of 'Low temperature thermal history of eastern Ireland: effects of fluid flow' by A. A. McCulloch (1994). *Marine and Petroleum Geology*, **12**, 773–776.

GREEN, P. F., DUDDY, I. R., & BRAY, R. J. 1997. Variation in thermal history styles around the Irish Sea and adjacent areas: implications for hydrocarbon occurrence and tectonic evolution. *In*: MEADOWS, N. S., TRUEBLOOD, S. P., HARDMAN, S. P. & COWAN, G. (eds) *Petroleum Geology of the Irish Sea and Adjacent Areas*. Geological Society, London, Special Publications, **124**, 73–93.

GREEN, P. F., DUDDY, I. R., & BRAY, R. J. & LEWIS, C. L. E. 1993*b*. Elevated palaeotemperatures prior to early Tertiary cooling throughout the UK region: implications for hydrocarbon generation. *In*: PARKER, J. R. (ed.) *Petroleum Geology of Northwest Europe: Proceedings of the 4th Conference*. Geological Society, London, 1067–1074.

GREEN, P. F., DUDDY, I. R., GLEADOW, A. J. W., TINGATE, P. R. & LASLETT, G. M. 1986. Thermal annealing of fission tracks in apatite 1. A qualitative description. *Chemical Geology (Isotope Geoscience Section)*, **59**, 237–253.

GREEN, P. F., DUDDY, I. R., HEGARTY, K. A. & BRAY, R. J. 1999. Early Tertiary heat flow along the UK Atlantic margin and adjacent areas. *In*: FLEET, A. J. & BOLDY, S. A. R. (eds) *Petroleum Geology of Northwest Europe: Proceedings of the 5th Conference*. Geological Society, London, 349–357.

GREEN, P. F., DUDDY, I. R., HEGARTY, K. A. & LOVERING, J. F. 1989*a*. Apatite fission track analysis as a palaeotemperature indicator for hydrocarbon exploration. *In*: NAESER, N. D. & MCCULLOCH, T. (eds) *Thermal History of Sedimentary Basins – Methods and Case Histories*. Springer-Verlag, New York, 181–195.

GREEN, P. F., DUDDY, I. R., LASLETT, G. M., HEGARTY, K. A., GLEADOW, A. J. W. & LOVERING, J. F. 1989*b*. Thermal annealing of fission tracks in apatite 4. Quantitative modelling techniques and extension to geological timescales. *Chemical Geology (Isotope Geoscience Section)*, **79**, 155–182.

GREEN, T. H. & RINGWOOD, A. E. 1968. Origin of garnet phenocrysts in calc-alkaline rocks. *Contributions to Mineralogy and Petrology*, **18**, 163–174.

GREENOUGH, G. B. (delineated by T. WEBSTER; engraved by S. NEELE & Son) 1819. *A Geological Map of England & Wales by G. B. Greenough Esq. President of the Geological Society FRS, FLS*. Longman, Hurst, Rees, Orme & Brown, London.

GREENWOOD, C. & GREENWOOD, J. 1823. *Map of the County of Cumberland from an Actual Survey Made in the Years 1821 & 1822 by C. & J. Greenwood*. George Pringle, London.

GREENWOOD, C. & GREENWOOD, J. 1824. *Map of the County of Westmorland, from an Actual Survey Made in the Years 1822 & 1823 by C. & J. Greenwood*. George Pringle, London.

GREGORY, J. W. 1909. Obituary notice of John George Goodchild, born 26th May 1844, died 21st Feb. 1906. *Transactions of the Edinburgh Geological Society*, **9**, 331–350 and plate.

GURNIS, M. 2001. Sculpting the Earth from inside out: powerful motions deep inside the planet do not merely shove fragments of the rocky shell horizontally round the globe – they also lift and lower entire continents. *Scientific American*, March, 34–41.

GUTMANIS, J. C., LANYON, G. W., WYNN, T. J. & WATSON, C. R. 1998. Fluid flow in faults: a study of fault hydrogeology in Triassic sandstone and Ordovician volcaniclastic rocks at Sellafield, North-west England. *Proceedings of the Yorkshire Geological Society*, **52**, 159–175.

H., R. G. S. 1950. In memoriam Edmund Johnston Garwood 1864–1949. *Proceedings of the Yorkshire Geological Society*, **28**, 1–4.

HACKETT, D. 1999. *Our Corporate History: Key Events Affecting the British Geological Survey, 1967–1998*. British Geological Survey Technical Report **WQ/99/1**.

HADFIELD, G. S. & WHITESIDE, H. C. M. 1936. The Borrowdale Volcanic Series of High Rigg and the adjoining Low Rigg microgranite. *Proceedings of the Geologists' Association*, **47**, 42–64 and plates.

HALL, J. 1865. *Graptolites of the Quebec Group*. Dawson Brothers, Montreal; Ballière, London and New York.

HAMILTON, B. M. 2001. Charles Lapworth's "The Moffat series", 1878. *Episodes*, **24**, 194–200.

HANCOX, E. G. 1934. The Haweswater Dolerite, Westmorland. *Proceedings of the Liverpool Geological Society*, **16**, 173–197.

HARKER, A. 1894*a*. Carrock Fell: a study in the variation of igneous rock-masses – Part I. The Gabbro. *Quarterly Journal of the Geological Society of London*, **50**, 311–317 and plates.

HARKER, A. 1894*b*. On some variolitic rocks on Carrock Fell. *Geological Magazine*, **1**(Decade 4), 551–553.

HARKER, A. 1895. Carrock Fell: a study in the variation of igneous rock-masses. Part II. The Carrock Fell granophyre. Part III. The Grainsgill greisen. *Quarterly Journal of the Geological Society of London*, **51**, 125–148, plate and map.

HARKER, A. 1902. Notes on the igneous rocks of the English Lake District. *Proceedings of the Yorkshire Geological Society*, **14**, 487–493.

HARKER, A. 1909. *The Natural History of Igneous Rocks*. Methuen, London.

HARKER, A. 1933. John Edward Marr. *The Eagle*, **48**, 64–65.

HARKER, A. & MARR, J. E. 1891. The Shap Granite and associated igneous and metamorphic rocks. *Quarterly Journal of the Geological Society of London*, **47**, 266–328 and plate.

HARKER, A. & MARR, J. E. 1893. Supplementary notes on the metamorphic rocks around the Shap Granite. *Quarterly Journal of the Geological Society of London*, **5**, 359–371 and plate.

HARKNESS, R. 1858 (for 1857). On the geology of the Caldbeck fells, and the lower sedimentary rocks of Cumberland. *Report of the British Association for the Advancement of Science*, 67–68.

HARKNESS, R. 1863. On the Skiddaw Slate Series with a note on the graptolites by J. W. Salter. *Quarterly Journal of the Geological Society of London*, **19**, 113–140.

HARKNESS, R. 1864 (for 1863). On the fossils of the Skiddaw Slates. *Report of the British Association for the Advancement of Science*, 69.

HARKNESS, R. 1865. On the Lower Silurian rocks of the south-east of Cumberland and the north-east of Westmorland. *Quarterly Journal of the Geological Society of London*, **21**, 235–249.

HARKNESS, R. 1870. On the distribution of the Wasdale Crag blocks, 'Shap Fell Granite boulders', in Westmorland. *Quarterly Journal of the Geological Society of London*, **26**, 517–528.

HARKNESS, R. & NICHOLSON, H. A. 1866. Additional observations on the geology of the Lake Country, with a note on two new species of trilobites by J. W. Salter. *Proceedings of the Geological Society*, **22**, 480–486.

HARKNESS, R. & NICHOLSON, H. A. 1868. On the Coniston Group. *Quarterly Journal of the Geological Society of London*, **24**, 296–303.

HARKNESS, R. & NICHOLSON, H. A. 1870 (for 1869). On the Green Slates and Porphyries of the Lake District. *Report of the British Association for the Advancement of Science*, 74.

HARKNESS, R. & NICHOLSON, H. A. 1877. On the strata and their fossil contents between the Borrowdale Series of the north of England and the Coniston Flags. *Quarterly Journal of the Geological Society of London*, **33**, 461–484.

HARLAND, W. B. 1971. Tectonic transpression in Caledonian Spitsbergen. *Geological Magazine*, **108**, 27–42.

HARLAND, W. B. & GAYER, R. A. 1972. The Arctic Caledonides and earlier oceans. *Geological Magazine*, **109**, 289–314.

HARLAND, W. B., ARMSTRONG, R. L., COX, A. V., CRAIG, L. E., SMITH, A. G. & SMITH, D. G. 1989. *A Geologic Time Scale 1989*. Cambridge University Press, Cambridge.

HARTLEY, J. J. 1924. *The geological structure of the Borrowdale Volcanic Series as developed in the area lying between Ambleside, Grasmere and Coniston*. MSc dissertation, London University external.

HARTLEY, J. J. 1925. The succession and structure of the Borrowdale Volcanic Series as developed in the area lying between the lakes of Grasmere, Windermere and Coniston. *Proceedings of the Geologists' Association*, **36**, 203–206.

HARTLEY, J. J. 1932. The volcanic and other igneous rocks of great and Little Langdale, Westmorland; with notes on the tectonics of the district. *Proceedings of the Geologists' Association*, **43**, 32–69.

HARTLEY, J. J. 1941. The geology of Helvellyn and the southern part of Thirlmere. *Quarterly Journal of the Geological Society of London*, **97**, 129–162 and plates.

HASZELDINE, R. S. & MCKEOWN, C. 1995. A model approach to radioactive waste disposal at Sellafield. *Terra Nova*, **7**, 87–95.

HASZELDINE, R. S. & SMYTHE, D. K. (eds) 1996. *Radioactive Waste Disposal at Sellafield, UK: Site Selection, Geological and Engineering Problems*. Department of Geology and Applied Geology, The University of Glasgow, Glasgow.

HASZELDINE, R. S. & SMYTHE, D. K. 1997. Why was Sellafield rejected as a disposal site for radioactive waste? *Geoscientist*, **7**(7), 18–20.

HEATHCOTE, J. A., JONES, M. A. & HERBERT, A. W. 1996. Modelling groundwater flow in the Sellafield area. *Quarterly Journal of Engineering Geology*, **29**, S59–S81.

HEER, O. 1858. *Die Schieferkohle von Utznach und Dürnten – Oeffentlicher Vortrag gehalten den 7. Januar 1858*. Orell, Füßli, Zurich.

HELM, D. G. 1968. *The stratigraphy and Caledonian tectonic history of the Black Combe area, Lake District*. PhD dissertation, Birkbeck College, London University.

HELM, D. G. 1970. Stratigraphy and structure in the Black Combe Inlier, English Lake District. *Proceedings of the Yorkshire Geological Society*, **38**, 105–148 and plate.

HELM, D. G. & ROBERTS, B. 1968. The Caledonian history of the northeastern Irish Sea region. *Scottish Journal of Geology*, **4**, 375–376.

HELM, D. G. & ROBERTS, B. 1971*a*. The relationship between the Skiddaw and Borrowdale Volcanic Groups in the English Lake District. *Nature Physical Science Series*, **232**(35), 181–183.

HELM, D. G. & ROBERTS, B. 1971*b*. Reply on Skiddaw Slates and Borrowdale Volcanics. *Nature Physical Science Series*, **234**(46), 59–60.

HELM, D. G. & SIDDANS, A. W. B. 1971. Deformation of slaty lapillar tuff in the English Lake District: discussion. *Bulletin of the Geological Society of America*, **82**, 523–531.

HELM, D. G., ROBERTS, B. & SIMPSON, A. 1963. Polyphase folding in the Caledonides south of the Scottish Highlands. *Naure*, **200**, 1060–1062.

HICKS, H. 1875. On the succession of the ancient rocks in the vicinity of St David's, Pembrokeshire, with special reference to those of the Arenig and Llandeilo groups. *Quarterly Journal of the Geological Society of London*, **31**, 171–195 and plates.

HICKS, H. 1876*a*. Some considerations on the probable conditions under which the Palaeozoic rocks were deposited over the northern hemisphere. *Geological Magazine*, **3**(Decade 2), 156–160, 215–218, 249–253.

HICKS, H. 1876*b*. Llandovery rocks in the Lake District. *Geological Magazine*, **3**(Decade 2), 335–336.

HICKS, H. 1876*c*. Llandovery rocks in the Lake District. *Geological Magazine*, **3**(Decade 2), 429–430.

HICKS, H. 1881. The classification of the Eozoic and Lower Palaeozoic rocks of the British Isles. *Popular Science Review*, **5**, New Series, 308–309.

HINDE, G. J. 1899. Prof. H. A. Nicholson, MD, DSc, FRS, FLS, FGS. *Geological Magazine*, **6**(Decade 4), 138–144 and plate.

HODGSON, E. 1863. On a deposit containing diatomaceae, leaves, &c., in the iron-ore mines near Ulverston. *Quarterly Journal of the Geological Society of London*, **19**, 19–31.

HOLLAND, C. H., AUDLEY-CHARLES, M. *et al.* 1978. *A Guide to Stratigraphic Procedure*. Geological Society, London.

HOLLIDAY, D. W. 1993. Mesozoic cover over northern England: interpretation of apatite fission data. *Journal of the Geological Society*, **150**, 657–660.

HOLLIDAY, D. W. 1994. Reply to A. A. Macculloch. *Journal of the Geological Society*, **151**, 736.

HOLLIDAY, D. W. 1999. Palaeotemperatures, thermal modelling and depth of burial studies in northern and eastern England. *Proceedings of the Yorkshire Geological Society*, **52**, 337–352.

HOLLIDAY, D. W. & REES, J. G. 1994. Advances in west Cumbrian geology resulting from the Nirex investigations around Sellafield: convenors' review. *Proceedings of the Yorkshire Geological Society*, **50**, 1–4.

HOLLIDAY, D. W. & REES, J. G. 1998. Recent advances in the understanding of the geology of West Cumbria: II. Convenors' review. *Proceedings of the Yorkshire Geological Society*, **52**, 121–124.

HOLLINGWORTH, S. E. 1936. Platforms around the Lake District. *Report of the British Association for the Advancement of Science*, 348–349.

HOLLINGWORTH, S. E. 1935–1937 (1937). High level erosional platforms in Cumberland and Furness. *Proceedings of the Yorkshire Geological Society*, **23**, 159–177 and plates.

HOLLINGWORTH, S. E. 1938. The recognition and correlation of high-level erosion surfaces in Britain: a statistical study. *Quarterly Journal of the Geological Society of London*, **94**, 55–84 and plates.

HOLLINGWORTH, S. E. 1964. Ernest Edward Leslie Dixon. *Proceedings of the Geologists' Association*, **75**, 112–113.

HOLLINGWORTH, S. E., ROSE, W. C. C., OLIVER, R. L. & FIRMAN, R. J.

1954. The geology of the Lake District – a review. *Proceedings of the Geologists' Association*, **65**, 385–402.

HOLMES, T. V. 1881. Notes on the geology of Cumberland north of the Lake District. *Proceedings of the Geologists' Association*, **7**, 404–419.

HOOYKAAS, R. 1970. *Catastrophism in Geology, Its Scientific Character in Relation to Actualism and Uniformitarianism*. North Holland, Amsterdam.

HOPKINS, W. 1842. On the elevation and denudation of the district of the Lakes of Cumberland and Westmorland. *Proceedings of the Geological Society*, **3**(2), No. 82, 757–766.

HOPKINS, W. 1848. On the elevation and denudation of the district of the Lakes of Cumberland and Westmoreland. *Quarterly Journal of the Geological Society of London*, **4**, 70–98 and plate.

HOUSMAN, J. 1800. *A Descriptive Tour, and Guide to the Lakes, Caves, Mountains, and other Natural Curiosities in Cumberland, Westmoreland, Lancashire, and a Part of the West Riding of Yorkshire*. F. Jollie, Carlisle & C. Law, London.

HOWARTH, R. J. 1999. Measurement, portrayal and analysis of orientation data and the origins of early modern structural geology (1670–1967). *Proceedings of the Geologists' Association*, **110**, 273–309.

HSÜ, K. J. 1974. Melanges and their distinction from olistostromes. *In*: DOTT, R. H. & SHAVER, R. H. (eds) *Modern and Ancient Geosynclinal Sedimentation*. Society of Economic Paleontologists and Mineralogists, Tulsa, 321–333.

HUDDART, D. 1970. *Aspects of glacial sedimentation in the Cumberland lowland*. PhD dissertation, Reading University.

HUDDART, D. 1970–1971. A relative glacial chronology from the tills of the Cumberland lowland. *Proceedings of the Cumberland Geological Society*, **3**(1), 21–32.

HUDDART, D. 1991. The glacial history and glacial deposits of the north and west Cumbrian lowlands. *In*: EHLARS, J., GIBBARD, P. L. & ROSE, J. (eds) *Glacial Deposits in Great Britain and Ireland*. Balkema, Rotterdam, 151–183.

HUDDART, D. & CLARK, R. 1992–1993 (1994). Conflicting interpretations of glacial sediments and landforms in Cumbria. *Proceedings of the Cumberland Geological Society*, **5**, 419–436.

HUDDART, D., TOOLEY, M. J. & CARTER, P. A. 1977. The coasts of north-west England. *In*: KIDSON, C. & TOOLEY, M. J. (eds) *The Quaternary History of the Irish Sea*. Seel House Press, Liverpool, 119–154.

HUDLESTON, W. H. 1881*a*. Excursion to the Lake District. Monday, July 18th, to Saturday, July 23rd. *Proceedings of the Geologists' Association*, **7**, 236–244.

HUDLESTON, W. H. 1881*b*. On the geology of the neighbourhood of Keswick (with reference to the Long Excursion). *Proceedings of the Geologists' Association*, **7**, 213–235.

HUGHES, R. A. 1990–1991 (1992). Recent advances in the understanding of the Skiddaw Group. *Proceedings of the Cumberland Geological Society*, **5**, 380–382.

HUGHES, R. A. 1995*a*. *The Skiddaw Group and its Contacts with the Borrowdale Volcanic Group in the Ullswater Inlier. Parts of the 1: 10,000 Sheets NY 31 NE, NY 32 SE, NY 41 NW, NY 42 SW & NY 42 SE Part of the 1: 50,000 Sheets 29 (Keswick) and 30 (Appleby)*. British Geological Survey Technical Report **WA/95/59**.

HUGHES, R. A. 1995*b*. The durations of Silurian graptolite zones. *Geological Magazine*, **132**, 113–115.

HUGHES, R. A. & FETTES, D. J. 1994. *The Geology of the 1: 10,000 Sheet NY 11 NW (Flautern Tarn)*. British Geological Survey Technical Report **WA/94/70**.

HUGHES, R. A. & KOKELAAR, B. P. 1993. The timing of Ordovician magmatism in the English Lake District and Cross Fell inliers. *Geological Magazine*, **130**, 369–377.

HUGHES, R. A., COOPER, A. H. & STONE, P. 1993. Structural evolution of the Skiddaw Group (English Lake District) on the northern margin of Eastern Avalonia. *Geological Magazine*, **130**, 621–629.

HUGHES, R. A., EVANS, J. A., NOBLE, S. R. & RUNDLE, C. C. 1996. U–Pb chronology of the Ennerdale and Eskdale intrusions supports sub-volcanic relationships with the Borrowdale Volcanic Group (Ordovician, English Lake District). *Journal of the Geological Society*, **153**, 33–38.

HUGHES, T. McK. 1867. On the break between the Upper and Lower Silurian rocks of the Lake District, as seen between Kirkby Lonsdale and Malham, near Settle. *Geological Magazine*, **4**(Decade 1), 346–356.

HULL, E. 1865. Additional observations on the drift-deposits and more recent gravels in the neighbourhood of Manchester. *MPLMPS*, **2**(3rd Series), 449–461.

HUMBOLDT, F. H. A. von 1880–1883. *Cosmos*, translated by Otté, E. C., Paul, B. H. & Dallas, W. S., 5 vols. George Bell & Sons, London.

HURST, J. M. 1979. Evolution, succession and replacement in the type Upper Caradoc (Ordovician) benthic faunas of England. *Bulletin of the British Museum (Natural History) Geology*, **27**, 189–246.

HUTT, J. E. 1974. *The Llandovery Graptolites of the English Lake District: Part I*. Palaeontographical Society Monographs, Publication 540, **128**, 1–56 and plates.

HUTT, J. E. 1975*b*. *The Llandovery Graptolites of the English Lake District: Part II*. Palaeontographical Society Monographs, Publication 542, **129**, 57–137 and plates.

HUTT, J. E. 1975*a*. *The Llandovery graptolites of the English Lake District*. PhD dissertation, Leicester University.

HUTTON, D. H. W. 1982. A tectonic model for the emplacement of the main Donegal Granite, NW Ireland. *Journal of the Geological Society*, **139**, 615–631.

HUTTON, D. H. W. 1987. Strike-slip terranes and a model for the evolution of the British and Irish Caledonides. *Geological Magazine*, **124**, 405–425.

HUTTON, D. H. W. & MURPHY, F. C. 1987. The Silurian of the Southern Uplands and Ireland as a successor basin to the end-Ordovician closure of Iapetus. *Journal of the Geological Society*, **14**, 765–772.

HUTTON, J. 1795. *Theory of the Earth, With Proofs and Illustrations*, 2 vols. Cadell, Junior & Davies, London; William Creech, Edinburgh.

INGHAM, J. K. 1966. The Ordovician rocks in the Cautley and Dent districts of Westmorland and Yorkshire. *Proceedings of the Yorkshire Geological Society*, **35**, 455–505 and plates.

INGHAM, J. K. 1970–1977. *The Upper Ordovician Trilobites from the Cautley and Dent Districts of Westmorland and Yorkshire*, 3 vols. The Palaeontographical Society, London.

INGHAM, J. K. & RICKARDS, R. B. 1974. Lower Palaeozoic rocks. *In*: RAYNER, D. H. & HEMINGWAY, J. E. (eds) *The Geology and Mineral Resources of Yorkshire*. The Yorkshire Geological Society, Leeds, 29–44.

INGHAM, J. K. & WRIGHT, A. D. 1970. A revised classification of the Ashgill Series. *Lethaia*, **3**, 233–242.

INGHAM, J. K., MCNAMARA, K. J. & RICKARDS, R. B. 1978. The Upper Ordovician and Silurian rocks. *In*: MOSELEY, F. (ed.) *The Geology of the Lake District*. Yorkshire Geological Society, Leeds, 121–145.

INSTITUTE OF GEOLOGICAL SCIENCES 1972*a*. *Aeromagnetic Map of Great Britain, 1: 625,000, Sheet 1*. Southampton, Ordnance Survey.

INSTITUTE OF GEOLOGICAL SCIENCES 1972*b*. *Kirkby Stephen Sheet 40. Solid Edition. One-Inch Series*. Geological Survey of Great Britain (England & Wales), Southampton.

INSTITUTE OF GEOLOGICAL SCIENCES 1972*c*. *The Cross Fell Inlier. Scale 1: 25,000 or About 2 1/2 Inches to the Mile. Resurveyed on the Six-Inch Scale in 1963–67 by I. C. Burgess, A. J. Wadge and R. S. Arthurton*. Geological Survey of Great Britain (England & Wales), Southampton.

INSTITUTE OF GEOLOGICAL SCIENCES 1977. *1:250,000 Series Bouguer Anomaly Map: Lake District Sheet*. Southampton, Ordnance Survey.

ISACKS, B., OLIVER, J. & SYKES, L. R. 1968. Seismology and the new global tectonics. *Journal of Geophysical Research*, **73**, 5855–5899.

JACKSON, D. E. 1956. *Geology of the Skiddaw Slates between the Buttermere Valley and Troutbeck, Cumberland*. PhD dissertation, King's College, University of Durham.

JACKSON, D. E. 1961. Stratigraphy of the Skiddaw Group between Buttermere and Mungrisdale, Cumberland. *Geological Magazine*, **98**, 515–528.

JACKSON, D. E. 1962. Graptolite zones in the Skiddaw Group. *JP*, **36**, 300–313.

JACKSON, D. E. 1978. The Skiddaw Group. *In*: MOSELEY, F. (ed.) *The Geology of the Lake District*. Yorkshire Geological Society, Leeds, 79–98.

JACKSON, D. I. & MULHOLLAND, P. 1993. Tectonic and stratigraphic aspects of the East Irish Sea Basin and adjacent areas: contrasts in their post-Carboniferous structural styles. *In*: PARKER, J. R. (ed.) *Petroleum Geology of Northwest Europe: Proceedings of the Fourth Conference held at the Barbican Centre, London 29 March–1 April 1992*. Geological Society, London, 791–808.

JACKSON, D. I., JACKSON, A. A., EVANS, D., WINGFIELD, R. T. R., BARNES, R. P. & ARTHUR, M. J. 1995. *The Geology of the Irish Sea*. HMSO, London.

JAMIESON, T. F. 1860. On the drift and rolled gravel of the north of Scotland. *Quarterly Journal of the Geological Society of London*, **16**, 347–371.

JAMIESON, T. F. 1863. On the Parallel Roads of Glen Roy, and their place in the history of the glacial period. *Quarterly Journal of the Geological Society of London*, **19**, 235–259

JAMIESON, T. F. 1865. On the history of the last geological changes in Scotland. *Quarterly Journal of the Geological Society of London*, **21**, 161–203.

JEANS, P. J. F. 1971. Relationship between the Skiddaw Slates and the Borrowdale Volcanics. *Nature Physical Science Series*, **234**, 46, p. 59.

JEANS, P. J. F. 1972. The junction between the Skiddaw Slates and Borrowdale Volcanics in Newlands Beck, Cumberland. *Geological Magazine*, **109**, 25–28.

JEANS, P. J. F. 1973*a*. *The structure, metamorphism, and stratigraphy of the Skiddaw Slates, east of Crummock Water, Cumberland*. PhD dissertation, The University of Birmingham.

JEANS, P. J. F. 1973*b*. Plate tectonic reconstruction of the southern Caledonides of Great Britain. *Nature Physical Science Series*, **245**, 120–122.

JOHNSON, E. W. 1992. *Geology of the Stoupdale Area of Black Combe, South-West Cumbria*. British Geological Survey Technical Report **WA/92/71**.

JOHNSON, E. W. & MILLWARD, D. 1987. Borrowdale Volcanic rocks in the Duddon Valley and Eskdale 8th–10th June, 1984. *Proceedings of the Yorkshire Geological Society*, **46**, 170–172.

JOHNSON, E. W., BRIGGS, D. E. G., SUTHREN, R. J., WRIGHT, J. L. & TUNNICLIFF, S. P. 1994. Non-marine arthropod traces from the subaerial Ordovician Borrowdale Volcanic Group, English Lake District. *Geological Magazine*, **131**, 395–406.

JOHNSON, E. W., BRIGGS, D. E. G. & WRIGHT, J. L. 1996. Lake District pioneers – the earliest footprints on land. *Geology Today*, **12**, 147–151.

JOHNSON, E. W., COOPER, A. H. & MILLWARD, D. n.d. Reply to: M. J. BRANNEY short paper: The subaerial setting of the Ordovician Borrowdale Volcanic Group, English Lake District. *Journal of the Geological Society*, **145**, 887–890, unpublished typescript.

JOHNSON, E. W., SOPER, N. J. & BURGESS, I. C., ET AL. 2001. *Geology of the Country around Ulverston: Memoir for 1:50 000 Geological Sheet **48*** (England & Wales). The Stationery Office, London.

JOHNSON, G. A. L. 1961. Skiddaw Slates proved in the Teesdale Inlier. *Nature*, **190**, 996–997.

JOHNSON, M. R. W. & STEWART, F. H. (eds) 1963. *The British Caledonides*. Oliver & Boyd, Edinburgh and London.

JOHNSON, M. R. W., SANDERSON, D. J. & SOPER, N. J. 1979. Deformation in the Caledonides of England, Ireland and Scotland. *In*: HARRIS, A. L., HOLLAND, C. H. & LEAKE, B. E. (eds) *The Caledonides of the British Isles – Reviewed*. Scottish Academic Press for the Geological Society, Edinburgh, 165–186.

JOHNSON, T. 2001. Sir Kingsley Dunham. *The Independent*, 16 April.

JOHNSTONE, S. G. 1976. George Hoole Mitchell (1902–1976). *Geological Society Annual Report 1976*. Geological Society, London, 32–33.

K., J. R. 1977. George Samuel Hadfield (1910–1976). *Geological Society: Annual Report*, 33.

KELLING, G. & KNILL, J. 1997. The Nirex story: a geological perspective. *Geoscientist*, **7**(7), 10–13.

KELLY, D. 1998–1999 (2000). John Bolton (1790–1873): pioneer geologist in the Furness area. *Proceedings of the Cumberland Geological Society*, **6**, 325–328.

KENDALL, J. D. 1875. The haematite deposits of Whitehaven and Furness. *Transactions of the Manchester Geological Society*, **13**, 231–283.

KENDALL, J. D. 1881. Interglacial deposits of West Cumberland and north Lancashire. *Quarterly Journal of the Geological Society of London*, **37**, 29–39 and plate.

KENDALL, J. D. 1881–1882. The haematite deposits of Furness. *TNEIMME*, **31**, 211–238 and plates.

KENDALL, J. D. 1884. The mineral veins of the Lake District. *Transactions of the Manchester Geological Society*, **17**, 292–341.

KENDALL, J. D. 1884–1885. The Carboniferous of Cumberland and north Lancashire, or Furness. *TNEIMME*, **34**, 125–137 and plates.

KENDALL, J. D. 1920*a*. Haematites of west Cumberland, Lancashire, and the Lake District [review of *Memoirs of the Geological Survey*, **8**]. *Mining Magazine*, **22**, 58–62.

KENDALL, J. D. 1920*b*. Lateral distribution of metallic minerals as illustrated by the lead, copper, and iron deposits of Cumberland and Lancashire. *Mining Magazine*, **23**, 75–80.

KENDALL, J. D. 1920*c*. Genesis of Cumberland iron ores [letter]. *Mining Magazine*, **24**, 96–97.

KENDALL, J. D. 1925. The hematites of West Cumberland and Furness. *Mining Magazine*, **32**, 137–144, 213–222.

KENDALL, J. D. 1929. The origin of Cumberland hematite. *Mining Magazine*, **40**, 141–147.

KENDALL, P. 1892. Glacial geology: old and new. *Geological Magazine*, **9**(Decade 3), 491–500.

KENDALL, P. F. & WROOT, H. E. 1924. *Geology of Yorkshire: An Illustration of the Evolution of Northern England*. Published privately, Vienna.

KEPPIE, J. D. 1989. Northern Appalachian Terranes and their Accretionary History. *In*: DALLMEYER (ed.) *Terranes in the Circum-Atlantic Paleozoic Orogens*. Geological Society of America, Boulder, Special Paper **230**, 159–192.

KIMBELL, G. S. 1994. The interpretation of gravity and aeromagnetic data for the geological investigation of the Sellafield area, west Cumbria. *Proceedings of the Yorkshire Geological Society*, **50**, 103–112.

KIMBELL, G. S. & STONE, P. 1995. Crustal magnetization variations across the Iapetus Suture. *Geological Magazine*, **132**, 599–609.

KIMBELL, G. S., CHADWICK, R. A., HOLLIDAY, D. W. & WERNGREN, O. C. 1989. The structure and evolution of the Northumberland Trough from new seismic reflection data and its bearing on modes of continental extension. *Journal of the Geological Society*, **146**, 775–787.

KING, L. M. 1992. *A basin study of the early Palaeozoic Windermere Group*. PhD dissertation, University of Cambridge.

KING, L. M. 1994*a*. Turbidite to storm transition in a migrating foreland basin: the Kendal Group (Upper Silurian), northwest England. *Geological Magazine*, **131**, 255–267.

KING, L. M. 1994*b*. Subsidence analysis of Eastern Avalonian sequences: implications for Iapetus closure. *Journal of the Geological Society*, **151**, 647–657.

KING, W. B. R. 1934. John Edward Marr. *Proceedings of the Geologists' Association*, **45**, 101–102.

KING, W. B. R. 1961. Gertrude Lilian Elles. *Proceedings of the Geologists' Association*, **72**, 168–170.

KING, W. B. R. & WILLIAMS, A. 1948. On the lower part of the Ashgillian Series in the north of England. *Geological Magazine*, **85**, 205–212 and plate.

KIRTON, S. R. & DONATO, J. A. 1985. Some buried Tertiary dykes of Britain and surrounding waters deduced by magnetic modelling and seismic reflection methods. *Journal of the Geological Society*, **142**, 1047–1058.

KNELL, S. J. 2000. *The Culture of English Geology, 1815–1851: A Science Revealed through its Collecting*. Ashgate, Aldershot, Burlington, Singapore and Sydney.

KNELLER, B. C. 1991. A foreland basin on the southern margin of the Iapetus Ocean. *Journal of the Geological Society*, **148**, 207–210.

KNELLER, B. C. & BELL, A. M. 1993. An Acadian mountain front in the English Lake District: the Westmorland Monocline. *Geological Magazine*, **130**, 203–213.

KNELLER, B. C. & SOPER, N. J. 1990. *Geology of SD39 SE and SD39 SW (Part of 1: 50,000 Geological Map Sheet 38, Ambleside*. British Geological Survey Technical Report **WA/90/65**.

KNELLER, B. C., KING, L. M. & BELL, A. M. 1993. Foreland basin development and tectonics on the northwest margin of Eastern Avalonia. *Geological Magazine*, **130**, 691–697.

KNELLER, B. C., SCOTT, R. W., SOPER, N. J., JOHNSON, E. W. & ALLEN, P. M. 1994. Lithostratigraphy of the Windermere Supergroup, Northern England. *Geological Journal*, **29**, 219–240.

KNILL, J. L. 1996. Discussion. *In*: The geology and hydrogeology of the Sellafield area (11 May 1994): chairman's remarks and discussion. *Quarterly Journal of Engineering Geology*, **29** (Supplement), S105–S107.

KOKELAAR, B. P. 1982. Fluidization of wet sediment during the emplacement and cooling of various igneous bodies. *Journal of the Geological Society*, **139**, 21–33.

KOKELAAR, B. P. & BRANNEY, M. J. 1999. *Inside Silicic Calderas (Snowdon, Scafell and Glencoe, UK): Interactions of Caldera*

Development, Tectonism and Hydrovolcanism IAVCEI Commission on Explosive Volcanism/IUGG (Field Workshop, July 7–18 1999).

KONIG, J. M. 1964. *A sedimentological, diagenetic, and structural study in the Middle Tuffs, Borrowdale Volcanics, English Lake District*. PhD dissertation, University College, London.

KUNO, H. 1966. Lateral variation in basalt magma type across continental margins and island arcs. *In*: CANADIAN GEOLOGICAL SURVEY, *Continental Margins and Island Arcs*. Paper **66–15**, 317–336.

LACROIX, A. 1903. L'éruption de la montagne Pelée en Janvier 1903. *Comptes rendus hebdomadaires des séances de l'Académie des Sciences,* **136**, 443–444.

LAMONT, A. 1941. Trinucleidae in Eire. *Annals and Magazine of Natural History*, **8**, 438–469 and plate.

LAMONT, A. 1946. The late Mr. B. B. Bancroft, MA (Cantab.) MSc (Manchester). *The Quarry Managers' Journal*, **29**, 319–321.

LAMONT, A. 1948. B. B. Bancroft's geological work in the Cross Fell Inlier. *The Quarry Managers' Journal*, **31**, 416–418, 466–469.

LAMPLUGH, G. W. 1903. *The Geology of the Isle of Man*. HMSO, London.

LAPWORTH, C. 1870. On the Lower Silurian Rocks of Galashiels. *Geological Magazine*, **7**(Decade 1), 204–209, 279–284.

LAPWORTH, C. 1873. Notes on the British graptolites and their allies. On an improved classification of the Rhabdophora. *Geological Magazine*, **10**(Decade 1), 500–504, 555–560 and table.

LAPWORTH, C. 1876. Llandovery rocks in the Lake District. *Geological Magazine*, **3**(Decade 2), 477–480.

LAPWORTH, C. 1878. The Moffat Series. *Quarterly Journal of the Geological Society of London*, **34**, 240–346.

LAPWORTH, C. 1879. On the tripartite classification of the Lower Palaeozoic rocks. *Geological Magazine*, **6**(Decade 2), 1–15.

LAPWORTH, C. 1879–1880. On the geological distribution of the Rhabdophora. *Annals and Magazine of Natural History*, **3** (5th Series), 245–257; **4**, 333–341; **5**, 45–62, 273–285, 358–369; **6**, 16–29, 185–207.

LAPWORTH, C. 1899. Professor Nicholson's work among the graptolites. *Alma Mater: Aberdeen University Magazine*, **16**, 176–178.

LAPWORTH, C. n.d. Section through Lake District [perhaps prepared for lecture at Keswick, *c*. 1870–1880, The Silurian System in Britain]. Lapworth archive, University of Birmingham.

LAUDAN, L. 1982. Science at the bar: causes for concern. *Science, Technology & Human Values*, **7**, 16–19.

LAWRENCE, D. J. D., WEBB, B. C., YOUNG, B. & WHITE, D. E. 1986. *The Geology of the Late Ordovician and Silurian Rocks (Windermere Group) in the Area around Kentmere and Crook*. British Geological Survey Report **18**, No. **5**, HMSO, London (with map).

LEAKE, B. E. & COBBING, J. 1993. Transient and long-term correspondence of erosion level and tops of granite plutons. *Scottish Journal of Geology*, **29**, 177–182.

LEAN, G. 2001. Public to lose voice in major planning rows. *The Independent on Sunday*, 17 June, 1.

LEE, M. K. 1984. *The Three-Dimensional Form of the Lake District Granite Batholith: Investigation of the Geothermal Potential of the UK*. British Geological Survey Report **WJ/GE/84/16**.

LEE, M. K. 1986. A new gravity survey of the Lake District and three-dimensional model of the granite batholith. *Journal of the Geological Society*, **143**, 425–435.

LEE, M. K. 1988. *Density Variations within Lake District Granites and Lower Palaeozoic Rocks*. British Geological Survey Technical Report **WK/88/9**.

LEE, M. K. 1989. *Upper Crustal Structure of the Lake District from Modelling and Image Processing of Potential Field Data*. British Geological Survey Technical Report **WK/89/1**.

LEE, M. K. 1990. *Geophysical investigations in the English Lake District*. PhD dissertation, The Open University.

LEGGETT, J. K., MCKERROW, W. S. & EALES, M. H. 1979. The Southern Uplands of Scotland: a Lower Palaeozoic accretionary prism. *Journal of the Geological Society*, **136**, 755–770.

LEGGETT, J. K., MCKERROW, W. S. & SOPER, N. J. 1983. A model for the crustal evolution of southern Scotland. *Tectonics*, **2**, 187–210.

LEITH, C. K. 1913. *Structural Geology*, H. Holt, New York.

LEONOV, M. G. 1978. Olistostromes and their origin. *Geotectonics*, **12**, 333–342.

LEWIS, C. L. E., GREEN, P. F., CARTER, A. & HURFORD, A. J. 1992 Elevated K/T palaeotemperatures throughout northwest England: three kilometres of Tertiary erosion? *Earth and Planetary Science Letters*, **112**, 131–145.

LIETCH, N. D. 1857. *A Memoir of Jonathan Otley: The Original Guide to the English Lakes* (publisher not stated).

LINNARSSON, G. 1876. On the vertical range of the graptolitic types in Sweden. *Geological Magazine*, **3**(Decade 2), 241–245 and plate.

LINTON, D. L. 1957. Radiating valleys in glaciated lands. *Tijdschrift van het Koninklijk Nederlansch Aardrijkskundig Genootschap*, **74**, 297–312.

LLEWELLYN, P. G. 1960. *The Middle and Upper Silurian rocks between Longsleddale and the Shap Granite, Westmorland*. PhD dissertation, University of Cambridge.

LLEWELLYN, P. G. 1965. The graptolitic mudstones of the Howgill Fells. *Geological Magazine*, **102**, 277–278.

LUNN, A. G. 1998–1999 (2000). John Goodchild (1844–1906) and Frederick Trotter (1897–1968): pioneering workers on the glaciation of Cumbria. *Proceedings of the Cumberland Geological Society*, **6**, 353–359.

LYELL, C. 1830–1833. *Principles of Geology, Being an Attempt to Explain the Former Changes of the Earth's Surface by Reference to Causes now in Operation*, 3 vols. John Murray, London.

LYELL, C. 1857 *Supplement to the Fifth Edition of A Manual of Elementary Geology*, 2nd edn. John Murray, London.

LYELL, C. 1863. *The Geological Evidences of the Antiquity of Man With Remarks on Theories of the Origin of Species by Variation*. John Murray, London.

MACKINTOSH, D. 1865. A tourist's notes on the surface-geology of the Lake-District. *Geological Magazine*, **2**(Decade 1), 299–306.

MACKINTOSH, D. 1869*a*. On the correlation, nature, and origin of the drifts of north-west Lancashire and a part of Cumberland, with remarks on denudation. *Quarterly Journal of the Geological Society of London*, **25**, 407–431.

MACKINTOSH, D. 1869*b*. *The Scenery of England and Wales*. Longmans, Green & Co., London.

MACKINTOSH, D. 1871*a*. On the drifts of the west and south borders of the Lake District, and on the three great granitic dispersions. *Geological Magazine*, **8**(Decade 1), 250–256, 302–312.

MACKINTOSH, D. 1871*b*. Mr. Croll on the dispersion of boulders. *Geological Magazine*, **8**(Decade 1), 94–95.

MACKINTOSH, D. 1879. Results of a systematic survey, in 1878, of the directions and limits of dispersion, mode of occurrence, and relation to drift-deposits, of the erratic blocks or boulders of the west of England and east of Wales, including a revision of many years' previous observations. *Quarterly Journal of the Geological Society of London*, **35**, 425–455 and plate.

MACLAREN, C. 1842. The glacial theory of Prof. Agassiz. *American Journal of Science*, **42**, 346–365.

MARR, J. E. 1878. On some well-defined life-zones in the lower part of the Silurian (Sedgwick) of the Lake-District. *Quarterly Journal of the Geological Society of London*, **34**, 871–885.

MARR, J. E. 1880*a*. On the predevonian rocks of Bohemia. *Quarterly Journal of the Geological Society of London*, **36**, 591–619.

MARR, J. E. 1880*b*. On the Cambrian (Sedg.) and Silurian beds of the Dee Valley, as compared with those of the Lake District. *Quarterly Journal of the Geological Society of London*, **36**, 277–284.

MARR, J. E. 1881*a*. On some sections in the Lower Palaeozoic rocks of the Craven district. *Proceedings of the Yorkshire Geological and Polytechnic Society*, **7**, 397–399.

MARR, J. E. 1881*b*. The classification of the Cambrian and Silurian rocks. *Geological Magazine*, **8**(Decade 2), 245–250.

MARR, J. E. 1882. On the Cambrian (Sedgw.) and Silurian rocks of Scandinavia. *Quarterly Journal of the Geological Society of London*, **38**, 313–327.

MARR, J. E. 1883. *The Classification of the Cambrian and Silurian Rocks*. Deighton & Bell, London; George Bell, Cambridge.

MARR, J. E. 1889. On the superimposed drainage of the English Lake District. *Geological Magazine*, **6**(Decade 3), 150–155.

MARR, J. E. 1892*a*. The Coniston Limestone Series. *Geological Magazine*, **9**(Decade 3), 97–110 and plate.

MARR, J. E. 1892*b*. Further remarks on the Coniston Limestone. *Geological Magazine*, **9**(Decade 3), 443–447.

MARR, J. E. 1892*c*. On the Wenlock and Ludlow strata of the Lake District. *Geological Magazine*, **9**(Decade 3), 534–541.

MARR, J. E. 1896*a*. The waterways of E. Lakeland. *Geographical Journal*, **7**, 602–605.

MARR, J. E. 1896*b*. On the lake basins of Lakeland. *Proceedings of the Geologists' Association*, **14**, 273–286.

MARR, J. E. 1900*a*. Notes on the geology of English Lake District. *Proceedings of the Geologists' Association*, **16**, 449–483.

MARR, J. E. 1900*b*. Long excursion to Keswick, Monday, August 20th, to Saturday, August 25th, 1900. *Proceedings of the Geologists' Association*, **16**, 526–531 and plates.

MARR, J. E. 1902. Exhibit of specimens from a metamorphosed metalliferous vein in basic andesites near the Shap Granite. *Quarterly Journal of the Geological Society of London*, **58**, lxxxi–lxxxii.

MARR , J. E. 1905*a*. Henry Alleyne Nicholson, *Proceedings of the Royal Society of London Containing Obituaries of Deceased Fellows . . .*, Harrison & Sons, London, 35–38.

MARR, J. E. 1905*b*. Anniversary address: classification of the sedimentary rocks. *Quarterly Journal of the Geological Society of London*, **61**, lxi–lxxxvi.

MARR, J. E. 1906*a*. On the stratigraphical relations of the Dufton Shales and the Keisley Limestone of the Cross Fell Inlier. *Geological Magazine*, **3**(Decade 5), 481–487.

MARR, J. E. 1906*b*. The influences of the geological structure of English Lakeland upon its present features – a study in physiography. *Quarterly Journal of the Geological Society of London*, **62**, lxvi–cxxviii.

MARR, J. E. 1907. On the Ashgillian Series. *Geological Magazine*, **4**(Decade 5), 59–69.

MARR, J. E. 1910. The Lake District and neighbourhood: Upper Palaeozoic and Neozoic times. *In*: MONCKTON, H. W. & HERRIES, R. S. (eds) *Geology in the Field: Jubilee Volume of the Geologists' Association (1858–1908)*. Stanford, London, **2**, 624–660.

MARR, J. E. 1913. The Lower Palaeozoic rocks of the Cautley district (Yorkshire). *Quarterly Journal of the Geological Society of London*, **69**, 1–18.

MARR, J. E. 1915. The Ashgillian succession in the tract to the west of Coniston Lake. *Quarterly Journal of the Geological Society of London*, **71**, 189–204.

MARR, J. E. 1916. *The Geology of the Lake District and the Scenery as influenced by the Geological Structure*. Cambridge University Press, Cambridge.

MARR, J. E. 1924. Notes on the glaciation of the Coniston Fells. *Geological Magazine*, **61**, 264–269.

MARR, J. E. & NICHOLSON, H. A. 1888. The Stockdale Shales. *Quarterly Journal of the Geological Society of London*, **44**, 654–732 and plate.

MARR, J. E. & ROBERTS, T. 1885 The Lower Palaeozoic rocks of the neighbourhood of Haverfordwest. *Quarterly Journal of the Geological Society of London*, **41**, 476–490.

MARSHALL , J. G. 1840 (for 1839). Description of a section across the Silurian rocks in Westmoreland, from the Shap Granite to Casterton Fell. *Report of the British Association for the Advancement of Science*, 67.

MARSHALL, J. G. 1858 (for 1857). On the geology of the Lake District, in reference especially to the metamorphic and igneous rocks. *Report of the British Association for the Advancement of Science*, 84–93.

MARSHALL, P. 1935 Acid rocks of the Taupo–Rotorua volcanic district. *Transactions and Proceedings of the Royal Society of New Zealand*, **64**, 323–366 and plates (read to the Wellington Philosophical Society, 25 October 1933).

MASON, A. P. 2001. Robin Langford Oliver 21 December 1921–1 February 2001. *Newsletter of the Historical Studies Group of the Geological Society of New Zealand*, **22**, 3–4.

MATHIESON, N. A. 1986. *Geology, structure and geochemistry of the Ordovician volcanic section in SW Cumbria*. PhD dissertation, Sheffield University.

MCCAFFREY, W. D. 1991. *Silurian turbidite provenance and basement development in northwest England*. PhD dissertation, Sheffield University.

MCCAFFREY, W. D. 1994. Sm–Nd isotopic characteristics of sedimentary provenance: the Windermere supergroup of NW England. *Journal of the Geological Society*, **151**, 1017–1021.

MCCAFFREY, W. D. & KNELLER, B. C. 1996. Silurian turbidite provenance on the northern Avalonian margin. *Journal of the Geological Society*, **153**, 437–450.

MCCONNELL, R. B. 1938–1941 (1939). Residual erosion surfaces in mountain ranges with special reference to the southern slopes of the Lake District. *Proceedings of the Yorkshire Geological Society*, **24**(2), 76–98 and plates.

MCCONNELL, R. B. 1938–1941 (1940). The relic surfaces of the Howgill Fells. *Proceedings of the Yorkshire Geological Society*, **24**(3), 152–164.

MCCOY, F. 1852. *Description of the British Palaeozoic Fossils in the Geological Museum of the University of Cambridge*. Cambridge University Press, Cambridge, Fascicule 2.

MCCULLOCH, A. A. 1994*a*. Discussion on Mesozoic cover over northern England: interpretation of apatite fission track data. *Journal of the Geological Society*, **151**, 735–736.

MCCULLOCH, A. A. 1994*b*. Low temperature thermal history of eastern Ireland: effects of fluid flow. *Marine and Petroleum Geology*, **11**, 389–399.

MCDONALD, C. S. 1996. Cumbria County Council: appeal by United Kingdom Nirex Limited, File APP/H0900/A/94/247019.

MCKENZIE, D. 1978. Some remarks on the development of sedimentary basins. *Earth and Planetary Science Letters*, **40**, 25–32.

MCKENZIE, D. 1984. A possible mechanism for epeirogenic uplift. *Nature*, **307**, 616–618.

MCKEOWN, C. 1997. *A model approach to radioactive waste disposal at Sellafield*. PhD dissertation, The University of Glasgow.

MCKEOWN, C., HASZELDINE, R. S. & COUPLES, G. D. 1999. Mathematical modelling of groundwater flow at Sellafield, U.K. *Engineering Geology*, **52**, 231–250.

MCKERROW, W. S. 1982. The northwest margin of the Iapetus Ocean during the early Palaeozoic. *Memoirs of the Association of Petroleum Geologists*, **34**, 521–533.

MCKERROW, W. S. & COCKS, L. R. M. 1976. Progressive faunal migration across the Iapetus Ocean. *Nature*, **263**, 304–306.

MCKERROW, W. S. & SOPER, N. J. 1989. The Iapetus Suture in the British Isles. *Geological Magazine*, **126**, 1–8.

MCKERROW, W. S., LEGGETT, J. K. & EALES, M. H. 1977. Imbricate thrust model of the Southern Uplands of Scotland. *Nature*, **267**, 237–239.

MCKERROW, W. S., LEGGETT, J. K. & EALES, M. H. 1978*a*. Caledonian plate tectonics and the place of the English Lake District: discussion and reply. *Bulletin of the Geological Society of America*, **89**, 1694.

MCKERROW, W. S., LEGGETT, J. K. & EALES, M. H. 1978*b*. Discussion of: MOSELEY, F., Caledonian plate tectonics and the place of the English Lake District (1977). *Bulletin of the Geological Society of America*, **89**, 1695.

MCNAMARA, K. J. 1975. *Upper Ordovician stratigraphy and trilobites of the southern Lake district: a geofantasy [sic] in four movements*. PhD dissertation, Cambridge University.

MCNAMARA, K. J. 1979*a*. The age, stratigraphy and genesis of the Coniston Limestone Group in the southern Lake District. *Geological Journal*, **14**, 41–68.

MCNAMARA, K. J. 1979*b*. Trilobites from the Coniston Limestone Group (Ashgill Series) of the Lake District, England. *Palaeontology*, **22**, 53–92 and plates.

MCQUILLIN, R., NICHOL, I. & RHODES, F. H. T. 1986. Sir Kingsley Dunham FRS – a biographical note. *In*: NESBITT, R. W. & NICHOL, I. (eds) *Geology in the Real World – Kingsley Dunham Volume*. Institute of Mining & Metallurgy, Basildon, vii–viii.

MENARD, H. W. 1971. *Science: Growth and Change*. Harvard University Press, Cambridge, Mass.

MERRIMAN, R. J., REX, D. C., SOPER, N. J. & PEACOR, D. R. 1995 The age of the Acadian cleavage in northern England, UK: K–Ar and TEM analysis of a Silurian metaturbidite. *Proceedings of the Yorkshire Geological Society*, **50**, 255–265.

MERTON, R. K. 1942. Science and technology in a democratic order, *Journal of Legal and Political* Sociology, **1**, 115–126 (republished as: The normative structure of science. *In*: MERTON, R. K. 1973 *The Sociology of Science: Theoretical and Empirical Investigations*. The University of Chicago Press, Chicago, 267–278).

MICHIE, U. McL 1998. Deep geological disposal of radioactive waste: a historical review of the UK experience. *Interdisciplinary Science Reviews*, **23**, 242–257.

MICHIE, U. McL & BOWDEN, R. A. 1994. UK Nirex geological investigations at Sellafield. *Proceedings of the Yorkshire Geological Society*, **50**, 5–9.

MILANKOVICH, M. 1920. *Théorie mathématique des phénomènes thermiques produits par la radiation solaire*. Gauthier-Villars, Paris.

MILANKOVICH, M. 1938. Astronomische Mittel zur Erforschungder erdgeschichtlichen Klimate. *In*: GUTENBERG, B. (ed.) *Handbuch der Geophysik*. Gerbrüder Bornträger, Berlin, **9**, 593–698.

MILL, H. R. 1895. Bathymetrical survey of the English Lakes. *Geological Journal*, **6**, 46–73, 135–166 with maps, 108, 204.

MILLER, J. A. 1961. The potassium–argon ages of the Skiddaw and Eskdale Granites. *Geophysical Journal*, **6**, 391–393.

MILLWARD, D. 1976. *The Borrowdale Volcanics of the English Lake District: a study of the petrology, vulcanicity and geochemistry of selected areas with particular reference to ignimbrites*. PhD dissertation, Birmingham University.

MILLWARD, D. 1978. Ordovician Lake District volcanoes: a review. *Amateur Geologist*, **8**(2), 4–14.

MILLWARD, D. 1979. Ignimbrite volcanism in the Ordovician Borrowdale Volcanics of the English Lake District. *In*: HARRIS, A. L., HOLLAND, C. H. & LEAKE, B. E. (eds) *The Caledonides of the British Isles – Reviewed*. Scottish Academic Press for the Geological Society, Edinburgh, 629–634.

MILLWARD, D. & LAWRENCE, D. J. D. 1985. The Stockdale (Yarlside) Rhyolite – a rheomorphic ignimbrite? *Proceedings of the Yorkshire Geological Society*, **45**, 299–306.

MILLWARD, D. & MOLYNEAUX, S. G. 1992. Field and biostratigraphic evidence for an unconformity at the base of the Eycott Volcanic Group in the English Lake District. *Geological Magazine*, **129**, 77–92.

MILLWARD, D., & SOPER, N. J. 1998–1999 (2000). George Hoole Mitchell (1902–1976): pioneer mapper of the Borrowdale Volcanic Group. *Proceedings of the Cumberland Geological Society*, **6**, 415–419.

MILLWARD, D., BEDDOE-STEPHENS, B., WILLIAMSON, I. T., YOUNG, S. R. & PETTERSON, M. G. 1994. Lithostratigraphy of a concealed caldera-related ignimbrite sequence within the Borrowdale Volcanic Group of west Cumbria. *Proceedings of the Yorkshire Geological Society*, **50**, 25–36.

MILLWARD, D., BEDDOE-STEPHENS, B. & YOUNG, B. 1999. Pre-Acadian copper mineralization in the English Lake District. *Geological Magazine*, **136**, 159–176.

MILLWARD, D., JOHNSON, E. W., BEDDOE-STEPHENS, B., YOUNG, B., KNELLER, B. C., LEE, M. K., FORTEY, N. J., *et al.* 2000. *Geology of the Ambleside District: Memoir for 1:50,000 Geological Sheet 38 (England and Wales)*. The Stationery Office, London.

MILLWARD, D., MOSELEY, F. & SOPER, N. J. 1978. The Eycott and Borrowdale Volcanic rocks. *In*: MOSELEY, F. (ed.) *The Geology of the Lake District*. Yorkshire Geological Society, Leeds, 99–120.

MILODOWSKI, A. E., GILLESPIE, M. R., NADEN, J., FORTEY, N. J., SHEPHERD, T. J., PEARCE, J. M. & METCALFE, R. 1998. The petrology and paragenesis of fracture mineralization in the Sellafield area, west Cumbria. *Proceedings of the Yorkshire Geological Society*, **52**, 215–214,

MITCHELL, A. H. G. & MCKERROW, W. S. 1975. Analogous evolution of the Burma Orogen and the Scottish Caledonides. *Bulletin of the Geological Society of America*, **86**, 305–315.

MITCHELL, G. F. 1960. The Pleistocene history of the Irish Sea. *Advancement of Science*, **17**, 313–325.

MITCHELL, G. F. 1963. Morainic ridges on the floor of the Irish Sea. *Irish Geography*, **4**, 335–344.

MITCHELL, G. F. 1972. The Pleistocene history of the Irish Sea: second approximation. *The Scientific Proceedings of the Royal Dublin Society*, Series A, **4**, 181–199.

MITCHELL, G. F., PENNY, L. F., SHOTTON, F. W. & WEST, R. G. 1973. *A Correlation of Quaternary Deposits in the British Isles*. Geological Society, London, Special Report **4**.

MITCHELL, G. H. 1925*a*. *The geology of the Borrowdale Volcanic Series of Troutbeck, Kentmere and the western part of Long Sleddale, Westmorland*. MSc dissertation, Liverpool University.

MITCHELL, G. H. 1925*b*. The Coniston Limestone Series of the Kentmere district. *Geological Magazine*, **62**, 264–267.

MITCHELL, G. H. 1929. The succession and structure of the Borrowdale Volcanic Series in Troutbeck, Kentmere, and the western part of Long Sleddale (Westmorland). *Quarterly Journal of the Geological Society of London*, **85**, 9–44 and plates.

MITCHELL, G. H. 1930. Notes on the petrography of the Borrowdale Volcanic Series of Kentmere (Westmorland). *Quarterly Journal of the Geological Society of London*, **86**, 1–8.

MITCHELL, G. H. 1934. The Borrowdale Volcanic Series and associated rocks in the country between Long Sleddale and Shap. *Quarterly Journal of the Geological Society of London*, **90**, 418–444 and plates.

MITCHELL, G. H. 1940. The Borrowdale Volcanic Series of Coniston, Lancashire. *Quarterly Journal of the Geological Society of London*, **96**, 301–319 and plate.

MITCHELL, G. H. 1956*a*. The Borrowdale Volcanic Series of the Dunnerdale Fells, Lancashire. *Liverpool and Manchester Geological Journal*, **1**, 428–449 and plates.

MITCHELL, G. H. 1956*b*. The geological history of the Lake District. *Proceedings of the Yorkshire Geological Society*, **30**, 407–463.

MITCHELL, G. H. 1957. Ordovician volcanoes. *Advancement of Science*, **14**, 34–47.

MITCHELL, G. H. 1963. The Borrowdale Volcanic Rocks of the Seathwaite Fells, Lancashire. *Liverpool and Manchester Geological Journal*, **3**, 289–300 and plate.

MITCHELL, G. H. 1967. The Caledonian orogeny in northern England. *Proceedings of the Yorkshire Geological Society*, **36**, 135–138.

MITCHELL, G. H. 1970 *Geologists' Association Guide, No. 2: The Lake District*. Geologists' Association, Colchester.

MITCHELL, G. H., MOSELEY, F. & De BOER, G. 1958. Haweswater and Ullswater: 5th and 6th July, 1958, excursion report. *Proceedings of the Yorkshire Geological Society*, **32**, 138–140.

MITCHELL, G. H., MOSELEY, F., FIRMAN, R. J., SOPER, N. J., ROBERTS, D. E., NUTT, M. J. & WADGE, A. J. 1972. Excursion to the northern Lake District 30 August–5 September 1970. *Proceedings of the Geologists' Association*, **83**, 443–470 and plates.

MOJSISOVICS EDLER VON MOJSVÁR, E. 1893. Die Cephalopoden der Hallstätter Kalke. *Abhandlungen der Kaiserliche-Königlichen Geologische Reichanstalt*. Geologische Reichanstalt. Vienna, **6**.

MOLYNEUX, S. G. 1979. New evidence for the age of the Manx Group, Isle of Man. *In*: HARRIS, A. L., HOLLAND, C. H. & LEAKE, B. E. (eds) *The Caledonides of the British Isles Reviewed*. Scottish Academic Press for the Geological Society, Edinburgh, 415–421.

MOLYNEUX, S. G. 1981. *The micropalaeontology and age of the Manx Group, Isle of Man*. PhD dissertation, Sheffield University.

MOLYNEUX, S. G. 1988. Micropalaeontological evidence for the age of the Borrowdale Volcanic Group (correspondence and notes). *Geological Magazine*, **135**, 541–542.

MOLYNEUX, S. G. & RUSHTON, A. W. A. 1984. Discovery of Tremadoc rocks in the Lake District. *Proceedings of the Yorkshire Geological Society*, **45**, 123–127.

MOLYNEUX, S. G. & RUSHTON, A. W. A. 1988. The age of the Watch Hill Grits (Ordovician), English Lake District: structural and palaeogeographical implications. *TRSE ES*, **79**, 43–69.

MOORE, R. M. 1992. *The Skiddaw Group of Cumbria: early Ordovician turbidite sedimentation and provenance on an evolving microcontinental margin*. PhD dissertation, The University of Leeds.

MORGAN, W. J. 1971. Convection plumes in the lower mantle. *Nature*, **230**, 42–43.

MORGAN, W. J. 1972. Plate motions and deep mantle convection. Geological Society of America, Memoir **132**, 7–22.

MORGAN, W. J. 1981. Hotspot tracks and the opening of the Atlantic and Indian Oceans. *In*: EMILIANI, C. (ed.) *The Sea: Ideas and Observations on Progress in the Study of the Seas. Volume 7. The Oceanic Lithosphere*. John Wiley, New York, 443–487.

MORLOT, C. A. 1855. On the post-Tertiary and Quaternary formations of Switzerland. *Edinburgh New Philosophical Journal*, **2**, New Series, 14–20 and plate.

MORRIS, W. A. 1976. Transcurrent motion determined palaeomagnetically in the northern Appalachians and Caledonides, and the Acadian orogeny. *Canadian Journal of Earth Sciences*, **13**, 1236–1243.

MOSELEY, F. 1952. *The Namurian of the Lancaster fells*. PhD dissertation, Sheffield University.

MOSELEY, F. 1960. The succession and structure of the Borrowdale Volcanic rocks south-east of Ullswater. *Quarterly Journal of the Geological Society of London*, **116**, 55–84 and plate.

MOSELEY, F. 1961. Erosion surfaces in the Forest of Bowland. *Proceedings of the Yorkshire Geological Society*, **33**, 173–196 and plate.

MOSELEY, F. 1964. The succession and structure of the Borrowdale

Volcanic rocks northwest of Ullswater. *Geological Journal*, **4**, 127–142.

Moseley, F. 1968. Joints and other structures in the Silurian rocks of the southern Shap fells, Westmorland. *Geological Journal*, **6**, 79–86.

Moseley, F. 1972. A tectonic history of northwest England. *Journal of the Geological Society*, **128**, 561–598 and plates.

Moseley, F. 1973–1974 (1975). Structural relations between the Skiddaw Slates and the Borrowdale Volcanics. *Proceedings of the Cumberland Geological Society*, **3**, 127–145.

Moseley, F. 1977. Caledonian plate tectonics and the place of the English Lake District. *Bulletin of the Geological Society of America*, **88**, 764–768.

Moseley, F. (ed.) 1978*a*. *The Geology of the Lake District*. Yorkshire Geological Society, Leeds.

Moseley, F. 1978*b*. Reply to discussion by McKerrow *et al*. *Bulletin of the Geological Society of America*, **89**, 1695–1696.

Moseley, F. 1981. Geological mapping and photogeology. *Geology Teaching*, **8**, 82–87.

Moseley, F. 1982. Lower Palaeozoic volcanic environments in the British Isles. *In*: Sutherland, D. S. (ed.) *Igneous Rocks of the British Isles*. John Wiley, London, 39–44.

Moseley, F. 1983*a*. Geological field mapping and photogeology. *Geology Teaching*, **8**, 82–87.

Moseley, F. 1983*b*. *The Volcanic Rocks of the Lake District: A Geological Guide to the Central Fells*. Macmillan Press, London and Basingstoke.

Moseley, F. 1984. Lower Palaeozoic lithostratigraphical classification in the English Lake District. *Geological Journal*, **19**, 239–247.

Moseley, F. 1984–1985 (1986). The use of aerial photography in Lake District geology. *Proceedings of the Cumberland Geological Society*, **4**, 361–375.

Moseley, F. 1990. *The Lake District: Geologists' Association Guide*. The Geologists' Association.

Moseley, F. & Millward, D. 1982. Ordovician volcanicity in the English Lake District. *In*: Sutherland, D. S. (ed.) *Igneous Rocks of the British Isles*. John Wiley, London, 93–111.

Murchison, R. I. 1839. *The Silurian System*, 3 vols. John Murray, London.

Murchison, R. I. 1851. On the distribution of the flint drift of the south-east of England, on the flanks of the Weald, and over the surface of the South and North Downs. *Quarterly Journal of the Geological Society of London*, **7**, 349–398.

Murchison, R. I. 1854. *Siluria: The History of the Oldest Rocks Containing Organic Remains*. John Murray, London.

Murchison, R. I. 1872. *Siluria: A History of the Oldest Rocks in the British Isles and Other Countries,* 5th edn. John Murray, London.

Murphy, F. C. & Hutton, D. H. W. 1986. Is the Southern Uplands of Scotland really an accretionary prism? *Geology*, **14**, 354–357.

Murphy, J. M. & Aalto, K. R. 1983. Flow mélanges: numerical modeling and geologic constraints on their origin in the Franciscan subduction complex, California, discussion and reply. *Bulletin of the Geological Society of America*, **94**, 1241–1244.

Nicholson, H. A. 1868*a*. *Geology of Cumberland and Westmorland*. Robert Hardwicke, London.

Nicholson, H. A. 1868*b*. The graptolites of the Skiddaw Series. *Quarterly Journal of the Geological Society of London*, **24**, 125–145 and plates.

Nicholson, H. A. 1868*c*. On the graptolites of the Coniston Flags; with notes on the British species of the genus Graptolites. *Quarterly Journal of the Geological Society of London*, **24**, 521–546 and plates.

Nicholson, H. A. 1868*d*. Notes on the distribution in time of the various British species of the genus graptolites. *Annals and Magazine of Natural History*, **2**, 347–357.

Nicholson, H. A. 1868*e* (for 1867). On the graptolites of the Skiddaw Slates. *Report of the British Association for the Advancement of Science*, 71–72.

Nicholson, H. A. 1868*f* (for 1867). On the Coniston Group of the Lake District. *Report of the British Association for the Advancement of Science*, 61.

Nicholson, H. A. 1868*g*. On the granite of Shap, in Westmoreland. *Transactions of the Edinburgh Geological Society*, **1**, 133–137.

Nicholson, H. A. 1869*a*. On the relations between the Skiddaw Slates and the Green Slates and Porphyries of the Lake-District. *Geological Magazine*, **6**(Decade 1), 105–108, 167–173.

Nicholson, H. A. 1869*b*. Notes on the Green Slates and Porphyries of the neighbourhood of Ingleton. *Geological Magazine*, **6**(Decade 1), 435–491.

Nicholson, H. A. 1869*c*. Notes on the geology of Derwentwater, in Cumberland. *Transactions of the Edinburgh Geological Society*, **1**, 274–278.

Nicholson, H. A. 1869*d*. On the occurrence of plants in the Skiddaw Slates. *Geological Magazine*, **6**(Decade 1), 494–498 and plate.

Nicholson, H. A. 1869*e*. Notes on certain of the intrusive igneous rocks of the Lake-District. *Quarterly Journal of the Geological Society of London*, **25**, 435–441.

Nicholson, H. A. 1869*f*. On some new species of graptolites. *Annals and Magazine of Natural History*, **4**(Series 4), 231–242 and plate.

Nicholson, H. A. 1870*a*. On the Coniston Limestone Series of Cumberland and Westmorland and its associated rocks. *Transactions of the Edinburgh Geological Society*, **2**, 84–94.

Nicholson, H. A. 1870*b*. Notes on the lower portion of the Green Slates and Porphyries of the Lake-District between Ulleswater and Keswick. *Quarterly Journal of the Geological Society of London*, **26**, 599–610.

Nicholson, H. A. 1872*a*. *A Monograph of the British Graptolitidae*. William Blackwood, Edinburgh and London.

Nicholson, H. A. 1872*b*. On the correlation of the Silurian deposits of the north of England with those of the south of Scotland. *Transactions of the Edinburgh Geological Society*, **2**, 105–113 (read 17 November 1870).

Nicholson, H. A. 1872*c*. Migrations of the graptolites. *Quarterly Journal of the Geological Society of London*, **28**, 217–232.

Nicholson, H. A. 1874 (for 1873). On the Silurian rocks of the English Lake District. *Proceedings of the Geologists' Association*, **3**, 105–114.

Nicholson, H. A. 1876. Notes on the correlation of the graptolitic deposits of Sweden with those of Britain. *Geological Magazine*, **3**(Decade 2), 245–249 and plate.

Nicholson, H. A. & Lapworth, C. 1876 (for 1875). On the central group of the Silurian Series of the north of England. *Report of the British Association for the Advancement of Science*, 78–79.

Nicholson, H. A. & Marr, J. E. 1887. On the occurrence of a new fossiliferous horizon in the Ordovician Series of the Lake District. *Geological Magazine*, **4**(Decade 3), 339–344.

Nicholson, H. A. & Marr, J. E. 1891. The Cross Fell Inlier. *Quarterly Journal of the Geological Society of London*, **47**, 500–529 and plates.

Nicholson, H. A. & Marr, J. E. 1895. Notes on the phylogeny of the graptolites. *Geological Magazine*, **2**(Decade 4), 529–539 and plates.

Nuclear Industry Radioactive Waste Executive 1989. Deep repository project: preliminary environmental and radiological assessment and preliminary safety report. *Nirex Report* **71**.

Nuclear Industry Radioactive Waste Executive 1993*a*. The geology and hydrogeology of the Sellafield area. Interim assessment December 1993. The geology (Volume 1); The geological structure of the Sellafield area (Volume 2); The hydrogeology (Volume 3); Investigation results (Volume 4). *Nirex Report* **524**.

Nuclear Industry Radioactive Waste Executive 1993*b*. Nirex deep waste repository project, scientific update 1993. *Nirex Report* **525**.

Nuclear Industry Radioactive Waste Executive 1995. A preliminary analysis of the groundwater pathway for a deep repository at Sellafield (Volume 1); Derivation of effective hydrogeological parameters for regional modelling (Volume 2); Calculations of risk (Volume 3). *Nirex Science Report* **S/95/012**.

Nuclear Industry Radioactive Waste Executive 1996. The borehole RCF3 pump test. *Nirex Science Report* **S/96/003** (issued 1997).

Nuclear Industry Radioactive Waste Executive 1997*a*. The Quaternary of the Sellafield area, Cumbria. *Nirex Science Report* **S/97/002**.

Nuclear Industry Radioactive Waste Executive 1997*b*. Planning the scientific programme for the rock characterisation facility. *Nirex Science Report* **S/97/006**.

Nuclear Industry Radioactive Waste Executive 1997*c*. The 3-D geological structure of the RCF south shaft location: summary report. *Nirex Science Report* **S/97/007**.

Nuclear Industry Radioactive Waste Executive 1997*d*. The hydrogeology of the Sellafield area: 1997 update. *Nirex Science Report* **S/97/008**.

Nuclear Industry Radioactive Waste Executive 1997*e*. The geology and hydrogeology of the borehole 8 and borehole 9 area in the vicinity of the Lake District Boundary fault zone and the Seascale–Gosforth fault zone. *Nirex Science Report* **SA/97/032**.

NUCLEAR INDUSTRY RADIOACTIVE WASTE EXECUTIVE 1997*f*. The petrology of fractures and fracture mineralisation associated with the flowing features in Sellafield boreholes 8A and 8B. *Nirex Science Report* **SA/97/036**.

NUCLEAR INDUSTRY RADIOACTIVE WASTE EXECUTIVE 1997*g*. The hydrogeology of the Sellafield area, Cumbria. *Nirex Science Report* **SA/97/002**.

NUCLEAR INDUSTRY RADIOACTIVE WASTE EXECUTIVE 1997*h*. Evaluation and sealing of fractures in the Borrowdale Volcanic Group in the Sellafield boreholes. *Nirex Science Report* **SA/97/031**.

NUCLEAR INDUSTRY RADIOACTIVE WASTE EXECUTIVE 2000. *Annual Report and Accounts 1999/2000*. United Kingdom Nirex Limited, Didcot.

NUMAN, N. M. S. 1974. *Structure and stratigraphy of the southern part of the Borrowdale Volcanic Group, English Lake District*. PhD dissertation, Sheffield University.

NUTT, M. J. C. 1966. Haweswater, 21st to 23rd May, 1965. *Proceedings of the Yorkshire Geological Society*, **35**, 429–433.

NUTT, M. J. C. 1968. Borrowdale Volcanic Series and associated rocks around Haweswater, Westmorland. *PGSL*, **1649**, 112–113.

NUTT, M. J. C. 1970. *The Borrowdale Volcanic Series and associated rocks around Haweswater, Westmorland*. PhD dissertation, Queen Mary College, London University.

OERTEL, G. 1970. Deformation of a slaty lapillar tuff in the Lake District, England. *Bulletin of the Geological Society of America*, **81**, 1173–1188.

OERTEL, G. 1971. Deformation of a slaty lapillar tuff in the Lake District, England: reply. *Bulletin of the Geological Society of America*, **82**, 533–536.

OERTEL, G. 1972. Deformation of a slaty lapillar tuff in the Lake District, England: another reply. *Bulletin of the Geological Society of America*, **83**, 549–550.

OLDHAM, R. D. 1900. The basal (Carboniferous) conglomerate of Ullswater and its mode of origin. *Report of the British Association for the Advancement of Science*, 764.

OLDROYD, D. R. 1990. *The Highlands Controversy: Constructing Geological Knowledge by Fieldwork in Nineteenth-Century Britain*. The University of Chicago Press, Chicago.

OLDROYD, D. R. 1991. The Archaean controversy in Britain. Part I – the rocks of St David's. *Annals of Science*, **48**, 407–452.

OLDROYD, D. R. 1992. The Archaean controversy in Britain. Part II – the Malverns and Shropshire. *Annals of Science*, **49**, 401–460.

OLDROYD, D. R. 1993. The Archaean controversy in Britain. Part III – the rocks of Anglesey and Caernarvonshire. *Annals of Science*, **50**, 523–584.

OLDROYD, D. R. 1994. The Archaean controversy in Britain. Part IV – some general theoretical and social issues. *Annals of Science*, **51**, 571–592.

OLDROYD, D. R. 1996. *Thinking About the Earth: A History of Ideas in Geology*. The Athlone Press, London; Harvard University Press, Cambridge, Mass.

OLDROYD, D. R. 1998–1999 (2000)*a*. Adam Sedgwick (1785–1873) and Lakeland geology (1822–1824): the first professional Lake District geologist. *Proceedings of the Cumberland Geological Society*, **6**, 315–324.

OLDROYD, D. R. 1998–1999 (2000)*b*. John Edward Marr (1857–1933): the foremost Lake District geologist of his era. *Proceedings of the Cumberland Geological Society*, **6**, 361–379.

OLDROYD, D. R. 1999*a*. Evidence that disappears: John Marr *et al.* and Lakeland geological sites. *The Geological Curator*, **7**, 17–25.

OLDROYD, D. R. 1999*b*. Early ideas about glaciation in the English Lake District: The problem of making sense of glaciation in a glaciated region. *Annals of Science*, **56**, 175–203.

OLDROYD, D. R. 2002. Adam Sedgwick: a confident mind in turmoil. *In*: P. HARMAN (ed.), *Cambridge Minds*. Cambridge University Press, Cambridge, 64–78.

OLDROYD, D. R. & HAMILTON, B. M. 1997. Geikie and Judd, and controversies about the igneous rocks of the Hebrides: theory, practice, and power in the geological community. *Annals of Science*, **54**, 221–268.

OLIVER, J. 1996. *Shocks and Rocks: Seismology in the Plate Tectonics Revolution*. American Geophysical Union, Washington, DC.

OLIVER, R. L. 1953. *The Borrowdale Volcanic and associated rocks of the Scafell area, Lake District (England)*. PhD dissertation, The University of Cambridge.

OLIVER, R. L. 1954*a*. Note on the succession in the region around the head of Borrrowdale, Cumberland. *Proceedings of the Geologists' Association*, **65**, 407–411.

OLIVER, R. L. 1954*b*. Welded tuffs in the Borrowdale Volcanic Series, English Lake District, with a note on similar rocks in Wales. *Geological Magazine*, **91**, 473–483.

OLIVER, R. L. 1956*a*. The origin of garnets in the Borrowdale Volcanic Series and associated rocks, English Lake District. *Geological Magazine*, **93**, 121–139.

OLIVER, R. L. 1956*b*. The origin of garnets in the Borrowdale Volcanic Series, *Geological Magazine*, **93**, 516–517.

OLIVER, R. L. 1961 The Borrowdale Volcanic and associated rocks of the Scafell Area, English Lake District. *Quarterly Journal of the Geological Society of London*, **117**, 377–417 and plates.

OPPEL, C. A. VON 1856–1858. *Die Juraformation Englands, Frankreichs und des Südwestlichen Deutshlands*. Verlag Von Ebner and Seubert, Stuttgart.

OTLEY, J. 1818. *A New Map of the Lakes in Cumberland, Westmorland, and Lancashire, with the Heights of its Principal Hills, the Stages and Favourite Excursions Distinctly Pointed Out*. J. Otley, Keswick.

OTLEY, J. 1820. Remarks on the succession of rocks in the district of the Lakes. *Lonsdale Magazine*, **1**, 433–438; *PM*, **56**, 257–261.

OTLEY, J. 1823. *A Concise Description of the English Lakes, the Mountains in their Vicinity, and the Roads by which they may be Visited; With Remarks on the Mineralogy and Geology of the District*. Published privately, Keswick, J. Richardson, London; A. Foster, Kirkby Lonsdale.

OTLEY, J. 1843. *A Descriptive Guide to the English Lakes, and Adjacent Mountains: With Notices of the Botany, Mineralogy, and Geology of the District*, 7th edn. Published privately, Keswick; Simpkin and Marshall, London; Arthur Foster, Kirkby Lonsdale.

OWEN, A. W. & RUSHTON, A. W. A. 1999. Arenig to Ashgill of Northern England. *In*: RUSHTON, A. W. A., OWEN, A. W., OWENS, R. M. & PRIGMORE, J. K., *British Cambrian to Ordovician Stratigraphy*. Joint Nature Conservation Committee, Peterborough, 258–292.

OWENS, B. & SARJEANT, W. A. S. 1999. Charles Downie (1923–1999). *Commission Internationale de Microflore du Paleozoique Newsletter*, **57**, 5–7.

PALMER, D. & RICKARDS, B. (eds) 1991. *Graptolites: Writing in the Rocks*. Boydell Press, Woodbridge and Rochester, NY.

PARLIAMENTARY OFFICE OF SCIENCE AND TECHNOLOGY 1997. *Radioactive Waste – Where Next?* POST, London.

PARRY, J. T. 1958. *The erosion surfaces of the south-western Lake District*. MSc dissertation, The University of Liverpool.

PARRY, J. T. 1960. The erosion surfaces of the south-western Lake District. *Transactions of the Institute of British Geographers*, **28**, 39–54.

PARSON, W. & WHITE, W. M. 1829. Geology of the Lake District. *In*: *History, Directory, and Gazetteer of the Counties of Westmorland and Cumberland, with That Part of the Lake District, forming the Lordships of Furness and Cartmel*. W. White, Leeds and Newcastle, 84–88.

PARSONS, V. (ed.) 1998. *A Brief History of the Westmorland Geological Society Based on Notes Left by the Late Walter Annis*. The Westmorland Geological Society, Carnforth.

PEACH, B. N. & HORNE, J. 1880. The glaciation of the Orkney islands. *Quarterly Journal of the Geological Society of London*, **36**, 648–663 and plates.

PENCK, F. C. A. 1882. *Die Vergletscherung der deutschen Alpen, ihre Ursachen, periodische Wiederkehr und ihr Einfluss auf die Bodengestaltung*. J. A. Barth, Leipzig.

PENCK, F. C. A. & BRÜCKNER, E. 1901–1909. *Die Alpen im Eiszeitalte*, 3 vols. Tauchnitz, Leipzig.

PENNINGTON, W. 1970. Vegetation history in north-west England. *In*: WALKER, D. & WEST, R. G. (eds) *Studies in the Vegetational History of the British Isles*. Cambridge University Press, Cambridge, 97–116.

PENNINGTON, W. 1978. Quaternary geology. *In*: MOSELEY, F. (ed.) *The Geology of the Lake District*. The Yorkshire Geological Society, Leeds, 207–225.

PERNER, J. 1937. Les "colonies de Barrande". *Bulletin de la Société Géologique de France*, **7**, Series 5, 513–526.

PETTERSON, M. G. 1988–1989 (1990). Recent developments concerning the Borrowdale Volcanic Group of the English Lake District. *Proceedings of the Cumberland Geological Society*, **5**, 151–168.

PETTERSON, M. G., BEDDOE-STEPHENS, B., MILLWARD, D. & JOHNSON, E. W. 1992. A pre-caldera plateau-andesite field in the Borrowdale Volcanic Group of the English Lake District. *Journal of the Geological Society*, **149**, 889–906.

PHILLIPS, J. 1827–1828. On a group of slate-rocks in Yorkshire between the rivers Lune and Wharfe from near Kirkby Lonsdale to Malham. *PGSL*, **1**, 39–40.

PHILLIPS, J. 1829. On a group of slate rocks ranging E.S.E. between the rivers Lune and Wharfe, from near Kirkby Lonsdale to near Malham; and on the attendant phenomena. *Transactions of the Geological Society of London*, Series 2, **3**, 1–19 (read 21 December 1827 and 4 January 1828).

PHILLIPS, J. 1836. *Illustrations of the Geology of Yorkshire, Or a Description of the Strata and Organic Remains. Part 2. The Mountain Limestone District.* John Murray, London.

PHILLIPS, J. 1837–1838. *A Treatise of Geology*, 2 vols. Longman, Brown, Green & Longmans & John Taylor, London.

PHILLIPS, J. 1844. *Memoirs of William Smith, LL.D., Author of the Strata of England and Wales.* John Murray, London.

PHILLIPS, J. 1849. Geology of the Lakes. *In*: BLACK, J. & BLACK, C. 203–243 (essay dated August 1846).

PHILLIPS, J. 1856 (for 1855). Report on cleavage and foliation in rocks and on the theoretical explanation of these phenomena. *Report of the British Association for the Advancement of Science*, 369–396.

PHILLIPS, J. 1873. Sedgwick. *Nature*, **7**, 257–259.

PHILLIPS, W. E. A., STILLMAN, C. J. & MURPHY, T. 1976. A Caledonian plate tectonic model. *Journal of the Geological Society*, **132**, 579–609.

PICKERING, K. T., BASSETT, M. G. & SIVETER, M. G. 1988. Late Ordovician–early Silurian destruction of the Iapetus Ocean: Newfoundland, British Isles and Scandinavia – a discussion. *TRSE ES*, **79**, 361–382.

PLAYFAIR, J. 1802. *Illustrations of the Huttonian Theory of the Earth.* Cadell & Davies, London; William Creech, Edinburgh.

POSTLETHWAITE, J. 1882–1883. Graptolites of the Skiddaw Slates. *Transactions of the Cumberland and Westmorland Association for the Advancement of Literature and Science*, **8**, 37–53 and plate.

POSTLETHWAITE, J. 1884–1885. Trilobites of the Skiddaw Slates. *Transactions of the Cumberland and Westmorland Association for the Advancement of Literature and Science*, **10**, 71–80.

POSTLETHWAITE, J. 1890–1891. The cleaved ashes and breccias of the Volcanic Series of Borrowdale. *Transactions of the Cumberland and Westmorland Association for the Advancement of Literature and Science*, **15**, 41–53.

POSTLETHWAITE, J. 1897. *The Geology of the English Lake District, with Notes on the Minerals.* T. Bakewell, Keswick.

POSTLETHWAITE, J. 1913. *Mines and Mining in the (English) Lake District*, 3rd edn. W. H. Moss, Whitehaven (1st edn 1877; reprinted Michael Moon, Whitehaven, 1987).

POWELL, D. W. 1971. A model for the Lower Palaeozoic evolution of the southern margin of the early Caledonides of Scotland and Ireland. *Scottish Journal of Geology*, **7**, 369–372.

RAISIN, C. 1889. On some nodular felstones of the Lleyn. *Quarterly Journal of the Geological Society of London*, **45**, 247–269.

RAMSAY, A. C. 1852. On the superficial accumulations and surface-markings of North Wales. *Quarterly Journal of the Geological Society of London*, **8**, 371–376

RAMSAY, A. C. 1862. On the glacial origin of certain lakes in Switzerland, the Black Forest, Great Britain, Sweden, North America, and elsewhere. *Quarterly Journal of the Geological Society of London*, **18**, 185–204.

RAMSAY, A. C. 1864. On the erosion of valleys and lakes. *PM*, **28**, Series 4, 293–311.

RAMSAY, A. C. 1866. *The Geology of North Wales, with Map and Sections. And an Appendix on the Fossils, with Plates, by J. W. Salter.* Longmans, Green, Reader & Dyer, London.

RAMSAY, A. C. 1881. *The Geology of North Wales, with an Appendix on the Fossils by J. W. Salter. Revised by Robert Etheridge.* HMSO, London.

RASTALL, R. H. 1906. The Buttermere and Ennerdale Granophyre. *Quarterly Journal of the Geological Society of London*, **62**, 253–274.

RASTALL, R. H. 1910. The Skiddaw Granite and its metamorphism. *Quarterly Journal of the Geological Society of London*, **66**, 116–140 and plates.

RATHORE, J. S. & KAFAFY, A. M. 1986. A magnetic fabric study of the Shap region in the English Lake District. *Journal of Structural Geology*, **8**, 69–77.

RAYSKA, S. 1994. *Through Thick and Thin: The History of Johnson, Poole and Bloomer 1844–1994.* Johnson, Poole and Bloomer, Stourbridge.

READ, H. H. 1957. *The Granite Controversy: Geological Addresses Illustrating the Evolution of a Disputant.* Interscience, New York.

READE, T. M. 1893. Glacial geology: old and new. *Geological Magazine*, **10**(Decade 3), 35–37.

REED, F. R. C. 1896. The fauna of the Keisley Limestone – Part I. *Quarterly Journal of the Geological Society of London*, **52**, 407–437 and plates.

REED, F. R. C. 1897. The fauna of the Keisley Limestone – Part II. Conclusion. *Quarterly Journal of the Geological Society of London*, **53**, 57–106 and plate.

RICHTER, R. 1871. Aus dem thüringischen Schiefergebirge. *Zeitschrift der Deutschen Geologischen Gesellschaft*, **23**, 231–256 and plate.

RICKARDS, R. B. 1963. *The Silurian strata of the Howgill Fells.* PhD dissertation, Hull University.

RICKARDS, R. B. 1964. The graptolitic mudstone and associated facies in the Silurian strata of the Howgill Fells. *Geological M gazine*, **101**, 435–451.

RICKARDS, R. B. 1967. The Wenlock and Ludlow succession in the Howgill Fells (north-west Yorkshire and Westmoreland). *Quarterly Journal of the Geological Society of London*, **123**, 215–251 and plate.

RICKARDS, R. B. 1969. Wenlock graptolite zones in the English Lake District. *PGSL*, **1654**, 61–65.

RICKARDS, R. B. 1970*a*. *The Llandovery (Silurian) Graptolites of the Howgill Fells, Northern England.* Palaeontographical Society, London.

RICKARDS, R. B. 1970*b*. Age of the Middle Coldwell Beds. *PGSL*, **1663**, 111–114.

RICKARDS, R. B. 1973. On some highest Llandovery red beds and graptolite assemblages in Britain and Eire. *Geological Magazine*, **110**, 70–72.

RICKARDS, R. B. 1978. Silurian. *In*: MOSELEY, F. (ed.) *The Geology of the Lake District.* Yorkshire Geological Society, Leeds, 121–145.

RICKARDS, R. B. 1999. A century of graptolite research in Cambridge. *The Geological Curator*, **7**, 71–76.

RICKARDS, R. B. n.d. *Sedgwick Geological Trail.* Yorkshire Dales National Park, Leyburn.

ROBERTS, D. E. 1971. Structures of the Skiddaw Slates in the Caldew Valley, Cumberland. *Geological Journal*, **7**, 225–238 and plates.

ROBERTS, D. E. 1973. *The structure of the Skiddaw Slates in the Blencathra–Mungrisdale area, Cumberland.* PhD dissertation, University of Birmingham.

ROBERTS, D. E. 1976. Cleavage formation in the Skiddaw Slates of the northern Lake District, England. *Geological Magazine*, **113**, 377–382.

ROBERTS, D. E. 1977*a*. Minor tectonic structures in the Skiddaw Slates of Raven Crags Mungrisdale, northern Lake District. *Proceedings of the Geologists' Association*, **88**, 117–124.

ROBERTS, D. E. 1977*b*. The structure of the Skiddaw Slates in the Blencathra–Mungrisdale Area, Cumbria. *Geological Journal*, **12**, 33–58.

ROBINS, N. S. 1980. *The Geology of some United Kingdom Nuclear Sites Related to the Disposal of Low and Medium-Level Radioactive Wastes Part I: UKAEA and BNFL Sites.* IGS Report **ENPU 80–5**, Environmental Protection Unit, Harwell.

ROBINSON, E. 1991. And did those feet, in ancient times? *Amateur Geologist*, **13/14**, 22–26.

ROBINSON, E. 1998–1999 (2000). Sydney Hollingworth (1899–1966): an exemplary field geologist. *Proceedings of the Cumberland Geological Society*, **6**, 407–413.

ROBINSON, T. 1709. *An Essay Towards a Natural History of Westmoreland and Cumberland.* W. Freeman, London.

ROSE, W. C. C. 1954. The sequence and structure of the Skiddaw Slates in the Keswick–Buttermere area. *Proceedings of the Geologists' Association*, **65**, 403–406.

ROSE, W. C. C. 1969*a*. Frederick Murray Trotter. *Proceedings of the Geologists' Association*, **80**, 126–127.

ROSE, W. C. C. 1969*b*. Stratigraphical sub-divisions of the Skiddaw Slates of the Lake District. *Geological Magazine*, **106**, 293–294.

ROSE, W. C. C. 1971. Obituary Tom Eastwood. *Proceedings of the Yorkshire Geological Society*, **38**, 295–297.

ROSE, W. C. C. & DUNHAM, K. C. 1969. Frederick Murray Trotter (1897–1968). *PGSL*, **1655**, 133–135.

ROSE, W. C. C. & DUNHAM, K. C. 1977. *Geology and Hematite Deposits of South Cumbria: Economic Memoir for 1:50 000 Sheet 58, and Southern Part of Sheet 48*. HMSO, London.

ROSENBUSCH, H. 1877. Die Steiger Stiefer, und ihre Contactzone an den Granititen von Barr-Andlau und Hohwald. *Abhandlungen zur geologischen Specialkarte von Elsass-Lothringen*, **1**, 79–393.

ROWNSLEY, H. D. 1894. *Literary Associations for the English Lakes*, vol. 1. James Maclehose, Glasgow.

ROYAL SOCIETY STUD GROUP 1994. *Disposal of Radioactive Wastes in Deep Repositories*. The Royal Society, London.

RUDWICK, M. J. S. 1974. Darwin and Glen Roy: a "great failure" in scientific method. *Studies in History and Philosophy of Science*, **5**, 97–185.

RUDWICK, M. J. S. 1985. *The Great Devonian Controversy: The Shaping of Scientific Knowledge among Gentlemanly Specialists*. The University of Chicago Press, Chicago, 1985.

RUNDLE, C. C. 1979. Ordovician intrusions in the English Lake District. *Journal of the Geological Society*, **136**, 29–38.

RUSHTON, A. W. A. 1985. A Lancefieldian graptolite from the Lake District. *Geological Magazine*, **122**, 329–333.

RUSHTON, A. W. A. 1988. Tremadoc trilobites from the Skiddaw Group in the English Lake District. *Palaeontology*, **31**, 677–698 and plates.

RUSHTON, A. W. A. 1993. Graptolites from the Manx Group. *Proceedings of the Yorkshire Geological Society*, **49**, 259–262.

RUSHTON, A. W. A. 1996. Graptolites from Aik Beck and the Bampton Inlier. British Geological Survey Technical Report **WH96/260/R**.

RUSHTON, A. W. A. & MOLYNEUX, S. G. 1989. The biostratigraphic age of the Ordovician Skiddaw Group in the Black Combe Inlier, English Lake District. *Proceedings of the Yorkshire Geological Society*, **47**, 267–276.

RUTHVEN, J. 1855. *Map of the English Lakes and Adjoining Country Geologically Coloured by John Ruthven of Kendal*. John Garnett, Windermere.

SALTER, J. W. 1861. New fossils from Skiddaw Slates. *The Geologist*, **4**, 74.

SALTER, J. W. 1864. General sketch [section] of the geology of the Lake Country. MS in Hughes papers, Cambridge University Library.

SARJEANT, W. A. S. 1992. Acritarchs. *In*: NILSSON, S. & PRAGLOWSKI, J. (eds) *Erdtman's Handbook of Palynology*. Munksgaard, Copenhagen, 486–494.

SARJEANT, W. A. S. 1999. Charles Downie (1923–1999). *Acritarch Newsletter*, **15**, 4–8.

SARJEANT, W. A. S. 2000. Charles Downie (1923–1999) and his work on fossil dinoflagellates. *Suifm@il*, Spring Issue, 33–44.

SAUNDERS, G. E. 1968. Glaciation of possible Scottish re-advance age in north west Wales. *Nature*, **218**, 76–78.

SCHARENBERG, W. 1851. *Ueber Graptolithen mit besonderer Berücksichtigung der bei Christiania vorkommenden Arten*. Trewendt & Granier, Breslau.

SCHRÖDER, H. 1888. Diluviale Süsswasser-Chonchylien auf primärer Lagerstatte in Ostpreussen. *Jahrbuch der Königlich Preussischen geologischen Landesanstalt und Bergakademie zu Berlin*, **8**, 349–362 and plate.

SCHWAB, F. L. 1986. Sedimentary 'signatures' of foreland basin assemblages: real or counterfeit? *In*: ALLEN, P. A. & HOMEWOOD, P. (eds) *Foreland Basins: Papers Presented at a Symposium held at Fribourg, Switzerland in 1986*. International Association of Sedimentologists, Special Publication **8**, 395–410.

SCOTT, R. W. 1992. *A palaeoenvironmental analysis of the Coniston Limestone Formation (Ashgill) in the north of England*. PhD dissertation, Sheffield University.

SCOTT, R. W. & KNELLER, B. C. 1990. *A Report on the Ashgill and Llandovery Age Rocks of Sheet 38 (Ambleside)*. British Geological Survey Technical Report **WA/90/63**.

SECORD, J. A. 1985. John W. Salter: the rise and fall of a Victorian palaeontological career. *In*: WHEELER, A. & PRICE, J. H. (eds) *From Linnaeus to Darwin: Commentaries on the History of Biology and Geology*. Society for the History of Natural History, London, 61–75.

SECORD, J. A. 1986. *Controversy in Geology: The Cambrian–Silurian Dispute*. Princeton University Press, Princeton.

SEDGWICK, A. 1825*a*. On diluvial formations. *Annals of Philosophy*, **9**, New Series, 18–37 (letter dated May 1825).

SEDGWICK, A. 1825*b*. On the origin of alluvial and diluvial formations. *Annals of Philosophy*, **9**, New Series, 241–257 (letter dated 11 March 1825).

SEDGWICK, A. 1826–1833 (1831*a*). On the general structure of the Lake Mountains of the north of England and on the great dislocations by which they have been separated from the neighbouring chains. *Proceedings of the Geological Society*, **1**(19), 247–249 (read 5 January 1831 and 16 March 1831).

SEDGWICK, A. 1831*b*. On the general structure of the Lake mountains of the north of England, and on the great dislocations by which they have been separated from the neighbouring chains. *PM*, **9**, 211–213, 377–379 (read 5 January 1831 and 16 March 1831).

SEDGWICK, A. 1826–1833 (1831*c*). Address to the Geological Society, delivered on the evening of the 18th of February by the Rev. Professor Sedgwick, MA, FRS on retiring from the President's chair 1831. *Proceedings of the Geological Society*, **1**(20), 281–316.

SEDGWICK, A. 1826–1833 (1831*d*). A description of longitudinal and transverse sections through a portion of the Carboniferous chain between Penigent and Kirkby Stephen. *Proceedings of the Geological Society*, **1**(21), 318–320 (read 2 March and 16 March 1831).

SEDGWICK, A. 1826–1833 (1832). On the geological relations of the stratified and unstratified groups of rocks composing the Cumbrian mountains. *Proceedings of the Geological Society*, **1**(27), 399–401 (read 16 May 1832).

SEDGWICK, A. 1833. *A Discourse on the Studies of the University*. Pitt Press, Cambridge; John W. Parker, London (republished Leicester University Press, Leicester & Humanities Press, New York, 1969).

SEDGWICK, A. 1833–1838 (1833). On a band of Transition Limestone, and on granite veins, appearing in the greywacke slate of Westmoreland, near Shap Wells and Wasdale Head. *Proceedings of the Geological Society*, **2**(33), 1–10 (read 6 November 1833).

SEDGWICK, A. 1835*a*. Introduction to the general structure of the Cumbrian mountains; with a description of the great dislocations by which they have been separated from the neighbouring Carboniferous chains. *Transactions of the Geological Society of London*, **4**, Series 2, 47–68 (read 5 January 1831 and 6 November 1833).

SEDGWICK, A. 1835*b*. A description of a series of longitudinal and transverse sections through a portion of the Carboniferous chain between Penigent and Kirkby Stephen. *Transactions of the Geological Society of London*, **4**, Series 2, 69–102 and plates (read 5 January 1831).

SEDGWICK, A. 1835*c*. Remarks on the structure of large mineral masses, and especially on the chemical changes produced in the aggregation of stratified rocks during different periods after their deposition. *Transactions of the Geological Society of London*, **3**, Series 3, 461–486 and plate (read 11 March 1835).

SEDGWICK, A. 1836. On the New Red Sandstones series in the basin of the Eden, and north-western coasts of Cumberland and Lancashire. *Transactions of the Geological Society of London*, **4**, Series 2, 383–407 and plates (read 1 February 1832).

SEDGWICK, A. 1833–1838 (1838). A synopsis of the English series of stratified rocks inferior to the Old Red Sandstone – with an attempt to determine the successive natural groups and formations. *Proceedings of the Geological Society*, **2**(58), 675–685 (read 21 March and 23 May 1838).

SEDGWICK, A. 1838–1842 (1841). Supplement to a 'Synopsis of the English series of stratified rocks inferior to the Old Red Sandstone,' with additional remarks on the relations of the Carboniferous Series and the Old Red Sandstone of the British Isles. *Proceedings of the Geological Society*, **3**(82), 545–558 (read 3 November and 17 November 1841).

SEDGWICK, A. 1842. Three letters on the geology of the Lake District. *In*: WORDSWORTH, W., *Description of the Scenery of the Lakes*. J. Hudson, Kendal.

SEDGWICK, A. 1843–1863 (1857). Description of a series of dislocations which have moved the Cambrian and Silurian rocks between Leven sands and Duddon sands, several miles out of their normal position in the geological map of the Lake mountains. *Proceedings of the Cambridge Philosophical Society*, **13**, 187–190 (read 9 November 1857).

SEDGWICK, A. 1844. On the older Palaeozoic (Protozoic) rocks of North Wales. *Proceedings of the Geological Society*, **4**, 255–256 (read 29 November 1843).

SEDGWICK, A. 1845*a*. On the comparative classification of the fossiliferous strata of North Wales, with corresponding deposits of Cumberland, Westmoreland, and Lancashire. *Quarterly Journal of the Geological Society of London*, **1**, 442–450.

SEDGWICK, A. 1845*b*. On the older Palaeozoic (Protozoic) rocks of North Wales. *Quarterly Journal of the Geological Society of London*, **1**, 5–22.

SEDGWICK, A. 1846. On the classification of the fossiliferous slates of Cumberland, Westmoreland, and Lancashire. *Quarterly Journal of the Geological Society of London*, **2**, 106–131.

SEDGWICK, A. 1847. On the classification of the fossiliferous slates of North Wales, Cumberland, Westmoreland and Lancashire (being a supplement to a paper read to the Society, March 12 1845). *Quarterly Journal of the Geological Society of London*, **3**, 133–166.

SEDGWICK, A. 1848. On the organic remains found in the Skiddaw slate with some remarks on the classification of the older rocks of Cumberland and Westmoreland, &c. *Quarterly Journal of the Geological Society of London*, **4**, 216–225.

SEDGWICK, A. 1851 (for 1850). On the geological structure and relations of the frontier chain of Scotland. *Report of the British Association for the Advancement of Science*, 103–107.

SEDGWICK, A. 1852*a*. On the Lower Palaeozoic rocks at the base of the Carboniferous chain between Ravenstonedale and Ribblesdale. *Quarterly Journal of the Geological Society of London*, **8**, 35–54.

SEDGWICK, A. 1852*b*. On the classification and nomenclature of the Lower Palaeozoic rocks of England and Wales. *Quarterly Journal of the Geological Society of London*, **8**, 136–168.

SEDGWICK, A. 1853*a*. *Geology of the Lake District, Westmorland, and Lancashire, in Letters Addressed to Wordsworth*. John Hudson, Kendal.

SEDGWICK, A. 1853*b* On a proposed separation of the so-called Caradoc Sandstone into two distinct groups; viz. (1) May Hill Sandstone, (2) Caradoc Sandstone. *Quarterly Journal of the Geological Society of London*, **9**, 215–230.

SEDGWICK, A. 1873. Letter concerning Bolton [bound in with some copies of BOLTON (1869)].

SEDGWICK, A. n.d. Geological Map of Cumberland, Sedgwick Museum Archives.

SEDGWICK, A. & MCCOY, F. 1855. *A Synopsis of the Classification of the British Palaeozoic Rocks, ... With a Systematic Description of the British Palaeozoic Fossils in the Geological Museum of the University of Cambridge by Frederick McCoy*. John W. Parker, London; Deighton & Bell and Macmillan, Cambridge.

SEDGWICK, A. & PEILE, W. 1833–1838 (1836). On the coal-fields on the north-western coast of Cumberland. *Proceedings of the Geological Society*, **2**(46), 419–422 (read 8 June 1836).

SEDGWICK, A. & SALTER, J. W. 1873. *A Catalogue of the Collection of Cambrian and Silurian Fossils Contained in the Geological Museum of the University of Cambridge, with a Preface by The Rev. Adam Sedgwick*. Cambridge University Press, Cambridge.

SHACKLETON, E. H. 1963. Jonathan Otley. *Proceedings of the Cumberland Geological Society*, **2**, 5–8.

SHACKLETON, E. H. 1966*a*. Robert Harkness. *Proceedings of the Cumberland Geological Society*, **2**(1), 19–23.

SHACKLETON, E. H. 1966*b*. *Lakeland Geology*. Dalesman, Clapham.

SHACKLETON, E. H. 1969. Henry Alleyne Nicholson. *Proceedings of the Cumberland Geological Society*, **2**(3), 105–107.

SHACKLETON, E. H. 1970–1971 (1971). Adam Sedgwick. *Proceedings of the Cumberland Geological Society*, **3**(1), 33–36.

SHACKLETON, E. H. 1973–1974 (1975). J. F. N. Green: amateur geologist. *Proceedings of the Cumberland Geological Society*, **3**(3), 147–149.

SHACKLETON, E. H. 1974. John Bolton of Ulverston. *Cumbria*, **23**, 627–628.

SHACKLETON, N. J. 1969. The last interglacial in the marine and terrestrial records. *Proceedings of the Royal Society*, Series B, **174**, 135–154.

SHACKLETON, N. J. & OPDYKE, N. D. 1973. Oxygen isotope and palaeomagnetic stratigraphy of Equatorial Pacific core V28-238: oxygen isotope temperatures and ice volumes on a 105 year and a 106 year scale. *Quaternary Research*, **3**, 39–55.

SHARPE, D. 1842. Sketch of the geology of the south of Westmoreland. *Proceedings of the Geological Society*, **3**, 602–608.

SHARPE, D. 1843*a*. On the Silurian rocks of the south of Westmoreland and north of Lancashire. *Proceedings of the Geological Society*, **4**, 23–29.

SHARPE, D. 1843*b*. On the Bala Limestone. *Proceedings of the Geological Society*, **4**, 10–14.

SHARPE, T. & MCCARTNEY, P. J. 1998. *The Papers of H. T. De la Beche (1796–1855) in the National Museum of Wales*. National Museums and Galleries of Wales, Cardiff.

SHAW, H. S. 1971. Charles Stansfield Hitchen (1904–1971). *Yearbook of the Geological Society for 1971*. Geological Society, London, 32–33.

SHAW, R. W. L. 1969. *The faunal stratigraphy of the Kirkby Moor Flags of the type area near Kendal, Westmorland*. PhD dissertation, The University of Newcastle-upon-Tyne.

SHAW, R. W. L. 1971. The faunal stratigraphy of the Kirkby Moor Flags of the type area near Kendal, Westmorland. *Geological Journal*, **7**, 359–380 and map.

SHIPP, T. 1990–1991 (1992). Edgar Howard Shackleton, FGS. *Proceedings of the Cumberland Geological Society*, **5**, 392–395.

SHOTTON, F. W. 1935. The stratigraphy and tectonics of the Cross Fell Inlier. *Quarterly Journal of the Geological Society of London*, **91**, 639–704 and plate.

SHOTTON, F. W. 1963. W. B. R. King – obituary. *Proceedings of the Yorkshire Geological Society*, **34**, 91–93.

SHOTTON, F. W. n.d. Section through Cross Fell Inlier. Lapworth archive, University of Birmingham.

SHOTTON, F. W., WADGE, A. J. & BURGESS, I. C. 1968. Field meeting. Cross Fell area. 14th to 16th July 1967. *Proceedings of the Yorkshire Geological Society*, **36**, 340–344.

SIMPSON, A. 1961. *The stratigraphy and tectonics of the Manx Slate Series, Isle of Man*. PhD dissertation, Birkbeck College, London University.

SIMPSON, A. 1963. The stratigraphy and tectonics of the Manx Slate Series, Isle of Man. *Quarterly Journal of the Geological Society of London*, **119**, 367–400 and plate.

SIMPSON, A. 1965. F_1 cross-folding in the Manx Slate Series, Isle of Man. *Geological Magazine*, **102**, 440–444.

SIMPSON, A. 1967. The stratigraphy and tectonics of the Skiddaw Slates and the relationship of the overlying Borrowdale Volcanic Series in part of the Lake District. *Geological Journal*, **5**, 391–418.

SIMPSON, A. 1968*a*. The Caledonian history of the north-eastern Irish Sea region, and its relation to surrounding areas. *Scottish Journal of Geology*, **4**, 135–163.

SIMPSON, A. 1968*b*. The Caledonian history of the North-Eastern Irish Sea region [letters of reply to Helm & Roberts, Wadge & Burgess, and Soper & Brown]. *Scottish Journal of Geology*, **4**, 380–385.

SIMPSON, B. 1934. The petrology of the Eskdale (Cumberland) Granite. *Proceedings of the Geologists' Association*, **45**, 17–34.

SISSONS, J. B. 1960. Erosion surfaces, cyclic slopes and drainage systems in southern Scotland and northern England. *Transactions of the Institute of British Geographers*, **28**, 23–38.

SISSONS, J. B. 1964. The glacial period. *In*: WREFORD WATSON, J. & SISSONS, J. B. (eds) *The British Isles: A Systematic Geography*. Nelson, London and Edinburgh, 131–152.

SISSONS, J. B. 1974. The Quaternary in Scotland: a review. *Scottish Journal of Geology*, **10**, 311–337.

SISSONS, J. B. 1980. The Loch Lomond advance in the Lake District, northern England. *TRSE, Earth Sciences,* **71**, 13–27.

SKEVINGTON, D. 1964. A new occurrence of the *Didymograptus bifidus* zone in the Skiddaw Group. *Nature*, **202**, 585.

SKEVINGTON, D. 1970. A Lower Llanvirn graptolite fauna from the Skiddaw Slates, Westmorland. *Proceedings of the Yorkshire Geological Society*, **37**, 395–444.

SKEVINGTON, D. 1974. Controls influencing the composition and distribution of Ordovician graptolite faunal provinces. *In*: RICKARDS, R. B., JACKSON, D. E. & HUGHES, C. P. (eds) *Graptolite Studies in Honour of O. M. B. BULMAN: A Memorial Volume of Twenty Papers*. Special Papers in Palaeontology, **13**, The Palaeontological Association, London, 59–73.

SMITH, B. 1912. The glaciation of the Black Combe district, Cumberland. *Quarterly Journal of the Geological Society of London*, **68**, 402–448 and plates.

SMITH, B. 1919. *Iron Ores: Haematites of West Cumberland, Lancashire and the Lake District*. HMSO, London.

SMITH, B. 1921 (for 1920). On borings for coal near Maryport, Cumberland. *Summary of Progress of the Geological Survey of Great Britain*, 85–91.

SMITH, B. 1922 (for 1921). Cumberland district. *Summary of Progress of the Geological Survey of Great Britain*, 44–68.

SMITH, B. 1923 (for 1922). Cumberland district. *Summary of Progress of the Geological Survey of Great Britain*, 58–69.

SMITH, B. 1924*a* (for 1923). Cumberland district [Whitehaven (Sheet 28) and Brampton (Sheet 18)]. *Summary of Progress of the Geological Survey of Great Britain*, 69–79.

SMITH, B. 1924*b*. *Iron Ores: Haematites of West Cumberland, Lancashire and the Lake District*, 2nd edn, HMSO, London.

SMITH, B. 1924*c*. The unconformable base of the Coniston Limestone Series in the Lake District. *Geological Magazine*, **61**, 163–167.

SMITH, B. 1925 (for 1924). Cumberland district [Whitehaven (Sheet 28) and Brampton (Sheet 18)]. *Summary of Progress of the Geological Survey of Great Britain*, 70–80.

SMITH, B. 1926 (for 1925). Cumberland district [Sheets 28 (Whitehaven), 22 (Maryport) and 18 (Brampton)]. *Summary of Progress of the Geological Survey of Great Britain*, 72–85.

SMITH, B. 1927 (for 1926). Cumberland district [Brampton, Whitehaven, and Maryport Sheets]. *Summary of Progress of the Geological Survey of Great Britain*, 51–53.

SMITH, B. 1928*a* (for 1927). The origin of Cumberland haematite (2). *Summary of Progress of the Geological Survey of Great Britain* (Part 2), 33–36.

SMITH, B. 1928*b* (for 1927). Cumberland district. *Summary of Progress of the Geological Survey of Great Britain* (Part 1), 51–53.

SMITH, B. 1930 (for 1929). Cumbrian district [Cockermouth Sheet 23]. *Summary of Progress of the Geological Survey of Great Britain*, 63–66.

SMITH, B., DIXON, E. E. L., EASTWOOD, T., TROTTER, F. M. & HOLLINGWORTH, S. E. 1929 (for 1928). Cumbrian district. *Summary of Progress of the Geological Survey of Great Britain*, 68–71.

SMITH, C. 1985. Geologists and mathematicians: the rise of mathematical geology. *In*: HARMAN, P. M. (ed.) *Wranglers and Physicists: Studies on Cambridge Physics in the Nineteenth Century*. Manchester University Press, Manchester, 49–83.

SMITH, C. 1989. William Hopkins and the shaping of dynamical geology: 1830–1860. *The British Journal for the History of Science*, **22**, 27–52.

SMITH, R. A. 1967. The deglaciation of south-west Cumberland: a reappraisal of some features in the Eskdale and Bootle Areas. *Proceedings of the Cumberland Geological Society*, **2**(2), 76–83.

SMITH, R. A. 2000. Jonathan Otley: a pioneer of Lakeland geology. *Geology Today*, **16**, 31–34.

SMITH, R. A. 1998–1999 (2000). Jonathan Otley (1788–1856): father of Lakeland geology. *Proceedings of the Cumberland Geological Society*, **6**, 309–314.

SMITH, R. A. 1998–1999 (2000). James Clifton Ward (1843–1880): an early populariser of geology. *Proceedings of the Cumberland Geological Society*, **6**, 341–347.

SMITH, R. A. n.d. *A Bibliography of the Geology and Geomorphology of Cumbria*. The Cumberland Geological Society, Penrith.

SMITH, R. A. n.d. *A Bibliography of the Geology and Geomorphology of Cumbria: Part 2 (1974–1990)*. The Cumberland Geological Society, Penrith.

SMITH, W. 1815. *A Delineation of the Strata of England and Wales with part of Scotland* [hand-coloured map].

SMYTHE, D. K., HOLT, J. M., ELSTO, M. & ROBSON, C. 1995. A high-resolution 3D vibroseis trial survey of a potential nuclear waste repository. *European Association of Exploration Geophysicists: Extended Abstracts of the 57th Conference and Technical Exhibition, Glasgow, 29 May–2 June 1995*, No. **C054**.

SOPER, N. J. 1970. Three critical localities on the junction of the Borrowdale Volcanic Rocks with the Skiddaw Slates in the Lake District. *Proceedings of the Yorkshire Geological Society*, **37**, 461–493 and plates.

SOPER, N. J. 1978. Structure. *In*: MOSELEY, F. (ed.) *The Geology of the Lake District*. Yorkshire Geological Society, Leeds, 45–67.

SOPER, N. J. 1986. The newer granite problem: a geotectonic view. *Geological Magazine*, **123**, 227–236.

SOPER, N. J. 1987. The Ordovician batholith of the English Lake District. *Geological Magazine*, **124**, 481–482.

SOPER, N. J. 1988. Timing and geometry of collision, terrane accretion and sinistral strike-slip events in the British Caledonides. *In*: HARRIS, A. L. & FETTES, D. J. (eds) *The Caledonian–Appalachian Orogen*. Blackwell Scientific, Oxford, 481–492.

SOPER, N. J. n.d. [2000]. Lake District: early Palaeozoic tectonostratigraphy [unpublished table].

SOPER, N. J. & ANDERTON, R. 1984. Did the Dalradian slides originate as extensional slides? *Nature*, **307**, 357–360.

SOPER, N. J. & BROWN, P. E. 1968. The Caledonian history of the northeastern Irish Sea region [letter]. *Scottish Journal of Geology*, **4**, 378–379.

SOPER, N. J. & HUTTON, D. H. W. 1984. Late Caledonian sinistral displacements in Britain: implications for a three-plate collision model. *Tectonics*, **3**, 781–794.

SOPER, N. J. & KNELLER, B. C. 1990. Cleaved microgranite dykes of the Shap swarm in the Silurian of NW England. *Geological Journal*, **25**, 161–170.

SOPER, N. J. & NUMAN, N. M. S. 1974. Structure and stratigraphy of the Borrowdale Volcanic Rocks of the Kentmere Area, English Lake District. *Geological Journal*, **9**, 147–166.

SOPER, N. J. & ROBERTS, D. E. 1971. Age of cleavage in the Skiddaw Slates in relation to the Skiddaw aureole. *Geological Magazine*, **108**, 293–302.

SOPER, N. J. & WOODCOCK, N. H. 1990. Silurian collision and sediment dispersal patterns in southern Britain. *Geological Magazine*, **127**, 527–542.

SOPER, N. J., BRANNEY, M. J., MATHIESON, N. A. & DAVIS, N. C. 1987*b*. Reply [to WEBB *et al.*]. *Geological Magazine*, **124**, 483–484.

SOPER, N. J., COOPER, A. H, HUGHES, R. A., MILLWARD, D. & MOLYNEUX, S. G 1993. Discussion on evidence for an unconformity at the base of the Eycott Volcanic Group. *Geological Magazine*, **130**, 271–273.

SOPER, N. J., ENGLAND, R. W., SNYDER, D. B. & RYAN, P. D. 1992*a*. The Iapetus Suture zone in England, Scotland and eastern Ireland: a reconciliation of geological and deep seismic data. *Journal of the Geological Society*, **149**, 697–700.

SOPER, N. J., STRAHAN, R. A., HOLDSWORTH, R. E., GAYER, R. A. & GREILING, R. O. 1992*b*. Sinistral transpression and the Silurian closure of Iapetus. *Journal of the Geological Society*, **149**, 871–880.

SOPER, N. J., WEBB, B. C. & WOODCOCK, N. H. 1987*a*. Late-Caledonian (Acadian) transgression in north-west England: timing, geometry and geotectonic significance. *Proceedings of the Yorkshire Geological Society*, **46**, 175–192.

SORBY, H. C. 1858. On the microscopical structure of crystals, indicating the origin of minerals and rocks, *Quarterly Journal of the Geological Society of London*, **14**, 453–500 and plates.

SORBY, H. C. 1880. [Slates of the English Lake District – part of Presidential address]. *Quarterly Journal of the Geological Society of London*, **36**, 74–76.

SORBY, H. C. 1908. On the application of quantitative methods to the study of the structure and history of rocks. *Quarterly Journal of the Geological Society of London*, **64**, 171–233 and plates.

SOUTHEY, R. 1909. *The Poems of Robert Southey*, edited by Maurice H. Fitzgerald. Oxford University Press, London.

SPARKS, R. S. J. & WILSON, L. 1976. A model for the formation of ignimbrite by gravitational column collapse. *Journal of the Geological Society*, **132**, 441–451.

SPEAKMAN, C. 1982. *Adam Sedgwick: Geologist and Dalesman, 1785–1873*. Broad Oak Press, The Geological Society of London & Trinity College, Cambridge, Broad Oak, Sussex.

STONE, P. & KIMBELL, G. S. 1995. Caledonian terrane relationships in Britain: an introduction. *Geological Magazine*, **132**, 461–464 and plate.

STONE, P., COOPER, A. H. & EVANS, J. A. 1999. The Skiddaw Group (English Lake District) reviewed: early Palaeozoic sedimentation and tectonism at the northern margin of Avalonia. *In*: WOODCOCK, N. H., QUIRK, D. G., FITCHES, W. R. & BARNES, R. P. (eds) *In Sight of the Suture: The Palaeozoic Geology of the Isle of Man in its Iapetus Ocean Context*. Geological Society, London, 325–336.

STONE, P., FLOYD, J. D., BARNES, R. P. & LINTERN, B. C. 1987. A sequential back-arc and foreland basin thrust duplex model for the Southern Uplands of Scotland. *Journal of the Geological Society*, **144**, 753–764.

STONE, P., PLANT, J. A., MENDUM, J. R. & GREEN, P. M. 1999. A regional geochemical assessment of some terrane relationships in the British Caledonides. *Scottish Journal of Geology*, **35**, 145–156.

STRENS, R. G. J. 1962. *The geology of the Borrowdale–Honister district (Cumberland) with special reference to the mineralisation*. PhD dissertation, Nottingham University.

STROTHERS, P. K. 1996. Acritarchs. *In*: JANSONIUS, J. & MCGREGOR, D. C. (eds) *Palynology: Principles and Applications*, 3 vols. American Association of Stratigraphic Palynologists Foundation, Salt Lake City, **1**, 81–109.

STUBBLEFIELD, J. & DUNHAM, K. C. 1977. George Hoole Mitchell 31 December 1902–11 March 1976. *Biographical Memoirs of the Fellows of the Royal Society*, **23**, 367–383 and plate.

STYCHE, P. C. 1997. Letter of 17 March 1997 on behalf of Secretary of State for the Environment to David Brown, Director Nirex. Ref APP/H0900/A/94/247019.

SUESS, E. 1906. *The Face of the Earth (Das Antlitz der* Erde). Translated by Hertha B. C. Sollas, Vol. II. Clarendon Press, Oxford.

SUTHREN, R. J. 1977*a*. *Volcanic and sedimentary facies of Part of the Borrowdale Volcanic Group, Cumbria*. PhD dissertation, Keele University.

SUTHREN, R. J. 1977*b*. Sedimentary processes in the Borrowdale Volcanic Group. *In*: STILLMAN C. J., Conference report – Palaeozoic volcanism in Great Britain and Ireland. *Journal of the Geological Society*, **133**, 410–411.

SUTHREN, R. J. & FURNES, H. 1980. Origin of some bedded welded tuffs. *Bulletin of Volcanology*, **43**, 61–71.

SUTTON, J. 1974. Floating on high [granite gravity anomalies in the Lake District]. *Nature*, **250**, 374–375.

SUTTON, J. S. 1996. Hydrogeological testing in the Sellafield area. *Quarterly Journal of Engineering Geology*, **29**, Supplement 1, S29–S38.

SWEETING, G. S. (ed.) 1958. *The Geologists' Association 1858–1958: A History of the First Hundred Years*. Benham & Co. for the Geologists' Association, London.

SYNGE, F. M. 1952. Retreat stages of the last ice-sheet in the British Isles. *Irish Geography*, **2**, 168–171.

TAPPIN, D. 2001. Local tsunamis. *Geoscientist*, **11**(5), 4–7.

TARR, R. S. 1906*a*. Watkins Glen and other gorges in the Finger Lake region of central New York. *Popular Science Monthly*, May 1906, 387–397.

TARR, R. S. 1906*b*. Glacial erosion in the Finger-Lake region of central New York. *Journal of Geology*, **14**, 215–228 and map.

TAYLOR, B. J. 1968–1970. F. M. Trotter. *Proceedings of the Yorkshire Geological Society*, **37**, 101–103.

TAYLOR, B. J., BURGESS, I. C., LAND, D. H., MILLS, D. A. C., SMITH, D. B. & WARREN, P. T. 1971. *British Regional Geology: Northern England*. HMSO, London.

TEALL, J. J. H. 1888. *British Petrography*. Dulau, London.

TEMPLE, J. T. 1968. *A Monograph of the Lower Llandovery (Silurian) Brachiopods from Keisley, Westmorland*. The Palaeontographical Society, London.

TEMPLE, J. T. 1969. Lower Llandovery (Silurian) trilobites from Keisley, Westmorland. *Bulletin of the British Museum (Natural History) Geology*, **18**, 47–65 and plates.

THIRLWALL, M. F. & FITTON, J. G. 1983. Sm–Nd garnet age for the Ordovician Borrowdale Volcanic Group, English Lake District. *Journal of the Geological Society*, **140**, 511–518.

THOMAS, L. J. 1985. *Low-grade metamorphism and the stable isotopic composition of alteration fluids, Lower Palaeozoic succession, English Lake District*. PhD dissertation, St Andrews University.

THOMSON, K. 1995. Discussion on a latest Cretaceous hotspot and the southesaterly tilt of Britain. *Journal of the Geological Society*, **152**, 729.

TIDDEMAN, R. H. 1872. On the evidence for the ice-sheet in north Lancashire and adjacent parts of Yorkshire and Westmoreland. *Quarterly Journal of the Geological Society of London*, **28**, 471–490 and plate.

TODHUNTER, I. 1876. *William Whewell, . . . An Account of his Writings with Selections from his Literary and Scientific Correspondence*, 2 vols. Macmillan, London.

TOOLEY, M. J. 1982. Sea-level changes in northern England. *Proceedings of the Geologists' Association*, **93**, 43–51.

TORRENS, H. S. 1994. Joseph Harrison Fryer (1777–1855): geologist and mining engineer, in England 1803–1825 and South America 1826–1828. *In*: FIGUEIRÔA, S. & LOPES, M. (eds) *Geological Sciences in Latin America: Scientific Relations and Exchanges*. Universidade Estadual de Campinas, Campinas, Brazil, 29–46.

TREAGUS, J. E. (ed.) 1992. *Caledonian Structures in Britain South of the Midland Valley*. Chapman & Hall, London.

TRECHMANN, C. O. 1882. Note on the so-called 'hypersthenite' of Carrock Fell, Cumberland. *Geological Magazine*, **9**(Decade 2), 210–212.

TRIMMER, J. 1834 (for 1831). On the diluvial deposits of Caernarvonshire, between the Snowdon chain of hills and the Menai Strait, and on the discovery of marine shells in diluvial sand and gravel on the summit of Moel Tryfane, near Caernarvon, 1,000 ft above the Level of the Sea. *Proceedings of the Geological Society*, **1**, 331–332.

TROTTER, F. M. 1929*a*. The Tertiary uplift and resultant drainage of the Alston block and adjacent areas. *Proceedings of the Yorkshire Geological Society*, **21**, 161–180.

TROTTER, F. M. 1929*b*. The glaciation of eastern Edenside, the Alston block, and the Carlisle plain. *Quarterly Journal of the Geological Society of London*, **85**, 599–612 and plates.

TROTTER, F. M. 1945. The origin of the west Cumberland haematites. *Geological Magazine*, **82**, 67–80.

TROTTER, F. M. 1952. The west Cumberland haematites. *XIXth International Geological Congress: Symposium sur les Gisements de Fer du Monde*, **2**, Algeria, 427–431.

TROTTER, F. M. & DUNHAM, K. C. 1952. Genesis of the west coast haematite deposits. *XIXth International Geological Congress, Symposium sur les Gisements de Fer du Monde*, **2**, Algeria, 435–438.

TROTTER, F. M. & HOLLINGWORTH, S. E. 1932*a*. *The Geology of the Brampton District*, HMSO, London.

TROTTER, F. M. & HOLLINGWORTH, S. E. 1932*b*. The glacial sequence in the North of England. *Geological Magazine*, **69**, 374–380.

TROTTER, F. M., HOLLINGWORTH, S. E., EASTWOOD, T. & ROSE, W. C. C. 1937. *Gosforth District (One-inch Geological Sheet 37 New Series)*. HMSO, London.

TULLBERG, S. A. T. 1883. *Skånes Graptoliter*, **2**, Section C, No. **55**.

TURNER, J. S. 1935. Structural geology of Stainmore, Westmorland, and notes on the Late-Palaeozoic (Late-Variscan) tectonics of the north of England. *Proceedings of the Geologists' Association*, **46**, 121–151 and plate.

TURNER, J. S. 1949. The deeper structure of central and northern England. *Proceedings of the Yorkshire Geological Society*, **27**, 280–297.

TURNER, R. E. & WADGE, A. J. 1979. Acritarch dating of Arenig volcanism in the Lake District. *Proceedings of the Yorkshire Geological Society*, **42**, 405–414 and plates.

VANN, I. R., GRAHAM, H. R. & HAYWARD, A. B. 1986. The structure of mountain fronts. *Journal of Structural Geology*, **8**, 215–228.

VICTORIA & ALBERT MUSEUM 1984. *The Discovery of the Lake District: A Northern Arcadia and its Uses*. Victoria & Albert Museum, London.

VON BUCH, L. 1810. Etwas über locale und allgemeine Gebirgsformationen. *Magazin der Gesellschaft Naturforschende Freunde zu Berlin*, **4**, 69–74.

VON SCHLOTHEIM, E. F. 1822–1823. *Nachträge zur Petrefaktenkunge*. Part **2**, Becker, Gotha.

WADGE, A. J. 1971. Skiddaw Slates and the Borrowdale Volcanics. *Nature Physical Science Series*, **234**(36), 58–59.

WADGE, A. J. 1972. Sections through the Skiddaw–Borrowdale unconformity in Eastern Lakeland. *Proceedings of the Yorkshire Geological Society*, **39**, 179–198.

WADGE, A. J. 1978*a*. Devonian. *In*: MOSELEY, F. (ed.) *The Geology of the Lake District*. Yorkshire Geological Society, Leeds, 164–167.

WADGE, A. J. 1978*b*. Classification and stratigraphical relationships of the Lower Ordovician rocks. *In*: MOSELEY, F. (ed.) *The Geology of the Lake District*. Yorkshire Geological Society, Leeds, 68–77.

WADGE, A. J., GALE, N. H., BECKINSALE, R. D. & RUNDLE, C. C. 1978. A Rb–Sr isochron for the Shap Granite. *Proceedings of the Yorkshire Geological Society*, **40**, 297–305.

WADGE, A. J., NUTT, M. J. C., LISTER, T. R. & SKEVINGTON, D. 1969. A probable *Didymograptus murchisoni* zone fauna from the Lake District. *Geological Magazine*, **106**, 595–598.

WADGE, A. J., NUTT, M. J. C. & SKEVINGTON, D. 1972. Geology of the Tarn Moor Tunnel in the Lake District. *Bulletin of the Geological Survey of Great Britain*, **41**, 55–73 and plates.

WADGE, A. J., OWENS, B. & DOWNIE, C. 1967. Microfossils from the Skiddaw Group. *Geological Magazine*, **104**, 506–507.

WALKER, E. E. 1904. Notes on the garnet-bearing and associated rocks of the Borrowdale Volcanic Series. *Quarterly Journal of the Geological Society of London*, **60**, 70–105 and plates (communicated by J. E. Marr).

WALLACE, W. 1861. *The Mineral Deposits of Alston Moor, The Laws which Regulate the Disposition of Lead Ore in Veins; Illustrated by an Examination of the Geological Structure of the Mining Districts of Alston Moor*. Edward Stanford, London.

WALTON, E. K. 1963. Sedimentation and structure in the Southern Uplands. *In*: JOHNSON, M. R. W. & STEWART, F. H. (eds) *The British Caledonides*. Oliver & Boyd, Edinburgh and London, 71–97.

WARD, J. C. 1873. The glaciation of the northern part of the Lake-District. *Quarterly Journal of the Geological Society of London*, **29**, 422–441.

WARD, J. C. 1875*a*. On the granitic, granitoid, and associated metamorphic rocks of the Lake-District. *Quarterly Journal of the Geological Society of London*, **31**, Parts I–II, 568–602 and plates.

WARD, J. C. 1875*b*. Notes on the comparative microscopic rock-structure of some ancient and modern volcanic rocks. *Quarterly Journal of the Geological Society of London*, **31**, 388–422.

WARD, J. C. 1875*c*. The glaciation of the southern part of the Lake-District and the glacial origin of the lake-basins of Cumberland and Westmoreland. *Quarterly Journal of the Geological Society of London*, **31**, 152–166.

WARD, J. C. 1876*a*. *The Geology of the Northern Part of the English Lake District*. Longmans & Edward Stanford, London.

WARD, J. C. 1876*b*. Absence of Llandovery rocks in the Lake District. *Geological Magazine*, **3**(Decade 2), 383.

WARD, J. C. 1876*c*. On the granitic, granitoid, and associated metamorphic rocks of the Lake-District. Parts III–V. *Quarterly Journal of the Geological Society of London*, **32**, 11–34 and plates.

WARD, J. C. 1877. Jonathan Otley, geologist and guide. *Transactions of the Cumberland and Westmorland Association for the Advancement of Literature and Science*, **2**, 125–169.

WARD, J. C. 1879. On the physical history of the English Lake District. With notes on the possible subdivisions of the Skiddaw Slates. *Geological Magazine*, **6**(Decade 2), 49–61, 110–125.

WATNEY, G. R. & WELCH, E. G. 1910. The graptolite zones of the Salopian rocks of the Cautley area near Sedbergh, Yorkshire. *Geological Magazine*, **7**(Decade 5), 473.

WATNEY, G. R. & WELCH, E. G. 1911*a* (for 1910). The graptolitic zones of the Salopian rocks of the Cautley area near Sedbergh, Yorkshire. *Report of the British Association for the Advancement of Science*, 603.

WATNEY, G. R. & WELCH, E. G. 1911*b*. The zonal classification of the Salopian rocks of the Cautley and Ravenstonedale. *Quarterly Journal of the Geological Society of London*, **67**, 215–237 and plate (communicated by J. E. Marr).

WEBB, B. C. 1972. North–south trending pre-cleavage folds in the Skiddaw Slate Group of the English Lake District. *Nature Physical Science Series*, **235**(59), 138–140.

WEBB, B. C. 1975. *The structure and stratigraphy of the Skiddaw Slates between Buttermere and Newlands, Cumbria, and their relationship to the overlying volcanic rocks*. PhD dissertation, Sheffield University.

WEBB, B. C. & COOPER, A. H. 1988. Slump and gravity slide structures in a Lower Palaeozoic marginal basin sequence (the Skiddaw Group), NW England. *Journal of Structural Geology*, **10**, 463–472.

WEBB, B. C., MILLWARD, D., JOHNSON, E. & COOPER, A. 1987. The Ordovician(?) batholith of the English Lake District. *Geological Magazine*, **124**, 482–483.

WEBB, P. C. & BROWN, G. C. 1984. *Lake District Granites: Heat Production and Related Geochemistry*. British Geological Survey Technical Report **WJ/GE/84/14**.

WELLS, M. K. 1967. Sydney Ewart Hollingworth, MA, DSc. *PGSL*, **1636**, 192–196.

WHEILDON, J., KING, J., CROOK, C. N. & THOMAS-BETTS, A. 1984. *The Lake District Granites – Heat Flow, Heat Production and Geothermal Model Studies*. British Geological Survey.

WHITAKER, W. 1899. Henry Alleyne Nicholson. *PGSL*, **55**, lxiv–lxvi.

WHITE, A. W. 1961. Gertrude Lilian Elles. *Newnham College Roll Letter, 1961*. W. Heffer, Cambridge, 45–49.

WHITE, N. & LOVELL, B. 1997. Measuring the pulse of a plume with the sedimentary record. *Nature*, **387**, 888–891.

WHITE, R. S. 1988. A hot-spot model for early Tertiary volcanism in the N Atlantic. *In*: MORTON, A. C. & PARSON, L. M. (eds) *Early Tertiary Volcanism and the Opening of the NE Atlantic*. Geological Society & Blackwell Scientific Publications, London, Special Publications, **39**, 3–13.

WHITE, R. S. & MCKENZIE, D. 1989. Magmatism at rift zones: the generation of volcanic continental margins and flood basalts. *Journal of Geophysical Research*, **94**, 7685–7729.

WHITTARD, W. F. 1952. A geology of south Shropshire. *Proceedings of the Geologists' Association*, **63**, 143–197.

WHITTARD, W. F. & SIMPSON, S. (eds) 1960. *Lexique stratigraphique international. Volume 1 Europe. Fascicule 3a: Angleterre Pays de Galle Écosse. Fascicule 3a IV: Ordovicien*. Centre Nationale de la Recherche Scientifique, Paris.

WHITTARD, W. F. & SIMPSON, S. (eds) 1961. *Lexique stratigraphique international. Volume 1 Europe. Fascicule 3a: Angleterre Pays de Galle Écosse. Fascicule 3a V: Silurien*. Centre Nationale de la Recherche Scientifique, Paris.

WIESHEU, R. & HEIN, U. F. 1998. The history of fluid inclusion studies. *In*: FRITSCHER, B. & HENDERSON, F. (eds) *Toward a History of Mineralogy, Petrology, and Geochemistry*. Institut für Geschichte der Naturwissenschaften, Munich, 309–325.

WILLIAMS, A. 1969. Ordovician of the British Isles. *In*: KAY, M. (ed.) *North Atlantic Geology and Continental Drift: A Symposium*. The American Association of Petroleum Geologists, Tulsa, 236–264.

WILLIAMS, A. 1976. Plate tectonics and biofacies evolution as factors in Ordovician correlation. *In*: BASSETT, M. G. (ed.) *The Ordovician System: Proceedings of a Palaeontological Association Symposium, Birmingham, September, 1974*. University of Wales Press and National Museum of Wales, Cardiff, 29–66.

WILLIAMS, G. D., BRABHAM, P. J., EATON, G. P. & HARRIS, C. 2001. Late Devensian glaciotectonic deformation at St Bees, Cumbria: a critical wedge model. *Journal of the Geological Society*, **158**, 125–135.

WILLIAMS, S. H. 1988. Dobs Linn – the Ordovician–Silurian boundary stratotype. *Bulletin of the British Museum (Natural History) Geology*, **43**, 17–30.

WILLS, L. J. 1937. Bernard Smith. *Quarterly Journal of the Geological Society of London*, **93**, cvii–cix.

WILSON, D. W. R. 1954. *The stratigraphy and palaeontology of Valentian rocks of Cautley (Yorks W. R.)*. PhD dissertation, Birmingham University.

WILSON, H. E. 1960. J. J. H.: an appreciation. *Irish Naturalists' Journal*, **13**, 134.

WILSON, J. T. 1963*a*. Evidence from islands on the spreading of ocean floors. *Nature*, **197**, 536–538.

WILSON, J. T. 1963*b*. A possible origin of the Hawaiian islands. *Canadian Journal of Physics*, **41**, 863–870.

WILSON, J. T. 1966. Did the Atlantic close and then re-open? *Nature*, **211**, 676–681.

WILSON, T. n.d. Jonathan Otley: clockmaker and geologist: the story of Keswick's "back-room" boy. Reprinted from *The Keswick Reminder* (1956).

WINCH, N. J. 1817. Observations on the geology of Northumberland and Durham. *Transactions of the Geological Society of London*, **4**, 1–101.

WOOD, E. M. R. 1900. The Lower Ludlow Formation and its graptolite fauna. *Quarterly Journal of the Geological Society of London*, **56**, 415–492 and plates.

WOODCOCK, N. H. 1990. Sequence stratigraphy of the Palaeozoic Welsh Basin. *Journal of the Geological Society*, **147**, 537–547.

WOODCOCK, N. H., QUIRK, D. G., FITCHES, W. R. & BARNES, R. P. (eds) 1999. *In Sight of the Suture: The Palaeozoic Geology of the Isle of Man in the Iapetus Ocean Context*. Geological Society, London, Special Publications, **160**.

WOODHALL, D. G. 2000. *Geology of the Keswick District*. British Geological Survey, Keyworth.

WOOLACOTT, D. 1921. The interglacial problem and the glacial and post-glacial sequence in Northumberland and Durham. *Geological Magazine*, **8**(Decade 6), 60–69.

WOOLDRIDGE, S. W. & LINTON, D. L. 1939. *Structure, Surface and*

Drainage in South East England. Institute of British Geographers, Publication **10**.

WORDSWORTH, W. 1810. Introduction to: Rev. Joseph Wilkinson. *Select Views in Cumberland, Westmorland, and Lancashire*, R. Ackerman for Joseph Wilkinson, London, i–xxxiv.

WORDSWORTH, W. 1820. Topographical description of the country of the Lakes. *In*: *Poems by William Wordsworth: Including the River Duddon; Vaudracour & Julia; Peter Bell; The Waggoner; A Thanksgiving Ode; and Miscellaneous Poems*, vol. 3. Longman, Hurst, Rees, Orme & Brown, London, 215–321.

WORDSWORTH, W. 1994. *The Works of William Wordsworth*. Wordsworth Editions, Ware, Hertfordshire.

WORSLEY, P. 1991. Obituary notice: Frederick William Shotton. *Proceedings of the Geologists' Association*, **102**, 322–323.

WRIGHT, A. D. 1985. The Ordovician–Silurian boundary at Keisley, northern England. Geological Magazine, **122**, 261–273.

WRIGHT, A. D. 1988. The Ordovician–Silurian boundary at Keisley, Cumbria. *Bulletin of the British Museum (Natural History) Geology*, **43**, 59–63.

WU, K. 1999. *Numerical modelling of groundwater flow and radioactive waste migration, Sellafield, England*. PhD dissertation, Edinburgh University.

WYATT, J. 1996. *Wordsworth and the Geologists*, Cambridge University Press, Cambridge, New York and Melbourne.

ZIEGLER, P. A. 1982. *Geological Atlas of Western and Central Europe*. Elsevier (for Shell Internationale Petroleum Maatschappil Bulletin of Volcanology, Amsterdam and New York.

ZIEGLER, P. A. 1984. Caledonian and Hercynian crustal consolidation of western and central Europe – a working hypothesis. *Geologie en Mijnbouw*, **63**, 93–108.

Index

Note: Page numbers in *italic* type refer to illustrations; those in **bold** type refer to tables.

WHAT'S UP IN THE AIR?

ANITA GANERI

SIMON & SCHUSTER
YOUNG BOOKS

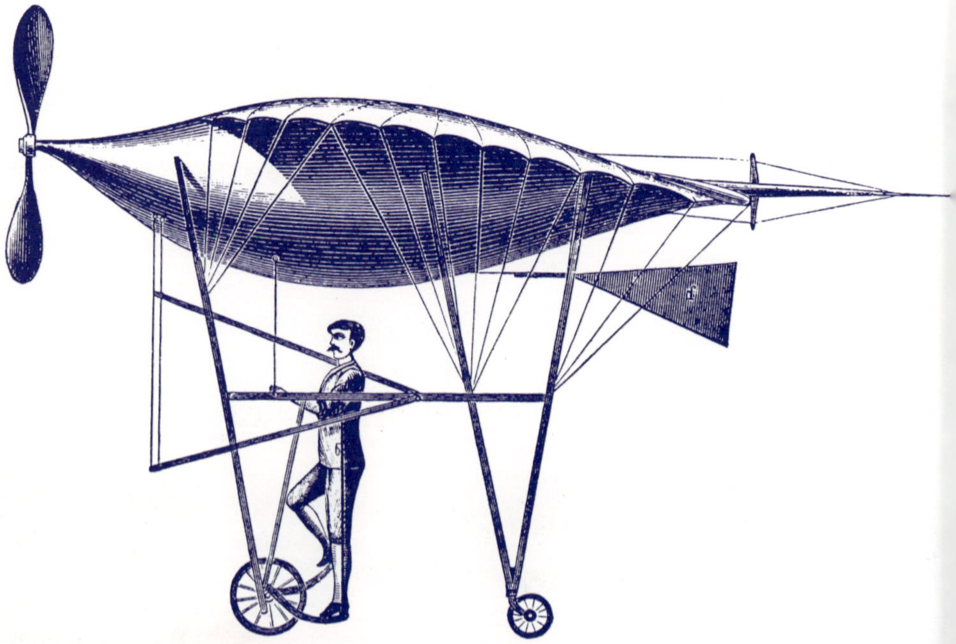

First published in 1993 by
Simon & Schuster Young Books

Simon & Schuster Young Books
Campus 400
Maylands Avenue
Hemel Hempstead
Herts
HP2 7EZ

Design: Jane Hannath

Illustrator: Brett Breckon

Commissioning editor: Debbie Fox

Editor: Jayne Booth

Picture research: Val Mulcahy

Photograph acknowledgements
Front cover top (t) & back cover: Zefa (Michael Wray); centre (c) The Image Bank (Mitchell Funk); bottom (b) Zefa; spine: Tony Stone Worldwide (John Chard); Allsport/Vandystadt, p.28(r) (Bernard Giani); Heather Angel, pp.12(tr), 13(t), 15(c); Art Directors Photo Library, pp.9(c), 38–39 (C.F.E. Smedley), 41(br); Chris Bonington Library, pp.29(r), 29(b) (Chris Bonington); Britstock-IFA, endpapers and p.19 (Fred); John Cleare/Mountain Camera, p.28(1); Bruce Coleman, pp.10(b) (Rod Williams), 11(cl) (Kim Taylor), 11(cr) (Jeff Foott Productions), 12(b) (Hans Reinhard), 13(br) (Udo Hirsch), 14(l) (Orion Service & Trading Co), 15(b) (Hans Reinhard), 18 (Jane Burton), 26(l) (Gerald Cubitt); Comstock, pp.31(r) (Adam Woolfitt), 33(cl); © Crown Copyright (National Meteorological Library), p.23(t); Eye Ubiquitous/TRIP, pp.17(t) (P.M. Fielding), 22 (R.D. Raby); Robert Harding Picture Library, pp.27(t), 27(c), 33(cr) (Paul van Riel); The Hulton Deutsch Collection, pp.32(br), 37(tl), 37(tr), 37(b), 41(bl) (William England); The Image Bank, pp.25(tr) (Alberto Incrocci), 37(c) (Mercury Archive); The ISE Photo Library, pp. 32(bl), 39(cl); Kunsthistorisches Museum, Vienna, p.30(t); Frank Lane Picture Agency, pp.11(t) (E & D Hosking), 15(tl) (R Tidman), 21(t) (Lucille Handberg); Robert Opie Collection p.33(b); Oxford Scientific Films, pp.12(tl) (Tony Allen), 15(tr) (Wendy Shattil & Bob Rozinski), 23(c) (Doug Allan); Picturepoint, pp.8(b), 9(b), 16(b), 17(cr), 23(b), 26(r), 30(c), 32(t), 34(t), 35(t), 35(b); Planet Earth, p.10(c) (Colin Pennycuick); Popperfoto, p.36; Royal Geographical Society, pp.9(t), 29(l); Royal Gorge Attractions Association, p.34(b); Brian Rogers/Biofotos, p.10(t); Science Photo Library, pp.13(bl) (John Mead), 21(cl) (Claud Nuridsany & Marie Perennou), 24(t) (Philippe Plailly), 25(tl) (Will McIntyre), 25(c) (Sheila Terry); South American Pictures, pp.16(t) (Hilary Bradt), 17(b) (Tony Morrison); Spectrum Colour Library, pp.8(c), 25(b); Tony Stone Worldwide, pp.20(l) (Hilarie Kavanagh), 21(cr) (Cosmo Condina), 24(b) (Keith Wood), 27(b) (Charles Thatcher), 31(l) (Alan Smith), 33(t); Sygma, pp.39(bl) 39(br) (J.P. Laffont), 41(t) (J.P. Bolle); TRIP, p.38(bl); Zefa, pp. 8(t) (Art Wolfe), 14(r), 17(cl) (Spichtinger), 20(r), 30(b) (K. Kerth), 31(b) (Sunak), 38(br) (Paul Freytag), 40, 41(c).

Typeset by: Goodfellow and Egan, Cambridge

Printed and bound: in Hong Kong by Wing King Tong

A CIP catalogue record for this book is available from the British Library

ISBN 0 7500 1327 3

CONTENTS

Introduction

YOU CANNOT SEE THE AIR but it is all around you. It is invisible but vital to the survival of life on Earth. Without it there would be no oxygen to breathe and no weather to warm us up or cool us down. There is plenty to see up in the air – clouds in the sky, birds and insects, tall trees, high buildings and hills. All these things reach up into the sky or live in the air itself. This book looks at the amazing world of the air and sky. There are chapters on living things, natural resources, natural features, exploration and history, buildings and services, and transport and sport.

LIVING THINGS

A wide variety of animals and plants live up in the air or in high places such as mountains or trees. The mountains are home to people too. These living things all have special features to help them survive. This Japanese macaque has thick fur to keep it warm in its mountain habitat.

NATURAL RESOURCES

The atmosphere around the Earth allows us to breathe, gives us the weather and a constant water supply. The energy of the wind and other natural resources are now being tapped to provide electricity. It is hoped that this will help to ease the serious problem of air pollution and its effects, such as acid rain and the thinning of the ozone layer.

NATURAL FEATURES

Towering, snow-capped mountains are among the world's most dramatic natural features. Some mountains are volcanic; others are formed by movements of the Earth's crust. Most are millions of years old. The highest point on Earth is Mount Everest in the Himalayas. The Matterhorn (right) is one of the most distinctive mountains in the Alps.

EXPLORATION AND HISTORY

Over the last 150 years, many daring attempts have been made to climb and explore the world's highest mountains. In 1953, Sir Edmund Hillary and Sherpa Tenzing Norgay (left) became the first people to stand on the top of Mount Everest. Many people enjoy the challenge of climbing despite the hazardous conditions, altitude sickness, frostbite and other problems they might face. There are ancient cities, temples and monasteries to be discovered in the mountains too.

BUILDINGS AND SERVICES

In many large cities, the skyline is crowded with skyscrapers. They have been built because the cities are getting too crowded. There is no space for buildings to reach outwards, only up. Other tall structures include radio and television masts, electricity pylons, bridges and aqueducts. Railways and cable cars have also been built on some mountainsides so that people can get up and down.

TRANSPORT AND SPORT

Today, aircraft carry people and goods all over the world. But the first powered flight was not made until 1903, although people had dreamt of flying for centuries. Flying machines range from hot-air balloons to helicopters and from hang-gliders to *Concorde*, the first supersonic plane to carry passengers. Sky sports such as sky-diving, gliding and aerobatics have also become very popular.

Flying creatures

THE FIRST FLYING creatures were insects which lived about 380 million years ago. They took to the prehistoric skies to find food and to escape from their enemies on the ground – the same reasons why animals fly today. Other creatures followed them into the air. The first bird, archaeopteryx, lived about 150 million years ago. Modern flying creatures include birds, insects and bats. Some other animals, such as frogs, snakes and fish, can glide through the air, but they can't really fly. Birds can fly further and faster than any other animals. Their bodies are light and their bones hollow – perfect adaptations for flight. They can flap their wings, soar, glide or even hover in the air.

Cape epauletted fruit bat

BATS

Bats are the only mammals that can fly. Their wings are flaps of leathery skin stretched over their very long fingers and hands. The scientific name for bats is *chiroptera* which means 'hand wings'. Most bats are nocturnal, coming out at night to feed. During the day, some hang upside-down from a branch with their wings wrapped round them.

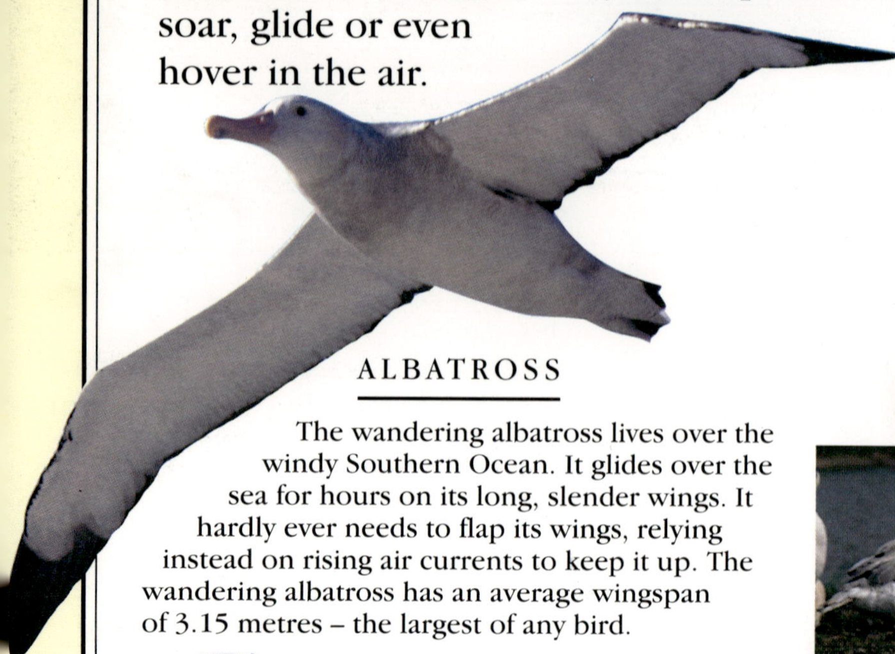

ALBATROSS

The wandering albatross lives over the windy Southern Ocean. It glides over the sea for hours on its long, slender wings. It hardly ever needs to flap its wings, relying instead on rising air currents to keep it up. The wandering albatross has an average wingspan of 3.15 metres – the largest of any bird.

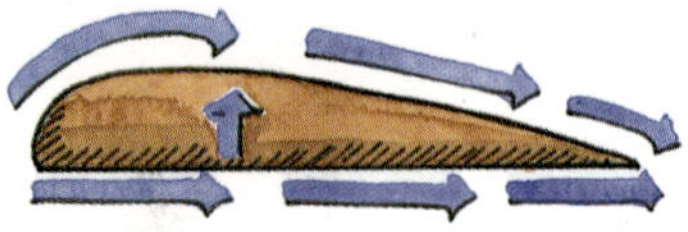

THE WING SHAPE

The aerofoil shape of a bird's wings is the key to its flying success. Air flows over the top of the wing faster than it flows underneath. This produces lift.

PEREGRINE FALCON

The peregrine falcon can fly faster than any other bird. It dives down on its prey of small birds from a great height, killing them with the sharp claws on its back toes. This dive is called 'stooping'. The peregrine reaches a speed of at least 200km/h as it stoops. Some experts think it may even reach an incredible 350km/h.

BUMBLE BEE

Bumble bees shouldn't be able to fly at all, according to the laws of physics. Their wings are too small for their bodies. But bees are good fliers, as they buzz from flower to flower collecting pollen and nectar.

MONARCH BUTTERFLIES

Very few insects make long migrations. But every year monarch butterflies travel some 3,000 kilometres from the USA to spend the winter in Mexico. They gather at the same tree roosts each year. A single site may contain over 14 million butterflies.

DID YOU KNOW

The Arctic tern makes the longest journey of any animal. Each year, it migrates from the Arctic to the Antarctic and back – a return trip of over 40,000 kilometres. It spends eight months of every year in flight.

FLYING SQUIRREL

Some mammals can glide through the treetops, although they cannot actually fly. The flying squirrel lives in North America. It has furry flaps of skin between its wrists and ankles. As it leaps, it spreads these flaps out like parachutes. It uses its bushy tail for steering. A flying squirrel can glide for about 100 metres in one go.

BAR-HEADED GEESE

Bar-headed geese (left) migrate from central Asia to northern India and Burma each year. On the way, they have to cross the world's highest mountains – the Himalayas. This means flying at a height of over 8,000 metres. Most migrating birds rarely fly above 100 metres. The highest flying bird seen so far was a Ruppell's griffon vulture at 11,277 metres.

Plants

GREEN PLANTS MAKE FOOD by a chemical process called photosynthesis. They use energy from sunlight to convert carbon dioxide from the air and water from the ground into a type of sugar. The plants trap sunlight using a green pigment, called chlorophyll, in their leaves. This is why they grow upwards into the air, to give their leaves a better chance of reaching the light. Some trees grow very tall indeed, supported by their strong, woody stems. Plants living high up on mountainsides have to cope with freezing temperatures and gale-force winds. Many have special survival features.

DID YOU KNOW

The snow on some mountain slopes is sometimes coloured red by millions of tiny plants, called algae. The red pigment they contain shields them from the harsh Sun.

DWARF WILLOW

Dwarf forms of willow and fir trees are found on the sides of some mountains. They grow very low to the ground to stop their branches being broken and battered by strong winds. The trees are only a few centimetres high – short enough for a person to step over.

LICHENS

No plants can live on the tops of mountains, but lichens come close. These hardy plants grow on rocks 7,000 metres up. They grow very slowly, taking 60 years to cover a patch the size of a postage stamp.

EDELWEISS

Edelweiss flowers grow in the Alps in Europe at heights of over 3,000 metres. They have thick, hairy petals which trap the Sun's heat and keep the plant warm. They also stop water escaping so that the plant does not dry out in the wind. Edelweiss are very rare. In some places, they are now protected by law.

RAINFOREST LAYERS

Rainforests are lush, green jungles which grow near to the Equator. They are hot and wet all year round. Rainforest trees and plants grow in layers, depending on their height above the ground. A few very tall trees, the emergents, reach heights of 45 metres. Below them is the canopy, with trees about 30 metres tall. They get the best of the sunlight. The understorey beneath consists of smaller trees and saplings. Shrubs and herbs form the lowest layers.

GIANT LOBELIAS

Giant lobelias and groundsels grow on the sides of Mount Kenya in Africa. To survive the frosty nights on the mountain, groundsels close their leaves over their delicate buds and shoots. Lobelias have downy leaves to keep them warm. Giant lobelias grow up to 8 metres high and groundsels over 6 metres high.

TOWERING TREES

The biggest living things in the world are trees. Giant sequoias can grow over 83 metres tall and measure 25 metres round their trunks. The tallest tree is another type of sequoia called a coast redwood. It stands almost 114 metres tall, higher than a 20-storey building. Both types of sequoia grow in California, USA.

FLOWER FLOURISH

It takes a *puya raimondii* plant up to 150 years to produce its first flowers. But the result is worth waiting for. The plant grows a huge flower spike, over 10 metres tall. It is covered in about 800 small, white flowers. The *puya raimondii* grows in the Andes mountains in South America. It is a type of herb.

Mountain Life

THE HIGHER UP a mountain you go, the colder it gets. For every 500 metres you climb, the temperature drops by about 3°C. On the highest mountain tops, the temperature is well below –20°C. It feels even colder because of the wind. There is another problem too – the higher you go, the less oxygen there is and the more difficult it is to breathe. Mountain animals have special ways of coping with the thin air and the cold. They have thick wool or fur for warmth. They also have compact body shapes which conserve heat. Some animals hibernate in winter.

ZONES OF LIFE

As the temperature changes up a mountainside, so do the plants and animals that live in each zone. Forests and meadows give way to hardier shrubs, mosses and lichens. Some animals migrate downhill in winter to find food and shelter.

SNOW LEOPARD

Snow leopards have thick, smokey-grey and black coats to keep them warm, and broad, furry feet which act as snowshoes. These beautiful and very rare animals live in the mountains of central Asia, feeding on marmots, wild sheep and ibex.

SNOW MONKEYS

Japanese macaques, or snow monkeys, live in the mountains of northern Japan. The winter months are very harsh in the mountains. Thick snow falls and it is bitterly cold. The macaques have an ingenious way of keeping warm. They take long, hot baths in the volcanic springs.

BONE BREAKERS

Lammergeiers soar above the mountains. They feed on dead animals, carrying the bones high into the air and dropping them on the rocks below to smash them open. Then they pick out the juicy marrow inside.

ALPINE SALAMANDER

The Alpine salamander is found as high as 3,000 metres. Most amphibians lay eggs in water. But Alpine salamanders give birth to live young on land. This gives the young a better chance of survival.

YAKS

Yaks live higher than any other mammal. They sometimes climb to over 6,000 metres in search of food. Their long, thick, matted coats protect them from the icy cold. Yaks live in Nepal, Tibet and China. They are kept by mountain villagers for their milk, wool and dung. The dung is burnt as fuel.

AMAZING BUT TRUE

Many people claim to have seen the yeti in the Himalayas. This huge, shaggy creature is reported to be half-man, half-ape. People have even produced photos of yeti footprints. But no one can prove the yeti exists. It may just be a case of 'seeing things'. On the other hand, it may be true . . .

MOUNTAIN GOATS

Keeping your footing on a steep mountain slope is not easy. But mountain goats are famous for their surefootedness. Their hooves have tough, rubbery pads which act like suction pads to stop the goats slipping. Mountain goats have no difficulty climbing near vertical cliffs or leaping along the narrowest ledges.

SPRINGTAILS

Springtails and other flightless insects survive much higher up than other animals because their bodies contain a type of anti-freeze. They feed on pollen and dead insects blown up by the wind.

The High Life

PEOPLE LIVING HIGH UP in the mountains face the same problems as mountain animals and plants. They have to survive the cold and the lack of oxygen. Some have physical adaptations which help them to breathe the thin air. They make the most of the oxygen available by making more red blood corpuscles. These contain the substance haemoglobin, which soaks up oxygen. Studies of people in the Peruvian Andes have shown that they live longer at higher altitudes than lower down. They also suffer less from illnesses such as high blood pressure and heart attacks. The culture of many mountain people is linked closely with their surroundings. The mountains also provide them with all they need to eat and live.

QUECHUA INDIANS

The Quechua Indians of Peru live over 3,500 metres up in the Andes mountains. They can breathe the thin mountain air easily because they have bigger hearts and lungs than normal. They also have such good blood circulation in their hands and feet that they can walk barefoot in the snow without suffering from frostbite.

AMAZING BUT TRUE

The tallest people in the world are the Watusi of Rwanda and Burundi in Africa. On average, men are 196 centimetres tall and women 178 centimetres tall. The Mbuti pygmies of Zaire are the shortest people. Men and women are 135 – 137 centimetres tall, on average.

LAKE TITICACA

Lake Titicaca in Peru and Bolivia is the highest large lake in the world. It is 3,811 metres above sea level. People have lived on and around the lake since the time of the Incas over 500 years ago. They make their homes on the great reed islands floating on the lake. They also use the reeds for boat building.

THE SHERPAS

The Sherpa people live in Nepal, high up in the Himalayas. They treat the mountains with great respect, believing them to be the homes of the gods. They live by farming and raising yaks. They are famous for their climbing skills and strength. Many have helped mountaineers on their expeditions. Sherpa Tenzing Norgay was one of the first people to climb Mount Everest (see page 29).

NORTH AMERICAN INDIANS

Banff National Park lies in the Rocky Mountains in Alberta, Canada. Its spectacular scenery includes snow-capped mountains, glaciers and valleys. Alberta is also home to some 45,000 North American Indians, including members of the Blood and Blackfoot tribes.

POTALA PALACE

The Potala Palace stands high above the city of Lhasa in Tibet. The city is 3,684 metres above sea level. The Potala's golden roofs and colourful walls can be seen all over the city. For centuries, the palace was the home of the Dalai Lama, the leader of Tibet and the head of the Buddhist religion in Tibet. The building was begun by the fifth Dalai Lama in 1645. It acted as a monastery as well as a palace.

LA PAZ

La Paz is the capital of Bolivia. It is the highest capital city in the world, situated 3,631 metres above sea level. Its airport is 4,080 metres above sea level, so you get a good view of the mountains as you come in to land!

The Air Around Us

YOU CANNOT SEE THE AIR, but it is all around you. The Earth is surrounded by a thick blanket of air called the atmosphere. It is the key to our survival. It contains oxygen for us to breathe and provides green plants with the carbon dioxide they need to make food. It shields the Earth from the Sun's harmful ultraviolet rays and stops it getting too hot. It traps some heat to stop the Earth getting too cold. The atmosphere also produces the weather (see pages 20–21). The atmosphere is divided into layers, reaching 8,000 kilometres above you. It presses down on the Earth and everything on it, including you. This force is called air, or atmospheric, pressure. You do not normally feel it because your breathing helps to balance out its effect.

ALGAE

When green plants make food, they use carbon dioxide from the air and produce bubbles of oxygen as waste. About 2,000 million years ago, algae were the first plants to make oxygen. Without them, people and animals could never have evolved. Algae make over two thirds of the oxygen we breathe today.

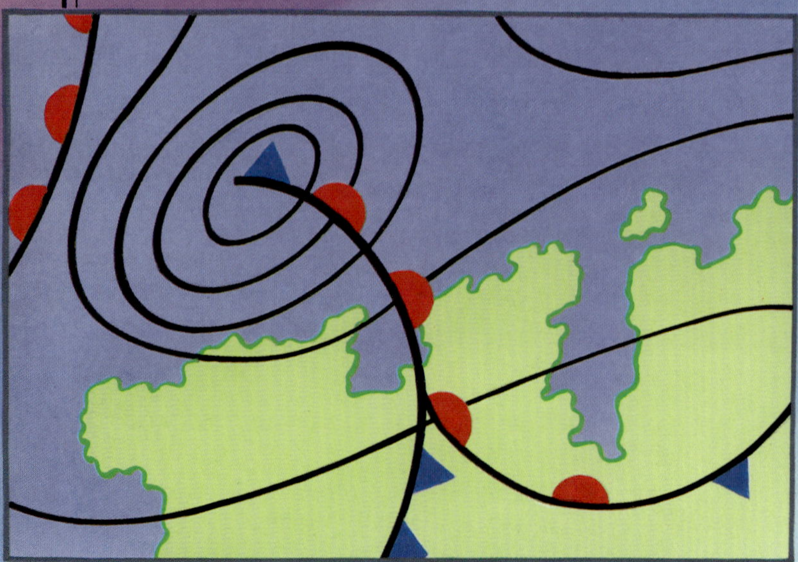

ISOBARS

The wiggly lines on a weather map are called isobars. They join areas of equal air pressure. The closer together they are, the stronger the wind will be. Areas of low pressure are marked as 'lows' on the map. They tend to bring clouds and rain. Areas of high pressure, marked as 'highs', bring sunnier weather.

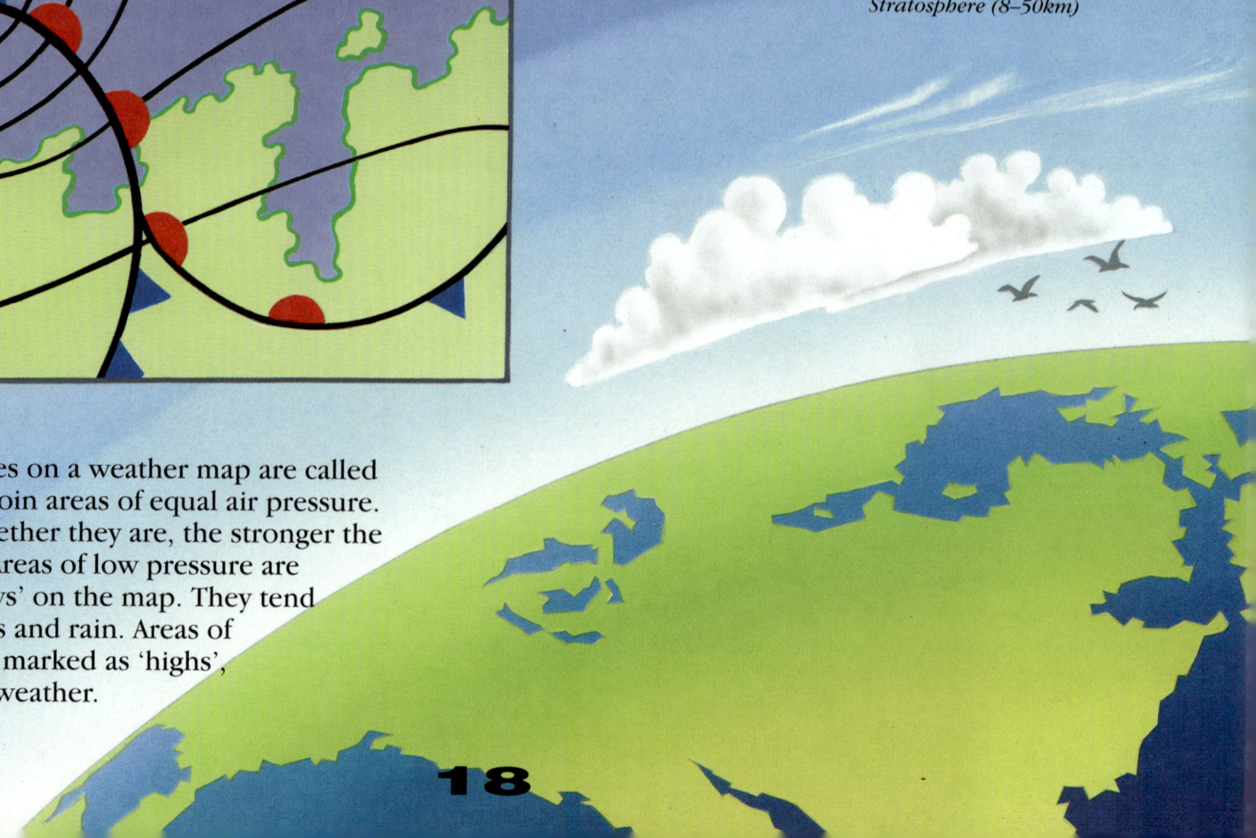

BLUE SKIES

Light coming from the Sun is called white light but it is made up of different colours – red, orange, yellow, green, blue, indigo and violet (see page 21). On a clear, sunny day the sky looks blue because the light is split and scattered by particles of dust and gases in the atmosphere. Blue is scattered more widely than the other colours so the sky looks blue.

Exosphere (500–8,000km)

Thermosphere (80–500km)

Mesosphere (50–80km)

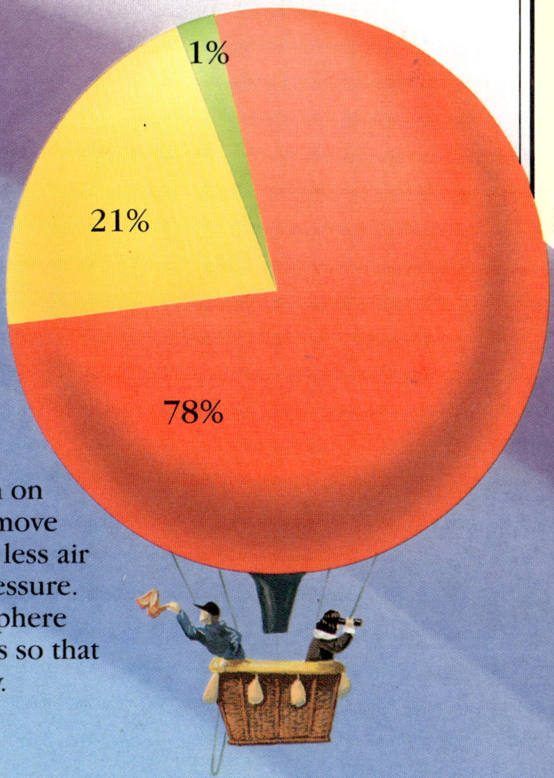

THINNING AIR

At sea level, air pressure is equal to 1.03 kilograms pressing down on each square centimetre. As you move up through the atmosphere, the less air there is and the lower the air pressure. Jet planes cruising in the stratosphere need specially pressurised cabins so that passengers can breathe normally.

Troposphere (0–8km above Poles)
(0–16km above Equator)

DID YOU KNOW

The atmosphere is held in place by the force of gravity. Otherwise the air would simply float off into space. Spacecraft need to travel at speeds of over 40,000km/h to break free of gravity.

WHAT IS AIR MADE OF?

Air is a mixture of different gases. Nitrogen makes up about 78 per cent of the air, oxygen about 21 per cent and argon about 1 per cent. There are also minute amounts of carbon dioxide, hydrogen, water vapour, ozone, methane, neon, xenon and krypton. Over three quarters of the air in the atmosphere is in the troposphere.

The Weather Machine

THE WEATHER HAPPENS in the layer of the atmosphere closest to the ground – the troposphere. Scientists define weather as the state of the air at any given time. The air, the Sun and water vapour (the gas form of water) are the main ingredients in weather. The way they react together from day to day gives us our many different types of weather – from wind and rain, to summer Sun. It affects us all wherever we live, dictating what we wear, what we eat and what we do. We might complain about the weather but we couldn't live without it. If it did not spread the Sun's heat more evenly around the world, some places would burn up and others freeze.

LIGHTNING

Inside towering cumulo-nimbus clouds, strong air currents toss ice crystals up and down. As they rub together, they become electrically charged. Lightning is a sudden flash of electricity inside the cloud or from the cloud to the ground. It heats the air in its path to 30,000°C. The air expands at high speed, making the booming sound of thunder.

CLOUDS

When the Sun heats the ground, the ground warms the air above it. This warm air rises and, as it rises, it cools. The water vapour it contains condenses and turns into tiny water droplets. A cloud is made up of millions and millions of these droplets. Different clouds bring different weather. Inside rain clouds, water droplets crash together to form larger, heavier drops.These fall as rain.

WIND

Wind is air moving from areas of high pressure to areas of low pressure. The air does not flow in a straight line. As the Earth spins on its axis, it makes the winds swing to the right (northern hemisphere) or left (southern hemisphere).

TORNADOES

Tornadoes are very violent tropical storms. They look like dark funnels of wind and cloud hanging down from storm clouds. As the funnel sucks up hot air, it starts to twist. Winds inside them can reach over 600km/h. Tornadoes seem to jump across the ground, destroying anything in their path. They are powerful enough to lift up trees, vehicles and even people.

RAINBOWS

When sunlight shines through a prism, it splits into its different colours (see page 19). Raindrops or the water droplets in spray also act as tiny prisms. They split the light, making the colours fan out as a rainbow. Red is always at the top of the rainbow.

SNOWFLAKES

Snowflakes form inside high, cold clouds when ice crystals collide and join together. All snowflakes have six sides but no two are ever exactly the same shape. Their shape depends on conditions inside the cloud – the temperature, humidity and movement of the air.

THE WATER CYCLE

The Earth's water supply is constantly being recycled and reused. In the water cycle, the Sun's heat causes millions of litres of water to evaporate from the oceans every day. As the water vapour rises, it cools and condenses to form clouds. Rain and snow fall from the clouds onto the land below. Some falls directly into the sea; some falls into rivers which carry the water to the sea.

AMAZING BUT TRUE

Showers of fish and frogs are probably caused by tornadoes. They suck the animals from ponds and drop them some way away. Some tornadoes are capable of even more amazing feats. In 1931, a tornado in the USA picked up an 83-tonne railway carriage with 117 passengers inside. It carried its load a short way then dropped it in a ditch.

Measuring the weather

SCIENTISTS WHO STUDY the weather are called meteorologists. They also prepare weather forecasts for the newspapers and television, and for people such as pilots, farmers and sailors. There is a network of about 10,000 weather stations all over the world, constantly measuring temperature, air pressure, humidity and so on. They use a variety of sophisticated instruments to collect information, including radar and satellites. This information is sent to meteorological offices all over the world where it is plotted on special maps called synoptic charts, and fed into huge computers. The most powerful of these supercomputers can do over 400 million calculations a second. They help the meteorologists to prepare forecasts.

SATELLITES

Weather satellites high above the Earth send back photographs of cloud and snow cover, storms and temperature patterns. There are two types of weather satellite. Geostationary satellites, such as Meteosat above, stay in a fixed place 36,000 kilometres above the Earth. Polar-orbiting satellites circle the Earth.

ANEROID BAROMETER

Air pressure is measured in units called millibars (mb) by instruments called barometers. An aneroid barometer is made of a metal box, with a vacuum inside. As the air pressure rises or falls, the box expands or gets smaller. The changes in pressure are shown on a dial. A rise in pressure usually means good weather ahead; a fall in pressure means stormy weather.

CUP ANEMOMETER

Meteorologists need to measure wind speed and direction. Wind speed is measured in kilometres per hour using an anemometer. The most common type consists of three cups attached to a central shaft. As the wind blows, the cups spin round and the wind speed is shown on a dial. Wind direction is measured with a wind vane or wind sock, using the points of the compass. A wind is named after the direction it blows from.

DID YOU KNOW

The amount of cloud covering the sky is measured in oktas, or eighths of the sky. So, 0 oktas means there is not a cloud in the sky, but 8 oktas means the sky is completely covered in cloud.

WEATHER AIRCRAFT

There are weather stations on land, at sea and on aircraft. The specially modified aircraft below has been fitted with radar and other instruments for measuring the weather high up in the sky. The instruments fit inside the plane's long nose cone. In the USA, weather aircraft are used to monitor hurricanes. They fly right into the middle or 'eye' of the storm.

WEATHER BALLOONS

To measure the weather in the upper atmosphere, meteorologists send up special balloons. These are filled with helium to make them rise. They carry instruments called radiosondes which radio information back to the weather station on the ground. The balloons usually reach a height of 25 kilometres before they burst. Then the instruments float safely down to Earth on small parachutes. Radiosonde instruments collect information about air pressure, temperature, humidity and wind speed.

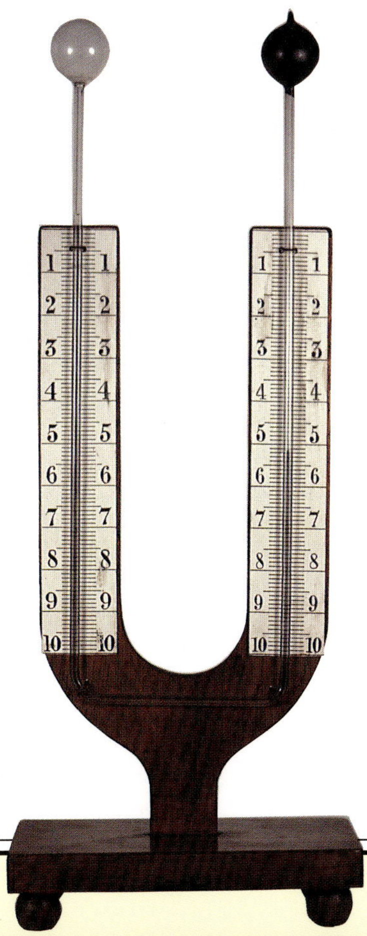

THERMOMETERS

Temperature is measured in degrees Centigrade (°C) or in degrees Fahrenheit (°F), using a thermometer. Maximum and minimum temperatures are recorded. Thermometers are often kept in a case called a Stevenson screen. This is placed in the shade and raised off the ground so that air can flow around the thermometers. The highest shade temperature ever recorded was 58°C in Libya; the lowest was –89.2°C in Antarctica.

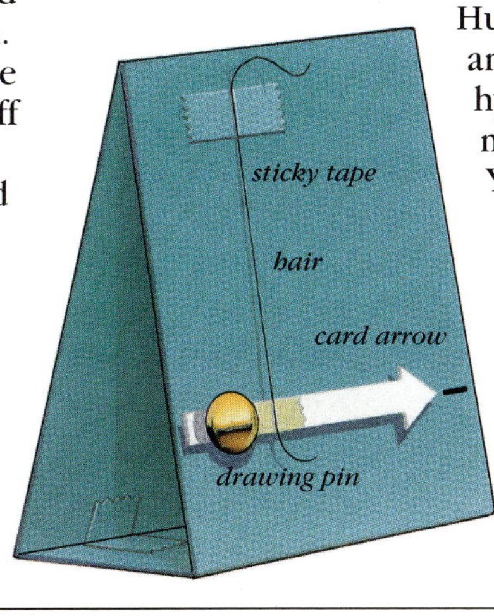

HUMAN HAIR HYGROMETER

Humidity is measured with an instrument called a hygrometer. High humidity makes the air feel very sticky. You can make your own hygrometer using a strand of hair. Human hair gets longer when the air is humid and shrinks when it is dry. The arrow on the hygrometer should go up or down as the hair shrinks or grows.

Air Pollution

ONE OF THE MOST serious problems facing planet Earth today is air pollution. We could not survive without the air, yet we are damaging it a great deal. The worst offenders are smoke and fumes from power stations which burn coal, gas and oil to produce electricity, and car exhaust fumes. Scientists are now trying to find cleaner ways of making electricity, using the Sun, the wind and water power. Many cars now use unleaded petrol which is less harmful to the air.

OZONE LAYER

About 20 kilometres up in the atmosphere is a layer of a gas called ozone. It shields the Earth from the Sun's harmful ultraviolet rays. Without it, the Earth would fry. In 1985, scientists discovered a hole the size of the USA in the ozone layer above Anarctica. It is being destroyed by gases called CFCs found in aerosols and fridges.

FOGS AND SMOGS

Some cities, including Tokyo and Los Angeles, suffer from thick smogs which can damage people's health. These smogs are caused by exhaust fumes reacting with sunlight. Until the 1960s, London had terrible smogs called 'pea soupers' because of their green colour. They were caused by coal smoke from fires and factories. In the worst smog in 1952, thousands of people died from pneumonia and bronchitis.

THE GREENHOUSE EFFECT

When coal, oil, gas or rainforest trees are burnt, they release carbon dioxide and other gases into the air. These gases may be stopping excess heat escaping from the Earth and making it warmer than it should be.

AMAZING BUT TRUE

Each year, an amazing 1.2 million tonnes of CFCs, 400,000 million tonnes of carbon dioxide, 130 million tonnes of sulphur dioxide and 36 million tonnes of nitrogen oxide are pumped into the atmosphere from factories, power stations, cars and so on. A car gives out four times its weight in carbon dioxide a year.

ACID RAIN

Acid rain is one of the worst types of air pollution in North America and Europe. Power stations pump huge amounts of sulphur dioxide and nitrogen oxide into the air where they form sulphuric and nitric acids. The acids fall with the rain, often hundreds of kilometres away from their source. Acid rain kills trees, poisons lakes and eats away at stone buildings.

SOLAR ENERGY

Solar energy is a much cleaner source of power than coal and oil. In some hot countries, people already have water-filled solar panels on their roofs. The water heats up in the Sun and is then piped around the houses. Solar energy can also be turned into electricity using solar cells. They can also be used to power cars, aircraft and space satellites.

WIND POWER

The power of the wind has been used for hundreds of years to push sailing ships along, grind corn and pump water. It can now be used to generate electricity. Modern windmills are called wind turbines. The wind turns their blades, powering a generator which makes electricity.

Mountains

ABOUT A QUARTER of the Earth's land surface is over 900 metres above sea level. Mountains are formed in several ways. Some are volcanoes. They may be extinct or still active. Others are outcrops of hard rock, left standing when the soft rock around them has been worn away. Ayers Rock in Australia formed in this way. The two major types of mountains, however, are built by movements of the Earth's crust. Fold mountains form when two plates of crust collide and the crust between them crumples and buckles up. Block mountains form when a gigantic block of rock is squeezed up between two cracks, or faults, in the crust.

THE MATTERHORN

The Matterhorn is part of the Alps in Switzerland. It is 4,477 metres high. Its distinctive shape is a result of millions of years of weathering by the wind, ice and frost. Glaciers have carved out sharp ridges on its sides. Mountains are constantly being worn down by the weather, but only by about 8.5 centimetres every 1,000 years.

THE HIMALAYAS

The Himalayas stretch for about 2,410 kilometres across the top of northern India and its neighbours. They are the highest mountains in the world, containing 95 of the world's 109 peaks over 7,500 metres high. At 8,848 metres, Mount Everest is the world's highest mountain. It stands in the Himalayas on the border of Nepal and Tibet.

MOUNT FUJI

Mount Fuji, or Fujiyama, is the highest mountain in Japan. It stands 3,776 metres tall. It is a volcanic mountain which last erupted in 1707. Its snow-capped cone can be seen for many kilometres around. The Japanese consider Mount Fuji to be a sacred mountain. It is a traditional place of pilgrimage for followers of Japan's Shinto religion.

AMAZING BUT TRUE

The Highlands of Scotland are some of the oldest mountains on Earth, at about 400 million years old. The Andes are about 80 million years old, the Rockies 70 million, the Himalayas 45 million and the Alps 15 million years old.

THE ROCKIES

The Rocky Mountains form the longest mountain chain in North America. They stretch for more than 4,800 kilometres from New Mexico in the south to Alaska in the north. The highest point in the Rockies is Mount Elbert at 4,399 metres. The Rockies are rich in timber and minerals. Their national parks and ski resorts attract many tourists.

SIERRA NEVADA

The Sierra Nevada in the USA are good examples of block mountains. The mountains run mainly through eastern California for a distance of about 640 kilometres. Their highest point is Mount Whitney which stands 4,418 metres high. Mount Whitney is the second highest mountain in the USA, next to Mount McKinley in Alaska.

MOUNT RUSHMORE

Mount Rushmore in South Dakota, USA, is 1,745 metres high. It is famous for the sculptures carved into its side. They show the heads of four past presidents – George Washington, Thomas Jefferson, Theodore Roosevelt and Abraham Lincoln. The heads are 18 metres high. They were carved in 1927–1941 by Gutzon Borglun and his son.

Exploring the mountains

THE FIRST PERSON known to have climbed a mountain for fun was a French courtier, Antoine de Ville, in 1492. He was ordered by King Charles VIII of France to lead a party up Mont Aiguille. De Ville was so pleased with his climb that he stayed at the summit for six days. People also climbed mountains to collect plants and rocks for scientific study. But the sport of mountaineering only really began about 150 years ago. Since then, thousands of climbers have braved the weather, altitude sickness, frostbite and hypothermia to climb up the world's highest and most difficult peaks. Climbers today take months preparing for expeditions, and new climbing techniques and gear are being developed all the time.

ROCK CLIMBING

Rock climbing skills are needed to ascend steep cliffs or sheer rock walls on a high mountain. But rock climbing has also become a popular sport in its own right. Climbers wear special 'rock boots' which have rubber soles for grip. They climb by jamming their hands and feet into small cracks in the rock face. Some climb with a rope for safety. Others climb without ropes but may cover their hands in chalk for grip.

BASE CAMP

A climbing expedition to a mountain such as Mount Everest sets up a base camp near the bottom of the mountain. Groups of climbers set off from here to try for the summit. Modern expeditions take hundreds of tonnes of equipment with them, including tents, food, clothing and radio equipment.

PETRARCH

The Italian poet, Petrarch, is credited with being the first known Alpine climber. Petrarch lived from 1304–1374. He climbed Mont Ventoux in the Alps on 26 April 1336. Mont Ventoux is 1,909 metres high.

REINHOLD MESSNER

The Italian, Reinhold Messner is one of the greatest climbers of recent times. In 1982, he became the first person to have climbed the three highest mountains in the world (Everest, K2 and Kanchenjunga). In 1986, he became the first person to have climbed all 14 of the world's mountains over 8,000 metres high.

DID YOU KNOW

In 1871, Lucy Walker became the first woman to climb the Matterhorn. She wore the fashion of the time, long skirts and a hat with a veil. She kept her energy up with an unusual diet of champagne and cake!

ON TOP OF THE WORLD

On 29 May 1953, Sir Edmund Hillary and Sherpa Tenzing Norgay became the first people to stand on the summit of Mount Everest. Ten previous expeditions had failed. On the final stage of their climb, weather conditions were so bad that it took them five minutes to cover every 1.5 metres. They eventually reached the summit at 11.30am.

OXYGEN SUPPLY

The lack of oxygen high up in the mountains can cause climbers to have difficulty with breathing, and hallucinations (visions). This is known as altitude sickness. It can be fatal. Some climbers wear oxygen masks to help them cope with the conditions. But some climbers manage without extra oxygen. In May 1978, Reinhold Messner and Peter Habeler became the first people to climb Mount Everest without using oxygen.

History and religion

THERE ARE LOTS of reasons why people in the past built their homes or places of worship high up on mountainsides or high into the sky. In some religions, people believed that the higher you built, the closer to heaven you got. In other religions, the sky itself was worshipped as a god or goddess. The sky, the weather and nature in general were so important in people's lives that they believed they must be controlled by the gods. Many mountains were, and still are, considered to be sacred. People also built high up on hillsides to look out for enemies and defend themselves more easily against attack.

TOWER OF BABEL

According to the Bible, the variety of languages spoken in the world was first heard in the Tower of Babel. This was supposed to be a building whose top reached heaven. Archaeologists think they may have discovered the tower's foundations in southern Iraq.

MACHU PICHU

The ruins of the ancient city of Machu Pichu lie high up in the Andes in Peru. The city was the last stronghold of the Incas during the Spanish invasion of the 1530s. Many tales had been told about such a city, but the ruins were not discovered until 1911. The city was big enough for 2,000 people. They grew food on huge terraces cut out of the rock and filled with soil.

LE PUY MONASTERY

The chapel of Saint-Michel-d'Aiguilhe was built in the 11th century on the top of an 80-metre high extinct volcano near the town of Le Puy in France. The men hired to build the chapel had to pull all their materials and tools up the mountain in baskets.

DID YOU KNOW

The Parsees in India follow the religion of Zoroastrianism. They do not believe in burying or burning their dead. So they lay the bodies of the dead on huge 'towers of silence' to be picked clean by vultures.

SKY GODDESS

The ancient Egyptians believed in many gods and goddesses connected with the world around them. The goddess of the sky was called Nut. She was held up by the god of the air, Shu, and formed an arch over Geb, the god of the Earth. The most important god was Re, the Sun god.

QUTAB MINAR

The Qutab Minar near Delhi, India, was built as a victory tower. Begun in 1193, it was designed to celebrate the Muslim victory over the last Hindu kingdom in north India. The tower is 73 metres high and tapers towards the top. It has five storeys. The first three storeys are red sandstone; the upper two, marble and sandstone.

CASTLE DEFENCES

In the Middle Ages, many castles were built on high ground to make them easier to defend. The castle guards had a good view of the surrounding countryside and of any attackers. They could raise the alarm before the enemy got too close. The castle above is Harlech Castle in Wales. It was built by King Edward I in 1283.

SWAYAMBUNATH

Swayambunath is a Buddhist temple built on top of a sacred hill on the outskirts of Kathmandu in Nepal. It was built about 2,000 years ago. Visitors can climb the 365 steps up the hillside to the temple, or travel up by road. The painted eyes on the top of the stupa (dome) represent the Buddha looking out on the world.

Skyscrapers

MANY BIG CITIES around the world have become very crowded and there is very little room left for building homes and offices. It was this lack of space that led architects to think about building upwards into the air, rather than outwards. The result was the skyscraper. The very first skyscrapers were built in Chicago, USA, in the 1890s but they were only 10-12 storeys high. Modern skyscrapers may have over 100 storeys. They can weigh over 100,000 tonnes so they need strong foundations. Architects also have to take into account how the building will withstand high winds and earthquakes.

CN TOWER

The CN Tower in Toronto, Canada, stands 553.34 metres high. It was built between 1973-1975 out of 130,000 tonnes of reinforced concrete. At 347.5 metres up the tower there is an area called the 'Sky Pod'. It contains a large, revolving restaurant.

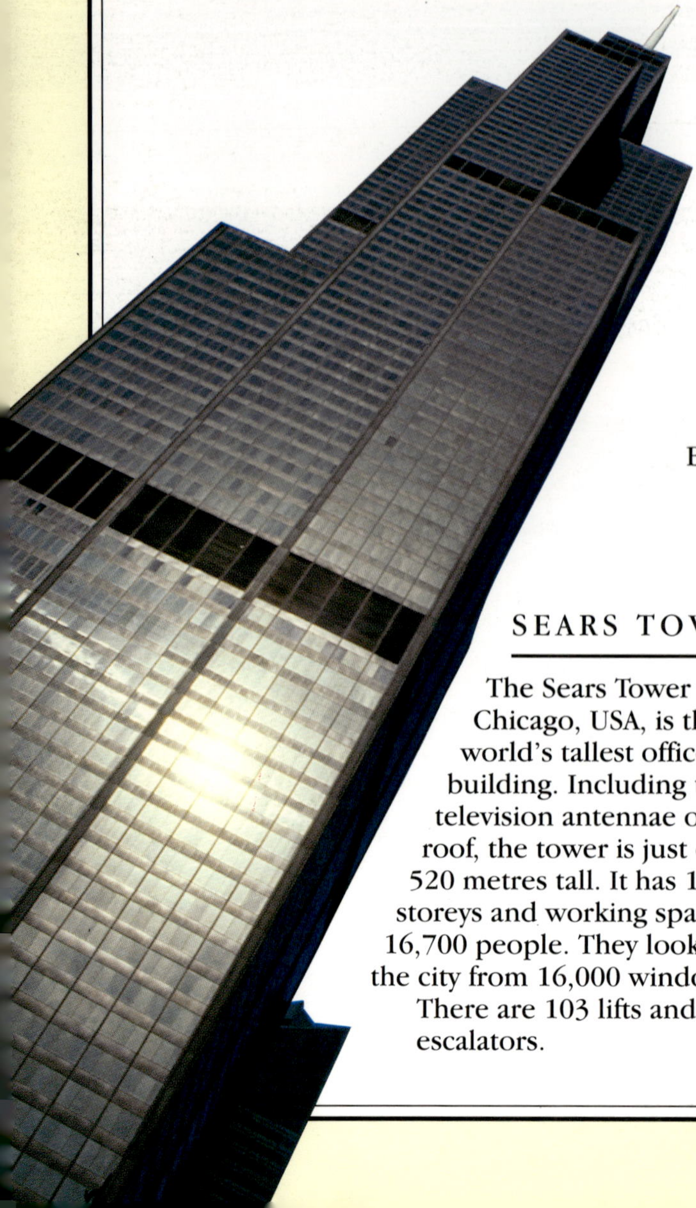

CONSTRUCTION WORKERS

Mohawk Indians from Montreal, Canada, spend much of their working day balanced on steel beams high above New York. Because they have such good heads for height, they are employed to put up the steel frames for many of the city's skyscrapers. A building rises by about one storey a week.

SEARS TOWER

The Sears Tower in Chicago, USA, is the world's tallest office building. Including the television antennae on its roof, the tower is just over 520 metres tall. It has 110 storeys and working space for 16,700 people. They look over the city from 16,000 windows. There are 103 lifts and 18 escalators.

THE TALLEST LIGHTHOUSE

At 106 metres, the steel lighthouse in Yokohama, Japan, is the tallest in the world. Its light shines with the power of 600,000 candles and can be seen by ships over 30 kilometres away. The earliest lighthouse was the Pharos of Alexandria off the coast of Egypt, built in about 270BC. It was even taller, at 122 metres.

HONG KONG

Hong Kong is one of the most densely populated places in the world. About 5,600 people live in each square kilometre, compared to 240 per square kilometre in Britain. Many huge skyscrapers have now been built in Hong Kong to make the best use of the small space available. They are used as offices, banks, hotels and apartment blocks.

HIGHEST CHURCH

The Chicago Temple of the First Methodist Church in the USA holds the record for the highest church spire. But it is given a bit of help. The church is part of a 22-storey skyscraper. There is a parsonage at about 100 metres and a chapel at 122 metres. The cross on top of the spire is 173 metres above street level.

THE EIFFEL TOWER

The Eiffel Tower was built in 1889 for the Paris Exposition by Alexandre Gustave Eiffel. He also designed the Statue of Liberty in New York. The Eiffel Tower is 300.5 metres tall and contains over 7,000 tonnes of steel. There are 1,792 steps leading to the top, together with lifts.

Bridges and services

PEOPLE BUILD BRIDGES, railways and roads for crossing rivers and valleys, and for getting up and down mountains. The first bridges were simply logs or stones for stepping across. The first people to build arch bridges were the Romans. Arches are still used in many bridges today, although concrete and steel have replaced stone and rope as the main building materials. The sky also provides a pathway for services such as electricity, telephone calls and even water.

ROMAN AQUEDUCT

Aqueducts are used to carry water. The Romans built many stone aqueducts with high arches to carry drinking water into towns and cities. The greatest aqueducts were over 140 kilometres long and could carry over 30 million litres of water a day. The Roman aqueduct at Segovia in Spain (left) was built in the 1st century BC. It has 148 arches and is about 800 metres long.

THE HIGHEST BRIDGE

The Royal Gorge Bridge in Colorado, USA, is the highest bridge in the world. It spans the Royal Gorge on the Arkansas River, 321 metres above the water. The gorge has been carved out of the rock over three million years by the river and the weather. The bridge was built in 1929. It is about 0.5 kilometres long, suspended from two 300-tonne steel cables. The world's steepest railway track runs down to the bottom of the gorge.

MOUNTAIN RAILWAY

The world's highest passenger railway runs between Lima and Huancayo in the Andes in Peru. The highest station on the route is Galera at 4,781 metres above sea level. It takes the train about four hours to climb up the mountainside to Galera. The track has to zig-zag as it climbs because the slope is so steep. At the top, some passengers have to be given oxygen to help them breathe.

PYLONS AND CABLES

Cables carrying electricity supplies and telephone calls are often held high in the air by tall pylons and poles. These keep the electricity, for example, off the ground where it could be dangerous. Cables and poles are designed to cope with strong winds and heavy snowfalls without snapping.

AMAZING BUT TRUE

The Tacoma Narrows Bridge in the USA was built to withstand winds of 190km/h. It collapsed four months after it opened in 1940, in winds of just 67km/h. The bridge was nicknamed 'Galloping Gertie' because of the way it shook up and down in the wind. Motorists drove over the bridge for the thrill. Fortunately, no one was hurt.

CABLE CARS

Cable cars and chair lifts are used to take sightseers and skiers up and down high mountains. Cable cars have also been used in San Francisco since 1873 to take passengers up and down the steep city streets. The first cable railway was opened in 1908. It carried people up the Wetterhorn in Switzerland. The longest chair lift ever climbed 5.6 kilometres up the Snowy Mountains in Australia. The journey up took between 45–75 minutes, depending on the weather.

Early Flights

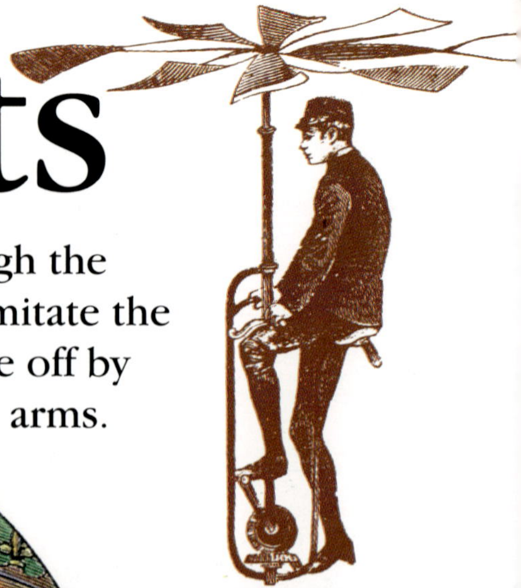

FOR CENTURIES, people have wanted to fly through the air. The first fliers made the mistake of trying to imitate the way birds fly. They built model wings and tried to take off by running fast downhill and flapping their arms. All these attempts were doomed to failure. People do not have flying muscles or the same body structure as birds. People experimented later with gliders, using rising air currents to soar through the sky. This was much more successful. In the 1800s, two great pioneers of flight, Sir George Cayley and Otto Lilienthal, made gliders which could carry a person.

CRAZY FLIGHTS

People were so obsessed by the idea of flying that they built all sorts of crazy machines to get them into the air. Trying to fly became a popular, if dangerous, hobby.

HOT-AIR BALLOONS

Two French brothers, Joseph and Etienne Montgolfier, pioneered hot-air balloon flights in the late 18th century. In 1783, they launched the first passengers into the air – a duck, a sheep and a cockerel. The flight was an experiment to see how well the animals survived. It was such a success that, later that year, the brothers became the first people to fly in a hot-air balloon.

FLYER 1

On 17 December 1903 in North Carolina, USA, Orville and Wilbur Wright made history by becoming the first people to achieve powered flight. Orville flew 36.5 metres in the brothers' bi-plane, *Flyer 1*, watched by Wilbur. The flight lasted for 12 seconds. Orville controlled the plane lying along its lower wing.

DID YOU KNOW

On 25 July 1909, French pilot Louis Blériot made the first aeroplane flight across the sea. His plane, Blériot XI, flew 37 kilometres across the English Channel from Dover to Calais in 36.5 minutes.

CHARLES LINDBERGH

In May 1927, an American pilot, Charles Lindbergh, became the first person to fly solo across the Atlantic Ocean. He flew 5,790 kilometres from New York to Paris in his plane, the *Spirit of St Louis*. The journey took nearly 33½ hours. It was a great achievement, particularly as Lindbergh was scared of heights as a child.

AMY JOHNSON

The first solo flight from Britain to Australia was made by a woman pilot, Amy Johnson. She made the journey in 1930 in a plane called a De Havilland Gypsy Moth. She had only learned to fly a year earlier. Amy Johnson died in 1941, flying in World War II.

WAX WINGS

In ancient Greek legend, the inventor Daedalus was imprisoned on the island of Crete. To escape, he made wings of bird feathers and wax for himself and his son, Icarus. But Icarus flew too close to the Sun, the wax melted and he fell into the sea.

AIRSHIPS

In the 1930s, airships were a popular form of transport for crossing the Atlantic. Airships were giant hydrogen-filled balloons, fitted with engines and propellers. Some of the largest and most famous were built by the German Zeppelin company. Among them was the 247-metre long *Hindenberg*. On 6 May 1937, the *Hindenberg* burst into flames as it landed in New Jersey, USA, and 35 people died. The golden age of airships had come to an end.

Modern Flying Machines

THE INVENTION OF the jet engine in the 1930s transformed the world of flight and flying machines. The first jet-powered plane was the Heinkel He 178, built by a German engineer, Ernst Heinkel, in 1939. Aircraft could now fly much faster. The first jet passenger plane, a De Havilland Comet, started making regular flights in 1952. The world was suddenly a much smaller place. Long-distance journeys that once took weeks or even months by ship could now be made in hours. In 1947, another milestone was reached. The American rocket plane, Bell X81, flew at a speed of 1,078km/h, faster than the speed of sound.

HELICOPTERS

Leonardo Da Vinci made a sketch of a helicopter in the 15th century. But the first helicopter was not built until 1936. Helicopters are VTOL (vertical take-off and landing) aircraft. They are often used for air and sea rescue, traffic control and on oil rigs.

PEDAL-POWERED PLANE

In 1979, a plane called the *Gossamer Albatross* was pedalled through the air across the English Channel. Its pilot was an American racing cyclist, Bryan Allen. The plane has a wingspan of 29 metres but was extremely lightweight. Allen pedalled for two hours and 49 minutes at a height of less than a metre above the sea.

CONCORDE

Concorde was the first supersonic passenger aircraft. It first flew in 1969 and was used on regular flights from 1976. It cruises at 2,333km/h, over twice the speed of sound.

In April 1990, Concorde crossed the Atlantic in a record time of two hours, 55 minutes and 15 seconds – 11 times faster than Charles Lindbergh had done in 1927.

AMAZING BUT TRUE

Today, thousands of people travel by plane every year. Heathrow Airport in London is the world's busiest international airport. Each year, over 30 million international passengers pass through the airport. The shortest scheduled flight is between two of the Orkney Islands near Scotland. The flight takes two minutes but can be cut to 58 seconds if the wind is right!

JUMBO JETS

The Boeing 747 'Jumbo' jet is the biggest passenger airliner. It is almost as long as a soccer pitch and as tall as a six-storey building. It can carry up to 500 passengers. Some 747s carry cargo only. Their nose cones swing upwards so that cargo can be loaded at the front of the plane.

STEALTH BOMBER

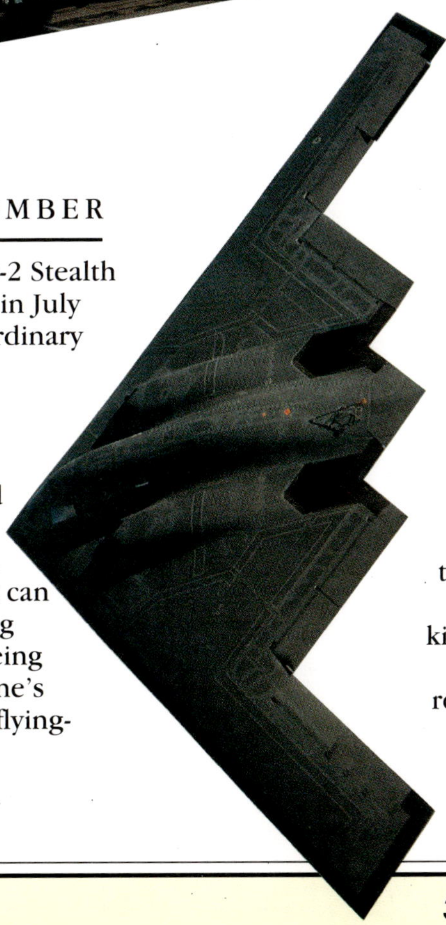

The US Airforce B-2 Stealth Bomber first flew in July 1989. This extraordinary plane is made of special, non-metallic materials and has a shape specially designed so that it is not detected by radar. This means that it can approach bombing targets without being detected. The plane's shape is called a 'flying-wing' shape.

VOYAGER

Voyager was the first plane to fly around the world without stopping. It flew from California, USA, on 14 December 1986 and returned there nine days later. It had flown 40,212.139 kilometres. Its two crew members worked and rested in cabins smaller than single beds.

Fun in the Air

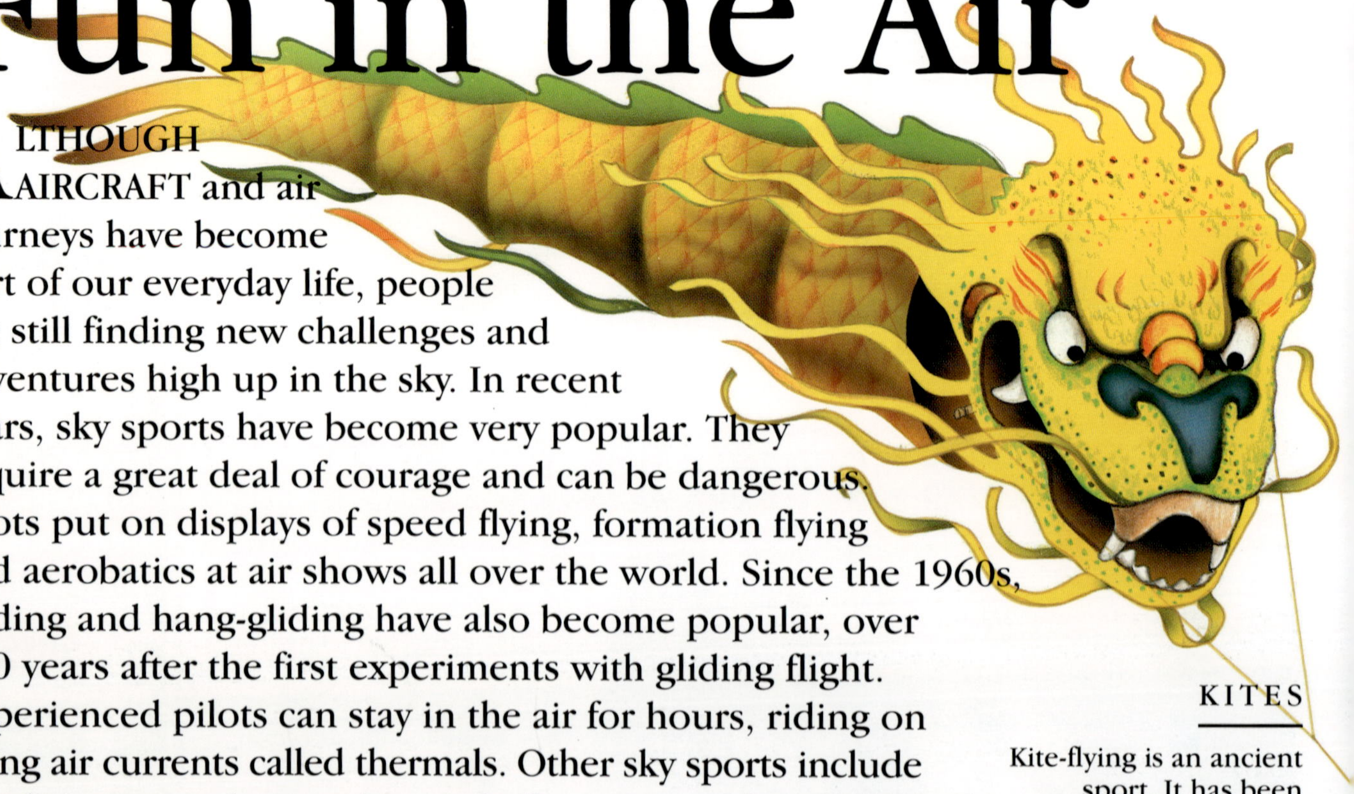

ALTHOUGH AIRCRAFT and air journeys have become part of our everyday life, people are still finding new challenges and adventures high up in the sky. In recent years, sky sports have become very popular. They require a great deal of courage and can be dangerous. Pilots put on displays of speed flying, formation flying and aerobatics at air shows all over the world. Since the 1960s, gliding and hang-gliding have also become popular, over 100 years after the first experiments with gliding flight. Experienced pilots can stay in the air for hours, riding on rising air currents called thermals. Other sky sports include sky-diving, parachuting, hot-air ballooning and kite-flying.

KITES

Kite-flying is an ancient sport. It has been popular in China and other Asian countries for over 3,000 years. Many kite-flying competitions were held and kites were also used for lifting and carrying things, including people. In parts of Asia, kites were also thought to ward off evil spirits.

DID YOU KNOW

A Frenchman, André Garnerin, gave an exhibition of parachute jumping in Paris in 1797. He jumped out of a hot-air balloon 680 metres above the ground, despite the fact he suffered from air sickness.

UP, UP AND AWAY

The first hot-air ballooning world championships were held in New Mexico, USA, in 1973. Since then, they have become an annual event. In the 18th century, hot-air ballooning was the latest craze. It has recently enjoyed a revival, with safer, less flammable gases used to lift the balloons. Hot-air balloons are also used on safari holidays to give a better view.

SKY-DIVING

Parachutes can be used as a means of escape, as brakes for space vehicles and to land equipment. They are also used in the sport of sky-diving in which people parachute down from planes for fun. Sky-divers 'free fall' through the air for thousands of metres, before opening their parachutes. They spread out their arms and legs to slow their fall.

WING-WALKING

Wing-walkers stand on the top of a plane's wings. Strong, steel safety ropes hold them in place. In 1980, Jaromir Wagner from Czechoslovakia crossed the Atlantic Ocean from Scotland on top of a plane. Wing-walking was a very popular attraction at airshows in America in the 1930s.

HANG-GLIDING

A hang-glider is like a giant kite with a harness underneath. Pilots take off from high hills or cliffs so that rising air currents push against the glider, keeping it up in the air. The pilot hangs underneath the canvas and steers by moving his body from one side to the other.

BLONDIN

In the 1850s and 1860s, a Frenchman known as Blondin crossed the Niagara Falls many times on a tightrope stretched 50 metres above the water. On one occasion, he crossed with his manager on his back; on another, he stopped halfway and cooked himself an omelette.

HEIGHT GUIDE

UP IN THE AIR

Look at the guide to discover what can be found at different heights in the air. Mount Everest is the highest point on Earth at 8,848 metres. The highest layer of the atmosphere, the exosphere, reaches about 8,000 kilometres into space.

Meteosat satellite
36,000 kilometres

The average height for parachute jumps is 3000 metres. The longest drop made is 25,816 metres from a hot-air balloon in New Mexico, USA, in 1960.

The average height for hot-air ballooning is about 1750 metres. The highest ascent is 19,811 metres, made over Texas, USA, in 1988.

Jet plane 15,000 metres

Cumulonimbus cloud
Top reaches 14,000 metres

Mount Everest 8,848 metres

Bar-headed geese
8,000 metres

Yak 6,100 metres
(highest living mammal)

Cirrus clouds 6,000 metres

Quechua Indians
5,200 metres

Mount Rushmore 1,745 metres

CN Tower, Canada
553.34 metres

World Trade Center
415.22 metres

Giant redwood 113.7 metres
(tallest tree)

ALGAE Very simple, often single-celled plants, found mainly in water.

AQUEDUCT A channel built to carry water from place to place.

ARCHAEOLOGIST A scientist who studies the past.

ATMOSPHERE The blanket of air around the Earth.

AXIS An imaginary line running through the centre of the Earth from North Pole to South Pole.

CFCs Short for gases called chlorofluorocarbons.

CHLOROPHYLL A green chemical in plant cells which traps the sunlight needed for photosynthesis.

FLAMMABLE Something which bursts into flame quite easily.

GRAVITY A force which pulls things down towards the centre of the Earth.

HAEMOGLOBIN A chemical in the blood which carries oxygen around the body and colours blood red.

HALLUCINATIONS Illusions or visions; things which are not really there.

HELIUM A colourless gas which is lighter than air.

HIBERNATE Spend the winter in a deep sleep. Some animals hibernate to avoid the winter weather and the lack of food. Their breathing and heart rates slow right down to save energy.

HUMIDITY The amount of water vapour in the air. When humidity is high, there is a lot of water vapour in the air and it feels hot and sticky.

HYPOTHERMIA A type of illness caused when the body gets too cold. A person loses consciousness and can die.

MAMMALS Warm-blooded animals with backbones which feed their young on milk.

MIGRATION A long journey made each year by some animals; they travel between their feeding and breeding grounds to take advantage of better weather and food supplies.

OZONE A gas which is form of oxygen.

PHOTOSYNTHESIS The chemical process by which green plants make their own food. They use sunlight to convert carbon dioxide and water into food.

PRISM A transparent object, usually triangular, which splits white light into its various colours. Raindrops act as prisms to create rainbows.

SAPLING Young tree.

SUPERSONIC Faster than the speed of sound.

THERMALS Rising currents of warm air.